ENGINEERING
APPLICATIONS
of
NONCOMMUTATIVE
HARMONIC ANALYSIS

With Emphasis on
Rotation and Motion Groups

ENGINEERING APPLICATIONS of NONCOMMUTATIVE HARMONIC ANALYSIS

With Emphasis on Rotation and Motion Groups

Gregory S. Chirikjian

Alexander B. Kyatkin

CRC Press
Taylor & Francis Group
Boca Raton London New York

CRC Press is an imprint of the
Taylor & Francis Group, an **informa** business

CRC Press
Taylor & Francis Group
6000 Broken Sound Parkway NW, Suite 300
Boca Raton, FL 33487-2742

Reissued 2019 by CRC Press

A Library of Congress record exists under LC control number:

Publisher's Note
The publisher has gone to great lengths to ensure the quality of this reprint but points out that some imperfections in the original copies may be apparent.

Disclaimer
The publisher has made every effort to trace copyright holders and welcomes correspondence from those they have been unable to contact.

ISBN 13: 978-0-367-25718-7 (hbk)
ISBN 13: 978-0-367-25720-0 (pbk)
ISBN 13: 978-0-429-28938-5 (ebk)

Visit the Taylor & Francis Web site at http://www.taylorandfrancis.com and the
CRC Press Web site at http://www.crcpress.com

To Atsuko, John, Margot, Brian, Pa, Baba-Marj, JiJi, Baba-Isako, and the Uncles

Authors

Gregory Chirikjian's research interests include applications of noncommutative harmonic analysis in engineering. Traditionally his research has focused on the study of highly articulated and modular self-reconfigurable robotic systems. Since the mid '90s, Dr. Chirikjian's interests have broadened to include applied mathematics, image analysis, and the conformational statistics of macromolecules. Dr. Chirikjian received a Ph.D. in applied mechanics from the California Institute of Technology in 1992. Since then, he has been on the faculty of the Department of Mechanical Engineering at The Johns Hopkins University.

Alexander Kyatkin's research interests include harmonic analysis on the motion group, computer vision and image processing problems, and numerical algorithm development. He also has conducted research in theoretical high energy particle physics where he studied non-perturbative aspects of field theory. Currently he develops machine vision systems for industrial applications. He received a Ph.D. in physics from The Johns Hopkins University in 1996.

Preface

This book is intended for engineers, applied mathematicians, and scientists who are interested in learning about the theory and applications of a generalization of Fourier analysis called *noncommutative harmonic analysis*. Existing books on this topic are written primarily for specialists in modern mathematics or theoretical physics. In contrast, we have made every attempt to make this book self-contained and readable for those with a standard background in engineering mathematics. Thus we present concepts in a heavily coordinate-dependent way without sacrificing the rigorous nature of the subject.

Fourier analysis is an invaluable tool to engineers and scientists. There are two reasons for this: a wide range of physical phenomena have been modeled using the classical theory of Fourier series and transforms; and the computational benefits that follow from fast Fourier transform (FFT) techniques have had tremendous impact in many fields including communications and signal and image processing.

While classical Fourier analysis is such a ubiquitous tool, it is relatively rare to find engineers who are familiar with the mathematical concept of a group. Even among the minority of engineers familiar with the group concept, it is rare to find engineers familiar with generalizations of Fourier analysis and orthogonal expansion of functions on groups. We believe that there are two reasons for this. Firstly, this generalization of Fourier analysis (noncommutative harmonic analysis) has historically been a subject developed by and for pure mathematicians and theoretical physicists. Hence, few with knowledge in this area have had access to the engineering problems that can be addressed using this area of mathematics. Secondly, without knowing of the existence of this powerful and beautiful area of mathematics, it has been impossible for engineers to formulate their problems within this framework, and thus impossible to apply it. Thus this book can be viewed as a bridge of sorts. On the one hand, mathematicians who are familiar with abstract formulations can look to this book and see group representation theory and harmonic analysis being applied to a number of problems of interest in engineering. Likewise, engineers completely unfamiliar with these concepts can look to this text to gain the vocabulary needed to express their problems in a new and potentially more efficient way.

Chapter 1 is an overview of several engineering problems to which this theory is applicable. In Chapter 2 we review commutative harmonic analysis (classical Fourier analysis) and associated computational issues that have become so important in all areas of modern engineering. Chapter 3 is a review of the orthogonal expansions of classical mathematical physics combined with more modern techniques such as Walsh functions, discrete polynomial transforms, and wavelets. Advanced engineering mathematics with emphasis on curvilinear coordinate systems and basic coordinate-dependent differential geometry of hyper-surfaces in $\mathbb{R}^N$ is the subject of Chapter 4. Chapters 5 and 6 are each devoted to the properties and parameterizations of the two groups of most importance in the application areas covered here: the rotation group and the group of rigid-body motions. For the most part, we do not use group theoretical notation in Chapters 5 and 6. In Chapter 7 we present ideas from group theory that are the most critical for the formulation in the remainder of the book. This includes many geometrical examples of groups and cosets. The distinction between discrete and continuous groups is made, and the classification of continuous groups using matrix techniques is examined. In Chapter 8 we discuss how to integrate functions on continuous (Lie) groups and present the general ideas of representation theory and harmonic analysis on finite and compact Lie groups. Chapter 9 is devoted to orthonormal expansions on the rotation group and related mathematical objects such as the unit sphere in four dimensions. In Chapter 10 we present the theory behind Fourier analysis on the Euclidean motion groups (groups of rigid-body motions in the plane and space). These are groups that we have found to be of great importance in engineering applications (though they are rarely discussed in any detail in mathematics textbooks and research monographs). Chapter 11 develops a fast Fourier transform for the motion groups. Chapters 12–18 are devoted exclusively to concrete applications of noncommutative harmonic analysis in engineering and applied science. A chapter is devoted to each application area: robotics; image analysis and tomography; pose determination and sensor calibration; estimation and control with emphasis on spacecraft attitude; rotational Brownian motion and diffusion (including the analysis of liquid crystals); conformational statistics of macromolecules; and orientational averaging in mechanics.

This book can serve as a supplement for engineering, applied science, and mathematics students at the beginning of their graduate careers, as well as a reference book for researchers in any field of engineering, applied science, or mathematics. Clearly there is too much material for this book to be used as a 1–year course. However, there are many ways to use parts of this book as reference material for semester-long (or quarter-long) courses. We list several options below:

- Graduate Engineering Mathematics: Chapters 2, 3, 4, Appendix.

- Robotics/Theoretical Kinematics: Chapters 5, 6, 9, 10, 12.

- Mathematical Image Analysis: Chapters 2, 10, 11, 13, 14.

- Introduction to Group Theory: Chapters 5, 6, 7, 8.

- Computational Harmonic Analysis on Groups: Chapters 7, 8, 9, 10, 11.

- Engineering Applications of Harmonic Analysis: Chapters 12, 13, 15, 18.

- Applications of Harmonic Analysis in Statistical and Chemical Physics: Chapters 14, 16, 17.

Writing a book such as this requires the efforts and input of many people. We would like to thank a number of people for their helpful comments and discussions which have been incorporated into this text in one way or another.

First, we would like to thank (in alphabetical order) those who suffered through the "Orientational Phenomena" course given by the first author based on an early version of this book: Richard Altendorfer, Commander Steve Chism, Lara Diamond, Vijay Kumar, Sang Yoon Lee, Subramanian Ramakrishnan, Ed Scheinerman, Arun Sripati, David Stein, Jackrit Suthakorn, Mag Tan, Yunfeng Wang, and Yu Zhou.

We would also like to thank (in alphabetical order) Jonas August, Troels Roussau Johansen, David Joyner, Omar Knio, David Maslen, Keizo Miyahara, Joshua Neuheisel, Frank Chongwoo Park, Bahram Ravani, Dan Rockmore, Wilson J. Rugh, Ramin Takloo-Bighash, Panagiotis Tsiotras, Louis Whitcomb, and Hemin Yang for proofreading and/or helpful suggestions.

Special thanks goes to Imme Ebert-Uphoff, whose initial work (under the supervision of the first author) led us to study noncommutative harmonic analysis, and to David Stein for preparing many of the figures in this book. We also acknowledge the support of the NSF, without which this work would not have been pursued.

G. Chirikjian
A. Kyatkin

Contents

Preface ix

1 Introduction and Overview of Applications 1
1.1 Noncommutative Operations 1
    1.1.1 Assembly Planning 1
    1.1.2 Rubik's Cube 3
    1.1.3 Rigid-Body Kinematics 4
1.2 Harmonic Analysis: Commutative and Noncommutative 5
1.3 Commutative and Noncommutative Convolution 7
1.4 Robot Workspaces, Configuration-space Obstacles, and Localization 9
1.5 Template Matching and Tomography 10
1.6 Attitude Estimation and Rotational Brownian Motion 11
1.7 Conformational Statistics of Macromolecules 12
1.8 Mechanics and Texture Analysis of Materials 12
    References 13

2 Classical Fourier Analysis 15
2.1 Fourier Series: Decomposition of Functions on the Circle 15
2.2 The Fourier Transform: Decomposition of Functions on the Line 19
2.3 Using the Fourier Transform to Solve PDEs and Integral Equations 23
    2.3.1 Diffusion Equations 23
    2.3.2 The Schrödinger Equation 23
    2.3.3 The Wave Equation 24
    2.3.4 The Telegraph Equation 25
    2.3.5 Integral Equations with Convolution Kernels 26
2.4 Fourier Optics 27
2.5 Discrete and Fast Fourier Transforms 30
    2.5.1 The Discrete Fourier Transform: Definitions and Properties 30
    2.5.2 The Fast Fourier Transform and Convolution 32
    2.5.3 Sampling, Aliasing, and Filtering 33
2.6 Summary 36
    References 36

3 Sturm-Liouville Expansions, Discrete Polynomial Transforms and Wavelets 39
3.1 Sturm-Liouville Theory 39
    3.1.1 Orthogonality of Eigenfunctions 39
    3.1.2 Completeness of Eigenfunctions 41

    3.1.3   Sampling Band-Limited Sturm-Liouville Expansions 42
3.2 Legendre Polynomials and Associated Legendre Functions 43
    3.2.1   Legendre Polynomials 43
    3.2.2   Associated Legendre Functions 44
    3.2.3   Fast Legendre and Associated Legendre Transforms 45
3.3 Jacobi, Gegenbauer, and Chebyshev Polynomials 46
3.4 Hermite and Laguerre Polynomials 47
3.5 Quadrature Rules and Discrete Polynomial Transforms 49
    3.5.1   Recurrence Relations for Orthogonal Polynomials 50
    3.5.2   Gaussian Quadrature 52
    3.5.3   Fast Polynomial Transforms 54
3.6 Bessel and Spherical Bessel Functions 55
    3.6.1   Bessel Functions 55
    3.6.2   Spherical Bessel Functions 57
    3.6.3   The Bessel Polynomials 58
    3.6.4   Fast Numerical Transforms 58
3.7 Hypergeometric Series and Gamma Functions 61
    3.7.1   Hahn Polynomials 62
    3.7.2   Charlier, Krawtchouk, and Meixner Polynomials 63
    3.7.3   Zernike Polynomials 64
    3.7.4   Other Polynomials with Continuous and Discrete Orthogonalities 65
3.8 Piecewise Constant Orthogonal Functions 65
    3.8.1   Haar Functions 66
    3.8.2   Rademacher Functions 67
    3.8.3   Walsh Functions and Transforms 67
3.9 Wavelets 69
    3.9.1   Continuous Wavelet Transforms 69
    3.9.2   Discrete Wavelet Transforms 72
3.10 Summary 74
    References 74

4 Orthogonal Expansions in Curvilinear Coordinates   81
4.1 Introduction to Curvilinear Coordinates and Surface Parameterizations 81
4.2 Parameterizations of the Unit Circle, Semi-Circle, and Planar Rotations 82
4.3 M-Dimensional Hyper-Surfaces in $\mathbb{R}^N$ 85
4.4 Gradients, Divergence, and the Laplacian 88
4.5 Examples of Curvilinear Coordinates 90
    4.5.1   Polar Coordinates 90
    4.5.2   Spherical Coordinates 91
    4.5.3   Stereographic Projection 93
    4.5.4   The Möbius Band 94
4.6 Topology of Surfaces and Regular Tessellations 95
    4.6.1   The Euler Characteristic and Gauss-Bonnet Theorem 96
    4.6.2   Regular Tessellations of the Plane and Sphere 97
4.7 Orthogonal Expansions and Transforms in Curvilinear Coordinates 98
    4.7.1   The 2-D Fourier Transform in Polar Coordinates 98
    4.7.2   Orthogonal Expansions on the Sphere 99
    4.7.3   The 3-D Fourier Transform in Spherical Coordinates 101
4.8 Continuous Wavelet Transforms on the Plane and $\mathbb{R}^n$ 102
    4.8.1   Multi-dimensional Gabor Transform 102

         4.8.2 Multi-Dimensional Continuous Scale-Translation Wavelet Transforms . . . 103
   4.9   Gabor Wavelets on the Circle and Sphere 103
         4.9.1 The Modulation-Shift Wavelet Transform for the Circle 103
         4.9.2 The Modulation-Shift Wavelet Transform for the Sphere 105
   4.10 Scale-Translation Wavelet Transforms for the Circle and Sphere 105
         4.10.1 The Circle 106
         4.10.2 The Sphere 106
         4.10.3 Discrete Wavelets and Other Expansions on the Sphere 107
   4.11 Summary 107
         References 108

5 **Rotations in Three Dimensions**   **111**
   5.1   Deformations of Nonrigid Objects 111
   5.2   Rigid-Body Rotations 112
         5.2.1 Eigenvalues and Eigenvectors of Rotation Matrices 114
         5.2.2 Relationships Between Rotation and Skew-Symmetric Matrices 115
         5.2.3 The Matrix Exponential 119
   5.3   Rules for Composing Rotations 121
   5.4   Parameterizations of Rotation 122
         5.4.1 Euler Parameters 122
         5.4.2 Cayley/Rodrigues Parameters 123
         5.4.3 Cartesian Coordinates in $\mathbb{R}^4$ 124
         5.4.4 Spherical Coordinates 125
         5.4.5 Parameterization of Rotation as a Solid Ball in $\mathbb{R}^3$ 125
         5.4.6 Euler Angles 125
         5.4.7 Parameterization Based on Stereographic Projection 126
   5.5   Infinitesimal Rotations, Angular Velocity, and Integration 127
         5.5.1 Jacobians Associated with Parameterized Rotations 128
         5.5.2 Rigid-Body Mechanics 132
   5.6   Other Methods for Describing Rotations in Three-Space 135
         5.6.1 Quaternions 135
         5.6.2 Expressing Rotations as 4×4 Matrices 137
         5.6.3 Special Unitary 2×2 Matrices: $SU(2)$ 138
         5.6.4 Rotations in 3D as Bilinear Transformations in the Complex Plane 141
   5.7   Metrics on Rotations 141
         5.7.1 Metrics on Rotations Viewed as Points in $\mathbb{R}^4$ 143
         5.7.2 Metrics Resulting from Matrix Norms 143
         5.7.3 Geometry-Based Metrics 144
         5.7.4 Metrics Based on Dynamics 144
   5.8   Integration and Convolution of Rotation-Dependent Functions 145
   5.9   Summary 147
         References 147

6 **Rigid-Body Motion**   **149**
   6.1   Composition of Motions 149
         6.1.1 Homogeneous Transformation Matrices 150
   6.2   Screw Motions 152
   6.3   Parameterization of Motions and Associated Jacobians 154
         6.3.1 The Matrix Exponential 155
         6.3.2 Infinitesimal Motions 156

6.4   Integration over Rigid-Body Motions   157
6.5   Assigning Frames to Curves and Serial Chains   159
    6.5.1   The Frenet-Serret Apparatus   159
    6.5.2   Frames of Least Variation   160
    6.5.3   Global Properties of Closed Curves   162
    6.5.4   Frames Attached to Serial Linkages   164
6.6   Dual Numbers   168
    6.6.1   Dual Orthogonal Matrices   169
    6.6.2   Dual Unitary Matrices   169
    6.6.3   Dual Quaternions   169
6.7   Approximating Rigid-Body Motions in $\mathbb{R}^N$ as Rotations in $\mathbb{R}^{N+1}$ 170
    6.7.1   Planar Motions   170
    6.7.2   Spatial Motions   171
6.8   Metrics on Motion   174
    6.8.1   Metrics on Motion Induced by Metrics on $\mathbb{R}^N$ 174
    6.8.2   Metrics on $SE(N)$ Induced by Norms on $\mathcal{L}^2(SE(N))$ 178
    6.8.3   Park's Metric   180
    6.8.4   Kinetic Energy Metric   181
    6.8.5   Metrics on $SE(3)$ from Metrics on $SO(4)$ 181
6.9   Summary   181
    References   182

7   Group Theory   **187**
7.1   Introduction   187
    7.1.1   Motivational Examples   187
    7.1.2   General Terminology   192
7.2   Finite Groups   195
    7.2.1   Multiplication Tables   195
    7.2.2   Permutations and Matrices   197
    7.2.3   Cosets and Orbits   199
    7.2.4   Mappings   203
    7.2.5   Conjugacy Classes, Class Functions, and Class Products   208
    7.2.6   Examples of Definitions: Symmetry Operations on the Equilateral Triangle
        (Revisited)   214
7.3   Lie Groups   216
    7.3.1   An Intuitive Introduction to Lie Groups   216
    7.3.2   Rigorous Definitions   218
    7.3.3   Examples   220
    7.3.4   Demonstration of Theorems with $SO(3)$ 220
    7.3.5   Calculating Jacobians   224
    7.3.6   The Killing Form   232
    7.3.7   The Matrices of $Ad(G)$, $ad(X)$, and $B(X, Y)$ 233
7.4   Summary   236
    References   236

8   Harmonic Analysis on Groups   **239**
8.1   Fourier Transforms for Finite Groups   240
    8.1.1   Representations of Finite Groups   241
    8.1.2   Characters of Finite Groups   245
    8.1.3   Fourier Transform, Inversion, and Convolution Theorem   248

      8.1.4   Fast Fourier Transforms for Finite Groups 252

  8.2 Differentiation and Integration of Functions on Lie Groups 254

      8.2.1   Derivatives, Gradients, and Laplacians of Functions on Lie Groups 254

      8.2.2   Integration Measures on Lie Groups and their Homogeneous Spaces . . . 255

      8.2.3   Constructing Invariant Integration Measures 259

      8.2.4   The Relationship between Modular Functions and the Adjoint 261

      8.2.5   Examples of Volume Elements 262

  8.3 Harmonic Analysis on Lie Groups 262

      8.3.1   Representations of Lie Groups 262

      8.3.2   Compact Lie Groups 263

      8.3.3   Noncommutative Unimodular Groups in General 266

  8.4 Induced Representations and Tests for Irreducibility 268

      8.4.1   Finite Groups 269

      8.4.2   Lie Groups 271

  8.5 Wavelets on Groups 273

      8.5.1   The Gabor Transform on a Unimodular Group 274

      8.5.2   Wavelet Transforms on Groups based on Dilation and Translation 274

  8.6 Summary 275

      References 276

9 Representation Theory and Operational Calculus for $SU(2)$ and $SO(3)$ 281

  9.1 Representations of $SU(2)$ and $SO(3)$ 281

      9.1.1   Irreducible Representations from Homogeneous Polynomials 281

      9.1.2   The Adjoint and Tensor Representations of $SO(3)$ 284

      9.1.3   Irreducibility of the Representations $U_i(g)$ of $SU(2)$ 287

  9.2 Some Differential Geometry of S^3 and $SU(2)$ 290

      9.2.1   The Metric Tensor 290

      9.2.2   Differential Operators and Laplacian for $SO(3)$ in Spherical Coordinates . 292

  9.3 Matrix Elements of $SU(2)$ Representations as Eigenfunctions of the Laplacian . . 294

  9.4 $SO(3)$ Matrix Representations in Various Parameterizations 298

      9.4.1   Matrix Elements Parameterized with Euler Angles 298

      9.4.2   Axis-Angle and Cayley Parameterizations 300

  9.5 Sampling and FFT for $SO(3)$ and $SU(2)$ 301

  9.6 Wavelets on the Sphere and Rotation Group 302

      9.6.1   Inversion Formulas for Wavelets on Spheres 302

      9.6.2   Diffusion-Based Wavelets for $SO(3)$ 305

  9.7 Helicity Representations 306

  9.8 Induced Representations 308

  9.9 The Clebsch-Gordan Coefficients and Wigner 3jm Symbols 309

  9.10 Differential Operators for $SO(3)$ 310

  9.11 Operational Properties 313

  9.12 Classification of Quantum States According to Representations of $SU(2)$ 315

  9.13 Representations of $SO(4)$ and $SU(2) \times SU(2)$ 317

  9.14 Summary 317

      References 318

10 Harmonic Analysis on the Euclidean Motion Groups   **321**

10.1   Introduction 321

10.2   Matrix Elements of IURs of $SE(2)$ 322

     10.2.1   Irreducibility of the Representations $U(g, p)$ of $SE(2)$ 323

10.3   The Fourier Transform for $SE(2)$ 324

10.4   Properties of Convolution and Fourier Transforms of Functions on $SE(2)$ 325

     10.4.1   The Convolution Theorem 325

     10.4.2   Proof of the Inversion Formula 327

     10.4.3   Parseval's Equality 328

     10.4.4   Operational Properties 329

10.5   Differential Operators for $SE(3)$ 330

10.6   Irreducible Unitary Representations of $SE(3)$ 331

10.7   Matrix Elements 334

10.8   The Fourier Transform for $SE(3)$ 336

10.9   Operational Properties 339

     10.9.1   Properties of Translation Differential Operators 340

     10.9.2   Properties of Rotational Differential Operators 342

     10.9.3   Other Operational Properties 343

10.10 Analytical Examples 344

10.11 Linear-Algebraic Properties of Fourier Transform Matrices for $SE(2)$ and $SE(3)$. 346

10.12 Summary 350

     References 350

11 Fast Fourier Transforms for Motion Groups   **353**

11.1   Preliminaries: Direct Convolution and the Cost of Interpolation 353

     11.1.1   Fast Exact Fourier Interpolation 354

     11.1.2   Approximate Spline Interpolation 355

11.2   A Fast Algorithm for the Fourier Transform on the 2D Motion Group 356

11.3   Algorithms for Fast $SE(3)$ Convolutions Using FFTs 358

     11.3.1   Direct Fourier Transform 358

     11.3.2   Inverse Fourier Transform 361

11.4   Fourier Transform on the Discrete Motion Group of the Plane 363

     11.4.1   Irreducible Unitary Representations of the Discrete Motion Group 363

     11.4.2   Fourier Transforms on the Discrete Motion Group 364

     11.4.3   Efficiency of Computation of $SE(2)$ Convolution Integrals Using the Discrete-Motion-Group Fourier Transform 365

11.5   Fourier Transform for the 3D "Discrete" Motion Groups 366

     11.5.1   Mathematical Formulation 367

     11.5.2   Computational Complexity 371

11.6   Alternative Methods for Computing $SE(D)$ Convolutions 371

     11.6.1   An Alternative Motion-Group Fourier Transform Based on Reducible Representations 372

     11.6.2   Computing $SE(D)$ Convolutions Using the Fourier Transform for $\mathbb{R}^D \times SO(D)$ 373

     11.6.3   Contraction of $SE(D)$ to $SO(D+1)$ 375

11.7   Summary 376

     References 376

12 Robotics **379**
12.1 A Brief Introduction to Robotics 379
12.2 The Density Function of a Discretely Actuated Manipulator 380
    12.2.1 Geometric Interpretation of Convolution of Functions on $SE(D)$ 382
    12.2.2 The Use of Convolution for Workspace Generation 384
    12.2.3 Computational Benefit of this Approach 385
    12.2.4 Computation of the Convolution Product of Functions on $SE(2)$ 386
    12.2.5 Workspace Generation for Planar Manipulators 387
    12.2.6 Numerical Results for Planar Workspace Generation 390
12.3 Inverse Kinematics of Binary Manipulators: the Ebert-Uphoff Algorithm . . . 399
12.4 Symmetries in Workspace Density Functions 400
    12.4.1 Discrete Manipulator Symmetries 400
    12.4.2 3D Manipulators with Continuous Symmetries:
           Simplified Density Functions 403
    12.4.3 Another Efficient Case: Uniformly Tapered Manipulators 405
12.5 Inverse Problems in Binary Manipulator Design 405
12.6 Error Accumulation in Serial Linkages 407
    12.6.1 Accumulation of Infinitesimal Spatial Errors 407
    12.6.2 Model Formulation for Finite Error 408
    12.6.3 Related Mathematical Issues 409
12.7 Mobile Robots: Generation of C-Space Obstacles 410
    12.7.1 Configuration-Space Obstacles of a Mobile Robot 411
12.8 Mobile Robot Localization 414
12.9 Summary 414
    References 414

13 Image Analysis and Tomography **419**
13.1 Image Analysis: Template Matching 419
    13.1.1 Method for Pattern Recognition 420
    13.1.2 Application of the FFT on $SE(2)$ to the Correlation Method 421
    13.1.3 Numerical Examples 425
13.2 The Radon Transform and Tomography 432
    13.2.1 Radon Transforms as Motion-Group Convolutions 433
    13.2.2 The Radon Transform for Finite Beam Width 437
    13.2.3 Computed Tomography Algorithm on the Motion Group 441
13.3 Inverse Tomography: Radiation Therapy Treatment Planning 446
13.4 Summary 450
    References 450

14 Statistical Pose Determination and Camera Calibration **455**
14.1 Review of Basic Probability 455
    14.1.1 Bayes' Rule 457
    14.1.2 The Gaussian Distribution 457
    14.1.3 Other Common Probability Density Functions on the Line 458
14.2 Probability and Statistics on the Circle 459
14.3 Metrics as a Tool for Statistics on Groups 461
    14.3.1 Moments of Probability Density Functions on Groups
           and their Homogeneous Spaces 461
14.4 PDFs on Rotation Groups with Special Properties 464
    14.4.1 The Folded Gaussian for One-Dimensional Rotations 464

14.4.2 Gaussians for the Rotation Group of Three-Dimensional Space 465
14.5 Mean and Variance for $SO(N)$ and $SE(N)$ 467
  14.5.1 Explicit Calculation for $SO(3)$ 467
  14.5.2 Explicit Calculation for $SE(2)$ and $SE(3)$ 468
14.6 Statistical Determination of a Rigid-Body Displacement 469
  14.6.1 Pose Determination without *A Priori* Correspondence 469
  14.6.2 Pose Determination with *A Priori* Correspondence 471
  14.6.3 Problem Statement Using Harmonic Analysis 473
14.7 Robot Sensor Calibration 474
  14.7.1 Solution with Two Sets of Exact Measurements 475
  14.7.2 Calibration with Two Noisy Measurements 478
14.8 Summary 478
  References 479

15 Stochastic Processes, Estimation, and Control   485
15.1 Stability and Control of Nonlinear Systems 485
15.2 Linear Systems Theory and Control 489
15.3 Recursive Estimation of a Static Quantity 491
15.4 Determining a Rigid-Body Displacement:
      Recursive Estimation of a Static Orientation or Pose 493
15.5 Gaussian and Markov Processes and Associated Probability Density Functions
      and Conditional Probabilities 493
15.6 Wiener Processes and Stochastic Differential Equations 495
15.7 Stochastic Optimal Control for Linear Systems 497
15.8 Deterministic Control on Lie Groups 498
  15.8.1 Proportional Derivative Control on $SO(3)$ and $SE(3)$ 498
  15.8.2 Deterministic Optimal Control on $SO(3)$ 500
  15.8.3 Euler's Equations as a Double Bracket 501
  15.8.4 Digital Control on $SO(3)$ 501
  15.8.5 Analogs of LQR for $SO(3)$ and $SE(3)$ 502
15.9 Dynamic Estimation and Detection of Rotational Processes 503
  15.9.1 Attitude Estimation Using an Inertial Navigation System 503
  15.9.2 Exponential Fourier Densities 505
15.10 Towards Stochastic Control on $SO(3)$ 506
15.11 Summary 507
  References 508

16 Rotational Brownian Motion and Diffusion   515
16.1 Translational Diffusion 516
  16.1.1 Einstein's Theory 516
  16.1.2 The Langevin, Smoluchowski, and Fokker-Planck Equations . . . 517
  16.1.3 A Special Fokker-Planck Equation 518
16.2 The General Fokker-Planck Equation 518
  16.2.1 Derivation of the Classical Fokker-Planck Equation 519
  16.2.2 The Fokker-Planck Equation on Manifolds 520
16.3 A Historical Introduction to Rotational Brownian Motion 522
  16.3.1 Theories of Debye and Perrin: Exclusion of Inertial Terms 522
  16.3.2 Non-Inertial Theory of Diffusion on $SO(3)$ 524
  16.3.3 Diffusion of Angular Velocity 525
16.4 Brownian Motion of a Rigid Body: Diffusion Equations on $SO(3) \times \mathbb{R}^3$ 526

16.5   Liquid Crystals 527
       16.5.1   The Nematic Phase 528
       16.5.2   The Smectic Phases 529
       16.5.3   Evolution of Liquid Crystal Orientation 530
       16.5.4   Explicit Solution of a Diffusion Equation for Liquid Crystals 531
16.6   A Rotational Diffusion Equation from Polymer Science 531
16.7   Other Models for Rotational Brownian Motion 535
       16.7.1   The Evolution of Estimation Error PDFs 535
       16.7.2   Rotational Brownian Motion Based on Randomized Angular Momentum . 535
       16.7.3   The Effects of Time Lag and Memory 536
16.8   Summary 537
       References 537

17 Statistical Mechanics of Macromolecules   545
17.1   General Concepts in Systems of Particles and Serial Chains 545
17.2   Statistical Ensembles and Kinematic Modeling of Polymers 548
       17.2.1   The Gaussian Chain (Random Walk) 548
       17.2.2   The Freely-Jointed Chain 549
       17.2.3   The Freely Rotating Chain 550
       17.2.4   Kratky-Porod Chain 551
17.3   Theories Including the Effects of Conformational Energy 553
       17.3.1   Chain with Hindered Rotations 553
       17.3.2   Rotational Isomeric State Model 554
       17.3.3   Helical Wormlike Models 556
       17.3.4   Other Computational and Analytical Models of Polymers 558
17.4   Mass Density, Frame Density, and Euclidean Group Convolutions 559
17.5   Generating Ensemble Properties for Purely Kinematical Models 562
       17.5.1   The Mathematical Formulation 562
       17.5.2   Generating $\mu_{i,i+1}$ from μ_i and μ_{i+1} 563
       17.5.3   Computationally Efficient Strategies for Calculating $f(g)$ and $\rho(\mathbf{x})$ . . . 564
17.6   Incorporating Conformational Energy Effects 564
       17.6.1   Two-State and Nearest-Neighbor Energy Functions 565
       17.6.2   Interdependent Potential Functions 565
       17.6.3   Ensemble Properties Including Long-Range Conformational Energy . . . 567
17.7   Statistics of Stiff Molecules as Solutions to PDEs on $SO(3)$ and $SE(3)$ 568
       17.7.1   Model Formulation 568
       17.7.2   Operational Properties and Solutions of PDEs 571
17.8   Summary 572
       References 573

18 Mechanics and Texture Analysis   579
18.1   Review of Basic Continuum Mechanics 579
       18.1.1   Tensors and How They Transform 580
       18.1.2   Mechanics of Elastic Solids 582
       18.1.3   Fluid Mechanics 584
18.2   Orientational and Motion-Group Averaging in Solid Mechanics 585
       18.2.1   Tensor-Valued Functions of Orientation 585
       18.2.2   Useful Formulas in Orientational Averaging of Tensor Components . . . 586
       18.2.3   Motion-Group Averaging: Modeling-Distributed Small Cracks
                and Dilute Composites 588

18.3 The Orientational Distribution of Polycrystals 589
    18.3.1 Averaging Single-Crystal Properties over Orientations 590
    18.3.2 Determining Single Crystal Strength Properties from Bulk Measurements
         of Polycrystalline Materials 592
18.4 Convolution Equations in Texture Analysis 592
18.5 Constitutive Laws in Solid Mechanics 594
    18.5.1 Isotropic Elastic Materials 594
    18.5.2 Materials with Crystalline Symmetry 596
18.6 Orientational Distribution Functions for Non-Solid Media 597
    18.6.1 Orientational Dynamics of Fibers in Fluid Suspensions 597
    18.6.2 Applications to Liquid Crystals 598
18.7 Summary 598
    References 598

A Computational Complexity, Matrices, and Polynomials   **607**
A.1 The Sum, Product, and Big-$\mathcal{O}$ Symbols 607
A.2 The Complexity of Matrix Multiplication 608
A.3 Polynomials 610
A.4 Efficient Multiplication of Polynomials 610
A.5 Efficient Division of Polynomials 611
A.6 Polynomial Evaluation 611
A.7 Polynomial Interpolation 613

B Set Theory   **615**
B.1 Basic Definitions 615
B.2 Maps and Functions 617
B.3 Invariant Measures and Metrics 619

C Vector Spaces and Algebras   **623**
C.1 Rings and Fields 626

D Matrices   **627**
D.1 Special Properties of Symmetric Matrices 627
D.2 Special Properties of the Matrix Exponential 630
D.3 Matrix Decompositions 632
    D.3.1 Decomposition of Complex Matrices 632
    D.3.2 Decompositions of Real Matrices 634

E Techniques from Mathematical Physics   **635**
E.1 The Dirac Delta Function 635
    E.1.1 Derivatives of Delta Functions 636
    E.1.2 Dirac Delta Functions with Functional Arguments 637
    E.1.3 The Delta Function in Curvilinear Coordinates: What to do at Singularities   638
E.2 Self-Adjoint Differential Operators 639
E.3 Functional Integration 640
E.4 Quantization of Classical Equations of Motion 642

F Variational Calculus   **645**
F.1 Derivation of the Euler-Lagrange Equation 646
F.2 Sufficient Conditions for Optimality 649

G   Manifolds and Riemannian Metrics   651

References   655

Index   659

Chapter 1

Introduction and Overview of Applications

In this chapter we provide a layman's introduction to the meaning of *noncommutative* operations. We also illustrate a number of applications in which the noncommutative convolution product arises. In subsequent chapters we provide the detail required to solve problems in each application area. Chapters 12–18 are each devoted to one of these applications.

1.1 Noncommutative Operations

The word *noncommutative* is used in mathematical contexts to indicate that the order in which some kind of operation is performed matters. For instance, the result of the cross product of two vectors depends on the order in which it is performed because

$$\mathbf{a} \times \mathbf{b} = -\mathbf{b} \times \mathbf{a}.$$

Likewise, the product of two square matrices generally depends on the order in which the matrices are multiplied. For instance if

$$A = \begin{pmatrix} 1 & 3 \\ 2 & 4 \end{pmatrix} \text{ and } B = \begin{pmatrix} 3 & 5 \\ 1 & 6 \end{pmatrix}$$

then

$$AB \neq BA.$$

Noncommutative operations are in contrast to commutative ones such as the addition of vectors or multiplication of real numbers. Both commutative and noncommutative operations arise outside the realm of pure mathematics. The following subsections review some common examples.

1.1.1 Assembly Planning

A common problem experienced by engineers is how to design a functional device so that it can be assembled either manually or by an automated factory. In order to make the assembly process as efficient as possible, it is necessary to know which parts need to be assembled in specific orders and for which assembly operations the order does not matter. Once this information is known, the best schemes for assembling a given device can be devised. The determination of such schemes is called *assembly planning* [5, 6].

An assembly plan for relatively simple devices can be expressed in an "exploded" schematic such as in Fig. 1.1.

1

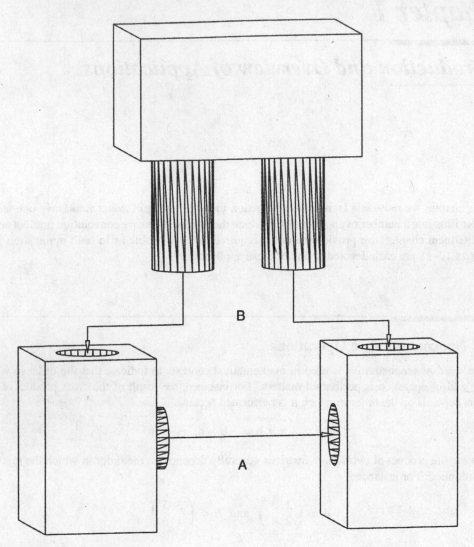

FIGURE 1.1
An assembly schematic.

Clearly the assembly operations A and B can be performed only in one order.

The importance of the order in which assembly operations are performed is common to our everyday experience outside of the technical world. For instance, when getting dressed in the morning, the order in which socks, shirt, and pants are put on does not matter (these operations, which we will denote as SX, ST, and PA, commute with each other). However, the operation of putting on shoes (which we will denote as SH) never commutes with SX, and rarely with PA (depending on the kind of pants). If we use the symbol ∘ to mean "followed by" then, for instance,

$$SX \circ ST = ST \circ SX$$

is a statement that the operations commute, and hence result in the same state of being dressed. Using this notation, $SX \circ SH$ works but $SH \circ SX$ does not.

Any number of operations that we perform on a daily basis can be categorized according to whether they are commutative or not. In the following subsections we review some examples with which everyone is familiar. These provide an intuitive introduction to the mathematics of noncommutative operations.

1.1.2 Rubik's Cube

An easy way to demonstrate noncommutativity is to play with a Rubik's cube. This toy is a cube with each of its six faces painted a different color. Each face consists of nine squares. There are three kinds of squares: corner, edge, and center. The corner squares are each rigidly attached to the two other corner squares of adjacent faces. The edge squares are each attached to one edge square of an adjacent face. The center squares are fixed to the center of the cube, and can rotate about their face. A whole face can be rotated about its center by increments of 90° resulting in a whole row in each of four faces changing color. Multiple rotations of this kind about each of the faces result in a scrambling of the squares.

In order to illustrate the noncommutativity of these operations, we introduce the following notation [8]. Let U, D, R, L, F, B each denote up, down, right, left, front, and back of the cube, respectively. An operation such as U^i for $i = 1, 2, 3, 4$ means rotation of the face U by an angle of $i \times \pi/2$ radians counterclockwise. Figure 1.2 illustrates U^1 and R^1. Since rotation by 2π brings a face back to its starting point, $U^4 = D^4 = R^4 = L^4 = F^4 = B^4 = I$, the identity (or "do nothing") operation.

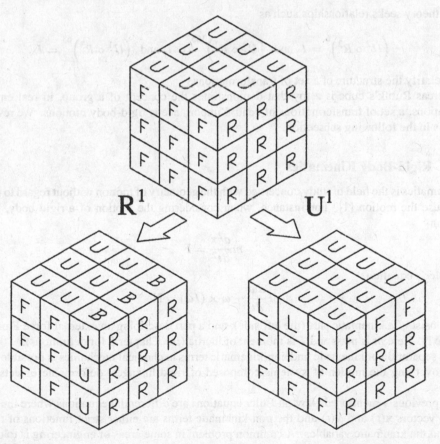

FIGURE 1.2
Rubik's cube.

The composition of two cube operations is simply one followed by another. These operations are noncommutative because, for instance,

$$U^1 \circ R^1 \neq R^1 \circ U^1.$$

This is not to say that all the cube operations are noncommutative. For instance

$$U^1 \circ U^2 = U^3 = U^2 \circ U^1.$$

However, since the whole set of cube operations contains at least some noncommutative elements, the whole set together with composition is called noncommutative.

The study of sets of transformations (or operations) that are either commutative or noncommutative and satisfy some additional properties is called *group theory*. This is the subject of Chapter 7. For now it suffices to say that group theory is a powerful tool in the analysis of a set of operations. It can answer questions such as how many distinct states Rubik's cube can attain (which happens to be approximately 4.3×10^{19}), and what strategies are efficient for restoring the cube to its original state without *a priori* knowledge of the moves that scrambled the cube. Using the notation $(\cdot)^n$ to mean that the contents of the parentheses is composed with itself n times, we see that

$$\left(U^1\right)^4 = U^1 \circ U^1 \circ U^1 \circ U^1 = U^4 = I.$$

Group theory seeks relationships such as

$$\left(U^2 \circ R^2\right)^6 = I \text{ and } \left(U^1 \circ R^1\right)^{105} = I \text{ and } \left(U^1 \circ R^3\right)^{63} = I$$

which clarify the structure of a set of transformations.

Whereas Rubik's cube is a toy that demonstrates the concept of a group, in real engineering applications, a set of transformations of central importance is rigid-body motions. We review such motions in the following subsection.

1.1.3 Rigid-Body Kinematics

Kinematics is the field of study concerned with the geometry of motion without regard to the forces that cause the motion [1]. For instance, when considering the motion of a rigid body, Newton's equation

$$m\frac{d^2\mathbf{x}}{dt^2} = \mathbf{F}$$

and Euler's equation

$$I\frac{d\omega}{dt} + \omega \times (I\omega) = \mathbf{N}$$

both consist of a kinematic part (the left side), and a part describing an external force $\mathbf{F}$ or moment of force $\mathbf{N}$. Here m is mass and I is moment of inertia (see Chapter 5 for definitions).

This separation of kinematic and non-kinematic terms in classical mechanics is possible for equations governing the motion of systems composed of rigid linkages, deformable objects, or fluid systems.

The previously mentioned Newton-Euler equations are differential equations where the variables are the vectors $\mathbf{x}(t)$ and $\omega(t)$ and the non-kinematic terms are either strict functions of time or of time and the kinematic variables. A common problem in some areas of engineering is to determine the value of the kinematic variables as a function of time. The inverse problem of determining forces and moments from observed motions is also of importance.

In the Newton-Euler equations, the variables are vector quantities (and hence are commutative when using the operation of vector addition). The only hint of the noncommutativity of rigid-body motion is given by the cross product in Euler's equations of motion.

In order to illustrate the noncommutative nature of rigid-body motion (without resorting to detailed equations), observe Fig. 1.3. This illustrates the composition of two rigid-body motions (or transformations) in two different orders. In Fig. 1.3a the frame of reference is translated and rotated from its initial pose to a new one by the motion g_1. Then g_2 is the relative motion of the frame from its intermediate pose to the final one. This composition of motions is denoted as $g_1 \circ g_2$. In contrast, in Fig. 1.3b g_2 is the motion from the initial pose to the intermediate one, and g_1 is the motion from intermediate to final. This results in the composite motion $g_2 \circ g_1$. The motions g_1 and g_2 in both figures are the same in the sense that if one were to cut out Fig. 1.3b it would be possible to align Fig. 1.3b so that the top two frames would coincide with the bottom two in Fig. 1.3a and vice versa. Needless to say, we do not recommend this experimental verification! The importance of Fig. 1.3 is that it illustrates

$$g_1 \circ g_2 \neq g_2 \circ g_1.$$

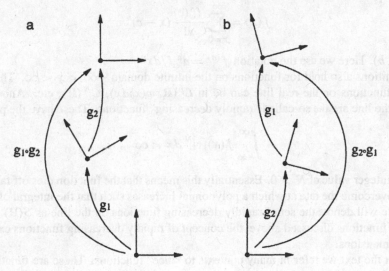

FIGURE 1.3
The frames $g_1 \circ g_2$ and $g_2 \circ g_1$.

1.2 Harmonic Analysis: Commutative and Noncommutative

Classical (pre-20th century) harmonic analysis is concerned with the expansion of functions into fundamental (usually orthogonal) components. Of particular interest in classical harmonic analysis is the expansion of functions on intervals of the form $a \leq x \leq b$ where x, a, and b are real numbers. Since the addition of real numbers is a commutative operation ($x + y = y + x$ for $x, y \in \mathbb{R}$), we will refer to the terms "classical" and "commutative" interchangeably when referring to harmonic analysis.

Essentially, the goal of commutative harmonic analysis is to expand a function $f(x)$ (that either assigns real or complex values to each value of x) in a series

$$f(x) \approx \sum_{i=1}^{N} a_i \phi_i(x). \tag{1.1}$$

The choice of the set of functions $\{\phi_i(x)\}$ as well as the meaning of approximate equality ($\approx$) depends on "what kind" of function $f(x)$ is.

For instance, if

$$\int_a^b |f(x)|^p w(x)\, dx < \infty$$

for some $w(x) > 0$ for all $a < x < b$, we say $f \in \mathcal{L}^p([a,b], w(x)dx)$. In the special case when $p = 1$, $f(x)$ is called "absolutely integrable," and when $p = 2$, $f(x)$ is called "square integrable." But these are not the only distinctions that can be made. For instance, a function that is continuous on the interval $[a, b]$ is called $C^0([a, b])$, and one for which the first N derivatives exist for all $x \in [a, b]$ is called $C^N([a, b])$. A function $f \in C^\infty([a, b])$ is called *analytic* for $x \in [a, b]$ if it has a convergent Taylor series

$$f(x) = \sum_{n=0}^{\infty} \frac{f^n(c)}{n!}(x - c)^n$$

for all $c \in (a, b)$. Here we use the notation $f^{(n)} = d^n f/dx^n$.

These definitions also hold for functions on the infinite domain $-\infty < x < \infty$. This is the real line, $\mathbb{R}$, and functions on the real line can be in $\mathcal{L}^p(\mathbb{R}, w(x)dx)$, $C^N(\mathbb{R})$, etc. Another class of functions on the line are the so-called "rapidly decreasing" functions. These have the property

$$\int_{-\infty}^{\infty} f(x)|x|^N\, dx < \infty$$

for any finite integer value of $N > 0$. Essentially this means that the function dies off fast enough as $|x| \to \infty$ to overcome the rate at which a polynomial increases such that the integral of the product converges. We will denote the set of rapidly decreasing functions on the line as $\mathcal{S}(\mathbb{R})$. As with all the classes of functions discussed above, the concept of rapidly decreasing functions extends easily to multiple dimensions.

Throughout the text we refer in many contexts to "nice" functions. These are functions that are analytic (infinitely differentiable at all points, and have convergent Taylor series), absolutely and square integrable, and rapidly decreasing. For functions defined on certain domains it is redundant to specify all of these conditions. For instance, a bounded function with compact support $[a, b] \subset \mathbb{R}$ will necessarily be rapidly decreasing and be contained in $\mathcal{L}^p([a, b], dx)$ for all finite positive integer values of p.

A typical set of nice functions on the real line is

$$f_m(x) = x^m e^{-x^2}$$

for $m = 0, 1, \ldots$

Whereas in pure mathematics the main goal is to find the broadest possible mathematical truths (and hence "weird" functions serve as the cornerstones of powerful theorems), in engineering and the macroscopic physical world the vast majority of functions vary smoothly and decay as desired for large values of their argument. For example, to an engineer working on a particular application, there may be no difference between the unit step function shown in Fig. 1.4 with solid lines [which is not continuous, rapidly decreasing, or even in $\mathcal{L}^p(\mathbb{R}, dx)$], and its approximation in Fig. 1.4 with dashed lines (which is a nice function).

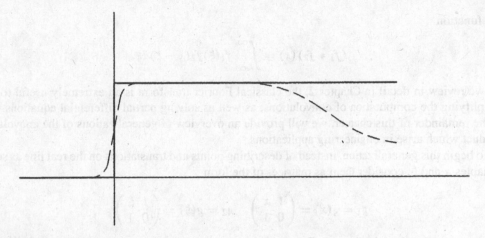

FIGURE 1.4
A step function and its nice approximation.

Hence, while many of the mathematical statements we will make for nice functions also hold for functions that are not so nice, we will not be concerned with the broadest possible mathematical statements. Rather we will focus on 20th century generalizations of classical harmonic analysis in the context of engineering applications with a noncommutative component.

Having said this, harmonic analysis is a useful tool because given a nice function, we can approximate it using a series of the form of Eq. (1.1). For a nice function $f(x)$ and a set of functions $\{\phi_i(x)\}$ satisfying certain properties, the values $\{a_i\}$ can be found as the solution to the minimization problem

$$\min_{\{a_i\}} \int_a^b \left| f(x) - \sum_{i=1}^{N} a_i \phi_i(x) \right|^2.$$

Not any set of nice functions $\{\phi_i\}$ will do if we want the series approximation to accurately capture the behavior of the function. Chapters 2 and 3 provide the requirements for such a set of functions to be acceptable. But before getting into the details of classical harmonic analysis, we review in each of the following sections (at a very coarse and nonmathematical level) the kinds of problems to which noncommutative harmonic analysis is applied in the later chapters of the book.

1.3 Commutative and Noncommutative Convolution

The concept of convolution of functions arises in an amazingly varied set of engineering scenarios. Convolution of functions which take their argument from the real line is an essential concept in signal/image processing and probability theory, just to name a few areas.

The classical convolution integral can be visualized geometrically as follows. One function, $f_2(x)$, is translated past another, $f_1(x)$. A translated version of $f_2(x)$ by amount ξ in the positive x direction takes the form $f_2(x - \xi)$. Then, the convolution is formed as the weighted sum of each of these translated copies, where the weight is $f_1(\xi)$ [value of $f_1(x)$ at the amount of shift $x = \xi$] and the sum, which is over all translations on the real line, becomes an integral. The result of this procedure is a function

$$(f_1 * f_2)(x) = \int_{-\infty}^{\infty} f_1(\xi) f_2(x - \xi) d\xi. \tag{1.2}$$

As we review in detail in Chapter 2, the classical Fourier transform is an extremely useful tool in simplifying the computation of convolutions, as well as solving partial differential equations. But in the remainder of this chapter, we will provide an overview of generalizations of the convolution product which arise in engineering applications.

To begin this generalization, instead of describing points and translations on the real line as scalar variables x and ξ, consider them as matrices of the form

$$g_1 = g(x) = \begin{pmatrix} 1 & x \\ 0 & 1 \end{pmatrix} \quad g_2 = g(\xi) = \begin{pmatrix} 1 & \xi \\ 0 & 1 \end{pmatrix}.$$

Then it is easy to verify that $g(x) \circ g(\xi) = g(x + \xi)$ and $[g(x)]^{-1} = g(-x)$ where g^{-1} is the inverse of the matrix g and $\circ$ stands for matrix multiplication[1].

Then the convolution integral is generalized to the form

$$(f_1 * f_2)(g_1) = \int_G f_1(g_2) f_2 \left(g_2^{-1} \circ g_1 \right) d\mu(g_2) \tag{1.3}$$

where G is the set of all matrices of the form $g(x)$ for all real numbers x, and $d\mu(g(\xi)) = d\xi$ is an appropriate integration measure for the set G. Henceforth, since we have fixed which measure is being used, the shorthand dg will be used in place of the rather cumbersome $d\mu(g)$.

A natural question to ask at this point is why we have written $f_2(g_2^{-1} \circ g_1)$ instead of $f_2(g_1 \circ g_2^{-1})$. The answer is that in this particular example both are acceptable since $g(x) \circ g(y) = g(y) \circ g(x)$, but in the generalizations to follow, this equality does not hold, and the definition in Eq. (1.3) remains a useful definition in applications.

Denoting the real numbers as $\mathbb{R}$, it is clear that G in this example is in some sense equivalent to $\mathbb{R}$, and we write $G \cong R$. The precise meaning of this equivalence will be given in Chapter 7.

We now examine a generalization of this concept which will be extremely useful later in the book. Consider, instead of points on the line and translations long the line, points in the plane and rigid motions in the plane. Any point in the plane can be described with matrices of the form

$$g_1 = \begin{pmatrix} 1 & 0 & x_1 \\ 0 & 1 & x_2 \\ 0 & 0 & 1 \end{pmatrix},$$

and any rigid-body motion in the plane (translation and rotation) can be described with matrices of the form

$$g_2 = g(a_1, a_2, \alpha) = \begin{pmatrix} \cos\alpha & -\sin\alpha & a_1 \\ \sin\alpha & \cos\alpha & a_2 \\ 0 & 0 & 1 \end{pmatrix}$$

with

$$g_2^{-1} = \begin{pmatrix} \cos\alpha & \sin\alpha & -a_1\cos\alpha - a_2\sin\alpha \\ -\sin\alpha & \cos\alpha & a_1\sin\alpha - a_2\cos\alpha \\ 0 & 0 & 1 \end{pmatrix}.$$

---

[1] It is standard to simply omit the $\circ$, in which case juxtaposition of two matrices stands for multiplication, but we will keep the $\circ$ for clarity.

By defining $dg_2 = da_1 da_2 d\alpha$, Eq. (1.3) now takes on a new meaning. In this context $f_1(g(a_1, a_2, \alpha)) = \tilde{f}_1(a_1, a_2, \alpha)$ is a function that has motions as its argument, and $f_2(g(x_1, x_2, 0)) = \tilde{f}_2(x_1, x_2)$ is a function of position. Both functions return real numbers. Explicitly we write $(f_1 * f_2)(g(x_1, x_2, 0))$ as

$$\left(\tilde{f}_1 * \tilde{f}_2\right)(x_1, x_2) = \int_{a_1=-\infty}^{\infty} \int_{a_2=-\infty}^{\infty} \int_{\alpha=-\pi}^{\pi} \tilde{f}_1(a_1, a_2, \alpha) \cdot$$
$$\tilde{f}_2((x_1 - a_1)\cos\alpha + (x_2 - a_2)\sin\alpha, -(x_1 - a_1)\sin\alpha$$
$$+ (x_2 - a_2)\cos\alpha)\, da_1 da_2 d\alpha,$$

which is interpreted as the function f_2 being swept by f_1. We can think of this as copies of f_2 which are translated and rotated and then deposited on the plane with an intensity governed by f_1. With a slight modification, so that $g_1 = g(x_1, x_2, \theta)$ and $f_2(g(x_1, x_2, \theta)) = \tilde{f}_2(x_1, x_2, \theta)$, the result is precisely a convolution of functions on the group of rigid-body motions of the plane. It is easy to check that in this case $g_1 \circ g_2 \neq g_2 \circ g_1$ in general and so $f_1 * f_a \neq f_2 * f_1$.

The middle third of this book is devoted to developing tools for the fast numerical computation of such generalized convolutions and the inversion of integral equations of the form $(f_1 * f_2)(g) = h(g)$ where $h(g)$ and $f_1(g)$ are known and $f_2(g)$ is sought.

The remaining sections of this chapter are each devoted to an overview of applications examined in detail in the last third of the book.

1.4 Robot Workspaces, Configuration-space Obstacles, and Localization

A robotic manipulator is a particular kind of device used to pick up objects from one place and put them in another. Often the manipulator resembles a mechanical arm and has a gripper or hand-like mechanism at the distal end (called the end effector). The proximal end of the manipulator is usually affixed to a torso, gantry, or moving platform. The *workspace* of a manipulator is the set of all positions and orientations that a frame of reference attached to the distal end can attain relative to a frame fixed to the proximal end. Figure 1.5 shows an arm that is highly articulated (has many joints). For such an arm we can ask how the workspace for the whole arm can be found if we know the workspaces of two halves. Superimposed on the arm in this figure are the positional boundaries of the workspaces of the proximal and distal halves (small ellipses), together with the workspace for the whole arm (large ellipse). One can imagine that a sweeping procedure involving the two halves generates the whole [3]. In Chapter 12 we show that this sweeping is, in fact, a generalized convolution, which can be performed efficiently using the computational and analytical techniques discussed in Chapters 2–11.

Manipulator arms are one of several classes of robotic devices. Another important class consists of mobile robots which can be viewed as single rigid bodies that can move in any direction. A central task in robotics is to navigate such objects through known environments. A popular technique to achieve this goal is to expand the obstacles by the girth of the robot and shrink the robot to a point [10]. Then the motion planning problem becomes one of finding the paths that a point robot is able to traverse in an environment of enlarged obstacles. The procedure of generating the enlarged obstacles is, again, a generalized convolution [9], as shown in Fig. 1.6.

As a practical matter, in motion planning the mobile robot does not always know exactly where it is. Using sensors, such as sonars, that bounce signals off walls to determine the robot's distance from the nearest walls, an internal representation of the local environment can be constructed. With this information, the estimated shape of the local free space (places where the robot can move) is

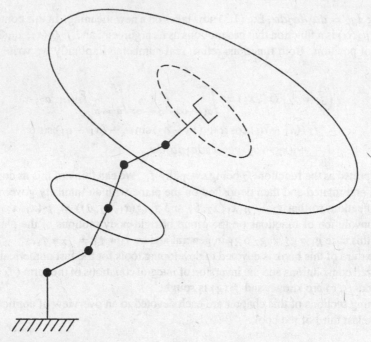

FIGURE 1.5
Workspace generation by sweeping.

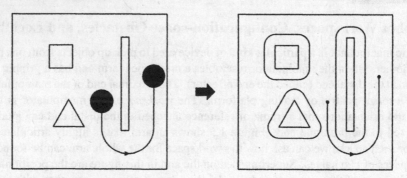

FIGURE 1.6
Configuration-space obstacles of mobile robots.

calculated. Given a global map of the environment, a generalized convolution of the local free space estimate with the global map will generate a number of likely locations of the robot.

1.5   Template Matching and Tomography

Template matching (also called pattern matching or image registration) is an important problem in many fields. For instance, in automated military surveillance, a problem of interest is to match a building or piece of military hardware with representative depictions in a database. Likewise, in medicine it is desirable to map anatomical features from a database onto images observed through medical imaging. In the case when the best rigid-body fit between the feature in the database and the observed data is desired, the generalized convolution technique is a natural tool.

Two problems in tomography can also be posed as generalized convolutions. The *forward* tomography problem is the way in which data about the three-dimensional structure of an object (typically a person) is reconstructed from multiple planar projections, as in Fig. 1.7. (See [4] for a detailed description and analysis of this problem.) The *inverse* tomography problem (also called the *tomotherapy* problem) is concerned with how much radiation a patient is given by way of X-ray (or other high energy) beams so as to maximize the damage to a tumor, while keeping exposure to surrounding tissues within acceptable limits. Both the forward and inverse tomography problems can be posed as integral equations that are of the form of generalized convolutions.

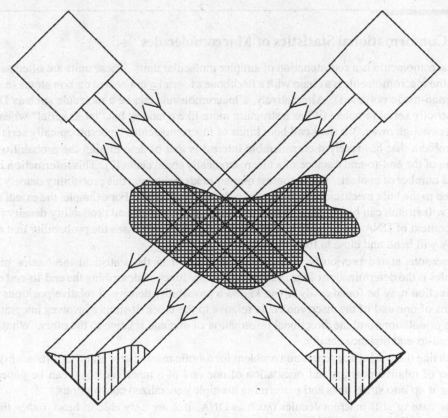

FIGURE 1.7
The forward tomography problem.

1.6 Attitude Estimation and Rotational Brownian Motion

A variety of problems in engineering, physics, and chemistry are modeled well as rigid bodies subjected to random disturbances [11]. This includes everything from satellites in space to the dynamics of liquid crystals composed of many molecules modeled as identical rigid bodies. Regardless of the particular physical situation, they have the common feature that they are described by linear partial differential equations describing processes which evolve on the group of rotations in three dimensional space.

In this context, noncommutative harmonic analysis provides a tool for converting partial differential equations into systems of ordinary differential equations in a different domain. In this other domain, the equations can be solved, and the solution can be converted back to the original domain. This is exactly analogous with the way the classical Fourier transform has been used in engineering science for more than a century. The only difference is that in the classical engineering context the equations evolve on a commutative object (i.e., the real line with the operation of addition).

1.7   Conformational Statistics of Macromolecules

A macromolecule is a concatenation of simpler molecular units. These units are often assembled so that the macromolecule is a chain with a backbone of serially connected carbon atoms (e.g., many of the man-made polymers). Alternatively, a macromolecule can be a molecule such as DNA that is not strictly serial in nature (it has a structure more like a ladder) but "looks serial" when viewed from far enough away. We shall call both kinds of macromolecules "macroscopically serial."

A problem that has received considerable interest is that of determining the probability density function of the end-to-end distance of a macroscopically serial chain [7]. This information is important in a number of contexts. In the context of man-made polymers, this probability density function is related to the bulk mechanical properties of polymeric materials. For example, the extent to which rubber will stretch can be explained with knowledge of the end-to-end probability density function. In the context of DNA, a vast body of literature exists which addresses the probability that a section of DNA will bend and close to form a ring.

For reasons stated previously, a fundamental problem in the statistical mechanics of macromolecules is the determination of the probability density function describing the end-to-end distance. This function may be found easily if one knows a probability density of relative positions and orientations of one end of the macromolecule relative to the other. It simply involves integrating over relative orientations and the directional information of one end relative to the other. What remains is the end-to-end distance density.

Much like the workspace generation problem for robotic manipulator arms, the probability density function of relative position and orientation of one end of a macromolecule can be generated by dividing it up into small units and performing multiple generalized convolutions.

In the case of stiff macromolecules (such as DNA) that are only able to bend, rather than kink, it is also possible to write diffusion-type partial differential equations which evolve on the group of rigid-body motions. Noncommutative harmonic analysis serves as a tool for converting such equations into a domain where they can be solved efficiently and transformed back.

1.8   Mechanics and Texture Analysis of Materials

Mechanics is the study of the behavior of solids and fluids under static and dynamic conditions. A useful approximation in mechanics is the *continuum* model, in which a solid or fluid is approximated as a continuous domain (rather than modeling the motion of each atom). Our discussion of applications of noncommutative harmonic analysis in mechanics will be restricted to solids. Applications in fluid mechanics are in some ways similar to rotational Brownian motion and diffusion.

The continuum model has its limitations. For instance, a real solid material may be composed of many tiny crystals, it may have many small cracks, or it may be a composite of two very different kinds of materials. A goal in the mechanics of solids is to find an equivalent continuum model that reflects the small scale structure without explicitly modeling the motion of every small component. This goal is depicted in Fig. 1.8.

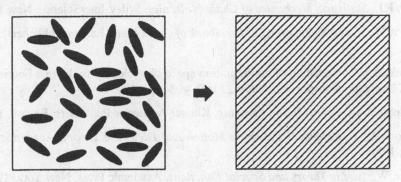

FIGURE 1.8
Averaging strength properties and continuum approximation.

To this end, if the orientation distribution of constituents (e.g., crystals) is known throughout the material, a goal is to somehow average the strength properties of each crystal over all orientations in order to determine the strength of the bulk material. The fields in which such problems have been studied are called *mathematical texture analysis* and *orientational averaging* in the mechanics of solids and materials [2, 13].

In Chapters 2–11, we provide the mathematical background required to pose all of the applications discussed in this introductory chapter in the language of noncommutative harmonic analysis (see for example [12]). In some instances harmonic analysis will serve the role of a language in which to describe the problems systematically. In other contexts, it will be clear that harmonic analysis is the only tool available to solve problems efficiently. The various applications are addressed in Chapters 12–18, together with numerous pointers to the literature.

References

[1] Bottema, O. and Roth, B., *Theoretical Kinematics*, Dover Publications, New York, reprint, 1990.

[2] Bunge, H.-J., *Mathematische Methoden der Texturanalyse*, (*Texture Analysis in Materials Science*), Akademie-Verlag, Berlin, 1969, translated by Morris, P.R., Butterworth Publishers, Stoneham, MA, 1982.

[3] Chirikjian, G.S. and Ebert-Uphoff, I., Numerical convolution on the Euclidean group with applications to workspace generation, *IEEE Trans. on Robotics and Autom.*, 14(1), 123–136, 1998.

[4] Deans, S.R., *The Radon Transform and Some of its Applications,* John Wiley & Sons, New York, 1983.

[5] DeFazio, T.L. and Whitney, D.E., Simplified generation of all mechanical assembly sequences, *IEEE J. Robotics and Autom.,* 3(6), 640–658, 1987.

[6] DeMello, L.S.H. and Sanderson, A.C., A correct and complete algorithm for the generation of mechanical assembly sequences, *IEEE Trans. on Robotics and Autom.,* 7(2), 228–240, 1991.

[7] Flory, P.J., *Statistical Mechanics of Chain Molecules,* Wiley-InterScience, New York, 1969.

[8] Frey, A.H., Jr. and Singmaster, D., *Handbook of Cubik Math,* Enslow Publishers, Hillside, NJ, 1982.

[9] Kavraki, L., Computation of configuration space obstacles using the fast Fourier transform, *IEEE Trans. on Robotics and Autom.,* 11(3), 408–413, 1995.

[10] Latombe, J.-C., *Robot Motion Planning,* Kluwer Academic Publishers, Boston, 1991.

[11] McConnell, J., *Rotational Brownian Motion and Dielectric Theory,* Academic Press, New York, 1980.

[12] Miller, W., Jr., *Lie Theory and Special Functions,* Academic Press, New York, 1968.

[13] Wenk, H.-R., Ed., *Preferred Orientation in Deformed Metals and Rocks: An Introduction to Modern Texture Analysis,* Academic Press, New York, 1985.

Chapter 2

Classical Fourier Analysis

In this chapter we review classical Fourier analysis, which is a tool for generating continuous approximations of a wide variety of functions on the circle and the line. We also illustrate how the Fourier transform can be calculated in an analog fashion using properties of optical systems, and how the discrete Fourier transform (DFT) lends itself to efficient numerical implementation by way of fast Fourier transform (FFT) techniques. The material in this chapter is known in varying levels of detail by most engineers. Good understanding of this material is critical for the developments in Chapters 8–11, which form the foundation on which the applications in Chapters 12–18 are built.

2.1 Fourier Series: Decomposition of Functions on the Circle

A Fourier series is a weighted sum of trigonometric basis functions used to approximate *periodic* functions.[1]

A periodic function of real-valued argument and period $L > 0$ is one that satisfies the constraint

$$f(x) = f(x + L). \tag{2.1}$$

We use the notation $\mathbb{R}$ for the real numbers, and say $x, L \in \mathbb{R}$ to denote the fact that x and L are real. The set of positive real numbers is denoted $\mathbb{R}^+$. When $L \in \mathbb{R}^+$ is the smallest positive number for which Eq. (2.1) holds for all $x \in \mathbb{R}$, then L is called the *fundamental period* of $f(x)$, and $f(x)$ is called *L-periodic*.

By definition, this means that $f(x + L) = f((x + L) + L) = f(x + 2L)$. Similarly, by the change of variables $y = x + L$, one observes that $f(y - L) = f(y)$ for all $y \in \mathbb{R}$. Thus it follows immediately from Eq. (2.1) that $f(x) = f(x + nL)$ for any $n \in \mathbb{Z}$ (the integers). That is, a periodic function with fundamental period L also has period nL.

Instead of viewing such functions as periodic on the line, it is often more convenient to view them as functions on a circle of circumference L. The mapping from line to circle can be viewed as "cutting" an L−length segment from the line and "pasting" the ends together. The mapping from circle to line can be viewed as the circle rolling on the line. As explained in Chapter 7, the group theoretical way to say this is that the 1-torus, T, is congruent to a quotient: $T \cong \mathbb{R}/L\mathbb{Z}$. Roughly speaking, this says that the circle with circular addition behaves like a representative piece of the line cut out with the ends glued.

---

[1]Jean Baptiste Joseph Fourier (1768–1830) initiated the use of such series in the study of heat conduction. The work was originally rejected by the Academy of Sciences in Paris when presented in 1807, and was finally published in 1822 [10].

In engineering it is common to encounter L-periodic functions which satisfy the inequality

$$\|f\|_2^2 \overset{\Delta}{=} \int_0^L |f(x)|^2 \, dx < \infty, \tag{2.2}$$

where for a number $c \in \mathbb{C}$ (the complex numbers), the squared modulus is $|c|^2 = c\bar{c}$, where $\bar{c}$ is the complex conjugate of c. Thus if $c = a + ib$ where $i = \sqrt{-1}$ and $a, b \in \mathbb{R}$, then $\bar{c} = a - ib$ and $|c|^2 = a^2 + b^2$. The set of all square integrable complex-valued functions with respect to integration measure dx and with argument in the interval $[0, L]$ is denoted as $\mathcal{L}^2([0, L], \mathbb{C}, dx)$ or simply $\mathcal{L}^2([0, L])$. Hence we should say that if Eq. (2.2) holds, then $f \in \mathcal{L}^2([0, L])$. It is also common to encounter functions whose modulus is absolutely integrable

$$\|f\|_1 \overset{\Delta}{=} \int_0^L |f(x)| \, dx < \infty.$$

Such functions are said to be in $\mathcal{L}^1([0, L])$. When L is finite, it may be shown that a function $f \in \mathcal{L}^2([0, L])$ is automatically in $\mathcal{L}^1([0, L])$ as well. This follows from the Cauchy-Schwarz inequality given in Appendix C. We will not distinguish between functions on $[0, L]$, functions on the circle of circumference L, and L-periodic functions on the line. In the following discussion, it is also assumed that the functions under consideration have, at most, a finite number of jump discontinuities.

Observe that for L-periodic functions, the following is true for any $c \in \mathbb{R}$

$$\int_0^L f(x) \, dx = \int_c^{c+L} f(x) \, dx.$$

That is, integration is independent of shifting the limits. Likewise, if we shift or reflect a periodic function, the value of its integral over a period is unchanged

$$\int_0^L f(x) \, dx = \int_0^L f(x + c) \, dx = \int_0^L f(-x) \, dx.$$

The *Fourier series* expansion of a complex-valued function with fundamental period L is defined as:

$$f(x) = \sum_{k=-\infty}^{+\infty} \hat{f}(k) e^{2\pi i k x / L}. \tag{2.3}$$

$\hat{f}(k)$ is called the kth *Fourier coefficient*. For the moment we will consider only functions of the form in Eq. (2.3). We shall see later in this section the conditions under which this expansion holds.

A function $f(x) \in \mathcal{L}^2([0, L])$ for which $\hat{f}(k) = 0$ for all $|k| > B$ for some fixed natural number (nonnegative integer) B is called a *band-limited* function. If $f(x)$ is not band-limited, it can be approximated with the *partial sum*

$$f(x) \approx f_B(x) \overset{\Delta}{=} \sum_{k=-B}^{B} \hat{f}(k) e^{2\pi i k x / L}.$$

Recall that the exponential functions above have the properties $d/d\theta(e^{i\theta}) = i e^{i\theta}$, $e^{i\theta} = \cos\theta + i\sin\theta$, $|e^{i\theta}| = 1$, and $e^{i\theta_1} e^{i\theta_2} = e^{i(\theta_1 + \theta_2)}$ for any $\theta, \theta_1, \theta_2 \in \mathbb{R}$. Furthermore, for integers k and n,

$$\int_0^L e^{2\pi i k x / L} \cdot e^{-2\pi i n x / L} \, dx = \int_0^L e^{2\pi i (k-n) x / L} \, dx = L \delta_{k,n}$$

where

$$\delta_{k,n} = \begin{cases} 1 & \text{for } k = n \\ 0 & \text{for } k \neq n \end{cases}$$

is the Kronecker delta function.

This means that the Fourier coefficients, $\hat{f}(n)$, can be calculated by multiplying both sides of Eq. (2.3) by $e^{-2\pi inx/L}$ and integrating x over the interval $[0, L]$. The result is

$$\mathcal{F}_k(f(x)) \triangleq \hat{f}(n) \triangleq \frac{1}{L} \int_0^L f(x) e^{-2\pi inx/L} \, dx. \tag{2.4}$$

Written as one expression, the original function is then

$$f(x) = \sum_{k=-\infty}^{+\infty} \left(\frac{1}{L} \int_0^L f(y) e^{-2\pi iky/L} \, dy \right) e^{2\pi ikx/L}. \tag{2.5}$$

It becomes obvious when written this way that there is freedom to insert an arbitrary constant C as follows

$$f(x) = \frac{1}{C} \sum_{k=-\infty}^{+\infty} \left(\frac{C}{L} \int_0^L f(y) e^{-2\pi iky/L} \, dy \right) e^{2\pi ikx/L}.$$

In other words, we could have defined the Fourier series as

$$f(x) = \frac{1}{C} \sum_{k=-\infty}^{+\infty} \tilde{f}(k) e^{2\pi ikx/L}$$

from the beginning [with $\tilde{f}(k) = C \hat{f}(k)$], and calculated the corresponding coefficients

$$\tilde{f}(n) = \frac{C}{L} \int_0^L f(x) e^{-2\pi inx/L} \, dx.$$

Some authors prefer the choice of $C = L$ to make the Fourier coefficients look neater. Others prefer to choose $C = \sqrt{L}$ to make the Fourier series and coefficients look more symmetrical. For simplicity, we will keep $C = 1$.

Fourier series are useful in engineering because they possess certain interesting properties. For instance, operations such as addition and scaling of functions result in addition and scaling of Fourier coefficients. If $f(x)$ is an even/odd function, $f(x) = \pm f(-x)$, then the Fourier coefficients will have the symmetry/antisymmetry $\hat{f}(n) = \pm \hat{f}(-n)$ (which reflects the fact that even functions can be expanded in a cosine series, while odd ones can be expanded in sine series). Similarly, if we define the function $f^*(x) = \overline{f(-x)}$, one observes that the Fourier coefficients of $f^*(x)$ are $\hat{f}^*(k) = \overline{\hat{f}(k)}$. In the case when $f(x)$ is real valued, $f(x) = \overline{f(x)} = f^*(-x)$, and this is reflected in the Fourier coefficients as $\hat{f}(k) = \overline{\hat{f}(-k)}$.

One of the most important properties of Fourier series (and Fourier analysis in general) is the *convolution property*. The (periodic) convolution of two $L-$periodic functions is defined as

$$(f_1 * f_2)(x) = \frac{1}{L} \int_0^L f_1(\xi) f_2(x - \xi) d\xi.$$

Expanding both functions in their Fourier series, one observes that

$$(f_1 * f_2)(x) = \frac{1}{L} \int_0^L \left(\sum_{k=-\infty}^{+\infty} \hat{f}_1(k) e^{2\pi ik\xi/L} \right) \left(\sum_{n=-\infty}^{+\infty} \hat{f}_2(n) e^{2\pi in(x-\xi)/L} \right) d\xi$$

$$= \frac{1}{L} \sum_{k,n=-\infty}^{+\infty} \hat{f}_1(k) \hat{f}_2(n) e^{2\pi inx/L} \int_0^L e^{2\pi i(k-n)\xi/L} d\xi. \tag{2.6}$$

The integral in the above expression has value $L\delta_{k,n}$, and so the following simplification results:

$$(f_1 * f_2)(x) = \sum_{k=-\infty}^{+\infty} \hat{f}_1(k)\hat{f}_2(k)e^{2\pi ikx/L}. \tag{2.7}$$

In other words, the kth Fourier coefficient of the convolution of two functions is simply the product of the kth Fourier coefficient of each of the two functions. Note that because the order of multiplication of two complex numbers does not matter, it follows from switching the order of $\hat{f}_1$ and $\hat{f}_2$ in Eq. (2.7) that $(f_1 * f_2)(x) = (f_2 * f_1)(x)$.

One can define the *Dirac delta function* (see Appendix E) as the function for which

$$(\delta * f)(x) = f(x).$$

In other words, all of the Fourier coefficients in the expansion of $\delta(x)$ must be $\hat{\delta}(k) = 1$. The obvious problem with this is that the Fourier coefficients of $\delta(x)$ do not diminish as k increases, and so this Fourier series does not converge. This reflects the fact that $\delta(x)$ is not a "nice" function [i.e., it is not in $\mathcal{L}^2([0, L])$]. It is, nevertheless, convenient to view $\delta(x)$ as the band-limited function

$$\delta_B(x) = \sum_{k=-B}^{B} e^{2\pi ikx/L}$$

as $B \to \infty$.

With the aid of $\delta(x)$, we can show that if $f(x)$ is L-periodic and the sequence $\{\hat{f}(k)\}$ for $k = 0, 1, \ldots$ converges, then $f(x)$ can be reconstructed from its Fourier coefficients. All that is required is to rearrange terms in Eq. (2.5), and to substitute the series representation of the Dirac delta function

$$f(x) = \frac{1}{L} \int_0^L f(y) \sum_{k=-\infty}^{+\infty} e^{2\pi ik(x-y)/L} \, dy = (f * \delta)(x). \tag{2.8}$$

Whereas the limits in the summation are to infinity, the change in order of summation and integration that was performed is really only valid if $\delta_B(x)$ is used in place of $\delta(x)$. Since B can be chosen to be an arbitrarily large finite number, the equality above can be made to hold to within any desired precision.

If we define $f_1(x) = f(x)$ and $f_2(x) = f^*(x) = \overline{f(-x)}$ where $f(x)$ is a "nice" function, then evaluating the convolution at $x = 0$ yields

$$(f_1 * f_2)(0) = \frac{1}{L} \int_0^L |f(\xi)|^2 d\xi.$$

The corresponding kth Fourier coefficient of this convolution is $\hat{f}(k)\overline{\hat{f}(k)} = |\hat{f}(k)|^2$. Evaluating the Fourier series of the convolution at $x = 0$ and equating, we get

$$\frac{1}{L} \int_0^L |f(\xi)|^2 d\xi = \sum_{k=-\infty}^{+\infty} \left|\hat{f}(k)\right|^2.$$

This is *Parseval's equality*. It is important because it equates the "power" of a signal (function) in the spatial and Fourier domains.

Evaluating the convolution of $f_1(x)$ and $f_2^*(x)$ at $x = 0$, one observes the generalized Parseval relationship

$$\frac{1}{L} \int_0^L f_1(\xi)\overline{f_2(\xi)}d\xi = \sum_{k=-\infty}^{+\infty} \hat{f}_1(k)\overline{\hat{f}_2(k)}.$$

Finally, we note that certain "operational" properties associated with the Fourier series exist. For instance, the Fourier coefficients of the function $f(x - x_0)$ can be calculated from those of $f(x)$ by making a change of variables $y = x - x_0$ and using the fact that integration over a period is invariant under shifts of the limits of integration

$$\begin{aligned} \mathcal{F}_k(f(x - x_0)) &= \frac{1}{L} \int_0^L f(x - x_0) e^{-2\pi i k x/L} \, dx \\ &= \frac{1}{L} \int_0^L f(y) e^{-2\pi i k(y+x_0)/L} \, dy \\ &= e^{-2\pi i k x_0/L} \hat{f}(k). \end{aligned}$$

Similarly, one can use integration by parts to calculate the Fourier coefficients of df/dx, assuming f is at least a once differentiable function. Or, simply taking the derivative of the Fourier series,

$$\frac{df}{dx} = \sum_{k=-\infty}^{+\infty} \hat{f}(k) \frac{d}{dx} \left(e^{2\pi i k x/L} \right)$$

one finds that

$$\frac{df}{dx} = \sum_{k=-\infty}^{+\infty} \left(\hat{f}(k) \cdot 2\pi i k/L \right) e^{2\pi i k x/L},$$

and so the Fourier coefficients of the derivative of $f(x)$ are

$$\mathcal{F}_k \left(\frac{df}{dx} \right) = \hat{f}(k) \cdot 2\pi i k/L.$$

If $f(x)$ can be differentiated multiple times, the corresponding Fourier coefficients are multiplied by one copy of $2\pi i k/L$ for each derivative. Note that since each differentiation results in an additional power of k in the Fourier coefficients, the tendency is for these Fourier coefficients to not converge as quickly as those for the original function, and they may not even converge at all.

2.2 The Fourier Transform: Decomposition of Functions on the Line

A function on the line, $f(x)$, is said to be in $\mathcal{L}^p(\mathbb{R})$ if

$$(\|f\|_p)^p \triangleq \int_{-\infty}^{\infty} |f(x)|^p \, dx < \infty.$$

Analogous with the way a periodic function (function on the circle) is expressed as a Fourier series, one can express an arbitrary "nice" function on the line [i.e., one in $\mathcal{L}^2(\mathbb{R}) \cap \mathcal{L}^1(\mathbb{R})$ which, in addition, has a finite number of finite discontinuities and oscillations in any finite interval of $\mathbb{R}$] as

$$f(x) = \mathcal{F}^{-1} \left(\hat{f} \right) = \frac{1}{2\pi} \int_{-\infty}^{\infty} \hat{f}(\omega) e^{i\omega x} \, d\omega \tag{2.9}$$

where the Fourier coefficients $\hat{f}(\omega)$ (now called the *Fourier transform*) depend on the continuous parameter $\omega \in \mathbb{R}$. The operation of reconstructing a function from its Fourier transform in Eq. (2.9) is called the *Fourier inversion* formula, or inverse Fourier transform. The Fourier transform of any nice function, $f(x)$, is calculated as

$$\hat{f}(\omega) = \mathcal{F}(f) = \int_{-\infty}^{\infty} f(x)e^{-i\omega x}\, dx. \tag{2.10}$$

The fact that a function is recovered from its Fourier transform is found by first observing that it is true for the special case of $g(x) = e^{-ax^2}$ for $a > 0$. One way to calculate

$$\hat{g}(\omega) = \int_{-\infty}^{\infty} e^{-ax^2} e^{-i\omega x}\, dx$$

is to differentiate both sides with respect to ω to yield

$$\frac{d\hat{g}(\omega)}{d\omega} = -i \int_{-\infty}^{\infty} x e^{-ax^2} e^{-i\omega x}\, dx = \frac{i}{2a} \int_{-\infty}^{\infty} \frac{dg}{dx} e^{-i\omega x}\, dx.$$

Integrating by parts, and observing that $e^{-i\omega x} g(x)$ vanishes at the limits of integration yields

$$\frac{d}{d\omega}\left(\hat{g}\right) = -\frac{\omega}{2a}\hat{g}(\omega).$$

The solution of this first-order ordinary differential equation is of the form

$$\hat{g}(\omega) = \hat{g}(0)e^{-\frac{\omega^2}{4a}},$$

$$\hat{g}(0) = \int_{-\infty}^{\infty} e^{-ax^2}\, dx = \sqrt{\frac{\pi}{a}}.$$

The last equality is well known, and can be calculated by observing that

$$\left(\int_{-\infty}^{\infty} e^{-ax^2}\, dx\right)^2 = \int_{-\infty}^{\infty}\int_{-\infty}^{\infty} e^{-a(x^2+y^2)}\, dx\, dy = \int_{0}^{2\pi}\int_{0}^{\infty} e^{-ar^2} r\, dr\, d\theta.$$

The integral in polar coordinates (r, θ) is calculated easily in closed form as π/a.

Having found the form of $\hat{g}(\omega)$, it is easy to see that $g(x)$ is reconstructed from $\hat{g}(\omega)$ using the inversion formula Eq. (2.9) (the calculation is essentially the same as for the forward Fourier transform). Likewise, the Gaussian function

$$h_\sigma(x) = \frac{1}{\sqrt{2\pi}\sigma} e^{-\frac{x^2}{2\sigma^2}}$$

has Fourier transform

$$\hat{h}_\sigma(\omega) = e^{-\frac{\sigma^2}{2}\omega^2},$$

and the reconstruction formula holds. The Gaussian function is a probability density function, i.e., it is non-negative everywhere and

$$\int_{-\infty}^{\infty} h_\sigma(x)\, dx = 1.$$

As σ becomes small, $h_\sigma(x)$ becomes like $\delta(x)$ (see Appendix E).[2] To emphasize this fact, we write $\delta_\epsilon(x) = h_\epsilon(x)$ for $0 < \epsilon << 1$. Since it is always possible to choose ϵ small enough that

$$\|f - \delta_\epsilon * f\|_2 < \nu$$

---

[2] Other "special" functions such as the Heaviside (unit) step function [17] defined as $u(x) = \int_{-\infty}^{x} \delta(y)dy$ play an important role in many fields including mechanics and systems theory.

for any given $\nu \in \mathbb{R}^+$, for engineering purposes $f(x) = (\delta_\epsilon * f)(x)$, and $\delta_\epsilon(x)$ is indistinguishable from $\delta(x)$.

As with the case of Fourier series, the Dirac δ-function is the one whose Fourier transform is defined as

$$\mathcal{F}(\delta) = 1,$$

and has the property that

$$\int_{-\infty}^{\infty} f(\xi)\delta(x - \xi)\, d\xi = f(x).$$

Since the Fourier inversion formula works for a Gaussian for all values of $\sigma > 0$, it also works for $\delta_\epsilon(x)$. We therefore write

$$\delta(x) = \frac{1}{2\pi} \int_{-\infty}^{\infty} e^{i\omega x}\, d\omega \quad \text{or} \quad \delta(x - \xi) = \frac{1}{2\pi} \int_{-\infty}^{\infty} e^{i\omega(x-\xi)}\, d\omega. \tag{2.11}$$

This formula is critical in proving the inversion formula in general, and it is often called a *completeness relation* because it proves that the functions $e^{i\omega x}$ for all $\omega \in \mathbb{R}$ are complete in $\mathcal{L}^2(\mathbb{R})$ [i.e., any function in $\mathcal{L}^2(\mathbb{R})$ can be expanded in terms of these basic functions]. Using the completeness relation, one easily sees

$$\begin{aligned}
\mathcal{F}^{-1}(\mathcal{F}(f)) &= \frac{1}{2\pi} \int_{-\infty}^{\infty} \left[\int_{-\infty}^{\infty} f(\xi)e^{-i\omega\xi}\, d\xi \right] e^{i\omega x}\, d\omega \\
&= \int_{-\infty}^{\infty} f(\xi) \left[\frac{1}{2\pi} \int_{-\infty}^{\infty} e^{i\omega(x-\xi)}\, d\omega \right] d\xi \\
&= \int_{-\infty}^{\infty} f(\xi)\delta(x - \xi)\, d\xi \\
&= f(x).
\end{aligned} \tag{2.12}$$

We note that the combination of Fourier transform and inversion formula can be written in the form

$$f(x) = \frac{C_1 C_2}{2\pi} \int_{-\infty}^{\infty} \left[\frac{1}{C_1} \int_{-\infty}^{\infty} f(\xi)e^{-iC_2\omega'\xi}\, d\xi \right] e^{iC_2\omega'x}\, d\omega' \tag{2.13}$$

where C_1 is an arbitrary constant (like C in the case of Fourier series) and C_2 is a second arbitrary constant by which we can change the frequency variable: $\omega = C_2\omega'$. We have chosen $C_1 = C_2 = 1$. Other popular choices in the literature are $(C_1, C_2) = (\sqrt{2\pi}, \pm 1)$, $(C_1, C_2) = (2\pi, \pm 1)$, and $(C_1, C_2) = (1, \pm 2\pi)$.

The convolution of "nice" functions on the line is defined by

$$(f_1 * f_2)(x) = \int_{-\infty}^{\infty} f_1(\xi)f_2(x - \xi)\, d\xi.$$

This operation produces a new well-behaved function from two old ones. For instance, we see that

$$\begin{aligned}
\|f_1 * f_2\|_1 &= \int_{-\infty}^{\infty} \left| \int_{-\infty}^{\infty} f_1(\xi)f_2(x - \xi)\, d\xi \right| dx \\
&\leq \int_{-\infty}^{\infty} \int_{-\infty}^{\infty} |f_1(\xi)| \cdot |f_2(x - \xi)|\, d\xi\, dx \\
&= \|f_1\|_1 \cdot \|f_2\|_1,
\end{aligned}$$

and when f_1 and f_2 are both differentiable, so too is $f_1 * f_2$ from the product rule.

Sometimes it is useful to consider the convolution of two shifted functions. This is defined as:

$$f_1(x - x_1) * f_2(x - x_2) \triangleq \int_{-\infty}^{\infty} f_1(\xi - x_1) f_2(x - \xi - x_2) \, d\xi = (f_1 * f_2)(x - x_1 - x_2).$$

The Fourier transform has the property that

$$\mathcal{F}((f_1 * f_2)(x)) = \hat{f}_1(\omega) \hat{f}_2(\omega).$$

This follows from the property $e^{i\omega(x_1 + x_2)} = e^{i\omega x_1} \cdot e^{i\omega x_2}$. More explicitly,

$$\mathcal{F}((f_1 * f_2)(x)) = \int_{-\infty}^{\infty} \left(\int_{-\infty}^{\infty} f_1(\xi) f_2(x - \xi) \, d\xi \right) e^{-i\omega x} \, dx.$$

By changing variables $y = x - \xi$, and observing that the limits of integration are invariant to finite shifts, this is rewritten as

$$\int_{-\infty}^{\infty} \int_{-\infty}^{\infty} f_1(\xi) f_2(y) e^{-i\omega(y+\xi)} \, dy \, d\xi = \left(\int_{-\infty}^{\infty} f_1(\xi) e^{-i\omega \xi} \, d\xi \right) \left(\int_{-\infty}^{\infty} f_2(y) e^{-i\omega y} \, dy \right)$$

$$= \hat{f}_1(\omega) \hat{f}_2(\omega).$$

Using exactly the same argument as in the previous section, we observe the Parseval equality

$$\int_{-\infty}^{\infty} |f(x)|^2 \, dx = \frac{1}{2\pi} \int_{-\infty}^{\infty} \left| \hat{f}(\omega) \right|^2 \, d\omega$$

and its generalization

$$\int_{-\infty}^{\infty} f_1(x) \overline{f_2(x)} \, dx = \frac{1}{2\pi} \int_{-\infty}^{\infty} \hat{f}_1(\omega) \overline{\hat{f}_2(\omega)} \, d\omega.$$

In complete analogy with the Fourier series, a number of symmetry and operational properties of the Fourier transform result from symmetries and operations performed on the original functions. The following is a summary of the most common operational properties:

$$f(x) = \pm f(-x) \longrightarrow \hat{f}(\omega) = \pm \hat{f}(-\omega) \tag{2.14}$$

$$f(x) = \overline{f(x)} \longrightarrow \hat{f}(\omega) = \overline{\hat{f}(-\omega)} \tag{2.15}$$

$$\frac{df}{dx} \longrightarrow i\omega \hat{f}(\omega) \tag{2.16}$$

$$f(x - x_0) \longrightarrow e^{-i\omega x_0} \hat{f}(\omega) \tag{2.17}$$

$$f(ax) \longrightarrow \frac{1}{|a|} \hat{f}(\omega/a) \tag{2.18}$$

$$(f_1 * f_2)(x) \longrightarrow \hat{f}_1(\omega) \hat{f}_2(\omega) \tag{2.19}$$

Finally, we note that the Fourier transform of functions with multiple domain dimensions follows in exactly the same way as in one dimension. For functions taking their argument in $\mathbb{R}^N$, the Fourier transform can be thought of as N one-dimensional Fourier transforms applied to the function with all but one Cartesian coordinate held fixed during each one-dimensional integration. Using vector notation

$$\hat{f}(\omega) = \int_{\mathbb{R}^N} f(\mathbf{x}) e^{-i\omega \cdot \mathbf{x}} \, d\mathbf{x}$$

where $d\mathbf{x} = dx_1 \ldots dx_N$. The inversion, viewed as N one-dimensional inversions, yields

$$f(\mathbf{x}) = \frac{1}{(2\pi)^N} \int_{\mathbb{R}^N} \hat{f}(\omega) e^{i\omega \cdot \mathbf{x}} \, d\omega$$

where $d\omega = d\omega_1 \ldots d\omega_N$.

2.3   Using the Fourier Transform to Solve PDEs and Integral Equations

The Fourier transform is a powerful tool when used to solve, or simplify, linear partial differential equations (PDEs) with constant coefficients. Several examples are given in Sections 2.3.1–2.3.4. The usefulness of the Fourier transform in this context is a direct result of the operational properties and convolution theorem discussed earlier in this chapter. The convolution theorem, Parseval equality, and operational properties are also useful tools in solving certain kinds of integral equations. Often such equations do not admit exact solutions, and Fourier techniques provide the tools for regularization (approximate solution) of such ill-posed problems.

2.3.1   Diffusion Equations

A one-dimensional linear diffusion equation with constant coefficients is one of the form

$$\frac{\partial u}{\partial t} = a \frac{\partial u}{\partial x} + b \frac{\partial^2 u}{\partial x^2} \tag{2.20}$$

where $a \in \mathbb{R}$ is called the drift coefficient and $b \in \mathbb{R}^+$ is called the diffusion coefficient. When modeling diffusion phenomena in an infinite medium, one often solves the above diffusion equation for $u(x, t)$ subject to given initial conditions $u(x, 0) = f(x)$. We note that Eq. (2.20) is a special case of the Fokker-Planck equation which we will examine in great detail in Chapter 16. When the drift coefficient is zero, the diffusion equation is called the *heat equation*.

Taking the Fourier transform of $u(x, t)$ for each value of t (i.e., treating time as a constant for the moment and x as the independent variable), one writes $\hat{u}(\omega, t)$. Applying the Fourier transform to both sides of Eq. (2.20), and the initial conditions, one observes that for each fixed frequency ω one has a linear first-order ordinary differential equation in the Fourier transform with initial conditions

$$\frac{d\hat{u}}{dt} = \left(ia\omega - b\omega^2\right)\hat{u} \text{ with } \hat{u}(\omega, 0) = \hat{f}(\omega).$$

The solution to this initial value problem is of the form

$$\hat{u}(\omega, t) = \hat{f}(\omega)e^{(ia\omega - b\omega^2)t}.$$

Application of the inverse Fourier transform yields a solution. One recognizes that in the above expression for $\hat{u}(\omega, t)$ is a Gaussian with phase factor, and on inversion this becomes a shifted Gaussian

$$\mathcal{F}^{-1}\left(e^{iat\omega}e^{-b\omega^2 t}\right) = \frac{1}{\sqrt{4\pi bt}} \exp\left(-\frac{(x + at)^2}{4bt}\right).$$

The convolution theorem then allows us to write

$$u(x, t) = \frac{1}{\sqrt{4\pi bt}} \int_{-\infty}^{\infty} f(\xi) \exp\left(-\frac{(x + at - \xi)^2}{4bt}\right) d\xi. \tag{2.21}$$

2.3.2   The Schrödinger Equation

In quantum mechanics, the Schrödinger equation for a free particle (i.e., when no potential field is present) is

$$\frac{\partial u}{\partial t} = ic^2 \frac{\partial^2 u}{\partial x^2}. \tag{2.22}$$

The corresponding initial and boundary conditions are $u(x, 0) = f(x)$ and $u(-\infty, t) = u(\infty, t) = 0$. At first sight this looks much like the heat equation. Only now, u is a complex-valued function, and $c^2 = \hbar/2m$ where m is the mass of the particle and $\hbar = 1.05457 \times 10^{-27} erg \cdot sec = 1.05457 \times 10^{-34} J \cdot sec$ is Planck's constant divided by 2π ($\hbar = h/2\pi$).

The solution of this equation progresses much like before. Taking the Fourier transform of both sides yields

$$\frac{d\hat{u}}{dt} = -ic^2\omega^2\hat{u}$$

for each fixed value of ω. The Fourier transform is then written in the form

$$\hat{u}(\omega, t) = \hat{f}(\omega)e^{-ic^2\omega^2 t}.$$

Applying the Fourier inversion formula yields the solution in the form of the integral

$$u(x, t) = \frac{1}{2\pi} \int_{-\infty}^{\infty} \hat{f}(\omega)e^{-ic^2\omega^2 t} e^{i\omega x} \, d\omega. \tag{2.23}$$

While this is not a closed-form solution, it does tell us important information about the behavior of the solution. For instance, if $\|f\|_2^2 = 1$, then from the above equation

$$\|u\|_2^2 = \int_{-\infty}^{\infty} u(x, t)\overline{u(x, t)} \, dx = 1.$$

This follows from direct substitution of Eq. (2.23) into the above equation as

$$\|u\|_2^2 = \int_{-\infty}^{\infty} \left(\frac{1}{2\pi} \int_{-\infty}^{\infty} \hat{f}(\omega)e^{-ic^2\omega^2 t} e^{i\omega x} \, d\omega \right) \left(\frac{1}{2\pi} \int_{-\infty}^{\infty} \overline{\hat{f}(v)}e^{+ic^2v^2 t} e^{-ivx} \, dv \right) dx.$$

Reversing the order of integration so that integration with respect to x is performed first, a closed-form solution of the x-integral is $2\pi\delta(\omega - v)$. Integration over v then results in

$$\|u\|_2^2 = \frac{1}{2\pi} \int_{-\infty}^{\infty} \hat{f}(\omega)\overline{\hat{f}(\omega)} \, d\omega = \|f\|_2^2.$$

2.3.3 The Wave Equation

The one-dimensional wave equation is

$$\frac{\partial^2 u}{\partial t^2} = c^2 \frac{\partial^2 u}{\partial x^2}. \tag{2.24}$$

The corresponding initial and boundary conditions are $u(x, 0) = f(x)$ and $u(-\infty, t) = u(\infty, t) = 0$. This equation is often used to model a long, taut string with uniform mass density per unit length ρ and tension T. The constant c is related to these physical quantities as $c^2 = T/\rho$.

One begins to solve this problem by applying the Fourier transform to the equation, initial conditions, and boundary conditions to yield

$$\frac{d^2\hat{u}}{dt^2} = -c^2\omega^2\hat{u} \qquad \hat{u}(\omega, 0) = \hat{f}(\omega) \qquad \hat{u}(0, t) = 0.$$

Again, ω is treated as a constant in this stage of the solution process. Clearly the solution to this ordinary differential equation (the simple harmonic oscillator) for the given initial conditions is of the form

$$\hat{u}(\omega, t) = \hat{f}(\omega) \cos c\omega t.$$

Using the identity $\cos\theta = (e^{i\theta} + e^{-i\theta})/2$ together with the operational property for shifts allows us to write

$$u(x) = \mathcal{F}^{-1}\left(\hat{f}(\omega)(e^{-ic\omega t} + e^{ic\omega t})/2\right) = \frac{1}{2}\{f(x - ct) + f(x + ct)\}. \tag{2.25}$$

This is known as d'Alembert's solution of the wave equation.

2.3.4 The Telegraph Equation

The telegraph (or transmission-line) equation models the flow of electrical current in a long cable. It is expressed as [23, 9, 25]

$$\frac{\partial^2 u}{\partial t^2} + 2b\frac{\partial u}{\partial t} + au = c^2\frac{\partial^2 u}{\partial x^2}. \tag{2.26}$$

The constants a, b, and c can be written in terms of the following cable properties (all per unit length): resistance, R, inductance, L, capacitance, C, and leakage, S. Explicitly,

$$b = \frac{1}{2}(R/L + S/C), \qquad c^2 = 1/CL, \qquad a = RS/CL.$$

It is straightforward to see that application of the Fourier transform yields

$$\frac{d^2\hat{u}}{dt^2} + 2b\frac{d\hat{u}}{dt} + \left(a + c^2\omega^2\right)\hat{u} = 0 \tag{2.27}$$

for each fixed value of ω. Initial conditions of the form $u(x, 0) = f_1(x)$ and $\partial u/\partial t(x, 0) = f_2(x)$ transform to $\hat{u}(\omega, 0) = \hat{f}_1(\omega)$ and $\partial\hat{u}/\partial t(\omega, 0) = \hat{f}_2(\omega)$.

Since Eq. (2.27) is a linear ordinary differential equation with constant coefficients, we seek solutions of the form $\hat{u} = e^{\lambda t}$ and find that the solutions of the corresponding characteristic equation

$$\lambda^2 + 2b\lambda + \left(a + c^2\omega^2\right) = 0$$

are of the form

$$\lambda = -b \pm \sqrt{b^2 - c^2\omega^2 - a}.$$

From the physical meaning of the parameters a and b, it must be that $b^2 - a \geq 0$. Hence we have two cases:

Case 1: $b^2 - a = 0, c \neq 0$.

In this case the solution for $\hat{u}$ is of the form

$$\hat{u}(\omega, t) = e^{-bt}\left\{C_1(\omega)e^{i\omega ct} + C_2(\omega)e^{-i\omega ct}\right\},$$

where $C_i(\omega)$ for $i = 1, 2$ are determined by the initial conditions by solving the equations

$$C_1(\omega) + C_2(\omega) = \hat{f}_1(\omega)$$

and

$$C_1(\omega)[i\omega c - b] - C_2(\omega)[b + i\omega c] = \hat{f}_2(\omega)$$

simultaneously.

Case 2: $b^2 - a > 0$.

In this case there are three subcases, corresponding to the sign of $b^2 - c^2\omega^2 - a$. Hence we write

$$\hat{u}(\omega, t) = \begin{cases} e^{-bt}\{C_1(\omega) + C_2(\omega)t\}; & b^2 - c^2\omega^2 - a = 0 \\ e^{-bt}\left\{C_1(\omega)\exp\left(it\sqrt{c^2\omega^2 - b^2 + a}\right) + C_2(\omega)\exp\left(-it\sqrt{c^2\omega^2 - b^2 + a}\right)\right\}; & b^2 - c^2\omega^2 - a < 0 \\ e^{-bt}\left\{C_1(\omega)\exp\left(t\sqrt{b^2 - c^2\omega^2 - a}\right) + C_2(\omega)\exp\left(-t\sqrt{b^2 - c^2\omega^2 - a}\right)\right\}; & b^2 - c^2\omega^2 - a > 0 \end{cases}$$

The constants $C_1(\omega)$ and $C_2(\omega)$ are again solved for given initial conditions represented in Fourier space.

Application of the Fourier inversion formula to the resulting Fourier transforms provides an integral representation of the solution.

2.3.5   Integral Equations with Convolution Kernels

Often when one takes a physical measurement, the quantity of interest is not found directly, and is obscured by a filter or measurement device of some sort. Examples of this are considered in Chapter 13 in the context of the tomography/tomotherapy problem.

In general, an *inverse problem*[3] is one in which the desired quantity (or function) is inside of an integral and the values of that integral are given (or measured). Extracting the desired function is the goal in solving the inverse problem.

Fourier analysis is a valuable tool in solving a particular kind of inverse problem which is posed as a Fredholm integral equation of the first kind with a convolution kernel

$$g(x) = cu(x) + \int_{-\infty}^{\infty} f(x - \xi)u(\xi)\,d\xi. \tag{2.28}$$

Here all the functions are real and c is either 1 or 0 since all other real coefficients can be absorbed in the definitions of the given "nice" functions $g(x)$ and $f(x)$. Our goal is, of course, to solve for $u(x)$.

Clearly the Fourier transform is a useful tool in this context because it yields

$$\hat{g}(\omega) = \left(c + \hat{f}(\omega)\right)\hat{u}(\omega).$$

If $c + \hat{f}(\omega) \neq 0$ for all values of ω, one can apply the Fourier inversion formula to find

$$u(x) = \mathcal{F}^{-1}\left(\frac{\hat{g}(\omega)}{c + \hat{f}(\omega)}\right).$$

In the case when $c + \hat{f}(\omega) = 0$ for one or more values of ω a *regularization* technique is required. Essentially, regularization is any method for generating approximate solutions where the singularity associated with division by zero is "smoothed out." One popular method is zeroth order *Tikhonov regularization* [15].

In this method, one seeks a solution to the modified problem of minimization of the functional[4]

$$F = \|cu + f * u - g\|_2^2 + \epsilon^2\|u\|_2^2$$

---

[3]The subject of inverse problems is more general than the solution of integral equations, but all of the inverse problems encountered in this book are integral equations.

[4]Higher order Tikhonov regularization methods add derivative terms to the functional. For instance, first order would add a term such as $\nu^2\|du/dx\|^2$ for a chosen parameter $\nu^2 << 1$ in an attempt to generate smoother solutions.

where ϵ is a small real parameter. In the limit as $\epsilon \to 0$, the solution to this modified problem approaches the exact solution to Eq. (2.28).

The regularized solution is generated using Fourier techniques by first using the Parseval equality to write

$$F = \frac{1}{2\pi} \left\{ \left\| \left(c + \hat{f} \right) \hat{u} - \hat{g} \right\|_2^2 + \epsilon^2 \left\| \hat{u} \right\|_2^2 \right\}.$$

Then treating $\hat{u}$ as a complex variable, one sets

$$2\pi \frac{\partial F}{\partial \hat{u}} = \left[\left(c + \hat{f} \right) \hat{u} - \hat{g} \right] \left(c + \overline{\hat{f}} \right) + \epsilon^2 \hat{u} = 0.$$

Solving for $\hat{u}$ and applying the Fourier inversion formula we write the regularized solution (which now depends on the choice of ϵ) as

$$u(x, \epsilon) = \mathcal{F}^{-1} \left(\frac{\hat{g}(\omega) \left(c + \overline{\hat{f}(\omega)} \right)}{\left| c + \hat{f}(\omega) \right|^2 + \epsilon^2} \right).$$

We conclude this section by stating that the examples considered here are only a few of the kinds of partial differential and integral equations solvable using classical Fourier methods. The interested reader is referred to [2, 3, 25, 30] for further reading.

2.4   Fourier Optics

In this section, we review how the behavior of light in some situations is described naturally using the Fourier transform, and conversely, how a combination of light and lenses can be used as a tool to calculate the Fourier transforms of certain kinds of functions.

Light is an electromagnetic wave, and is thus described by the equations governing electromagnetic phenomena. The electromagnetic state in a homogeneous isotropic medium (including a vacuum) without sources is described by Maxwell's equations [16]:

$$\nabla \times \mathbf{E} = -\mu \frac{\partial \mathbf{H}}{\partial t}$$

$$\nabla \times \mathbf{H} = \epsilon \frac{\partial \mathbf{E}}{\partial t}$$

$$\nabla \cdot \mathbf{E} = 0$$

$$\nabla \cdot \mathbf{H} = 0,$$

where $\mathbf{E} = \mathbf{E}(\mathbf{x}, t)$ is the electric field vector, and $\mathbf{H} = \mathbf{H}(\mathbf{x}, t)$ is the magnetic field vector. $\mathbf{x} \in \mathbb{R}^3$ is the position in the medium, ∇ is the gradient with respect to $\mathbf{x}$, $\nabla \times$ is the curl operation, and $\nabla \cdot$ is the divergence. The constants μ and ϵ are called *permeability* and *permittivity* in the medium, respectively. In the case of a vacuum, these are denoted μ_0 and ϵ_0, and the speed of light in a vacuum is given by $c = (\mu_0 \epsilon_0)^{-\frac{1}{2}} \approx 3 \times 10^8 m/sec$.

Observe that for an arbitrary vector-valued $\mathbf{K}$ that is at least twice differentiable in each component of $\mathbf{x}$,

$$\nabla \times (\nabla \times \mathbf{K}) = \nabla(\nabla \cdot \mathbf{K}) - \nabla^2(\mathbf{K}) \qquad (2.29)$$

where

$$\nabla^2 = \nabla \cdot \nabla = \sum_{i=1}^{3} \frac{\partial^2}{\partial x_i^2}$$

is the Laplacian.

Taking the curl of both sides of the first two of Maxwell's equations, and using Eq. (2.29), one can reduce the first two of Maxwell's equations in a vacuum to the form

$$\nabla^2 \mathbf{E} = \frac{1}{c^2} \frac{\partial^2 \mathbf{E}}{\partial t^2}, \quad \nabla^2 \mathbf{H} = \frac{1}{c^2} \frac{\partial^2 \mathbf{H}}{\partial t^2}. \tag{2.30}$$

This is because for the case when $\mathbf{K}$ is $\mathbf{E}$ or $\mathbf{H}$, we see that Eq. (2.29) simplifies from the last two of Maxwell's equations: $\nabla \cdot \mathbf{E} = \nabla \cdot \mathbf{H} = 0$. Let us now define the function $u(\mathbf{x}, t)$ to represent any component of the electric or magnetic field. The function $u(\mathbf{x}, t)$ must then satisfy the scalar *wave equation*

$$\nabla^2 u = \frac{1}{c^2} \frac{\partial^2 u}{\partial t^2}.$$

Solutions of this equation are of the form

$$u(\mathbf{x}, t) = U(\mathbf{x}) e^{-i\omega t}.$$

Substitution of this solution into the wave equation yields the Helmholtz equation

$$\left(\nabla^2 + k^2 \right) U = 0$$

where $k = \omega/c = 2\pi/\lambda$ is called the wave number, and λ is the wave length of the light.

A particular solution to the Helmholtz equation for the case of a spherical point source at $\mathbf{x} = \mathbf{0}$ is given by

$$U(\mathbf{x}) = \frac{e^{ik|\mathbf{x}|}}{|\mathbf{x}|}.$$

We now may consider what happens to light from this point source as it passes through an opening, or aperture, in an otherwise opaque screen, as illustrated in Fig. 2.1. According to the Huygens-Fresnel principle, the observable light on the other side of the screen is the same as if a continuum of spherical sources had been placed at the opening. Only now, the intensity of these "secondary" spherical waves is modulated by a factor of the form $\cos(\mathbf{n} \cdot \mathbf{x})/i\lambda$, where $\mathbf{n}$ is the normal to the screen. We will not go into the details of the physics behind this modulation. Discussions can be found in [13, 11].

What is important is that the amplitude of the light at a point on the opposite side of the screen from the source, $U(\mathbf{x}_0)$, can be found from the amplitude at the aperture, $U(\mathbf{x}_1)$, by adding the contributions from all of the secondary waves. This is performed with the superposition integral

$$U(\mathbf{x}_0) = \int_{\mathbb{R}^2} \frac{1}{i\lambda} \frac{e^{ik|\mathbf{x}_{01}|}}{|\mathbf{x}_{01}|} \cos(\mathbf{n} \cdot \mathbf{x}_{01}) U(\mathbf{x}_1) \, d\mathbf{x}_1, \tag{2.31}$$

which is valid when the size of the aperture is small in comparison to the distance of the source from the aperture. The notation $\mathbf{x}_{01} = \mathbf{x}_0 - \mathbf{x}_1$ is shorthand for the relative position vector of point 0 with respect to point 1. The integral over the aperture area is written as being over the whole plane because $U(\mathbf{x}_1)$ has compact support (i.e., the aperture area) and $d\mathbf{x}_1 = dx_1 dy_1$ represents a differential area element.

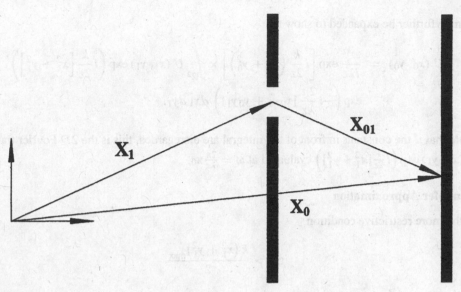

FIGURE 2.1
Diffraction of light traveling through an aperture.

Fresnel Approximation

Assuming that the point x_0 is far enough from the aperture, and its first and second components are small compared to its third, one has the following approximations

$$\cos(\mathbf{n} \cdot \mathbf{x}_{01}) \approx 1 \quad \text{and} \quad |\mathbf{x}_{01}| \approx z \left[1 + \frac{1}{2z^2} \left((x_0 - x_1)^2 + (y_0 - y_1)^2 \right) \right], \quad (2.32)$$

where z is the component of $\mathbf{x}_{01}$ in the direction normal to the plane of the aperture.

The second approximation just mentioned is acceptable for the denominator of the integrand in Eq. (2.31) provided

$$z^2 >> (x_0 - x_1)^2 + (y_0 - y_1)^2.$$

This just ensures that the Taylor series expansion truncated at the first term is an accurate approximation. However, it can lead to a poor result when substituted into the numerator of the integrand in Eq. (2.31) because $k = 2\pi/\lambda$, and the wavelength of visible light is extremely small (and hence its reciprocal is huge).

To ensure that $e^{ik|\mathbf{x}_{01}|}$ can be approximated well with the substitution Eq. (2.32), one assumes that the stronger condition

$$z^3 >> \frac{\pi}{4\lambda} \left[(x_0 - x_1)^2 + (y_0 - y_1)^2 \right]_{\max}^2 \quad (2.33)$$

is satisfied.

All of these approximations [and particularly Eq. (2.33)] are called the *Fresnel approximations*. Assuming these are valid, the superposition integral Eq. (2.31) becomes

$$U(x_0, y_0) = \frac{e^{ikz}}{i\lambda z} \int_{\mathbb{R}^2} U(x_1, y_1) \exp\left(i\frac{k}{2z} \left[(x_0 - x_1)^2 + (y_0 - y_1)^2 \right] \right) dx_1 \, dy_1.$$

This may further be expanded to show that

$$U(x_0, y_0) = \frac{e^{ikz}}{i\lambda z} \exp\left[i\frac{k}{2z}\left(x_0^2 + y_0^2\right)\right] \times \int_{\mathbb{R}^2} U(x_1, y_1) \exp\left(i\frac{k}{2z}\left[x_1^2 + y_1^2\right]\right)$$

$$\exp\left(-i\frac{2\pi}{\lambda z}[x_0 x_1 + y_0 y_1]\right) dx_1 \, dy_1.$$

Note that if the constants in front of the integral are disregarded, this is the $2D$ Fourier transform of $U(x_1, y_1) \exp\left(i\frac{k}{2z}[x_1^2 + y_1^2]\right)$ evaluated at $\omega = \frac{2\pi}{\lambda z}x_0$.

Fraunhofer Approximation

If the more restrictive condition

$$z >> \frac{k\left(x_1^2 + y_1^2\right)_{\max}}{2} \tag{2.34}$$

is valid, then $\exp\left(i\frac{k}{2z}[x_1^2 + y_1^2]\right) \approx 1$ and the superposition integral Eq. (2.31) is reduced further to

$$U(x_0, y_0) = \frac{e^{ikz}}{i\lambda z} \exp\left[i\frac{k}{2z}\left(x_0^2 + y_0^2\right)\right] \int_{\mathbb{R}^2} U(x_1, y_1) \exp\left(-i\frac{2\pi}{\lambda z}[x_0 x_1 + y_0 y_1]\right) dx_1 \, dy_1.$$

This shows that light traveling through an aperture is, in fact, performing a Fourier transform with modified phase and amplitude (due to the constant in front of the integral).

In practice, the approximation in Eq. (2.34) is extremely limiting. However, one can observe the same effect by use of lenses, and the physical phenomena discussed in this section can be used as an analog method for computing Fourier transforms (see, for example, [20]).

2.5   Discrete and Fast Fourier Transforms

Many physical phenomena are approximated well as being continuous, and the Fourier transform is a natural tool to model or aid in the analytical solution of equations describing these phenomena. However, when it comes time to perform calculations on a digital computer, discretization and truncation of these continuous descriptions of the world are required. In the subsections that follow, we present the general definitions and properties of the discrete Fourier transform, or DFT; show how it can be used to exactly reconstruct and convolve band-limited functions sampled at finite points; and sketch the basic ideas of the fast (discrete) Fourier transform, or FFT.

2.5.1   The Discrete Fourier Transform: Definitions and Properties

A digital computer cannot numerically perform the continuous integrations required in the definitions of the forward and inverse Fourier transform. However, a computer can calculate very well what is called the *discrete Fourier transform,* or DFT, of a function that is defined on a finite number, N, of equally spaced points. These points can be viewed as being on the unit circle, corresponding to the vertices of a regular $N-$gon. If the points are labeled $x_0, \ldots, x_{N-1}$, then the values of the discrete function, which are generally complex, are $f_j = f(x_j)$ for $j \in [0, N-1]$.

The DFT of such a function is defined as[5]

$$\hat{f}_k = \mathcal{F}(f_j) = \frac{1}{N} \sum_{j=0}^{N-1} f_j e^{-i2\pi jk/N}. \tag{2.35}$$

This can be viewed as an N-point approximation in the integral defining the Fourier coefficients $\hat{f}(k)$, where the sample points are $x_j = jL/N$, $f_j = f(x_j)$, and dx is approximated as $\Delta x = L/N$. Or it can simply be viewed as the definition of the Fourier transform of a function whose arguments take values in a discrete set.

As with the Fourier series and transform, the DFT inherits its most interesting properties from the exponential function. In this case, we have

$$\left[e^{-i2\pi/N}\right]^N = 1.$$

Using this property and the definition of the DFT it is easy to show that $\hat{f}_{k+N} = \hat{f}_k$. The following completeness relation can also be shown

$$\frac{1}{N} \sum_{k=0}^{N-1} e^{i2\pi(n-m)k/N} = \delta_{n,m} \tag{2.36}$$

by observing that for a geometric sum with $r \neq 1$ and $|r| \leq 1$,

$$\sum_{k=0}^{N-1} r^k = \frac{1-r^N}{1-r}.$$

Setting $r = e^{i2\pi(n-m)/N}$ for $n \neq m$ yields the property that $r^N = 1$, and so the numerator in the above equation is zero in this case. When $n = m$, all the exponentials in the sum reduce to the number 1, and so summing N times and dividing by N yields 1.

Equipped with Eq. (2.36), one observes that

$$\sum_{k=0}^{N-1} \hat{f}_k e^{i2\pi jk/N} = \sum_{k=0}^{N-1} \left(\frac{1}{N} \sum_{n=0}^{N-1} f_n e^{-i2\pi nk/N} \right) e^{i2\pi jk/N}$$

$$= \sum_{n=0}^{N-1} f_n \left(\frac{1}{N} \sum_{k=0}^{N-1} e^{i2\pi(j-n)k/N} \right)$$

$$= \sum_{n=0}^{N-1} f_n \delta_{j,n}$$

and thus the DFT inversion formula

$$f_j = \sum_{k=0}^{N-1} \hat{f}_k e^{i2\pi jk/N}. \tag{2.37}$$

This is much like the Fourier series expansion of a band-limited function $f(x)$ sampled at $x = x_j$. Only in the case of the Fourier series, the sum contains $2B + 1$ terms (which is always an odd

---

[5]As with the Fourier transform and Fourier series, one finds a variety of normalizations in the literature, the most common of which are $1/N$ and 1.

number), and in the case of the DFT, N is usually taken to be a power of 2 (which is always an even number). This discrepancy between the DFT and sampled band-limited Fourier series is rectified if the constraint $\hat{f}(B) = \hat{f}(-B)$ is imposed on the definition of a band-limited Fourier series. Then there are $N = 2B$ free parameters in both.

The convolution of two discrete functions f_j and g_j for $j = 0, \ldots, N - 1$ is defined as

$$(f \star g)_j = \frac{1}{N} \sum_{k=0}^{N-1} f_k g_{j-k},$$

and the corresponding convolution theorem is

$$\mathcal{F}(f \star g) = \hat{f}_j \hat{g}_j.$$

The proofs for Parseval's equality,

$$\frac{1}{N} \sum_{j=0}^{N-1} |f_j|^2 = \sum_{k=0}^{N-1} |\hat{f}_k|^2,$$

and the more general relationship,

$$\frac{1}{N} \sum_{j=0}^{N-1} f_j \overline{g_j} = \sum_{k=0}^{N-1} \hat{f}_k \overline{\hat{g}_k},$$

follow in an analogous way to those for the continuous Fourier series.

2.5.2   The Fast Fourier Transform and Convolution

Direct evaluation of the DFT using Eq. (2.35) can be achieved with N multiplications and $N - 1$ additions for *each* of the N values of k. Since for $k = 0$ and $j = 0$ the multiplications reduce to a product of a number with $e^0 = 1$, these need not be calculated explicitly, and a total of $(N - 1)^2$ multiplications and $N(N - 1)$ additions are sufficient. The values $e^{i2\pi jk/N}$ can be stored in advance and need not be calculated. Thus $\mathcal{O}(N^2)$ arithmetic operations are sufficient to compute the DFT, and because of the symmetry of the definition, the inverse DFT is computed in this order as well.[6] One likewise reasons that the discrete convolution of two functions with values at N points can be computed in $N - 1$ additions and N multiplications for each value of k in the convolution product $(f \star g)_k$, and so N^2 multiplications and $N(N - 1)$ additions are used for all values of k. Thus computation of the convolution product can also be performed in $\mathcal{O}(N^2)$ computations.

The *fast Fourier transform* refers to a class of algorithms, building on the result of Cooley and Tukey from the 1960s [7], that the DFT can actually be computed in $\mathcal{O}(N \log_2 N)$ computations. By the convolution theorem, this means that convolutions can be computed in this order as well. For large values of N, the relative amount of savings is dramatic, since the rate at which the function $N / \log N$ increases continues to increase with N.

The core of this somewhat remarkable result actually goes back hundreds of years. It is based on the fact that

$$\left(e^{-i2\pi jk/N} \right)^N = 1.$$

---

[6] See Appendix A for the meaning of the $\mathcal{O}(\cdot)$ notation.

Gauss[7] is purported to have used DFT and FFT ideas, and some reports indicate that others may have even preceded him [18]. In the modern era, the result is attributed to the 1965 paper of Cooley and Tukey, though many elements of the FFT had been known previously [18, 6]. Over the past few decades, a number of different FFT algorithms have been developed, but they all have the same $\mathcal{O}(N \log N)$ computational complexity [though the notation $\mathcal{O}(\cdot)$ can hide quite significant differences in the actual time the algorithms take to execute]. For an examination of how the FFT became a ubiquitous computational tool, see [5].

The core idea of the FFT is simple enough. The number N is restricted to be of the form $N = K^m$ for positive integers K and m. Then one can rewrite

$$\hat{f}_k = \sum_{j=0}^{K^m-1} f_j e^{-i2\pi jk/K^m} = \sum_{l=0}^{K-1} \left[\sum_{j=0}^{K^{m-1}-1} f_{K \cdot j + l} \left(e^{-i2\pi/K^{m-1}} \right)^{jk} \right] \left(e^{i2\pi/K^m} \right)^{lk}.$$

Each of the inner summations for each of the K values of l can be viewed as independent Fourier transforms for sequences with K^{m-1} points, and thus can be computed for all values of k in order $(N/K)^2$ time using the DFT. To compute all of the inner sums and to combine them all together to generate $\hat{f}_k$ for all values of k requires on the order of $(N/K)^2 \cdot K = N^2/K$ computations. Thus splitting the sum once reduces the computation time by a factor of K. In fact, we can keep doing this to each of the remaining smaller sequences, until we get K^m sequences each with $K^0 = 1$ point. In this case there are $m = \log_K N$ levels to the recursion, each requiring no more than $\mathcal{O}(N)$ multiplications and additions, so the total computational requirements are $\mathcal{O}(N \log_K N)$. Usually the most convenient choice is $K = 2$, and the trickiest part of the implementation of an FFT algorithm is the bookkeeping. Since a number of good books and FFT software packages are available (see for example, [4, 8, 14, 22, 28, 29, 30]) we will not discuss the particular methods in any detail. We note also that in recent years, efforts devoted to FFTs for nonuniformly spaced data have been investigated [1].

2.5.3 Sampling, Aliasing, and Filtering

It is well known in many engineering disciplines that an arbitrary signal (function) cannot be reconstructed from a finite number of samples. Any attempt to do so leads to a phenomenon known as *aliasing*. One physical example of aliasing is when the wheels of an automobile in a movie appear to be turning slowly clockwise when the wheels are actually rotating rapidly counterclockwise. This is because the sampling rate of the movie camera is not fast enough to capture the true motion of the wheels. In general, aliasing results because the amount of information in an arbitrary signal is infinite and cannot be captured with a finite amount of collected data.

However, a signal (or function) on the circle with Fourier coefficients $\hat{f}(k)$ for $k \in \mathbb{Z}$ can be *band-pass filtered* to band-limit B by essentially throwing away (or zeroing) all Fourier coefficients for which $|k| > B$. In this way, $2B + 1$ Fourier coefficients completely define the filtered signal.

Since the Fourier coefficients enter linearly in the expansion of a function in a Fourier series, one can write at each sample time

$$f_B(\theta(t_k)) = \sum_{j=-B}^{B} \hat{f}_j e^{ij\theta(t_k)}$$

for each of $2B + 1$ sample values of t_k. In principle, all that is then required to calculate the $2B + 1$ independent Fourier coefficients from $2B+1$ samples of the function is to invert a $(2B+1) \times (2B+1)$

---

[7]Carl Friedrich Gauss (1777–1855), one of the greatest mathematicians of all time, was responsible for the divergence theorem, and many results in number theory and differential geometry. His unpublished work on the DFT and FFT was not collected until after his death [12].

dimensional matrix with k, jth entry $e^{ij\theta(t_k)}$ (though the computation is not actually preformed as a matrix inversion). However, in practice two issues arise: (1) real filters only approximate band-pass filters; and (2) measurements always contain some error. Hence it is safest to sample at a rate above the minimal rate to obtain a more robust estimate of a filtered signal.

Sampling and Reconstruction of Functions on the Line

Similar ideas apply for functions on the line. In this context a band-limited function is one whose Fourier transform vanishes outside of a finite range of frequencies. The formal criteria for sampling and reconstruction of functions on the line are often collectively referred to as the *Shannon sampling theorem*.[8] If samples are taken at no less than twice the band-limit frequency, the band-limited function can be exactly reconstructed. In the case of the line, this is referred to as the *Nyquist criterion*.[9]

Clear descriptions of the Shannon sampling theorem can be found in [19, 24]. Our proof proceeds along similar lines as follows. Assume that $f \in L^2(\mathbb{R})$ is a band-limited function with band-limit Ω. That is, $\hat{f}(\omega) = 0$ for $|\omega| > \Omega$. We will view f as a function of time, t. The product of $f(t)$ and an infinite sum of equally spaced shifted delta functions (called a periodic pulse train) yields

$$f_s(t) = f(t) \sum_{n=-\infty}^{\infty} \delta(t - nT) = \sum_{n=-\infty}^{\infty} f(nT)\delta(t - nT)$$

where T is the sample interval. Since the impulse train is T-periodic, it can be viewed as a single delta function on the circle of circumference T, and can be expanded in a Fourier series as

$$\sum_{n=-\infty}^{\infty} \delta(t - nT) = \sum_{n=-\infty}^{\infty} e^{2\pi nit/T}.$$

Taking the Fourier transform of both sides of this equality yields

$$\mathcal{F}\left(\sum_{n=-\infty}^{\infty} \delta(t - nT) \right) = \sum_{n=-\infty}^{\infty} \mathcal{F}\left(e^{2\pi nit/T} \right) = 2\pi \sum_{n=-\infty}^{\infty} \delta(\omega - 2\pi n/T).$$

Since the product of functions in the time domain corresponds to a convolution in the frequency domain multiplied by a factor of $1/2\pi$, we find that

$$\hat{f}_s(\omega) = \sum_{n=-\infty}^{\infty} \hat{f}(\omega) * \delta(\omega - 2\pi n/T) = \sum_{n=-\infty}^{\infty} \hat{f}(\omega - 2\pi n/T).$$

This is a periodic function with period T, and hence can be expanded in a Fourier series. As long as $2\pi/T \geq 2\Omega$, $\hat{f}_s(\omega)$ [which is $\hat{f}(\omega)$ "wrapped around the circle"] will be identical to $\hat{f}(\omega)$ for $|\omega| < \Omega$. If $\pi/T < \Omega$, the wrapped version of the function will overlap itself. This phenomenon, called aliasing, results in the destruction of some information in the original function, and makes it impossible to perform an exact reconstruction.

Since $\hat{f}_s(\omega)$ is $2\pi/T$-periodic, and is identical to $\hat{f}(\omega)$ for $|\omega| < \Omega$, it can be expanded in a Fourier series as

$$\hat{f}_s(\omega) = \sum_{n=-\infty}^{\infty} \left(\frac{T}{2\pi} \int_{-\Omega}^{\Omega} \hat{f}(\sigma)e^{-inT\sigma}d\sigma \right) e^{inT\omega} \tag{2.38}$$

---

[8]This was first proved in the context of communication theory by Shannon (See [27, pp. 160–162]).

[9]Nyquist reported on the importance of this sample frequency in the context of the telegraph [23].

where σ is the variable of integration.

For the band-limited function $f(t)$, the Fourier inversion formula is written as

$$f(t) = \frac{1}{2\pi} \int_{-\Omega}^{\Omega} \hat{f}(\omega) e^{i\omega t} \, d\omega. \tag{2.39}$$

It then follows that the Fourier coefficients (quantities in parentheses) in Eq. (2.38) are equal to $T \cdot f(-nT)$ for each $n \in \mathbb{Z}$. Hence we can rewrite Eq. (2.39) as

$$f(t) = \frac{1}{2\pi} \int_{-\Omega}^{\Omega} \left[\sum_{n=-\infty}^{\infty} T \cdot f(-nT) e^{inT\omega} \right] e^{i\omega t} \, d\omega.$$

After changing the order of integration and summation, and making the substitution $n \to -n$, one finds Shannon's reconstruction

$$f(t) = T/\pi \sum_{n=-\infty}^{\infty} f(nT) \frac{\sin \Omega(t - nT)}{(t - nT)}.$$

When $f(t)$ is sampled at the Nyquist rate ($\Omega = \pi/T$) this becomes

$$f(t) = \sum_{n=-\infty}^{\infty} f(n\pi/\Omega) \frac{\sin \Omega(t - n\pi)}{\Omega(t - n\pi)}.$$

In addition to communication and information theory, the sampling theorem is of general usefulness in linear systems theory and control (see, for example, [26]). In the applications we will explore, it suffices to view functions on the line with finite support as functions on the circle provided the function is scaled properly, i.e., so that a sufficient portion of the circle maps to function values that are zero.

Commutativity of Sampling and Convolution

We now make one final note regarding the connection between classical continuous and discrete Fourier analysis. Let us assume that two band-limited functions on the unit circle ($L = 2\pi$) with the same band limit, B, are sampled at the points $\theta_k = 2\pi k/N$. Here $N \geq 2B + 1$. We will explicitly prove a point that is often used, but not always stated, in texts on Fourier analysis and numerical methods. Namely, the discrete convolution of evenly sampled band-limited functions on the circle yields exactly the same result as first convolving these functions, and then sampling.

To begin, consider band-limited functions of the form

$$f(x) = \sum_{k=-B}^{B} \hat{f}(k) e^{ikx} \tag{2.40}$$

where

$$\hat{f}(n) = \frac{1}{2\pi} \int_{0}^{2\pi} f(x) e^{-inx} \, dx.$$

As we discussed in the previous subsection, if we sample at $N \geq 2B + 1$ points, enough information exists to reconstruct the function from its samples. We denote the set of sampled values as $S[f]$, with $S[f]_n$ denoting the single sample $f(2\pi n/N)$.

The convolution of two band-limited functions of the form of Eq. (2.40) is

$$(f_1 * f_2)(x) = \sum_{k=-B}^{B} \hat{f}_1(k) \hat{f}_2(k) e^{ikx}.$$

Sampling at points $x = 2\pi n/N$ for $n \in [0, N-1]$ yields the set of sampled values $S[f_1 * f_2]$. Performing the convolution of the two sets of discrete values $S[f_1]$ and $S[f_2]$ yields

$$
\begin{aligned}
(S[f_1] \star S[f_2])_j &= \frac{1}{N} \sum_{k=0}^{N-1} S[f_1]_k \, S[f_2]_{j-k} \\
&= \frac{1}{N} \sum_{k=0}^{N-1} \left(\sum_{l=-B}^{B} \hat{f}_1(l) e^{2\pi i l k/N} \right) \left(\sum_{m=-B}^{B} \hat{f}_2(m) e^{2\pi i m(j-k)/N} \right) \\
&= \frac{1}{N} \sum_{l=-B}^{B} \sum_{m=-B}^{B} \hat{f}_1(l) \hat{f}_2(m) \left(e^{2\pi i m j/N} \sum_{k=0}^{N-1} e^{2\pi i k(l-m)/N} \right) \\
&= \sum_{l=-B}^{B} \hat{f}_1(l) \hat{f}_2(l) e^{2\pi i j l/N} \\
&= S[f_1 * f_2]_j \, .
\end{aligned}
$$

This follows from Eq. (2.36).

Hence, we write

$$
S[f_1 * f_2] = S[f_1] \star S[f_2] \, . \tag{2.41}
$$

2.6 Summary

Classical Fourier analysis is ubiquitous in engineering and science. Its applications range from optics and X-ray crystallography, to the solution of linear partial differential equations with constant coefficients, and the theory of sound and vibration. We know of no discipline in the physical sciences and engineering which does not make use of either Fourier series, the Fourier transform, or the DFT/FFT. For a complete overview of applications and the history of Fourier analysis see [21].

In contrast to classical Fourier analysis, which is built on the arithmetic of scalar addition and multiplication (commutative operations), the more modern area of mathematics referred to as noncommutative harmonic analysis is unknown to the vast majority of engineers. In fact it is unknown to most scientists outside of theoretical physics.

In subsequent chapters we explain the basic theory behind noncommutative harmonic analysis in a way which we hope is much more accessible to the various engineering communities than the pure mathematics books on the subject. We then illustrate a number of concrete applications and numerical implementations from our own research in the last seven chapters of the book.

References

[1] Bagchi, S. and Mitra, S., *The Nonuniform Discrete Fourier Transform and Its Applications in Signal Processing*, Kluwer Academic Publishers, Boston, 1999.

[2] Benedetto, J.J., *Harmonic Analysis and Applications*, CRC Press, Boca Raton, FL, 1997.

[3] Bracewell, R.N., *The Fourier Transform and Its Applications*, 2nd ed., McGraw-Hill, New York, 1986.

[4] Burrus, C.S. and Parks, T.W., *DFT/FFT and Convolution Algorithms*, John Wiley & Sons, New York, 1985.

[5] Cooley, J.W., How the FFT gained acceptance, Proc. ACM Conf. on the History of Scientific and Numeric Computation, Princeton, NJ, May 13–15, 1987, 133–140.

[6] Cooley, J.W., The re-discovery of the fast Fourier transform algorithm, *Mikrochimica Acta III*, 33–45, 1987.

[7] Cooley, J.W. and Tukey, J., An algorithm for the machine calculation of complex Fourier series, *Math. of Comput.*, **19**, 297–301, 1965.

[8] Elliott, D.F. and Rao, K.R., *Fast Transforms: Algorithms, Analyses, Applications*, Academic Press, New York, 1982.

[9] Farlow, S.J., *Partial Differential Equations for Scientists and Engineers*, Dover, New York, 1993.

[10] Fourier, J.B.J., *Théorie Analytique de la Chaleur*, F. Didot, Paris, 1822.

[11] Fowles, G.R., *Introduction to Modern Optics*, 2nd ed., Dover, New York, 1975.

[12] Gauß, C.F., Nachlass, Theoria Interpolationis Methodo Nova Tractata, in *Carl Friedrich GaußWerke, Band 3*, Königliche Gesellschaft der Wissenschaften: Göttingen, 265–330, 1866.

[13] Goodman, J.W., *Introduction to Fourier Optics*, McGraw-Hill, New York, 1968.

[14] Gray, R.M. and Goodman, J.W., *Fourier Transforms: An Introduction for Engineers*, Kluwer Academic Publishers, Boston, 1995.

[15] Groetsch, C.W., *The Theory of Tikhonov Regularization for Fredholm Equations of the First Kind*, Pitman, Boston, 1984.

[16] Hayt, W.H., Jr., *Engineering Electromagnetics*, 5th ed., McGraw-Hill, New York, 1989.

[17] Heaviside, O., On operators in physical mathematics, *Proc. Royal Soc.*, 52, 1893, 504–529, and 54, 105–143, 1894.

[18] Heideman, M.T., Johnson, D.H., and Burrus, C.S., Gauss and the history of the fast Fourier transform, *Arch. for Hist. of Exact Sci.*, 34(3), 265–277, 1985.

[19] Hsu, H.P., *Applied Fourier Analysis*, Harcourt Brace Jovanovich College Outline Series, New York, 1984.

[20] Karim, M.A. and Awwal, A.A.S., *Optical Computing, An Introduction*, John Wiley & Sons, New York, 1992.

[21] Körner, T.W., *Fourier Analysis*, Cambridge University Press, 1988, (reprint, 1993).

[22] Nussbaumer, H.J., *Fast Fourier Transform and Convolution Algorithms*, Springer-Verlag, New York, 1982.

[23] Nyquist, H., Certain topics in telegraph transmission theory, *AIEE Trans.*, 47, 617–644, 1928.

[24] Oppenheim, A.V. and Willsky, A.S., *Signals and Systems*, Prentice-Hall, Englewood Cliffs, NJ, 1983.

[25] Pinsky, M.A., *Introduction to Partial Differential Equations with Applications*, McGraw-Hill, New York, 1984.

[26] Rugh, W.J., *Johns Hopkins University Signals, Systems, and Control Demonstrations*, http://www.jhu.edu/~signals/, 1996-present.

[27] Sloane, N.J.A. and Wyner, A.D., Eds., *Claude Elwood Shannon: Collected Papers*, IEEE Press, New York, 1993.

[28] Tolimieri, R., An, M., and Lu, C., *Algorithms for Discrete Fourier Transform and Convolution*, Springer-Verlag, New York, 1989.

[29] Van Loan, C., *Computational Frameworks for the Fast Fourier Transform*, SIAM, Philadelphia, 1992.

[30] Walker, J.S., *Fast Fourier Transforms*, 2nd ed., CRC Press, Boca Raton, FL, 1996.

Chapter 3

Sturm-Liouville Expansions, Discrete Polynomial Transforms and Wavelets

This chapter reviews some generalizations of Fourier analysis. We begin with Sturm-Liouville theory, which provides a tool for generating functions that are orthogonal with respect to a given weight function on an interval. We also examine modern orthogonal polynomials and wavelet expansions.

3.1 Sturm-Liouville Theory

Consider the closed interval $a \le x \le b$ and real-valued differentiable functions $s(x)$, $w(x)$, and $q(x)$ defined on this interval, where $s(x) > 0$ and $w(x) > 0$ on the open interval $a < x < b$. The values of these functions may be positive at $x = a$ and $x = b$, or they may be zero. It has been known since the mid-19th century that the solutions of differential equations of the form

$$\frac{d}{dx}\left(s(x)\frac{dy}{dx}\right) + (\lambda w(x) - q(x))y = 0 \tag{3.1}$$

for different *eigenvalues*, λ, and appropriate boundary conditions produce sets of orthogonal *eigenfunctions* $y(x)$. Each eigenfunction-eigenvalue pair is denoted as $(y_i(x), \lambda_i)$ for $i = 1, 2, \ldots$

3.1.1 Orthogonality of Eigenfunctions

To observe the orthogonality of solutions, multiply Eq. (3.1) evaluated at the solution $(y_i(x), \lambda_i)$ by the eigenfunction $y_j(x)$ and then integrate

$$\int_a^b y_j \left(sy_i'\right)' dx + \int_a^b y_j(\lambda_i w - q)y_i \, dx = 0.$$

The notation y' is used in this context to denote dy/dx. The first integral in the above equation is written using integration by parts as

$$\int_a^b y_j \left(sy_i'\right)' dx = y_j sy_i' \big|_a^b - \int_a^b y_j' sy_i' \, dx.$$

Hence,

$$y_j sy_i' \big|_a^b - \int_a^b y_j' sy_i' \, dx + \int_a^b y_j \left(\lambda_i w - q\right) y_i \, dx = 0.$$

39

If we were to repeat the same procedure with the roles of (y_i, λ_i) and (y_j, λ_j) reversed, then

$$y_i s y_j' \big|_a^b - \int_a^b y_i' s y_j' \, dx + \int_a^b y_i \left(\lambda_j w - q \right) y_j \, dx = 0.$$

Subtracting the second equation from the first, we see that if

$$y_i s y_j' \big|_a^b - y_j s y_i' \big|_a^b = 0 \tag{3.2}$$

then

$$\left(\lambda_i - \lambda_j \right) \int_a^b y_i(x) y_j(x) w(x) \, dx = 0. \tag{3.3}$$

In other words, y_i and y_j are orthogonal with respect to the weight function w if $\lambda_i \neq \lambda_j$. If $s(a) \cdot w(a) > 0$ and $s(b) \cdot w(b) > 0$, then this is called a *regular* Sturm-Liouville problem. Otherwise it is called a *singular* Sturm-Liouville problem.[1]

Ways for $y_i s y_j' \big|_a^b - y_j s y_i' \big|_a^b = 0$ to be satisfied include the following boundary conditions for Eq. (3.1)

$$y'(a) \cos \alpha - y(a) \sin \alpha = 0 \quad \text{and} \quad y'(b) \cos \beta - y(b) \sin \beta = 0$$

for any $\alpha, \beta \in [0, 2\pi]$ (called *separable* boundary conditions);

$$y(a) = y(b) \quad \text{and} \quad y'(a) = y'(b) \quad \text{when} \quad s(a) = s(b)$$

(called *periodic* boundary conditions); and for a singular Sturm-Liouville problem with $s(a) = s(b) = 0$ the condition Eq. (3.2) is automatically satisfied.

One of the key features of Sturm-Liouville theory is that any "nice" function $f(x)$ defined on $a \leq x \leq b$ that satisfies the boundary conditions in Eq. (3.2) can be expanded in a series of eigenfunctions [10]:

$$f(x) = \sum_{n=0}^{\infty} c_n y_n(x) \tag{3.4}$$

where

$$c_n = \frac{\int_a^b f(x) y_n(x) w(x) \, dx}{\int_a^b y_n^2(x) w(x) \, dx}.$$

In this context we say $f(x)$ is band-limited with band-limit B if the sum in Eq. (3.4) is truncated at $n = B - 1$. The existence and uniqueness of the eigenfunctions up to an arbitrary scale factor is addressed in a number of works (see, for example, [77, 57]).

What is sometimes not stated is the relationship between different normalizations. Using the notation

$$\left(y_i, y_j \right)_w = \int_a^b y_i(x) y_j(x) w(x) \, dx,$$

orthogonality is stated as

$$\left(y_i, y_j \right)_w = C_i \delta_{ij}$$

where C_i is a constant. We can define orthonormal eigenfunctions as $\tilde{y}_i = y_i / \sqrt{C_i}$ so that

$$\left(\tilde{y}_i, \tilde{y}_j \right)_w = \delta_{ij}.$$

---

[1] It is named after Jacques Charles François Sturm (1803–1855) and Joseph Liouville (1809–1882).

However, when a and b are finite, it is common for the first eigenfunction, y_0, to be constant, and to be chosen as $y_0 = 1$. It is desirable to retain this normalization of the first eigenfunction. This may be achieved together with orthonormality if we normalize the weighting function as

$$\tilde{w}(x) = w(x) / \int_a^b w(x)\, dx$$

(the functions s and q must be rescaled as well) and then normalize the eigenfunctions a second time as

$$\tilde{\tilde{y}}_i = \tilde{y}_i \sqrt{C_0} = y_i \sqrt{C_0/C_i}.$$

In this way, when $y_0 = 1$ so is $\tilde{\tilde{y}}_0$. Since by definition in the case when $y_0 = 1$ we have

$$(1, 1)_w = \int_a^b w(x)\, dx = C_0,$$

it follows that

$$\left(\tilde{\tilde{y}}_i, \tilde{\tilde{y}}_j\right)_{\tilde{w}} = C_0 \left(\tilde{y}_i, \tilde{y}_j\right)_{\tilde{w}} = \left(\tilde{y}_i, \tilde{y}_j\right)_w = \delta_{ij}.$$

Hence the two conditions of orthonormality of all the eigenfunctions and normalization of the first eigenfunction to unity are simultaneously achievable.

Having said this, we will use whatever normalization is the most convenient in any given context, with the knowledge that the eigenfunctions *can* be normalized in the variety of ways described above.

Clearly, the Fourier series expansion of functions on the circle is one example of a solution to a Sturm-Liouville problem with $s(x) = w(x) = 1$, $q(x) = 0$, $\lambda = n^2$ for $n \in \mathbb{Z}$, $[a, b] = [0, 2\pi]$, and periodic boundary conditions. Throughout this chapter we examine others.

3.1.2 Completeness of Eigenfunctions

Let

$$f_B(x) = \sum_{n=0}^{B-1} c_n y_n(x) \tag{3.5}$$

be a band-limited approximation of a real-valued function $f(x)$ with the eigenfunction expansion in Eq. (3.4). The square of the norm of a function in this context is defined as

$$\|f(x)\|^2 = \int_a^b [f(x)]^2 w(x)\, dx.$$

The sequence of functions $f_1, f_2, \ldots, f_B$ is called *mean-square convergent* if

$$\lim_{B \to \infty} \|f_B(x) - f(x)\|^2 = 0. \tag{3.6}$$

The set of functions $\{y_n(x)\}$ is called *complete in a set of functions* S if for any $f \in S$ it is possible to approximate f to arbitrary precision in the mean-squared sense with a band-limited series of the form in Eq. (3.5). That is, $\{y_n(x)\}$ is complete in S if

$$\|f_B(x) - f(x)\| < \epsilon$$

for any $\epsilon > 0$, $f \in S$, and sufficiently large B. Here $\| \cdot \|$ is shorthand for $\| \cdot \|_2$.

When $\{y_n(x)\}$ is orthonormal and complete in $\mathcal{L}^2([a, b], w(x)\,dx)$, expanding

$$\|f_B(x) - f(x)\|^2 = \|f_B(x)\|^2 + \|f(x)\|^2 - 2\int_a^b f(x)f_B(x)w(x)\,dx,$$

and substituting Eq. (3.5) into the above expanded form of Eq. (3.6) yields *Bessel's inequality*

$$\sum_{n=0}^{B-1} c_n^2 \le \|f\|^2.$$

As $B \to \infty$, this becomes *Parseval's equality*

$$\sum_{n=0}^{\infty} c_n^2 = \|f\|^2.$$

Parseval's equality holds for eigenfunctions of Sturm-Liouville equations (i.e., the eigenfunctions form a complete orthonormal set in $\mathcal{L}^2([a, b], w(x)\,dx)$). To our knowledge, only case-by-case proofs of completeness were known for the century following the development of the theory by Sturm and Liouville.

A paper by Birkoff and Rota [10] from the latter half of the 20th century provides a general proof of the completeness of eigenfunctions of Sturm-Liouville equations. The essential point of this proof is that given one orthonormal basis $\{\phi_n\}$ for $\mathcal{L}^2([a, b], w(x)\,dx)$, then another infinite sequence of orthonormal functions $\{\psi_n\}$ for $n = 0, 1, 2, \ldots$ will also be a basis if

$$\sum_{n=0}^{\infty} \|\psi_n - \phi_n\|^2 < \infty.$$

Hence, starting with a known orthonormal basis (such as the trigonometric functions $\{\sqrt{2/\pi}\cos nx\}$ for $[a, b] = [0, \pi]$ and $w(x) = 1$) and relating the asymptotic behavior of other eigenfunctions on the same interval to this basis provides a way to prove completeness.

The interested reader can see [10] for details. Throughout the remainder of the text, the completeness of eigenfunctions is assumed.

3.1.3 Sampling Band-Limited Sturm-Liouville Expansions

Let $y_k(x)$ for $k = 1, 2, \ldots$ be normalized eigenfunctions that solve a Sturm-Liouville problem. Rewriting Eq. (3.5), recall that the function

$$f(x) = \sum_{k=0}^{B-1} c_k y_k(x) \tag{3.7}$$

is said to have band-limit B. If any such band-limited function is sampled at B distinct points $a < x_0 < x_1 < \cdots < x_{B-1} < b$, then Eq. (3.7) implies

$$f_j = \sum_{k=0}^{B-1} Y_{jk} c_k$$

where $f_j = f(x_j)$ and $Y_{jk} = y_k(x_j)$ are the entries of a $B \times B$ matrix Y. Assuming that the points $\{x_j\}$ can be chosen so that $\det(Y) \ne 0$ (where Y is the matrix with elements Y_{jk}), and the matrix

Y^{-1} can be calculated exactly, then it is possible to calculate

$$c_j = \int_a^b f(x) y_j(x) w(x) \, dx = \sum_{k=0}^{B-1} Y_{jk}^{-1} f_k. \tag{3.8}$$

That is, when $f(x)$ is band-limited and $\det(Y) \neq 0$, the integral defining the coefficients $\{c_j\}$ can be computed exactly as a finite sum.

At first glance it might appear that the above sums defining f_j and c_j require $\mathcal{O}(B^2)$ arithmetic operations for each value of j (when Y^{-1} is precomputed), and hence $\mathcal{O}(B^3)$ in total.[2] However, more efficient algorithms exist. When the functions $\{y_k\}$ satisfy the properties described below following, the previous computation can be performed *much* more efficiently.

For all of the Sturm-Liouville systems considered in this book, the functions $\{y_k\}$ satisfy *three-term recurrence relations* of the form

$$y_{k+1}(x) = \mathcal{C}_1(x, k) y_k(x) + \mathcal{C}_2(x, k) y_{k-1}(x). \tag{3.9}$$

In the special case when $\{y_k\}$ is a set of polynomials that solve a Sturm-Liouville problem, the coefficient $\mathcal{C}_1(x, k)$ is of the form $a_k x + b_k$, and $\mathcal{C}_2$ is a constant that depends on k (see Section 3.5.1). These recurrence relations allow for much more efficient evaluation of the above sums [4, 19]. Polynomials are also convenient because a function that is band-limited in one system of complete orthonormal polynomials will be band-limited with the same band-limit in every such system.

The use of recurrence relations for the fast evaluation of a weighted sum of orthogonal polynomials has a long history (see [31, 32, 35, 36, 48, 92] and references therein), as do algorithms for the conversion from one polynomial basis to another (see, for example, [86, 87]).

Appendix A reviews computational tools used in connection with polynomials that are independent of recurrence relations. The subsequent sections of this chapter provide concrete examples of solutions to Sturm-Liouville problems, as well as complete orthonormal systems of functions that are not generated as the eigenfunctions of such expansions.

3.2 Legendre Polynomials and Associated Legendre Functions

3.2.1 Legendre Polynomials

Legendre polynomials,[3] $P_l(x)$, are defined as solutions of the singular Sturm-Liouville equation

$$\left(\left(1 - x^2 \right) y' \right)' + l(l+1)y = 0,$$

or equivalently

$$\left(1 - x^2 \right) y'' - 2xy' + l(l+1)y = 0,$$

for $x \in [-1, 1]$ where $l = 0, 1, 2, \ldots$. The Legendre polynomials can be calculated analytically using the *Rodrigues formula*

$$P_l(x) = \frac{1}{2^l l!} \frac{d^l}{dx^l} \left[\left(x^2 - 1 \right)^l \right],$$

---

[2]See Appendix A for an explanation of the $\mathcal{O}(\cdot)$ notation.

[3]These are named after Adrien Marie Legendre (1752–1833).

or the integral

$$P_l(x) = \frac{1}{2\pi} \int_0^{2\pi} \left(x + i\sqrt{1-x^2} \cos\phi \right)^l d\phi.$$

The first few Legendre polynomials are $P_0(x) = 1$, $P_1(x) = x$, $P_2(x) = (3x^2 - 1)/2$. All the others can be calculated using the recurrence relation

$$(l+1)P_{l+1}(x) - (2l+1)xP_l(x) + (l)P_{l-1}(x) = 0.$$

This is sometimes called *Bonner's recursion*.

By the general Sturm-Liouville theory, the Legendre polynomials satisfy the orthogonality conditions

$$\int_{-1}^{1} P_{l_1}(x) P_{l_2}(x)\, dx = \frac{2}{2l_1 + 1} \delta_{l_1, l_2}.$$

Note that either l_1 or l_2 can be used in the denominator of the normalization constant in the above equation because δ_{l_1, l_2} is only nonzero when $l_1 = l_2$.

3.2.2   Associated Legendre Functions

The *associated Legendre functions*, P_l^m, are solutions of the equation

$$\left(\left(1 - x^2\right) y' \right)' + \left(l(l+1) - \frac{m^2}{1-x^2} \right) y = 0,$$

which is equivalent to

$$\left(1 - x^2\right) y'' - 2xy' + \left(l(l+1) - \frac{m^2}{1-x^2} \right) y = 0.$$

Here $l = 0, 1, 2, \ldots$ and $0 \leq m \leq l$. The associated Legendre functions are calculated from the Legendre polynomials as

$$\begin{aligned}
P_l^m(x) &= (-1)^m \left(1 - x^2\right)^{m/2} \frac{d^m}{dx^m} P_l(x) \\
&= \frac{(l+m)!}{2\pi l!} i^m \int_0^{2\pi} \left(x + i\sqrt{1-x^2} \cos\phi \right)^l e^{im\phi} d\phi.
\end{aligned} \tag{3.10}$$

We note that some texts define the associated Legendre functions without the $(-1)^m$ factor. Either way, it is easy to see that $P_l(x) = P_l^0$.

The associated Legendre functions follow the recurrence relations

$$(l - m + 1)P_{l+1}^m(x) - (2l+1)xP_l^m(x) + (l+m)P_{l-1}^m(x) = 0, \tag{3.11}$$

and are extended to negative values of the superscript as

$$P_l^{-m}(x) = (-1)^m \frac{(l-m)!}{(l+m)!} P_l^m(x)$$

for all $0 \leq m \leq l$.

The orthogonality of the associated Legendre functions is

$$\int_{-1}^{1} P_{l_1}^m(x) P_{l_2}^m(x)\, dx = \frac{2}{2l_1 + 1} \frac{(l_1 + m)!}{(l_1 - m)!} \delta_{l_1, l_2}.$$

For any fixed $m \in [-l, l]$, the associated Legendre functions form an orthonormal basis for $\mathcal{L}^2([-1, 1])$. That is, any "nice" function on the interval $-1 \leq x \leq 1$ can be expanded in a series of normalized associated Legendre functions as

$$f(x) = \sum_{l=|m|}^{\infty} \hat{f}(l, m) \tilde{P}_l^m(x).$$

Due to the orthonormality of these functions, we have

$$\hat{f}(l, m) = \int_{-1}^{1} f(x) \tilde{P}_l^m(x) \, dx,$$

where

$$\tilde{P}_l^m(x) = \sqrt{\frac{2l + 1}{2} \frac{(l - m)!}{(l + m)!}} \, P_l^m(x).$$

The coefficient $\hat{f}(l, m)$ is called the associated Legendre transform.

3.2.3 Fast Legendre and Associated Legendre Transforms

Using recurrence relations to efficiently evaluate a weighted sum of Legendre polynomials is an idea that has been explored for some time (see, for example, [27, 82, 102] and references therein). In recent years there has been some very interesting work on the development of sampling and "fast" discrete Legendre (and associated Legendre) transforms. A fast approximate Legendre transform is presented in [3]. The remainder of this section reviews the work of Driscoll, Healy, and Rockmore [28, 29]. In particular, they have obtained the following result:

THEOREM 3.1 [28]
The associated Legendre transform, $\hat{f}(l, m)$, of a band-limited function, $f(x)$, with band limit $B = 2^K$, can be computed exactly (in exact arithmetic) for a fixed value of $m \leq l$ and all values of $l \leq B$ as

$$\hat{f}(l, m) = \sqrt{2} \sum_{j=0}^{2B-1} a_j f(x_j) P_l^m(x_j)$$

in $\mathcal{O}(B(\log B)^2)$ operations, where $x_j = \cos \pi j / 2B$, and all the a_j are determined by solving the linear system of equations

$$\sum_{k=0}^{2B-1} a_k P_l(x_k) = \delta_{l,0}.$$

Furthermore, these coefficients can be written in closed form as

$$a_j = \frac{1}{B} \sin(\pi j / 2B) \sum_{l=0}^{B-1} \frac{1}{2l + 1} \sin\left([2l + 1]\frac{\pi j}{2B}\right) \qquad (3.12)$$

for $j = 0, \ldots, 2B - 1$.

This is recent work which can be found in the literature, and we will not review the derivations for these results. It suffices to say that the recurrence relations like those in Eq. (3.11) provide a tool for performing fast implementations.

As a special case, when $m = 0$, the Legendre transform can be computed in a fast way as well.

Throughout the text we will use asymptotic complexity bounds established for exact computation of problems using exact arithmetic. In practice, numerical stability issues and the tradeoff between numerical error and running time must be taken into account. For the Legendre transform, this has been addressed in [3].

3.3   Jacobi, Gegenbauer, and Chebyshev Polynomials

The Legendre polynomials form one complete orthogonal basis for functions on the interval $[-1, 1]$. There are several other important examples of polynomials that satisfy

$$(p_k, p_m) = \int_{-1}^{1} p_k(x) p_m(x) w(x)\, dx = C_k \delta_{k,m} \tag{3.13}$$

for some normalization C_k and weight $w(x)$. One very broad class of orthogonal polynomials is the *Jacobi polynomials*, $\{P_n^{(\alpha, \beta)}(x)\}$, which, for each fixed $\alpha, \beta > -1$ and $n = 0, 1, 2, \ldots$, form an orthogonal basis with respect to the weight

$$w(x) = (1 - x)^{\alpha} (1 + x)^{\beta}.$$

The Jacobi polynomials are calculated using the Rodrigues formula

$$P_n^{(\alpha, \beta)}(x) = \frac{(-1)^n}{2^n n!} (1 - x)^{-\alpha} (1 + x)^{-\beta} \frac{d^n}{dx^n} \left[(1 - x)^{\alpha+n} (1 + x)^{\beta+n} \right]. \tag{3.14}$$

In the case when $\alpha = \beta = \lambda - \frac{1}{2}$, the set of functions $\{P_n^{(\lambda)}(x)\}$ for fixed λ and $n = 0, 1, 2, \ldots$ are called the *Gegenbauer polynomials*. If $\lambda = 0$, the Gegenbauer polynomials reduce to the Chebyshev (or Tschebysheff) polynomials of the second kind, which are defined as

$$T_n(x) = \cos \left(n \cos^{-1} x \right) = \frac{1}{2} \left[\left(x + \sqrt{x^2 - 1} \right)^n + \left(x - \sqrt{x^2 - 1} \right)^n \right]$$

for $0 \le \cos^{-1} x \le \pi$. They are solutions of the equation

$$\left(1 - x^2 \right) y'' - xy' + n^2 y = 0$$

where $n = 0, 1, 2, \ldots$. It may be shown that $T_n(-x) = (-1)^n T_n(x)$ and

$$\int_{-1}^{1} T_k(x) T_m(x) \frac{dx}{\sqrt{1 - x^2}} = F_m \delta_{k,m}$$

where

$$F_m = \begin{cases} \frac{\pi}{2} & m \ne 0 \\ \pi & m = 0. \end{cases}$$

See [84] for more properties. Some computational aspects are addressed in [19].

A band-limited function in this context is one for which

$$f(x) = \sum_{n=0}^{B-1} c_n T_n(x)$$

for finite B where

$$c_n = \frac{1}{F_n} \int_{-1}^{1} f(x) T_n(x) \frac{dx}{\sqrt{1-x^2}}.$$

We note that by setting $x = \cos y$, the function may be evaluated as $f(x) = f(\cos y) = \tilde{f}(y)$. Then the band-limited expansion above becomes

$$\tilde{f}(y) = \sum_{n=0}^{B-1} c_n \cos ny \qquad (3.15)$$

where

$$c_n = \frac{1}{F_n} \int_0^{\pi} \tilde{f}(y) \cos ny \, dy.$$

Sampling this integral at the points $y = \pi j / B$ for $j = 0, 1, 2, \ldots, B - 1$ produces the exact result

$$c_n = \frac{1}{B} \sum_{j=0}^{B-1} \tilde{f}(\pi j / B) \cos(\pi n j / B).$$

This *discrete cosine transformation* (DCT) may be viewed as a variant of the DFT, and together with Eq. (3.15) sampled at $y = \pi j / B$ may be computed for all $j = 0, \ldots, B - 1$ in $\mathcal{O}(B \log B)$ computations. Therefore, a fast discrete (sampled) Chebyshev transform results as well.

3.4 Hermite and Laguerre Polynomials

Hermite polynomials[4] are solutions to the differential equation

$$y'' - 2xy' + 2ny = 0$$

or equivalently,

$$\left(e^{-x^2} y'\right)' + 2ne^{-x^2} y = 0,$$

for $n = 0, 1, 2, \ldots$ and $x \in \mathbb{R}$. They are generated by the Rodrigues formula

$$H_n(x) = (-1)^n e^{x^2} \frac{d^n}{dx^n} \left(e^{-x^2}\right),$$

and they satisfy the recurrence formulas

$$H_{n+1} = 2x H_n(x) - 2n H_{n-1}(x); \quad H_n'(x) = 2n H_{n-1}(x).$$

The first few Hermite polynomials are $H_0(x) = 1$, $H_1(x) = 2x$, and $H_2(x) = 4x^2 - 2$. The Hermite polynomials satisfy the orthogonality conditions

$$\int_{-\infty}^{\infty} H_m(x) H_n(x) e^{-x^2} dx = 2^n n! \sqrt{\pi} \delta_{m,n}.$$

---

[4]These are named after Charles Hermite (1822–1901).

The *Hermite functions* are defined as

$$h_n(x) = \frac{1}{2^{n/2}\sqrt{n!}\sqrt{\pi}} H_n(x) e^{-\frac{x^2}{2}}.$$

Note that $h_0(x)$ is a Gaussian scaled so that $\|h_0^2\| = 1$.

The set $\{h_n(x)\}$ for $n = 0, 1, 2, \ldots$ forms a complete orthonormal basis for the set of square-integrable functions on the line, $\mathcal{L}^2(\mathbb{R})$ (with unit weighting function). The Hermite functions are also very special because they are eigenfunctions of the Fourier transform. That is, the Fourier transform of a Hermite function is a scalar multiple of the same Hermite function evaluated at the frequency parameter. More specifically, it can be shown that [96]

$$\int_{-\infty}^{\infty} e^{-i\omega x} h_n(x)\, dx = \sqrt{2\pi}(-i)^n h_n(\omega).$$

Hence, a function expanded in a series of Hermite functions in the spatial domain will have a Fourier transform that is expanded in a series of Hermite functions in the frequency domain.

The *Laguerre polynomials*[5] are solutions of the differential equation

$$xy'' + (1 - x)y' + ny = 0,$$

or

$$\left(xe^{-x}y'\right)' + ne^{-x}y = 0,$$

for $x \in \mathbb{R}^+$. They satisfy the Rodrigues formula

$$L_n(x) = e^x \frac{d^n}{dx^n}\left(x^n e^{-x}\right).$$

The first few Laguerre polynomials are $L_0(x) = 1$, $L_1(x) = 1 - x$, $L_2(x) = x^2 - 4x + 2$, and the rest can be generated using the recurrence formulas

$$\begin{aligned} L_{n+1}(x) &= (2n + 1 - x)L_n(x) - n^2 L_{n-1}(x); \\ L_n'(x) &= nL_{n-1}'(x) - nL_{n-1}(x); \\ xL_n'(x) &= nL_n(x) - n^2 L_{n-1}(x). \end{aligned}$$

The Laguerre polynomials satisfy the orthogonality conditions

$$\int_0^{\infty} L_m(x)L_n(x)e^{-x}\, dx = (n!)^2 \delta_{m,n}.$$

In analogy with the Legendre polynomials, we can define *associated Laguerre polynomials* as

$$L_n^m(x) = \frac{d^m}{dx^m}\left(L_n(x)\right)$$

for $0 \le m \le n$. They satisfy the equation

$$xy'' + (m + 1 - x)y' + (n - m)y = 0,$$

---

[5]These are named after Edmond Laguerre (1834–1886).

and they satisfy the orthogonality conditions

$$\int_0^\infty L_n^m(x) L_p^m(x) e^{-x} x^m \, dx = \frac{(n!)^3}{(n-m)!} \delta_{p,n}.$$

Often in the literature the alternate definitions

$$\tilde{L}_n(x) = \frac{1}{n!} L_n(x) \text{ and } \tilde{L}_n^m(x) = (-1)^m \frac{d^m}{dx^m} \left(\tilde{L}_{n+m}(x) \right) = \frac{(-1)^m}{(n+m)!} L_{n+m}^m(x) \qquad (3.16)$$

are used, in which case

$$(n+1)\tilde{L}_{n+1}^k = (2n+k+1-x)\tilde{L}_n^k - (n+k)\tilde{L}_{n-1}^k.$$

Corresponding changes to the orthogonality relation follow also.

3.5   Quadrature Rules and Discrete Polynomial Transforms

We saw earlier that exact fast discrete transforms exist for the Fourier series, Legendre transform, and Chebyshev series. A natural question to ask at this point is whether or not fast discrete polynomial transforms exist for the other polynomials orthogonal on the interval $[-1, 1]$. (We shall not consider discrete transforms on an infinite domain such as those corresponding to the Hermite and Laguerre polynomials, and we restrict the discussion to functions on $[-1, 1]$ because any closed finite interval can be mapped to $[-1, 1]$ and vice versa).

We first need a systematic way to exactly convert integrals of band-limited functions into sums

$$\int_{-1}^1 f(x) w(x) \, dx = \sum_{i=0}^n f(x_i) w_i. \qquad (3.17)$$

Equation (3.17) can be viewed as a special case of Eq. (3.8). In this context, a band-limited polynomial of band-limit $B + 1$ is one of the form

$$f(x) = a_B x^B + a_{B-1} x^{B-1} + \cdots + a_1 x + a_0.$$

The set of all polynomials with band-limit $B + 1$ (or smaller) is denoted $\mathbb{P}_B$.

Any choice of the set of ordered pairs $\{(x_i, w_i) | i = 0, \ldots, n < \infty\}$ for which Eq. (3.17) holds for a given (fixed) weighting function $w(x)$ and *all* band-limited functions of band-limit $B + 1$ is called a *quadrature rule*. An optimal quadrature rule is one which minimizes the required number of points, n, while satisfying Eq. (3.17) for a given B. On the other hand, this may require locating the points $x_0, \ldots, x_n$ in irregular ways. Therefore, we should expect a tradeoff between regular location of points and the minimal number of points required for exact computation.

We now make the discussion concrete by reviewing some of the most common quadrature rules. For modern generalizations see [63]. Two of the most commonly used elementary rules are the *trapezoid rule* and *Simpson's rule*. These are exact for $w(x) = 1$ and small values of B. The even more elementary *centered difference rule* for numerical integration is only guaranteed to be exact when the integrand is a constant. The trapezoid and Simpson's rules are both based on evenly spaced samples and are respectively

$$\int_{-1}^1 f(x) \, dx = \frac{1}{n} \left[f(-1) + 2 \sum_{k=1}^{n-1} f(-1 + 2k/n) + f(1) \right]$$

and

$$\int_{-1}^{1} f(x)\, dx = \frac{2}{3n} \left[f(-1) + 4 \sum_{k=1}^{n/2} f(-1 + 2(2k-1)/n) + 2 \sum_{k=1}^{n/2-1} f(-1 + 4k/n) + f(1) \right].$$

In the case of Simpson's rule, n is assumed even. While both of these rules usually behave well numerically for polynomial functions f and large enough values of n, the trapezoid rule provides an *exact* answer only when $f(x) = a_1 x + a_0$, and Simpson's rule provides an exact answer only when $f(x) = a_3 x^3 + a_2 x^2 + a_1 x + a_0$ for finite values of a_i. In the case of the trapezoid rule it is assumed that $n \geq 1$, while for Simpson's rule $n \geq 2$.

Clearly these rules will not suffice for our purposes. The optimal quadrature rule (in the sense of requiring the fewest sample points) is presented in Section 3.5.2. But first we review an important general property of orthogonal polynomials.

3.5.1 Recurrence Relations for Orthogonal Polynomials

All of the orthogonal polynomials encountered thus far have had three-term recurrence relations of the form

$$p_n(x) = (a_n x + b_n) p_{n-1}(x) + c_n p_{n-2}(x) \tag{3.18}$$

for $n = 1, 2, 3, \ldots$ and $p_{-1}(x) \overset{\Delta}{=} 0$. It may be shown that for appropriate choices of $a_n, b_n,$ and c_n, Eq. (3.18) is *always* true for orthogonal polynomials [24, 93].

To prove this inductively, assume $p_0, \ldots, p_n$ are orthogonal to each other, and are generated by recurrence. By defining

$$p_{n+1} = \left(\frac{x}{(p_n, p_n)^{\frac{1}{2}}} - \frac{(xp_n, p_n)}{(p_n, p_n)^{\frac{3}{2}}} \right) p_n - \frac{(p_n, p_n)^{\frac{1}{2}}}{(p_{n-1}, p_{n-1})^{\frac{1}{2}}} p_{n-1}, \tag{3.19}$$

we seek to show that $(p_{n+1}, p_k) = 0$ for all $k = 0, \ldots, n$ given that $(p_r, p_s) = 0$ for all $r \neq s \in [0, n]$. Here $(\cdot, \cdot)$ is defined as in Eq. (3.13).

Since $xp_j \in \mathbb{P}_{j+1}$ for $p_j \in \mathbb{P}_j$ we can write $xp_j = \sum_{k=0}^{j+1} \alpha_k p_k(x)$ for some set of constants $\{\alpha_k\}$. For all $k \leq j + 1 < n$, it follows from $(p_n, p_k) = 0$ that $(xp_n, p_j) = (p_n, xp_j) = 0$. We therefore conclude that $(p_{n+1}, p_j) = 0$ for all $j = 0, \ldots, n - 2$.

Similarly, one finds that

$$(p_{n+1}, p_{n-1}) = \frac{(xp_n, p_{n-1})}{(p_n, p_n)^{\frac{1}{2}}} - (p_n, p_n)^{\frac{1}{2}} (p_{n-1}, p_{n-1})^{\frac{1}{2}},$$

which reduces to zero because

$$(xp_n, p_{n-1}) = (p_n, xp_{n-1}) = (p_n, p_n)(p_{n-1}, p_{n-1})^{\frac{1}{2}}. \tag{3.20}$$

To see this, evaluate Eq. (3.19) at $n - 1$ in place of n

$$p_n = \left(\frac{x}{(p_{n-1}, p_{n-1})^{\frac{1}{2}}} - \frac{(xp_{n-1}, p_{n-1})}{(p_{n-1}, p_{n-1})^{\frac{3}{2}}} \right) p_{n-1} - \frac{(p_{n-1}, p_{n-1})^{\frac{1}{2}}}{(p_{n-2}, p_{n-2})^{\frac{1}{2}}} p_{n-2}.$$

Next, take the inner product with p_n and rearrange terms to result in Eq. (3.20). The equality $(p_{n+1}, p_n) = 0$ is verified by inspection.

Thus far we have shown that $(p_{n+1}, p_k) = 0$ for $k = 0, \ldots, n$ under the inductive hypothesis that $p_0, \ldots, p_n$ are orthogonal and generated by recurrence. By starting with $p_{-1}(x) = 0$, $p_0(x) = C$, and calculating $p_1(x)$ from Eq. (3.19), one finds $(p_0, p_1) = 0$. This justifies our hypothesis up to $n = 1$, and this justification is carried to arbitrary n by the discussion above.

Evaluating Eq. (3.19) with $n - 1$ instead of n and comparing with Eq. (3.18), one finds

$$a_n = \frac{1}{(p_{n-1}, p_{n-1})^{\frac{1}{2}}}; \quad b_n = -\frac{(xp_{n-1}, p_{n-1})}{(p_{n-1}, p_{n-1})^{\frac{3}{2}}}; \quad c_n = -\frac{(p_{n-1}, p_{n-1})^{\frac{1}{2}}}{(p_{n-2}, p_{n-2})^{\frac{1}{2}}}.$$

With appropriate normalization, p_i can be assumed to be orthonormal. We may also set

$$p_{-1}(x) = 0 \quad \text{and} \quad p_0(x) = 1 \tag{3.21}$$

by normalizing $w(x)$. In this case, the coefficients a_n, b_n, and c_n are related to the leading coefficients $p_n(x) = k_n x^n + s_n x^{n-1} + \cdots$ as [24, 94]

$$a_n = \frac{k_n}{k_{n-1}}; \quad b_n = -a_n \left(\frac{s_n}{k_n} - \frac{s_{n-1}}{k_{n-1}} \right); \quad c_n = -a_n \frac{k_{n-2}}{k_{n-1}}. \tag{3.22}$$

The previous expressions for a_n and b_n follow from expanding p_n, p_{n-1}, and p_{n-2} out term by term in Eq. (3.18) and equating coefficients of powers x^n and x^{n-1}. The expression for c_n follows by observing that

$$(p_n, p_{n-2}) = a_n (xp_{n-1}, p_{n-2}) + c_n$$

and

$$(xp_{n-1}, p_{n-2}) = (p_{n-1}, xp_{n-2}) = \left(p_{n-1}, k_{n-2}x^{n-1} + s_{n-2}x^{n-2} + \cdots \right) = \left(p_{n-1}, k_{n-2}x^{n-1} \right)$$

since p_{n-1} is orthogonal to all polynomials in $\mathbb{P}_{n-2}$. Likewise,

$$\left(p_{n-1}, k_{n-2}x^{n-1} \right) = \frac{k_{n-2}}{k_{n-1}} \left(p_{n-1}, k_{n-1}x^{n-1} \right) = \frac{k_{n-2}}{k_{n-1}} (p_{n-1}, p_{n-1}) = \frac{k_{n-2}}{k_{n-1}}$$

since p_{n-1} is normalized.

We note that since Eq. (3.18) holds, so too does the *Christoffel-Darboux formula* [93, 94]

$$\sum_{i=0}^{n} p_i(x)p_i(y) = \frac{k_n}{k_{n+1}} \frac{p_{n+1}(x)p_n(y) - p_n(x)p_{n+1}(y)}{x - y}. \tag{3.23}$$

This follows since by using the substitution $p_{n+1}(x) = (a_{n+1}x + b_{n+1})p_n(x) + c_{n+1}p_{n-1}(x)$ and rearranging terms one finds

$$\begin{aligned} p_{n+1}(x)p_n(y) - p_n(x)p_{n+1}(y) &= a_{n+1}(x - y)p_n(x)p_n(y) \\ &\quad -c_{n+1} \left(p_n(x)p_{n-1}(y) - p_{n-1}(x)p_n(y) \right). \end{aligned}$$

Recursively expanding the term in parenthesis on the right, and using Eq. (3.22) to rewrite the resulting coefficients, we arrive at Eq. (3.23). As $y \to x$, a straightforward application of l'Hôpital's rule to Eq. (3.23) yields

$$\sum_{i=0}^{n} (p_i(x))^2 = \frac{k_n}{k_{n+1}} \left[p_n(x)p_{n+1}'(x) - p_{n+1}(x)p_n'(x) \right]. \tag{3.24}$$

3.5.2   Gaussian Quadrature

Let $p_k(x)$ for $k = 0, 1, 2, \ldots$ be a complete system of orthonormal polynomials in $\mathcal{L}^2([-1, 1], w(x)\, dx)$. Let $x_0, \ldots, x_n$ be the roots of p_{n+1}, and let $w_0, \ldots, w_n$ be the solution of the system of equations

$$\sum_{j=0}^{n} (x_j)^k w_j = \int_{-1}^{1} x^k w(x)\, dx \tag{3.25}$$

for $k = 0, \ldots, n$. Then [16, 34]

$$\sum_{j=0}^{n} p_k\left(x_j\right) w_j = \int_{-1}^{1} p_k(x) w(x)\, dx$$

for any $p_k \in \mathbb{P}_{2n+1}$, or more generally

$$\sum_{j=0}^{n} f\left(x_j\right) w_j = \int_{-1}^{1} f(x) w(x)\, dx \tag{3.26}$$

for any $f \in \mathbb{P}_{2n+1}$. In order to verify Eq. (3.26), it is convenient to define the *Lagrange fundamental polynomials* for any set of distinct points $\xi_0, \ldots, \xi_n$ and $k \in \{0, 1, 2, \ldots, n\}$ as

$$l_k^n(x) = \frac{(x - \xi_0)(x - \xi_1) \cdots (x - \xi_{k-1})(x - \xi_{k+1}) \cdots (x - \xi_n)}{(\xi_k - \xi_0)(\xi_k - \xi_1) \cdots (\xi_k - \xi_{k-1})(\xi_k - \xi_{k+1}) \cdots (\xi_k - \xi_n)}.$$

From this definition $l_k^n(x) \in \mathbb{P}_n$ and $l_k^n(\xi_j) = \delta_{kj}$. The Lagrange fundamental polynomials can be written in the alternate form

$$l_k^n(x) = \frac{l^{n+1}(x)}{(x - \xi_k)\left(l^{n+1}\right)'(\xi_k)}$$

where

$$l^{n+1}(x) = \prod_{k=0}^{n} (x - \xi_k) \in \mathbb{P}_{n+1}.$$

It follows from these definitions that

$$\sum_{k=0}^{n} l_k^n(x) = 1$$

and the function

$$\mathcal{L}_n(x) = \sum_{k=0}^{n} f_k l_k^n(x) \in \mathbb{P}_n$$

has the property $\mathcal{L}_n(\xi_j) = f_j$ for $j = 0, \ldots, n$. $\mathcal{L}_n(x)$ is called a *Lagrange interpolation polynomial*.

In the special case where $\xi_i = x_i$ for $i = 0, \ldots, n$, where x_i satisfies $p_{n+1}(x_i) = 0$, it follows that $l^{n+1}(x) = c_{n+1} p_{n+1}(x)$ for some nonzero constant $c_{n+1} \in \mathbb{R}$ because every polynomial in $\mathbb{P}_{n+1}$ is determined up to a constant by its zeros. In this case we write

$$\mathcal{L}_n(x) = \sum_{k=0}^{n} f\left(x_k\right) \frac{p_{n+1}(x)}{(x - x_k)\, p_{n+1}'\left(x_k\right)} = \sum_{k=0}^{n} f\left(x_k\right) l_k^n(x).$$

For any $f(x) \in \mathbb{P}_{2n+1}$ it then may be shown that [94]

$$f(x) - \mathcal{L}_n(x) = p_{n+1}(x) r(x)$$

for some $r(x) \in \mathbb{P}_n$. Therefore,

$$
\begin{aligned}
\int_{-1}^{1} f(x)w(x) \, dx &= \int_{-1}^{1} \mathcal{L}_n(x)w(x) \, dx + \int_{-1}^{1} p_{n+1}(x)r(x)w(x) \, dx \\
&= \int_{-1}^{1} \mathcal{L}_n(x)w(x) \, dx + 0 \qquad (3.27) \\
&= \sum_{k=0}^{n} f(x_k) \int_{-1}^{1} l_k^n(x)w(x) \, dx.
\end{aligned}
$$

This is Eq. (3.26) with

$$
w_k = \int_{-1}^{1} l_k^n(x)w(x) \, dx. \qquad (3.28)
$$

It may be shown (see [93]) that these weights satisfy Eq. (3.25).

Given Eq. (3.26), it follows that any polynomial of the form $f(x) = \sum_{k=0}^{2n+1} c_k p_k(x)$ can be integrated exactly with this rule. In particular, the product $p_j(x) p_k(x) \in \mathbb{P}_{2n+1}$ is true when $j, k \leq n$, and so

$$
\sum_{k=0}^{n} p_i(x_k) \, p_j(x_k) \, w_i = \int_{-1}^{1} p_i(x) p_j(x) w(x) \, dx = \delta_{ij} \qquad (3.29)
$$

(the second equality holding by the definition of orthonormality). This means that orthonormality of discrete polynomials follows naturally from the orthonormality of continuous ones when using Gaussian quadrature.

Consider a band-limited polynomial function with band-limit $B + 1$

$$
f(x) = \sum_{k=0}^{B} c_k p_k(x).
$$

When sampled at the $n + 1$ Gaussian quadrature points $x_0, \ldots, x_n$ this becomes

$$
f(x_i) = \sum_{k=0}^{B} c_k p_k(x_i). \qquad (3.30)
$$

We may therefore write

$$
\sum_{i=0}^{n} f(x_i) \, p_j(x_i) \, w_i = \sum_{i=0}^{n} \left(\sum_{k=0}^{B} c_k p_k(x_i) \right) p_j(x_i) \, w_i = \sum_{k=0}^{B} c_k \left(\sum_{i=0}^{n} p_k(x_i) \, p_j(x_i) \, w_i \right).
$$

The last term in parenthesis may be simplified to δ_{jk} using Eq. (3.29) *only if* $B \leq n$. Taking $n = B$ then ensures that the coefficients corresponding to the discrete polynomial expansion in Eq. (3.30) are calculated as

$$
a_j = \sum_{i=0}^{n} f(x_i) \, p_j(x_i) \, w_i. \qquad (3.31)
$$

Hence, when using Gaussian quadrature, the number of sample points is equal to the bandlimit. This comes, however, at the price of not having the freedom to choose the location of sample points.

Using the Christoffel-Darboux formula Eq. (3.23), it may be shown (see [25, 58, 93]) that the weights that solve Eq. (3.25) are determined as

$$w_i = -\frac{k_{n+2}}{k_{n+1}}\frac{1}{p_{n+2}(x_i)\,p_{n+1}'(x_i)} = \frac{k_{n+1}}{k_n}\frac{1}{p_n(x_i)\,p_{n+1}'(x_i)} = \frac{1}{\sum_{j=0}^n (p_j(x_i))^2}$$

where k_n is the leading coefficient of $p_n(x) = k_n x^n + \cdots$.

To verify the first two equalities, substitute $y = x_i$ into Eq. (3.23), integrate with respect to $w(x)\,dx$ over $[-1, 1]$, and simplify with Eq. (3.28). Combining either of the first two equalities with Eq. (3.24), and evaluating at $x = x_i$ yields the last equality.

3.5.3 Fast Polynomial Transforms

In the context of discrete polynomial transforms it is often convenient to specify the sample points in advance. The sample points in the cases of the DFT and discrete Legendre transform are examples of this. In general, when $n + 1$ distinct sample points are chosen, Eq. (3.17) will still hold, but instead of holding for all $f \in \mathbb{P}_{2n+1}$, it will only hold for $f \in \mathbb{P}_n$. This makes sense because there are $n + 1$ free weights determined by the $n + 1$ equations resulting from Eq. (3.25) for $k = 0, \ldots, n$.

The implication of this for discrete polynomial transforms is that when the set of sample points are fixed *a priori*, the number must be related to the band-limit as $n \geq 2B$. Thus by choosing $n = 2B$, Eqs. (3.30)–(3.31) remain true. We already saw this for the discrete Legendre transform. A natural choice is to evenly space the sample points. This is the assumption in the following theorem due to Driscoll, Healy, and Rockmore.

THEOREM 3.2 [29]
Let $B = 2^K$ for positive integer K, and $\phi_i(x)$ for $i = 0, \ldots, B - 1$ be a set of functions satisfying Eqs. (3.18) and (3.21). Then for any set of positive weights $w_0, \ldots, w_{n-1}$ (where $n = 2B$)

$$\hat{f}(k) = \sum_{j=0}^{n-1} f(j)\phi_k(j)w_j$$

can be computed (exactly in exact arithmetic) for all $k < B$ in $\mathcal{O}(B(\log B)^2)$ arithmetic operations.

Because of the similarity of form between Eqs. (3.30) and (3.31), this theorem indicates the existence of fast algorithms for both the computation of a_j for all $j = 0, \ldots, n - 1$, as well as the fast reconstruction of $f(x_i)$ for all i.

For points not evenly spaced, the following is useful:

THEOREM 3.3 [66]
Let $B = 2^k$ and $\{p_0, \ldots, p_{B-1}\}$ be a set of real orthogonal polynomials. Then for any sequence of real numbers, $f_0, \ldots, f_{B-1}$, the discrete polynomial transform defined by the sums

$$\hat{f}(l) = \frac{1}{B}\sum_{j=0}^{B-1} f_j p_j\left(\cos^{-1}[(2j+1)\pi/2B]\right)$$

may be calculated exactly in $\mathcal{O}(B(\log B)^2)$ exact arithmetic operations.

Both of the above theorems depend on the fast polynomial evaluation and interpolation algorithms reviewed in Appendix A and assume exact arithmetic. In practice, modifications to account for numerical errors are required. Recent work on fast discrete polynomial transforms can be found in [78].

The following section reviews the Bessel functions, for which we know of no exact fast discrete transform. This is problematic when considering the fast numerical evaluation of the Fourier transform in polar coordinates discussed in Chapter 4, and will require us to interpolate back and forth between polar and Cartesian coordinates in order to benefit from FFT methods. This step, while usually numerically acceptable, does not preserve the exactness of the results when linear or spline interpolation is used.

3.6 Bessel and Spherical Bessel Functions

Bessel functions[6] are solutions, $(y_i(x), \lambda_i)$, to the equation

$$y'' + (d-1)\frac{y'}{x} + \left(\lambda - \frac{m^2}{x^2}\right)y = 0. \tag{3.32}$$

This equation may be written in the form of the singular Sturm-Liouville equation

$$\left(x^{d-1}y'\right)' + \left(\lambda x^{d-1} - m^2 x^{d-3}\right)y = 0.$$

The parameter d is called the dimension and m is called the angular frequency or order.

One feature of the above equations is that the alternate definition of the independent variable as $\xi = x\sqrt{\lambda}$ and corresponding function $z(\xi) = y(x)$ results in the equation

$$\frac{d^2z}{d\xi^2} + (d-1)\frac{1}{\xi}\frac{dz}{d\xi} + \left(1 - \frac{m^2}{\xi^2}\right)z = 0, \tag{3.33}$$

which is independent of λ.

This follows from the chain rule

$$\frac{d}{dx} = \frac{d}{d\xi}\frac{d\xi}{dx} = \sqrt{\lambda}\frac{d}{d\xi}.$$

Hence, in the absence of boundary conditions, we can take $\lambda = 1$ and consider solutions to Eq. (3.33) without loss of generality since the solution to Eq. (3.32) will then be $y(x) = z(x\sqrt{\lambda})$.

3.6.1 Bessel Functions

Consider first the most common case when $d = 2$. For each fixed $m \in \{0, 1, 2, \dots\}$ there is a solution corresponding to $\lambda = 1$ of the form [11]:

$$J_m(x) = \frac{x^m}{2^m m!}\left(1 + \sum_{n=1}^{\infty} \frac{(-1)^n x^{2n}}{2^{2n} n! \prod_{k=1}^{n}(k+m)}\right). \tag{3.34}$$

It may be shown that an equivalent expression for these *Bessel functions of the first kind* is

$$J_m(x) = \frac{i^{-m}}{2\pi}\int_{-\pi}^{\pi} e^{ix\cos\theta}e^{-im\theta}\,d\theta.$$

---

[6]These are named after Friedrich Wilhelm Bessel (1784–1846).

That is, $J_m(x)$ is the mth Fourier coefficient of the function $i^{-m}e^{ix\cos\theta}$.

By taking derivatives of either of the above expressions for $J_m(x)$, it is easy to see that

$$J'_m(x) = \frac{1}{2}\left[J_{m-1}(x) - J_{m+1}(x)\right], \quad \left[x^m J_m(x)\right]' = x^m J_{m-1}(x).$$

We also have the recurrence formula

$$J_m(x) = \frac{x}{2m}\left[J_{m-1}(x) + J_{m+1}(x)\right]$$

and the symmetry formula

$$J_{-m}(x) = (-1)^m J_m(x),$$

that extends the definition of the Bessel functions from the nonnegative integers to all the integers.

The Bessel functions, $\{J_m(x)\}$, satisfy several kinds of orthogonality conditions. Namely, if $\{x_n\}$ for $n = 1, 2, \ldots$ are nonnegative solutions of the equation

$$x_n J'_m(x_n)\cos\beta + J_m(x_n)\sin\beta = 0 \qquad (3.35)$$

for $m \geq 0$ and $0 \leq \beta \leq \pi/2$, then [77, 100]

$$\int_0^1 J_m\left(xx_{n_1}\right) J_m\left(xx_{n_2}\right) x\, dx = C_{m,n_1}\delta_{n_1,n_2}$$

for the constant

$$C_{m,n} = \begin{cases} \frac{1}{2}J_{m+1}^2(x_n) & \text{for } \beta = \pi/2 \\[2mm] \frac{(x_n^2 - m^2 + \tan^2\beta)J_m^2(x_n)}{2x_n^2} & \text{for } 0 \leq \beta < \pi/2. \end{cases}$$

Equation (3.35) results from a separable boundary condition on $y_n(x) = J_m(xx_n)$ at $x = 1$. The scaling of the dependent variable used here is consistent with the eigenvalue $\lambda_n = x_n^2$. From Sturm-Liouville theory it follows that the functions y_{n_1} and y_{n_2} corresponding to $\lambda_{n_1} \neq \lambda_{n_2}$ must be orthogonal as shown above.

The above implies that functions on the interval $[0, 1]$ that are square-integrable with respect to $x\,dx$ can be expanded in a *Fourier-Bessel series*[7] as

$$f(x) = \sum_{n=1}^{\infty} c_n J_m\left(xx_n\right),$$

where

$$c_n = \frac{1}{C_{m,n}}\int_0^1 f(x)J_m\left(xx_n\right) x\, dx.$$

In analogy with the way the concept of the Fourier series of functions on the circle is extended to the Fourier transform of functions on the line, a transform based on Bessel functions can be defined. It is called the mth *order Hankel transform*[8]

$$\hat{f}_m(p) = \int_0^{\infty} f(x)J_m(px)x\, dx, \qquad (3.36)$$

---

[7] This is also called *Dini's series*.
[8] This is named after Hermann Hankel (1839–1873).

and its inverse is of the same form

$$f(x) = \int_0^\infty \hat{f}_m(p) J_m(px) p \, dp. \tag{3.37}$$

The completeness relation associated with this transform pair is

$$\int_0^\infty J_m(px) J_m\left(p'x\right) x \, dx = \frac{1}{p} \delta\left(p - p'\right).$$

3.6.2 Spherical Bessel Functions

For the case when $d = 3$, Eq. (3.32) becomes

$$y'' + \frac{2}{x} y' + \left(\lambda - \frac{l(l+1)}{x^2}\right) y = 0$$

where m^2 is replaced with $l(l+1)$. The reason for this becomes clear in Chapter 4 in the context of orthogonal expansions in spherical coordinates. If we let $y = x^{-\frac{1}{2}} z$, then we see that

$$z'' + \frac{1}{x} z' + \left(\lambda - \frac{(m')^2}{x^2}\right) z = 0,$$

and hence $z(x) = c J_{m'}(x)$ for some normalizing constant c where $(m')^2 = l(l+1) + \frac{1}{4} = (l+\frac{1}{2})^2$. When $\lambda = 1$, solutions $y(x)$ are then of the form

$$j_l(x) = \left(\frac{\pi}{2x}\right)^{\frac{1}{2}} J_{l+\frac{1}{2}}(x)$$

for $l = 0, 1, 2, \ldots$ where c is chosen as $(\pi/2)^{1/2}$. Unlike the functions $J_m(x)$ for integer m, the *spherical Bessel functions* $j_l(x)$ can always be written as finite combinations of elementary trigonometric and rational functions for finite integers l. For instance, $j_0(x) = (1/x) \sin x$, $j_1(x) = (1/x^2) \sin x - (1/x) \cos x$, and $j_2(x) = (3/x^3 - 1/x) \sin x - (3/x^2) \cos x$.

Here again, as a special case of Sturm-Liouville theory, the functions $y_n(x) = j_l(x\sqrt{\lambda_n})$ are orthogonal on the finite domain $0 \le x \le 1$ where λ_n is determined by the separable boundary condition $y_n'(1) \cos \beta + y_n(1) \sin \beta = 0$. For $0 \le \beta \le \pi/2$, this orthogonality is [77]

$$\int_0^1 j_l\left(x\sqrt{\lambda_{n_1}}\right) j_l\left(x\sqrt{\lambda_{n_2}}\right) x^2 \, dx = D_{n_1,l} \delta_{n_1,n_2}$$

where

$$D_{n,l} = \begin{cases} \frac{\pi}{4} \left(J'_{l+1/2}\right)^2 \left(\sqrt{\lambda_n}\right) & \text{for } \beta = \pi/2 \\[2ex] \pi \left(\lambda_n + \tan^2 \beta - (m')^2\right) J^2_{l+1/2}\left(\sqrt{\lambda_n}\right) & \text{for } 0 \le \beta < \pi/2 \end{cases}$$

and $m' = l + \frac{1}{2}$.

Again, a continuous transform on $\mathbb{R}^+$ exists. It is the *spherical Hankel transform* [5]

$$\hat{f}_l(p) = \int_0^\infty f(x) j_l(px) x^2 \, dx, \tag{3.38}$$

and it is its own inverse,

$$f(x) = \int_0^\infty \hat{f}_l(p) j_l(px) p^2 \, dp. \tag{3.39}$$

The corresponding completeness relation is [5]

$$\int_0^\infty j_l(px) j_l\left(p'x\right) x^2 \, dx = \frac{\pi}{2p^2} \delta\left(p - p'\right).$$

3.6.3   The Bessel Polynomials

The *Bessel polynomials* [56], $y_n(x)$, are the functions that satisfy the Sturm-Liouville equation

$$\left(x^2 e^{-2/x} y_n'\right)' = n(n+1) e^{-2/x} y_n.$$

This is equivalent to the equation

$$x^2 y_n'' + 2(x+1) y_n' = n(n+1) y_n.$$

These polynomials satisfy the three-term recurrence relation

$$y_{n+1} = (2n+1) x y_n + y_{n-1},$$

and the first few are of the form $y_0(x) = 1$, $y_1(x) = 1 + x$, $y_2(x) = 1 + 3x + 3x^2$. They can be generated with the Rodrigues formula

$$y_n(x) = \frac{e^{2/x}}{2^n} \frac{d^n}{dx^n} \left(x^{2n} e^{-2/x}\right).$$

The orthogonality relation for these polynomials is [56]

$$\frac{1}{2\pi i} \oint_{S^1} y_m(z) y_n(z) e^{-2/z} \, dz = (-1)^{n+1} \frac{2}{2n+1} \delta_{m,n}$$

where S^1 is the unit circle in the complex plane. (This can be replaced with any other closed curve containing the origin.) Choosing S^1 and $z = e^{i\theta}$, we write

$$\frac{1}{2\pi} \int_0^{2\pi} y_m\left(e^{i\theta}\right) y_n\left(e^{i\theta}\right) \exp\left(i\theta - 2e^{-i\theta}\right) d\theta = (-1)^{n+1} \frac{2}{2n+1} \delta_{m,n}.$$

These polynomials are connected to the spherical Bessel functions through the relationship [56]

$$j_n(x) = (1/2x) \left[i^{-n-1} e^{ix} y_n(-1/ix) + i^{n+1} e^{-ix} y_n(1/ix)\right]. \tag{3.40}$$

For further reading on Bessel polynomials see [40].

3.6.4   Fast Numerical Transforms

In many applications it is desirable to compute either the Hankel or the spherical-Hankel transform as efficiently and accurately as possible (see, for example, [1, 17, 21, 30, 41, 44, 47, 49, 50, 59, 60, 61, 64, 70, 71, 76, 83, 89, 90]). Unfortunately, it seems that unlike the FFT for the exact computation of Fourier series sampled at a finite number of evenly spaced points, a fast, exact Hankel transform is yet to be formulated. A number of efficient and accurate (though not exact) numerical approximations have been reported in the literature. A review can be found in [22].

We are aware of four main categories of fast numerical Hankel transform algorithms. These are: (1) approximation of a compactly supported function in a Fourier-Bessel series and use of approximate quadrature rules; (2) conversion of the Hankel transform to the Fourier transform by a change of variables and interpolation; (3) expansion of a function in a band-limited series of Laguerre polynomials and closed-form analytical computation of Hankel transform; and (4) approximation of Bessel functions using asymptotic expansions. We review these methods in some detail below.

Sampling the Fourier-Bessel Series

If we assume that the independent variable is normalized so that $f(x) = 0$ for $x \geq 1$, then the Hankel transform becomes the equation for the coefficients of the Fourier-Bessel (Dini) series where $\beta = \pi/2$ and $J_m(x_n) = 0$ in Eq. (3.35).

Sampling the integral at quadrature points results in a sum. Since the Bessel functions are not polynomials, this will not result in an exact quadrature. Using Eq. (3.8) is also possible, but this is slow. Unfortunately, the numerical stability is poor when using the recurrence relations for Bessel functions [35, 37, 102], and we know of no exact fast algorithm.

Converting to the Fourier Transform

A number of methods use tricks that convert the Hankel transform to a form where the Fourier transform and/or the convolution theorem can be used.

Candel [12, 13, 14] uses the relationship

$$e^{ir\sin\theta} = \sum_{k=-\infty}^{+\infty} e^{ik\theta} J_k(r)$$

with $r = px$ to write

$$\int_0^\infty e^{ipx\sin\theta} f(x)x\,dx = \sum_{k=-\infty}^{+\infty} e^{ik\theta} \hat{f}_k(p)$$

where $\hat{f}_k(p)$ denotes the kth-order Hankel transform of $f(x)$. The left side of the above is the Fourier transform of the function

$$F(x) = \begin{cases} xf(x) & \text{for } x > 0 \\ 0 & \text{for } x \leq 0 \end{cases}$$

evaluated at frequency $p\sin\theta$. The right side indicates that the kth order Hankel transform of $f(x)$ is the kth Fourier series coefficient of the expansion on the right. Multiplying both sides by $e^{-ik\theta}$, integrating over $0 \leq \theta < 2\pi$ and dividing by 2π isolates $\hat{f}_k(p)$. Discretizing the above steps provides a numerical procedure for approximate calculation of the Hankel transforms. A similar result is obtained by using the projection-slice and back-projection theorems [73, 74] discussed in Chapter 13.

Another trick is to change variables so that the independent variables are of the form [91, 95]

$$x = x_0 e^{\alpha\xi} \quad p = p_0 e^{\alpha\eta} \tag{3.41}$$

where x_0, p_0, and α are constants chosen for good numerical performance [91]. Substituting Eq. (3.41) into the kth order Hankel transform results in the expression

$$\tilde{g}(\eta) = \int_{-\infty}^\infty \tilde{f}(\xi)\tilde{h}(\xi + \eta)d\xi \tag{3.42}$$

where $\tilde{f}(\xi) = xf(x)$, $\tilde{g}(\eta) = pg(p)$, and $\tilde{h}(\xi + \eta) = \alpha xp J_k(xp)$. Since Eq. (3.42) is of the form of a correlation (a convolution with a sign difference), the FFT can be used when $\tilde{f}$ and $\tilde{h}$ are approximated as band-limited functions. The drawbacks of this method are that the band-limited approximations are not exact, and the use of unequally (exponentially) spaced sample points in the original domain is awkward.

Using the Laguerre Series Expansion

The following method was developed in [15].

Recall that the associated Laguerre polynomials are an orthogonal basis for $\mathcal{L}^2(\mathbb{R}^+, x^m e^{-x} dx)$, and when the normalization in Eq. (3.16) is used,

$$\int_0^\infty \tilde{L}_n^m(x)\tilde{L}_p^m(x)x^m e^{-x} dx = \frac{(n+m)!}{n!}\delta_{p,n}.$$

The *associated-Laguerre functions* (also called *Gauss-Laguerre functions*) are defined as

$$\varphi_{n,m}(x) = N_{n,m} x^{m/2} \tilde{L}_n^m(x) e^{-x/2},$$

where $N_{n,m}$ is the normalization so that

$$\int_0^\infty \varphi_{n,m}(x)\varphi_{p,m}(x)\, dx = \delta_{n,p}.$$

These functions constitute an orthonormal basis for $\mathcal{L}^2(\mathbb{R}^+)$. They have the property [15, 88]

$$\int_0^\infty (\alpha x)^m \tilde{L}_n^m\left(\alpha^2 x^2\right) e^{-\alpha^2 x^2/2} J_m(px)x\, dx = (-1)^n \alpha^{-m-2}(k/\alpha)^m m \tilde{L}_n^m\left(p^2/\alpha_2\right) e^{p^2/2\alpha^2}$$

for an arbitrary positive constant α. Therefore if $f(x) \in \mathcal{L}^2(\mathbb{R}^+)$ is expanded in a Gauss-Laguerre series as

$$f(x) = \sum_{m=0}^\infty C_n^m \varphi_{n,m}\left(\alpha^2 x^2\right)$$

where

$$C_n^m = N_{n,m} \int_0^\infty f(x) L_n^m\left(\alpha^2 x^2\right) e^{-\alpha^2 x^2/2}(\alpha x)^m d\left(\alpha^2 x^2\right),$$

then

$$\hat{f}_m(p) = \int_0^\infty f(x) J_m(px)x\, dx = \sum_{m=0}^\infty (-1)^n \alpha^{-m-2} C_n^m \varphi_{n,m}\left(p^2/\alpha^2\right).$$

This is also a Gauss-Laguerre series. And if f has band limit B (i.e., $C_n^m = 0$ for all $m \geq B$) then $\hat{f}_m(p)$ will also have band-limit B.

The approximation required in this method is to calculate the coefficients C_n^m numerically. A recursive numerical procedure for doing this is explained in [15].

For functions that are not band-limited, the parameter α is chosen to minimize the number of terms in the series expansion for a given error criterion.

Asymptotic Expansions

In this approximate method, the asymptotic form of Bessel functions for large values of their argument is used to gain simplifications. We shall not consider this method in detail, and refer the reader to [75] for descriptions of this method.

3.7   Hypergeometric Series and Gamma Functions

The generalized *hypergeometric series* is defined as

$$_pF_q\left(\alpha_1, \alpha_2, \ldots, \alpha_p; \rho_1, \rho_2, \ldots, \rho_q; z\right) = \sum_{k=0}^{\infty} \frac{z^k}{k!}\left[\frac{\prod_{h=1}^{p}(\alpha_h)_k}{\prod_{l=1}^{q}(\rho_l)_k}\right] \tag{3.43}$$

where for $k \in \mathbb{Z}^+$ and for $a = \alpha$ or ρ,

$$(a)_k \overset{\Delta}{=} \prod_{j=0}^{k-1}(a+j) = a(a+1)\cdots(a+k-1) \text{ and } (a)_0 \overset{\Delta}{=} 1.$$

The term $(a)_n$ is called the *Pochhammer symbol*. The series in Eq. (3.43) terminates if one of the α_i is zero or a negative integer.

Using the previous notation, the *binomial coefficient* is written as

$$C_n^k = \binom{n}{k} = \frac{n!}{k!(n-k)!} = \frac{(-1)^k(-n)_k}{k!}.$$

The compact notation

$$_pF_q\left(\alpha_1, \alpha_2, \ldots, \alpha_p; \rho_1, \rho_2, \ldots, \rho_q; z\right) = \ _pF_q\left(\begin{matrix}\alpha_1, \alpha_2, \ldots, \alpha_p \\ \rho_1, \rho_2, \ldots, \rho_q\end{matrix}\ \middle|\ z\right)$$

is often used. The most commonly encountered values are $(p, q) = (2, 1)$ and $(p, q) = (1, 1)$

$$_2F_1(a, b; c; z) = \ _2F_1\left(\begin{matrix}a, b \\ c\end{matrix}\ \middle|\ z\right) = \sum_{k=0}^{\infty}\frac{(a)_k(b)_k}{(c)_k}\frac{z^k}{k!}$$

$$_1F_1(a; c; z) = \ _1F_1\left(\begin{matrix}a \\ c\end{matrix}\ \middle|\ z\right) = \sum_{k=0}^{\infty}\frac{(a)_k}{(c)_k}\frac{z^k}{k!}.$$

The former is called the *hypergeometric function*, and the latter is called the *confluent hypergeometric function*.

The *gamma function* is defined for complex argument z when $Re(z) > 0$ as [62]

$$\Gamma(z) = \int_0^{\infty} e^{-t}t^{z-1}\,dt.$$

It satisfies the difference equation

$$\Gamma(z+1) = z\Gamma(z),$$

and for positive integer argument

$$\Gamma(n) = (n-1)!$$

The gamma function is related to the coefficients in the hypergeometric series as

$$(a)_k = \Gamma(a+k)\Gamma(a).$$

The hypergeometric and confluent hypergeometric functions can be written as the integrals [97]

$$_2F_1(a, b; c; z) = \frac{\Gamma(c)}{\Gamma(a)\Gamma(c-a)}\int_0^1 (1-tz)^{-b}t^{a-1}(1-t)^{c-a-1}\,dt$$

and

$$_1F_1(a; c; z) = \frac{\Gamma(c)}{\Gamma(a)\Gamma(c-a)} \int_0^1 e^{zt} t^{a-1} (1-t)^{c-a-1} \, dt.$$

The classical functions of mathematical physics such as Bessel functions and orthogonal polynomials can be written in terms of hypergeometric functions and gamma functions. Following are some of these expressions. Many more such relationships can be found in [62, 97].

$$e^z = {}_0F_0(z);$$

$$(1+z)^a = {}_1F_0(-a; -z) \text{ for } |z| < 1;$$

$$\ln(1+z) = (z{}_2F_1)(1, 1; 2; -z) \text{ for } |z| < 1;$$

$$\text{Erf}(z) \triangleq \int_0^z e^{-x^2} \, dx = z \, {}_1F_1\left(\frac{1}{2}; \frac{3}{2}; -z^2\right);$$

$$J_\nu(z) = \frac{(z/2)^\nu e^{\pm iz}}{\Gamma(\nu+1)} \, {}_1F_1\left(\frac{1}{2}+\nu; 1+2\nu; \mp 2iz\right);$$

$$P_\nu^\mu(z) = \frac{[(z+1)/(z-1)]^{\mu/2}}{\Gamma(1-\mu)} \, {}_2F_1(-\nu, \nu+1; 1-\mu; (1-z)/2);$$

$$P_n^{(\alpha,\beta)}(z) = \frac{\Gamma(n+\alpha+1)}{n!\Gamma(\alpha+1)} \, {}_2F_1(-n, n+\alpha+\beta+1; \alpha+1; (1-z)/2);$$

$$\tilde{L}_n^\alpha(z) = \frac{\Gamma(n+\alpha+1)}{n!\Gamma(\alpha+1)} \, {}_1F_1(-n; \alpha+1; z).$$

The hypergeometric functions are convenient for defining the polynomials in the subsections that follow.

3.7.1 Hahn Polynomials

Whereas quadrature rules induce discrete orthogonality relations from continuous ones, a number of 20th-century polynomials are defined with discrete orthogonality in mind from the beginning.

The *Hahn polynomials* [43, 52] are defined for $\alpha > -1$ and $\beta > -1$ and for positive integer N as

$$Q_n(x; \alpha, \beta, N) = {}_3F_2(-n, -x, n+\alpha+\beta+1; \alpha+1, -N+1; 1). \tag{3.44}$$

For fixed values of α and β, Q_n is a polynomial in the variable x that satisfies two discrete orthogonalities. These are

$$\sum_{x=0}^{N-1} Q_n(x; \alpha, \beta, N) Q_m(x; \alpha, \beta, N) \mu(x; \alpha, \beta, N) = \delta_{m,n}/v_n(\alpha, \beta, N)$$

and

$$\sum_{n=0}^{N-1} Q_n(x; \alpha, \beta, N) Q_n(y; \alpha, \beta, N) v_n(\alpha, \beta, N) = \delta_{x,y}/\mu(x; \alpha, \beta, N)$$

where

$$\mu(x; \alpha, \beta, N) = \frac{\dbinom{\alpha+x}{x}\dbinom{\beta+N-1-x}{N-1-x}}{\dbinom{N+\alpha+\beta}{N-1}}$$

and

$$v_n(\alpha, \beta, N) = \frac{\binom{N-1}{n}}{\binom{N+\alpha+\beta+n}{n}} \times \frac{\Gamma(\beta+1)}{\Gamma(\alpha+1)\Gamma(\alpha+\beta+1)}$$

$$\times \frac{\Gamma(n+\alpha+1)\Gamma(n+\alpha+\beta+1)}{\Gamma(n+\beta+1)\Gamma(n+1)} \times \frac{(2n+\alpha+\beta+1)}{(\alpha+\beta+1)}.$$

The Hahn polynomials satisfy a three-term recurrence relation and a difference equation that can be found in [52].

3.7.2 Charlier, Krawtchouk, and Meixner Polynomials

In this subsection, we review three modern polynomials with discrete orthogonalities. The functions

$$C_n(x, a) = \frac{\Gamma(x+1)(-a)^{-n}}{\Gamma(x-n+1)} \, {}_1F_1(-n; x-n+1; a)$$

are called *Charlier polynomials*. When the argument x is chosen from the numbers $0, 1, 2, \ldots$ the symmetry

$$C_n(x, a) = C_x(n, a) \tag{3.45}$$

is observed. They satisfy two discrete orthogonalities:

$$\sum_{x=0}^{\infty} \frac{a^x}{x!} C_k(x, a) C_p(x, a) = k! e^a a^{-k} \delta_{kp}$$

and

$$\sum_{n=0}^{\infty} \frac{a^n}{n!} C_n(x, a) C_n(y, a) = x! e^a a^{-x} \delta_{xy}.$$

The second orthogonality results from the first orthogonality and the symmetry Eq. (3.45).

The *Krawtchouk polynomials* of degree s and argument x are defined as

$$K_s(x; p; N) = \frac{(-1)^s}{p^s C_N^s} P_s^{(x-s, N-s-x)}(1+2p) = \, {}_2F_1\left(-x; -s; -N; p^{-1}\right).$$

They satisfy the discrete orthogonality [97]

$$\sum_{x=0}^{N} C_N^x p^x (1-p)^{N-x} K_s(x; p; N) K_q(x; p; N) = \frac{1}{C_N^s} \left(\frac{1-p}{p}\right)^s \delta_{sq}$$

(where C_N^x is the binomial coefficient) and the symmetry relation

$$K_s(x; p; N) = K_x(s; p; N).$$

Combining the previous two equations, we find the second orthogonality

$$\sum_{s=0}^{N} C_N^s p^s (1-p)^{N-s} K_s(x; p; N) K_s(y; p; N) = \frac{1}{C_N^x} \left(\frac{1-p}{p}\right)^x \delta_{xy}.$$

The *Meixner polynomials* [67, 68, 97] are defined as

$$M_s(x; \gamma; c) = s! P_s^{(\gamma-1,-s-x-\gamma)} \left(\frac{2}{c} - 1 \right) = \frac{\Gamma(\gamma + s)}{\Gamma(\gamma)} \, _1F_1(-x; -s; \gamma; 1 - 1/c)$$

for $0 < c < 1$ and $\gamma > 0$. The symmetry relation

$$M_s(x; \gamma; c) = \frac{\Gamma(\gamma + s)}{\Gamma(\gamma + x)} M_x(s; \gamma; c)$$

holds, as does the discrete orthogonality relation

$$\sum_{x=0}^{\infty} \frac{c^x (\gamma + x - 1)!}{x!} M_s(x; \gamma; c) M_q(x; \gamma; c) = s! (\gamma + s - 1)! c^{-s} (1 - c)^{-\gamma} \delta_{sq}.$$

Similar to the Charlier and Krawtchouk polynomials, a second orthogonality results from the symmetry relation

$$\sum_{s=0}^{\infty} \frac{c^s (\gamma + s - 1)!}{s!} M_s(x; \gamma; c) M_s(y; \gamma; c) = x! (\gamma + x - 1)! c^{-x} (1 - c)^{-\gamma} \delta_{xy}.$$

3.7.3   Zernike Polynomials

The *Zernike polynomials* [103] are defined as

$$Z_n^m(x) = \sum_{s=0}^{(n-|m|)/2} (-1)^s \frac{(n - s)!}{s! \left(\frac{n+|m|}{2} - s \right)! \left(\frac{n-|m|}{2} - s \right)!} x^{n-2s}$$

for $0 \leq m \leq n$ with $n \equiv m \, \mathrm{mod} 2$. They satisfy the differential equation

$$r \left(1 - r^2 \right) y'' + \left(1 - 3r^2 \right) y' + \left[n(n + 2)r - m^2/r \right] y = 0,$$

and can be written in terms of the hypergeometric function as

$$Z_n^m(x) = (-1)^{\frac{n-m}{2}} \begin{pmatrix} \frac{1}{2}(n + m) \\ m \end{pmatrix} x^m \, _2F_1 \left(\frac{n+m+1}{2}, \frac{m-n}{2}; m + 1, x^2 \right).$$

They are orthogonal on the interval $[0, 1]$ with respect to the weight $w(x) = x$

$$\int_0^1 Z_n^m(x) Z_{n'}^m(x) x \, dx = \frac{1}{2n + 2} \delta_{n,n'}.$$

They satisfy the Rodrigues formula

$$Z_n^m(x) = \frac{x^{-m}}{\left(\frac{n-m}{2} \right)!} \left(\frac{d}{d(x^2)} \right)^{\frac{n-m}{2}} \left[x^{n+m} \left(x^2 - 1 \right)^{\frac{n-m}{2}} \right]$$

and can be related to the Jacobi polynomials as

$$Z_n^m(x) = x^m \frac{P_{\frac{n-m}{2}}^{(0,m)} \left(2x^2 - 1 \right)}{P_{\frac{n-m}{2}}^{(0,m)}(1)}.$$

The normalization is chosen so that $Z_n^m(1) = 1$. The first few are $Z_0^0(x) = 1$, $Z_1^1(x) = x$, $Z_2^0(x) = 2x^2 - 1$, $Z_2^2(x) = x^2$, $Z_3^1(x) = 3x^2 - 2x$.
Zernike showed that

$$\int_0^1 Z_n^m(x) J_m(px)x \, dx = (-1)^{\frac{n-m}{2}} p^{-1} J_{n+1}(p).$$

The Zernike polynomials are often used in combination with the functions $e^{\pm im\theta}$ to form the orthogonal basis elements

$$V_n^{\pm m}(r, \theta) = Z_n^m(r)e^{\pm im\theta}$$

for the set of square-integrable functions on the disc of unit radius. For recent examples of Zernike polynomials being applied in the field of pattern analysis, see [54] (and references therein).

3.7.4   Other Polynomials with Continuous and Discrete Orthogonalities

A set of functions that generalizes the Zernike polynomials has been developed by Koornwinder (see [55] and references therein) which are orthogonal on the interval [0, 1] with respect to the weight $(1 - x^2)^\alpha$ for $\alpha > -1$.

A collection of polynomials of complex argument (called *Wilson polynomials*) defined as

$$\mathcal{W}_n\left(z^2; a, b, c, d\right) = (a+b)_n(a+c)_n(a+d)_n \cdot$$

$$_4F_3\left(\begin{array}{c} -n, a+b+c+d+n-1, a-z, a+z \\ a+b, a+c, a+d \end{array} \middle| 1\right)$$

and satisfying certain orthogonalities can be found in [101, 97].

Sets of polynomials with discrete orthogonality related to topics covered in Chapter 9 can be found in [7, 81].

3.8   Piecewise Constant Orthogonal Functions

Sturm-Liouville theory is over a century old. The recent advances in discrete polynomial transforms reviewed in Section 3.5.2 have deep historical roots as well. In contrast, the focus of this section is on non-polynomial orthonormal expansions and transforms developed in the 20th century. In the subsections that follow, we review the Haar and Walsh functions as well as the basics of wavelets.

In the 20th century, a number of new classes of non-polynomial orthogonal expansions on intervals and the real line were developed. One such class consists of functions that are piecewise constant over intervals. Three examples of this class are the Haar, Rademacher, and Walsh functions. All of these functions are defined for the independent variable $x \in [0, 1]$, are indexed by the number $m = 0, 1, 2, \ldots,$ and return values in the set $\{0, \pm C_m\}$ for a real constant C_m. They are all real-valued and orthonormal with respect to the inner product

$$(f_1, f_2) = \int_0^1 f_1(x)f_2(x) \, dx.$$

3.8.1   Haar Functions

The *Haar function* har$(0, 1, x)$ is defined as [42]

$$
\text{har}(0, 1, x) = \begin{cases} 1 & \text{for } x \in (0, 1/2) \\ -1 & \text{for } x \in (1/2, 1) \\ 0 & \text{elsewhere.} \end{cases}
$$

Higher order Haar functions are defined as

$$
\text{har}(r, m, x) = \begin{cases} 2^{r/2} & \text{for } \frac{m-1}{2^r} < x < \frac{m-1/2}{2^r} \\ -2^{r/2} & \text{for } \frac{m-1/2}{2^r} < x < \frac{m}{2^r} \\ 0 & \text{elsewhere.} \end{cases}
$$

Here $1 \leq m \leq 2^r$, and the only Haar function not defined by the previous rule, is har$(0, 0, x) = 1$.

We note that the Haar system of functions is orthonormal and complete, and the definition of har(r, m, x) for $r = 0, 1, 2, \ldots$ and $m = 1, \ldots, 2^r$ can be written in terms of the recurrence relations:

$$
\text{har}(r, 1, x) = \sqrt{2}\,\text{har}(r - 1, 1, 2x)
$$

and

$$
\text{har}(r, m, x) = \text{har}\left(r, m - 1, x - 2^{-r}\right).
$$

The Haar functions form a complete orthonormal system, and so any function $f(x) \in \mathcal{L}^2([0, 1])$ can be expanded in a Haar series as

$$
f(x) = \sum_{n=0}^{\infty} h_n \text{har}(n, x)
$$

where

$$
h_n = \int_0^1 f(x) \text{har}(n, x)\,dx.
$$

If $f(x)$ is piecewise constant over $N = 2^r$ evenly spaced sub-intervals of $[0, 1]$, then the *finite Haar transform* pair results from truncating the sum and evaluating the previous integral

$$
h_n = \frac{1}{N} \sum_{i=0}^{N-1} f(i/N) \text{har}(n, i/N)
$$

and

$$
f(i/n) = \sum_{n=0}^{N-1} h_n \text{har}(n, i/N).
$$

A fast algorithm exists for evaluating the finite Haar transform. It can be performed in $\mathcal{O}(N)$ arithmetic operations [85].

3.8.2 Rademacher Functions

The *Rademacher function* [80] rad(m, x) is defined for $x \in [0, 1]$ and returns values ± 1 at all but a finite number of points. The Rademacher functions are generated using the equality

$$\text{rad}(m, x) = \text{rad}\left(1, 2^{m-1}x\right),$$

or equivalently, the one-term recurrence relation

$$\text{rad}(m, x) = \text{rad}(m - 1, 2x),$$

where

$$\text{rad}(1, x) = \begin{cases} 1 & \text{for } x \in (0, 1/2) \\ -1 & \text{for } x \in (1/2, 1) \\ 0 & \text{elsewhere.} \end{cases}$$

The exception to the above rule for calculating rad(m, x) is rad$(0, x) = 1$. They observe the periodicity [2]

$$\text{rad}\left(m, t + n/2^{m-1}\right) = \text{rad}(m, t)$$

for $m = 1, 2, \ldots$ and $n = \pm 1, \pm 2, \ldots$. These periodicities are apparent when writing the Rademacher functions in the form

$$\text{rad}(m, x) = \text{sign}\left[\sin 2^m \pi x\right].$$

While the Rademacher functions are orthonormal, they are not complete in the sense that an arbitrary $f \in \mathcal{L}^2([0, 1])$ cannot be approximated to any desired accuracy in the $\mathcal{L}^2$ sense.

3.8.3 Walsh Functions and Transforms

In order to understand Walsh functions [98], we first need to introduce some notation for binary numbers. The value of the sth bit of an r-bit binary number b is denoted $b_s \in \{0, 1\}$ for $s = 0, 1, \ldots, r - 1$. Here $s = 0$ and $s = r - 1$ correspond to the most and least significant bit, respectively. Hence, in the binary number system we would write

$$b = b_0 b_1 \cdots b_{r-1} = \sum_{s=0}^{r-1} b_s 2^{r-1-s}.$$

Given the integers $n, k \in [0, 2^r - 1]$, their sth bit values when written as binary numbers are n_s and k_s. We define the coefficient [53]

$$W_k\left(n/2^r\right) = (-1)^{\sum_{s=0}^{r-1} k_{r-1-s} n_s} = \exp\left(\pi i \sum_{s=0}^{r-1} k_{r-1-s} n_s\right).$$

This can be related to the Rademacher functions as

$$W_k\left(n/2^r\right) = \prod_{s=0}^{r-1} \left(\text{rad}_{s+1}\left(n/2^r\right)\right)^{k_{r-1-s}}.$$

It is also convenient for all $x, a, b \in \mathbb{R}$ with $a < b$ to define the *window function*

$$\text{win}(x, a, b) = \begin{cases} 1 & \text{for } x \in (a, b) \\ 0 & \text{for } x \notin [a, b] \\ \frac{1}{2} & \text{for } x \in \{a, b\} \end{cases}$$

This function has the property

$$\text{win}(x, c \cdot a, c \cdot b) = \text{win}(x/c, a, b)$$

for any $c \in \mathbb{R}^+$.

The *Walsh function* $\text{wal}(k, x)$ is defined on the interval $[0, 1]$ as

$$\text{wal}(k, x) = \sum_{n=0}^{2^r - 1} W_k \left(n/2^r \right) \cdot \text{win} \left(x \cdot 2^r, n, n+1 \right). \tag{3.46}$$

When $x = m/N$ where $N = 2^r$ for some integer $0 < m < N$ we see that

$$\text{wal}(k, m/N) = W_k(m/N)\delta_{m,n} = W_k(n/N)\delta_{n,m} = \text{wal}(m, k/N).$$

In addition to the orthonormality

$$(\text{wal}(m, x), \text{wal}(n, x)) = \delta_{m,n},$$

the Walsh functions satisfy the discrete orthogonality [9]

$$\sum_{i=0}^{N-1} \text{wal}(m, i/N)\text{wal}(n, i/N) = N\delta_{m,n}$$

where $N = 2^r$.

The Walsh functions form a complete orthonormal system, and so any function $f(x) \in \mathcal{L}^2([0, 1])$ can be expanded in a Walsh series as

$$f(x) = \sum_{n=0}^{\infty} w_n \text{wal}(n, x)$$

where

$$w_n = \int_0^1 f(x)\text{wal}(n, x)\,dx.$$

If f is piecewise constant over $N = 2^r$ evenly spaced sub-intervals of $[0, 1]$, then the *finite Walsh transform* pair results from truncating the sum and evaluating the previous integral

$$w_n = \frac{1}{N} \sum_{i=0}^{N-1} f(i/N)\text{wal}(n, i/N) = \frac{1}{N} \sum_{i=0}^{N-1} f(i/N)W_k(n/N)$$

and

$$f(i/n) = \sum_{n=0}^{N-1} w_n\text{wal}(n, i/N) = \sum_{n=0}^{N-1} w_n W_k(n/N).$$

Walsh functions can be generated by a recurrence relation, and fast algorithms exist for evaluating these [9, 45]. One of these algorithms is called the *Walsh-Hadamard transform*, and can be performed in $\mathcal{O}(N \log_2 N)$ arithmetic operations.

See [2, 9], and [53] for modern reviews of applications and algorithms for fast implementations.

3.9 Wavelets

3.9.1 Continuous Wavelet Transforms

The basic reason for the popularity of wavelets since the mid-1980s is that certain functions can be expressed very efficiently (in the sense of requiring few terms for good approximation) as a weighted sum (or integral) of scaled, shifted, and modulated versions of a single function ("mother wavelet") with certain properties. The concept of wavelets initially entered the recent literature as a technique for the analysis of seismic measurements as described by Morlet et al. [72]. The connection between the concept of wavelets and group representations (the latter of which is a large part of Chapter 8) was established shortly thereafter by Grossmann, Morlet, and Paul [38, 39]. Daubechies (see, for example, [23]), Meyer (see, for example, [69]), Mallat (see, for example, [65]), and others contributed further to a rigorous theory behind the concept of wavelets, connected the idea to other techniques of classical harmonic analysis, and popularized the concept. Since then, there has been an explosion of work in both the theory and applications of wavelets. While wavelets can be superior to other expansions (in the sense or requiring fewer terms to achieve the same least-squares error), when a high degree of self-similarity exists in the signal or function being decomposed, they do have some drawbacks. These include poor performance under traditional operations such as convolution and differentiation.

The three basic operations upon which wavelet analysis is based are scale (dilation), translation (shift), and modulation of functions. These are applied to a function f as

$$(\sigma_s f)(x) = s^{-\frac{1}{2}} f(x/s) \tag{3.47}$$

$$(\tau_t f)(x) = f(x - t) \tag{3.48}$$

$$(\mu_m f)(x) = e^{-imx} f(x), \tag{3.49}$$

respectively. These are *unitary operations* in the sense that $U \in \{\sigma_s, \tau_t, \mu_m\}$ preserves the inner product of two complex-valued functions

$$(Uf_1, Uf_2) = (f_1, f_2) \tag{3.50}$$

where

$$(f_1, f_2) = \int_{-\infty}^{\infty} \overline{f_1(x)} f_2(x) \, dx.$$

As a special case, we see

$$\|Uf\|^2 = \|f\|^2$$

where $\|f\|^2 = (f, f)$. We note that operators unitary with respect to (f_1, f_2) are unitary with respect to $\overline{(f_1, f_2)}$ (and vice versa).

Continuous wavelet transforms are based on combinations of continuous translational and scale changes to a primitive function $\varphi \in \mathcal{L}^2(\mathbb{R})$. This is in contrast to the continuous Fourier transform which is based on continuous modulations of the function $\varphi(x) = 1$.

Continuous Scale-Translation-Based Wavelet Transforms

Let $\varphi \in \mathcal{L}^2(\mathbb{R})$ and $\hat{\varphi}$ be its Fourier transform. φ is chosen so that

$$0 < \int_0^{\infty} |\hat{\varphi}(\pm\omega)|^2 \frac{d\omega}{|\omega|} = C_{\pm} < \infty \tag{3.51}$$

(for reasons that will become clear shortly). It follows that

$$0 < \int_{-\infty}^{\infty} |\hat{\varphi}(\pm\omega)|^2 \frac{d\omega}{|\omega|} = C = C_+ + C_- < \infty.$$

A typical (but not the only) choice that satisfies this constraint is

$$\varphi(x) = \frac{2}{\pi^{1/4}\sqrt{3}} \left(1 - x^2\right) e^{-x^2/2}.$$

As a second example of a suitable function φ, consider the sinc function

$$\varphi(x) = \text{sinc}(x) = \frac{\sin \pi x}{\pi x}.$$

This is also called the *Shannon wavelet*, and its orthogonality and completeness as a wavelet system follow from the properties expressed in our discussion of the Shannon sampling theorem in Chapter 2. In particular Shannon observed the orthogonality

$$\int_{-\infty}^{\infty} \text{sinc}(2Wx - m)\text{sinc}(2Wx - n)\,dx = \frac{1}{2W}\delta_{m,n}.$$

In general, let

$$\varphi_{s,t}^p(x) = |s|^{-p}\varphi\left(\frac{x - t}{s}\right). \tag{3.52}$$

The range of values for scale changes and translations for the independent variable is $s \in \mathbb{R} - \{0\}$ and $t \in \mathbb{R}$. The number $p \in \mathbb{R}^+$ is chosen differently in various papers on wavelets. We follow the notation in [51] and leave p undetermined for the moment.

Since $\varphi \in \mathcal{L}^2(\mathbb{R})$, so is $\varphi_{s,t}^p$; since

$$\|\varphi_{s,t}^p\|^2 = |s|^{-2p} \int_{-\infty}^{\infty} \left|\varphi\left(\frac{x - t}{s}\right)\right|^2 dx = |s|^{1-2p}\|\varphi\|^2.$$

(The choice $p = 1/2$ is common and natural given the previous equation but we will not fix the value of p).

The function $\varphi_{s,t}^p$ is called a *wavelet*, and $\varphi_{1,0}^p = \varphi$ is called the *mother wavelet*. The Fourier transform of $\varphi_{s,t}^p$ is

$$\hat{\varphi}_{s,t}^p(\omega) = \mathcal{F}\left(\varphi_{s,t}^p\right) = \int_{-\infty}^{\infty} |s|^{-p}\varphi\left(\frac{x - t}{s}\right) e^{-i\omega x}\,dx = |s|^{1-p}e^{-i\omega t}\hat{\varphi}(\omega s).$$

(The choice $p = 1$ is also natural to simplify the previous equation.)

The *continuous wavelet transform* of the function $f \in \mathcal{L}^2(\mathbb{R})$ associated with the wavelet $\varphi_{s,t}^p$ is

$$\tilde{f}_p(s, t) = (\varphi_{s,t}^p, f) = \int_{-\infty}^{\infty} \overline{\varphi_{s,t}^p(x)} f(x)\,dx = \overline{\varphi_{-s,0}^p(t)} * f(t). \tag{3.53}$$

The derivation of the inverse of the continuous wavelet transform proceeds as follows [51]. Applying the Fourier transform to both sides of Eq. (3.53) and using the convolution theorem, we see that

$$\mathcal{F}\left(\overline{\varphi_{-s,0}^p(t)} * f(t)\right) = \mathcal{F}\left(\overline{\varphi_{-s,0}^p(t)}\right)\hat{f}(\omega).$$

Explicitly,

$$\mathcal{F}\left(\overline{\varphi_{-s,0}^p(t)}\right) = \int_{-\infty}^{\infty} |s|^{-p}\overline{\varphi(-t/s)}e^{-i\omega t}\,dt$$

$$= \int_{\mathrm{sgn}(s)\cdot\infty}^{-\mathrm{sgn}(s)\cdot\infty} |s|^{-p}\overline{\varphi(\tau)}e^{-i\omega t}(-sdt)$$

$$= |s|^{1-p}\overline{\hat{\varphi}(-s\omega)}.$$

Therefore, $f(x)$ is extracted from Eq. (3.53) by first observing that

$$\int_0^\infty \mathcal{F}\left(\tilde{f}_p(s,t)\right)\hat{\varphi}(s\omega)s^{p-2}\,ds = \int_0^\infty \mathcal{F}\left(\overline{\varphi_{-s,0}^p(t)} * f(t)\right)\hat{\varphi}(s\omega)s^{p-2}\,ds$$

$$= C_+\hat{f}(\omega)$$

where C_+ is defined in Eq. (3.51). Dividing by C_+ and writing the result using the relationships given previously,

$$\hat{f}(\omega) = \frac{1}{C_+}\int_0^\infty \left(\int_{-\infty}^\infty e^{-i\omega t}\hat{\varphi}(s\omega)\tilde{f}(s,t)dt\right)s^{p-2}\,ds$$

$$= \frac{1}{C_+}\int_0^\infty \left(\int_{-\infty}^\infty \hat{\varphi}_{s,t}^p(\omega)\tilde{f}(s,t)dt\right)s^{2p-3}\,ds.$$

Applying the inverse Fourier transform to both sides yields

$$f(x) = \frac{1}{C_+}\int_{-\infty}^\infty \int_0^\infty \varphi_{s,t}^p(x)\tilde{f}(s,t)s^{2p-3}\,ds\,dt.$$

Sometimes it is more natural to integrate over negative as well as positive values of s. Since $s = 0$ constitutes a set of measure zero (i.e., a set of no area) in the (s,t) plane, it need not be excluded from the integral defining the inverse wavelet transform, and we write

$$f(x) = \frac{1}{C}\int_{\mathbb{R}^2} \varphi_{s,t}^p(x)\tilde{f}(s,t)|s|^{2p-3}\,ds\,dt. \tag{3.54}$$

From the previous equation, another "natural" choice for the value of p is $3/2$.

Continuous Modulation-Translation-Based Wavelet Transforms

In 1946, Gabor [33] introduced the idea of expanding functions in terms of translated and modulated versions of a single window-like function. Continuous transforms based on this idea can be found in the literature (see [8] and references therein). To begin, assume that a complex-valued function $g \in \mathcal{L}^2(\mathbb{R})$ is given, and has the additional property that it is either supported on a closed interval, or at least decreases rapidly to zero outside of such an interval. Further, assume that $g(x)$ is normalized so that

$$\|g\|^2 = (g, g) = \int_{-\infty}^\infty |g(x)|^2\,dx = 1.$$

Then the *continuous Gabor transform* (or *complex spectrogram* [8]) of a function $f \in \mathcal{L}^2(\mathbb{R})$ is defined as

$$\tilde{f}_g(t, \omega) = \int_{\mathbb{R}} f(x)\overline{g(x-t)}e^{-i\omega x}\,dx. \tag{3.55}$$

If we multiply by $g(x - t)$ and integrate both sides over $t \in \mathbb{R}$, the result is

$$\int_{\mathbb{R}} \tilde{f}_g(t, \omega) g(x - t) dt = \int_{\mathbb{R}} f(x) \|g(x - t)\|^2 e^{-i\omega x} dx = \|g\|^2 \int_{\mathbb{R}} f(x) e^{-i\omega x} dx.$$

In the previous manipulations, we used the fact that the value of $\|g(x - t)\|^2$ is independent of t. Since $\|g\|^2 = 1$, all that remains is the Fourier transform on f. Hence, application of the inverse Fourier transform recovers f. Combining all the steps, we obtain the inversion formula

$$f(x) = \int_{\mathbb{R}^2} \tilde{f}_g(t, \omega) g(x - t) e^{i\omega x} d\omega dt. \tag{3.56}$$

3.9.2 Discrete Wavelet Transforms

Analogous to complete series expansions on the real line (like the Hermite functions), orthonormal wavelet series expansions (as opposed to transforms) are possible. As with the continuous wavelet transform where translation-scale and translation-modulation combinations are possible, we find these combinations for the discrete wavelet transform.

Discrete Scale-Translation Wavelet Series

Let $\psi \in \mathcal{L}^2(\mathbb{R})$ be normalized so that $\|\psi\|_2 = 1$, and let $\hat{\psi}$ be its Fourier transform. We seek conditions on the function ψ so that the set of functions

$$\psi_{j,k}(x) = 2^{j/2} \psi \left(2^j x - k\right) \tag{3.57}$$

for all $j, k \in \mathbb{Z}$ is orthonormal and complete. Following [46], and using the notation

$$(\psi, \phi) = \int_{\mathbb{R}} \psi \overline{\phi} \, dx,$$

a straightforward change of variables $x = 2^{-n}(y + m)$ shows that

$$\left(\psi_{j,k}, \psi_{n,m}\right) = \int_{\mathbb{R}} 2^{j/2} \psi \left(2^j x - k\right) \cdot 2^{n/2} \overline{\psi \left(2^n x - m\right)} \, dx \tag{3.58}$$

$$= \int_{\mathbb{R}} 2^{(j-n)/2} \overline{\psi \left(2^{(j-n)} y - \left(k - 2^{(j-n)} m\right)\right)} \cdot \psi(y) \, dy \tag{3.59}$$

$$= \left(\psi_{l,p}, \psi\right) \tag{3.60}$$

where $l = j - n$ and $p = k - 2^{j-n} m$. A similar calculation shows that

$$\left(\psi_{j,k}, \psi_{j,l}\right) = \left(\psi_{0,k}, \psi_{0,l}\right).$$

The orthonormality condition $(\psi_{j,k}, \psi_{j,l}) = \delta_{k,l}$ can, therefore, be examined in the context of the simpler expression $(\psi_{0,k}, \psi_{0,l}) = (\psi, \psi_{0,l-k}) = \delta_{k,l} = \delta_{0,l-k}$. It is convenient to define $r = l - k$. Again, following [46], necessary and sufficient conditions for this orthogonality to hold are

$$\delta_{0,r} = \int_{\mathbb{R}} \psi(x) \overline{\psi(x - r)} \, dx = \frac{1}{2\pi} \int_{\mathbb{R}} \left\|\hat{\psi}\right\|^2 e^{ir\xi} \, d\xi.$$

We have used the shift operational property and Parseval's equality previously. Since the integrand on the right is oscillatory, it makes sense to break the integration up as

$$\frac{1}{2\pi} \sum_{r=-\infty}^{+\infty} \int_{2r\pi}^{2(r+1)\pi} \left|\hat{\psi}(\xi)\right|^2 e^{ir\xi} \, d\xi = \frac{1}{2\pi} \int_0^{2\pi} \left(\sum_{r=-\infty}^{+\infty} \left|\hat{\psi}(\xi + 2r\pi)\right|^2\right) e^{ir\xi} \, d\xi.$$

This last integral is equal to $\delta_{0,r}$ when the term in parenthesis satisfies

$$\sum_{r \in \mathbb{Z}} \left| \hat{\psi}(\xi + 2r\pi) \right|^2 = 1. \tag{3.61}$$

A similar calculation follows from the condition $(\psi, \psi_{j,k}) = 0$. Namely, observing that the Fourier transform of Eq. (3.57) is

$$\mathcal{F}(\psi_{j,k}) = 2^{-j/2} \hat{\psi}\left(2^{-j}\xi\right) e^{-ik\xi/2^j},$$

Parseval's equality together with the change of scale in frequency $\nu = 2^{-j}\xi$ yields [46]

$$0 = (\psi, \psi_{j,k}) = \frac{1}{2\pi} \int_{\mathbb{R}} 2^{j/2} \hat{\psi}\left(2^j \nu\right) \overline{\hat{\psi}(\nu)} e^{ik\nu} \, d\nu$$

for $j, k \neq 0$. Breaking up the integral into a sum of integrals over intervals of length 2π as before and extracting the factor multiplying $e^{ik\nu}$, we see that the previous equation holds when

$$\sum_{k \in \mathbb{Z}} \hat{\psi}\left(2^j(\nu + 2k\pi)\right) \overline{\hat{\psi}(\nu + 2k\pi)} = 0. \tag{3.62}$$

The *discrete wavelet transform* (or discrete wavelet coefficient j, k) is defined as

$$\tilde{f}_{j,k} = (\psi_{j,k}, f).$$

It may be shown [46] that the conditions

$$\sum_{j \in \mathbb{Z}} \left| \hat{\psi}\left(2^j \xi\right) \right|^2 = 1 \tag{3.63}$$

and

$$\sum_{j=0}^{\infty} \hat{\psi}\left(2^j \xi\right) \overline{\hat{\psi}\left(2^j(\xi + 2m\pi)\right)} = 0 \tag{3.64}$$

for all $\xi \in \mathbb{R}$ (except possibly for a set of measure zero) and all odd integers $m \in 2\mathbb{Z} + 1$ ensure the completeness of the set of functions $\{\psi_{j,k} | j, k \in \mathbb{Z}\}$. It may also be shown that the orthonormality conditions Eqs. (3.61) and (3.62) follow from Eqs. (3.63) and (3.64), and so these conditions completely characterize functions ψ that give rise to systems of complete orthonormal wavelets.

When these conditions hold, the Parseval (Plancherel) equality

$$\sum_{j,k \in \mathbb{Z}} \left| (\psi_{j,k}, f) \right|^2 = \|f\|_2^2,$$

and the inversion formula

$$f = \sum_{j,k \in \mathbb{Z}} (\psi_{j,k}, f) \psi_{j,k}$$

do as well.

A concrete example of a discrete wavelet basis is the *Haar system* discussed in Section 3.8.1.

Discrete Modulation-Translation Wavelet Series

Discrete wavelet transforms based on modulations and translations (instead of dilations and translations) are possible. For example, the system

$$g_{m,n}(x) = e^{2\pi i m x} g(x - n)$$

for $m, n \in \mathbb{Z}$ introduced by Gabor in 1946 can be shown to be an orthonormal basis for certain classes of $g \in \mathcal{L}^2(\mathbb{R})$. See [33, 46] for a discussion.

For further reading on wavelets see [18, 20, 99].

3.10   Summary

In this chapter we reviewed a number of orthogonal expansions on the real line and various intervals. The topics ranged in chronology from Sturm-Liouville theory and the 19th century functions of mathematical physics, to the beginning of the 20th century (Haar, Walsh, Zernike, Meixner, etc.). Our discussion of orthonormal wavelets and associated transforms is work that can be found in the literature of the late 1980s, and the discussion of fast polynomial transforms dates to the late 1990s. Of course we have not considered all possible orthogonal functions and transforms. Our goal here was to provide the necessary background for the reader to put the developments of the later chapters of this book in perspective. For those readers interested in more exhaustive treatments of orthogonal functions and transforms, [6, 26, 79] may be useful.

References

[1] Agnesi, A., Reali, G.C., Patrini, G., and Tomaselli, A., Numerical evaluation of the Hankel transform — remarks, *J. of the Opt. Soc. of Am. A-Opt. Image Sci. and Vision,* 10(9), 1872–1874, 1993.

[2] Ahmed, N. and Rao, K.R., *Orthogonal Transforms for Digital Signal Processing,* Springer-Verlag, New York, 1975.

[3] Alpert, B. and Rokhlin, V., A fast algorithm for the evaluation of Legendre expansions, *SIAM J. Sci. and Stat. Computing,* 12(1), 158–179, 1991.

[4] Amos, D.E. and Burgmeier, J.W., Computation with three-term linear nonhomogeneous recurrence relations, *SIAM Rev.,* 15, 335–351, 1973.

[5] Arfken, G.B. and Weber, H.J., *Mathematical Methods for Physicists,* 4th ed., Academic Press, San Diego, 1995.

[6] Askey, R., *Orthogonal Polynomials and Special Functions,* SIAM, Philadelphia, 1975.

[7] Askey, R. and Wilson, J., Set of orthogonal polynomials that generalize the Racah coefficients or 6-j symbols, *SIAM J. on Math. Anal.,* 10(5), 1008–1016, 1979.

[8] Bastiaans, M.J., A sampling theorem for the complex spectrogram, and Gabor's expansion of a signal in Gaussian elementary signals, *Opt. Eng.*, 20(4), 594–598, 1981.

[9] Beauchamp, K.G., *Applications of Walsh and Related Functions*, Academic Press, New York, 1984.

[10] Birkhoff, G. and Rota, G.-C., On the completeness of Sturm-Liouville expansions, *Am. Math. Mon.*, 67(9), 835–841, 1960.

[11] Bowman, F., *Introduction to Bessel Functions*, Dover Publications, New York, 1968.

[12] Candel, S.M., An algorithm for the Fourier-Bessel transform, *Comput. Phys. Commun.*, 23(4), 343–353, 1981.

[13] Candel, S.M., Simultaneous calculation of Fourier-Bessel transforms up to order n, *J. of Computational Phys.*, 44(2), 243–261, 1981.

[14] Candel, S.M., Dual algorithms for fast calculation of the Fourier-Bessel transform, *IEEE Trans. on Acoust. Speech and Signal Process.*, 29(5), 963–972, 1981.

[15] Cavanagh, E. and Cook, B.D., Numerical evaluation of Hankel transforms via Gaussian-Laguerre polynomial expansions, *IEEE Trans. on Acoust. Speech and Signal Process.*, 27(4), 361–366, 1979.

[16] Canuto, C., Hussaini, M.Y., Quarteroni, A., and Zang, T.A., *Spectral Methods in Fluid Dynamics*, Springer-Verlag, Berlin, 1988.

[17] Christensen, N.B., The fast Hankel transform — comment, *Geophys. Prospecting*, 44(3), 469–471, 1996.

[18] Chui, C.K., *An Introduction to Wavelets*, Academic Press, Boston, 1992.

[19] Clenshaw, C.W., A note on the summation of Chebyshev series, *Math. Tables and other Aids to Computation*, 9, 118–120, 1955.

[20] Cody, M.A., The fast wavelet transform, *Dr. Dobb's J.*, 16(4), 16–28, 1992.

[21] Corbató, F.J. and Uretsky, J.L., Generation of spherical Bessel functions in digital computers, *J. of the ACM*, 6(3), 366–375, 1959.

[22] Cree, M.J. and Bones, P.J., Algorithms to numerically evaluate the Hankel transform, *Comput. & Math. with Appl.*, 26(1), 1–12, 1993.

[23] Daubechies, I., Orthonormal bases of compactly supported wavelets, *Commun. on Pure and Applied Math.*, 49, 909–996, 1988.

[24] Davis, P.J., *Interpolation and Approximation*, Ginn (Blaisdell), Boston, 1963.

[25] Davis, P.J. and Rabinowitz, P., *Methods of Numerical Integration*, Academic Press, New York, 1975.

[26] Debnath, L., *Integral Transforms and Their Applications*, CRC Press, Boca Raton, FL, 1995.

[27] Deprit, A., Note on the summation of Legendre series, *Celestial Mech.*, 20, 319–323, 1979.

[28] Driscoll, J.R. and Healy, D., Computing Fourier transforms and convolutions on the 2-sphere, *Adv. in Appl. Math.*, 15, 202–250, 1994.

[29] Driscoll, J.R., Healy, D., and Rockmore, D., Fast discrete polynomial transforms with applications to data analysis for distance transitive graphs, *SIAM J. Computing*, 26, 1066–1099, 1997.

[30] Ferrari, J.A., Fast Hankel transform of order zero, *J. of the Opt. Soc. of Am. A-Opt. Image Sci. and Vision*, 12(8), 1812–1813, 1995.

[31] Fino, B.J. and Algazi, V.R., A unified treatment of discrete fast transforms, *SIAM J. Computing*, 6, 700–717, 1977.

[32] Forsythe, G.E., Generation and use of orthogonal polynomials for data fitting with a digital computer, *J. SIAM*, 5, 74–88, 1957.

[33] Gabor, D., Theory of communication, *J. Inst. Electr. Eng.*, 93(III), 429–457, 1946.

[34] Gauß, C.F., Methodus nova integralium valores per approximationen inveniendi, in *Carl Friedrich Gauß Werke, Band 3*, Königliche Gesellschaft der Wissenschaften: Göttingen, 163–196, 1866.

[35] Gautschi, W., Computational aspects of three-term recurrence relations, *SIAM Rev.*, 9(1), 24–82, 1967.

[36] Gautschi, W., Minimal solutions of three-term recurrence relations and orthogonal polynomials, *Math. of Computation*, 36, 547–554, 1981.

[37] Goldstein, M. and Thaler, R.M., Recurrence techniques for the calculation of Bessel functions, *Math. Tables and other Aids to Computation*, 13(66), 102–108, 1959.

[38] Grossmann, A., Morlet, J., and Paul, T., Transforms associated to square integrable group representations. I. General results, *J. of Math. Phys.*, 26(10), 2473–2479, 1985.

[39] Grossmann, A., Morlet, J., and Paul, T., Transforms associated to square integrable group representations. II. Examples, *Annales de l'Institut Henri Poincaré-Physique théorique*, 45(3), 293–309, 1986.

[40] Grosswald, E., *Bessel Polynomials*, Lecture Notes in Mathematics No. 698, Springer-Verlag, New York, 1978.

[41] Gueron, S., Methods for fast computation of integral-transforms, *J. of Computational Phys.*, 110(1), 164–170, 1994.

[42] Haar, A., Zur Theorie Der Orthogonal Funktionensysteme, *Math. Ann.*, 69, 331–371, 1910.

[43] Hahn, W., Über orthogonalpolynome, die q-differenzengleichungen genügen, *Math. Nachrichten*, 2, 4–34, 1949.

[44] Hansen, E.W., Fast Hankel transform algorithm, *IEEE Trans. on Acoust. Speech and Signal Process.*, 33(3), 666–671, 1985.

[45] Harmuth, H.F., *Transmission of Information by Orthogonal Functions*, 2nd ed., Springer-Verlag, Berlin, 1972.

[46] Hernández, E. and Weiss, G., *A First Course on Wavelets*, CRC Press, Boca Raton, FL, 1996.

[47] Higgins, W.E. and Munson, D.C., An algorithm for computing general integer-order Hankel-transforms, *IEEE Trans. on Acoust. Speech and Signal Process.*, 35(1), 86–97, 1987.

[48] Higham, N.J., Fast solution of Vandermonde-like systems involving orthogonal polynomials, *IMA J. Numerical Anal.*, 8, 473–486, 1988.

[49] Jerri, A.J., Towards a discrete Hankel transform and its applications, *Appl. Anal.*, 7, 97–109, 1978.

[50] Johansen, H.K. and Sorensen, K., Fast Hankel-transforms, *Geophys. Prospecting*, 27(4), 876–901, 1979.

[51] Kaiser, G., *A Friendly Guide to Wavelets*, Birkhäuser, Boston, 1994.

[52] Karlin, S. and McGregor, J.L., The Hahn polynomials, formulas and an application, *Scripta Math.*, 26(1), 33–46, 1961.

[53] Karpovsky, M.G., *Finite Orthogonal Series in the Design of Digital Devices*, Halsted Press, John Wiley & Sons, New York, 1975.

[54] Kim, W.-Y. and Kim, Y.-S., Robust rotation angle estimator, *IEEE Trans. on Pattern Anal. and Machine Intelligence*, 21(8), 768–773, 1999.

[55] Koornwinder, T., The addition formula for Laguerre polynomials, *SIAM J. Math. Anal.*, 8, 535–540, 1977.

[56] Krall, H.L. and Frink, O., A new class of orthogonal polynomials: the Bessel polynomials, *Trans. of the Am. Math. Soc.*, 65, 100–115, 1949.

[57] Kreyszig, E., *Advanced Engineering Mathematics*, 6th ed., John Wiley & Sons, New York, 1988.

[58] Krylov, V.I., *Approximate Calculation of Integrals*, MacMillan, New York, 1962.

[59] Lemoine, D., The discrete Bessel transform algorithm, *J. of Chem. Phys.*, 101(5), 3936–3944, 1994.

[60] Liu, Q.H., Applications of the conjugate-gradient fast Hankel, Fourier transfer method with an improved fast Hankel transform algorithm, *Radio Sci.*, 30(2), 469–479, 1995.

[61] Liu, Q.H. and Chew, W.C., Applications of the conjugate-gradient fast Fourier Hankel transfer method with an improved fast Hankel transform algorithm, *Radio Sci.*, 29(4), 1009–1022, 1994.

[62] Luke, Y.L., *Mathematical Functions and their Approximations*, Academic Press, New York, 1975.

[63] Ma, J., Rokhlin, V., and Wandzura, S., Generalized Gaussian quadrature rules for systems of arbitrary functions, *SIAM J. Numerical Anal.*, 33(3), 971–996, 1996.

[64] Magni, V., Cerullo, G., and Desilvestri, S., High–accuracy fast Hankel transform for optical beam propagation, *J. of the Opt. Soc. of Am. A-Opt. Image Sci. and Vision*, 9(11), 2031–2033, 1992.

[65] Mallat, S., Multiresolution approximations and wavelet orthonormal bases of $\mathcal{L}^2(\mathbb{R})$, *Trans. of the Am. Math. Soc.*, 315, 69–87, 1989.

[66] Maslen, D.K. and Rockmore, D.N., Generalized FFTs — a survey of some recent results, *DIMACS Ser. in Discrete Math. and Theor. Comput. Sci.*, 28, 183–237, 1997.

[67] Meixner, J., Orthogonale polynomsysteme mit einer besonderen gestalt der erzeugenden funktion, *J. of the London Math. Soc.*, 9, 6–13, 1934.

[68] Meixner, J., Symmetric systems of orthogonal polynomials, *Arch. for Ration. Mech. and Anal.*, 44, 69–75, 1972.

[69] Meyer, Y., *Ondelettes Et Opérateurs I: Ondelettes*, Hermann, 1989.

[70] Mohsen, A.A. and Hashish, E.A., The fast Hankel transform, *Geophys. Prospecting,* 42(2), 131–139, 1994.

[71] Mook, D.R., An algorithm for the numerical evaluation of the Hankel and Abel transforms, *IEEE Trans. on Acoust. Speech and Signal Process.,* 31(4), 979–985, 1983.

[72] Morlet, J., Arens, G., Fourgeau, I., and Giard, D., Wave propagation and sampling theory, *Geophys.,* 47, 203–236, 1982.

[73] Oppenheim, A.V., Frisk, G.V., and Martinez, D.R., Algorithm for numerical evaluation of Hankel transform, *Proc. of the IEEE,* 66(2), 264–265, 1978.

[74] Oppenheim, A.V., Frisk, G.V., and Martinez, D.R., Computation of the Hankel transform using projections, *J. Acoust. Soc. Am.,* 68(2), 523–529, 1980.

[75] Orszag, S.A., Fast Eigenfunction Transforms, in *Sci. and Comput., Adv. in Math. Suppl. Stud.,* 10, 23–30, Academic Press, New York, 1986.

[76] Piessens, R. and Branders, M., Modified Clenshaw-Curtis method for the computation of Bessel-function integrals, *BIT,* 23(3), 370–381, 1983.

[77] Pinsky, M.A., *Introduction to Partial Differential Equations with Applications,* McGraw-Hill, New York, 1984.

[78] Potts, D., Steidl, G., and Tasche, M., Fast algorithms for discrete polynomial transforms, *Math. of Computation,* 67(224), 1577–1590, 1998.

[79] Poularikas, A., Ed., *The Transforms and Applications Handbook,* 2nd ed., CRC Press, Boca Raton, FL, 1999.

[80] Rademacher, H., Einige sätze von allgemeinen orthogonalfunktionen, *Math. Annalen,* 87, 122–138, 1922.

[81] Rao, K.S., Santhanam, T.S., and Gustafson, R.A., Racah polynomials and a 3-term recurrence relation for the Racah coefficients, *J. of Phys. A-Math. and General,* 20(10), 3041–3045, 1987.

[82] Renault, O., A new algorithm for computing orthogonal polynomials, *J. of Computational and Appl. Math.,* 75(2), 231–248, 1996.

[83] Rijo, L., 'The fast Hankel transform' — Comment, *Geophys. Prospecting,* 44(3), 473–477, 1996.

[84] Rivlin, T.J., *Chebyshev Polynomials,* 2nd ed., John Wiley & Sons, New York, 1990.

[85] Roeser, P.R. and Jernigan, M.E., Fast Haar transform algorithms, *IEEE Trans. Comput.,* C-31, 175–177, 1982.

[86] Salzer, H.E., A recurrence scheme for converting from one orthogonal expansion into another, *Commun. of the ACM,* 16, 705–707, 1973.

[87] Salzer, H.E., Computing interpolation series into Chebyshev series, *Math. of Computation,* 30, 295–302, 1976.

[88] Sanzone, G., *Orthogonal Functions,* Wiley InterScience, New York, 1958.

[89] Secada, J.D., Numerical evaluation of the Hankel transform, *Comput. Phys. Commun.,* 116(2-3), 278–294, 1999.

[90] Sharafeddin, O.A., Bowen, H.F., Kouri, D.J., and Hoffman, D.K., Numerical evaluation of spherical Bessel transforms via fast Fourier-transforms, *J. of Computational Phys.,* 100(2), 294–296, 1992.

[91] Siegman, A.E., Quasi fast Hankel transform, *Opt. Lett.,* 1(1), 13–15, 1977.

[92] Smith, F.J., An algorithm for summing orthogonal polynomial series and their derivatives with application to curve fitting and interpolation, *Math. of Computation,* 19, 33–36, 1965.

[93] Stroud, A.H. and Secrest, D., *Gaussian Quadrature Formulas,* Prentice-Hall, Englewood Cliffs, NJ, 1996.

[94] Szegö G., *Orthogonal Polynomials,* Vol. 23, American Mathematical Society Colloquium Publications, Providence, RI, 1939, (4th ed., 1975).

[95] Talman, J.D., Numerical Fourier and Bessel transforms in logarithmic variables, *J. of Computational Phys.,* 29(1), 35–48, 1978.

[96] Thangavelu, S., *Lectures on Hermite and Laguerre Expansions,* Princeton University Press, Princeton, NJ, 1993.

[97] Vilenkin, N.J. and Klimyk, A.U., *Representation of Lie Group and Special Functions,* Vol. 1-3, Kluwer Academic Publishers, The Netherlands, 1991.

[98] Walsh, J.L., A closed set of orthogonal functions, *Am. J. of Math.,* 45, 5–24, 1923.

[99] Walter, G.G., *Wavelets and Other Orthogonal Systems with Applications,* CRC Press, Boca Raton, FL, 1994.

[100] Watson, G.N., *A Treatise on the Theory of Bessel Functions,* Cambridge University Press, New York, 1995.

[101] Wilson, J.A., Some hypergeometric orthogonal polynomials, *SIAM J. on Math. Anal.,* 11(4), 690–701, 1980.

[102] Wimp, J., *Computation with Recurrence Relations,* Pitman Press, Boston, 1984.

[103] Zernike, F., Beugungstheorie des schneidenverfahrens und seiner verbesserten form, der phasenkontrastmethode, *Physica,* 1, 689–704, 1934.

Chapter 4

Orthogonal Expansions in Curvilinear Coordinates

In this chapter we review orthogonal expansions for functions on $\mathbb{R}^N$ expressed in curvilinear coordinates. We also examine orthogonal expansions on surfaces in $\mathbb{R}^3$ and in higher dimensions.

4.1 Introduction to Curvilinear Coordinates and Surface Parameterizations

While Cartesian coordinates are the most common method for identifying positions in space, they are by no means the only one. In general, positions in space are parameterized by an array of parameters $(u_1, \ldots, u_N)$, which we will denote as $\mathbf{u}$. Then $\mathbf{x} = \tilde{\mathbf{x}}(\mathbf{u})$, where $\mathbf{x}, \mathbf{u} \in \mathbb{R}^N$.

The relationship between differential elements of the arclength, $dL = (dL/dt)dt$, of any curve $\mathbf{x}(t) = \tilde{\mathbf{x}}(\mathbf{u}(t))$ in the different curvilinear coordinates is related to those in Cartesian coordinates as

$$\left(\frac{dL}{dt}\right)^2 = \frac{d\mathbf{x}}{dt} \cdot \frac{d\mathbf{x}}{dt} = \left(J(\mathbf{u})\frac{d\mathbf{u}}{dt}\right)^T \left(J(\mathbf{u})\frac{d\mathbf{u}}{dt}\right) = \frac{d\mathbf{u}}{dt} \cdot \left(G(\mathbf{u})\frac{d\mathbf{u}}{dt}\right),$$

where

$$G(\mathbf{u}) = J^T(\mathbf{u})J(\mathbf{u}),$$

and

$$J(\mathbf{u}) = \left(\frac{\partial \mathbf{x}}{\partial u_1}, \cdots, \frac{\partial \mathbf{x}}{\partial u_N}\right).$$

G is called the *metric tensor* and J is called the *Jacobian matrix*.

If for some reason the units used to measure distance in each of the coordinate directions are different from each other, this must be rectified if some meaningful measure of arc length and volume are to be defined. In such cases, we have

$$G(\mathbf{u}) = J^T(\mathbf{u})WJ(\mathbf{u}), \tag{4.1}$$

where W is a diagonal matrix with elements of the form $w_{ij} = w_i \delta_{ij}$. The constant weights w_i are measures of distance or ratios of distance units used in each coordinate direction. For instance, if x_1 through x_{N-1} are measured in centimeters and x_N is measured in inches, then $w_1 = \cdots = w_{N-1} = 1$ and $w_N = 2.54$ will provide the required uniformity.

The volume of an infinitesimal element in $\mathbb{R}^N$ written in the curvilinear coordinates $\mathbf{u}$ is

$$\begin{aligned}
dx^N &= \sqrt{\det(G(\mathbf{u}))}du^N \\
&= w_1 \cdots w_N \sqrt{\det\left(J^T(\mathbf{u})J(\mathbf{u})\right)}du^N \\
&= w_1 \cdots w_N |\det(J(\mathbf{u}))|du^N.
\end{aligned}$$

We use the notation $da^N = da_1 da_2 \ldots da_N$ for any $\mathbf{a} \in \mathbb{R}^N$.

In the special case when

$$\sqrt{\det(G(\mathbf{u}))} = f_1(u_1) \cdots f_N(u_N),$$

we can construct orthogonal expansions in curvilinear coordinates as N separate one-dimensional problems (one in each parameter) provided the weighting functions $f_i(u_i)$ all satisfy the conditions imposed by Sturm-Liouville theory.

4.2 Parameterizations of the Unit Circle, Semi-Circle, and Planar Rotations

The unit circle, S^1, is the set of all points in $\mathbb{R}^2$ that are of unit distance from the origin

$$x_1^2 + x_2^2 = 1.$$

A parameterization of S^1 that is very natural (in the sense that uniform increments of the curve parameter correspond to equal increments of arc length) is

$$\mathbf{x}(\theta) = \begin{pmatrix} \cos\theta \\ \sin\theta \end{pmatrix}. \tag{4.2}$$

Here θ is the counter-clockwise measured angle that the vector $\mathbf{x}$ makes with the x_1-axis. In this parameterization of the circle, the Fourier series is a natural way to expand functions $f(\mathbf{x}(\theta))$.

Rotations in the plane are identified with the circle by mapping the angle θ to a particular rotation matrix

$$R(\theta) = \begin{pmatrix} \cos\theta & -\sin\theta \\ \sin\theta & \cos\theta \end{pmatrix}. \tag{4.3}$$

If we identify points in the plane $\mathbb{R}^2$ with points in the complex plane $\mathbb{C}$ by the rule

$$\mathbf{x} \leftrightarrow x_1 + ix_2,$$

the effect of the rotation matrix Eq. (4.3) applied to a vector $\mathbf{x}$ is equivalent to the product $e^{i\theta}(x_1 + ix_2)$

$$R(\theta)\mathbf{x} \leftrightarrow e^{i\theta}(x_1 + ix_2).$$

That which is most natural from a geometric perspective can sometimes be inefficient from a computational perspective. In the case of the circle and rotation parameterizations, it is computationally attractive to write

$$\mathbf{x}(z) = \begin{pmatrix} \frac{1-z^2}{1+z^2} \\ \frac{2z}{1+z^2} \end{pmatrix} \quad \text{and} \quad R(z) = \begin{pmatrix} \frac{1-z^2}{1+z^2} & \frac{-2z}{1+z^2} \\ \frac{2z}{1+z^2} & \frac{1-z^2}{1+z^2} \end{pmatrix}. \tag{4.4}$$

where

$$z = \tan\frac{\theta}{2} \quad \text{and} \quad \theta = 2\tan^{-1}z.$$

The reason why this parameterization is attractive is that the evaluation of Eq. (4.4) requires that only the most basic arithmetic operations (addition, subtraction, multiplication, and division) be performed, without the need to compute trigonometric functions. In computer graphics, these descriptions (called "algebraic" or "rational") are often preferred to the transcendental (trigonometric) descriptions.

One computationally attractive way to implement rotations is as a succession of shears. A linear shear transformation of the plane that maps vertical lines into vertical lines (preserving their x_1 value) is of the form $\mathbf{x}' = A\mathbf{x}$ where

$$A = \begin{pmatrix} 1 & 0 \\ \beta & 1 \end{pmatrix},$$

and $\beta \in \mathbb{R}$ describes the amount of shear. A linear shear transformation of the plane that maps horizontal lines into horizontal lines (preserving their x_2 value) is of the form

$$A = \begin{pmatrix} 1 & \alpha \\ 0 & 1 \end{pmatrix}.$$

It may be shown that the arbitrary planar rotation Eq. (4.3) can be implemented as concatenated shears [27]

$$\begin{pmatrix} \cos\theta & -\sin\theta \\ \sin\theta & \cos\theta \end{pmatrix} = \begin{pmatrix} 1 & \alpha(\theta) \\ 0 & 1 \end{pmatrix} \begin{pmatrix} 1 & 0 \\ \beta(\theta) & 1 \end{pmatrix} \begin{pmatrix} 1 & \alpha(\theta) \\ 0 & 1 \end{pmatrix}$$

where

$$\alpha(\theta) = -\tan\frac{\theta}{2} \quad \text{and} \quad \beta(\theta) = \sin\theta.$$

In rational form, one writes

$$\begin{pmatrix} \frac{1-z^2}{1+z^2} & \frac{-2z}{1+z^2} \\ \frac{2z}{1+z^2} & \frac{1-z^2}{1+z^2} \end{pmatrix} = \begin{pmatrix} 1 & -z \\ 0 & 1 \end{pmatrix} \begin{pmatrix} 1 & 0 \\ 2z/(1+z^2) & 1 \end{pmatrix} \begin{pmatrix} 1 & -z \\ 0 & 1 \end{pmatrix}.$$

This is a particularly useful tool for the rotation of images stored in pixelized form since the operations are performed on whole rows and columns of pixels.

In principle, one can generate an infinite number of different parameterizations of the circle and planar rotations by an appropriate choice of a monotonically increasing function $\theta(t)$ with $\theta(-\pi) = -\pi$ and $\theta(\pi) = \pi$. Geometrically, one class of parameterizations results from *stereographic projection* between the circle and line. This is illustrated in Fig. 4.1.

We refer to the horizontal line bisecting the circle as the *equatorial line* and use the variable x to denote the distance measured along this line from the center of the circle (with positive sense to the right). All lines passing through the south pole (called *projectors*) assign a unique point on the circle to the equatorial line and vice versa (except the south pole). Then, from elementary geometry,[1] one finds that the relationship between the point $(v_1, v_2)^T \in S^1$ and $x \in \mathbb{R}$ is

$$v_1 = \frac{2x}{1+x^2} \qquad v_2 = \frac{1-x^2}{1+x^2}; \qquad x = \frac{1-v_2}{v_1} = \frac{v_1}{1+v_2}. \tag{4.5}$$

In the preceding expressions for x as a function of v_1 and v_2, the first one breaks down when $v_1 = 0$, and the second breaks down when $v_2 = -1$.

This stereographic projection is the same as the rational description in Eq. (4.4) when the north pole is associated with $\theta = 0$ and x is made positive, in the sense of counter-clockwise rotation.

Of course, this is not the only kind of projection possible between the line and circle. Figure 4.1(b) illustrates a projection where a line tangent to the north pole replaces the role of the equatorial line.

---

[1] This kind of projection, together with planar and spherical trigonometry, is attributed to Hipparchus (circa 150 BC) [31].

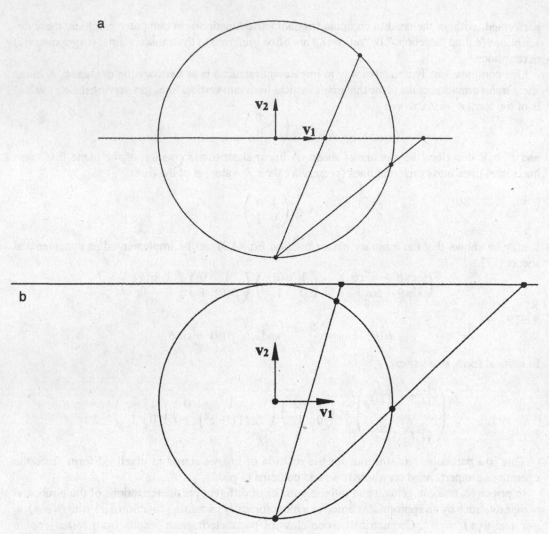

FIGURE 4.1

(a) Projection onto the equatorial line; (b) projection onto the tangent to the north pole.

Denoting the distance along this line as measured from the north pole (positive to the right) as y, the relationship between $(v_1, v_2)^T \in S^1$ and $y \in \mathbb{R}$ is

$$v_1 = \frac{4y}{4 + y^2} \qquad v_2 = \frac{4 - y^2}{4 + y^2}; \qquad y = \frac{2(1 - v_2)}{v_1} = \frac{2v_1}{1 + v_2}. \tag{4.6}$$

The preceding expressions for y as a function of v_1 and v_2 break down when $v_1 = 0$ and $v_2 = -1$, respectively.

We note that when using projectors through the south pole, a similar relationship can be established between the circle and any line lying between the tangent to the north pole and equatorial line and parallel to both.

The upper semi-circle (S^1 with $x_2 \geq 0$ and denoted here as $(S^1)^+$) can be parameterized using any of the previously mentioned circle parameterizations. We may also project the upper semi-circle to the line in additional ways not possible for the whole circle. For instance, points can be projected straight up and down yielding the relationships

$$(v_1, v_2) \to v_1; \qquad x \to \left(x, \sqrt{1 - x^2}\right).$$

And a projection between the tangent to the north pole and the circle through its center yields

$$(v_1, v_2) \to \frac{v_1}{v_2}; \qquad x \to \left(x/\sqrt{1 + x^2}, 1/\sqrt{1 + x^2}\right).$$

These projections are shown in Fig. 4.2.

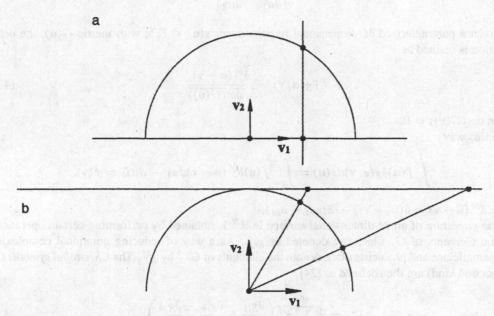

FIGURE 4.2

(a) Vertical projection; (b) projection through the center.

The extensions of these simple examples to higher dimensions will be valuable in Chapter 5 in the context of parameterizing rotations in three dimensions.

4.3 M-Dimensional Hyper-Surfaces in $\mathbb{R}^N$

In the case when we have an M-dimensional hyper-surface $\mathcal{S} \subset \mathbb{R}^N$, the parameterization follows in exactly the same way as for curvilinear coordinates, only now $\mathbf{u} \in \mathbb{R}^M$ where $M < N$. We note that $\mathbf{u}$ is not a vector in the physical sense, but rather an array of M real numbers.

The Jacobian matrix $J(\mathbf{u})$ is no longer square, and we may no longer compute $|\det(J(\mathbf{u}))|$. Thus in the calculation of a hyper-surface volume element $ds(\mathbf{u})$, one usually writes

$$ds(\mathbf{u}) = \sqrt{\det(G(\mathbf{u}))}\, du^M. \qquad (4.7)$$

The volume element for an M dimensional hyper-surface in $\mathbb{R}^N$ can also be computed without introducing the metric tensor G. This follows since the columns of J form a basis for the tangent hyper-plane at any point on the hyper-surface. By using the Gram-Schmidt orthogonalization procedure (see Appendix C), an orthonormal basis for the hyper-plane is constructed from the columns of J. We denote the elements of this basis as $\mathbf{v}_1, \ldots, \mathbf{v}_M$. Then an $M \times M$ matrix with entry i, j corresponding to the projection of $\partial \mathbf{x}/\partial u_i$ onto $\mathbf{v}_j$ can be constructed. The determinant of this matrix then yields the same result as $\sqrt{\det(G(\mathbf{u}))}$. This is perhaps the most direct generalization of the use of the triple product in three dimensions for computing volume in higher dimensions.

In the special case when $N = 3$ and $M = 2$, both of the preceding approaches are the same as

$$ds(\mathbf{u}) = \left| \frac{\partial \mathbf{x}}{\partial u_1} \times \frac{\partial \mathbf{x}}{\partial u_2} \right| du_1 du_2.$$

Given a parameterized M-dimensional hypersurface, $\mathbf{x}(\mathbf{u}) \in \mathbb{R}^N$, with metric $G(\mathbf{u})$, the delta function is defined as

$$\delta_S(\mathbf{u}, \mathbf{v}) = \frac{\delta^M(\mathbf{u} - \mathbf{v})}{\sqrt{\det(G(\mathbf{u}))}} \tag{4.8}$$

when $\det(G(\mathbf{v})) \neq 0$.

In this way

$$\int_S f(\mathbf{u}) \delta_S(\mathbf{u}, \mathbf{v}) ds(\mathbf{u}) = \int_{\mathbb{R}^N} f(\mathbf{u}) \delta^M(\mathbf{u} - \mathbf{v}) du_1 \cdots du_M = f(\mathbf{v}).$$

Here $\delta^M(\mathbf{u} - \mathbf{v}) = \delta(u_1 - v_1) \cdots \delta(u_M - v_M)$.

The *curvature* of an M-dimensional surface in $\mathbb{R}^N$ is obtained by performing certain operations on the elements of G, which are denoted as g_{ij}. As a way of reducing notational complexity, mathematicians and physicists often denote the elements of G^{-1} by g^{ij}. The *Christoffel symbols* (of the second kind) are then defined as [24]

$$\Gamma^i_{jk} = \frac{1}{2} \sum_{l=1}^M g^{il} \left(\frac{\partial g_{lj}}{\partial u_k} + \frac{\partial g_{lk}}{\partial u_j} - \frac{\partial g_{jk}}{\partial u_l} \right). \tag{4.9}$$

In the case of a two-dimensional surface in $\mathbb{R}^3$, the previous definition of the Christoffel symbols follows from the expression

$$\frac{\partial^2 \mathbf{x}}{\partial u_i \partial u_j} = L_{ij} \mathbf{n} + \sum_{k=1}^2 \Gamma^k_{ij} \frac{\partial \mathbf{x}}{\partial u_k}$$

where

$$\mathbf{n} = \frac{\frac{\partial \mathbf{x}}{\partial u_1} \times \frac{\partial \mathbf{x}}{\partial u_2}}{\left| \frac{\partial \mathbf{x}}{\partial u_1} \times \frac{\partial \mathbf{x}}{\partial u_2} \right|}$$

is the unit normal to the surface.

It follows that

$$L_{ij} = \frac{\partial^2 \mathbf{x}}{\partial u_i \partial u_j} \cdot \mathbf{n},$$

where $\cdot$ is the usual dot product in $\mathbb{R}^3$. If L is the matrix with coefficients L_{ij}, the *Gaussian curvature* of the surface at $\mathbf{x}(u_1, u_2)$ is

$$K = \det(L)/\det(G) = \frac{1}{\sqrt{\det(G)}} \mathbf{n} \cdot \left(\frac{\partial \mathbf{n}}{\partial u_1} \times \frac{\partial \mathbf{n}}{\partial u_2} \right). \tag{4.10}$$

The *Riemannian curvature tensor*[2] is defined as [24]

$$R^i_{jkl} = -\frac{\partial \Gamma^i_{jk}}{\partial u_l} + \frac{\partial \Gamma^i_{jl}}{\partial u_k} + \sum_{m=1}^M \left(-\Gamma^m_{jk}\Gamma^i_{ml} + \Gamma^m_{jl}\Gamma^i_{mk} \right). \tag{4.11}$$

The *Ricci* and *scalar curvature* formulas are defined relative to this as [24]

$$R_{jl} = \sum_{i=1}^M R^i_{jil} \quad \text{and} \quad R = \sum_{j,l=1}^M g^{jl} R_{jl},$$

respectively. The scalar curvature for $\mathbb{R}^N$ is zero. This is independent of what curvilinear coordinate system is used. For a sphere in any dimension, the scalar curvature is constant. In the case when $M = 2$ and $N = 3$, the scalar curvature is twice the *Gaussian* curvature.

As an example of these definitions, consider the 2-torus

$$\mathbf{x}(\theta, \phi) = \begin{pmatrix} (R + r\cos\theta)\cos\phi \\ (R + r\cos\theta)\sin\phi \\ r\sin\theta \end{pmatrix}.$$

The metric tensor is written in this parameterization as

$$G(\theta, \phi) = \begin{pmatrix} (R + r\cos\theta)^2 & 0 \\ 0 & r^2 \end{pmatrix}.$$

We find that

$$\Gamma^1_{11} = \Gamma^1_{22} = \Gamma^2_{12} = \Gamma^2_{21} = \Gamma^2_{22} = 0,$$
$$\Gamma^1_{12} = \Gamma^1_{21} = -\frac{r\sin\theta}{R + r\cos\theta},$$

and

$$\Gamma^2_{11} = \frac{(R + r\cos\theta)\sin\theta}{r}.$$

We also calculate

$$R^2_{121} = -R^2_{112} = \frac{R\cos\theta}{r} + \cos^2\theta,$$
$$R^1_{212} = -R^1_{221} = \frac{r\cos\theta}{R + r\cos\theta},$$

with all other $R^i_{jkl} = 0$.

In an effort to extend the concepts of length, volume, and curvature to the most general scenarios, mathematicians toward the end of the 19th century formulated the geometry of higher dimensional curved spaces without regard to whether or not these spaces fit in a higher dimensional Euclidean space.[3] Roughly speaking, the only properties that were found to be important were: (1) that the curved space "locally looks like" $\mathbb{R}^M$; (2) that it is possible to construct a set of overlapping patches in the curved space in such a way that coordinates can be imposed on each patch; and (3) that the description of points in the overlapping patches are consistent with respect to the different

---

[2] This is named after Bernhard Riemann (1826–1866).

[3] Albert Einstein used these mathematical techniques as a tool to express the general theory of relativity.

parameterizations that are defined in the overlapping regions. In mathematics, the study of curved spaces is called differential geometry. Smooth curved spaces that for all points look locally like Euclidean space are called *manifolds*.[4] Each manifold can be covered with a collection of overlapping *patches,* in which a local *coordinate chart* relates a point in the manifold to one in a copy of $\mathbb{R}^M$. The collection of all of the coordinate charts is called an *atlas*. The inverse of a coordinate chart gives the parameterization of a patch in the manifold. Precise definitions of these terms can be found in any of a number of excellent books on differential geometry, and a short review is provided in Appendix G. It is assumed from the beginning that all of the manifolds of interest in this book can be embedded in $\mathbb{R}^N$ for sufficiently large N. Hence, we need not use formal methods of coordinate-free differential geometry. Rather, it suffices to use the methods described previously for surfaces in higher dimensions, with the one modification described below.

We associate matrices $X \in \mathbb{R}^{N \times N}$ with vectors $\mathbf{x} \in \mathbb{R}^{N^2}$ by simply taking each column of the matrix and sequentially stacking the columns one on top of the other, until an N^2-dimensional vector results. The M^2 elements of the matrix $G(\mathbf{u})$ defined in Eq. (5.66) (called a *Riemannian metric tensor*) can be written in this case as

$$g_{ij} = \left(\frac{\partial \mathbf{x}}{\partial u_i} \right)^T W \frac{\partial \mathbf{x}}{\partial u_j} = \text{trace} \left(\frac{\partial X}{\partial u_i} W_0 \frac{\partial X^T}{\partial u_j} \right), \tag{4.12}$$

where $\mathbf{u} \in \mathbb{R}^M$ and $\mathbf{x}(\mathbf{u}) \in \mathbb{R}^{N^2}$ are the vectors associated with the matrices $X(\mathbf{u}) \in \mathbb{R}^{N \times N}$, and the matrix W_0 is defined relative to W in a way that makes the equality hold.

Note that it is always the case that $g_{ij} = g_{ji}$. In the special case when $G(\mathbf{u})$ is diagonal, the coordinates $\mathbf{u}$ are called *orthogonal* curvilinear coordinates.

4.4   Gradients, Divergence, and the Laplacian

In $\mathbb{R}^N$, the concept of the gradient of a scalar function of Cartesian coordinates $f(x_1, \ldots, x_N)$ is defined as the vector ∇f, where $\nabla = \sum_{i=1}^{N} \mathbf{e}_i \partial/\partial x_i$. Recall that $\mathbf{e}_i$ is the ith natural basis vector for $\mathbb{R}^N$ and $(\mathbf{e}_i)_j = \delta_{ij}$. The divergence of a vector field $\mathbf{F} = \sum_{i=1}^{N} \mathbf{e}_i F_i$ is $\nabla \cdot \mathbf{F} = \sum_{i=1}^{N} \partial F \cdot /\partial x_i$. The Laplacian[5] of a scalar function is $\nabla^2 = \nabla \cdot \nabla f = \sum_{i=1}^{N} \partial^2 f/\partial x_i^2$.

The concepts of gradient, divergence, and the Laplacian of smooth functions, $\tilde{f}(\mathbf{x}(\mathbf{u})) \stackrel{\Delta}{=} f(\mathbf{u})$, on a hyper-surface $\mathcal{S} \subset \mathbb{R}^N$ with points $\mathbf{x}(\mathbf{u}) \in \mathcal{S}$, are naturally generalized using the definitions given earlier in this section. Before continuing, we add that our notation of using subscripts to denote components is different than in physics, where subscripts and superscripts each have special meaning. Basis vectors for the tangent hyper-plane (which are also vectors in $\mathbb{R}^N$) are

$$\mathbf{v}_i = \frac{\partial \mathbf{x}}{\partial u_j}.$$

These are generally not orthogonal with respect to the natural inner product in $\mathbb{R}^N$ because for $W = \mathbb{I}$,

$$\mathbf{v}_i \cdot \mathbf{v}_j = g_{ij}.$$

---

[4]See Appendix G for a precise definition.
[5]This is named after Pierre Simon Marquis de Laplace (1749–1827).

However, when G is diagonal,

$$e_{u_i} = \mathbf{v}_i / (g_{ii})^{\frac{1}{2}} \tag{4.13}$$

are orthonormal with respect to this inner product.

We define the gradient of f as

$$\text{grad}(f) = \sum_{i=1}^{M} \text{grad}(f)_i \mathbf{v}_i$$

where

$$\text{grad}(f)_i = \sum_{j=1}^{M} g^{ij} \frac{\partial f}{\partial u_j},$$

and again we use the shorthand g^{ij} to denote $(G^{-1})_{ij}$. We define the divergence of a vector field $\mathbf{F} = \sum_{i=1}^{M} F_i \mathbf{v}_i$ on a hyper-surface as

$$\text{div}(\mathbf{F}) = \sum_{i=1}^{M} \frac{1}{\sqrt{\det G}} \frac{\partial}{\partial u_i} \left(\sqrt{\det G} F_i \right).$$

The gradient and divergence are the "dual" (or adjoint) of each other in the following sense. For real-valued functions f and g that take their argument in $\mathcal{S}$, define the inner product

$$(f, g) = \int_S f(\mathbf{x}) g(\mathbf{x}) ds(\mathbf{u}) = \int_{\mathbb{R}^M} f(\mathbf{x}(\mathbf{u})) g(\mathbf{x}(\mathbf{u})) \sqrt{\det G(\mathbf{u})} du_1 \cdots du_M.$$

Here we have assumed for convenience that the whole surface can be considered as one coordinate patch, and either $\mathcal{S}$ is closed and bounded [and $f(\mathbf{x}(\mathbf{u}))$ and $g(\mathbf{x}(\mathbf{u}))$ therefore have compact support], or $\mathcal{S}$ has infinite extent (and f and g decay rapidly to zero outside of a compact region).

If $\mathbf{F}$ and $\mathbf{K}$ are two real vector fields on $\mathcal{S}$ with components F_i and K_i, we define

$$\langle \mathbf{F}, \mathbf{K} \rangle = \int_S \mathbf{F}(\mathbf{x}) \cdot \mathbf{K}(\mathbf{x}) ds(\mathbf{u}).$$

It follows that

$$\langle \mathbf{F}, \mathbf{K} \rangle = \left\langle \sum_{i=1}^{m} F_i \mathbf{v}_i, \sum_{j=1}^{m} K_j \mathbf{v}_j \right\rangle = \sum_{i,j=1}^{m} \langle F_i \mathbf{v}_i, K_j \mathbf{v}_j \rangle.$$

Equivalently,

$$\langle \mathbf{F}, \mathbf{K} \rangle = \sum_{i,j=1}^{m} \int_{\mathbb{R}^M} F_i g_{ij} K_j \sqrt{\det G} du_1 \cdots du_M.$$

Then

$$\langle \text{grad}(f), \mathbf{F} \rangle = \sum_{i,j,k=1}^{M} \int_{\mathbb{R}^M} g^{ki} \frac{\partial f}{\partial u_i} g_{kj} F_j \sqrt{\det G(\mathbf{u})} du_1 \cdots du_M.$$

Using the fact that

$$\sum_{k=1}^{M} g_{kj} g^{ki} = \delta_{ij}$$

we write

$$\langle \text{grad}(f), \mathbf{F} \rangle = \sum_{i=1}^{M} \int_{\mathbb{R}^M} \frac{\partial f}{\partial u_i} F_i \sqrt{\det G(\mathbf{u})} du_1 \cdots du_M.$$

Rearranging terms and using integration by parts,

$$\sum_{i=1}^{M} \int_{\mathbb{R}^M} \frac{\partial f}{\partial u_i} F_i \sqrt{\det G(\mathbf{u})} du_1 \cdots du_M = -\sum_{i=1}^{M} \int_{\mathbb{R}^M} f \frac{\partial}{\partial u_i} \left(F_i \sqrt{\det G(\mathbf{u})} \right) du_1 \cdots du_M.$$

Dividing the integrand by $\sqrt{\det G(\mathbf{u})}$ and multiplying by the same factor to restore the integration measure $ds(\mathbf{u}) = \sqrt{\det G(\mathbf{u})} du_1 \cdots du_M$, we observe the duality

$$\langle \text{grad}(f), \mathbf{F} \rangle = -(f, \text{div}(\mathbf{F})). \tag{4.14}$$

The Laplacian (or *Laplace-Beltrami operator*) of a smooth real-valued function is defined as the divergence of the gradient

$$\text{div}(\text{grad } f) = \sum_{i=1}^{M} \frac{1}{\sqrt{\det G}} \frac{\partial}{\partial u_i} \left(\sqrt{\det G} \, g^{ij} \frac{\partial f}{\partial u_j} \right)$$

$$= \sum_{i,j=1}^{M} g^{ij} \left(\frac{\partial^2 f}{\partial u_i \partial u_j} - \sum_{k=1}^{M} \Gamma_{ij}^k \frac{\partial f}{\partial u_k} \right). \tag{4.15}$$

In the case of $\mathbb{R}^N$ with Cartesian coordinates, $\mathbf{v}_i = \mathbf{e}_i$, $g^{ij} = g_{ij} = \delta_{ij}$, and $\Gamma_{ij}^k = 0$ for all i, j, k.

In general, the eigenfunctions of the Laplacian on a surface (or Riemannian manifold) form a complete orthonormal basis for the set of square-integrable complex-valued functions on that surface [20, 22, 28].

___

4.5   Examples of Curvilinear Coordinates

In this section we illustrate the general definitions given previously with polar and spherical coordinates. See [5, 8] as references. For examples of other curvilinear coordinate systems see [26].

4.5.1   Polar Coordinates

Consider the polar (cylindrical) coordinates

$$x_1 = r \cos \phi \tag{4.16}$$
$$x_2 = r \sin \phi \tag{4.17}$$
$$x_3 = z. \tag{4.18}$$

Here $u_1 = r$, $u_2 = \phi$, and $u_3 = z$. In this case the Jacobian matrix is

$$J(r, \phi, z) = \left[\mathbf{v}_r, \mathbf{v}_\phi, \mathbf{v}_z \right] = \left[\frac{\partial \mathbf{x}}{\partial r}, \frac{\partial \mathbf{x}}{\partial \phi}, \frac{\partial \mathbf{x}}{\partial z} \right] = \begin{pmatrix} \cos \phi & -r \sin \phi & 0 \\ \sin \phi & r \cos \phi & 0 \\ 0 & 0 & 1 \end{pmatrix}.$$

The metric tensor is

$$G(r, \phi, z) = \begin{pmatrix} 1 & 0 & 0 \\ 0 & r^2 & 0 \\ 0 & 0 & 1 \end{pmatrix}.$$

According to Eq. (4.13),

$$\mathbf{v}_1 = \mathbf{e}_r = \begin{pmatrix} \cos\phi \\ \sin\phi \\ 0 \end{pmatrix}; \qquad \mathbf{v}_2/r = \mathbf{e}_\phi = \begin{pmatrix} -\sin\phi \\ \cos\phi \\ 0 \end{pmatrix}; \qquad \mathbf{v}_3 = \mathbf{e}_z = \begin{pmatrix} 0 \\ 0 \\ 1 \end{pmatrix}$$

and the vector $\mathbf{F}$ is expanded as

$$\mathbf{F} = F_1\mathbf{v}_1 + F_2\mathbf{v}_2 + F_3\mathbf{v}_3 = F_r\mathbf{e}_r + F_\phi\mathbf{e}_\phi + F_z\mathbf{e}_z.$$

The above expression serves as a definition. That is, $F_1 = F_r$, $F_2 = F_\phi/r$, and $F_3 = F_z$.

By the general definitions given previously, the gradient, divergence, and Laplacian are given in polar (cylindrical) coordinates as

$$\operatorname{grad} f = \frac{\partial f}{\partial r}\mathbf{v}_1 + \frac{1}{r^2}\frac{\partial f}{\partial \phi}\mathbf{v}_2 + \frac{\partial f}{\partial z}\mathbf{v}_3 = \frac{\partial f}{\partial r}\mathbf{e}_r + \frac{1}{r}\frac{\partial f}{\partial \phi}\mathbf{e}_\phi + \frac{\partial f}{\partial z}\mathbf{e}_z,$$

$$\operatorname{div}\mathbf{F} = \frac{1}{r}\frac{\partial(rF_1)}{\partial r} + \frac{\partial F_2}{\partial \phi} + \frac{\partial F_3}{\partial z} = \frac{1}{r}\frac{\partial(rF_r)}{\partial r} + \frac{1}{r}\frac{\partial F_\phi}{\partial \phi} + \frac{\partial F_z}{\partial z},$$

and

$$\operatorname{div}(\operatorname{grad} f) = \frac{1}{r}\frac{\partial}{\partial r}\left(r\frac{\partial f}{\partial r}\right) + \frac{1}{r^2}\frac{\partial^2 f}{\partial \phi^2} + \frac{\partial f^2}{\partial z^2}.$$

Note that the eigenfunctions of the Laplacian in polar coordinates involve Bessel functions, and for fixed ϕ and z, the eigenfunction problem reduces to the Sturm-Liouville equation defining the Bessel functions.

4.5.2 Spherical Coordinates

Consider the orthogonal curvilinear coordinates

$$\begin{aligned} x_1 &= r\cos\phi\sin\theta \\ x_2 &= r\sin\phi\sin\theta \\ x_3 &= r\cos\theta. \end{aligned} \tag{4.19}$$

These are the familiar spherical coordinates, where $0 \le \theta \le \pi$ and $0 \le \phi \le 2\pi$. r is the radial distance from the origin, θ is called the polar angle (or colatitude), and ϕ is called the azimuthal angle (or longitude).[6] These are denoted in Fig. 4.3.

In this case, the Jacobian matrix with $\mathbf{u} = [r, \phi, \theta]^T$ is

$$J(r, \phi, \theta) = \left[\frac{\partial\mathbf{x}}{\partial r}, \frac{\partial\mathbf{x}}{\partial \phi}, \frac{\partial\mathbf{x}}{\partial \theta}\right] = \begin{pmatrix} \cos\phi\sin\theta & -r\sin\phi\sin\theta & r\cos\phi\cos\theta \\ \sin\phi\sin\theta & r\cos\phi\sin\theta & r\sin\phi\cos\theta \\ \cos\theta & 0 & -r\sin\theta \end{pmatrix}.$$

---

[6] Often in engineering the angles ϕ and θ in Fig. 4.3 are reversed (see [19]). We use the notation more common in physics and mathematics to be consistent with the classical references (see [5]).

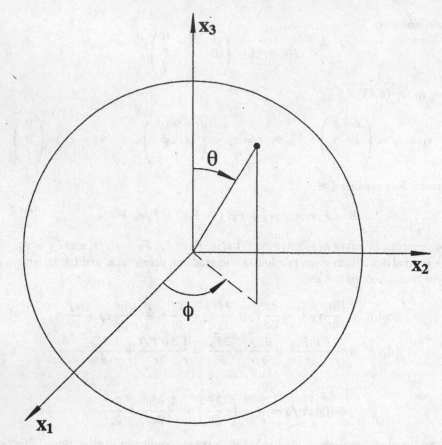

FIGURE 4.3
Spherical coordinates.

The metric tensor is

$$G(r, \phi, \theta) = \begin{pmatrix} 1 & 0 & 0 \\ 0 & r^2 \sin^2 \theta & 0 \\ 0 & 0 & r^2 \end{pmatrix}.$$

By the general definitions given previously, the gradient, divergence, and Laplacian in spherical coordinates are

$$\operatorname{grad} f = \frac{\partial f}{\partial r} \mathbf{e}_r + \frac{1}{r} \frac{\partial f}{\partial \theta} \mathbf{e}_\theta + \frac{1}{r \sin \theta} \frac{\partial f}{\partial \phi} \mathbf{e}_\phi, \tag{4.20}$$

$$\operatorname{div} \mathbf{F} = \frac{1}{r^2} \frac{\partial (r^2 F_r)}{\partial r} + \frac{1}{r \sin \theta} \frac{\partial F_\phi}{\partial \phi} + \frac{1}{r \sin \theta} \frac{\partial (F_\theta \sin \theta)}{\partial \theta}, \tag{4.21}$$

and

$$\operatorname{div}(\operatorname{grad} f) = \frac{1}{r^2} \frac{\partial}{\partial r} \left(r^2 \frac{\partial f}{\partial r} \right) + \frac{1}{r^2 \sin^2 \theta} \frac{\partial^2 f}{\partial \phi^2} + \frac{1}{r^2 \sin \theta} \frac{\partial}{\partial \theta} \left(\sin \theta \frac{\partial f}{\partial \theta} \right), \tag{4.22}$$

where

$$\mathbf{e}_r = \begin{pmatrix} \cos \phi \sin \theta \\ \sin \phi \sin \theta \\ \cos \theta \end{pmatrix}; \qquad \mathbf{e}_\phi = \begin{pmatrix} -\sin \phi \\ \cos \phi \\ 0 \end{pmatrix}; \qquad \mathbf{e}_\theta = \begin{pmatrix} \cos \phi \cos \theta \\ \sin \phi \cos \theta \\ -\sin \theta \end{pmatrix}$$

and the vector $\mathbf{F} = \sum_{i=1}^{3} F_i \mathbf{e}_i$ is expanded as $\mathbf{F} = F_r \mathbf{e}_r + F_\phi \mathbf{e}_\phi + F_\theta \mathbf{e}_\theta$.

The Dirac delta function for the sphere is

$$\delta_{S^2}(\mathbf{u}, \mathbf{u}_0) = \begin{cases} \delta(\theta - \theta_0)\delta(\phi - \phi_0)/\sin\theta & \theta_0 \notin \{0, \pi\} \\[2ex] \delta(\theta - \theta_0)/(2\pi \sin\theta) & \theta_0 \in \{0, \pi\}. \end{cases}$$

4.5.3   Stereographic Projection

Stereographic projections from the unit sphere to the plane can be defined in an analogous manner to those described earlier in the case of the projection of the circle onto the line.

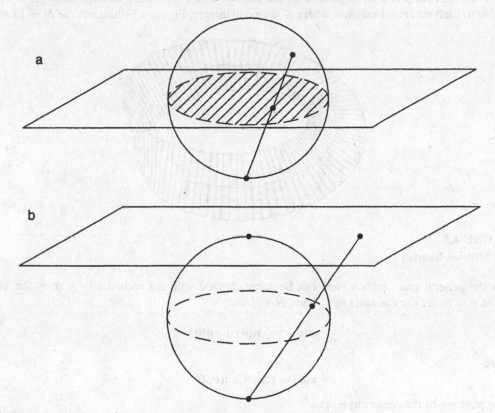

FIGURE 4.4
Stereographic projections: (a) equatorial; (b) tangent to north pole.

In the particular case of the projection when the projecting line originates at the south pole and the plane intersects the sphere at the equator [Fig. 4.4(a)], the relationship between planar coordinates (x_1, x_2) and $\mathbf{v} \in S^2$ is

$$x = x_1 + i x_2 = \frac{v_1 + i v_2}{1 + v_3}. \tag{4.23}$$

The inverse relationship is

$$v_1 = \frac{2x_1}{1 + x_1^2 + x_2^2} \qquad v_2 = \frac{2x_2}{1 + x_1^2 + x_2^2} \qquad v_3 = \frac{1 - x_1^2 - x_2^2}{1 + x_1^2 + x_2^2}. \tag{4.24}$$

One can verify that this projection has the following properties [10]: (1) circles on the sphere are mapped to circles or lines in the plane; and (2) angles between tangents to intersecting curves on the sphere are preserved under projection to the plane. The other stereographic projection shown in Fig. 4.4(b) is less common.

Using the general theory, it is possible to formulate the gradient, divergence, and Laplacian of functions on the sphere using the projected coordinates, though this is much less common than the (θ, ϕ) coordinates discussed earlier.

4.5.4   The Möbius Band

The Möbius band is a rectangular strip, the ends of which are turned through an angle of $N\pi$ relative to each other and attached, where N is an odd integer. Figure 4.5 illustrates the $N = 1$ case.

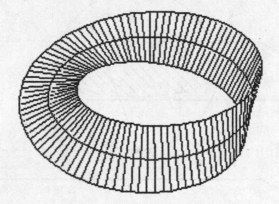

FIGURE 4.5
The Möbius band.

In the general case, such a band can be parameterized with the variables $0 \le \theta \le 2\pi$ and $-r_0 < r < r_0$ for the constants $r_0 \le 1$ and $N \in \mathbb{Z}$ as

$$\mathbf{x}(\theta, r) = \mathbf{n}(\theta) + r\mathbf{u}(\theta)$$

where

$$\mathbf{n}(\theta) = [\cos\theta, \sin\theta, 0]^T$$

is the position of the center circle, and

$$\mathbf{u}(\theta) = \begin{pmatrix} \cos\theta \sin N\theta/2 \\ \sin\theta \sin N\theta/2 \\ \cos N\theta/2 \end{pmatrix}$$

is the unit tangent to the surface in the direction orthogonal to $\mathbf{n}$.

We note that the values of the parameters (r, θ) and $((-1)^N r, \theta + 2\pi)$ define the same point on the strip

$$\mathbf{x}(r, \theta) = \mathbf{x}\left((-1)^N r, \theta + 2\pi\right).$$

However, when one computes the Jacobian

$$J(\theta, r) = \left[\frac{\partial \mathbf{x}}{\partial \theta}, \frac{\partial \mathbf{x}}{\partial r}\right],$$

one finds that

$$J(\theta, r) = J\left((-1)^N r, \theta + 2\pi\right) \begin{pmatrix} 1 & 0 \\ 0 & (-1)^N \end{pmatrix}.$$

This suggests that for odd N there is not a well-defined *orientation* for the Möbius band, since the direction of the normal $\partial x/\partial\theta \times \partial x/\partial r$ differs depending on whether it is evaluated at the parameters (r, θ) or $((-1)^N r, \theta + 2\pi)$.

In order to clarify what is meant by the word orientation, the following definitions are helpful.

DEFINITION 4.1 A Gauss map *is a function that assigns to each point on a surface a unit normal vector. Since this may be viewed as a mapping between the surface and the unit sphere, S^2, it is also called a* normal spherical image.

DEFINITION 4.2 A surface (2-manifold) is orientable *if the Gauss map is continuous at all points.*

The orientation of any orientable surface is fixed when one specifies the orientation of the normal at any point on the surface (e.g., up or down, in or out).

The Möbius band is the standard example of a manifold that is not *orientable*. Throughout the rest of the text, we will deal exclusively with orientable manifolds for which well-defined normals and concepts of volume and integration exist.

4.6 Topology of Surfaces and Regular Tessellations

Topology is a field of mathematics which considers only those properties of a manifold that remain invariant under certain kinds of invertible transformations. Topological properties include connectedness and compactness.

Connectedness means that it is possible to form a path between any two points in the object, space, or surface under consideration, with the whole path being contained in the object. We say the object is simply connected if any closed path in the object can be continuously "shrunk" to a point, while always remaining in the object. Compactness means that the object is closed and bounded.

An object can have more than one component, each of which is connected (such as a hyperboloid of two sheets or the set of all orthogonal matrices of dimension n), in which case the whole object is not connected; an object can be simply connected (such as the sphere, the surface of a cube, or the plane); or an object can be connected (but not simply) such as the torus or the set of rotations $SO(3)$.

Even though the plane and sphere are both simply connected, the former is noncompact while the latter is compact. Therefore, it is not possible to find an invertible mapping between the sphere and plane (recall that stereographic projection has a singularity). On the other hand, the sphere and surface of the cube can be invertibly mapped to each other (for instance, by lining up their centers and projecting along all lines passing through this point) and are considered to be the same from the topological perspective.

In Section 4.6.1 we provide the oldest and perhaps the most illustrative example of a topological property of surfaces. In Section 4.6.2 we apply this to the determination of decompositions of the sphere and plane into congruent regular components.

4.6.1   The Euler Characteristic and Gauss-Bonnet Theorem

Consider a polygon in the plane. Figure 4.6(a) shows what happens if we divide it in half by introducing a line segment (edge) terminating at two points (vertices). Figure 4.6(b) shows additional finer division with edges and vertices. Enclosed within some set of edges is always an area (called a face). If we count the number of vertices (V), edges (E), and faces (F), we find that no matter how we tessellate the polygon, the formula $V - E + F = 1$ holds.

In the plane with no boundary, if we start forming a larger and larger collection of polygons we observe the condition $V - E + F = 2$ where the whole plane, minus the collection of inscribed shapes, is counted as one face.

Likewise, consider a rectangular strip. If we imagine tessellating the strip in such a way that is consistent with the two ends being twisted by π radians relative to each other and glued, we observe the result $V - E + F = 0$. If without twisting, we glue parallel edges together, the same formula holds. This is the case of a torus.

The quantity

$$V - E + F = \chi(\mathcal{S}) \tag{4.25}$$

is called the *Euler characteristic* of the surface $\mathcal{S}$. See [12] for a more detailed discussion. Euler discovered it more than 200 years ago. From the previous arguments, we see $\chi(\mathbb{R}^2) = 2$, $\chi(T^2) = \chi(Mo) = 0$. Here T^2 and Mo denote the 2-torus and Möbius strip, respectively. Similar arguments for the sphere show that $\chi(S^2) = 2$.

a
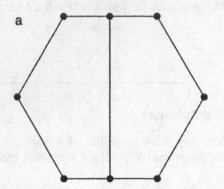

b
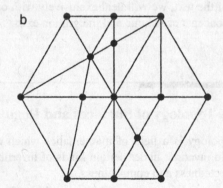

FIGURE 4.6
Successive tessellations of a polygon.

Another remarkable result is that the total curvature of a surface (curvature integrated over the whole surface) is a topological (as opposed to geometrical) property. The *Gauss-Bonnet* theorem says that the total curvature is related to the Euler characteristic as

$$\int_S K\,ds = 2\pi\chi(\mathcal{S}) \tag{4.26}$$

where K is the Gaussian curvature defined in Eq. (4.10) and $ds = \sqrt{\det G}\ du_1 du_2$ is the area element for the surface. This is easy to see by direct calculation for the sphere or the torus.

To see that it really is invariant under smooth topological transformations, start with the surface parameterized as $\mathbf{x}(u_1, u_2)$, and apply an arbitrary invertible transformation $f : \mathbb{R}^3 \to \mathbb{R}^3$ that is smooth with a smooth inverse. The surface then becomes $\mathbf{f}(\mathbf{x}(u_1, u_2))$. Going through the calculation of the volume element and curvature of this new surface, we find that the integral of total curvature remains unchanged.

In the next section we use the Euler characteristic to determine constraints that regular tessellations of the plane and sphere must obey.

4.6.2 Regular Tessellations of the Plane and Sphere

A *regular tessellation* is a division of a surface (or space) into congruent regular polygons (polyhedra in the three-dimensional case, and polytopes in higher dimensional cases). Our current discussion is restricted to tessellations of surfaces, in particular the plane and sphere.

The Case of the Plane

For a regular planar N-gon, the exterior angle between adjacent edges will be $\theta = 2\pi/N$. The interior angle will, therefore, be $\pi - \theta$.

Suppose M such N-gons meet at each vertex. Then it must be the case that

$$M \cdot (\pi - \theta) = 2\pi.$$

Dividing by π, this is rewritten as

$$M \cdot (1 - 2/N) = 2,$$

or solving for N in terms of M,

$$N = 2M/(M - 2).$$

Since M and N must both be positive integers, and $N \geq 3$, and $M - 2$ does not divide $2M$ without remainder for $M > 6$, we see immediately that

$$(M, N) \in \{(3, 6), (4, 4), (6, 3)\}.$$

These correspond to hexagonal, square, and equilateral triangular tessellations, respectively.

The Case of the Sphere

Given a regular tessellation, if m faces meet at each vertex and each face is a regular n-gon (polygon with n edges), the tessellation (as well as the constituent polygons) is said to be of type $\{n, m\}$. It follows that for this type of tessellation,

$$V \cdot m = 2E = n \cdot F. \tag{4.27}$$

The above equation simply says that each edge joins two vertices and each edge lies at the interface of two faces.

Substituting Eq. (4.27) into Eq. (4.25) for the case of a sphere, we have

$$2E/m - E + 2E/n = 2,$$

or

$$\frac{2n - mn + 2m}{2mn} = \frac{1}{E}.$$

Since $m, n, E > 0$, we have $2n - mn + 2m > 0$, or equivalently

$$(n - 2)(m - 2) < 4.$$

For positive integer values of m and n, it is easy to see that a necessary (though not sufficient) condition for this equation to hold is that both the numbers m and n must be less than six. Direct enumeration of all pairs $\{n, m\}$ weeded out by conditions such as that every face must be surrounded by at least three edges and vertices, yields

$$\{3, 3\}, \{4, 3\}, \{3, 4\}, \{5, 3\}, \{3, 5\}.$$

These correspond to the tetrahedron, cube, octahedron, dodecahedron, and icosahedron.

4.7   Orthogonal Expansions and Transforms in Curvilinear Coordinates

The definition of orthogonal functions on curved surfaces in higher dimensions is analogous in many ways to Sturm-Liouville theory. Given an M−dimensional hyper-surface S, or a finite-volume section of the surface $\Omega \subset S$, orthonormal real-valued functions are those for which

$$(\phi_i, \phi_j) = \int_{\Omega} \phi_i(\mathbf{u})\phi_j(\mathbf{u})\sqrt{\det(G(\mathbf{u}))}du^M = \delta_{ij}.$$

Hence, the square root of the determinant of the Riemannian metric tensor serves as a weighting function. If this determinant can be written as the product of functions of only one variable, the classical Sturm-Liouville theory may be used to expand functions on the surface in an orthonormal basis.

A complete orthogonal set of functions on any orientable hyper-surface can with finite volume be generated as eigenfunctions of the Laplacian operator. For example, the Laplacian of functions on the circle can be defined as $d^2/d\theta^2$. Eigenfunctions of the Laplacian are those which satisfy $d^2f/d\theta^2 = \lambda f$ and the periodic boundary conditions $f(0) = f(2\pi)$ and $f'(0) = f'(2\pi)$. The result is the set of functions $f_n(\theta) = e^{in\theta}$ with $\lambda = -n^2$ for $n \in \mathbb{Z}$. These functions form the starting point for classical Fourier analysis.

More generally, it has been observed [29] that the eigenfunctions of the Laplacian operator on a smooth compact Riemannian manifold also form an orthogonal basis for functions that are square integrable on the manifold. The integration measure is the one defined relative to the Riemannian metric in Eq. (4.7).

4.7.1   The 2-D Fourier Transform in Polar Coordinates

The Fourier transform in two dimensions is written in Cartesian coordinates as

$$\hat{f}(\omega) = \hat{f}(\omega_1, \omega_2) = \int_{-\infty}^{\infty} \int_{-\infty}^{\infty} f(x_1, x_2)e^{-i(\omega_1 x_1 + \omega_2 x_2)}dx_1 dx_2 = \int_{\mathbb{R}^2} f(\mathbf{x})e^{-i\boldsymbol{\omega}\cdot\mathbf{x}}d\mathbf{x}.$$

Changing to polar coordinates in both the spatial and Fourier domains

$$x_1 = r\cos\theta, \qquad x_2 = r\sin\theta,$$

and

$$\omega_1 = p\cos\alpha, \qquad \omega_2 = p\sin\alpha,$$

the original function is $f(x_1, x_2) = \tilde{f}(r, \theta)$, and the Fourier transform is rewritten as

$$\hat{\tilde{f}}(\rho, \alpha) = \int_0^{\infty} \int_0^{2\pi} \tilde{f}(r, \theta)e^{-ipr\cos(\theta-\alpha)}r\,dr\,d\theta$$

$$= \sum_{n=-\infty}^{\infty} e^{in\alpha}i^{-n} \int_0^{\infty} \int_0^{2\pi} \tilde{f}(r, \theta)J_n(pr)e^{-in\theta}r\,dr\,d\theta.$$

If we assume that

$$\tilde{f}(r, \theta) = \sum_{n=-\infty}^{\infty} f_n(r)e^{in\theta},$$

then

$$\int_0^\infty \int_0^{2\pi} \tilde{f}(r,\theta)e^{-ipr\cos(\theta-\alpha)}r\,dr\,d\theta = \sum_{n=-\infty}^{\infty} \int_0^\infty f_n(r)\left(\int_0^{2\pi} e^{in\theta}e^{-ipr\cos(\theta-\alpha)}d\theta\right)r\,dr.$$

The inner integral reduces to $e^{in\alpha}J_n(pr)$, and so the Fourier transform in polar coordinates reduces to a combination of a Fourier series in the angle variable and a Hankel transform in the radial variable. Another way to rearrange terms is

$$\hat{f}_n(p) = \int_0^{2\pi}\int_0^\infty f(r,\theta)e^{-in\theta}J_n(pr)r\,dr\,d\theta$$

$$f(r,\theta) = \sum_{n=-\infty}^{\infty} e^{in\theta}\int_0^\infty \hat{f}_n(p)J_n(pr)p\,dp. \tag{4.28}$$

We note also the expansion

$$e^{i\mathbf{p}\cdot\mathbf{r}} = \sum_{n=-\infty}^{\infty} i^n J_n(pr)e^{in\theta}e^{-in\phi} \tag{4.29}$$

and its complex conjugate

$$e^{-i\mathbf{p}\cdot\mathbf{r}} = \sum_{n=-\infty}^{\infty} i^{-n} J_n(pr)e^{-in\theta}e^{in\phi}$$

could have been substituted into the Fourier transform and its inverse to derive Eq. (4.28).

4.7.2 Orthogonal Expansions on the Sphere

When restricted to the surface of a sphere, the result of orthogonal expansions in spherical coordinates is the spherical harmonics. A more general theory of orthogonal expansions on higher dimensional manifolds follows in the same way.

The unit sphere in $\mathbb{R}^3$ is a two-dimensional surface denoted as S^2 defined by the constraint $x_1^2 + x_2^2 + x_3^2 = 1$. The spherical coordinates (θ, ϕ) are the most common.

Other parameterizations include stereographic projection [10] and other angular measurements. In the (θ, ϕ) parameterization, integration of a function on the sphere is performed as

$$\int_{S^2} f\,ds = c \int_{\theta=0}^{\pi}\int_{\phi=0}^{2\pi} f(\theta,\phi)\sin\theta\,d\theta\,d\phi \tag{4.30}$$

where c is a constant that is chosen to be either 1 or 4π depending on the context. Henceforth we choose $c = 1$ unless otherwise stated.

Using the spherical coordinates introduced in the previous section, we may expand functions on the sphere into a series approximation much like Fourier series. The volume element can be viewed as the product of the volume elements for S^1 and $[0, \pi]$ (with weight $\sin\theta$). Functions on the sphere are then viewed as functions on $S^1 \times [0, \pi]$ with the weighting factor $w(\theta, \phi) = w_1(\theta)w_2(\phi) = \sin\theta \cdot 1$. This allows us to use Sturm-Liouville theory to construct orthonormal functions in these two domains separately. We already know that an orthogonal basis for $\mathcal{L}^2(S^1)$ is given by $\{e^{im\phi}\}$ for $m \in \mathbb{Z}$. Likewise for $\mathcal{L}^2([-1, 1], dx)$, we have the Legendre polynomials $P_l(x)$. It follows that by the change of coordinates $x = \cos\theta$, the functions $\sqrt{(2l+1)/2}P_l(\theta)$ form an orthonormal basis for $\mathcal{L}^2([0, \pi], d\theta\sin\theta)$, and the collection of all products $\{\sqrt{(2l+1)/4\pi}e^{im\phi}P_l(\theta)\}$ for $l = 0, 1, 2, \ldots$

and $m \in \mathbb{Z}$ form a complete orthonormal set of functions on the sphere. This is, however, not the only choice. In fact, it is much more common to choose the *associated Legendre functions* $\{P_l^m(\theta)\}$ for any fixed $m \in \mathbb{Z}$ satisfying $|m| \leq l$ as the basis for $\mathcal{L}^2([0, \pi], d\theta \sin \theta)$. These functions were defined in Section 3.2. The orthonormal basis that is most commonly used to expand functions on the sphere has elements of the form

$$Y_l^m(\theta, \phi) = \sqrt{\frac{(2l+1)(l-m)!}{4\pi(l+m)!}} \, P_l^m(\cos \theta) e^{im\phi}. \tag{4.31}$$

These are called *spherical harmonics*[7] for reasons we will see shortly.

Any function in $\mathcal{L}^2(S^2)$ can be expanded in a *(spherical) Fourier series* as

$$f(\theta, \phi) = \sum_{l=0}^{\infty} \sum_{m=-l}^{l} \hat{f}(l, m) Y_l^m(\theta, \phi) \qquad \text{where} \qquad \hat{f}(l, m) = \int_{S^2} f \overline{Y_l^m} \, ds. \tag{4.32}$$

Here the convention in Eq. (4.30) is used with $c = 1$.

Each of the coefficients $\hat{f}(l, m)$ is called the *(spherical) Fourier transform* of $f(\theta, \phi)$, whereas the collection $\{\hat{f}(l, m)\}$ is called the *spectrum*. In order to avoid a proliferation of names of various transforms, we will henceforth refer to these simply as the Fourier series, transform and spectrum for functions on the sphere.

The spherical harmonics, $Y_l^m(\theta, \phi)$, are the most commonly used expansion on the sphere because they are eigenfunctions of the Laplacian operator defined below.

Spherical Harmonics as Eigenfunctions of the Laplacian

The gradient of a differentiable function on the sphere results from setting $r = 1$ in the expression for the gradient in spherical coordinates in Eq. (4.20)

$$\text{grad}(f) = \frac{\partial f}{\partial \theta} \mathbf{e}_\theta + \frac{1}{\sin \theta} \frac{\partial f}{\partial \phi} \mathbf{e}_\phi.$$

Likewise, the divergence of a differentiable vector field $\mathbf{F}(\theta, \phi) = F_\theta \mathbf{e}_\theta + F_\phi \mathbf{e}_\phi$ on the sphere results from setting $r = 1$ in Eq. (4.21):

$$\text{div}(\mathbf{F}) = \frac{1}{\sin \theta} \frac{\partial (F_\theta \sin \theta)}{\partial \theta} + \frac{1}{\sin \theta} \frac{\partial F_\phi}{\partial \phi}.$$

The Laplacian of a smooth function on the sphere is given by

$$\text{div}(\text{grad } f) = \frac{1}{\sin \theta} \frac{\partial}{\partial \theta} \left(\sin \theta \frac{\partial f}{\partial \theta} \right) + \frac{1}{\sin^2 \theta} \frac{\partial^2 f}{\partial \phi^2}.$$

In analogy with Sturm-Liouville theory, the eigenfunctions of the Laplacian operator are defined as

$$\text{div}(\text{grad } f) = \lambda f$$

for eigenvalues λ. The boundary conditions resulting from the spherical geometry are periodic in ϕ: $f(\theta, \phi + 2\pi) = f(\theta, \phi)$. One finds that one possible set of eigenfunctions are the spherical harmonics defined in Eq. (4.31), with $\lambda = -l(l+1)$.

---

[7]Two variants on this form are common. Often a factor of $(-1)^m$ is present, which we have absorbed in the definition of the associated Legendre functions in Eq. (3.10). Also, it is common to define harmonics $\tilde{Y}_l^m$ with the property $\overline{\tilde{Y}_l^m} = \tilde{Y}_l^{-m}$. In contrast $\overline{Y_l^m} = (-1)^m Y_l^{-m}$.

FFTs for the Sphere

Recall that the DFT is a regularly sampled version of the Fourier transform of a continuous band-limited function. In the case of the sphere, all regular samplings are very coarse, for example, the vertices of the platonic solids. Regular sampling in ϕ and θ is appealing, but yields closely packed sample points near the poles of the sphere, and sparsely packed points near the equator. In recent work, Driscoll and Healy [11] formulated an efficient technique for calculating the Fourier transform of band-limited functions on the sphere based on this sampling. Namely, if $\hat{f}(l, m)$ is the Fourier transform of $f(\theta, \phi)$, and $\hat{f}(l, m) = 0$ for all $l \geq B$ (where $B = 2^K$ is the band limit), then in exact arithmetic it is possible to exactly calculate the whole spectrum of $f(\theta, \phi)$ by the discrete computations

$$\hat{f}(l, m) = \frac{\sqrt{\pi}}{B} \sum_{j=0}^{2B-1} \sum_{k=0}^{2B-1} a_j f(\theta_j, \phi_k) \overline{Y_l^m(\theta_j, \phi_k)}, \tag{4.33}$$

where $\theta_j = \pi j / 2B$ and $\phi_k = \pi k / B$ and a_j are calculated as in Eq. (3.12). The total number of samples in this case is $N = (2B)^2$.

Using the recursion relations for Legendre polynomials, and writing the discrete spherical Fourier transform as

$$\hat{f}(l, m) = \frac{1}{2B} \sqrt{\frac{(2l+1)(l-m)!}{(l+m)!}} \sum_{j=0}^{2B-1} a_j P_l^m (\cos \theta_j) \left(\sum_{k=0}^{2B-1} e^{-im\phi_k} f(\theta_j, \phi_k) \right), \tag{4.34}$$

it is possible to use both the usual FFT and the fast associated Legendre transform to realize an FFT of functions on the sphere. There are $N = \mathcal{O}(B^2)$ sample points, and the whole spectrum of f can be computed in $\mathcal{O}(N(\log N)^2)$ operations. Using uniform sampling in $x = \cos \theta$ rather than uniform sampling in θ also results in an $\mathcal{O}(N(\log N)^2)$ transform as a result of the general theory reviewed in Section 3.5. The inversion formula may be computed at the N sample points in $\mathcal{O}(N(\log N)^2)$ as well [14]. We note that when the discrete Legendre transform in θ is used in the traditional way (without the fast implementation) together with the FFT in ϕ, an $\mathcal{O}(B^3) = \mathcal{O}(N^{3/2})$ algorithm results.

In contrast to this algorithm, which is exact in exact arithmetic, a number of approximate numerical methods for computing spherical Fourier transforms have been proposed [9, 17, 23]. While from a theoretical perspective these algorithms do not provide exact results, they have been shown to work well in practice.

4.7.3   The 3-D Fourier Transform in Spherical Coordinates

In analogy with the expansion Eq. (4.29) in the planar case, one makes the expansion [21]

$$\exp(i\mathbf{p} \cdot \mathbf{r}) = 4\pi \sum_{l=0}^{\infty} \sum_{k=-l}^{l} i^l j_l(pr) Y_l^k (\theta_{\mathbf{r}}, \phi_{\mathbf{r}}) \overline{Y_l^k (\theta_{\mathbf{p}}, \phi_{\mathbf{p}})}. \tag{4.35}$$

The direct Fourier transform is written in spherical coordinates as

$$\hat{f}(p, \theta_p, \phi_p) = \sum_{l=0}^{\infty} \sum_{k=-l}^{l} \hat{f}_l^k(p) Y_l^k (\theta_p, \phi_p)$$

(the inverse spherical Fourier transform), where

$$\hat{f}_l^k(p) = 4\pi i^l f_l^k(p)$$

and

$$f_l^k(p) = \int_0^\infty A_l^k(r) j_l(pr) r^2 dr$$

(the spherical Hankel transform).

The functions $A_l^k(r)$ are defined as

$$A_l^k(r) = \int_0^\pi \int_0^{2\pi} f(r, \theta_r, \phi_r) \overline{Y_l^k(\theta_r, \phi_r)} \sin \theta_r d\theta_r d\phi_r$$

(the direct spherical Fourier transform). Note that the complex conjugate has been interchanged between $Y_l^k(\theta_p, \theta_p)$ and $Y_l^k(\theta_r, \theta_r)$ in the expansion Eq. (4.35) of a plane wave in spherical harmonics.

The inverse Fourier transform is written as

$$\hat{f}(r, \theta_r, \phi_r) = \sum_{l=0}^\infty \sum_{k=-l}^l \hat{g}_l^k(r) Y_l^k(\theta_r, \phi_r)$$

where

$$\hat{g}_l^k(r) = \frac{1}{2\pi^2} i^{-l} g_l^k(r)$$

and

$$g_l^k(r) = \int_0^\infty B_l^k(p) j_l(pr) p^2 dp$$

(the spherical Hankel transform), and the functions $B_l^k(p)$ are defined as

$$B_l^k(p) = \int_0^\pi \int_0^{2\pi} f(p, \theta_p, \phi_p) \overline{Y_l^k(\theta_p, \phi_p)} \sin \theta_p d\theta_p d\phi_p.$$

4.8   Continuous Wavelet Transforms on the Plane and $\mathbb{R}^n$

The idea of describing a function as a weighted sum (or integral) of translated and scaled (or modulated) versions of a given function extends to domains other than the real line. In this section, we review the work of a number of researchers who have examined continuous wavelet transforms in multi-dimensional Cartesian space.

4.8.1   Multi-dimensional Gabor Transform

The continuous Gabor transform for $\mathbb{R}^n$ is defined in exact analogy as in the one-dimensional case

$$G_f(\mathbf{b}, \omega) = \int_{\mathbb{R}^n} f(\mathbf{x}) \overline{g_{(\mathbf{b}, \omega)}(\mathbf{x})} dx_1 \ldots dx_n$$

where

$$g_{(\mathbf{b}, \omega)} = e^{i\omega \cdot (\mathbf{x} - \mathbf{b})} g(\mathbf{x} - \mathbf{b}).$$

Using the same calculations as in the one-dimensional case with the n-dimensional Fourier transform replacing the one-dimensional version, the inversion formula

$$f(\mathbf{x}) = \frac{1}{(2\pi)^n \|g\|^2} \int_{\mathbb{R}^{2n}} G_f(\mathbf{b}, \omega) e^{i\omega \cdot (\mathbf{x} - \mathbf{b})} g(\mathbf{x} - \mathbf{b}) db_1 \ldots db_n d\omega_1 \ldots d\omega_n.$$

4.8.2 Multi-Dimensional Continuous Scale-Translation Wavelet Transforms

In a series of papers, the continuous wavelet transform based on translations and dilations has been extended to $\mathbb{R}^n$ by considering dilations and rigid-body motions (translations and rotations) [25, 4]. We review here only the planar case (the higher dimensional cases follow analogously, provided one knows how to describe rotations in $\mathbb{R}^n$).

Let $\varphi \in \mathcal{L}^1(\mathbb{R}^2) \cap \mathcal{L}^2(\mathbb{R}^2)$ and its 2-D Fourier transform, $\hat{\varphi}(\omega)$, satisfy the admissibility condition

$$0 < k_\varphi = 2\pi \int_{\mathbb{R}^2} |\hat{\varphi}(\omega)|^2 \frac{d\omega_1 d\omega_2}{\|\omega\|^2} < \infty.$$

Then the *scale-Euclidean* wavelet transform is defined as

$$T_f(\mathbf{b}, a, R(\theta)) = \int_{\mathbb{R}^2} f(\mathbf{x}) \overline{\varphi_{(\mathbf{b}, a, R(\theta))}(x)} dx_1 dx_2$$

where

$$\varphi_{(\mathbf{b}, a, R(\theta))}(x) = \frac{1}{a} \varphi \left(R^{-1}(\theta)(\mathbf{x} - \mathbf{b})/a \right).$$

Here $R(\theta) \in SO(2)$ is the 2×2 rotation matrix in Eq. (4.3).

The inversion follows in a similar way as for the one-dimensional case and is written explicitly as [25]

$$f(\mathbf{x}) = \frac{1}{k_\varphi} \int_{a \in \mathbb{R}^+} \int_{\mathbf{b} \in \mathbb{R}^2} \int_{\theta=0}^{2\pi} T_f(\mathbf{b}, a, R(\theta)) \varphi_{\mathbf{b}, a, R(\theta)}(x) \frac{da}{a^3} db_1 db_2 d\theta.$$

We also note that continuous wavelet transforms based on all three operations (rigid-body motion, modulation, and dilation) have also been presented in the literature. These are called *Weil-Heisenberg wavelets* [18].

4.9 Gabor Wavelets on the Circle and Sphere

In a sense, the basis $\{e^{in\theta}\}$ is a wavelet basis, and Fourier series expansions are Gabor wavelet expansions with mother wavelet $\varphi = 1$. Each $e^{in\theta}$ for $n \neq 0$ may also be viewed as a scaled version of $e^{i\theta}$. The set of Haar functions can also be used to expand functions on the circle, and so this is another natural wavelet series expansion on a domain other than the real line.

Here we explore a number of ways to extend the concept of modulation on the circle and sphere. Together with the appropriate definition of a shift (which in this case is simply a rotation), a number of different Gabor-like transforms can be defined. As always, we are considering only transforms of functions on the sphere that are well behaved (continuous, and hence square integrable).

4.9.1 The Modulation-Shift Wavelet Transform for the Circle

We examine two kinds of modulation-shift wavelets in this subsection. We refer to these as Gabor and Torresani wavelets, respectively.

Gabor Wavelets for the Circle

Given a mother wavelet $\psi \in \mathcal{L}^2(S^1)$, we define

$$\psi_{n,\alpha}(\theta) = e^{in\theta} \psi(\theta - \alpha).$$

Then the Gabor transform of a function $f \in \mathcal{L}^2(S^1)$ is

$$\tilde{f}(n, \alpha) = \int_0^{2\pi} f(\theta)\overline{\psi_{n,\alpha}(\theta)}d\theta.$$

The function $f(\theta)$ can be recovered from $\tilde{f}(n, \alpha)$ with the inverse transform

$$f(\theta) = \frac{1}{2\pi\|\psi\|_2^2} \sum_{n=-\infty}^{\infty} \int_0^{2\pi} \tilde{f}(n, \alpha)\psi_{n,\alpha}(\theta)d\alpha,$$

where $\|\psi\|^2 = (\psi, \psi)$. This follows by direct substitution and using the formula for the delta function on the circle

$$\sum_{n=-\infty}^{\infty} e^{in\theta} = 2\pi\delta(\theta).$$

The Torresani Wavelets for the Circle

This subsection describes a transform introduced in [32]. The starting point is a mother wavelet $\psi(\theta)$ with support on the circle in the interval $|\theta| \leq \pi/2$ and which satisfies the admissibility condition

$$0 \neq c_\psi = 2\pi \int_{-\pi/2}^{\pi/2} \frac{|\psi(\gamma)|^2}{\cos\gamma}d\gamma < \infty. \tag{4.36}$$

The corresponding wavelet transform is

$$\tilde{f}(\alpha, p) = \int_0^{2\pi} f(\theta)\overline{\psi_{\alpha,p}(\theta)}d\theta \tag{4.37}$$

where

$$\psi_{\alpha,p}(\theta) = e^{ip\sin(\theta-\alpha)}\psi(\theta - \alpha)$$

is a modulated and shifted version of $\psi(\theta)$ with p and α specifying the amount of modulation and rotation, respectively.

The inversion formula is

$$f(\theta) = \frac{1}{c_\psi} \int_0^{2\pi} \int_{-\infty}^{\infty} \tilde{f}(\alpha, p)\psi_{\alpha,p}(\theta)dp\,d\alpha. \tag{4.38}$$

In order to see that this does in fact reproduce the function f from the transform $\tilde{f}$, we substitute

$$\int_0^{2\pi} \int_{-\infty}^{\infty} \left(\int_0^{2\pi} f(\theta)\overline{\psi_{\alpha,p}(\theta)}\psi_{\alpha,p}(\gamma)d\theta \right) dp\,d\alpha$$

$$= \int_0^{2\pi} \int_0^{2\pi} f(\theta)\frac{\delta(\gamma - \theta)}{\cos(\theta - \alpha)}\overline{\psi(\theta - \alpha)}\psi(\gamma - \alpha)d\alpha\,d\theta = c_\psi f(\gamma). \tag{4.39}$$

In the first preceding equation, the integral over p is performed first, resulting in the delta function

$$\int_{-\infty}^{\infty} \exp(ip(\sin(\gamma - \alpha) - \sin(\theta - \alpha)))\,dp = 2\pi\delta(\sin(\gamma - \alpha) - \sin(\theta - \alpha))$$

$$= \frac{2\pi\delta(\gamma - \theta)}{\cos(\theta - \alpha)}.$$

The final step in Eq. (4.39) is the observation that integration on the circle is invariant under shifts.

4.9.2 The Modulation-Shift Wavelet Transform for the Sphere

In this subsection we examine Gabor and Torresani wavelets for the sphere that are analogous to the definitions for the circle.

Gabor Wavelets for the Sphere

The concepts of modulation and translation are generalized to the sphere as multiplication by a spherical harmonic and rotation, respectively. Given a mother wavelet function $\psi \in \mathcal{L}^2(S^2)$, the spherical Gabor wavelets are then

$$\psi_l^m(\mathbf{u}(\theta, \phi), R) = Y_l^m(\mathbf{u}(\theta, \phi))\psi\left(R^{-1}\mathbf{u}(\theta, \phi)\right)$$

where R is a 3×3 rotation matrix (see Chapter 5). The spherical Gabor transform is then

$$\tilde{f}_l^m(R) = \int_{S^2} f(\mathbf{u})\overline{\psi_l^m(\mathbf{u}(\theta, \phi), R)}ds(\mathbf{u}). \tag{4.40}$$

In order to determine an inversion formula to recover f from $\tilde{f}_l^m(R)$, we need to know how to integrate over the set of all rotations. This is addressed in Chapters 5 and 9. We, therefore, postpone the discussion of the inversion formula until Chapter 9.

Torresani Wavelets for the Sphere

Here a mother wavelet $\psi(\mathbf{u})$, with support on the northern hemisphere, is used to generate a family of wavelets of the form

$$\psi(\mathbf{u}; \mathbf{p}, R) = e^{i(R\mathbf{p})\cdot\mathbf{u}}\psi\left(R^{-1}\mathbf{u}\right)$$

where R is a rotation matrix and $\mathbf{p} = [p_1, p_2, 0]^T \in \mathbb{R}^3$. The transform is

$$\tilde{f}(\mathbf{p}, R) = \int_{S^2} f(\mathbf{u})\overline{\psi(\mathbf{u}; \mathbf{p}, R)}ds(\mathbf{u}). \tag{4.41}$$

As with the spherical Gabor transform, the inversion formula corresponding to Eq. (4.41) requires properties of integrals over the set of rotations. Such topics are addressed in Chapters 5 and 9. See [32] for the inversion formula.

4.10 Scale-Translation Wavelet Transforms for the Circle and Sphere

Whereas modulation of functions is a natural operation on the circle and sphere, the dilation of functions on these compact manifolds is somewhat problematic. This is because we are immediately confronted with the problem of what to do with the "extra" part of the dilated version of a function $\psi(\theta)$ that is "pushed outside" of the finite domain $[0, 2\pi]$ or $[0, 2\pi] \times [0, \pi]$ for the circle and sphere, respectively.

The two techniques for handling this problem are *periodization* and *stereographic projection*. Periodization is essentially a way to "wrap" a function on the line (plane) around the circle (sphere). Stereographic projection maps all points of finite value from the line (plane) onto all but one point on the circle (sphere), and vice versa. Hence these approaches convert a function on the circle (sphere) to one on the line (plane), perform the scale change, and then map the function back to the circle (sphere). An approach to wavelets on the sphere based on diffusion is presented in Chapter 9.

4.10.1   The Circle

Continuous dilation-translation-based wavelet transforms are perhaps less natural than the Gabor wavelets because of the issue of what to do when the support of the dilated function is wider than 2π. To this end, Holschneider [15] defined a continuous wavelet transform for the circle based on dilations and translations by "wrapping" the "leftover part" around the circle. The technical word for this wrapping is *periodization*. Here, we shall review the formulation in [15].

Given a function $\varphi \in \mathcal{L}^2(\mathbb{R})$ that satisfies the same conditions as a mother wavelet for the continuous wavelet transform on the line, the dilated, translated, and periodized version of φ (with period 2π) is denoted

$$\left[\varphi_{t,a}(x)\right]_{2\pi} = \sum_{n \in \mathbb{Z}} \frac{1}{a}\varphi\left(\frac{x - t + 2\pi n}{a}\right). \tag{4.42}$$

The periodization operation

$$\varphi \to \left[\varphi_{0,1}(x)\right]_{2\pi} = p_{2\pi}(\varphi)$$

commutes with modulations and translations

$$\begin{aligned}
(p_{2\pi}(\mu_m f))(x) &= (\mu_m (p_{2\pi} f))(x) \\
(p_{2\pi}(\tau_t f))(x) &= (\tau_t (p_{2\pi} f))(x).
\end{aligned}$$

Recall that

$$(\sigma_s f)(x) = s^{-\frac{1}{2}} f(x/s), \quad (\tau_t f)(x) = f(x - t) \quad \text{and} \quad (\mu_m f)(x) = e^{-imx} f(x),$$

where $s \in \mathbb{R}^+$, $t \in \mathbb{R}$ and $m \in \mathbb{Z}$.

The continuous wavelet transform of a complex-valued function on the circle, $f(x)$, is defined as

$$\left[\tilde{f}\right]_{2\pi}(t, a) = \int_0^{2\pi} f(x)\overline{\left[\varphi_{t,a}(x)\right]_{2\pi}}\,dx. \tag{4.43}$$

The function f is recovered from $[\tilde{f}]_{2\pi}$ as [15]

$$f(x) = \int_{S^1 \times \mathbb{R}^+} \left[\tilde{f}\right]_{2\pi}(t, a)\left[\varphi_{t,a}(x)\right]_{2\pi}\frac{da\,dt}{a}.$$

Our discussion here has been for a wavelet transform for the circle using periodization. Stereographic projection of the continuous translation-scale wavelet transform for the line onto the circle can also be used, and this is discussed in [4]. It follows in much the same way as the case for the sphere discussed in the next subsection.

4.10.2   The Sphere

Here we review the results of [2, 3] where the mother wavelet is assumed to satisfy the admissibility conditions

$$\int_0^{2\pi} \int_0^{\pi} \frac{\psi(\theta, \phi)}{1 + \cos\theta} \sin\theta\,d\theta\,d\phi = 0 \quad \text{and} \quad \int_0^{2\pi} \psi(\theta, \phi)d\phi \neq 0.$$

The spherical wavelet transform defined using such mother wavelets is based on stereographic projection of the continuous translation-scale wavelet transform for the plane.

The function $\psi(\theta, \phi)$ is projected onto the plane using stereographic projection. The resulting function on the plane is dilated by a and then projected back. The dilation operator defined in this way, together with a rotation operator, produces the family of wavelets

$$\psi(\mathbf{u}(\theta, \phi); a, R) = \frac{2a}{[(a^2 - 1)(R\mathbf{e}_3 \cdot \mathbf{u}) + (a^2 + 1)]} \psi\left(\left(R^{-1}\mathbf{u}(\theta, \phi)\right)_{1/a}\right)$$

where the notation $(\cdot)_{1/a}$ means

$$(\mathbf{u}(\theta, \phi))_{1/a} = \mathbf{u}\left(\theta_{1/a}, \phi\right)$$

where

$$\tan \frac{\theta_{1/a}}{2} = \frac{1}{a} \tan \frac{\theta}{2}.$$

The associated wavelet transform is

$$\tilde{f}(a, R) = \int_0^{2\pi} \int_0^\pi f(\mathbf{u}(\theta, \phi)) \overline{\psi(\mathbf{u}(\theta, \phi); a, R)} \sin\theta \, d\theta \, d\phi. \tag{4.44}$$

Again, in order to invert, we need techniques that are presented in subsequent chapters. See [2, 3] for the inversion formula.

4.10.3 Discrete Wavelets and Other Expansions on the Sphere

A number of other wavelet-like expansions on spheres have been presented in the recent literature.

Discrete wavelet series expansions on the sphere have been presented [7]. At the core of this approach is the use of a subset of the Haar functions on the product of intervals $[0, 1] \times [0, 1]$ that satisfy certain boundary conditions together with a mapping

$$[0, 1] \times [0, 1] \leftrightarrow S^2.$$

In this way, functions on the interior and boundary of the unit square are used to describe functions on the sphere.

An expansion based on pulse functions with support in successively smaller subdivisions of the sphere is presented in [30]. In that work, the triangular regions generated by the projection of an icosahedron onto the sphere serve as the starting point.

Other methods for describing functions on spheres based on splines can be found in the literature (see, for example, [6, 1]). For a complete review of many different methods for expanding functions on spheres, see [13].

4.11 Summary

In this chapter we examined various expansions of functions on curved surfaces (and manifolds) with emphasis on spheres. Some of the methods we reviewed here have been known for well over 100 years, while others have been introduced in the last few years. Knowledge of these methods provides a context for our discussion of Fourier expansions on functions of rotation and motion in later chapters. Chapters 5 and 6 review rotations and rigid-body motions in great detail.

References

[1] Alfred, P., Neamtu, M., and Schumaker, L.L., Bernstein-Bézier polynomials on spheres and sphere-like surfaces, *Comput. Aided Geometric Design*, 13, 333–349, 1996.

[2] Antoine, J.-P. and Vandergheynst, P., Wavelets on the n-sphere and related manifolds, *J. of Math. Phys.*, 39(8), 3987–4008, 1998.

[3] Antoine, J.-P. and Vandergheynst, P., Wavelets on the 2-sphere and related manifolds, *Rep. on Math. Phys.*, 43(1/2), 13–24, 1999.

[4] Antoine, J.-P., Carrette, P., Murenzi, R., and Piette, B., Image analysis with two-dimensional continuous wavelet transform, *Signal Process.*, 31(3), 241–272, 1993.

[5] Arfken, G.B. and Weber, H.J., *Mathematical Methods for Physicists*, 4th ed., Academic Press, San Diego, 1995.

[6] Dahlke, S. and Maass, P., A continuous wavelet transform on tangent bundles of spheres, *J. Fourier Anal. and Appl.*, 2, 379–396, 1996.

[7] Dahlke, S., Dahmen, W., and Weinreich, I., Multiresolution analysis and wavelets on S^2 and S^3, *Numerical Functional Anal. and Optimization*, 16(1-2), 19–41, 1995.

[8] Davis, H.F. and Snider, A.D., *Introduction to Vector Analysis*, 4th ed., Allyn and Bacon, Boston, 1979.

[9] Dilts, G.A., Computation of spherical harmonic expansion coefficients via FFTs, *J. of Computational Phys.*, 57, 439–453, 1985.

[10] Donnay, J.D.H., *Spherical Trigonometry: After the Cesàro Method*, InterScience Publishers, New York, 1945.

[11] Driscoll, J.R. and Healy, D.M., Jr., Computing Fourier transforms and convolutions on the 2-sphere, *Adv. in Appl. Math.*, 15, 202–250, 1994.

[12] Firby, P.A. and Gardiner, C.F., *Surface Topology*, 2nd ed., Ellis Horwood, New York, 1991.

[13] Freeden, W., Gervens, T., and Schreiner, M., *Constructive Approximation on the Sphere, With Applications to Geomathematics*, Clarendon Press, Oxford, 1998.

[14] Healy, D.M., Jr., Rockmore, D., Kostelec, P.J., and Moore, S.S.B., FFTs for the 2-sphere-improvements and variations, *Adv. in Appl. Math.*, in press.

[15] Holschneider, M., Wavelet analysis on the circle, *J. Math. Phys.*, 31(1), 39–44, 1990.

[16] Holschneider, M., Continuous wavelet transforms on the sphere, *J. Math. Phys.*, 37(8), 4156–4165, 1996.

[17] Jakob-Chien, R. and Alpert, B.K., A fast spherical filter with uniform resolution, *J. of Computational Phys.*, 136, 580–584, 1997.

[18] Kalisa, C. and Torrésani, B., N-dimensional affine Weyl-Heisenberg wavelets, *Annales de l'Institut Henri Poincaré — Physique théorique*, 59(2), 201–236, 1993.

[19] Kreyszig, E., *Advanced Engineering Mathematics*, 6th ed., John Wiley & Sons, New York, 1988.

[20] McKean, H.P. and Singer, I.M., Curvature and the eigenvalues of the Laplacian, *J. Differential Geometry,* 1, 43–69, 1967.

[21] Miller, W., Jr., *Lie Theory and Special Functions,* Academic Press, New York, 1968.

[22] Minakshisundaram, S. and Pleijel, Å., Some properties of the eigenfunctions of the Laplace-operator on Riemannian manifolds, *Can. J. Math.,* 1, 242–256, 1949.

[23] Mohlenkamp, M.J., A fast transform for spherical harmonics, *J. Fourier Anal. and Appl.,* 5(2/3), 159–184, 1999.

[24] Morgan, F., *Riemannian Geometry: A Beginner's Guide,* Jones and Bartlett, Boston, 1993.

[25] Murenzi, R., Wavelet transforms associated to the n-dimensional Euclidean group with dilations: signal in more than one dimension, in *Wavelets: Time-Frequency Methods and Phase Space,* Combes, J.M., Grossmann, A., and Tchamitchian. Ph., Eds., Springer-Verlag, New York, 1989.

[26] Neutsch, W., *Coordinates,* Walter de Gruyter, Berlin, 1996.

[27] Paeth, A.W., A fast algorithm for general Raster rotation, in *Graphics Gems,* Glassner, A.S., Ed., Academic Press, Boston, 179–195, 1990.

[28] Patodi, V.K., Curvature and the eigenforms of the Laplace operator, 5, 233–249, 1971.

[29] Rosenberg, S., *The Laplacian on a Riemannian Manifold: An Introduction to Analysis on Manifolds,* London Mathematical Society Student Texts, 31, Cambridge University Press, London, 1997.

[30] Schroeder, P. and Swelden, W., Spherical wavelets: texture processing, Tech. Report, University of South Carolina, Columbia, SC, 1995.

[31] Sohon, F.W., *The Stereographic Projection,* Chemical Publishing, Brooklyn, NY, 1941.

[32] Torresani, B., Position-frequency analysis for signals defined on spheres, *Signal Process.,* 43, 341–346, 1995.

Chapter 5

Rotations in Three Dimensions

5.1 Deformations of Nonrigid Objects

In order to fully understand the constraints that rigidity imposes, we begin our discussion with motion of nonrigid media. A general motion (or deformation) of a continuous medium (including a rigid body) is a transformation of the form

$$\mathbf{x} = \tilde{\mathbf{x}}(\mathbf{X}, t)$$

where $\mathbf{X}$ is a vector that parameterizes the position of any material point in the medium at some reference time $t = t_0$. $\mathbf{x}$ is the position of the same material point at time t. By definition, we then have $\mathbf{x}(\mathbf{X}, t_0) = \mathbf{X}$. The orthonormal basis $\{\mathbf{e}_1, \mathbf{e}_2, \mathbf{e}_3\}$ is used as a frame of reference fixed in space, and the vectors $\mathbf{x}$ and $\mathbf{X}$ are expressed in components as $x_i = \mathbf{x} \cdot \mathbf{e}_i$ and $X_i = \mathbf{X} \cdot \mathbf{e}_i$ for $i = 1, 2, 3$. In the case of a non-rigid continuous medium, many possible transformations exist. For instance, it may be possible to shear the medium in one direction

$$
\begin{aligned}
x_1 &= X_1 + k\,(t - t_0)\,X_2 \qquad k \in \mathbb{R} \\
x_2 &= X_2 \\
x_3 &= X_3.
\end{aligned}
\tag{5.1}
$$

Or, one could stretch the medium along one axis

$$
\begin{aligned}
x_1 &= e^{ct} X_1 \\
x_2 &= e^{-ct} X_2 \\
x_3 &= X_3
\end{aligned}
\tag{5.2}
$$

for $c \in \mathbb{R}$. These deformations are depicted in Fig. 5.1.

Researchers in the mechanics of solid media make great use of this kind of description of motion, which is called a *referential* or *Lagrangian* description. A subset of such motions that preserves the volume of any closed surface in $\mathbb{R}^3$ has been applied by one of the authors as a design tool in CAD and geometric modelling [4].

In contrast, one of the main aims of solid mechanics is to understand the relationship between forces applied to a continuous medium and the resulting deformation. We address applications of noncommutative harmonic analysis in solid mechanics in Chapter 18, which deals with the particular case of materials whose macroscopic strength properties depend on the orientational distribution of microscopic particles. An essential first step is to fully understand how orientations, or rotations, are quantified.

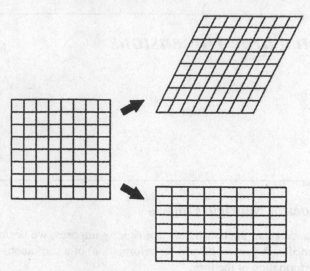

FIGURE 5.1
Geometric interpretation of shear and stretch.

5.2 Rigid-Body Rotations

Spatial rigid-body rotations are defined as motions that preserve the distance between points in a body before and after the motion and leave one point fixed under the motion. By definition a motion must be physically realizable, and so reflections are not allowed. If $\mathbf{X}_1$ and $\mathbf{X}_2$ are any two points in a body before a rigid motion, then $\mathbf{x}_1$ and $\mathbf{x}_2$ are the corresponding points after rotation, and

$$d\,(\mathbf{x}_1, \mathbf{x}_2) = d\,(\mathbf{X}_1, \mathbf{X}_2)$$

where

$$d(\mathbf{x}, \mathbf{y}) = ||\mathbf{x} - \mathbf{y}|| = \sqrt{(x_1 - y_1)^2 + (x_2 - y_2)^2 + (x_3 - y_3)^2}$$

is the *Euclidean* distance. Note that rotations do not generally preserve other kinds of distances between points.

By appropriately choosing our frame of reference in space, it is possible to make the pivot point (the point which does not move under rotation) the origin. Therefore $\mathbf{x}(\mathbf{0}, t) = \mathbf{0}$. With this choice, it can be shown that a necessary condition for a motion to be a rotation is

$$\mathbf{x}(\mathbf{X}, t) = A(t)\mathbf{X},$$

where $A(t) \in \mathbb{R}^{3 \times 3}$ is a time-dependent matrix. Note that this is *not* a sufficient condition, as the stretch and shear we saw earlier in Eqs. (5.1) and (5.2) can be written in this form.

Constraints on the form of $A(t)$ arise from the distance-preserving properties of rotations. If $\mathbf{X}_1$ and $\mathbf{X}_2$ are vectors defined in the frame of reference attached to the pivot, then the triangle with sides of length $||\mathbf{X}_1||$, $||\mathbf{X}_2||$, and $||\mathbf{X}_1 - \mathbf{X}_2||$ is congruent to the triangle with sides of length $||\mathbf{x}_1||$, $||\mathbf{x}_2||$, and $||\mathbf{x}_1 - \mathbf{x}_2||$. Hence, the angle between the vectors $\mathbf{x}_1$ and $\mathbf{x}_2$ must be the same as the angle between $\mathbf{X}_1$ and $\mathbf{X}_2$. In general $\mathbf{x} \cdot \mathbf{y} = ||\mathbf{x}||\,||\mathbf{y}|| \cos \theta$ where θ is the angle between $\mathbf{x}$ and $\mathbf{y}$. Since $||\mathbf{x}_i|| = ||\mathbf{X}_i||$ in our case, it follows that

$$\mathbf{x}_1 \cdot \mathbf{x}_2 = \mathbf{X}_1 \cdot \mathbf{X}_2.$$

Observing that $\mathbf{x} \cdot \mathbf{y} = \mathbf{x}^T \mathbf{y}$ and $\mathbf{x}_i = A\mathbf{X}_i$, we see that

$$(A\mathbf{X}_1)^T (A\mathbf{X}_2) = \mathbf{X}_1^T \mathbf{X}_2. \tag{5.3}$$

Moving everything to the left side of the equation, and using the transpose rule for matrix vector multiplication, Eq. (5.3) is rewritten as

$$\mathbf{X}_1^T \left(A^T A - \mathbb{I} \right) \mathbf{X}_2 = 0.$$

Since $\mathbf{X}_1$ and $\mathbf{X}_2$ where arbitrary points to begin with, this holds *for all* possible choices. The only way this can hold is if

$$A^T A = \mathbb{I} \tag{5.4}$$

where $\mathbb{I}$ is the 3×3 identity matrix.

An easy way to see this is to choose $\mathbf{X}_1 = \mathbf{e}_i$ and $\mathbf{X}_2 = \mathbf{e}_j$ for $i, j \in \{1, 2, 3\}$. This forces all the components of the matrix $A^T A - \mathbb{I}$ to be zero.

Equation (5.4) says that a rotation matrix is one whose inverse is its transpose. Taking the determinant of both sides of this equation yields $(\det A)^2 = 1$. There are two possibilities: $\det A = \pm 1$. The case $\det A = -1$ is a reflection and is not physically realizable in the sense that a rigid body cannot be reflected (only its image can be). A rotation is what remains

$$\det A = +1. \tag{5.5}$$

Thus a rotation matrix A is one which satisfies both Eqs. (5.4) and (5.5). The set of all real matrices satisfying both Eqs. (5.4) and (5.5) is called the set of *special orthogonal*[1] matrices. In general, the set of all $N \times N$ special orthogonal matrices is called $SO(N)$, and the set of all rotations in three-dimensional space is referred to as $SO(3)$.

We note that neither of the conditions in the definition is enough by itself. Equation (5.4) is satisfied by reflections as well as rotations, and Eq. (5.5) is satisfied by a matrix such as

$$S(t) = \begin{pmatrix} 1 & k(t-t_0) & 0 \\ 0 & 1 & 0 \\ 0 & 0 & 1 \end{pmatrix},$$

which corresponds to the shear deformation $\mathbf{x} = S(t)\mathbf{X}$ in Eq. (5.1). The equations defining a rotation impose constraints on the behavior of the nine elements of a rotation matrix. Equation (5.4) imposes six scalar constraints. To see this, first write $A = [\mathbf{a}_1, \mathbf{a}_2, \mathbf{a}_3]$, and rewrite Eq. (5.4) as the nine constraints $\mathbf{a}_i \cdot \mathbf{a}_j = \delta_{i,j}$ for $i, j \in \{1, 2, 3\}$. Since the order of the dot product does not matter, $\mathbf{a}_i \cdot \mathbf{a}_j = \mathbf{a}_j \cdot \mathbf{a}_i$, and so three of the nine constraints are redundant. Six independent constraints remain. These six constraints and Eq. (5.5) are equivalent to saying that $A = [\mathbf{u}, \mathbf{v}, \mathbf{u} \times \mathbf{v}]$ where $\mathbf{u}$ and $\mathbf{v}$ are unit vectors. Hence, the six components of $\mathbf{u}$ and $\mathbf{v}$, together with the constraints that $\mathbf{u}$ and $\mathbf{v}$ be unit vectors, and the fact that $\mathbf{u} \cdot \mathbf{v} = 0$, means that Eq. (5.5) does not reduce the number of degrees of freedom to less than three for spatial rotation.

---

[1] This is also called proper orthogonal.

In the special case of rotation about a fixed axis by an angle ϕ, the rotation only has one degree of freedom. In particular, for counter-clockwise rotations about the e_3, e_2, and e_1 axes

$$R_3(\phi) = \begin{pmatrix} \cos\phi & -\sin\phi & 0 \\ \sin\phi & \cos\phi & 0 \\ 0 & 0 & 1 \end{pmatrix}; \tag{5.6}$$

$$R_2(\phi) = \begin{pmatrix} \cos\phi & 0 & \sin\phi \\ 0 & 1 & 0 \\ -\sin\phi & 0 & \cos\phi \end{pmatrix}; \tag{5.7}$$

$$R_1(\phi) = \begin{pmatrix} 1 & 0 & 0 \\ 0 & \cos\phi & -\sin\phi \\ 0 & \sin\phi & \cos\phi \end{pmatrix}. \tag{5.8}$$

5.2.1   Eigenvalues and Eigenvectors of Rotation Matrices

Recall that an eigenvalue/eigenvector pair $(\lambda, \mathbf{x})$ for any square matrix A is defined as the solution to the equation

$$A\mathbf{x} = \lambda\mathbf{x}.$$

Since a rotation matrix consists of only real entries, complex conjugation of both sides of this equation yields $\overline{A\mathbf{x}} = A\overline{\mathbf{x}} = \overline{\lambda}\overline{\mathbf{x}}$. We may therefore equate the dot product of vectors on the left and right sides of these equations as

$$(A\mathbf{x}) \cdot (A\overline{\mathbf{x}}) = (\lambda\mathbf{x}) \cdot (\overline{\lambda}\overline{\mathbf{x}}).$$

Using the transpose rule on the left we have

$$(A\mathbf{x}) \cdot (A\overline{\mathbf{x}}) = \mathbf{x} \cdot \left(A^T A\overline{\mathbf{x}}\right) = \mathbf{x} \cdot \overline{\mathbf{x}}.$$

On the right, things simplify to $\lambda\overline{\lambda}\mathbf{x} \cdot \overline{\mathbf{x}}$. Since $A^T A = \mathbb{I}$ and $\|\mathbf{x}\| \neq 0$, we have

$$\lambda\overline{\lambda} = 1.$$

This can be satisfied for the eigenvalues $\lambda_1 = 1$ and $\lambda_{2,3} = e^{\pm i\phi}$ for $\phi \in [0, \pi]$. Since the trace of a matrix is equal to the sum of its eigenvalues, we have in this case

$$\text{trace}(A) = 1 + e^{i\phi} + e^{-i\phi} = 1 + 2\cos\phi,$$

or equivalently, the value of ϕ can be computed from the expression

$$\cos\phi = \frac{\text{trace}(A) - 1}{2}.$$

The eigenvalue $\lambda = 1$ has a corresponding eigenvector $\mathbf{n}$, which is unchanged under application of the rotation matrix. Furthermore, $A\mathbf{n} = \mathbf{n}$ implies $A^T\mathbf{n} = \mathbf{n}$ and $A^m\mathbf{n} = \mathbf{n}$ for any integer power m. The eigenvalues $e^{\pm i\phi}$ have corresponding eigenvectors $\mathbf{c}_{\pm}$. By taking the complex conjugate of both sides of the equation $A\mathbf{c}_+ = e^{i\phi}\mathbf{c}_+$, it is clear that $\mathbf{c}_- = \overline{\mathbf{c}_+}$. Taking the dot product of the eigenvector/eigenvalue equations corresponding to $(1, \mathbf{n})$ and $(e^{\pm i\phi}, \mathbf{c}_{\pm})$ we see that

$$(A\mathbf{n}) \cdot (A\mathbf{c}_{\pm}) = e^{\pm i\phi}\mathbf{n} \cdot \mathbf{c}_{\pm},$$

and so

$$\mathbf{n} \cdot \mathbf{c}_{\pm} = 0$$

when $e^{\pm i\phi} \neq 1$. From the complex vectors $\mathbf{c}_\pm$, we construct the real vectors

$$\mathbf{c}_1 = (\mathbf{c}_+ + \mathbf{c}_-)/2 \qquad \text{and} \qquad \mathbf{c}_2 = i\,(\mathbf{c}_+ - \mathbf{c}_-)/2.$$

The vectors $\{\mathbf{n}, \mathbf{c}_1, \mathbf{c}_2\}$ can be taken to be unit vectors without loss of generality, and due to their mutual orthonormality, they represent an orthonormal basis for $\mathbb{R}^3$. For further reading see [1, 2, 16]. For surveys on different parameterizations of rotation see [23, 25, 27].

5.2.2   Relationships Between Rotation and Skew-Symmetric Matrices

There is a deep relationship between rotation matrices and *skew-symmetric* matrices. Recall that a real matrix is skew-symmetric if its transpose is its negative. Any 3×3 skew-symmetric matrix, $S = -S^T$, can be written as

$$S = \begin{pmatrix} 0 & -s_3 & s_2 \\ s_3 & 0 & -s_1 \\ -s_2 & s_1 & 0 \end{pmatrix}, \tag{5.9}$$

where s_1, s_2, and s_3 can be viewed as the components of a vector $\mathbf{s} \in \mathbb{R}^3$, called the *dual vector* of S. The set of all 3×3 skew-symmetric matrices is denoted $so(3)$. The product of a skew-symmetric matrix with an arbitrary vector is the same as the cross product of the dual vector with the same arbitrary vector

$$S\mathbf{x} = \mathbf{s} \times \mathbf{x}. \tag{5.10}$$

We use the notation $\text{vect}(S) = \mathbf{s}$ and $S = \text{matr}(\mathbf{s})$ to express this relationship throughout the rest of the book.

It is easy to check that the eigenvalues of a 3×3 skew-symmetric matrix satisfy

$$-\det(S - \lambda\mathbb{I}) = \lambda^3 + (\mathbf{s} \cdot \mathbf{s})\lambda = \left(\lambda^2 + \|\mathbf{s}\|^2\right)\lambda = 0$$

and so the eigenvalues of S are 0, $\pm i\,\|\mathbf{s}\|$.

It is also useful to know that the eigenvectors and eigenvalues of real symmetric matrices $M = M^T$ are real, and it is always possible to construct eigenvectors of M which form an orthonormal set. It follows that for any real symmetric matrix we can write

$$M = Q\Lambda Q^T; \qquad MQ = Q\Lambda; \qquad \Lambda = Q^T M Q \tag{5.11}$$

where Λ is a real diagonal matrix with nonzero entries corresponding to the eigenvalues of M, and Q is a rotation matrix whose columns are the eigenvectors of M. A proof of this statement is provided in Appendix D.

Cayley's Formula

Let $\mathbf{X}$ be an arbitrary vector in $\mathbb{R}^3$, and let $\mathbf{x} = A\mathbf{X}$ be a rotated version of this vector. Then it is clear from Fig. 5.2 that the vectors $\mathbf{x} + \mathbf{X}$ and $\mathbf{x} - \mathbf{X}$ are orthogonal, and so one writes

$$(\mathbf{x} + \mathbf{X}) \cdot (\mathbf{x} - \mathbf{X}) = 0. \tag{5.12}$$

Since by definition $\mathbf{x} = A\mathbf{X}$, it follows that

$$\mathbf{x} + \mathbf{X} = (A + \mathbb{I})\mathbf{X} \qquad \text{and} \qquad \mathbf{x} - \mathbf{X} = (A - \mathbb{I})\mathbf{X}. \tag{5.13}$$

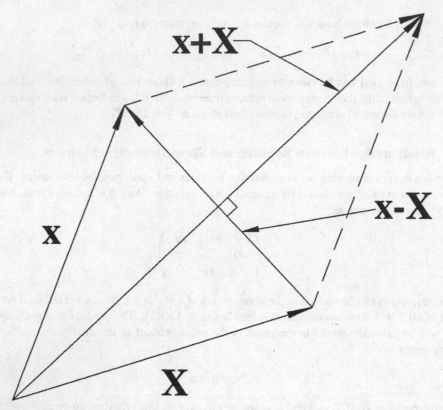

FIGURE 5.2
Orthogonality of x + X and x − X.

For all $\phi \neq \pm\pi$, the eigenvalues of a rotation matrix are not equal to -1, and so $A+\mathbb{I}$ is invertible. We will disregard the case when $\phi = \pm\pi$ for the moment and examine this special case more closely later. Thus we write $\mathbf{X} = (A+\mathbb{I})^{-1}(\mathbf{x}+\mathbf{X})$, and substitute back into the second equation in Eq. (5.13) to yield

$$\mathbf{x} - \mathbf{X} = (A - \mathbb{I})(A + \mathbb{I})^{-1}(\mathbf{x} + \mathbf{X}).$$

From this statement and the condition in Eq. (5.12), we see that the matrix

$$S = (A - \mathbb{I})(A + \mathbb{I})^{-1} \tag{5.14}$$

must satisfy

$$\mathbf{y}^T S \mathbf{y} = 0 \tag{5.15}$$

where $\mathbf{y} = \mathbf{x} + \mathbf{X}$. Since $\mathbf{X}$ was arbitrary to begin with, so is $\mathbf{y}$.

We now prove that S in Eq. (5.15) must be skew-symmetric. As with any real matrix, we can make the following decomposition: $S = S_{sym} + S_{skew}$ where $S_{sym} = \frac{1}{2}(S + S^T)$ and $S_{skew} = \frac{1}{2}(S - S^T)$ are respectively the symmetric and skew-symmetric parts of S. Hence

$$\mathbf{y}^T S \mathbf{y} = \mathbf{y}^T S_{sym} \mathbf{y} + \mathbf{y}^T S_{skew} \mathbf{y}.$$

The second term on the right hand side of the preceding equation must be zero because of the vector identity $\mathbf{y} \cdot (\mathbf{s}_{skew} \times \mathbf{y}) = 0$. As for the first term, since $\mathbf{y}$ is arbitrary, it is always possible to choose

$\mathbf{y} = Q\mathbf{e}_i$ for any $i = 1, 2, 3$, in which case from Eq. (5.11) we have $\mathbf{y}^T S_{sym} \mathbf{y} = \lambda_i$ where λ_i are the eigenvalues of S_{sym} and $Q\mathbf{e}_i$ are the corresponding eigenvectors. Therefore $\mathbf{y}^T S_{sym} \mathbf{y}$ can only be zero for all choices of $\mathbf{y}$ if $\lambda_i = 0$ for $i = 1, 2, 3$, which means that $S_{sym} = 0$. This means that S satisfying Eq. (5.15) must be skew-symmetric. The equation

$$S = (A - \mathbb{I})(A + \mathbb{I})^{-1}$$

can be inverted to find A by multiplying by $A + \mathbb{I}$ on the right and recollecting terms to get

$$\mathbb{I} + S = (\mathbb{I} - S)A.$$

Since the number 1 is not an eigenvalue of S, $\mathbb{I} - S$ is invertible, and we arrive at *Cayley's formula* [3] for a rotation matrix

$$A = (\mathbb{I} - S)^{-1}(\mathbb{I} + S). \tag{5.16}$$

This is the first of several examples we will see that demonstrate the relationship between skew-symmetric matrices and rotations. Cayley's formula provides one technique for *parameterizing* rotations. Each vector $\mathbf{s}$ contains the information required to specify a rotation, and given a rotation, a unique $\mathbf{s}$ can be extracted from the rotation matrix using Eq. (5.14).

Expanding out Eq. (5.16) in components, one finds that

$$A = \frac{1}{1 + \|\mathbf{s}\|^2} \begin{pmatrix} 1 + s_1^2 - s_2^2 - s_3^2 & 2(s_1 s_2 - s_3) & 2(s_1 s_3 + s_2) \\ 2(s_1 s_2 + s_3) & 1 - s_1^2 + s_2^2 - s_3^2 & 2(s_2 s_3 - s_1) \\ 2(s_1 s_3 - s_2) & 2(s_2 s_3 + s_1) & 1 - s_1^2 - s_2^2 + s_3^2 \end{pmatrix}. \tag{5.17}$$

From this expression it is clear that this description of rotation breaks down, for example, for rotation matrices of the form

$$R_1(\pi) = \begin{pmatrix} 1 & 0 & 0 \\ 0 & -1 & 0 \\ 0 & 0 & -1 \end{pmatrix};$$

$$R_2(\pi) = \begin{pmatrix} -1 & 0 & 0 \\ 0 & 1 & 0 \\ 0 & 0 & -1 \end{pmatrix};$$

$$R_3(\pi) = \begin{pmatrix} -1 & 0 & 0 \\ 0 & -1 & 0 \\ 0 & 0 & 1 \end{pmatrix}$$

because there are no positive values of s_1^2, s_2^2, s_3^2 in Eq. (5.17) that will allow A to attain these values. All of the matrices mentioned above, which are of the form $R_i(\pi)$ for $i = 1, 2, 3$, are similar to each other under the appropriate similarity transformation $Q R_i(\pi) Q^T$ for $Q \in SO(3)$. More generally, any rotation matrix similar to these will also not lend itself to description using Cayley's formula. These all correspond to the case $\phi = \pi$ discussed earlier.

Transformation of Cross Products under Rotation

Given two vectors in a rotated frame of reference, $\mathbf{x}$ and $\mathbf{y}$, their cross product in the same rotated frame of reference will be $\mathbf{x} \times \mathbf{y}$. In the unrotated (inertial) frame, this result will appear as $A(\mathbf{x} \times \mathbf{y})$ where A is the rotation that performs motion from the unrotated to rotated frame (and hence, converts vectors described in the rotated frame to their description in the unrotated frame). On the other hand, the vectors that appear as $\mathbf{x}$ and $\mathbf{y}$ in the rotated frame will be viewed as $A\mathbf{x}$ and $A\mathbf{y}$ in the inertial frame. Taking their cross product must then give

$$(A\mathbf{x}) \times (A\mathbf{y}) = A(\mathbf{x} \times \mathbf{y}). \tag{5.18}$$

The preceding geometrical argument may be verified by algebraic manipulations.

We note also that if $X\mathbf{y} = \mathbf{x} \times \mathbf{y}$ then

$$\left(AXA^T\right)\mathbf{y} = (A\mathbf{x}) \times \mathbf{y}. \tag{5.19}$$

That is, the skew-symmetric matrix corresponding to a rotated vector is the similarity-transformed version of the skew-symmetric matrix corresponding to the unrotated vector.

Euler's Theorem

Euler's theorem [6, 7][2] states that the general displacement of a rigid body with one point fixed is a rotation about some axis. That is not to say that every rotational *motion* $A(t) \in SO(3)$ for $t \in \mathbb{R}$ is a rotation about a fixed axis. Rather, the *result* of any rotational motion between two times $t = t_1$ and $t = t_2$ is a rotational *displacement*. This resulting displacement can be described as a rotation about an axis fixed in space that depends on the times t_1 and t_2.

By geometric construction, we can find the relationship between the axis of rotation, $\mathbf{n}$, and the angle of rotation about that axis, θ, which together describe an arbitrary spatial displacement. Observing an arbitrary point $\mathbf{X}$ in the body before rotation we break this into components as follows: $(\mathbf{X} \cdot \mathbf{n})\mathbf{n}$ is the component along the axis of rotation, and $\mathbf{X} - (\mathbf{X} \cdot \mathbf{n})\mathbf{n}$ is the component orthogonal to the axis of rotation. The plane containing the point $(\mathbf{X} \cdot \mathbf{n})\mathbf{n}$ and orthogonal to the axis of rotation is spanned by the two vectors $\mathbf{X} - (\mathbf{X} \cdot \mathbf{n})\mathbf{n}$ and $\mathbf{X} \times \mathbf{n}$. (See Fig. 5.3(a), (b).) These vectors are orthogonal to each other and have the same magnitude. Rotation about $\mathbf{n}$ by θ may be viewed as a planar rotation in this plane, and so the rotated version of the point $\mathbf{X}$ is

$$\mathbf{x} = (\mathbf{X} \cdot \mathbf{n})\mathbf{n} + (\mathbf{X} - (\mathbf{X} \cdot \mathbf{n})\mathbf{n})\cos\theta + (\mathbf{n} \times \mathbf{X})\sin\theta. \tag{5.20}$$

To each unit vector $\mathbf{n}$ describing the axis of rotation, we assign a skew-symmetric matrix N. Then Eq. (5.20) is written as

$$\mathbf{x} = A(\mathbf{n}, \theta)\mathbf{X} \quad \text{where} \quad A(\mathbf{n}, \theta) = \mathbb{I} + \sin\theta\, N + (1 - \cos\theta)N^2 \tag{5.21}$$

where $\mathbf{n} \in S^2$ and $\theta \in [-\pi, \pi]$. Equations (5.20) and (5.21) are often called *Rodrigues'* equations for a rotational displacement, and (θ, n_1, n_2, n_3) is a four-parameter description of rotation called the *Rodrigues parameters* (after [22]).

Given any rotation matrix, these parameters can be extracted by observing that

$$A^T(\mathbf{n}, \theta) = A(\mathbf{n}, -\theta) = \mathbb{I} - \sin\theta\, N + (1 - \cos\theta)N^2,$$

and so

$$2\sin\theta N = A - A^T.$$

Then $\mathbf{n}$ is the normalized dual vector of $N = (A - A^T)/(2\sin\theta)$, and θ is determined by solving $2\sin\theta = \|\text{vect}(A - A^T)\|$.

---

[2]Leonard Euler (1707–1783) was one of the most prolific mathematicians in history. This is but one of the many theorems and formulae attributed to him.

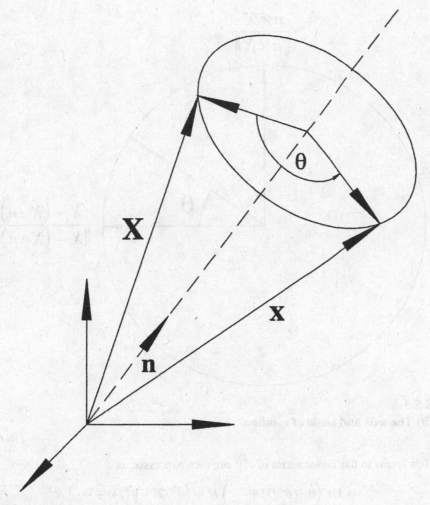

FIGURE 5.3
(a) The axis and angle of rotation. *(Continued.)*

5.2.3   The Matrix Exponential

The result of Euler's theorem can be viewed in another way using the concept of a matrix exponential. Recall that the Taylor series expansion of the scalar exponential function is

$$e^x = 1 + \sum_{k=1}^{\infty} \frac{x^k}{k!}.$$

The matrix exponential is the same formula evaluated at a square matrix

$$e^X = \mathbb{I} + \sum_{k=1}^{\infty} \frac{X^k}{k!}.$$

In the case when $X = \theta N \in \mathbb{R}^{3\times3}$ where $N = -N^T$ is the matrix whose dual is the unit vector **n**, it may be shown that $N^2 = \mathbf{n}\mathbf{n}^T - \mathbb{I}$ and all higher powers of N can be related to either N or N^2 as

$$N^{2k+1} = (-1)^k N \quad \text{and} \quad N^{2k} = (-1)^{k+1} N^2. \tag{5.22}$$

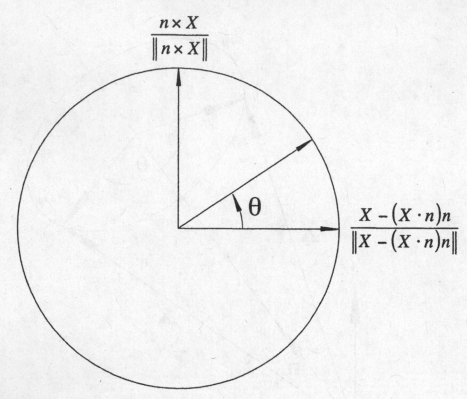

FIGURE 5.3
(Cont.) **(b) The axis and angle of rotation.**

The first few terms in the Taylor series of $e^{\theta N}$ are then expressed as

$$e^{\theta N} = \mathbb{I} + \left(\theta - \theta^3/3! + \cdots\right) N + \left(\theta^2/2! - \theta^4/4! + \cdots\right) N^2.$$

Hence Eq. (5.21) results, and we can write any rotational displacement as

$$A(\theta, \mathbf{n}) = e^{\theta N}.$$

This form clearly illustrates that $(\theta, \mathbf{n})$ and $(-\theta, -\mathbf{n})$ correspond to the same rotation.

Since $\theta = \|\mathbf{x}\|$ where $\mathbf{x} = \text{vect}(X)$ and $N = X/\|\mathbf{x}\|$, one often writes the alternative form

$$e^X = \mathbb{I} + \frac{\sin \|\mathbf{x}\|}{\|\mathbf{x}\|} X + \frac{(1 - \cos \|\mathbf{x}\|)}{\|\mathbf{x}\|^2} X^2. \tag{5.23}$$

We note that in the special case when considering the rotation which transforms a unit vector $\mathbf{a}$ into the unit vector $\mathbf{b}$,

$$\mathbf{b} = R(\mathbf{a}, \mathbf{b})\mathbf{a},$$

then

$$R(\mathbf{a}, \mathbf{b}) = e^C = \mathbb{I} + C + \frac{(1 - \mathbf{a} \cdot \mathbf{b})}{\|\mathbf{a} \times \mathbf{b}\|^2} C^2 \tag{5.24}$$

where $C = \text{matr}(\mathbf{a} \times \mathbf{b})$. This follows easily from the facts that $\|\mathbf{a} \times \mathbf{b}\| = \sin \theta_{ab}$ and $\mathbf{a} \cdot \mathbf{b} = \cos \theta_{ab}$ where $0 \leq \theta_{ab} \leq \pi$ is the counterclockwise measured angle from $\mathbf{a}$ to $\mathbf{b}$ as measured in the direction defined by $\mathbf{c}$.

The matrix logarithm is defined as the inverse of the exponential

$$X = \log(\exp X).$$

When $A = \exp X$ is a rotation, one explicitly calculates the logarithm as

$$\log(A) = \frac{1}{2} \frac{\theta(A)}{\sin \theta(A)} \left(A - A^T \right)$$

where

$$\theta(A) = \cos^{-1} \left(\frac{\text{trace}(A) - 1}{2} \right)$$

is the angle of rotation. From this definition it follows that for $\theta = 0$ we have $\log(\mathbb{I}_{3\times3}) = 0_{3\times3}$. The preceding definitions break down when $\theta = \pi$.

5.3 Rules for Composing Rotations

Consider three frames of reference A, B, and C, which all have the same origin. The vectors $\mathbf{x}^A$, $\mathbf{x}^B$, $\mathbf{x}^C$ represent the position of *the same* arbitrary point in space, $\mathbf{x}$, as it is viewed in the three different frames. With respect to some common frame fixed in space with axes defined by $\{e_1, e_2, e_3\}$ where $(e_i)_j = \delta_{ij}$, the rotation matrices describing the basis vectors of the frames A, B, and C are

$$R_A = \left[e_1^A, e_2^A, e_3^A \right] \qquad R_B = \left[e_1^B, e_2^B, e_3^B \right] \qquad R_C = \left[e_1^C, e_2^C, e_3^C \right],$$

where the vectors e_i^A, e_i^B, and e_i^C are unit vectors along the ith axis of frame A, B, and C, respectively. The "absolute" coordinates of the vector $\mathbf{x}$ are then given by

$$\mathbf{x} = R_A \mathbf{x}^A = R_B \mathbf{x}^B = R_C \mathbf{x}^C.$$

In this notation, which is often used in the field of robotics (see, for example, [5, 18]), there is effectively a "cancellation" of indices along the upper right to lower left diagonal.

Given the rotation matrices R_A, R_B, and R_C, it is possible to define rotations of one frame *relative* to another by observing that, for instance, $R_A \mathbf{x}^A = R_B \mathbf{x}^B$ implies $\mathbf{x}^A = (R_A)^{-1} R_B \mathbf{x}^B$. Therefore, given any vector $\mathbf{x}^B$ as it looks in B, we can find how it looks in A, $\mathbf{x}^A$, by performing the transformation

$$\mathbf{x}^A = R_B^A \mathbf{x}^B \qquad \text{where} \qquad R_B^A = (R_A)^{-1} R_B. \tag{5.25}$$

It follows from substituting the analogous expression $\mathbf{x}^B = R_C^B \mathbf{x}^C$ into $\mathbf{x}^A = R_B^A \mathbf{x}^B$ that concatenation of rotations is calculated as

$$\mathbf{x}^A = R_C^A \mathbf{x}^C \qquad \text{where} \qquad R_C^A = R_B^A R_C^B. \tag{5.26}$$

Again there is effectively a cancellation of indices, and this propagates through for any number of relative rotations. Note that the order of multiplication is critical.

In addition to changes of basis, rotation matrices can be viewed as descriptions of motion. Multiplication of a rotation matrix Q (which represents a frame of reference) by a rotation matrix R (representing motion) on the left, RQ, has the effect of moving Q by R relative to the *base frame*.

Multiplying by the same rotation matrix on the right, QR, has the effect of moving by R relative to the frame Q.

To demonstrate the difference, consider again the result of Euler's theorem. Let us define a frame of reference $Q = [\mathbf{a}, \mathbf{b}, \mathbf{n}]$ where $\mathbf{a}$ and $\mathbf{b}$ are unit vectors orthogonal to each other, and $\mathbf{a} \times \mathbf{b} = \mathbf{n}$. By first rotating from the identity $\mathbb{I} = [\mathbf{e}_1, \mathbf{e}_2, \mathbf{e}_3]$ fixed in space to Q and then rotating relative to Q by $R_3(\theta)$, results in $QR_3(\theta)$. On the other hand, a rotation about the vector $\mathbf{e}_3^Q = \mathbf{n}$ as viewed in the fixed frame is $A(\theta, \mathbf{n})$. Hence, shifting the frame of reference Q by multiplying on the left by $A(\theta, \mathbf{n})$ has the same effect as $QR_3(\theta)$, and so we write

$$A(\theta, \mathbf{n})Q = QR_3(\theta) \qquad \text{or} \qquad A(\theta, \mathbf{n}) = QR_3(\theta)Q^T. \tag{5.27}$$

This is yet another way to view the expression of $A(\theta, \mathbf{n})$ given previously. We note that in component form this is expressed as

$$A(\theta, \mathbf{n}) = \begin{pmatrix} n_1^2 v\theta + c\theta & n_2 n_1 v\theta - n_3 s\theta & n_3 n_1 v\theta + n_2 s\theta \\ n_1 n_2 v\theta + n_3 s\theta & n_2^2 v\theta + c\theta & n_3 n_2 v\theta - n_1 s\theta \\ n_1 n_3 v\theta - n_2 s\theta & n_2 n_3 v\theta + n_1 s\theta & n_3^2 v\theta + c\theta \end{pmatrix}$$

where $s\theta = \sin\theta$, $c\theta = \cos\theta$, and $v\theta = 1 - \cos\theta$.

Note that $\mathbf{a}$ and $\mathbf{b}$ do not appear in the final expression. There is nothing magical about $\mathbf{e}_3$, and we could have used the same construction using any other basis vector, $\mathbf{e}_i$, and we would get the same result so long as $\mathbf{n}$ is in the ith column of Q.

5.4   Parameterizations of Rotation

In this section, several parameterizations other than the axis-angle description of rotation are reviewed.

5.4.1   Euler Parameters

Euler parameters, $\{u_1, u_2, u_3, u_4\}$, are defined relative to the axis-angle parameters $\mathbf{n}, \theta$ as

$$\mathbf{u} = \mathbf{n}\sin(\theta/2) \qquad u_4 = \cos(\theta/2) \tag{5.28}$$

where $\mathbf{u} = (u_1, u_2, u_3)^T$. With this definition, and using the fact that $\mathbf{n} \cdot \mathbf{n} = 1$, we see that

$$u_1^2 + u_2^2 + u_3^2 + u_4^2 = 1,$$

and so a natural relationship clearly exists between $SO(3)$ and the 3–sphere, S^3. In particular, because $-\pi \le \theta \le \pi$ and $\cos(\theta/2) \ge 0$, u_4 is limited to the "upper half space" of $\mathbb{R}^4$ and so we associate $SO(3)$ with the "upper hemisphere" of S^3 corresponding to $u_4 \ge 0$.

The rotation matrix when using Euler parameters takes the form

$$A_E(u_1, u_2, u_3, u_4) = \begin{pmatrix} u_1^2 - u_2^2 - u_3^2 + u_4^2 & 2(u_1 u_2 - u_3 u_4) & 2(u_3 u_1 + u_2 u_4) \\ 2(u_1 u_2 + u_3 u_4) & u_2^2 - u_3^2 - u_1^2 + u_4^2 & 2(u_2 u_3 - u_1 u_4) \\ 2(u_3 u_1 - u_2 u_4) & 2(u_2 u_3 + u_1 u_4) & u_3^2 - u_1^2 - u_2^2 + u_4^2 \end{pmatrix}. \tag{5.29}$$

The Euler parameters can be extracted from any rotation matrix A as

$$u_4 = \frac{1}{2}(1 + \text{trace}(A))^{\frac{1}{2}}$$

$$u_1 = \frac{a_{32} - a_{23}}{4u_4}$$

$$u_2 = \frac{a_{13} - a_{31}}{4u_4}$$

$$u_3 = \frac{a_{21} - a_{12}}{4u_4}$$

[where a_{ij} is the (i, j)th entry of A] so long as $u_4 \neq 0$. In this special case, we have

$$A_E(u_1, u_2, u_3, 0) = \begin{pmatrix} u_1^2 - u_2^2 - u_3^2 & 2u_1u_2 & 2u_3u_1 \\ 2u_1u_2 & u_2^2 - u_3^2 - u_1^2 & 2u_2u_3 \\ 2u_3u_1 & 2u_2u_3 & u_3^2 - u_1^2 - u_2^2 \end{pmatrix},$$

and we can usually find the remaining parameters using the formula

$$\mathbf{u} = \frac{1}{\sqrt{(a_{12}a_{13})^2 + (a_{12}a_{23})^2 + (a_{13}a_{23})^2}} \begin{pmatrix} a_{13}a_{12} \\ a_{12}a_{23} \\ a_{13}a_{23} \end{pmatrix}.$$

However, this, too, can be degenerate if $a_{12} = a_{23} = 0$ or $a_{13} = a_{23} = 0$ or $a_{12} = a_{13} = 0$. These cases correspond to the rotations about natural basis vectors in Eqs. (5.6)–(5.8), and can be detected in advance by counting the number of zeros in the matrix.

5.4.2 Cayley/Rodrigues Parameters

If we set $u_4 = 1$ and normalize the resulting matrix $A_E(u_1, u_2, u_3, 1)$ by $1/(1 + u_1^2 + u_2^2 + u_3^2)$, the three parameters u_1, u_2, u_3 can be viewed as the elements of the vector $\mathbf{s}$ corresponding to the skew-symmetric matrix in Cayley's formula, Eq. (5.16). In this case, u_1, u_2, u_3 are called *Cayley's parameters.* As with all 3−parameter descriptions of rotation, the Cayley parameters are singular at some points.

Rodrigues parameters may also be defined relative to the Euler and axis-angle parameterization. They are defined as

$$\mathbf{r} = \mathbf{n} \tan \theta/2 = \mathbf{u}/u_4.$$

$\mathbf{r}$ is often called the *Rodrigues vector* or the *Gibbs vector.*

In terms of the exponential coordinates, $\mathbf{x}$, in Eq. (5.23), one finds the relationship

$$\mathbf{r} = \frac{\mathbf{x}}{\|\mathbf{x}\|} \tan\left(\frac{\|\mathbf{x}\|}{2}\right) \qquad \leftrightarrow \qquad \mathbf{x} = 2\tan^{-1}(\|\mathbf{r}\|)\frac{\mathbf{r}}{\|\mathbf{r}\|}.$$

The resulting rotation matrix is the same as that for the Cayley parameters (and so the Cayley and Rodrigues parameters are equivalent)

$$A_R(\mathbf{r}) = \frac{1}{1 + r_1^2 + r_2^2 + r_3^2} \begin{pmatrix} 1 + r_1^2 - r_2^2 - r_3^2 & 2(r_1r_2 - r_3) & 2(r_3r_1 + r_2) \\ 2(r_1r_2 + r_3) & 1 + r_2^2 - r_3^2 - r_1^2 & 2(r_2r_3 - r_1) \\ 2(r_3r_1 - r_2) & 2(r_2r_3 + r_1) & 1 + r_3^2 - r_1^2 - r_2^2 \end{pmatrix}. \qquad (5.30)$$

We note that

$$A_R(\mathbf{r}) = \frac{\left(1 - \|\mathbf{r}\|^2\right)\mathbb{I} + 2\mathbf{r}\mathbf{r}^T + [\text{matr}(\mathbf{r})]}{1 + \|(\mathbf{r})\|^2}$$

where $[\text{matr}(\mathbf{r})]$ is the skew-symmetric matrix with

$$\text{matr}(\mathbf{r})\mathbf{x} = \mathbf{r} \times \mathbf{x}.$$

We note that when

$$A_R(\mathbf{r})A_R(\mathbf{r}') = A_R(\mathbf{r}''),$$

the Rodrigues vector $\mathbf{r}''$ is calculated in terms of $\mathbf{r}$ and $\mathbf{r}'$ as

$$\mathbf{r}'' = \frac{\mathbf{r} + \mathbf{r}' - \mathbf{r}' \times \mathbf{r}}{1 - \mathbf{r} \cdot \mathbf{r}'}.$$

The Rodrigues parameters can be extracted from any rotation matrix as

$$r_1 = \frac{a_{32} - a_{23}}{1 + \text{trace}(A)}$$

$$r_2 = \frac{a_{13} - a_{31}}{1 + \text{trace}(A)}$$

$$r_3 = \frac{a_{21} - a_{12}}{1 + \text{trace}(A)}$$

so long as $\text{trace}(A) \neq -1$. This is the singularity in this description of rotation.

As explained by Kane and Levinson [13], when one knows the position of two points $(\mathbf{X}_1, \mathbf{X}_2)$ in the rigid body prior to rotation and the same points after rotation $(\mathbf{x}_1, \mathbf{x}_2)$, then the relationship

$$\mathbf{x}_i - \mathbf{X}_i = \mathbf{r} \times (\mathbf{x}_i + \mathbf{X}_i)$$

(which follows from our discussion of Cayley's formula) is used to extract the Rodrigues vector as

$$\mathbf{r} = -\frac{(\mathbf{x}_1 - \mathbf{X}_1) \times (\mathbf{x}_2 - \mathbf{X}_2)}{(\mathbf{x}_1 + \mathbf{X}_1) \cdot (\mathbf{x}_2 - \mathbf{X}_2)}. \tag{5.31}$$

This formula follows by observing that

$$(\mathbf{x}_1 - \mathbf{X}_1) \times (\mathbf{x}_2 - \mathbf{X}_2) = (\mathbf{r} \times (\mathbf{x}_1 + \mathbf{X}_1)) \times (\mathbf{x}_2 - \mathbf{X}_2)$$

and expanding the right side using the vector identity

$$(\mathbf{a} \times \mathbf{b}) \times \mathbf{c} = (\mathbf{a} \cdot \mathbf{c})\mathbf{b} - (\mathbf{b} \cdot \mathbf{c})\mathbf{a}$$

to isolate $\mathbf{r}$.

5.4.3 Cartesian Coordinates in $\mathbb{R}^4$

Identifying $SO(3)$ with the upper hemisphere in $\mathbb{R}^4$, the Cayley parameters correspond to a description of the surface in Cartesian coordinates as

$$x_1 = u_1$$
$$x_2 = u_2$$
$$x_3 = u_3$$
$$x_4 = \sqrt{1 - u_1^2 - u_2^2 - u_3^2}.$$

5.4.4 Spherical Coordinates

Parameterizing rotations using spherical coordinates is essentially the same as parameterizing S^3 in spherical coordinates. We associate the set of all rotations with the upper hemisphere, which is parameterized as

$$
\begin{aligned}
x_1 &= \cos \lambda \sin \nu \sin \theta/2 \\
x_2 &= \sin \lambda \sin \nu \sin \theta/2 \\
x_3 &= \cos \nu \sin \theta/2 \\
x_4 &= \cos \theta/2.
\end{aligned}
$$

λ and ν are the polar and azimuthal angles for the axis of rotation, and θ is the angle of rotation.

The corresponding rotation matrix results from the spherical-coordinate parameterization of the vector of the axis of rotation, $\mathbf{n} = \mathbf{n}(\lambda, \nu)$, and is given by

$$
A_S(\lambda, \nu, \theta) = \begin{pmatrix} v\theta\, s^2\nu\, c^2\lambda + c\theta & v\theta\, s^2\nu\, c\lambda\, s\lambda - s\theta\, c\nu & v\theta\, s\nu\, c\nu\, c\lambda + s\theta\, s\nu\, s\lambda \\ v\theta\, s^2\nu\, c\lambda\, s\lambda + s\theta\, c\nu & v\theta\, s^2\nu\, s^2\lambda + c\theta & v\theta\, s\nu\, c\nu\, s\lambda - s\theta\, s\nu\, c\lambda \\ v\theta\, s\nu\, c\nu\, c\lambda - s\theta\, s\nu\, s\lambda & v\theta\, s\nu\, c\nu\, s\lambda + s\theta\, s\nu\, c\lambda & v\theta\, c^2\nu + c\theta \end{pmatrix}, \quad (5.32)
$$

where $s\theta = \sin \theta$, $c\theta = \cos \theta$, and $v\theta = 1 - \cos \theta$. It is clear from Eq. (5.32) that (θ, λ, ν) and $(-\theta, \lambda \pm \pi, \pi - \nu)$ correspond to the same rotation.

5.4.5 Parameterization of Rotation as a Solid Ball in $\mathbb{R}^3$

Another way to view the set of rotations is as a solid ball in $\mathbb{R}^3$ with radius π. In this description, each concentric spherical shell within the ball represents a rotation by an angle which is the radius of the shell. That is, given any point in the solid ball,

$$
\mathbf{x}(\theta, \lambda, \nu) = \theta \begin{pmatrix} \sin \nu \cos \lambda \\ \sin \nu \sin \lambda \\ \cos \nu \end{pmatrix}.
$$

The ranges of these angles are $-\pi \leq \theta \leq \pi, 0 \leq \nu \leq \pi, 0 \leq \lambda \leq 2\pi$. In this model, antipodal points on the surface of the ball, $\theta = \pm \pi$, represent the same rotations.

Exponentiating the skew-symmetric matrix corresponding to $\mathbf{x}(\theta, \lambda, \nu)$ results in the matrix in Eq. (5.32). Similarly, the parameters (x_1, x_2, x_3) defining the matrix exponential Eq. (5.23) can be viewed as Cartesian coordinates to the interior of this same ball.

5.4.6 Euler Angles

Euler angles are by far the most widely known parameterization of rotation. They are generated by three successive rotations about independent axes. Three of the most common choices are the ZXZ, ZYZ, and ZYX Euler angles. We will denote these as

$$
\begin{aligned}
A_{ZXZ}(\alpha, \beta, \gamma) &= R_3(\alpha) R_1(\beta) R_3(\gamma), & (5.33) \\
A_{ZYZ}(\alpha, \beta, \gamma) &= R_3(\alpha) R_2(\beta) R_3(\gamma), & (5.34) \\
A_{ZYX}(\alpha, \beta, \gamma) &= R_3(\alpha) R_2(\beta) R_1(\gamma). & (5.35)
\end{aligned}
$$

Of these, the ZXZ and ZYZ Euler angles are the most common, and the corresponding matrices are explicitly

$$A_{ZXZ}(\alpha, \beta, \gamma)$$
$$= \begin{pmatrix} \cos\gamma\cos\alpha - \sin\gamma\sin\alpha\cos\beta & -\sin\gamma\cos\alpha - \cos\gamma\sin\alpha\cos\beta & \sin\beta\sin\alpha \\ \cos\gamma\sin\alpha + \sin\gamma\cos\alpha\cos\beta & -\sin\gamma\sin\alpha + \cos\gamma\cos\alpha\cos\beta & -\sin\beta\cos\alpha \\ \sin\beta\sin\gamma & \sin\beta\cos\gamma & \cos\beta \end{pmatrix},$$

and

$$A_{ZYZ}(\alpha, \beta, \gamma)$$
$$= \begin{pmatrix} \cos\gamma\cos\alpha\cos\beta - \sin\gamma\sin\alpha & -\sin\gamma\cos\alpha\cos\beta - \cos\gamma\sin\alpha & \sin\beta\cos\alpha \\ \sin\alpha\cos\gamma\cos\beta + \sin\gamma\cos\alpha & -\sin\gamma\sin\alpha\cos\beta + \cos\gamma\cos\alpha & \sin\beta\sin\alpha \\ -\sin\beta\cos\gamma & \sin\beta\sin\gamma & \cos\beta \end{pmatrix}.$$

The ranges of angles for these choices are $0 \le \alpha \le 2\pi$, $0 \le \beta \le \pi$, and $0 \le \gamma \le 2\pi$. When ZYZ Euler angles are used,

$$\begin{aligned} R_3(\alpha)R_2(\beta)R_3(\gamma) &= R_3(\alpha)\left(R_3(\pi/2)R_1(\beta)R_3(-\pi/2)\right)R_3(\gamma) \\ &= R_3(\alpha + \pi/2)R_1(\beta)R_3(-\pi/2 + \gamma), \end{aligned}$$

and so

$$R_{ZYZ}(\alpha, \beta, \gamma) = R_{ZXZ}(\alpha + \pi/2, \beta, \gamma - \pi/2).$$

In recent times, the Euler angles have been considered by some to be undesirable as compared to the axis-angle or Euler parameters because of the singularities that they (and all) three-parameter descriptions possess. However, for our purposes, where integral quantities that are not sensitive to singularities are of interest, the Euler angles will serve well.

5.4.7 Parameterization Based on Stereographic Projection

The Cayley/Rodrigues parameters can be viewed geometrically as points in $\mathbb{R}^3$ that are mapped onto half of the unit hypersphere in $\mathbb{R}^4$ via the four-dimensional analog of the two-dimensional stereographic projection shown in Fig. 4.1(a). This is not the only way stereographic projection can be used to generate parameterizations of rotation. This section describes parameterizations developed in the spacecraft attitude literature based on stereographic projection.

Modified Rodrigues Parameters

If the stereographic projection in $\mathbb{R}^4$ analogous to the one described in Fig. 4.1(b) is used, the Euler parameters $\{u_i\}$ are mapped to values [24, 15]

$$\sigma_i = \frac{u_i}{1 + u_4} = n_i \tan(\theta/4)$$

for $i = 1, 2, 3$. The benefit of this is that the singularity at $\theta = \pm\pi$ is pushed to $\theta = \pm 2\pi$. The corresponding rotation matrix is [24]

$$R(\sigma) = \frac{1}{(1+\|\sigma\|^2)^2}$$

$$\times \begin{pmatrix} 4\left(\sigma_1^2 - \sigma_2^2 - \sigma_3^2\right) + \Sigma^2 & 8\sigma_1\sigma_2 - 4\sigma_3\Sigma & 8\sigma_3\sigma_1 + 4\sigma_2\Sigma \\ 8\sigma_1\sigma_2 + 4\sigma_3\Sigma & 4\left(-\sigma_1^2 + \sigma_2^2 - \sigma_3^2\right) + \Sigma^2 & 8\sigma_2\sigma_3 - 4\sigma_1\Sigma \\ 8\sigma_3\sigma_1 - 4\sigma_2\Sigma & 8\sigma_2\sigma_3 + 4\sigma_1\Sigma & 4\left(-\sigma_1^2 - \sigma_2^2 + \sigma_3^2\right) + \Sigma^2 \end{pmatrix} \quad (5.36)$$

where $\Sigma = 1 - \|\sigma\|^2$.

The Parameterization of Tsiotras and Longuski

Consider the concatenation of the transformation $R(\mathbf{e}_3, \mathbf{v})$ (the rotation that moves $\mathbf{e}_3$ to $\mathbf{v}$) and $R_3(\varphi)$ (the rotation about the local z axis). The result is a four-parameter description of rotation with matrix

$$R(\mathbf{e}_3, \mathbf{v}) R_3(\varphi) = \begin{pmatrix} \frac{v_3 \cos\varphi - v_1 v_2 \sin\varphi + (v_2^2 + v_3^2) \cos\varphi}{1 + v_3} & -\frac{v_3 \sin\varphi + v_1 v_2 \cos\varphi + (v_2^2 + v_3^2) \sin\varphi}{1 + v_3} & v_1 \\ \frac{v_3 \sin\varphi - v_1 v_2 \cos\varphi + (v_1^2 + v_3^2) \sin\varphi}{1 + v_3} & \frac{v_3 \cos\varphi + v_1 v_2 \sin\varphi + (v_1^2 + v_3^2) \cos\varphi}{1 + v_3} & v_2 \\ -v_2 \sin\varphi - v_1 \cos\varphi & v_1 \sin\varphi - v_2 \cos\varphi & v_3 \end{pmatrix} \qquad (5.37)$$

where $\|\mathbf{v}\| = 1$.

Tsiotras and Longuski [28] have used stereographic projection Eq. (4.23) of the $\|\mathbf{v}\| = 1$ sphere to the plane (viewed as the complex plane $\mathbb{C}$). In this way, two parameters, $w_1 = x_2$ and $w_2 = -x_1$, are used to parameterize the unit vector $\mathbf{v}$. This allows one to write [28]

$$R(w, \varphi) = \frac{1}{1 + |w|^2} \begin{pmatrix} Re[(1 + w^2) e^{-i\varphi}] & Im[(1 + w^2) e^{-i\varphi}] & -2Im(w) \\ Im[(1 - \overline{w}^2) e^{i\varphi}] & Re[(1 - \overline{w}^2) e^{i\varphi}] & 2Re(w) \\ 2Im(we^{-i\varphi}) & -2Re(we^{-i\varphi}) & 1 - |w|^2 \end{pmatrix}. \qquad (5.38)$$

Modified Axis-Angle Parameterization

The four-parameter axis-angle parameterization $R(\mathbf{n}, \theta)$ resulting from Euler's theorem and written as $\exp[\theta N]$ was reduced to a three-dimensional parameterization earlier in this chapter by expressing $\mathbf{n}$ in spherical coordinates. A different reduction to three parameters is achieved by sterographic projection of the sphere $\|\mathbf{n}\| = 1$ onto the plane in the same way discussed in the previous subsection. We then write

$$R(\mathbf{n}(w), \theta) = \frac{1}{(k_+(w))^2}$$
$$\times \begin{pmatrix} 4w_2^2 v\theta + (k_+(w))^2 c\theta & -4w_1 w_2 v\theta - (1 - |w|^4) s\theta & -2w_2(k_-(w)) v\theta + 2w_1(k_+(w)) s\theta \\ -4w_1 w_2 v\theta + (1 - |w|^4) s\theta & 4w_1^2 v\theta + (k_+(w))^2 c\theta & 2w_1(k_-(w)) v\theta + 2w_2(k_+(w)) s\theta \\ -2w_2(k_-(w)) v\theta - 2w_1(k_+(w)) s\theta & 2w_1(k_-(w)) v\theta - 2w_2(k_+(w)) s\theta & (k_-(w))^2 v\theta + (k_+(w))^2 c\theta \end{pmatrix}$$

where $k_+(w) = 1 + |w|^2$ and $k_-(w) = 1 - |w|^2$. One can go further and make the substitution

$$\cos\theta = \frac{1 - z^2}{1 + z^2} \qquad \sin\theta = \frac{2z}{1 + z^2}$$

so that $R(\mathbf{n}(w), \theta(z))$ is a rational expression.

5.5 Infinitesimal Rotations, Angular Velocity, and Integration

It is clear from Euler's theorem that when $|\theta| \ll 1$, a rotation matrix reduces to the form

$$A_E(\theta, \mathbf{n}) = \mathbb{I} + \theta N.$$

This means that for small rotation angles θ_1 and θ_2, rotations commute

$$A_E(\theta_1, \mathbf{n}_1) A_E(\theta_2, \mathbf{n}_2) = \mathbb{I} + \theta_1 N_1 + \theta_2 N_2 = A_E(\theta_2, \mathbf{n}_2) A_E(\theta_1, \mathbf{n}_1).$$

Given two frames of reference, one of which is fixed in space and the other of which is rotating relative to it, a rotation matrix describing the orientation of the rotating frame as seen in the fixed frame is written as $R(t)$ at each time t. One connects the concepts of small rotations and angular velocity by observing that if $\mathbf{x}_0$ is a fixed (constant) position vector in the rotating frame of reference, the position of the same point as seen in a frame of reference fixed in space with the same origin as the rotating frame is related to this as

$$\mathbf{x} = R(t)\mathbf{x}_0 \quad \leftrightarrow \quad \mathbf{x}_0 = R^T(t)\mathbf{x}.$$

The velocity as seen in the frame fixed in space is then

$$\mathbf{v} = \dot{\mathbf{x}} = \dot{R}\mathbf{x}_0 = \dot{R}R^T(\dot{R}R^T\mathbf{x})\mathbf{x}. \tag{5.39}$$

Observing that since R is a rotation matrix,

$$\frac{d}{dt}\left(RR^T\right) = \frac{d}{dt}(\mathbb{I}_{3\times3}) = 0_{3\times3},$$

and one writes

$$\dot{R}R^T = -R\dot{R}^T = -\left(\dot{R}R^T\right)^T.$$

Due to the skew-symmetry of this matrix, we can rewrite Eq. (5.39) in the form most familiar to engineers

$$\mathbf{v} = \boldsymbol{\omega}_L \times \mathbf{x},$$

where $\boldsymbol{\omega}_L = \text{vect}(\dot{R}R^T)$. The vector $\boldsymbol{\omega}_L$ is the angular velocity as seen in the space-fixed frame of reference (i.e., the frame in which the moving frame appears to have orientation given by R). The subscript L is not conventional. It simply denotes that $\dot{R}$ is on the left of R^T in $\text{vect}(\dot{R}R^T)$. In contrast, the angular velocity as seen in the rotating frame of reference (where the orientation of the moving frame appears to have orientation given by the identity rotation) is the dual vector of $R^T\dot{R}$, which is also a skew-symmetric matrix. This is denoted as $\boldsymbol{\omega}_R$. Therefore we have

$$\boldsymbol{\omega}_R = \text{vect}\left(R^T\dot{R}\right) = \text{vect}\left(R^T\left(\dot{R}R^T\right)R\right) = R^T\boldsymbol{\omega}_L.$$

Another way to write this is

$$\boldsymbol{\omega}_L = R\boldsymbol{\omega}_R.$$

In other words, the angular velocity as seen in the frame of reference fixed in space is obtained from the angular velocity as seen in the rotating frame in the same way in which the absolute position is obtained from the relative position.

5.5.1 Jacobians Associated with Parameterized Rotations

When a time-varying rotation matrix is parameterized as

$$R(t) = A\left(q_1(t), q_2(t), q_3(t)\right) = A(\mathbf{q}(t)),$$

then by the chain rule from calculus, one has

$$\dot{R} = \frac{\partial A}{\partial q_1}\dot{q}_1 + \frac{\partial A}{\partial q_2}\dot{q}_2 + \frac{\partial A}{\partial q_3}\dot{q}_3.$$

Multiplying on the right by R^T and extracting the dual vector from both sides, one finds that

$$\boldsymbol{\omega}_L = J_L(A(\mathbf{q}))\dot{\mathbf{q}} \tag{5.40}$$

where

$$J_L(A(\mathbf{q})) = \left[\text{vect}\left(\frac{\partial A}{\partial q_1} A^T \right), \text{vect}\left(\frac{\partial A}{\partial q_2} A^T \right), \text{vect}\left(\frac{\partial A}{\partial q_3} A^T \right) \right].$$

Similarly,

$$\omega_R = J_R(A(\mathbf{q}))\dot{\mathbf{q}} \tag{5.41}$$

where

$$J_R(A(\mathbf{q})) = \left[\text{vect}\left(A^T \frac{\partial A}{\partial q_1} \right), \text{vect}\left(A^T \frac{\partial A}{\partial q_2} \right), \text{vect}\left(A^T \frac{\partial A}{\partial q_3} \right) \right].$$

These two Jacobian matrices are related as

$$J_L = A J_R. \tag{5.42}$$

It is easy to verify from the above expressions that for an arbitrary constant rotation $R_0 \in SO(3)$

$$J_L(R_0 A(\mathbf{q})) = R_0 J_L(A(\mathbf{q})); \qquad J_L(A(\mathbf{q})R_0) = J_L(A(\mathbf{q}))$$
$$J_R(R_0 A(\mathbf{q})) = J_R(A(\mathbf{q})); \qquad J_R(A(\mathbf{q})R_0) = R_0^T J_R(A(\mathbf{q})).$$

In the following subsections, we provide the explicit forms for the Jacobians of several of the parameterizations discussed earlier in this chapter. As a notational shorthand, we will write $J_L(\mathbf{q})$ in place of $J_L(A(\mathbf{q}))$, and similarly for J_R.

The Jacobians for ZXZ Euler Angles

In this subsection we explicitly calculate the Jacobian matrices J_L and J_R for the ZXZ Euler angles. In this case, $A(\alpha, \beta, \gamma) = R_3(\alpha)R_1(\beta)R_3(\gamma)$, and the skew-symmetric matrices whose dual vectors form the columns of the Jacobian matrix J_L are given as[3]

$$\frac{\partial A}{\partial \alpha} A^T = \left(R_3'(\alpha)R_1(\beta)R_3(\gamma) \right)(R_3(-\gamma)R_1(-\beta)R_3(-\alpha)) = R_3'(\alpha)R_3(-\alpha),$$

$$\frac{\partial A}{\partial \beta} A^T = \left(R_3(\alpha)R_1'(\beta)R_3(\gamma) \right)(R_3(-\gamma)R_1(-\beta)R_3(-\alpha))$$

$$= R_3(\alpha)\left(R_1'(\beta)R_1(-\beta) \right) R_3(-\alpha),$$

$$\frac{\partial A}{\partial \gamma} A^T = R_3(\alpha)R_1(\beta)(R_3'(\gamma)(R_3(-\gamma))R_1(-\beta)R_3(-\alpha).$$

Noting that $\text{vect}(R_i' R_i^T) = e_i$ regardless of the value of the parameter, and using the rule $\text{vect}(RXR^T) = R\text{vect}(X)$, one finds

$$J_L(\alpha, \beta, \gamma) = [e_3, R_3(\alpha)e_1, R_3(\alpha)R_1(\beta)e_3].$$

This is written explicitly as

$$J_L(\alpha, \beta, \gamma) = \begin{pmatrix} 0 & \cos\alpha & \sin\alpha\sin\beta \\ 0 & \sin\alpha & -\cos\alpha\sin\beta \\ 1 & 0 & \cos\beta \end{pmatrix}.$$

---

[3]For one-parameter rotations we use the notation "$'$" to denote differentiation with respect to the parameter.

The Jacobian J_R can be derived similarly, or one can calculate it easily from J_L as

$$J_R = A^T J_L = \left[R_3(-\gamma) R_1(-\beta) \mathbf{e}_3, \; R_3(-\gamma) \mathbf{e}_1, \; \mathbf{e}_3 \right].$$

Explicitly, this is

$$J_R = \begin{pmatrix} \sin\beta \sin\gamma & \cos\gamma & 0 \\ \sin\beta \cos\gamma & -\sin\gamma & 0 \\ \cos\beta & 0 & 1 \end{pmatrix}.$$

These Jacobian matrices are important in understanding the differential operators and integration measures used in Chapter 8. These, in turn, provide the tools for defining orthogonal expansions of functions of rotation-valued arguments.

It is easy to see that

$$J_L^{-1} = \begin{pmatrix} -\cot\beta \sin\alpha & \cos\alpha \cot\beta & 1 \\ \cos\alpha & \sin\alpha & 0 \\ \csc\beta \sin\alpha & -\cos\alpha \csc\beta & 0 \end{pmatrix}.$$

$$J_R^{-1} = \begin{pmatrix} \csc\beta \sin\gamma & \cos\gamma \csc\beta & 0 \\ \cos\gamma & -\sin\gamma & 0 \\ -\cot\beta \sin\gamma & -\cos\gamma \cot\beta & 1 \end{pmatrix}. \tag{5.43}$$

$$\det(J_L) = \det(J_R) = -\sin\beta.$$

The Jacobians for the Matrix Exponential

Relatively simple analytical expressions have been derived by Park [19] for the Jacobian J_L and its inverse when rotations are parameterized as in Eq. (5.23). These expressions are

$$J_L(\mathbf{x}) = \mathbb{I}_{3\times 3} + \frac{1 - \cos\|\mathbf{x}\|}{\|\mathbf{x}\|^2} X + \frac{\|\mathbf{x}\| - \sin\|\mathbf{x}\|}{\|\mathbf{x}\|^3} X^2 \tag{5.44}$$

and

$$J_L^{-1}(\mathbf{x}) = \mathbb{I}_{3\times 3} - \frac{1}{2} X + \left(\frac{1}{\|\mathbf{x}\|^2} - \frac{1 + \cos\|\mathbf{x}\|}{2\|\mathbf{x}\| \sin\|\mathbf{x}\|} \right) X^2.$$

The corresponding Jacobian J_L and its inverse are then calculated as $J_L = A J_R$ and $J_L^{-1} = J_R^{-1} A^T$ to yield

$$J_R(\mathbf{x}) = \mathbb{I}_{3\times 3} - \frac{1 - \cos\|\mathbf{x}\|}{\|\mathbf{x}\|^2} X + \frac{\|\mathbf{x}\| - \sin\|\mathbf{x}\|}{\|\mathbf{x}\|^3} X^2$$

and

$$J_R^{-1}(\mathbf{x}) = \mathbb{I}_{3\times 3} + \frac{1}{2} X + \left(\frac{1}{\|\mathbf{x}\|^2} - \frac{1 + \cos\|\mathbf{x}\|}{2\|\mathbf{x}\| \sin\|\mathbf{x}\|} \right) X^2.$$

Note that unlike the Euler angles,

$$J_L = J_R^T.$$

The determinants are

$$\det(J_L) = \det(J_R) = \frac{2(1 - \cos\|\mathbf{x}\|)}{\|\mathbf{x}\|^2}.$$

The Jacobians for the Cayley/Rodrigues Parameters

We note that

$$J_R = \frac{2}{1 + r_1^2 + r_2^2 + r_3^2} \begin{pmatrix} 1 & r_3 & -r_2 \\ -r_3 & 1 & r_1 \\ r_2 & -r_1 & 1 \end{pmatrix}$$

and

$$J_R^{-1} = \frac{1}{2} \begin{pmatrix} 1 + r_1^2 & r_1 r_2 - r_3 & r_1 r_3 + r_2 \\ r_2 r_1 + r_3 & 1 + r_2^2 & r_2 r_3 - r_1 \\ r_3 r_1 - r_2 & r_3 r_2 + r_1 & 1 + r_3^2 \end{pmatrix}.$$

We also observe the interesting property of this parameterization

$$J_L = J_R^T,$$

and so

$$J_L^{-1} = \left(J_R^T \right)^{-1} = \left(J_R^{-1} \right)^T.$$

The Jacobian determinant is

$$\det(J_L) = \det(J_R) = \frac{8}{\left(1 + r_1^2 + r_2^2 + r_3^2 \right)^2}.$$

The Jacobians for Spherical Coordinates

By ordering columns with variables in the order θ, λ, ν, the Jacobians in spherical coordinates are

$$J_L = \begin{pmatrix} c\lambda s\nu & -c\lambda s^2(\theta/2)s(2\nu) - s\lambda s\nu s\theta & -2s^2(\theta/2)s\lambda + c\lambda c\nu s\theta \\ s\lambda s\nu & -s\lambda s(2\nu)s^2(\theta/2) + c\lambda s\nu s\theta & 2c\lambda s^2(\theta/2) + c\nu s\lambda s\theta \\ c\nu & 2s^2\nu s^2(\theta/2) & -s\nu s\theta \end{pmatrix}$$

and

$$J_R = \begin{pmatrix} c\lambda s\nu & c\lambda s(2\nu)s^2(\theta/2) - s\lambda s\nu s\theta & 2s\lambda s^2(\theta/2) + c\lambda c\nu s\theta \\ s\lambda s\nu & s\lambda s(2\nu)s^2(\theta/2) + c\lambda s\nu s\theta & -2s^2(\theta/2)c\lambda + c\nu s\lambda s\theta \\ c\nu & -2s^2\nu s^2(\theta/2) & -s\nu s\theta \end{pmatrix}.$$

The Jacobian determinant is then

$$\det(J_L) = \det(J_R) = -4s^2(\theta/2)s\nu.$$

Note that while changing $(\theta, \lambda, \nu) \to (-\theta, \lambda \pm \pi, \pi - \nu)$ changes the sign of the first and third columns of the Jacobians, the sign of the Jacobian determinant does not change.

The Jacobians for the Modified Rodrigues Parameters

The Jacobians for this parameterization have some interesting properties. First we observe that

$$J_L = \frac{4}{1 + \|\sigma\|^2} A_R(\sigma) =$$

$$\frac{4}{(1 + \|\sigma\|^2)^2} \begin{pmatrix} 1 + \sigma_1^2 - \sigma_2^2 - \sigma_3^2 & 2(\sigma_1\sigma_2 - \sigma_3) & 2(\sigma_1\sigma_3 + \sigma_2) \\ 2(\sigma_2\sigma_1 + \sigma_3) & 1 - \sigma_1^2 + \sigma_2^2 - \sigma_3^2 & 2(\sigma_2\sigma_3 - \sigma_1) \\ 2(\sigma_3\sigma_1 - \sigma_2) & 2(\sigma_3\sigma_2 + \sigma_1) & 1 - \sigma_1^2 - \sigma_2^2 + \sigma_3^2 \end{pmatrix},$$

where A_R is the Rodrigues rotation matrix in Eq. (5.30), evaluated at σ instead of $\mathbf{r}$. Since this Jacobian is a scalar multiple of a rotation matrix (as was observed in [24]), it follows that its inverse is easy to calculate

$$J_L^{-1} = \frac{1}{4}\begin{pmatrix} 1 + \sigma_1^2 - \sigma_2^2 - \sigma_3^2 & 2(\sigma_1\sigma_2 + \sigma_3) & 2(\sigma_1\sigma_3 - \sigma_2) \\ 2(\sigma_2\sigma_1 - \sigma_3) & 1 - \sigma_1^2 + \sigma_2^2 - \sigma_3^2 & 2(\sigma_2\sigma_3 + \sigma_1) \\ 2(\sigma_3\sigma_1 + \sigma_2) & 2(\sigma_3\sigma_2 - \sigma_1) & 1 - \sigma_1^2 - \sigma_2^2 + \sigma_3^2 \end{pmatrix}.$$

As with the Rodrigues parameters, we have

$$J_R = J_L^T,$$

and so

$$J_R^{-1} = \left(J_L^{-1}\right)^T.$$

The Jacobian determinants are

$$\det(J_L) = \det(J_R) = \frac{64}{\left(1 + \|\sigma\|^2\right)^3}.$$

The Jacobians for the Euler Parameters

The relationship between the time derivatives of the four Euler parameters and the angular velocity ω_L can be expressed using a 4×4 Jacobian matrix J_L as

$$\begin{pmatrix} \omega \\ 0 \end{pmatrix} = J_L(\mathbf{u})\dot{\mathbf{u}}$$

or

$$\dot{\mathbf{u}} = J_L^{-1}(\mathbf{u})\left[\omega^T, 0\right]^T.$$

Here the last row of J_L corresponds to the constraint $\|\mathbf{u}\|^2 = 1$. The Jacobian matrix J_L is explicitly

$$J_L = 2\begin{pmatrix} u_4 & u_3 & -u_2 & -u_1 \\ -u_3 & u_4 & u_1 & -u_2 \\ u_2 & -u_1 & u_4 & -u_3 \\ u_1 & u_2 & u_3 & u_4 \end{pmatrix}$$

(from [24] with first row and column moved) and its inverse is

$$J_L^{-1} = \frac{1}{2}\begin{pmatrix} u_4 & -u_3 & u_2 & u_1 \\ u_3 & u_4 & -u_1 & u_2 \\ -u_2 & u_1 & u_4 & u_3 \\ -u_1 & -u_2 & -u_3 & u_4 \end{pmatrix}.$$

5.5.2 Rigid-Body Mechanics

In the case of a rigid body, the angular momentum $\mathbf{L}$ has a special form. This is because the absolute position to the ith particle can be decomposed into a component due to the position of the center of mass and a component due to rotation about the center of mass. That is, we have

$$\mathbf{x}_i = \mathbf{x}_{cm} + R\mathbf{y}_i \tag{5.45}$$

where $\mathbf{y}_i$ is the position of the ith particle as measured in a frame attached to the center of mass of the rigid body. The angular momentum is then

$$L = \sum_{i=1}^{n} m_i \left(\mathbf{x}_i \times \dot{\mathbf{x}}_i\right) = \sum_{i=1}^{n} m_i \left((\mathbf{x}_{cm} + R\mathbf{y}_i) \times (\dot{\mathbf{x}}_{cm} + \dot{R}\mathbf{y}_i)\right).$$

Expanding this, we get

$$L = \sum_{i=1}^{n} m_i \left(\mathbf{x}_{cm} \times \dot{\mathbf{x}}_{cm} + \mathbf{x}_{cm} \times \dot{R}\mathbf{y}_i + R\mathbf{y}_i \times \dot{\mathbf{x}}_{cm} + R\mathbf{y}_i \times \dot{R}\mathbf{y}_i\right).$$

Now, by the definition of the center of mass, if we multiply Eq. (5.45) by m_i and sum both sides, we find that $\sum_{i=1}^{n} m_i R\mathbf{y}_i = 0$, which means that $\sum_{i=1}^{n} m_i \mathbf{y}_i = 0$ (because the rotation matrix can come out of the summation sign and we can multiply both sides of the equation by its inverse on the left).

This means that all terms with a sum of this form will drop out, and so the angular momentum expression simplifies to

$$L = \left(\sum_{i=1}^{n} m_i\right) (\mathbf{x}_{cm} \times \dot{\mathbf{x}}_{cm}) + \sum_{i=1}^{n} m_i \left(R\mathbf{y}_i \times \dot{R}\mathbf{y}_i\right),$$

which is a component due to the motion of the center of mass and a component due to rotation about the center of mass. In cases where the center of mass is fixed or travels in a straight line (with constant or variable velocity) passing through the origin of a coordinate system, the first part of this expression will be zero.

In order to describe the second term in the angular momentum expression more easily, we can define the position of the ith particle relative to the center of mass (defined in a frame parallel to the inertial/fixed frame) as $\mathbf{z}_i = R\mathbf{y}_i$. This means that we have

$$R\mathbf{y}_i \times \dot{R}\mathbf{y}_i = \mathbf{z}_i \times \dot{R}R^T \mathbf{z}_i = \mathbf{z}_i \times (\boldsymbol{\omega}_L \times \mathbf{z}_i).$$

Using the identity $\mathbf{a} \times (\mathbf{b} \times \mathbf{a}) = (\mathbf{a} \cdot \mathbf{a})\mathbf{b} - (\mathbf{a} \cdot \mathbf{b})\mathbf{a}$, we can write for the case when $\mathbf{x}_{cm} \times \dot{\mathbf{x}}_{cm} = 0$

$$L_L = \sum_{i=1}^{n} m_i \left((\mathbf{z}_i \cdot \mathbf{z}_i) \boldsymbol{\omega} - (\mathbf{z}_i \cdot \boldsymbol{\omega}) \mathbf{z}_i\right) = I_L \boldsymbol{\omega}_L,$$

where

$$I_L \stackrel{\Delta}{=} \sum_{i=1}^{n} m_i \left(\mathbf{z}_i \cdot \mathbf{z}_i \mathbb{I} - \mathbf{z}_i \mathbf{z}_i^T\right).$$

Note that the angular momentum vector, just like any vector, can be represented in any frame. Therefore, if instead of representing it in the fixed frame we represent it in the rotated frame as L_R, then $L_L = RL_R$. Likewise, recall that angular velocity can then be represented in any frame, and so $\boldsymbol{\omega}_L = R\boldsymbol{\omega}_R$. Therefore, we can write

$$L_R = R^T L_L = R^T I_L R \boldsymbol{\omega}_R = I_R \boldsymbol{\omega}_R,$$

where

$$I_R = \sum_{i=1}^{n} m_i \left(\mathbf{y}_i \cdot \mathbf{y}_i \mathbb{I} - \mathbf{y}_i \mathbf{y}_i^T\right).$$

In other words, the moment of inertia matrix is different when represented in different frames. It is a constant matrix I_R, when represented in the frame fixed to the body, and time-dependent in any frame which moves relative to the body. Any physical quantity that can be represented by a matrix $A \in R^{3 \times 3}$, which transforms under any rotation $Q \in SO(3)$ as $A_L = Q A_R Q^T$, is called a *2nd order tensor*. Vectors are often called *1st order tensors* because they transform under the rule $a_L = Q a_R$ under rotations.

Parallel Axis Theorem

For a rigid body with one fixed point, we can attach a frame at that point which is fixed in space and a second frame at that point which is fixed in the body. The vector to the center of mass is then $x_{cm} = R y_{cm}$, where y_{cm} is a *constant* vector. In this case, we can write

$$\left(\sum_{i=1}^{n} m_i \right) (x_{cm} \times \dot{x}_{cm}) = \left(\sum_{i=1}^{n} m_i \right) ((R y_{cm}) \times \dot{R} y_{cm})$$

$$= \left(\sum_{i=1}^{n} m_i \right) R \left((y_{cm}) \times \left(R^T \dot{R} y_{cm} \right) \right)$$

$$= M \cdot R \left((y_{cm}) \times \left(R^T \dot{R} y_{cm} \right) \right)$$

$$= R \left(M (y_{cm}) \times (\omega_R \times y_{cm}) \right)$$

where the rule $(Qa) \times (Qb) = Q(a \times b)$ has been used (for $Q \in SO(3)$) and M is the total mass.

This can be simplified in one final step as

$$R (M (y_{cm}) \times (\omega_R \times y_{cm})) = R \left(M \left(y_{cm} \cdot y_{cm} \mathbb{I} - y_{cm} y_{cm}^T \right) \right) \omega_R.$$

In other words, if we define a constant matrix

$$P_R = (y_{cm} \cdot y_{cm}) \mathbb{I} - y_{cm} y_{cm}^T,$$

the angular momentum vector described in the rotating frame with origin at the fixed point is

$$L_R = (I_R + M P_R) \omega_R.$$

It is often easier to write these two matrices together as one matrix

$$I_R' = I_R + M P_R.$$

This is the *parallel axis theorem*. It says that if a rigid body has one fixed point, and that point is not the center of mass, the angular momentum vector can still be written as $L_R = I_R' \omega_R$, or $L_L = I_L' \omega_L$.

Euler's Equations of Motion

We can use this to derive compact equations of motion for any rigid body with one point fixed, e.g., a spatial pendulum, or spinning top. Since

$$N_L = \frac{dL_L}{dt} = \frac{d}{dt} (R L_R) = \dot{R} L_R + R \dot{L}_R,$$

we can define $N_R = R^T N_L$ to be the resultant moment represented in the rotated frame, and

$$N_R = R^T N_L = R^T \dot{R} L_R + \dot{L}_R.$$

Recognizing that $\omega_R = \text{vect}(R^T \dot{R})$, and substituting $\mathbf{L}_R = I'_R \omega_R$, we get

$$\mathbf{N}_R = \omega_R \times I'_R \omega_R + I'_R \dot{\omega}_R.$$

This is called *Euler's equation of motion*. In particular, if I'_R is diagonal with components I_1, I_2, and I_3 (which can always be achieved by choosing the body-fixed frame as the eigenvectors of the moment of inertia matrix), ω_R has components ω_1, ω_2, ω_3, and $\mathbf{N}_R$ has components N_1, N_2, N_3, then these equations have the form

$$I_1 \dot{\omega}_1 + (I_3 - I_2)\,\omega_2 \omega_3 = N_1$$
$$I_2 \dot{\omega}_2 + (I_1 - I_3)\,\omega_3 \omega_1 = N_2$$
$$I_3 \dot{\omega}_3 + (I_2 - I_1)\,\omega_1 \omega_2 = N_3.$$

Kinetic Energy of a Rotating Rigid Body

The kinetic energy of a rigid body consists of a part due to translation of the center of mass, and a part due to rotation about the center of mass

$$T = T_{trans} + T_{rot}.$$

The rotational kinetic energy is usually given by

$$T_{rot} = \frac{1}{2} \omega_R^T I_R \omega_R$$

where I_R and ω_R are the moment of inertia tensor and angular velocity, respectively, as seen in the frame of reference attached to the rigid body at its mass center. The same equation for T_{rot} could be used without the subscripts on I and ω. However, for convenience, the frame fixed in the body is usually chosen because in this frame I_R is a constant matrix. For a mass distribution that is continuous over a finite body, we write

$$I_R = \text{trace}\,(\mathcal{I}_R)\,\mathbb{I} - \mathcal{I}_R$$

where

$$\mathcal{I}_R = \int_{\mathbb{R}^3} \mathbf{z}\mathbf{z}^T \rho(\mathbf{z})\,dz_1 dz_2 dz_3$$

and $\mathbf{z}$ is position as defined with respect to the frame attached to the rigid body at its center of mass, and $\rho(\mathbf{z})$ is the mass density of the rigid body.

An equivalent expression for rotational kinetic energy in terms of the rotation matrix R is

$$T_{rot} = \frac{1}{2}\text{trace}\left(\dot{R}\mathcal{I}_R\dot{R}^T\right).$$

5.6 Other Methods for Describing Rotations in Three-Space

5.6.1 Quaternions

Quaternions[4] were developed in an attempt to generalize the concept of the complex numbers from $2D$ to $3D$ in the following sense. Each complex number $x = x_1 + i x_2$ is associated with a

---

[4]These were developed by Sir William Rowan Hamilton (1805–1865).

position in the plane $\mathbf{x} = (x_1, x_2)^T \in \mathbb{R}^2$. While it does not make sense to talk about dividing by vectors, it is perfectly legitimate to find for any complex number $x \neq 0$, its inverse x^{-1} such that $xx^{-1} = x^{-1}x = 1$. This is given explicitly as $x^{-1} = (x_1 - ix_2)/(x_1^2 + x_2^2)$ and is easily verified by writing the rule for multiplication of complex numbers as $xy = (x_1 y_1 - x_2 y_2) + i(x_1 y_2 + x_2 y_1)$. It is interesting to note that the operation on vectors $\mathbf{x}, \mathbf{y} \in \mathbb{R}^2$ defined as

$$\mathbf{x} \wedge \mathbf{y} = (x_1 y_1 - x_2 y_2)\,\mathbf{e}_1 + (x_1 y_2 + x_2 y_1)\,\mathbf{e}_2$$

endows the pair $(\mathbb{R}^2, \wedge)$ with the properties of an algebra (see Appendix C). Rotations in the complex plane are achieved as $x' = e^{i\theta}x$ for $\theta \in [0, 2\pi]$. This yields exactly the same result as the matrix-vector operation

$$\begin{pmatrix} x_1' \\ x_2' \end{pmatrix} = \begin{pmatrix} \cos\theta & -\sin\theta \\ \sin\theta & \cos\theta \end{pmatrix} \begin{pmatrix} x_1 \\ x_2 \end{pmatrix}.$$

Hamilton attempted for over 10 years to find a spatial analog of the complex numbers, i.e., one of the form $x = x_1 + ix_2 + jx_3$ where x^{-1} is well defined. It is rumoured that during this time his son would often ask, "Father, have you learned how to divide vectors?"[9]. Needless to say, progress was slow until Hamilton had the inspiration to expand his thinking to four dimensions. Analogous to the complex numbers which have the basis $\{1, i\}$ satisfying $i^2 = -1$, the *quaternions* (or hyper-complex numbers) have a basis $\{1, \tilde{\mathbf{i}}, \tilde{\mathbf{j}}, \tilde{\mathbf{k}}\}$ satisfying the conditions

$$\tilde{\mathbf{i}}^2 = \tilde{\mathbf{j}}^2 = \tilde{\mathbf{k}}^2 = -1 \tag{5.46}$$

$$\tilde{\mathbf{i}}\tilde{\mathbf{j}} = -\tilde{\mathbf{j}}\tilde{\mathbf{i}} = \tilde{\mathbf{k}} \tag{5.47}$$

$$\tilde{\mathbf{k}}\tilde{\mathbf{i}} = -\tilde{\mathbf{i}}\tilde{\mathbf{k}} = \tilde{\mathbf{j}} \tag{5.48}$$

$$\tilde{\mathbf{j}}\tilde{\mathbf{k}} = -\tilde{\mathbf{k}}\tilde{\mathbf{j}} = \tilde{\mathbf{i}}. \tag{5.49}$$

Multiplication by 1 on either side of $\tilde{\mathbf{i}}, \tilde{\mathbf{j}}$, or $\tilde{\mathbf{k}}$ leaves it unchanged. The quaternions form an algebra over $\mathbb{R}^4$, when viewing the multiplication of two quaternions $p = p_4 + \tilde{\mathbf{i}}p_1 + \tilde{\mathbf{j}}p_2 + \tilde{\mathbf{k}}p_3$ and $q = q_4 + \tilde{\mathbf{i}}q_1 + \tilde{\mathbf{j}}q_2 + \tilde{\mathbf{k}}q_3$ as an operation on vectors in $\mathbb{R}^4$ with components (q_1, q_2, q_3, q_4) and (p_1, p_2, p_3, p_4). They also almost satisfy the definition of a *field* (see Appendix C for definition), except for the fact that $pq \neq qp$ in general, and so the quaternions are referred to as a *skew field*. Often a quaternion q is written as the pair $(q_4, \mathbf{q})$ where $\mathbf{q} = \tilde{\mathbf{i}}q_1 + \tilde{\mathbf{j}}q_2 + \tilde{\mathbf{k}}q_3$ is called a "pure quaternion" or the "vector part" of q, and q_4 is called the "real part" of q. Using the rules in Eqs. (5.46)–(5.49), the product of two quaternions is written as

$$(q_4, \mathbf{q})\,(p_4, \mathbf{p}) = (q_4 p_4 - \mathbf{q} \cdot \mathbf{p},\ q_4 \mathbf{p} + p_4 \mathbf{q} + \mathbf{q} \times \mathbf{p})$$

where $\cdot$ and $\times$ are the scalar and cross products for $\mathbb{R}^3$.

Analogous to the complex numbers, the conjugate of a quaternion $q = (q_4, \mathbf{q})$ is $q^* = (q_4, -\mathbf{q})$. The norm (or modulus) of a quaternion is the square root of

$$|q|^2 = qq^* = q^*q = q_1^2 + q_2^2 + q_3^2 + q_4^2.$$

The inverse of a quaternion is

$$q^{-1} = \frac{q^*}{qq^*}.$$

In addition to being an algebra (and hence a vector space and an Abelian group under addition), the quaternions form a group under multiplication, with the small modification that $q = 0$ be excluded.[5]

---

[5] See Chapter 7 for an introduction to group theory.

Unit quaternions are those satisfying $|q| = 1$. Henceforth we use the notation u to denote unit quaternions instead of q. As with rotations, the unit quaternions can be viewed as lying on the surface of S^3. In fact, for $x_4 = 0$ the transformation

$$x' = uxu^* \tag{5.50}$$

for $u = u_4 + \tilde{i}u_1 + \tilde{j}u_2 + \tilde{k}u_3$ implements exactly the same rotation of the vector $\mathbf{x} = (x_1, x_2, x_3)^T$ as multiplication by the matrix $A(u_1, u_2, u_3, u_3)$ of Euler parameters in Eq. (5.29). The unit quaternion corresponding to rotation about $\mathbf{n}$ by angle θ is $q = (\cos\theta/2, \mathbf{n}\sin\theta/2)$. Whereas the set of rotations is viewed as the upper hemisphere of S^3, two unit quaternions, u and $-u$, represent the same rotation. If u is in the upper hemisphere, then $-u$ is in the lower one, and the set of all unit quaternions is viewed as the whole of S^3 and is called the *quaternion sphere*.

For further reading on quaternions, see Hamilton's original paper [10], or any thorough book on kinematics or group theory.

As a final note to this subsection, we observe that generalizations of the Fourier transform based on unit quaternions, instead of unimodular complex numbers, have been investigated in recent years. The motivation goes something like this: $e^{i\phi} = \cos\phi + i\sin\phi$ and satisfies $(e^{i\phi})^k = \cos k\phi + i\sin k\phi$. A unit quaternion $u = (\cos\phi, \mathbf{n}\sin\phi)$ satisfies the analogous expression $u^k = (\cos k\phi, \mathbf{n}\sin k\phi)$. For real and complex-valued functions, the Fourier transform is defined based on the properties of $e^{i\phi}$, so an analogous Fourier transform for quaternion-valued functions should follow based on the properties of unit quaternions. Several recent research papers and books in the field of "hyper-complex analysis" have defined this kind of generalization of the Fourier transform [26, 14, 9].

To our knowledge, there is no associated generalization of the convolution theorem. It is also difficult to imagine that one is possible, given the noncommutative nature of multiplication of quaternion-valued functions. In later chapters we will explore noncommutative harmonic analysis, which is a generalization of Fourier analysis when the *argument* of a function is a quantity such as a rotation or quaternion and the *value* of the function is real or complex.

5.6.2 Expressing Rotations as 4×4 Matrices

The rules for quaternion multiplication reviewed in the previous subsection may be observed in several kinds of matrices. For instance, we may define

$$\mathbb{I}_{4\times4} = \begin{pmatrix} 1 & 0 & 0 & 0 \\ 0 & 1 & 0 & 0 \\ 0 & 0 & 1 & 0 \\ 0 & 0 & 0 & 1 \end{pmatrix}; \quad \tilde{i}_{4\times4} = \begin{pmatrix} 0 & -1 & 0 & 0 \\ 1 & 0 & 0 & 0 \\ 0 & 0 & 0 & 1 \\ 0 & 0 & -1 & 0 \end{pmatrix};$$

$$\tilde{j}_{4\times4} = \begin{pmatrix} 0 & 0 & -1 & 0 \\ 0 & 0 & 0 & -1 \\ 1 & 0 & 0 & 0 \\ 0 & 1 & 0 & 0 \end{pmatrix}; \quad \tilde{k}_{4\times4} = \begin{pmatrix} 0 & 0 & 0 & 1 \\ 0 & 0 & -1 & 0 \\ 0 & 1 & 0 & 0 \\ -1 & 0 & 0 & 0 \end{pmatrix}. \tag{5.51}$$

It is easy to see that $\tilde{i}_{4\times4}\tilde{i}_{4\times4} = -\mathbb{I}_{4\times4}$, $\tilde{i}_{4\times4}\tilde{j}_{4\times4} = \tilde{k}_{4\times4}$, etc. This choice is not unique. For example, if the matrices we have assigned to the symbols $\tilde{i}_{4\times4}$, $\tilde{j}_{4\times4}$, $\tilde{k}_{4\times4}$ are instead assigned to a cyclic reordering of these symbols, the quaternion multiplication rules will still hold. This is also the case if the assignment of these matrices is made to an acyclic reordering of the symbols, provided that certain signs are changed. For example, $i_{4\times4} = \tilde{k}_{4\times4}$, $j_{4\times4} = -\tilde{j}_{4\times4}$, $k_{4\times4} = \tilde{i}_{4\times4}$ is another choice that is found in the literature.

Using the assignment in Eq. (5.51), the rotation encoded by a quaternion can be expressed as

$$Q(u_1, u_2, u_3, u_4) = u_4 \mathbb{I}_{4 \times 4} + u_1 \tilde{i}_{4 \times 4} + u_2 \tilde{j}_{4 \times 4} + u_3 \tilde{k}_{4 \times 4}$$

$$= \begin{pmatrix} u_4 & -u_1 & -u_2 & u_3 \\ u_1 & u_4 & -u_3 & -u_2 \\ u_2 & u_3 & u_4 & u_1 \\ -u_3 & u_2 & -u_1 & u_4 \end{pmatrix}. \tag{5.52}$$

Note that since $u_1^2 + u_2^2 + u_3^2 + u_4^2 = 1$, all of the columns are orthonormal, and $\det Q = +1$. Therefore, $Q \in SO(4)$. However, four parameters are not sufficient to parameterize $SO(4)$, and hence it is *not* true that all 4×4 rotation matrices are of the form in Eq. (5.52).

Quaternion multiplication is implemented in this notation as 4×4 real-matrix multiplication, with the action of rotation of a vector $\mathbf{x} = (x_1, x_2, x_3)^T \in \mathbb{R}^3$ implemented as

$$X' = QXQ^\dagger \tag{5.53}$$

where

$$X = x_1 \tilde{i}_{4 \times 4} + x_2 \tilde{j}_{4 \times 4} + x_3 \tilde{k}_{4 \times 4} = \begin{pmatrix} 0 & -x_1 & -x_2 & x_3 \\ x_1 & 0 & -x_3 & -x_2 \\ x_2 & x_3 & 0 & x_1 \\ -x_3 & x_2 & -x_1 & 0 \end{pmatrix},$$

and

$$Q^\dagger(u_1, u_2, u_3, u_4) = \begin{pmatrix} u_4 & u_1 & u_2 & -u_3 \\ -u_1 & u_4 & u_3 & u_2 \\ -u_2 & -u_3 & u_4 & -u_1 \\ u_3 & -u_2 & u_1 & u_4 \end{pmatrix} = Q(-u_1, -u_2, -u_3, u_4).$$

Note that in the present context, $Q^\dagger$ is simply Q^T, and Eq. (5.53) is somewhat like Eq. (5.19). More generally, $\dagger$ is the Hermitian conjugate. One difference is that in the 4×4 description, taking the determinant of both sides yields $\|\mathbf{x}'\|^4 = \|\mathbf{x}\|^4$, which is a check that length is preserved, whereas in the 3×3 case the determinant is zero, since 3×3 skew-symmetric matrices are singular. However, in both cases, $\text{trace}(XX^T)$ (which is a multiple of $\|\mathbf{x}\|^2$) is preserved under the transformation.

Finally, we note that the axis-angle description of $3D$-rotations using 4×4 matrices takes the form

$$Q(n_1 \sin \theta/2, n_2 \sin \theta/2, n_3 \sin \theta/2, \cos \theta/2) = e^{\frac{\theta}{2}N} = \cos \frac{\theta}{2} \mathbb{I}_{4 \times 4} + \sin \frac{\theta}{2} N \tag{5.54}$$

where

$$N = n_1 \tilde{i}_{4 \times 4} + n_2 \tilde{j}_{4 \times 4} + n_3 \tilde{k}_{4 \times 4}$$

and $\mathbf{n} \cdot \mathbf{n} = 1$.

5.6.3 Special Unitary 2×2 Matrices: $SU(2)$

The rules of quaternion multiplication can also be realized with matrices of the form

$$\mathbb{I}_{2 \times 2} = \begin{pmatrix} 1 & 0 \\ 0 & 1 \end{pmatrix}; \qquad \tilde{i}_{2 \times 2} = \begin{pmatrix} 0 & -1 \\ 1 & 0 \end{pmatrix};$$

$$\tilde{j}_{2 \times 2} = \begin{pmatrix} -i & 0 \\ 0 & i \end{pmatrix}; \qquad \tilde{k}_{2 \times 2} = \begin{pmatrix} 0 & -i \\ -i & 0 \end{pmatrix}. \tag{5.55}$$

As with the case of the 4×4 matrices in the previous subsection, this assignment of matrices to the symbols $\tilde{i}_{2\times2}, \tilde{j}_{2\times2}, \tilde{k}_{2\times2}$ is not unique. Within the convention defined in Eq. (5.55), 2×2 matrices of the form

$$U(u_1, u_2, u_3, u_4) = u_4 \mathbb{I}_{2\times2} + u_1 \tilde{i}_{2\times2} + u_2 \tilde{j}_{2\times2} + u_3 \tilde{k}_{2\times2}$$
$$= \begin{pmatrix} u_4 - iu_2 & -u_1 - iu_3 \\ u_1 - iu_3 & u_4 + iu_2 \end{pmatrix} \quad (5.56)$$

encode rotations. Composition of rotations is achieved by multiplication of these complex matrices, and rotation of a vector $\mathbf{x} = (x_1, x_2, x_3)^T \in \mathbb{R}^3$ is achieved as

$$X' = UXU^\dagger$$

where

$$X = \begin{pmatrix} -ix_2 & -x_1 - ix_3 \\ x_1 - ix_3 & ix_2 \end{pmatrix} = -X^\dagger \quad (5.57)$$

is a traceless *skew-Hermitian* matrix corresponding to the pure quaternion representing $\mathbf{x}$. It is easy to verify that

$$\det(X) = \|\mathbf{x}\|^2 = \frac{1}{2}\text{trace}\left(XX^\dagger\right)$$

is preserved under the transformation, as should be the case for a rotation.

The matrix exponential of a skew-Hermitian matrix S, of the same form as X in Eq. (5.57), corresponds to a unitary matrix

$$e^S = \mathbb{I}_{2\times2} \cos \|\mathbf{s}\| + S\frac{\sin \|\mathbf{s}\|}{\|\mathbf{s}\|}.$$

Likewise, the axis-angle description becomes

$$e^{\frac{\theta}{2}N} = \mathbb{I}_{2\times2} \cos \frac{\theta}{2} + N \sin \frac{\theta}{2}$$

where $N = S/\|\mathbf{s}\|$.

In physics it is common to describe rotations using the *Pauli spin matrices*[6]

$$\sigma_1 = i \cdot \tilde{k}_{2\times2} = \begin{pmatrix} 0 & 1 \\ 1 & 0 \end{pmatrix};$$

$$\sigma_2 = i \cdot \tilde{i}_{2\times2} = \begin{pmatrix} 0 & -i \\ i & 0 \end{pmatrix};$$

$$\sigma_3 = i \cdot \tilde{j}_{2\times2} = \begin{pmatrix} 1 & 0 \\ 0 & -1 \end{pmatrix}.$$

In this notation, a position in space is represented as a traceless *Hermitian* matrix

$$Y = y_1\sigma_1^T + y_2\sigma_2^T + y_3\sigma_3^T = \begin{pmatrix} y_3 & y_1 + iy_2 \\ y_1 - iy_2 & -y_3 \end{pmatrix},$$

(here T denotes transpose), and a rotation is given by a matrix of the form

$$U(a, b) = \begin{pmatrix} a & b \\ -\bar{b} & \bar{a} \end{pmatrix}$$

---

[6]These are named after the theoretical physicist Wolfgang Pauli (1900–1958).

where $a = a_1 + ia_2$, $b = b_1 + ib_2$, and $a\bar{a} + b\bar{b} = 1$. These are called the *Cayley-Klein* parameters [8], which are essentially a variant of the Euler parameters when rotation is performed as

$$Y' = UYU^\dagger. \tag{5.58}$$

In the most general case, the rotation matrix $R(a, b) \in SO(3)$ to which the special unitary matrices $\pm U(a, b)$ correspond can be found using the formula

$$U(a, b) \left(\sum_{j=1}^{3} y_j \sigma_j^T \right) U^\dagger(a, b) = \sum_{j=1}^{3} (R(a, b)\mathbf{y})_j \, \sigma_j^T \tag{5.59}$$

which must hold for an arbitrary vector $\mathbf{y} = [y_1, y_2, y_3]^T \in \mathbb{R}^3$.

Explicitly, the rotation matrix can be written as

$$R(a, b) = \begin{pmatrix} \frac{a^2 - b^2 + \bar{a}^2 - \bar{b}^2}{2} & \frac{-i\left(\bar{a}^2 + \bar{b}^2 - a^2 - b^2\right)}{2} & -\left(\bar{a}b + ab\right) \\ \frac{-i\left(a^2 - b^2 - \bar{a}^2 + \bar{b}^2\right)}{2} & \frac{\left(\bar{a}^2 + \bar{b}^2 + a^2 + b^2\right)}{2} & i\left(-\bar{a}b + ab\right) \\ \bar{a}b + a\bar{b} & i\left(-\bar{a}b + a\bar{b}\right) & a\bar{a} - b\bar{b} \end{pmatrix}. \tag{5.60}$$

It may be easily shown by direct calculation for particular cases that rotations of $\mathbf{y} \in \mathbb{R}^3$ about $\mathbf{e}_1$, $\mathbf{e}_2$, $\mathbf{e}_3$ are achieved by substituting the following matrices in for U in Eq. (5.58)

$$U_1(\phi) = \begin{pmatrix} \cos(\phi/2) & i\sin(\phi/2) \\ i\sin(\phi/2) & \cos(\phi/2) \end{pmatrix} = \mathbb{I}\cos(\phi/2) + i\sigma_1^T \sin(\phi/2);$$

$$U_2(\phi) = \begin{pmatrix} \cos\phi/2 & -\sin\phi/2 \\ \sin\phi/2 & \cos\phi/2 \end{pmatrix} = \mathbb{I}\cos(\phi/2) + i\sigma_2^T \sin(\phi/2);$$

$$U_3(\phi) = \begin{pmatrix} e^{i\phi/2} & 0 \\ 0 & e^{-i\phi/2} \end{pmatrix} = \mathbb{I}\cos(\phi/2) + i\sigma_3^T \sin(\phi/2).$$

It is also easy to show by direct calculation that for any traceless 2×2 Hermitian matrix T, that $A = e^{iT}$ is a 2×2 unitary matrix of the form

$$e^{iT} = \mathbb{I}_{2\times 2} \cos\|\mathbf{t}\| + iT \frac{\sin\|\mathbf{t}\|}{\|\mathbf{t}\|}.$$

Likewise, the axis-angle description becomes

$$e^{i\theta N} = \mathbb{I}_{2\times 2} \cos\frac{\theta}{2} + iN\sin\frac{\theta}{2}$$

where $N = T/\|\mathbf{t}\|$.

In particular,

$$U_k(\phi) = e^{i(\phi/2)\sigma_k^T} \tag{5.61}$$

for $k = 1, 2, 3$.

Finally, we note that other conventions for the definitions of the Pauli spin matrices lead to similar-looking results. For instance, Miller [17] uses $\tilde{\sigma}_i = U_1(\pi)\sigma_i^T U_1(-\pi)$. Substitution of this into Eq. (5.60) results in a matrix $R'(a, b)$ which, in our notation, corresponds to the unitary matrices $U'(a, b) = U_1(-\pi)U(a, b)U_1(\pi)$ and is calculated relative to ours by the similarity transformation $R'(a, b) = R_1(-\pi)R(a, b)R_1(\pi)$.

5.6.4   Rotations in 3D as Bilinear Transformations in the Complex Plane

A *bilinear* (or *fractional linear*) transformation is one of the form

$$x' = \frac{ax + b}{cx + d} \tag{5.62}$$

where $x, a, b, c, d \in \mathbb{C}$. When $ad - bc = 1$, this is called a *Möbius transformation*.

Henceforth the real part of these variables will be denoted with a subscript 1 and the complex part with a subscript 2, e.g., $a = a_1 + ia_2$. Expanding this expression and separating real and imaginary parts, one finds

$$x_1' = \frac{(a_1 x_1 - a_2 x_2 + b_1)(c_1 x_1 - c_2 x_2 + d_1) + (b_2 + a_2 x_1 + a_1 x_2)(d_2 + c_2 x_1 + c_1 x_2)}{(c_1 x_1 - c_2 x_2 + d_1)^2 + (d_2 + c_2 x_1 + c_1 x_2)^2} \tag{5.63}$$

and

$$x_2' = \frac{(b_2 + a_2 x_1 + a_1 x_2)(c_1 x_1 - c_2 x_2 + d_1) - (a_1 x_1 - a_2 x_2 + b_1)(d_2 + c_2 x_1 + c_1 x_2)}{(c_1 x_1 - c_2 x_2 + d_1)^2 + (d_2 + c_2 x_1 + c_1 x_2)^2}. \tag{5.64}$$

Clearly this is a ratio of two second order polynomials in the variables x_1 and x_2.

Referring back to the equations of stereographic projection, Eqs. (4.23)–(4.24), we write

$$\mathbf{x} = \mathbf{f}(\mathbf{v}) \qquad \text{and} \qquad \mathbf{v} = \mathbf{g}(\mathbf{x})$$

where $\mathbf{v} = [v_1, v_2, v_3]^T$ and $\mathbf{x} = [x_1, x_2]^T$.

It may be verified that the composed operation

$$\mathbf{x}' = \mathbf{f}(R\mathbf{g}(\mathbf{x}))$$

also results in a ratio of two second order polynomials in the variables x_1 and x_2, where $R \in \mathbb{R}^{3 \times 3}$ is a matrix independent of $\mathbf{x}$ and $\mathbf{v}$. Expanding this expression and equating the components of $\mathbf{x}'$ defined in this way with those defined in Eqs. (5.63)–(5.64) imposes a system of constraint equations. When $|a|^2 + |b|^2 = 1$, and

$$x' = \frac{ax + b}{-\bar{b}x + \bar{a}} \tag{5.65}$$

(which is a special kind of Möbius transformation), the matrix R may be solved as

$$R = \begin{pmatrix} a_1^2 - a_2^2 - b_1^2 + b_2^2 & -2(a_1 a_2 + b_1 b_2) & 2(a_1 b_1 - a_2 b_2) \\ 2(a_1 a_2 - b_1 b_2) & a_1^2 - a_2^2 + b_1^2 - b_2^2 & 2(a_1 b_2 + a_2 b_1) \\ -2(a_1 b_1 + a_2 b_2) & -2(a_1 b_2 - a_2 b_1) & a_1^2 + a_2^2 - b_1^2 - b_2^2 \end{pmatrix}.$$

This is clearly related to the rotation matrices with Cayley-Klein parameters and Euler parameters.

5.7   Metrics on Rotations

In the literature, several metrics (distance functions) for rotations have been presented. Before examining these, we review the standard techniques for measuring distance between positions.

Definitions and Examples of Metrics

DEFINITION 5.1   *Given a set $\mathcal{S}$, a metric is a real-valued function, $d : \mathcal{S} \times \mathcal{S} \to \mathbb{R}$, which has the following properties for all $A, B, C \in \mathcal{S}$*

$$d(A, B) \geq 0 \quad \text{and} \quad d(A, B) = 0 \Longleftrightarrow A = B$$
$$d(A, B) = d(B, A) \tag{5.66}$$
$$d(A, B) + d(B, C) \geq d(A, C).$$

We refer to these as positive definiteness, symmetry, and the triangle inequality, respectively. The pair $(\mathcal{S}, d)$ is called a *metric space*.

The most common examples of metrics on $\mathbb{R}^N$ are of the form

$$d_p(\mathbf{x}, \mathbf{y}) = \sqrt[p]{\sum_{i=1}^{N} |x_i - y_i|^p} \quad \text{for} \ \ p \in \{1, 2, \dots\},$$

and for the case $p = 1$ the root sign disappears.

The above metrics are usually denoted $d_p(\mathbf{x}, \mathbf{y}) = \|\mathbf{x} - \mathbf{y}\|_p$ where $\|\mathbf{x}\|_p$ is the p-norm of $\mathbf{x} \in \mathbb{R}^N$. Note that $d_1(\mathbf{x}, \mathbf{y})$ is the Manhattan or "taxicab" metric, $d_2(\mathbf{x}, \mathbf{y})$ is the Euclidean metric, and $d_\infty(\mathbf{x}, \mathbf{y}) = \max_i |x_i - y_i|$ is the infinity metric. Clearly, evaluation of each of these metrics requires $\mathcal{O}(N)$ computations for fixed p.

The following metric (called the *discrete* or *trivial* metric) can also be thought of as a degenerate case of $d_p(\cdot, \cdot)$ defined above (for $p = 0$ if we allow the nonstandard definition $0^0 \overset{\Delta}{=} 0$) when $\mathcal{S} \subset \mathbb{R}^N$,

$$d_0(A, B) = \begin{cases} 0 & A = B \\ 1 & A \neq B. \end{cases}$$

However, this metric is much more general. In fact, this metric can be defined on *any* set. Clearly it satisfies the metric properties independent of the structure of $\mathbb{R}^N$.

THEOREM 5.1
Given metric spaces $(\mathcal{S}, d_1)$ and $(\mathcal{S}, d_2)$, the following are metric spaces: (a) $(\mathcal{S}, \alpha_1 d_1 + \alpha_2 d_2)$ for $\alpha_1, \alpha_2 \in \mathbb{R}^+$; (b) $(\mathcal{S}, \max(d_1, d_2))$ where $\max(\cdot, \cdot)$ takes the value of the greater of the two arguments; (c) $(\mathcal{S}, f(d_1))$ where $f(0) = 0$, $f'(0) > 0$, $f'(x) > 0$ and $f''(x) \leq 0$ for all $x \in \mathbb{R}^+$.

PROOF   Let $A, B, C \in \mathcal{S}$.
(a) Clearly the properties of scalar multiplication and addition cause positive definiteness, symmetry, and the triangle inequality to be preserved when $\alpha_1, \alpha_2 \in \mathbb{R}^+$.
(b) Again, positive definiteness and symmetry follow trivially. The triangle inequality follows because $d_1(A, C) \leq d_1(A, B) + d_1(B, C)$ and $d_2(A, C) \leq d_2(A, B) + d_2(B, C)$, and so

$$\max\left(d_1(A, C), d_2(A, C)\right) \leq \max\left(d_1(A, B) + d_1(B, C), d_2(A, B) + d_2(B, C)\right)$$
$$\leq \max\left(d_1(A, B), d_2(A, B)\right) + \max\left(d_1(B, C), d_2(B, C)\right).$$

(c) Symmetry and positive definiteness hold from the fact that d is a metric and $f(d(A, A)) = f(0) = 0$. The triangle inequality holds because

$$f(d(A, B)) + f(d(B, C)) \geq f(d(A, B) + d(B, C)) \geq f(d(A, C))$$

where the first inequality is due to the fact that $f''(x) \le 0$ and the second is due to the fact that $f'(x) > 0$ and $f(x) > 0$ for $x > 0$ [which follows from $f'(0) > 0$ and the other conditions imposed on $f(x)$ in the statement of the theorem]. ∎

In fact, $f(x)$ need not even be differentiable as long as it is positive and nondecreasing with nonincreasing slope. Concrete examples of acceptable functions $f(x)$ are $f(x) = \log(1 + x)$, $f(x) = x/(1 + x)$, $f(x) = \min(x, T)$ for some fixed threshold $T \in \mathbb{R}^+$, and $f(x) = x^r$ for $0 < r < 1$.

5.7.1 Metrics on Rotations Viewed as Points in $\mathbb{R}^4$

Using the description of rotations as unit vectors in $\mathbb{R}^4$, the vector 2-norm of the previous subsection can be used to define [21]

$$d(q, r) = \sqrt{(q_1 - r_1)^2 + (q_2 - r_2)^2 + (q_3 - r_3)^2 + (q_4 - r_4)^2}.$$

Since the length of a vector calculated using the 2-norm is invariant under rotations, it follows that for any unit quaternion s,

$$d(s\,q, s\,r) \overset{\triangle}{=} d(R(s)q, R(s)r) = d(q, r).$$

5.7.2 Metrics Resulting from Matrix Norms

When rotations are described using 3×3 matrices, an intuitive metric takes advantage of the matrix 2-norm

$$d(R_1, R_2) = \|R_1 - R_2\| \tag{5.67}$$

where

$$\|A\| = \sqrt{\text{trace}\left(AA^T\right)} = \left(\sum_{i,j=1}^{3} a_{ij}^2\right)^{\frac{1}{2}}.$$

This satisfies the general properties of a metric, and it is easy to see that

$$d(R_1 R, R_2 R) = d(R_1, R_2).$$

Because of the general property $\text{trace}(RAR^T) = \text{trace}(A)$ for any $A \in \mathbb{R}^{3\times3}$ and $R \in SO(3)$, it follows that

$$d(RR_1, RR_2) = d(R_1, R_2).$$

These properties are called *right* and *left* invariance, respectively. They are useful because they allow us to write the distance between orientations as a function of the relative rotation from one orientation to the other, without regard to the order

$$d(R_1, R_2) = d\left(\mathbb{I}, R_1^{-1} R_2\right) = d\left(\mathbb{I}, R_2^{-1} R_1\right).$$

This is similar to the result when the norm of the difference of two unit quaternions is considered.

By definition, we can write Eq. (5.67) in the different form

$$d(R_1, R_2) = \sqrt{tr\left[(R_1 - R_2)(R_1 - R_2)^T\right]} = \sqrt{6 - 2\text{trace}\left(R_1^T R_2\right)}$$

[or equivalently $d(R_1, R_2) = \sqrt{6 - 2\mathrm{trace}(R_2^T R_1)}$]. Since $\mathrm{trace}(R) = 2\cos\theta + 1$, where $R = e^{\theta V}$ for some $V \in so(3)$, we can write

$$d(R_1, R_2) = 2\sqrt{1 - \cos\theta_{1,2}}$$

where $|\theta_{1,2}| \le \pi$ is the relative angle of rotation.

5.7.3 Geometry-Based Metrics

The metrics discussed thus far are, in a sense, not natural because they do not take advantage of the geometric structure of the space of the rotations. For example, if the space of rotations is identified with the upper hemisphere of a unit sphere in $\mathbb{R}^4$, the above metrics measure the straight-line distance between points on the sphere, rather than along great arcs on its surface.

Using the unit-quaternion notation, the distance as measured on the unit hypersphere is given by the angle between the corresponding rotations described as unit vectors

$$d(q, r) = \cos^{-1}(q_1 r_1 + q_2 r_2 + q_3 r_3 + q_4 r_4).$$

When q is the north pole, this gives

$$d(1, r) = \theta/2$$

where θ is the angle of rotation.

Likewise, one can simply define the distance between orientations to be the angle of rotation from one to the other

$$d(R_1, R_2) = |\theta_{1,2}|,$$

where the angle of rotation is extracted from a rotation as

$$\theta(R) = \|\mathrm{vect}(\log(R))\|.$$

5.7.4 Metrics Based on Dynamics

In some situations, such as spacecraft attitude control,[7] the geometric definition of distance between orientations needs to be augmented so as to reflect the effort (power or fuel consumption) required to get to a final orientation from an initial one. In this case the inertial properties of the satellite should enter in the definition of distance. For instance, one can define distance between two orientations based on the minimum of the integral of kinetic energy required to move a rigid body from one orientation to another (where both orientations are static). This is written mathematically as

$$d(R_1, R_2) = \sqrt{\min_{R(t) \in SO(3)} \frac{1}{2} \int_{t_1}^{t_2} \mathrm{trace}\left(\dot{R}\mathcal{I}_R \dot{R}^T\right) dt} \tag{5.68}$$

subject to the boundary conditions $R(t_1) = R_1$ and $R(t_2) = R_2$. This is a variational calculus problem that can be handled using the methods described in Appendix F (and their generalizations). Problems like this arise in planning spacecraft attitude maneuvers [12]. The numerical calculation of $d(R_1, R_2)$ in Eq. (5.68) involves solving Euler's equations of motion together with the kinematic constraint relating angular velocity and the rotation matrix, subject to the boundary conditions stated previously. In practice, a numerical "shooting" technique can be used.

---

[7] See [11] for an introduction to this subject.

5.8 Integration and Convolution of Rotation-Dependent Functions

In many of the applications that follow, starting in Chapter 12, we will be considering functions that take rotations as their arguments and return real values. It is natural to ask how the standard concepts of differentiation and integration defined in the context of curvilinear coordinates are extended to the case of the set of rotations.

We present an informal discussion now, with a more formal treatment following in Chapter 7. To begin, recall that the Jacobian matrix relates the rate of change of the parameters used to describe the rotation to the angular velocity of a rigid body

$$\omega = J(R(\mathbf{q}))\dot{\mathbf{q}}.$$

Depending on whether body or spatial coordinates are used, the angular velocity and Jacobian would be subscribed with R or L, respectively. The angular velocity may be viewed as existing in the tangent to the set of rotations, analogous to the way a velocity vector of a particle moving on a surface moves instantaneously in the tangent plane to the surface. Analogous with the way area (or volume) is defined for a surface, volume for the set of rotations is defined relative to a metric tensor $G = J^T I_0 J$ for some positive definite I_0 by defining the volume element $\sqrt{\det(G)}\,dq_1 dq_2 dq_3$. Since J is 3×3, this reduces to a positive multiple of $|\det(J)|dq_1 dq_2 dq_3$. Hence we write the integral over the set of rotations as

$$\int_{SO(3)} f(R)d(R) = \frac{1}{V}\int_{\mathbf{q}\in Q} f(R(\mathbf{q}))|\det(J(R(\mathbf{q}))|dq_1 dq_2 dq_3.$$

In this equation it does not matter if J_R or J_L is used, since they are related by a rotation matrix, and their determinants are therefore the same. Q denotes the parameter space (which can be viewed as a subset of $\mathbb{R}^3$) and

$$V = \int_{\mathbf{q}\in Q} |\det(J(R(\mathbf{q}))|dq_1 dq_2 dq_3$$

is the volume as measured in this parameterization. The integral of a function on $SO(3)$ is normalized by the volume V so that the value of the integral does not depend on the parameterization or the choice of I_0, and

$$\int_{SO(3)} d(R) = 1.$$

As a concrete example, we have for the case of ZXZ Euler angles

$$\int_{SO(3)} f(R)d(R) = \frac{1}{8\pi^2}\int_0^{2\pi}\int_0^{\pi}\int_0^{2\pi} f(\alpha,\beta,\gamma)\sin\beta d\alpha d\beta d\gamma. \tag{5.69}$$

For the other 3-parameter descriptions we have

$$d(R) = \frac{1-\cos\|\mathbf{x}\|}{4\pi^2\|\mathbf{x}\|^2}dx_1 dx_2 dx_2 = \frac{dr_1 dr_2 dr_3}{\pi^2\left(1+\|\mathbf{r}\|^2\right)^2} \tag{5.70}$$

$$= \frac{1}{4\pi^2}s^2(\theta/2)svd\theta d\lambda dv = \frac{4d\sigma_1 d\sigma_2 d\sigma_3}{\pi^2\left(1+\|\sigma\|^2\right)^3}. \tag{5.71}$$

The bounds of integration for these variables are: $\mathbf{r}, \sigma \in \mathbb{R}^3$, $\mathbf{x}$ inside the ball of radius π in $\mathbb{R}^3$, and $-\pi \le \theta \le \pi, 0 \le v \le \pi, 0 \le \lambda \le 2\pi$.

The four-parameter descriptions will not be used in the context of integration in other chapters, but for completeness we write

$$d(R) = \frac{1}{8\pi^2}\delta\left(\|\mathbf{u}\|^2 - 1\right)du_1du_2du_3du_4$$

$$= \frac{1}{8\pi^2}(1 - \cos\theta)\delta\left(\|\mathbf{n}\|^2 - 1\right)dn_1dn_2dn_3d\theta$$

where $\delta(\cdot)$ is the Dirac delta function, and the bounds of integration are $\mathbf{u} \in \mathbb{R}^4$ and $(\mathbf{n}, \theta) \in \mathbb{R}^3 \times [0, 2\pi]$.

This definition of the integral of a function over the set of rotations has some useful properties. For instance, the integral of a function is invariant under "shifts" from the left and right by an arbitrary rotation. It is clear from the change of variables $A = R_1RR_2$ that

$$\int_{SO(3)} f(R_1RR_2)\,d(R) = \int_{SO(3)} f(A)d\left(R_1^{-1}AR_2^{-1}\right).$$

Using the fact that

$$\left|J\left(R_1^{-1}AR_2^{-1}\right)\right| = |J(A)|$$

one may write

$$\int_{SO(3)} f(R_1RR_2)\,d(R) = \int_{\mathbf{q}\in Q} f(A(\mathbf{q}))\left|\det\left(J\left(R_1^{-1}A(\mathbf{q})R_2\right)\right)\right|dq_1dq_2dq_3$$

$$= \int_{\mathbf{q}\in Q\subset\mathbb{R}^3} f(A(\mathbf{q}))|\det(J(A(\mathbf{q}))|dq_1dq_2dq_3 = \int_{SO(3)} f(A)d(A).$$

Given the ability to integrate functions on the set of rotations, a natural question to ask is if the concept of convolution also generalizes. The answer is most definitely yes, provided the types of functions one convolves are "nice" in the sense that

$$\int_{SO(3)} |f(R)|^2d(R) < \infty.$$

Analogous to the case of functions on the circle or line, the set of such square-integrable functions is denoted as $\mathcal{L}^2(SO(3))$. Any two functions $f_1, f_2 \in \mathcal{L}^2(SO(3))$ are convolved as

$$(f_1 * f_2)(R) = \int_{SO(3)} f_1(Q)f_2\left(Q^{-1}R\right)d(Q). \tag{5.72}$$

It is easy to see with the change of variables $Q \to RQ^{-1}$ that

$$(f_1 * f_2)(R) = \int_{SO(3)} f_1\left(RQ^{-1}\right) f_2(Q)d(Q).$$

While these two forms are always equal, it is rarely the case that the convolution of functions on $SO(3)$ commute. That is, in general

$$(f_1 * f_2)(R) \neq (f_2 * f_1)(R).$$

However, for certain special functions, convolutions will commute.

5.9 Summary

In this chapter we presented a variety of different ways to describe rotations in three-dimensional space. These methods include 3×3 special orthogonal matrices parameterized with three or four variables, 2×2 special unitary matrices, unit quaternions, and 4×4 special orthogonal matrices. The relationships between these descriptions of rotation were also examined. In addition we addressed topics such as how to define distance between two rotations, and how to integrate functions of rotation (or orientation) over all rotations.

References

[1] Angeles, J., *Rational Kinematics*, Springer-Verlag, New York, 1988.

[2] Bottema, O. and Roth, B., *Theoretical Kinematics*, Dover Publications, New York, reprint, 1990.

[3] Cayley, A., On the motion of rotation of a solid body, *Cambridge Math. J.*, 3, 224–232, 1843.

[4] Chirikjian, G.S., Closed-form primitives for generating volume preserving deformations, *ASME J. of Mech. Design*, 117, 347–354, 1995.

[5] Craig, J.J., *Introduction to Robotics, Mechanics and Control*, Addison-Wesley, Reading, MA, 1986.

[6] Euler, L., Du mouvement de rotation des corps solides autour d'un axe variable, *Mémoires de l'Académie des Sciences de Berlin*, 14, 154–193, 1758.

[7] Euler, L., Nova methodus motum corporum rigidorum determinandi, *Novii Comentarii Academiæ Scientiarum Petropolitanæ*, 20, 208–238, 1775/76.

[8] Goldstein, H., *Classical Mechanics, 2nd ed.*, Addison-Wesley, Reading, MA, 1980.

[9] Gürlebeck, K. and Sprössig, W., *Quaternionic and Clifford Calculus for Physicists and Engineers*, John Wiley & Sons, New York, 1997.

[10] Hamilton, Sir W.R., On a new species of imaginary quantities connected with a theory of quaternions, *Proc. of the Royal Irish Academy*, 2, 424–434, 1843.

[11] Hughes, P.C., *Spacecraft Attitude Dynamics*, John Wiley & Sons, New York, 1986.

[12] Junkins, J.L. and Turner, J.D., *Optimal Spacecraft Rotational Maneuvers*, Elsevier, New York, 1986.

[13] Kane, T.R. and Levinson, D.A., *Dynamics: Theory and Applications*, McGraw-Hall, New York, 1985.

[14] Li, C., McIntosh, A., and Qian, T., Clifford algebras, Fourier transforms, and singular convolution operators on Lipschitz surfaces, *Revista Matemática Iberoamericana*, 10, 665–721, 1994.

[15] Marandi, S.R. and Modi, V.J., A preferred coordinate system and the associated orientation representation in attitude dynamics, *Acta Astronautica,* 15, 833–843, 1987.

[16] McCarthy, J.M., *Introduction to Theoretical Kinematics,* MIT Press, 1990.

[17] Miller, W., Jr., *Lie Theory and Special Functions,* Academic Press, New York, 1968.

[18] Murray, R.M., Li, Z., and Sastry, S.S., *A Mathematical Introduction to Robotic Manipulation,* CRC Press, Boca Raton, FL, 1994.

[19] Park, F.C., *The Optimal Kinematic Design of Mechanisms,* Ph.D. thesis, Division of Engineering and Applied Sciences, Harvard University, Cambridge, MA, 1991.

[20] Park F.C. and Ravani, B., Smooth invariant interpolation of rotations, *ACM Trans. on Graphics,* 16(3), 277–295, 1997.

[21] Ravani, B. and Roth, B., Motion synthesis using kinematic mapping, *ASME J. of Mechanisms, Transmissions and Automation in Design,* 105, 460–467, 1983.

[22] Rodrigues, O., Des lois géométriques qui régissent les déplacements d'un système solide dans l'espace, et de la variation des coordonnées provenant de ces déplacements considérés independamment des causes qui peuvent les produire, *J. Mathématique Pures et Appliquées,* 5, 380–440, 1840.

[23] Rooney, J., A survey of representations of spatial rotation about a fixed point, *Environ. and Plann.,* B4, 185–210, 1977.

[24] Schaub, H., Tsiotras, P., and Junkins, J.L., Principal rotation representations of proper $N \times N$ orthogonal matrices, *Int. J. of Eng. Sci.,* 33(15), 2277–2295, 1995.

[25] Shuster, M.D., A survey of attitude representations, *J. of the Astronaut. Sci.,* 41(4), 439–517, 1993.

[26] Sommen, F., Plane waves, biregular functions and hypercomplex Fourier analysis, *Proc. of the 13th Winter School on Abstr. Anal.,* 1985.

[27] Stuelpnagel, J.H., On the parameterization of the three-dimensional rotation group, *SIAM Rev.,* 6, 422–430, 1964.

[28] Tsiotras, P. and Longuski, J.M., A new parameterization of the attitude kinematics, *The J. of the Astronaut. Sci.,* 43(3), 243–262, 1995.

Chapter 6

Rigid-Body Motion

Thus far we have considered spatial rotations, the subset of rigid-body motions which leave one point fixed. Clearly these are not the only rigid-body motions since a translation of the form $\mathbf{x}' = \mathbf{x} + \mathbf{b}$ preserves distance between points: $||\mathbf{x}'_1 - \mathbf{x}'_2||_p = ||\mathbf{x}_1 - \mathbf{x}_2||_p$. Note that unlike rotations, translations leave many more general measures of distance invariant than $d_2(\mathbf{x}, \mathbf{y}) = ||\mathbf{x} - \mathbf{y}||_2$, although $|| \cdot ||_2$ is the one of greatest interest.

The following statements address what comprises the complete set of rigid-body motions.

THEOREM 6.1 (Chasles [11])[1]

(1) Every motion of a rigid body can be considered as a translation in space and a rotation about a point; (2) every spatial displacement of a rigid body can be equivalently affected by a single rotation about an axis and translation along the same axis.

In modern notation, (1) is expressed by saying that every point $\mathbf{x}$ in a rigid body may be moved as

$$\mathbf{x}' = R\mathbf{x} + \mathbf{b} \tag{6.1}$$

where $R \in SO(3)$ is a rotation matrix, and $\mathbf{b} \in \mathbb{R}^3$ is a translation vector.

The pair $g = (R, \mathbf{b}) \in SO(3) \times \mathbb{R}^3$ describes both motion of a rigid body *and* the relationship between frames fixed in space and in the body. Furthermore, motions characterized by a pair $(R, \mathbf{b})$ could either describe the behavior of a rigid body or of a deformable object undergoing a rigid-body motion during the time interval for which this description is valid.

6.1 Composition of Motions

Consider a rigid-body motion which moves a frame originally coincident with the "natural" frame $(\mathbb{I}, \mathbf{0})$ to $(R_1, \mathbf{b}_1)$. Now consider a relative motion of the frame $(R_2, \mathbf{b}_2)$ with respect to the frame $(R_1, \mathbf{b}_1)$. That is, given any vector $\mathbf{x}$ defined in the terminal frame, it will look like $\mathbf{x}' = R_2\mathbf{x} + \mathbf{b}_2$ in the frame $(R_1, \mathbf{b}_1)$. Then the same vector will appear in the natural frame as

$$\mathbf{x}'' = R_1(R_2\mathbf{x} + \mathbf{b}_2) + \mathbf{b}_1 = R_1R_2\mathbf{x} + R_1\mathbf{b}_2 + \mathbf{b}_1.$$

---

[1] This was named after Michel Chasles (1793–1880).

The net effect of composing the two motions (or changes of reference frame) is equivalent to the definition

$$(R_3, \mathbf{b}_3) = (R_1, \mathbf{b}_1) \circ (R_2, \mathbf{b}_2) \stackrel{\triangle}{=} (R_1 R_2, R_1 \mathbf{b}_2 + \mathbf{b}_1). \tag{6.2}$$

From this expression, we can calculate the motion $(R_2, \mathbf{b}_2)$ that for any $(R_1, \mathbf{b}_1)$ will return the floating frame to the natural frame. All that is required is to solve $R_1 R_2 = \mathbb{I}$ and $R_1 \mathbf{b}_2 + \mathbf{b}_1 = \mathbf{0}$ for the variables R_2 and $\mathbf{b}_2$ and given R_1 and $\mathbf{b}_1$. The result is $R_2 = R_1^T$ and $\mathbf{b}_2 = -R_1^T \mathbf{b}_1$. Thus we denote the inverse of a motion as

$$(R, \mathbf{b})^{-1} = \left(R^T, -R^T \mathbf{b} \right). \tag{6.3}$$

This inverse, when composed either on the left or the right side of $(R, \mathbf{b})$, yields $(\mathbb{I}, \mathbf{0})$.

The set of all pairs $(R, \mathbf{b})$ together with the operation $\circ$ is denoted as $SE(3)$ for reasons that are explained in Chapter 7.

Note that every rigid-body motion [element of $SE(3)$] can be decomposed into a pure translation followed by a pure rotation as

$$(R, \mathbf{b}) = (\mathbb{I}, \mathbf{b}) \circ (R, \mathbf{0}),$$

and every translation *conjugated*[2] by a rotation yields a translation

$$(R, \mathbf{0}) \circ (\mathbb{I}, \mathbf{b}) \circ \left(R^T, \mathbf{0} \right) = (\mathbb{I}, R\mathbf{b}). \tag{6.4}$$

6.1.1 Homogeneous Transformation Matrices

It is of great convenience in many fields to *represent*[3] each rigid-body motion with a transformation matrix instead of a pair of the form $(R, \mathbf{b})$ and to use matrix multiplication in place of a composition rule.

This is achieved by assigning to each pair $(R, \mathbf{b})$ a unique 4×4 matrix

$$H(R, \mathbf{b}) = \begin{pmatrix} R & \mathbf{b} \\ \mathbf{0}^T & 1 \end{pmatrix}. \tag{6.5}$$

This is called a homogeneous transformation matrix, or simply a *homogeneous transform*. It is easy to see by the rules of matrix multiplication and the composition rule for rigid-body motions that

$$H \left((R_1, \mathbf{b}_1) \circ (R_2, \mathbf{b}_2) \right) = H \left(R_1, \mathbf{b}_1 \right) H \left(R_2, \mathbf{b}_2 \right).$$

Likewise, the inverse of a homogeneous transformation matrix represents the inverse of a motion

$$H \left((R, \mathbf{b})^{-1} \right) = [H(R, \mathbf{b})]^{-1}.$$

In this notation, vectors in $\mathbb{R}^3$ are augmented by appending a "1" to form a vector

$$\mathbf{X} = \begin{pmatrix} \mathbf{x} \\ 1 \end{pmatrix}.$$

---

[2]Conjugation of a motion $g = (R, \mathbf{x})$ by a motion $h = (Q, \mathbf{y})$ is defined as $h \circ g \circ h^{-1}$.

[3]The word "representation" will take on a very precise and important mathematical meaning in subsequent chapters. For now the usual meaning of the word will suffice.

The following are then equivalent expressions

$$\mathbf{Y} = H(R, \mathbf{b})\mathbf{X} \quad \leftrightarrow \quad \mathbf{y} = R\mathbf{x} + \mathbf{b}.$$

In the case of planar motion (viewed as a subset of spatial motion), we have

$$H\left(R\left(\mathbf{e}_3, \theta\right), \mathbf{b}\right) = \begin{pmatrix} \cos\theta & -\sin\theta & 0 & b_1 \\ \sin\theta & \cos\theta & 0 & b_2 \\ 0 & 0 & 1 & 0 \\ 0 & 0 & 0 & 1 \end{pmatrix}.$$

These are a lot of "0"s and "1"s to be carrying around, so it is common to reduce the transformation matrices for planar motion by removing the third row and column to yield

$$H\left(\theta, b_1, b_2\right) = \begin{pmatrix} \cos\theta & -\sin\theta & b_1 \\ \sin\theta & \cos\theta & b_2 \\ 0 & 0 & 1 \end{pmatrix}. \tag{6.6}$$

There is usually no confusion in distinguishing 3×3 and 4×4 transformation matrices, since the former is used exclusively in the case of planar motions, and the latter is used in the spatial case.

Analogous to rotations, one can express planar motions using the matrix exponential parameterization

$$\exp \begin{pmatrix} 0 & -\theta & v_1 \\ \theta & 0 & v_2 \\ 0 & 0 & 0 \end{pmatrix} = \begin{pmatrix} \cos\theta & -\sin\theta & [v_2(-1 + \cos\theta) + v_1\sin\theta]/\theta \\ \sin\theta & \cos\theta & [v_1(1 - \cos\theta) + v_2\sin\theta]/\theta \\ 0 & 0 & 1 \end{pmatrix}.$$

The logarithm of a planar motion is defined so that

$$\log \exp \begin{pmatrix} 0 & -\theta & v_1 \\ \theta & 0 & v_2 \\ 0 & 0 & 0 \end{pmatrix} = \begin{pmatrix} 0 & -\theta & v_1 \\ \theta & 0 & v_2 \\ 0 & 0 & 0 \end{pmatrix}.$$

As $\theta \to 0$, this expression is not singular from a mathematical perspective (since l'Hôpital's rule provides a clear limit), but care must be taken in numerical computations since there is a division by zero.

Alternative complex-number descriptions of planar motion exist as well, i.e., it is easy to see that the matrices

$$K_1\left(\theta, b_1, b_2\right) = \begin{pmatrix} e^{i\theta} & b_1 + ib_2 \\ 0 & 1 \end{pmatrix}$$

or

$$K_2\left(\theta, b_1, b_2\right) = \begin{pmatrix} e^{i\theta} & 0 & (b_2 - ib_1)/2 \\ 0 & e^{-i\theta} & (-b_2 - ib_1)/2 \\ 0 & 0 & 1 \end{pmatrix}$$

contain the same information as $H(\theta, b_1, b_2)$. It is easy to verify that the product of these matrices with matrices of the same kind results in matrices containing the information of the corresponding composed motions.

6.2   Screw Motions

The axis in the second part of Chasles' theorem is called the *screw axis*. It is a line in space about which a rotation is performed and along which a translation is performed.[4] As with any line in space, it is specified completely by a direction $\mathbf{n} \in S^2$ and the position of any point $\mathbf{p}$ on the line. Hence, a line is parameterized as

$$\mathbf{L}(t) = \mathbf{p} + t\mathbf{n}, \quad \forall \quad t \in \mathbb{R}.$$

Since there are an infinite number of vectors $\mathbf{p}$ on the line that can be chosen, the one which is "most natural" is that which has the smallest magnitude. This is the vector beginning at the origin of the coordinate system and terminating at the line to which it intersects orthogonally. Hence the condition $\mathbf{p} \cdot \mathbf{n} = 0$ is satisfied. Since $\mathbf{n}$ is a unit vector and $\mathbf{p}$ satisfies a constraint equation, a line is uniquely specified by only four parameters. Often, instead of the pair of line coordinates $(\mathbf{n}, \mathbf{p})$, the pair $(\mathbf{n}, \mathbf{p} \times \mathbf{n})$ is used to describe a line because this implicitly incorporates the constraint $\mathbf{p} \cdot \mathbf{n} = 0$. That is, when $\mathbf{p} \cdot \mathbf{n} = 0$, $\mathbf{p}$ can be reconstructed as $\mathbf{p} = \mathbf{n} \times (\mathbf{p} \times \mathbf{n})$, and it is clear that for unit $\mathbf{n}$, the pair $(\mathbf{n}, \mathbf{p} \times \mathbf{n})$ has four degrees of freedom. Such a description of lines is called the Plücker coordinates. For more on this subject, and kinematics in general, see [1, 6, 7, 8, 29, 30, 33, 41, 52, 55, 56, 57, 58, 59].

Given an arbitrary point $\mathbf{x}$ in a rigid body, the transformed position of the same point after translation by d units along a screw axis with direction specified by $\mathbf{n}$ is $\mathbf{x}' = \mathbf{x} + d\mathbf{n}$. Rotation about the same screw axis is given as $\mathbf{x}'' = \mathbf{p} + e^{\theta N}(\mathbf{x}' - \mathbf{p})$.

Since $e^{\theta N}\mathbf{n} = \mathbf{n}$, it does not matter if translation along a screw axis is performed before or after rotation. Either way, $\mathbf{x}'' = \mathbf{p} + e^{\theta N}(\mathbf{x} - \mathbf{p}) + d\mathbf{n}$. This is illustrated in Fig. 6.1.

Another way to view this is that the homogeneous transforms

$$\text{trans}(\mathbf{n}, d) = \begin{pmatrix} \mathbb{I} & d\mathbf{n} \\ \mathbf{0}^T & 1 \end{pmatrix}$$

and

$$\text{rot}(\mathbf{n}, \mathbf{p}, \theta) = \begin{pmatrix} e^{\theta N} & (\mathbb{I} - e^{\theta N})\mathbf{p} \\ \mathbf{0}^T & 1 \end{pmatrix}$$

commute, and the homogeneous transform for a general rigid-body motion along screw axis $(\mathbf{n}, \mathbf{p})$ is given as

$$\begin{aligned} \text{rot}(\mathbf{n}, \mathbf{p}, \theta)\text{trans}(\mathbf{n}, d) &= \text{trans}(\mathbf{n}, d)\text{rot}(\mathbf{n}, \mathbf{p}, \theta) \\ &= \begin{pmatrix} e^{\theta N} & (\mathbb{I} - e^{\theta N})\mathbf{p} + d\mathbf{n} \\ \mathbf{0}^T & 1 \end{pmatrix}. \end{aligned} \tag{6.7}$$

A natural question to ask at this point is how the screw axis parameters $(\mathbf{n}, \mathbf{p})$ and motion parameters (θ, d) can be extracted from a given rigid displacement $(R, \mathbf{b})$. Since we have already done this for pure rotations, half the problem is already solved, i.e., $\mathbf{n}$ and θ are calculated from $R = e^{\theta N}$ as described in Chapter 5. What remains is to find for given R, $\mathbf{n}$, and $\mathbf{b}$ the variables $\mathbf{p}$ and d satisfying

$$(\mathbb{I} - R)\mathbf{p} + d\mathbf{n} = \mathbf{b} \quad \text{and} \quad \mathbf{p} \cdot \mathbf{n} = 0.$$

This is achieved by first taking the dot product of the left of the above equations with $\mathbf{n}$ and observing that $\mathbf{n} \cdot e^{\theta N}\mathbf{p} = \mathbf{n} \cdot \mathbf{p} = 0$, and so

$$d = \mathbf{b} \cdot \mathbf{n}.$$

---

[4]The theory of screws was developed by Sir Robert Stawell Ball (1840–1913) [1].

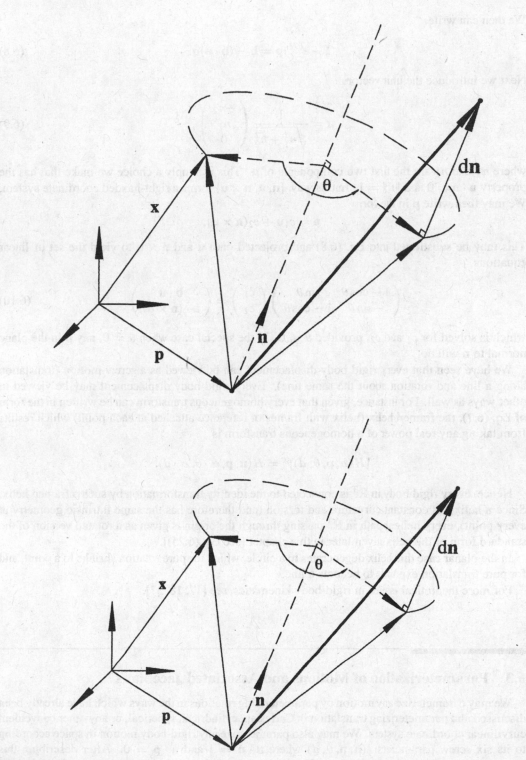

FIGURE 6.1
A general screw transformation.

We then can write

$$(\mathbb{I} - e^{\theta N})\mathbf{p} = \mathbf{b} - (\mathbf{b} \cdot \mathbf{n})\mathbf{n}. \tag{6.8}$$

Next we introduce the unit vector

$$\mathbf{u} = \frac{1}{\sqrt{n_1^2 + n_2^2}} \begin{pmatrix} -n_2 \\ n_1 \\ 0 \end{pmatrix}, \tag{6.9}$$

where n_1 and n_2 are the first two components of $\mathbf{n}$. This is simply a choice we make that has the property $\mathbf{u} \cdot \mathbf{n} = 0$ and $\|\mathbf{u}\| = 1$. In this way, $\{\mathbf{n}, \mathbf{u}, \mathbf{n} \times \mathbf{u}\}$ forms a right-handed coordinate system. We may then write $\mathbf{p}$ in the form

$$\mathbf{p} = c_1 \mathbf{u} + c_2(\mathbf{n} \times \mathbf{u}).$$

This may be substituted into Eq. (6.8) and projected onto $\mathbf{u}$ and $\mathbf{n} \times \mathbf{u}$ to yield the set of linear equations

$$\begin{pmatrix} 1 - \cos\theta & \sin\theta \\ -\sin\theta & 1 - \cos\theta \end{pmatrix} \begin{pmatrix} c_1 \\ c_2 \end{pmatrix} = \begin{pmatrix} \mathbf{b} \cdot \mathbf{u} \\ \mathbf{b} \cdot (\mathbf{n} \times \mathbf{u}) \end{pmatrix}, \tag{6.10}$$

which is solved for c_1 and c_2, provided $\theta \neq 0$. In the special case when $\theta = 0$, any $\mathbf{p}$ in the plane normal to $\mathbf{n}$ will do.

We have seen that every rigid body displacement can be viewed as a screw motion (translation along a line and rotation about the same line). Every rigid-body displacement may be viewed in other ways as well. For instance, given that every homogeneous transform can be written in the form of Eq. (6.7), the framed helix (helix with frames of reference attached at each point) which results from taking any real power of a homogeneous transform is

$$[H(\mathbf{n}, \mathbf{p}, \theta, d)]^\alpha = H(\mathbf{n}, \mathbf{p}, \alpha \cdot \theta, \alpha \cdot d).$$

Hence, every rigid body in $\mathbb{R}^3$ is connected to the identity transformation by such a framed helix. Since a helix has constant curvature and torsion (and therefore has the same intrinsic geometry at every point), every helical path in $\mathbb{R}^3$ passing through the origin is given as a rotated version of the standard form of the helix given later in this chapter [see Eq. (6.25)].

In the planar case this helix degenerates to a circle, which for pure rotation shrinks to a point, and for pure translation expands to become a line.

For more theoretical issues in rigid-body kinematics, see [17, 18, 37].

6.3   Parameterization of Motions and Associated Jacobians

We may parameterize any motion by parameterizing rotations in the ways which have already been discussed and parameterizing translation in Cartesian, cylindrical, spherical, or any other convenient curvilinear coordinate system. We may also parameterize any rigid-body motion in space according to its six screw parameters: $(\mathbf{n}, \mathbf{p}, \theta, d)$ where $\mathbf{n} \cdot \mathbf{n} = 1$ and $\mathbf{n} \cdot \mathbf{p} = 0$. After describing this parameterization, we examine the relationship between differential changes in parameters and the corresponding infinitesimal spatial motions. Two other ways of describing rigid body motions in space which do not use homogeneous transforms are addressed in the final sections of this chapter.

6.3.1 The Matrix Exponential

Analogous to the way skew-symmetric matrices are exponentiated to result in rotations, we may exponentiate scalar multiples of "screw matrices" of the form

$$\Sigma = \begin{pmatrix} N & h\mathbf{n} \\ \mathbf{0}^T & 0 \end{pmatrix}$$

to generate rigid body motions. Here $\mathbf{n} = \text{vect}(N)$ and $\mathbf{n} \cdot \mathbf{n} = 1$, and $h \in \mathbb{R}$ is a parameter called the *pitch* of the screw displacement. Analogous to a physical screw, the pitch of a screw displacement is the ratio of translation along the screw axis to the rotatation about it, i.e., $h = d/\theta$.

The idea of exponentiating screw matrices of a more genral form than those above to parameterize rigid-body motions was introduced into the engineering literature by Brockett [9]. It is easy to observe that for $m \geq 2$ [44]

$$\Sigma^m = \begin{pmatrix} N^m & \mathbf{0} \\ \mathbf{0}^T & 0 \end{pmatrix},$$

since $N_{\mathbf{n}} = \mathbf{0}$. Therefore, computing the exponential

$$e^{\theta\Sigma} = 1 + \theta\Sigma + \frac{1}{2!}\theta^2\Sigma^2 + \cdots$$

results in

$$e^{\theta\Sigma} = \begin{pmatrix} e^{\theta N} & h\theta\mathbf{n} \\ \mathbf{0}^T & 1 \end{pmatrix}. \tag{6.11}$$

While this certainly represents a rigid body motion, it does not parameterize all such motions. Analogous to rotations, where a general rotation $A(\mathbf{n}, \theta)$ is generated by similarity transformation of a rotation about e_3, the matrix exponential in Eq. (6.11) may be generalized to describe arbitrary rigid-body motions using a similarity transformation of the form

$$Be^{\theta\Sigma}B^{-1} = e^{\theta B\Sigma B^{-1}}.$$

Since the rotation part of the homogeneous transform in Eq. (6.11) is completely general, it is only the translational part which needs to be "fixed." This may be acheived by using the choice [44]

$$B = \begin{pmatrix} \mathbb{I} & \mathbf{n} \times \mathbf{r} \\ \mathbf{0}^T & 1 \end{pmatrix}.$$

The result is that any matrix

$$B(\theta\Sigma)B^{-1} = \begin{pmatrix} N & \mathbf{r} \\ \mathbf{0}^T & 0 \end{pmatrix},$$

when exponentiated yields [44]

$$B\left[\exp\theta\begin{pmatrix} N & h\mathbf{n} \\ \mathbf{0}^T & 0 \end{pmatrix}\right]B^{-1} = \exp\theta\begin{pmatrix} N & \mathbf{r} \\ \mathbf{0}^T & 0 \end{pmatrix}$$

$$= \begin{pmatrix} e^{\theta N} & (\mathbb{I} - e^{\theta N})(\mathbf{n} \times \mathbf{r}) + h\theta\mathbf{n} \\ \mathbf{0}^T & 1 \end{pmatrix}, \tag{6.12}$$

where the pitch is related to both $\mathbf{n}$ and $\mathbf{r}$ as $h = \mathbf{n}^T\mathbf{r}$.

It is possible to uniquely solve

$$\left(\mathbb{I} - e^{\theta N}\right)(\mathbf{n} \times \mathbf{r}) + h\theta\mathbf{n} = \mathbf{b} \tag{6.13}$$

for $\mathbf{r}$ and h for any given $\mathbf{b}$, θ, and $\mathbf{n}$ by performing a construction like in Section 6.2. In the current case, we write $\mathbf{r} = r_1\mathbf{n} + r_2\mathbf{u} + r_3(\mathbf{n} \times \mathbf{u})$ where $r_1 = h = \mathbf{b} \cdot \mathbf{n}/\theta$. Substitution into Eq. (6.13) and projection onto $\mathbf{u}$ and $\mathbf{n} \times \mathbf{u}$ yields

$$\begin{pmatrix} \sin\theta & \cos\theta - 1 \\ 1 - \cos\theta & \sin\theta \end{pmatrix} \begin{pmatrix} r_2 \\ r_3 \end{pmatrix} = \begin{pmatrix} \mathbf{b} \cdot \mathbf{u} \\ \mathbf{b} \cdot (\mathbf{n} \times \mathbf{u}) \end{pmatrix}, \tag{6.14}$$

provided $\theta \neq 0$. Note that the matrix in Eq. (6.14) is different than that in Eq. (6.10) due to the $\mathbf{n}\times$ term in Eq. (6.13) which is not present in Eq. (6.8). In the special case when $\theta \to 0$, the motion approaches being a pure translation, and $h \to \infty$. Letting $h = ||\mathbf{b}||/\theta$ and $\mathbf{r} = \mathbf{b}/||\mathbf{b}||$ solves the problem in this special case. If h is finite and $\theta \to 0$, the identity (null) motion results. Thus Eq. (6.12) is a parameterization of the entire set of rigid-body motions.

6.3.2 Infinitesimal Motions

For "small" motions the matrix exponential description is approximated well when truncated at the first two terms

$$\exp\left[\begin{pmatrix} \Omega & \mathbf{v} \\ \mathbf{0}^T & 0 \end{pmatrix} \Delta t\right] \approx \mathbb{I} + \begin{pmatrix} \Omega & \mathbf{v} \\ \mathbf{0}^T & 0 \end{pmatrix} \Delta t. \tag{6.15}$$

Here $\Omega = -\Omega^T$ and $\mathrm{vect}(\Omega) = \boldsymbol{\omega}$ describe the rotational part of the displacement. Since the second term in Eq. (6.15) consists mostly of zeros, it is common to extract the information necessary to describe the motion as

$$\begin{pmatrix} \Omega & \mathbf{v} \\ \mathbf{0}^T & 0 \end{pmatrix}^{\vee} = \begin{pmatrix} \boldsymbol{\omega} \\ \mathbf{v} \end{pmatrix}.$$

This six-dimensional vector is called an *infinitesimal* screw motion or *infinitesimal twist*.

Given a homogeneous transform

$$H(\mathbf{q}) = \begin{pmatrix} R(\mathbf{q}) & \mathbf{b}(\mathbf{q}) \\ \mathbf{0}^T & 1 \end{pmatrix}$$

parameterized with $(q_1, \ldots, q_6)$, which we write as a vector $\mathbf{q} \in \mathbb{R}^6$, one can express the homogeneous transform corresponding to a slightly changed set of parameters as the truncated Taylor series

$$H(\mathbf{q} + \Delta\mathbf{q}) = H(\mathbf{q}) + \sum_{i=1}^{6} \Delta q_i \frac{\partial H}{\partial q_i}(\mathbf{q}).$$

This result can be shifted to the identity transformation by multiplying on the right or left by H^{-1} to define an equivalent relative infinitesimal motion. In this case we write

$$\begin{pmatrix} \boldsymbol{\omega}_L \\ \mathbf{v}_L \end{pmatrix} = \mathcal{J}_L(\mathbf{q})\dot{\mathbf{q}} \quad \text{where} \quad \mathcal{J}_L(\mathbf{q}) = \left[\left(\frac{\partial H}{\partial q_1} H^{-1}\right)^{\vee}, \cdots, \left(\frac{\partial H}{\partial q_6} H^{-1}\right)^{\vee}\right] \tag{6.16}$$

where $\boldsymbol{\omega}_L$ is defined as before in the case of pure rotation and

$$\mathbf{v}_L = -\boldsymbol{\omega}_L \times \mathbf{b} + \dot{\mathbf{b}}. \tag{6.17}$$

Similarly,

$$\begin{pmatrix} \boldsymbol{\omega}_R \\ \mathbf{v}_R \end{pmatrix} = \mathcal{J}_R(\mathbf{q})\dot{\mathbf{q}} \quad \text{where} \quad \mathcal{J}_R(\mathbf{q}) = \left[\left(H^{-1}\frac{\partial H}{\partial q_1}\right)^{\vee}, \cdots, \left(H^{-1}\frac{\partial H}{\partial q_6}\right)^{\vee}\right]. \tag{6.18}$$

Here

$$\mathbf{v}_R = R^T \mathbf{b}.$$

The left and right Jacobian matrices are related as

$$\mathcal{J}_L(H) = [Ad(H)]\mathcal{J}_R(H) \tag{6.19}$$

where the matrix $[Ad(H)]$ is called the *adjoint*[5] and is written as [41, 44]

$$[Ad(H)] = \begin{pmatrix} R & 0 \\ BR & R \end{pmatrix}. \tag{6.20}$$

The matrix B is skew-symmetric, and vect$(B) = \mathbf{b}$.

Jacobians when Rotations and Translations are Parameterized Separately

When the rotations are parameterized as $R = R(q_1, q_2, q_3)$ and the translations are parameterized using Cartesian coordinates $\mathbf{b}(q_4, q_5, q_6) = [q_4, q_5, q_6]^T$, one finds that

$$\mathcal{J}_R = \begin{pmatrix} J_R & 0 \\ 0 & R^T \end{pmatrix} \quad \text{and} \quad \mathcal{J}_L = \begin{pmatrix} J_L & 0 \\ B J_L & \mathbb{I} \end{pmatrix}, \tag{6.21}$$

where J_L and J_R are the left and right Jacobians for the case of rotation.

6.4   Integration over Rigid-Body Motions

In general, the volume element with which to integrate over motions will be of the form

$$d(H) = |\det(\mathcal{J})| dq_1 \cdots dq_N$$

where $N = 3$ in the case of planar motion, and $N = 6$ in the case of spatial motion, and q_i are the parameters used to describe the motion. Here $\mathcal{J}$ is either $\mathcal{J}_L$ or $\mathcal{J}_R$; it does not matter which one is used. In both cases, $\det(\mathcal{J})$ reduces to a product of Jacobian determinants for the rotations and translations. When translations are described using Cartesian coordinates, $\det(\mathcal{J}) = \det(J)$ (the Jacobian determinant for rotations).

In particular, for the planar case we get within an arbitrary constant factor

$$d\left(H\left(x_1, x_2, \theta\right)\right) = \frac{1}{2\pi} dx_1 dx_2 d\theta,$$

which is essentially the same as the volume element for $\mathbb{R}^3$. It is trivial to show that this is left and right invariant by direct calculation.

For the spatial case, we see the invariance of the volume element as follows. Right invariance follows from the fact that for any constant homogeneous transform

$$H_0 = \begin{pmatrix} R_0 & \mathbf{b}_0 \\ \mathbf{0}^T & 1 \end{pmatrix},$$

---

[5]We will be seeing this in a more general context in Chapter 7.

$$\mathcal{J}_L(HH_0) = \left[\left(\frac{\partial H}{\partial q_1}H_0\,(HH_0)^{-1}\right)^{\vee} \cdots \left(\frac{\partial H}{\partial q_6}H_0\,(HH_0)^{-1}\right)^{\vee}\right].$$

Since $(HH_0)^{-1} = H_0^{-1}H^{-1}$, and $H_0 H_0^{-1} = \mathbb{I}$, we have that $\mathcal{J}_L(HH_0) = \mathcal{J}_L(H)$.

The left invariance follows from the fact that

$$\mathcal{J}_L(H_0 H) = \left[\left(H_0\frac{\partial H}{\partial q_1}H^{-1}H_0^{-1}\right)^{\vee} \cdots \left(H_0\frac{\partial H}{\partial q_6}H^{-1}H_0^{-1}\right)^{\vee}\right],$$

where

$$\left(H_0\frac{\partial H}{\partial q_i}H^{-1}H_0^{-1}\right)^{\vee} = [Ad\,(H_0)]\left(\frac{\partial H}{\partial q_i}H^{-1}\right)^{\vee}.$$

Therefore,

$$\mathcal{J}_L(H_0 H) = [Ad\,(H_0)]\,\mathcal{J}_L(H).$$

But since $\det(Ad(H_0)) = 1$,

$$\det\left(J_L(H_0 H)\right) = \det\left(J_L(H)\right).$$

Explicitly, the volume element in the case of $SE(3)$ is found when the rotation matrix is parameterized using ZXZ or ZYZ Euler angles (α, β, γ) as

$$d\left(H(x_1, x_2, x_3, \alpha, \beta, \gamma)\right) = \frac{1}{8\pi^2}\sin\beta d\alpha d\beta d\gamma dx_1 dx_2 dx_3,$$

which is the product of the volume elements for $\mathbb{R}^3$ ($dx = dx_1 dx_2 dx_3$), and for $SO(3)$ ($dR = \sin\beta d\alpha d\beta d\gamma$). Since $\beta \in [0, \pi]$, this is positive, except at the two points $\beta = 0$ and $\beta = \pi$, which constitute a set of zero measures and therefore does not contribute to the integral of singularity-free functions.

If the above derivation is too mathematical, there is a physically intuitive way to show the invariance of the integration measure. Starting with the definition

$$d(H) = d(R)d(\mathbf{x}),$$

the left shift by the constant $H_0 = (R_0, \mathbf{b}_0)$ gives

$$d(H_0 H) = d(R_0 R)\,d(R_0\mathbf{x} + \mathbf{b}_0) = d(R)d(R_0\mathbf{x} + \mathbf{b}_0) = d(R)d(\mathbf{x}). \qquad (6.22)$$

The first equality follows from the invariance of the $SO(3)$ volume element. The second equality follows from the invariance of integration of functions of spatial position under rigid-body motions. That is, whenever $\int_{\mathbb{R}^N} f(\mathbf{x})d\mathbf{x}$ exists, it is the case that

$$\int_{\mathbb{R}^N} f(R_0\mathbf{x} + \mathbf{b}_0)\,d\mathbf{x} = \int_{\mathbb{R}^N} f(\mathbf{x})d\mathbf{x}$$

for any $R_0 \in SO(3)$ and $\mathbf{b}_0 \in \mathbb{R}^3$ that do not depend on $\mathbf{x}$. Likewise, for right shifts,

$$d(HH_0) = d(RR_0)\,d(R\mathbf{b}_0 + \mathbf{x}) = d(R)d(R\mathbf{b}_0 + \mathbf{x}) = d(R)d(\mathbf{x}). \qquad (6.23)$$

The first equality again follows because of the invariance of the $SO(3)$ integration measure under shifts. And while $R\mathbf{b}_0$ is not constant, it is independent of $\mathbf{x}$, and so the same reasoning as before applies.

We note that unlike the case of the integration measure for the rotations, there is no unique way to scale the integration measure for rigid-body motions. This is due to the fact that translational motions can extend to infinity, and so it does not make sense to set the volume of the whole set of motions to unity as it did in the case of pure rotations. In the coming chapters, we will use different normalizations depending on the context so that equations take their simplest form.

Our motivation for studying the mathematical techniques reviewed in this section is that all of the problems stated in Chapter 1 involve convolutions of functions of rigid-body motion. The concept of convolution for rigid-body motions is exactly analogous to that discussed in the previous chapter for the case of rotations. For other applications of integration over rigid-body motions, see [47].

6.5 Assigning Frames to Curves and Serial Chains

In a variety of applications it is important to assign frames of reference (described using homogeneous transforms) to points along a curve or cascade of rigid links. Several methods for doing this, and their resulting properties, are explored in the subsections that follow.

6.5.1 The Frenet-Serret Apparatus

To every smooth curve $\mathbf{x}(s) \in \mathbb{R}^3$, it is possible to define two scalar functions, termed *curvature* and *torsion*, which completely specify the shape of the curve without regard to how it is positioned and oriented in space. That is, curvature and torsion are intrinsic properties of the curve, and are the same for $\mathbf{x}(s)$ and $\mathbf{x}'(s) = R\mathbf{x} + \mathbf{b}$ for any $R \in SO(3)$ and $\mathbf{b} \in \mathbb{R}^3$. The most natural curve parameterization is by arclength s. The *tangent* of a curve $\mathbf{x}(s)$ parameterized by arclength is

$$\mathbf{t}(s) = \frac{d\mathbf{x}}{ds}.$$

This is a unit vector satisfying $\mathbf{t}(s) \cdot \mathbf{t}(s) = 1$ because by the definition of arclength, the equality

$$s = \int_0^s \|\mathbf{t}(\sigma)\| d\sigma, \qquad (6.24)$$

can only hold *for all* values of $s \in \mathbb{R}$ if the integrand is equal to unity. By defining the *normal* vector as

$$\mathbf{n} = \frac{1}{\kappa(s)} \frac{d\mathbf{t}}{ds}$$

when $\kappa(s) = \|d\mathbf{t}/ds\| \neq 0$, one observes that $\frac{d}{ds}(\mathbf{t}(s) \cdot \mathbf{t}(s)) = \frac{d}{ds}(1) = 0$ implies $\mathbf{t}(s) \cdot \frac{d\mathbf{t}}{ds} = 0$ and so

$$\mathbf{t}(s) \cdot \mathbf{n}(s) = 0.$$

Thus the tangent and normal vectors define two orthonormal vectors in three-dimensional space at each s for which $\kappa(s) \neq 0$. A right-handed frame of reference may be defined in these cases by defining a third vector, termed the *binormal*, as

$$\mathbf{b}(s) = \mathbf{t}(s) \times \mathbf{n}(s).$$

The frames of reference given by the positions $\mathbf{x}(s)$ and orientations $[\mathbf{t}(s), \mathbf{n}(s), \mathbf{b}(s)] \in SO(3)$ for all values of s parameterizing the curve are called the *Frenet frames* attached to the curve.

The *torsion* of the curve is defined as

$$\tau(s) = -\frac{d\mathbf{b}(s)}{ds} \cdot \mathbf{n}(s)$$

and is a measure of how much the curve bends out of the $(\mathbf{t}, \mathbf{n})$- plane at each s.

A *right-circular helix* is a curve with constant curvature and torsion. Helices may be parameterized as

$$\mathbf{x}(s) = (r\cos as, r\sin as, has)^T \qquad (6.25)$$

where $a = (r^2 + h^2)^{-1/2}$, and r and h are constants.

The information contained in the collection of Frenet frames, the curvature, and the torsion, is termed the Frenet-Serret apparatus.[6] Since the curvature and torsion completely specify the intrinsic geometry of a curve, it should be no surprise that the way the Frenet frames change along the curve depends on these functions. In particular, it may be shown that

$$\frac{d}{ds}\begin{pmatrix} \mathbf{t}(s) \\ \mathbf{n}(s) \\ \mathbf{b}(s) \end{pmatrix} = \begin{pmatrix} 0 & \kappa(s) & 0 \\ -\kappa(s) & 0 & \tau(s) \\ 0 & -\tau(s) & 0 \end{pmatrix} \begin{pmatrix} \mathbf{t}(s) \\ \mathbf{n}(s) \\ \mathbf{b}(s) \end{pmatrix}. \qquad (6.26)$$

The vectors $\mathbf{t}$, $\mathbf{n}$, and $\mathbf{b}$ are treated like scalars when performing the matrix-vector multiplication on the right hand side of Eq. (6.26). This may be written in the different form

$$\frac{d}{ds}[\mathbf{t}(s), \mathbf{n}(s), \mathbf{b}(s)] = [\kappa(s)\mathbf{n}(s), -\kappa(s)\mathbf{t}(s) + \tau(s)\mathbf{b}(s), -\tau(s)\mathbf{n}(s)]$$
$$= -[\mathbf{t}(s), \mathbf{n}(s), \mathbf{b}(s)]\Lambda,$$

or

$$\frac{dQ_{FS}}{ds} = -Q_{FS}\Lambda$$

where Λ is the skew-symmetric matrix in Eq. (6.26) and $Q_{FS} = [\mathbf{t}(s), \mathbf{n}(s), \mathbf{b}(s)]$. The Frenet frame at each value of arclength is then $(Q_{FS}(s), \mathbf{x}(s))$.

6.5.2   Frames of Least Variation

The classical Frenet frames are only one of a number of techniques that can be used to assign frames of reference to space curves [2]. We now examine ways to assign frames which vary as little as possible along a curve segment. Two variants of this idea are presented here. First we examine how the Frenet frames should be "twisted" along the tangent to the curve for each value of arclength so as to result in a set of frames of least variation subject either to end constraints or no end constraints. Next, we examine how frames of reference should be distributed along the length of the curve in an optimal way. This constitutes a reparameterization of a space curve. Finally, we combine the optimal set of frames and reparameterization. The tools from variational calculus used throughout this section are reviewed in Appendix F.

Optimally Twisting Frames

In order to determine the twist about the tangent of the Frenet frames which yields a minimally varying set of frames with orientation $Q(s)$ with $Q(s)\mathbf{e}_1 = \mathbf{t}(s)$ for $s \in [0, 1]$, we seek an arclength-dependent angle, $\theta(s)$, such that

$$Q = ROT[\mathbf{t}, \theta]Q_{FS} = Q_{FS}ROT[\mathbf{e}_1, \theta]$$

---
[6]This was published independently by Frenet (1852) and Serret (1851).

minimizes the functional

$$J = \frac{1}{2} \int_0^1 tr\left(\frac{dQ}{ds}\frac{dQ^T}{ds}\right) ds. \tag{6.27}$$

Explicitly,

$$\frac{1}{2}tr\left(\frac{dQ}{ds}\frac{dQ^T}{ds}\right) = \kappa^2 + \tau^2 + 2\frac{d\theta}{ds}\mathbf{e}_1 \cdot \boldsymbol{\omega}_{FS} + \left(\frac{d\theta}{ds}\right)^2.$$

Here we have used the definition

$$\boldsymbol{\omega}_{FS} = \text{vect}(-\Lambda) = \begin{pmatrix} \tau \\ 0 \\ \kappa \end{pmatrix},$$

and the properties of the trace, including the fact that for skew-symmetric matrices, Ω_1 and Ω_2,

$$\frac{1}{2}tr(\Omega_1\Omega_2) = -\boldsymbol{\omega}_1 \cdot \boldsymbol{\omega}_2$$

where $\omega_i = \text{vect}(\Omega_i)$. Clearly,

$$\frac{1}{2}tr\left(\frac{dQ}{ds}\frac{dQ^T}{ds}\right) = \kappa^2 + \left(\tau + \frac{d\theta}{ds}\right)^2, \tag{6.28}$$

and the cost functional is minimized when

$$\frac{d\theta}{ds} = -\tau.$$

If, in addition, a specified twist relative to the Frenet frames is required at the end points, the optimal solution is obtained by substituting Eq. (6.28) into the Euler-Lagrange equations with θ as the generalized coordinate. The solution is then

$$\theta(s) = c_1 + c_2 s - \int_0^s \tau(\sigma)d\sigma$$

where c_1 and c_2 are determined by the end conditions $\theta(0) = \theta_0$ and $\theta(1) = \theta_1$.

Optimal Reparameterization for Least Variation

In the optimal reparameterization problem, we seek to replace the arclength, s, with a curve parameter t along a unit length of curve such that $s(t)$ satisfies $s(0) = 0$ and $s(1) = 1$ and the cost functional

$$J = \int_0^1 \left\{ \frac{1}{2}r^2 tr\left(\frac{dQ(s(t))}{dt}\frac{dQ^T(s(t))}{dt}\right) + \left(\frac{ds}{dt}\right)^2 \right\} dt$$

is minimized. Here r is a length constant introduced to define the trade off between the cost of bending and extending. The integrand in this problem is clearly of the form $g(s)(s')^2$ where $s' = ds/dt$ and

$$g(s) = \frac{1}{2}r^2 tr\left(\frac{dQ}{ds}\frac{dQ^T}{ds}\right) + 1,$$

and so the optimal solution to this reparemetrization is that given in Eq. (F.7) of Appendix F with $y = s$ and $x = t$.

An Alternative to the Intrinsic Approach

Instead of describing curve properties (and twist about a curve) based on the Frenet frames, it is often convenient to have a description that does not degenerate when $\kappa = 0$. One way to do this is to consider arclength-parameterized curves that evolve as

$$\mathbf{x}(s) = \int_0^s Q(\sigma)\mathbf{e}_3 d\sigma.$$

Here $Q(s) \in SO(3)$ is the orientation of the curve frame at s. When $Q(s) = R_{ZXZ}(\alpha(s), \beta(s), \gamma(s))$, then α and β describe the orientation of the unit tangent vector, and γ specifies the twist of the frame about the tangent. The least varying twist [in the sense of minimizing the cost functional Eq. (6.27)] subject to end constraints is of the form

$$\gamma(s) = c_1 + c_2 s - \int_0^s \cos \beta(\sigma)\alpha'(\sigma)d\sigma,$$

where α' is the derivative of α and c_1 and c_2 provide freedom to match the end constraints.

6.5.3   Global Properties of Closed Curves

We now present, without proof, some theorems relating the integrals of curvature and torsion of "nice" closed curves in $\mathbb{R}^3$ to global topological properties. The curves are all assumed to be smooth and self avoiding. The curve parameter, s, is taken to be arclength.

THEOREM 6.2 (Frenchel [26])

$$\oint \kappa(s)ds \geq 2\pi \tag{6.29}$$

with equality holding only for some kinds of planar ($\tau(s) = 0$) curves.

In contrast to the preceding theorem, one observes that for any closed smooth planar curve

$$\oint k(s)ds \cong 0 \bmod 2\pi$$

where $k(s)$ is the *signed curvature* of the curve, such that $|k(s)| = \kappa(s)$ with $k(s) > 0$ for counterclockwise bending and $k(s) < 0$ for clockwise bending.

Furthermore, any planar curve can be parameterized, up to a rigid-body displacement, as

$$\mathbf{x}(s) = \begin{pmatrix} \int_0^s \cos\left(\int_0^\sigma k(\nu)d\nu\right)d\sigma \\ \int_0^s \sin\left(\int_0^\sigma k(\nu)d\nu\right)d\sigma \end{pmatrix}. \tag{6.30}$$

THEOREM 6.3 (Fary-Milnor [23, 43])
For closed space curves forming a knot

$$\oint \kappa(s)ds \geq 4\pi. \tag{6.31}$$

THEOREM 6.4 (see [42, 53, 54])

If a closed curve is contained in the surface of the sphere S^2, then

$$\oint \tau(s)ds = \oint \frac{\tau(s)}{\kappa(s)}ds = 0. \tag{6.32}$$

Given two closed curves, $x_1(s)$ and $x_2(s)$, then the *Gauss integral* is a functional defined as

$$G(x_1, x_2) = \frac{1}{4\pi} \oint_{C_1} ds_1 \oint_{C_2} ds_2 \, [\dot{x}_1(s_1) \times \dot{x}_2(s_2)] \cdot \frac{x_1(s_1) - x_2(s_2)}{\|x_1(s_1) - x_2(s_2)\|^3} \tag{6.33}$$

where $\dot{}$ is shorthand for d/ds.

Gauss showed that this integral is a topological invariant in the sense that its value only depends on the degree to which the curves intertwine. It is also called the *linking number* of the two curves, and the notation $Lw = G(x_1, x_2)$ is common.

Suppose we are given a closed curve of unit length, $x(s)$. Then a *ribbon* (or strip) associated with this backbone curve is any smoothly evolving set of line segments of fixed length $2r$ for $0 \le s \le 1$ with centers at $x(s)$, such that the line segments point in a direction in the plane normal to $\dot{x}(s)$, and such that the tips of the line segments trace out closed curves. The tips of the ribbon can be described using the Frenet-Serret apparatus as the two curves

$$x_\pm(s) = s \pm rv(s)$$

where

$$v(s) = n(s)\cos\theta(s) + b(s)\sin\theta(s)$$

and $\theta(0) = \theta(1)$.

When r is sufficiently small, it is useful to represent the linking number into the sum of two quantities: the *writhe* (or writhing number), denoted as Wr, and the *twist* (or twisting number), denoted as Tw. It has been shown [10, 50, 49, 63] that the linking number of x and x_+ is

$$Lw(x, x + rv) = Wr(x) + Tw(x, v) \tag{6.34}$$

(or more simply $Lw = Wr + Tw$) where

$$Wr = \frac{1}{4\pi} \oint ds_1 \oint ds_2 \, [\dot{x}(s_1) \times \dot{x}(s_2)] \cdot \frac{x(s_1) - x(s_2)}{\|x(s_1) - x(s_2)\|^3}$$

and

$$Tw = \frac{1}{2\pi} \oint \dot{x}(s) \cdot [v(s) \times \dot{v}(s)]/\|\dot{x}(s)\| ds.$$

We note that in the case of an arc-length-parameterized curve $\|\dot{x}(s)\| = 1$. When $\theta(s) = 0$, one finds

$$Tw = \frac{1}{2\pi} \oint \tau(s)ds.$$

When $\theta(s) = \alpha(s)$ (the optimal twist distribution), then $Tw = 0$. For a simple (non-self-intersecting) closed curve in the plane or on the surface of a sphere, it has been shown [27] that $Wr = 0$.

For any fixed unit vector u not parallel to the tangent to the curve $x(s)$ for any value of s, one defines the *directional writhing number* [27] as

$$Wr(x, u) = Lk(x, x + \epsilon u)$$

where for sufficiently small ϵ, the value of the directional writhing number is independent of ϵ. The writhe can be calculated from the directional writhing number by integrating over all directions not parallel to the tangent of **x**. This amounts to integration over the sphere (except at the set of measure zero where the tangent traces out a curve on the surface of the sphere) and so [27]

$$Wr(\mathbf{x}) = \int_{S^2} Wr(\mathbf{x}, \mathbf{u}) d\mathbf{u}.$$

(Here the integral is normalized so that $\int_{S^2} d\mathbf{u} = 1$.)

We mention these relationships because they play an important role in the study of DNA topology [60]. Chapter 17 discusses statistical mechanics of macromolecules in some detail.

6.5.4 Frames Attached to Serial Linkages

It is interesting to note that while the Frenet-Serret apparatus has been known for over 150 years, methods for assigning frames of reference to chains with discrete links is a far newer problem. Two examples of this are the Denavit-Hartenberg framework in robotics, and the analogous formulation in polymer science and biophysics. We provide a short review of these formulations below. See references [15, 16], and [25] for detailed explanations as to how frames are uniquely attached to serial-chain robot arms and molecules.

The Denavit-Hartenberg Parameters in Robotics

The Denavit-Hartenberg (D-H) framework is a method for assigning frames of reference to a serial robot arm constructed of sequential rotary joints connected with rigid links. If the robot arm is imagined at any fixed time, the axes about which the joints turn are viewed as lines in space. In the most general case, these lines will be skew, and in degenerate cases they can be parallel or intersect.

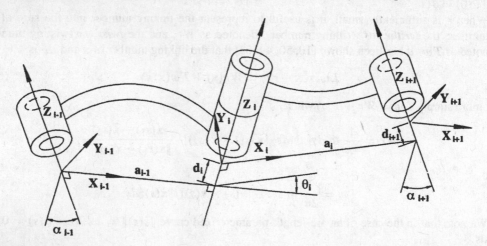

FIGURE 6.2
The Denavit-Hartenberg frames.

In the D-H framework (Fig. 6.2), a frame of reference is assigned to each link of the robot at the joint where it meets the previous link. The z-axis of the ith D-H frame points along the ith joint axis. Since a robot arm is usually attached to a base, there is no ambiguity in terms of which of the two ($\pm$) directions along the joint axis should be chosen, i.e., the "up" direction for the first joint is chosen. Since the $(i + 1)^{st}$ joint axis in space will generally be skew relative to axis i, a unique x-axis is assigned to frame i, by defining it to be the unit vector pointing in the direction of the shortest line segment from axis i to axis $i + 1$. This segment intersects both axes orthogonally. In addition to completely defining the relative orientation of the ith frame relative to the $(i-1)^{st}$, it also provides the relative position of the origin of this frame.

The D-H parameters, which completely specify this model, are:

- The distance from joint axis i to axis $i+1$ as measured along their mutual normal. This distance is denoted as a_i.

- The angle between the projection of joint axes i and $i+1$ in the plane of their common normal. The sense of this angle is measured counterclockwise around their mutual normal originating at axis i and terminating at axis $i+1$. This angle is denotes α_i.

- The distance between where the common normal of joint axes $i-1$ and i, and that of joint axes i and $i+1$ intersect joint axis i, as measured along joint axis i. This is denoted as d_i.

- The angle between the common normal of joint axes $i-1$ and i, and the common normal of joint axes i and $i+1$. This is denoted as θ_i, and has positive sense when rotation about axis i is counterclockwise.

Hence, given all the parameters $\{a_i, \alpha_i, d_i, \theta_i\}$ for all the links in the robot, together with how the base of the robot is situated in space, one can completely specify the geometry of the arm at any fixed time. Generally, θ_i is the only parameter that depends on time.

In order to solve the *forward kinematics* problem, which is to find the position and orientation of the distal end of the arm relative to the base, the homogeneous transformations of the relative displacements from one D-H frame to another are multiplied sequentially. This is written as

$$H_N^0 = H_1^0 H_2^1 \cdots H_N^{N-1}.$$

It is clear from Fig. 6.2 that the relative transformation, H_i^{i-1}, from frame $i-1$ to frame i is performed by first rotating about the x-axis of frame $i-1$ by α_{i-1}, then translating along this same axis by a_{i-1}. Next we rotate about the z-axis of frame i by θ_i and translate along the same axis by d_i. Since all these trasformations are relative, they are multiplied sequentially on the right as rotations (and translations) about (and along) natural basis vectors. Furthermore, since the rotations and translations appear as two screw motions (translations and rotations along the same axis), we write

$$H_i^{i-1} = \text{Screw}(e_1, a_{i-1}, \alpha_{i-1}) \, \text{Screw}(e_3, d_i, \theta_i),$$

where in this context

$$\text{Screw}(v, c, \gamma) = \text{rot}(v, 0, \gamma)\text{trans}(v, c).$$

Explicitly,

$$H_i^{i-1}(a_{i-1}, \alpha_{i-1}, d_i, \theta_i) =$$

$$\begin{pmatrix} \cos\theta_i & -\sin\theta_i & 0 & a_{i-1} \\ \sin\theta_i\cos\alpha_{i-1} & \cos\theta_i\cos\alpha_{i-1} & -\sin\alpha_{i-1} & -d_i\sin\alpha_{i-1} \\ \sin\theta_i\sin\alpha_{i-1} & \cos\theta_i\sin\alpha_{i-1} & \cos\alpha_{i-1} & d_i\cos\alpha_{i-1} \\ 0 & 0 & 0 & 1 \end{pmatrix}.$$

Polymer Kinematics

A polymer is a long chain of repeated chemical units, where the bonds between units can be modeled as a rotary joint. The $(i+1)^{st}$ joint axis intersects axis i at a distance l_i from where joint axis i is intersected by joint axis $i-1$. Polymer chains can be modeled using the D-H framework (as a degenerate case where the joint axes intersect). In polymer science, the joint (bond) axes are labeled as the x-axis, and the angle between joint axis i and $i + 1$ is denoted as θ_i. The z-axis is defined normal to the plane of joint axes i and $i + 1$, and the angle of rotation about bond i is denoted as ϕ_i. The corresponding homogeneous transformation matrix for one such convention is

$$H_{i+1}^i = \text{Screw}(\mathbf{e}_1, l_i, \phi_i)\text{Screw}(\mathbf{e}_3, 0, \theta_i).$$

Explicitly,

$$H_{i+1}^i (l_i, \phi_i, 0, \theta_i) = \begin{pmatrix} \cos\theta_i & -\sin\theta_i & 0 & l_i \\ \sin\theta_i \cos\phi_i & \cos\theta_i \cos\phi_i & -\sin\phi_i & 0 \\ \sin\theta_i \sin\phi_i & \cos\theta_i \sin\phi_i & \cos\phi_i & 0 \end{pmatrix}. \tag{6.35}$$

This assumes a convention for the sense of angles and axes consistent with the D-H framework outlined in the previous subsection. The preceding convention is the same (to within a shifting of axis labels by one place) as those presented in [45] and [21].

Other conventions for the signs of angles and directions of the axes in the context of polymers are explained in [25](p. 387), and [39](p. 112). In order to define angles and coordinates so as to be consistent with [25], one transforms Eq. (6.35) by setting $\phi_i \to \phi_i - \pi$ and performs a similarity transformation with respect to the matrix

$$S = \text{rot}(\mathbf{e}_1, \pi) = \begin{pmatrix} 1 & 0 & 0 & 0 \\ 0 & -1 & 0 & 0 \\ 0 & 0 & -1 & 0 \\ 0 & 0 & 0 & 1 \end{pmatrix}.$$

Therefore,

$$\mathcal{H}_i = SH_{i+1}^i (l_i, \phi_i - \pi, 0, \theta_i) S^{-1} = \begin{pmatrix} \cos\theta_i & \sin\theta_i & 0 & l_i \\ \sin\theta_i \cos\phi_i & -\cos\theta_i \cos\phi_i & \sin\phi_i & 0 \\ \sin\theta_i \sin\phi_i & -\cos\theta_i \sin\phi_i & -\cos\phi_i & 0 \end{pmatrix}.$$

The notation

$$\mathcal{H}_i = \begin{pmatrix} T_i & \mathbf{l}_i \\ \mathbf{0}^T & 1 \end{pmatrix}$$

is used in Chapter 17.

The angle ϕ_i is called the ith *torsion angle,* and the angle θ_i is related to the ith *bond angle* as $\pi - \theta_i$.

Several relationships exist between continuous curves and discrete chains with zero offset (such as polymer chains). Consider the planar N-link chain shown in Fig. 6.3 where the total length is normalized so that $\sum_{i=1}^N L_i = 1$.

The end position and orientation of this chain with respect to its base are given as

$$\mathbf{x}_{end} = \begin{pmatrix} \sum_{i=1}^N L_i \cos\left(\sum_{j=1}^i \theta_j\right) \\ \sum_{i=1}^N L_i \sin\left(\sum_{j=1}^i \theta_j\right) \end{pmatrix}$$

and

$$\theta_{end} = \sum_{i=1}^N \theta_i.$$

This example can be modeled using either the D-H or polymer frames as a degenerate case. What is perhaps less obvious is that it can also be described as the limiting case of a curve with signed

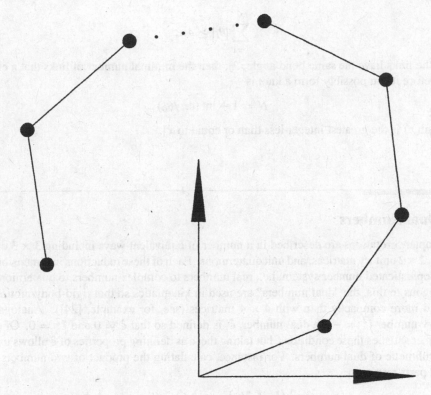

FIGURE 6.3
A planar N-link serial chain.

curvature

$$k(s) = \sum_{i=1}^{N} \theta_i \delta \left(s - \sum_{j=0}^{i-1} L_j \right)$$

where by definition $L_0 = 0$. It may be verified that substitution of this signed curvature into Eq. (6.30) results in $\mathbf{x}_{end} = \mathbf{x}(1)$ and $\theta_{end} = \int_0^1 k(s)ds$. From another perspective, as $N \to \infty$, and as $\theta_i \to 0$, the discrete linkage becomes a smooth curve with $\theta_i = (1/N) \cdot k(i/N)$. Both of these observed relationships between discrete chains and continuous curves have been useful in the kinematics and motion planning of snake-like ("hyper-redundant") manipulator arms [12].

In the case of a polymer, one expects a similar extension of the Frenet-Serret apparatus to be

$$\kappa(s) = \sum_{i=1}^{N} |\theta_i| \delta \left(s - \sum_{j=0}^{i-1} L_j \right),$$

and

$$|\tau(s)| = \sum_{i=1}^{N} |\phi_i| \delta \left(s - \sum_{j=0}^{i-1} L_j \right).$$

Hence, relationships such as the Fary-Milnor theorem should provide the necessary conditions for a discrete linkage to form a knot

$$\sum_{i=1}^{N} |\theta_i| \geq 4\pi,$$

or if all the links have the same bond angle, θ_0, then the minimal number of links that a chain must posess before it can possibly form a knot is

$$N = 1 + \text{int}\,(4\pi/\theta_0)$$

[where $\text{int}(x)$ is the greatest integer less than or equal to x].

6.6   Dual Numbers

In Chapter 5 rotations are described in a number of equivalent ways including 3×3 orthogonal matrices, 2×2 unitary matrices, and unit quaternions. Each of these reductions in dimension requires a more sophisticated number system, i.e., real numbers to complex numbers to quaternions.

Analogous to this, the "dual numbers" are used in kinematics so that rigid-body motions can be expressed more compactly than with 4×4 matrices (see, for example, [24]). Analogous to the imaginary number $i^2 = -1$, a dual number, $\hat{\epsilon}$, is defined so that $\hat{\epsilon} \neq 0$ and $\hat{\epsilon}^2 = 0$. Of course no real number satisfies these conditions, but taking these as defining properties of $\hat{\epsilon}$ allows us to define a new arithmetic of dual numbers. For instance, calculating the product of two numbers with real and dual parts results in

$$\left(a_1 + \hat{\epsilon} a_2\right)\left(b_1 + \hat{\epsilon} b_2\right) = a_1 b_1 + \hat{\epsilon}\left(b_1 a_2 + a_1 b_2\right).$$

The term $\hat{\epsilon}^2 (a_2 b_2)$ that would appear in usual arithmetic disappears by the definition of $\hat{\epsilon}$. Note that scalar multiplication using dual numbers is equivalent in some sense to multiplying 2×2 matrices with special structure

$$\begin{pmatrix} a_1 & a_2 \\ 0 & a_1 \end{pmatrix} \begin{pmatrix} b_1 & b_2 \\ 0 & b_1 \end{pmatrix} = \begin{pmatrix} a_1 b_1 & b_1 a_2 + a_1 b_2 \\ 0 & a_1 b_1 \end{pmatrix}.$$

One can "dualize" real, complex, or quaternion-valued vectors and matrices by adding another of the same kind of quantity multiplied by a dual number.

We may also discuss real-valued functions of a dual-number-valued argument. For instance, if $f(x)$ is an analytic function and hence has a Taylor series that exists and is convergent for all $x \in \mathbb{R}$, we can evaluate the same function at $\hat{x} = x_0 + \hat{\epsilon} x_1$ for $x_0, x_1 \in \mathbb{R}$ by using its Taylor series about x_0

$$
\begin{aligned}
f(\hat{x}) = f\left(x_0 + \hat{\epsilon} x_1\right) &= f(x_0) + \hat{\epsilon} x_1 f'(x_0) + \left(\hat{\epsilon} x_1\right)^2 f''(x_0)/2 + \cdots \\
&= f(x_0) + \hat{\epsilon} x_1 f'(x_0).
\end{aligned} \tag{6.36}
$$

The preceding equality holds because $\hat{\epsilon}^2 = 0$, and so all higher powers of $\hat{\epsilon}$ are also zero. As we will see later, an example of this with particular importance in kinematics is

$$\sin \hat{\theta} = \sin \theta + \hat{\epsilon} d \cos \theta \tag{6.37}$$
$$\cos \hat{\theta} = \cos \theta - \hat{\epsilon} d \sin \theta \tag{6.38}$$

where $\hat{\theta} = \theta + \hat{\epsilon} d$.

By using the concept of dual numbers, we can describe a rigid-body motion in space using a 3×3 matrix of real and dual numbers, a 2×2 matrix of complex and dual numbers, or a dual quaternion. Each of these different ways of describing motion in space is examined in the following subsections.

6.6.1 Dual Orthogonal Matrices

A dual orthogonal matrix is constructed to contain all the information concerning a rigid-body displacement in a 3×3 matrix instead of a homogeneous transform. In particular, given a rigid-body displacement $(R, \mathbf{b})$, the corresponding dual orthogonal matrix is

$$\hat{R} = R + \hat{\epsilon} BR$$

where B is defined to be the skew-symmetric matrix satisfying $B\mathbf{x} = \mathbf{b} \times \mathbf{x}$ for any $\mathbf{x} \in \mathbb{R}^3$. It is clear from the usual rules of matrix multiplication, the fact that $B^T = -B$, and the properties of dual-number arithmetic, that

$$\hat{R}\hat{R}^T = (R + \hat{\epsilon} BR)\left(R^T - \hat{\epsilon} R^T B\right) = \mathbb{I}.$$

Furthermore, the product of dual orthogonal matrices results in a dual orthogonal matrix as

$$\hat{R}_1 \hat{R}_2 = R_1 R_2 + \hat{\epsilon}\left(R_1 B_2 R_1^T + B_1\right) R_1 R_2.$$

Since $R_1 B_2 R_1^T$ is the skew-symmetric matrix corresponding to the vector $R_1 \mathbf{b}_2$, we see that the skew-symmetric matrix $R_1 B_2 R_1^T + B_1$ corresponds to the translation $R_1 \mathbf{b}_2 + \mathbf{b}_1$ resulting from concatenated rigid motions. Hence the product of dual orthogonal matrices captures the property of composition of motions.

6.6.2 Dual Unitary Matrices

If one desires to express motions using 2×2 matrices with entries that are both complex and dual numbers, this can be acheived by using dual unitary matrices defined as

$$\hat{U} = U + \frac{1}{2}\hat{\epsilon} BU$$

where

$$B = \begin{pmatrix} -ib_2 & -b_1 - ib_3 \\ b_1 - ib_3 & ib_2 \end{pmatrix}$$

is a matrix containing all the translational information, and $U \in SU(2)$ contains all of the rotational information. Composition of motions is achieved by the multiplication $\hat{U}_1 \hat{U}_2$.

6.6.3 Dual Quaternions

Starting with 4×4 homogeneous transform matrices to represent rigid-body displacements, we reduced the dimensions to 3×3 by introducing dual numbers, and to 2×2 by, in addition, introducing complex numbers. It should therefore be no surprise that the lowest dimensional representation of rigid-body motion uses the most complicated combination of elements — dual numbers and quaternions.

The dual unit quaternions are written as

$$\hat{u} = u + \frac{1}{2}\hat{\epsilon} bu$$

where $b = b_1 \tilde{i} + b_2 \tilde{j} + b_3 \tilde{k}$ is the vector quaternion representing translation vector $\mathbf{b}$ in the displacement $(R, \mathbf{b})$.

As with the case of pure rotation, where there was a one-to-one correspondence between elements of $SU(2)$ and unit quaternions, there is also a one-to-one correspondence between dual special unitary matrices and dual unit quaternions. It follows then that

$$\hat{u}\hat{u}^* = 1.$$

6.7   Approximating Rigid-Body Motions in $\mathbb{R}^N$ as Rotations in $\mathbb{R}^{N+1}$

In many ways it is more convenient to deal with pure rotations than the combination of rotations and translations which comprise rigid-body motions. The set of rotations have a well-defined measure of distance that is invariant under shifts from the left and right, and the set of rotations does not extend to infinity in any direction. Neither of these are characteristics of the set of rigid-body motions.

This fact has led some to investigate approximating rigid-body motions with pure rotations. To our knowledge, this concept was introduced into the engineering literature by McCarthy [40]. More generally, the concept of "contraction" of a set of transformations of infinite extent to one of finite extent has been known for many years in the physics literature (see, for example, [31, 64]).

Methods for performing this contraction are geometrically intuitive. We have already seen in previous chapters how stereographic projection can be used to map points on the sphere to points in the plane, and vice versa. One way to approximate a rigid-body motion with a pure rotation is to examine how an arbitrary moving point in the plane is mapped to the sphere. The rotation that approximates the motion is then calculated as that which rotates the image of the projected points on the sphere. Given a homogeneous transform, H, another approach is to calculate the closest rotation using the polar decomposition as $R = H(H^T H)^{-\frac{1}{2}}$. Both of these methods for contracting the set of rigid-body motions to the set of rotations are examined in detail for the planar and spatial cases in the following subsections.

6.7.1   Planar Motions

The set of planar rigid-body motions is parameterized with three variables (x_1, x_2, θ). We seek an explicit mapping between planar motions and three-dimensional rotations.

McCarthy's Approach

Any rigid-body transformation can be decomposed into the product of a translation followed by a rotation about the x_3-axis. Any rotation in space can be described by first rotating about the x_2 axis by an angle ϵ, then about the x_1 axis by $-\nu$, followed by a rotation about the x_3 axis by θ. These two scenarios are written as

$$R = R_2(\epsilon) R_1(-\nu) R_3(\theta)$$

and

$$H = \begin{pmatrix} 1 & 0 & b_1/L \\ 0 & 1 & b_2/L \\ 0 & 0 & 1 \end{pmatrix} \begin{pmatrix} \cos\theta & -\sin\theta & 0 \\ \sin\theta & \cos\theta & 0 \\ 0 & 0 & 1 \end{pmatrix},$$

where L is the measure of length used to normalize the translations.

The approximation

$$R \approx H$$

is then valid with $\mathcal{O}(1/L^2)$ error when the plane of the rigid motion is taken to be the one that kisses the sphere at the north pole when one sets

$$\tan \epsilon = b_1/L \qquad \text{and} \qquad \tan \nu = b_2/L.$$

This idea is presented in [40].

We note that rigid-body motions in the plane can also be approximated as pure rotations in $\mathbb{R}^3$ using stereographic projection or the polar decomposition.

6.7.2 Spatial Motions

The set of spatial rigid-body motions is parameterized with six variables (three for translation and three for rotation). We seek an explicit mapping between spatial motions and rotations in four dimensions. Since, in general, a rotation in $\mathbb{R}^N$ is described with $N(N-1)/2$ parameters, it should be no surprise that rotations in $\mathbb{R}^4$ require the same number of parameters as rigid-body motion in three-dimensional space.

Before going into the details of the mapping between $SE(3)$ and $SO(4)$ (the set of rotations in $\mathbb{R}^4$), we first review some important features of $SO(4)$.

Rotations in $\mathbb{R}^4$

Rotations in four-space can be parameterized in ways very similar to those reviewed in Chapter 5. For instance, any $SO(4)$-rotation can be described as the exponential of a 4×4 skew-symmetric matrix [28, 20]

$$M(\mathbf{z}, \mathbf{w}) = \begin{pmatrix} 0 & -z_3 & z_2 & w_1 \\ z_3 & 0 & -z_1 & w_2 \\ -z_2 & z_1 & 0 & w_3 \\ -w_1 & -w_2 & -w_3 & 0 \end{pmatrix}, \tag{6.39}$$

$$R = \exp M(\mathbf{z}, \mathbf{w}). \tag{6.40}$$

Following [28], two new matrices M^+ and M^- are formed as

$$M^{\pm} = (M(\mathbf{z}, \mathbf{w}) \pm M(\mathbf{w}, \mathbf{z}))/2.$$

These matrices have the form

$$M^+(\mathbf{s}) = \begin{pmatrix} 0 & -s_3 & s_2 & s_1 \\ s_3 & 0 & -s_1 & s_2 \\ -s_2 & s_1 & 0 & s_3 \\ -s_1 & -s_2 & -s_3 & 0 \end{pmatrix} = \|\mathbf{s}\| M^+(\hat{\mathbf{s}}).$$

and

$$M^-(\mathbf{s}) = \begin{pmatrix} 0 & -s_3 & s_2 & -s_1 \\ s_3 & 0 & -s_1 & -s_2 \\ -s_2 & s_1 & 0 & -s_3 \\ s_1 & s_2 & s_3 & 0 \end{pmatrix} = \|\mathbf{s}\| M^-(\hat{\mathbf{s}}).$$

where $\mathbf{s} = \|\mathbf{s}\| \hat{\mathbf{s}}$ and $\hat{\mathbf{s}} \in S^2$.

Vectors $\mathbf{s}_{\pm}$ are defined such that

$$M(\mathbf{z}, \mathbf{w}) = M^+(\mathbf{s}_+) + M^-(\mathbf{s}_-).$$

The reason for performing this decomposition is that

$$M^+(\mathbf{s}) M^-(\mathbf{t}) = M^-(\mathbf{s}) M^+(\mathbf{t})$$

for any $\mathbf{s}, \mathbf{t} \in \mathbb{R}^3$. Since, in general, the exponential of commuting matrices can be written as the product of each matrix exponentiated separately, one can rewrite Eq. (6.40) as

$$R = \exp M^+(\mathbf{s}_+) \exp M^-(\mathbf{s}_-).$$

This is convenient because each of these simpler matrices has the closed form

$$\exp M^{\pm}(\mathbf{s}) = \cos \|\mathbf{s}\| \mathbb{I}_{4\times 4} + \sin \|\mathbf{s}\| M^{\pm}(\hat{\mathbf{s}}). \tag{6.41}$$

It has been observed that $\exp M^+(\mathbf{s})$ and $\exp M^-(\mathbf{s})$ each correspond to unit quaternions describing rotation in three dimensions, with axis of rotation $\mathbf{s}$, and angle $\pm 2\|\mathbf{s}\|$, with sign corresponding to $M^{\pm}$. Hence the name *biquaternions* [14] is sometimes used.

This becomes particularly clear when the basis $i_{4\times 4} = \tilde{k}_{4\times 4}$, $j_{4\times 4} = -\tilde{j}_{4\times 4}$, $k_{4\times 4} = \tilde{i}_{4\times 4}$ is used. Then given a point $\mathbf{x} \in \mathbb{R}^3$, three dimensional rotations are implemented as

$$M^+\left(\mathbf{x}'\right) = \exp M^+(\mathbf{s}) M^+(\mathbf{x}) \exp M^+(-\mathbf{s}).$$

An interesting and well-known result is that by defining the matrix

$$X = x_4 \mathbb{I}_{4\times 4} + M^+(\mathbf{x}),$$

rotations in four dimensions are achieved as

$$X' = \exp M^+(\mathbf{s}) X \exp M^-(\mathbf{t}). \tag{6.42}$$

The fact that this is a rotation is easy to verify because

$$\det(X) = \left(x_1^2 + x_2^2 + x_3^2 + x_4^2\right)^2$$

and

$$\mathrm{tr}(XX^T) = 4\left(x_1^2 + x_2^2 + x_3^2 + x_4^2\right)$$

are preserved under the transformation.

One may also show that the transformation of the vector $\tilde{\mathbf{x}} = [x_1, x_2, x_3, x_4]^T$ is performed as the product [19]

$$\tilde{\mathbf{x}}' = \exp M^+(\mathbf{s}) \exp M^-(\mathbf{t}) \tilde{\mathbf{x}}, \tag{6.43}$$

and so it is clear that this transformation is both orientation and length preserving, and hence must be a rotation. Equation (6.43) follows from the relationship

$$X' = \exp M^{\pm}(\mathbf{s}) X \exp M^{\pm}(-\mathbf{s}) \leftrightarrow \tilde{\mathbf{x}}' = \exp M^{\pm}(\mathbf{s}) \tilde{\mathbf{x}},$$

and the commutivity of $M^+(\mathbf{s})$ and $M^-(\mathbf{t})$.

If instead of 4×4 real rotation matrices one prefers 2×2 special unitary matrices or unit quaternions, the equations analogous to Eqs. (6.41) and (6.42) follow naturally with the elements of $SU(2)$ or the unit quaternions corresponding to $M^{\pm}$ and

$$X = \begin{pmatrix} x_4 - ix_2 & -x_1 - ix_3 \\ x_1 - ix_3 & x_4 + ix_2 \end{pmatrix}$$

or $x = x_1 i + x_2 j + x_3 k + x_4$.

Since both of the pairs $(\exp M^+(\mathbf{s}), \exp M^-(\mathbf{t}))$ and $(-\exp M^+(\mathbf{s}), -\exp M^-(\mathbf{t}))$ describe the same rotation in $\mathbb{R}^4$, one says that the biquaternions and $SU(2) \times SU(2)$ [which is notation for two independent copies of $SU(2)$] both have two-to-one mappings to $SO(4)$. One calls the biquaternions and $SU(2) \times SU(2)$ the double covers of $SO(4)$.

Contraction of $SE(3)$ to $SO(4)$

We now consider the approximation of rigid-body motions in three space with rotations in four dimensions. The same three approaches presented in Section 6.7 are applicable here.

McCarthy's Approach

To begin, McCarthy's approach using Euler angles requires us to extend the concept of Euler angles to four dimensions. In four dimensions one no longer talks of rotations "about an axis k," but rather of rotations "in the plane of axes i and j." We denote such rotations as $R_{ij}(\theta)$, with the convention that $R_{ij}(\pi/2)$ is the rotation that takes axis i into axis j, leaving the other two axes unchanged. It then follows that

$$R_{12}(\theta) = \begin{pmatrix} \cos\theta & -\sin\theta & 0 & 0 \\ \sin\theta & \cos\theta & 0 & 0 \\ 0 & 0 & 1 & 0 \\ 0 & 0 & 0 & 1 \end{pmatrix} \qquad R_{31}(\theta) = \begin{pmatrix} \cos\theta & 0 & \sin\theta & 0 \\ 0 & 1 & 0 & 0 \\ -\sin\theta & 0 & \cos\theta & 0 \\ 0 & 0 & 0 & 1 \end{pmatrix}$$

$$R_{23}(\theta) = \begin{pmatrix} 1 & 0 & 0 & 0 \\ 0 & \cos\theta & -\sin\theta & 0 \\ 0 & \sin\theta & \cos\theta & 0 \\ 0 & 0 & 0 & 1 \end{pmatrix} \qquad R_{42}(\theta) = \begin{pmatrix} 1 & 0 & 0 & 0 \\ 0 & \cos\theta & 0 & \sin\theta \\ 0 & 0 & 1 & 0 \\ 0 & -\sin\theta & 0 & \cos\theta \end{pmatrix}$$

$$R_{34}(\theta) = \begin{pmatrix} 1 & 0 & 0 & 0 \\ 0 & 1 & 0 & 0 \\ 0 & 0 & \cos\theta & -\sin\theta \\ 0 & 0 & \sin\theta & \cos\theta \end{pmatrix} \qquad R_{14}(\theta) = \begin{pmatrix} \cos\theta & 0 & 0 & -\sin\theta \\ 0 & 1 & 0 & 0 \\ 0 & 0 & 1 & 0 \\ \sin\theta & 0 & 0 & \cos\theta \end{pmatrix}.$$

One of the many ways rotations in $\mathbb{R}^4$ can be parameterized using Euler angles is as follows [20]. Start with any of the usual Euler angles, e.g.,

$$K(\alpha, \beta, \gamma) = R_{12}(\alpha) R_{23}(\beta) R_{12}(\gamma) = \begin{pmatrix} A(\alpha, \beta, \gamma) & \mathbf{0} \\ \mathbf{0}^T & 1 \end{pmatrix}.$$

This particular choice is a 3×3 rotation matrix corresponding to ZXZ Euler angles embedded in a 4×4 matrix.

Recall that any rigid-body motion with translation measured in units of L can be decomposed into a translation followed by a rotation, as

$$H = \begin{pmatrix} \mathbb{I}_{3\times 3} & \mathbf{b}/L \\ \mathbf{0}^T & 1 \end{pmatrix} \begin{pmatrix} A(\alpha, \beta, \gamma) & \mathbf{0} \\ \mathbf{0}^T & 1 \end{pmatrix}.$$

We therefore seek a second 4×4 rotation matrix, J, to approximate the translation. so that

$$R = J(\epsilon, \nu, \eta) K(\alpha, \beta, \gamma) \approx H.$$

One possible choice is

$$J(\epsilon, \nu, \eta) = R_{43}(\eta) R_{42}(\nu) R_{41}(\epsilon) = \begin{pmatrix} c\epsilon & 0 & 0 & s\epsilon \\ -s\nu s\epsilon & c\nu & 0 & s\nu c\epsilon \\ -s\eta c\nu s\epsilon & -s\eta s\nu & c\eta & s\eta c\nu c\epsilon \\ -c\eta c\nu s\epsilon & -s\nu c\eta & -s\eta & c\eta c\nu c\epsilon \end{pmatrix}.$$

In the limit as $1/L \to 0$, the approximation $H \approx R$ has $\mathcal{O}(1/L^2)$ error with

$$\tan\epsilon = b_1/L \qquad \tan\nu = b_2/L \qquad \tan\eta = b_3/L.$$

We note that rigid-body motions in three-dimensional space can also be approximated as pure rotations in $\mathbb{R}^4$ using stereographic projection or the polar decomposition of a homogeneous transform.

6.8   Metrics on Motion

In this section we present two general methods for generating metrics on the set of rigid-body motions. The basic idea behind the two classes of metrics presented here is to use metrics on $\mathbb{R}^N$ and on function spaces on the set of motions to induce metrics on the set of motions itself. The formulation in this section follows those in [13, 22, 34, 38]. Applications of metrics on motion are discussed in [48, 51, 65].

6.8.1   Metrics on Motion Induced by Metrics on $\mathbb{R}^N$

Perhaps the most straightforward way to define metrics on the set of motions is to take advantage of the well-known metrics on $\mathbb{R}^N$. Namely, if $\rho(\mathbf{x})$ is a continuous real-valued function on $\mathbb{R}^N$ which satisfies the properties

$$0 \le \rho(\mathbf{x}) \quad \text{and} \quad 0 < \int_{\mathbb{R}^N} \|\mathbf{x}\|^m \rho(\mathbf{x}) dx_1 \dots dx_N < \infty,$$

for any $m \ge 0$, then

$$d(g_1, g_2) = \int_{\mathbb{R}^N} \|g_1 \circ \mathbf{x} - g_2 \circ \mathbf{x}\| \rho(\mathbf{x}) dx_1 \dots dx_N$$

is a metric when $\| \cdot \|$ is any norm for vectors in $\mathbb{R}^N$ (in particular, the p-norm is denoted $\| \cdot \|_p$). Here $g = (A, \mathbf{b})$ is an affine transformation with corresponding homogeneous transform $H(A, \mathbf{b})$. The fact that this is a metric was observed in [38] for the case of rigid-body motion, i.e., when $A \in SO(N)$. We use the notation $g \circ \mathbf{x} = A\mathbf{x} + \mathbf{b}$.

Natural choices for $\rho(\mathbf{x})$ are either the mass density of the object, or a function which is equal to one on the object and zero otherwise. However, it is also possible to choose a function such as $\rho(\mathbf{x}) = e^{-a^2 \mathbf{x} \cdot \mathbf{x}}$ (for any $a \in \mathbb{R}^+$) which is positive everywhere yet decreases rapidly enough for $d(g_1, g_2)$ to be finite.

The fact that this is a metric on the set of rigid-body transformations is observed as follows. The symmetry property $d(g_1, g_2) = d(g_2, g_1)$ results from the symmetry of vector addition and the properties of vector norms. The triangle inequality also follows from properties of norms. Positive definiteness of this metric follows from the fact that for rigid-body transformations

$$\|g_1 \circ \mathbf{x} - g_2 \circ \mathbf{x}\| = \|(A_1 - A_2)\mathbf{x} + (\mathbf{b}_1 - \mathbf{b}_2)\|,$$

and because of the positive definiteness of $\| \cdot \|$, the only time this quantity can be zero is when

$$(A_1 - A_2)\mathbf{x} = \mathbf{b}_2 - \mathbf{b}_1.$$

If $(A_1 - A_2)$ is invertible, this only happens at one point, i.e., $\mathbf{x} = (A_1 - A_2)^{-1}(\mathbf{b}_2 - \mathbf{b}_1)$.

In any case, the set of all $\mathbf{x}$ for which this equation is satisfied will have dimension less than N when $g_1 \ne g_2$, and so the value of the integrand at these points does not contribute to the integral.

Thus, because $\rho(\mathbf{x})$ satisfies the properties listed previously, and $\|g_1 \circ \mathbf{x} - g_2 \circ \mathbf{x}\| > 0$ for $g_1 \neq g_2$ except on a set of measure zero, the integral in the definition of the metric must satisfy $d(g_1, g_2) > 0$ unless $g_1 = g_2$, in which case $d(g_1, g_1) = 0$.

While this does satisfy the properties of a metric and *could* be used for CAD and robot design and path planning problems, it has the significant drawback that the integral of a pth root (or absolute value) must be taken. This means that numerical computations are required. For practical problems, devoting computer power to the computation of the metric detracts significantly from other aspects of the application in which the metric is being used. Therefore, this is not a practical metric. On the other hand, we may modify this approach slightly so as to generate metrics which are calculated in closed form. This yields tremendous computational advantages.

Namely, we observe that

$$d^{(p)}(g_1, g_2) = \sqrt[p]{\int_{\mathbb{R}^N} \|g_1 \circ \mathbf{x} - g_2 \circ \mathbf{x}\|_p^p \rho(\mathbf{x}) dx_1 \ldots dx_N}$$

is a metric. Clearly, this is symmetric and positive definite for all of the same reasons as the metric presented earlier. In order to prove the triangle inequality, we must use Minkowski's inequality. That is, if $a_1, a_2, \ldots a_n$ and $b_1, b_2, \ldots b_n$ are non-negative real numbers and $p > 1$, then

$$\left(\sum_{k=1}^{n} (a_k + b_k)^p \right)^{\frac{1}{p}} \leq \left(\sum_{k=1}^{n} a_k^p \right)^{\frac{1}{p}} + \left(\sum_{k=1}^{n} b_k^p \right)^{\frac{1}{p}}. \tag{6.44}$$

In our case, $a = \|g_1 \circ \mathbf{x} - g_2 \circ \mathbf{x}\|_p [\rho(\mathbf{x})]^{\frac{1}{p}}$, $b = \|g_2 \circ \mathbf{x} - g_3 \circ \mathbf{x}\|_p [\rho(\mathbf{x})]^{\frac{1}{p}}$, summation is replaced by integration, and because $\|g_1 \circ \mathbf{x} - g_2 \circ \mathbf{x}\|_p + \|g_2 \circ \mathbf{x} - g_3 \circ \mathbf{x}\|_p \geq \|g_1 \circ \mathbf{x} - g_3 \circ \mathbf{x}\|_p$, then

$$d^{(p)}(g_1, g_2) + d^{(p)}(g_2, g_3)$$
$$\geq \sqrt[p]{\int_{\mathbb{R}^N} (\|g_1 \circ \mathbf{x} - g_2 \circ \mathbf{x}\|_p + \|g_2 \circ \mathbf{x} - g_3 \circ \mathbf{x}\|_p)^p \rho(\mathbf{x}) dx_1 \ldots dx_N}$$
$$\geq \sqrt[p]{\int_{\mathbb{R}^N} \|g_1 \circ \mathbf{x} - g_3 \circ \mathbf{x}\|_p^p \rho(\mathbf{x}) dx_1 \ldots dx_N}$$
$$= d^{(p)}(g_1, g_3).$$

From the preceding argument, we can conclude that $d^p(\cdot, \cdot)$ is a metric. Likewise, it is easy to see that

$$d^{(p')}(g_1, g_2) = \sqrt[p]{\frac{\int_{\mathbb{R}^N} \|g_1 \circ \mathbf{x} - g_2 \circ \mathbf{x}\|_p^p \rho(\mathbf{x}) dx_1 \ldots dx_N}{\int_{\mathbb{R}^N} \rho(\mathbf{x}) dx_1 \ldots dx_N}}$$

is also a metric, since division by a positive real constant has no effect on metric properties.

The obvious benefit of using $d^{(p)}(\cdot, \cdot)$ or $d^{(p')}(\cdot, \cdot)$ is that the pth root is now outside of the integral, and so the integral can be calculated in closed form.

A particularly useful case is when $p = 2$. In this case it is easy to see that all of the metrics presented in this section satisfy the property

$$d(h \circ g_1, h \circ g_2) = d(g_1, g_2)$$

where $h, g_1, g_2 \in SE(N)$, which is the set of rigid-body motions of $\mathbb{R}^N$.

This is because if $h = (R, \mathbf{b}) \in SE(N)$, then

$$\|h \circ g_1 \circ \mathbf{x} - h \circ g_2 \circ \mathbf{x}\|_2 = \|R[g_1 \circ \mathbf{x}] + \mathbf{b} - R[g_2 \circ \mathbf{x}] - \mathbf{b}\|_2 = \|g_1 \circ \mathbf{x} - g_2 \circ \mathbf{x}\|_2.$$

It is also interesting to note that there is a relationship between this kind of metric for $SE(N)$ and the Hilbert-Schmidt norm of $N \times N$ matrices. That is, for $g \in SE(N)$

$$d^{(2)}(g, e) = \sqrt{\int_V \|g \circ \mathbf{x} - \mathbf{x}\|_2^2 \rho(\mathbf{x}) dV}$$

is the same as a weighted norm

$$\|g - e\|_W = \sqrt{\text{tr}((g - e)^T W(g - e))},$$

where $W = W^T \in \mathbb{R}^{4 \times 4}$ and g and e are expressed as 4×4 homogeneous transformations matrices. To show that this is true, we begin our argument by expanding as follows (let $g = (R, \mathbf{b})$)

$$\begin{aligned}
\|g \circ \mathbf{x} &- \mathbf{x}\|_2^2 \\
&= (R\mathbf{x} + \mathbf{b} - \mathbf{x})^T (R\mathbf{x} + \mathbf{b} - \mathbf{x}) \\
&= \text{tr}\left((R\mathbf{x} + \mathbf{b} - \mathbf{x})(R\mathbf{x} + \mathbf{b} - \mathbf{x})^T\right) \\
&= \text{tr}\left((R\mathbf{x} + \mathbf{b} - \mathbf{x})\left(\mathbf{x}^T R^T + \mathbf{b}^T - \mathbf{x}^T\right)\right) \\
&= \text{tr}\left(R\mathbf{x}\mathbf{x}^T R^T + R\mathbf{x}\mathbf{b}^T - R\mathbf{x}\mathbf{x}^T + \mathbf{b}\mathbf{x}^T R^T + \mathbf{b}\mathbf{b}^T - \mathbf{b}\mathbf{x}^T - \mathbf{x}\mathbf{x}^T R^T - \mathbf{x}\mathbf{b}^T + \mathbf{x}\mathbf{x}^T\right) \\
&= \text{tr}\left(\mathbf{x}\mathbf{x}^T + R\mathbf{x}\mathbf{b}^T - R\mathbf{x}\mathbf{x}^T + R\mathbf{x}\mathbf{b}^T + \mathbf{b}\mathbf{b}^T - \mathbf{x}\mathbf{b}^T - R\mathbf{x}\mathbf{x}^T - \mathbf{x}\mathbf{b}^T + \mathbf{x}\mathbf{x}^T\right) \\
&= 2\text{tr}\left(\mathbf{x}\mathbf{x}^T + R\mathbf{x}\mathbf{b}^T - R\mathbf{x}\mathbf{x}^T - \mathbf{x}\mathbf{b}^T + \frac{1}{2}\mathbf{b}\mathbf{b}^T\right).
\end{aligned}$$

Note that

$$\int_V \text{tr}[F(\mathbf{x})]\rho(\mathbf{x})dV = \text{tr}\left[\int_V F(\mathbf{x})\rho(\mathbf{x})dV\right],$$

where $F(\mathbf{x})$ is any matrix function. Under the assumption that

$$\int_V \mathbf{x}\rho(\mathbf{x})dV = \mathbf{0},$$

we then have

$$\|g \circ \mathbf{x} - \mathbf{x}\|_2^2 = 2\text{tr}\left[(\mathbb{I} - R)\mathbf{x}\mathbf{x}^T + \frac{1}{2}\mathbf{b}\mathbf{b}^T\right],$$

and the metric $d^{(2)}(g, e)$ satisfies

$$\left(d^{(2)}(g, e)\right)^2 = 2\text{tr}\left[(\mathbb{I} - R)\int_V \mathbf{x}\mathbf{x}^T \rho(\mathbf{x})dV\right] + \mathbf{b} \cdot \mathbf{b}\int_V \rho(\mathbf{x})dV$$

or

$$d^{(2)}(g, e) = \sqrt{2\text{tr}[(\mathbb{I} - R)\mathcal{I}] + \mathbf{b} \cdot \mathbf{b}M} \tag{6.45}$$

where $\mathbb{I}$ is the 3×3 identity matrix, $M = \int_V \rho(\mathbf{x})dV$ is the mass, and $\mathcal{I} = \int_V \mathbf{x}\mathbf{x}^T \rho(\mathbf{x})dV$ has a simple relationship with the moment of inertia matix of the rigid body

$$I = \int_V \left(\mathbf{x}^T \mathbf{x}\mathbb{I} - \mathbf{x}\mathbf{x}^T\right)\rho(\mathbf{x})dV = \text{tr}(\mathcal{I})\mathbb{I} - \mathcal{I}.$$

Now we compare this to the weighted norm of $\|g - e\|_W$, defined by $\|g - e\|_W^2 = \text{tr}((g-e)W(g-e)^T)$, where $W = \begin{pmatrix} W_{3\times3} & 0 \\ 0^T & w_{44} \end{pmatrix}$:

$$
\begin{aligned}
\|g - e\|_W^2 &= \text{tr}\left((g-e)W(g-e)^T\right) \\
&= \text{tr}\left((g-e)^T(g-e)W\right) \\
&= \text{tr}\left(\begin{pmatrix} R^T - \mathbb{I} & 0 \\ \mathbf{b}^T & 0 \end{pmatrix}\begin{pmatrix} R - \mathbb{I} & \mathbf{b} \\ 0^T & 0 \end{pmatrix}\begin{pmatrix} W_{3\times3} & 0 \\ 0^T & w_{44} \end{pmatrix}\right) \\
&= \text{tr}\left(\begin{pmatrix} (R^T - \mathbb{I})(R - \mathbb{I}) & (R^T - \mathbb{I})\mathbf{b} \\ \mathbf{b}^T(R - \mathbb{I}) & \mathbf{b}^T\mathbf{b} \end{pmatrix}\begin{pmatrix} W_{3\times3} & 0 \\ 0^T & w_{44} \end{pmatrix}\right) \\
&= \text{tr}\begin{pmatrix} (R^T - \mathbb{I})(R - \mathbb{I})W_{3\times3} & (R^T - \mathbb{I})\mathbf{b}w_{44} \\ \mathbf{b}^T(R - \mathbb{I})W_{3\times3} & \mathbf{b}^T\mathbf{b}w_{44} \end{pmatrix} \\
&= \text{tr}\left((R^T - \mathbb{I})(R - \mathbb{I})W_{3\times3}\right) + w_{44}\mathbf{b}^T\mathbf{b} \\
&= 2\text{tr}\left((\mathbb{I} - R)W_{3\times3}\right) + w_{44}\mathbf{b}^T\mathbf{b}.
\end{aligned}
$$

It is exactly in the same form as Eq. (6.45), with $J = W_{3\times3}$ and $M = w_{44}$. Therefore, we conclude that

$$
d^{(2)}(g, e) = \|g - e\|_W.
$$

Furthermore, we can prove that $d(g_1, g_2) = \|g_1 - g_2\|_W$. First note that

$$
g_1 - g_2 = \begin{pmatrix} R_1 - R_2 & \mathbf{b}_1 - \mathbf{b}_2 \\ 0^T & 0 \end{pmatrix},
$$

so

$$
\begin{aligned}
\|g_1 - g_2\|_W^2 &= \text{tr}\left(\begin{pmatrix} R_1^T - R_2^T & 0 \\ \mathbf{b}_1{}^T - \mathbf{b}_2{}^T & 0 \end{pmatrix}\begin{pmatrix} R_1 - R_2 & \mathbf{b}_1 - \mathbf{b}_2 \\ 0^T & 0 \end{pmatrix}\begin{pmatrix} W_{3\times3} & 0 \\ 0^T & w_{44} \end{pmatrix}\right) \\
&= \text{tr}\left(\begin{pmatrix} (R_1^T - R_2^T)(R_1 - R_2) & (R_1^T - R_2^T)(\mathbf{b}_1 - \mathbf{b}_2) \\ (\mathbf{b}_1{}^T - \mathbf{b}_2{}^T)(R_1 - R_2) & (\mathbf{b}_1{}^T - \mathbf{b}_2{}^T)(\mathbf{b}_1 - \mathbf{b}_2) \end{pmatrix}\begin{pmatrix} W_{3\times3} & 0 \\ 0^T & w_{44} \end{pmatrix}\right) \\
&= \text{tr}\left((R_1^T - R_2^T)(R_1 - R_2)W_{3\times3}\right) + w_{44}(\mathbf{b}_1 - \mathbf{b}_2)^T(\mathbf{b}_1 - \mathbf{b}_2) \\
&= \text{tr}\left((2\mathbb{I} - R_1^T R_2 - R_2^T R_1)W_{3\times3}\right) + w_{44}(\mathbf{b}_1 - \mathbf{b}_2)^T(\mathbf{b}_1 - \mathbf{b}_2).
\end{aligned}
$$

Note that

$$
\text{tr}\left(R_2^T R_1 W_{3\times3}\right) = \text{tr}\left(W_{3\times3}R_1^T R_2\right) = \text{tr}\left(R_1^T R_2 W_{3\times3}\right).
$$

We get

$$
\|g_1 - g_2\|_W^2 = 2\text{tr}\left((\mathbb{I} - R_1^T R_2)W_{3\times3}\right) + w_{44}\|\mathbf{b}_1 - \mathbf{b}_2\|^2.
$$

On the other hand, we already know that

$$
d\left(g_2^{-1} \circ g_1, e\right) = d\left(g_1^{-1} \circ g_2, e\right) = d(g_1, g_2).
$$

Noting that

$$
g_1^{-1} \circ g_2 = \begin{pmatrix} R_1^T & -R_1^T\mathbf{b}_1 \\ 0^T & 1 \end{pmatrix}\begin{pmatrix} R_2 & \mathbf{b}_2 \\ 0^T & 1 \end{pmatrix} = \begin{pmatrix} R_1^T R_2 & R_1^T(\mathbf{b}_2 - \mathbf{b}_1) \\ 0^T & 1 \end{pmatrix},
$$

we can get

$$
\left(d^{(2)}\left(g_1^{-1} \circ g_2, e\right)\right)^2 = \left\| g_1^{-1} \circ g_2 - e \right\|_W^2
$$

$$
= 2\mathrm{tr}\left(\left(\mathbb{I} - R_1^T R_2\right) W_{3\times 3}\right) + w_{44} \left\| R_1^T \left(\mathbf{b}_2 - \mathbf{b}_1\right) \right\|^2
$$

$$
= 2\mathrm{tr}\left(\left(\mathbb{I} - R_1^T R_2\right) W_{3\times 3}\right) + w_{44} \left\| \left(\mathbf{b}_2 - \mathbf{b}_1\right) \right\|^2 .
$$

This is exactly identical to $\|g_1 - g_2\|_W^2$. So we may conclude that for $g_1, g_2 \in SE(N)$

$$
d^{(2)}(g_1, g_2) = \|g_1 - g_2\|_W, \tag{6.46}
$$

where $W = \begin{pmatrix} J & 0 \\ 0^T & M \end{pmatrix}$.

This property of these metrics is very convenient since we can use many well-developed theories of matrix norms, and the only explicit integration that is required to compute the metric is the computation of moments of inertia (which are already tabulated for most common engineering shapes).

6.8.2   Metrics on $SE(N)$ Induced by Norms on $\mathcal{L}^2(SE(N))$

It is always possible to define a continuous real-valued function $f : SE(N) \to \mathbb{R}$ that decays rapidly outside of a neighborhood of the identity motion. Furthermore, if $f \in \mathcal{L}^p(SE(N))$, by definition it is possible to integrate f^p over all motions. In this case, the measure of the function,

$$
\mu(f) = \int_G f(g) d(g),
$$

(where $G = SE(N)$) is finite where $d(g)$ is an integration measure on the set of rigid-body motions discussed earlier, and g denotes a motion.

Given this background, it is possible to define left and right invariant metrics on the set of motions in the following way. Let $f(g)$ be a continuous nonperiodic p-integrable function [i.e., $\mu(|f|^p)$ is finite]. Then the following are metrics

$$
d_L^{(p)}(g_1, g_2) = \left(\int_G \left| f\left(g_1^{-1} \circ g\right) - f\left(g_2^{-1} \circ g\right) \right|^p d(g) \right)^{\frac{1}{p}},
$$

$$
d_R^{(p)}(g_1, g_2) = \left(\int_G \left| f(g \circ g_1) - f(g \circ g_2) \right|^p d(g) \right)^{\frac{1}{p}}.
$$

The fact that these are metrics follow easily. The triangle inequality holds from the Minkowski inequality, symmetry holds from the symmetry of scalar addition, and positive definiteness follows from the fact that we choose $f(g)$ to be a nonperiodic continuous function. That is, we choose $f(g)$ such that the equalities $f(g) = f(g_1 \circ g)$ and $f(g) = f(g \circ g_1)$ do not hold for $g_1 \neq e$ except possibly on sets of measure zero. Thus there is no way for the integral of the difference of shifted versions of this function to be zero other than when $g_1 = g_2$, where it must be zero.

We prove the left-invariance of $d_L^{(p)}(g_1, g_2)$ in the following. The proof for right-invariance for $d_R^{(p)}(g_1, g_2)$ follows analogously.

$$d_L^{(p)} (h \circ g_1, h \circ g_2) = \left(\int_G \left| f \left((h \circ g_1)^{-1} \circ g \right) - f \left((h \circ g_2)^{-1} \circ g \right) \right|^p d(g) \right)^{\frac{1}{p}}$$

$$= \left(\int_G \left| f \left((g_1^{-1} \circ h^{-1}) \circ g \right) - f \left((g_2^{-1} \circ h^{-1}) \circ g \right) \right|^p d(g) \right)^{\frac{1}{p}}$$

$$= \left(\int_G \left| f \left(g_1^{-1} \circ (h^{-1} \circ g) \right) - f \left(g_2^{-1} \circ (h^{-1} \circ g) \right) \right|^p d(g) \right)^{\frac{1}{p}} .$$

Because of the left-invariance of the integration measure, we then have

$$= \left(\int_G \left| f \left(g_1^{-1} \circ g' \right) - f \left(g_2^{-1} \circ g' \right) \right|^p d (h \circ g') \right)^{\frac{1}{p}}$$

$$= d_L^{(p)} (g_1, g_2),$$

where the change of variables $g' = h^{-1} \circ g$ has been made.

If it were possible to find a square-integrable function with the property $f(g \circ h) = f(h \circ g)$ for all $g, h \in SE(N)$, it would be possible to define a bi-invariant metric on $SE(N)$. This is clear as follows, by illustrating the right invariance of a left-invariant metric.

$$d_L^{(p)} (g_1 \circ h, g_2 \circ h) = \left(\int_G \left| f \left((g_1 \circ h)^{-1} \circ g \right) - f \left((g_2 \circ h)^{-1} \circ g \right) \right|^p d(g) \right)^{\frac{1}{p}}$$

$$= \left(\int_G \left| f \left(h^{-1} \circ g_1^{-1} \circ g \right) - f \left(h^{-1} \circ g_2^{-1} \circ g \right) \right|^p d(g) \right)^{\frac{1}{p}}$$

For such a function called (a class function), $f(g_1 \circ (g_2 \circ g_3)) = f((g_2 \circ g_3) \circ g_1)$ for any $g_1 \in G$, and so

$$d_L^{(p)} (g_1 \circ h, g_2 \circ h) = \left(\int_G \left| f \left(g_1^{-1} \circ g \circ h^{-1} \right) - f \left(g_2^{-1} \circ g \circ h^{-1} \right) \right|^p d(g) \right)^{\frac{1}{p}}$$

$$= \left(\int_G \left| f \left(g_1^{-1} \circ g' \right) - f \left(g_2^{-1} \circ g' \right) \right|^p d (h \circ g') \right)^{\frac{1}{p}}$$

$$= d_L^{(p)} (g_1, g_2),$$

where the substitution $g' = g \circ h^{-1}$ has been made and the right invariance of the integration has been assumed.

Unfortunately, for $SE(N)$ no such class functions exist [36]. However, the previous construction can be used to generate invariant metrics for $SO(N)$, where functions of the form $f(RQ) = f(QR)$ do exist.

This is consistent with results reported in the literature [46]. However, it does not rule out the existence of anomalous metrics such as the trivial metric, which can be generated for $p = 1$ when $f(g)$ is a delta function on $SE(N)$.

As a practical matter, $p = 1$ is a difficult case to work with since the integration must be performed numerically. Likewise, $p > 2$ does not offer computational benefits, and so we concentrate on the case $p = 2$. In this case we get

$$d_L^{(2)} (g_1, g_2) = \sqrt{\frac{1}{2} \int_G \left| f \left(g_1^{-1} \circ g \right) - f \left(g_2^{-1} \circ g \right) \right|^2 d(g)} .$$

Note that introducing the factor of $1/2$ does not change the fact that this is a metric. Expanding the square, we see

$$2\left(d_L^{(2)}(g_1, g_2)\right)^2 = \int_G f^2\left(g_1^{-1} \circ g\right) d(g) + \int_G f^2\left(g_2^{-1} \circ g\right) d(g)$$
$$- 2\int_G f\left(g_1^{-1} \circ g\right) f\left(g_2^{-1} \circ g\right) d(g). \qquad (6.47)$$

Because of the left invariance of the measure, the first two integrals are equal. Furthermore, if we define $f^*(g) = f(g^{-1})$, the last term may be written as a convolution. That is,

$$d_L^{(2)}(g_1, g_2) = \sqrt{\|f\|_2^2 - (f * f^*)(g_1^{-1} \circ g_2)}, \qquad (6.48)$$

where

$$\|f\|_2^2 = \int_G f^2(g) d(g),$$

and convolution on the set of rigid-body motions was defined earlier.

Since the maximum value of $(f * f^*)(g)$ occurs at $g = e$ and has the value $\|f\|^2$, we see that $d_L(g_1, g_1) = 0$, as must be the case for it to be a metric. The left invariance is clearly evident when written in the form of Eq. (6.48), since $(h \circ g_1)^{-1} \circ (h \circ g_2) = g_1^{-1} \circ (h^{-1} \circ h) \circ g_2 = g_1^{-1} \circ g_2$. We also recognize that unlike the class of metrics in the previous section, this one has a bounded value. That is

$$\max_{g_1, g_2 \in G} \sqrt{\|f\|_2^2 - (f * f^*)(g_1^{-1} \circ g_2)} \leq \|f\|_2.$$

By restricting the choice of f to symmetric functions, i.e., $f(g) = f(g^{-1})$, this metric takes the form

$$d_L^{(2)}(g_1, g_2) = d_L^{(2)}\left(e, g_1^{-1} \circ g_2\right) = \sqrt{\|f\|_2^2 - (f * f)\left(g_1^{-1} \circ g_2\right)}.$$

One reason why the left invariance of metrics is useful is because as a practical matter it can be more convenient to calculate $d_L^{(2)}(e, g)$ and evaluate it at $g = g_1^{-1} \circ g_2$ than to calculate $d_L^{(2)}(g_1, g_2)$ directly.

6.8.3   Park's Metric

Distance metrics for $SE(3)$ can be constructed from those for $\mathbb{R}^3$ and $SO(3)$. If we denote $d_{SO(3)}(R_1, R_2)$ to be any metric on $SO(3)$ and $d_{\mathbb{R}^3}(\mathbf{b}_1, \mathbf{b}_2)$ to be any metric on $\mathbb{R}^3$, the following will be metrics on $SE(3)$ for $g_i = (R_i, \mathbf{b}_i) \in SE(3)$

$$d_{SE(3)}^{(1)}(g_1, g_2) = L \cdot d_{SO(3)}(R_1, R_2) + d_{\mathbb{R}^3}(\mathbf{b}_1, \mathbf{b}_2);$$

and

$$d_{SE(3)}^{(2)}(g_1, g_2) = \sqrt{L^2\left(d_{SO(3)}(R_1, R_2)\right)^2 + \left(d_{\mathbb{R}^3}(\mathbf{b}_1, \mathbf{b}_2)\right)^2}.$$

Here L is a measure of length that makes the units of orientational and translational displacements compatable. If the metric is normalized *a priori* by some meaningful length, we can choose $L = 1$. In the special case when $d_{SO(3)}(R_1, R_2) = \theta(R_1^{-1} R_2)$, the metric $d_{SE(3)}^{(1)}(g_1, g_2)$ previously mentioned was introduced into the mechanical design community by Park [46].

6.8.4 Kinetic Energy Metric

A metric can be constructed for $SE(3)$ that is analogous to the one in Eq. (5.68). Namely, we can define

$$d\left(g_1, g_2\right) = \sqrt{\min_{g(t) \in SE(3)} \frac{1}{2} \int_{t_1}^{t_2} \left[M\dot{\mathbf{b}} \cdot \dot{\mathbf{b}} + \text{trace}(\dot{R}\mathcal{I}_R\dot{R}^T)\right] dt} \tag{6.49}$$

subject to the boundary conditions $g(t_1) = g_1$ and $g(t_2) = g_2$. For a nonspherical rigid body with mass M and moment of inertia in the body-fixed frame $\mathcal{I}_R$, the path $g(t) = (R(t), \mathbf{b}(t))$ that satisfies the above conditions must usually be generated numerically. This makes it less attractive than some of the other metrics discussed earlier.

6.8.5 Metrics on $SE(3)$ from Metrics on $SO(4)$

In Subsection 6.7.2 we reviewed how rigid-body motions in three dimensions can be approximated as rotations in four dimensions, and how rotations in four dimensions decompose into two copies of rotations (unit quaternions) in three dimensions. This fact has been used by Etzel and McCarthy [19, 20] to define approximate metrics on $SE(3)$. We now generalize this formulation and review their results in this context.

Any given pair of 4×4 rotation matrices $(\exp M^+(\mathbf{s}), \exp M^-(\mathbf{t}))$ can be mapped to a pair of 3×3 rotation matrices $(\exp S, \exp -T)$. We have metrics between rotations in three dimensions that can be easily extended to define distance between pairs of pairs of matrices. For instance, given the two pairs (R_1, Q_1) and (R_2, Q_2), it is easy to see that

$$D_1\left((R_1, Q_1), (R_2, Q_2)\right) = d\left(R_1, R_2\right) + d\left(Q_1, Q_2\right)$$

and

$$D_2\left((R_1, Q_1), (R_2, Q_2)\right) = \left([d\left(R_1, R_2\right)]^2 + [d\left(Q_1, Q_2\right)]^2\right)^{\frac{1}{2}}$$

satisfy the metric properties whenever $d(R, Q)$ does. Here $d(R, Q)$ can be any of the metrics for $SO(3)$ discussed earlier. In addition, it is clear that these metrics $D_i(\cdot, \cdot)$ for $SO(4)$ inherit the bi-invarance of $d(\cdot, \cdot)$. That is, given any $R_0, Q_0 \in SO(3)$,

$$D_i\left(R_0 R, Q_0 Q\right) = D_i\left(R R_0, Q Q_0\right) = D_i(R, Q).$$

Likewise for right shits.

6.9 Summary

In this chapter we have reviewed a number of different ways to describe rigid-body motions, to measure distance between motions, and to assign frames of reference to curves and discrete linkages. Applications of these techniques from rigid-body kinematics can be found in a number of areas including manufacturing, robotics, computer-aided design (CAD), polymer science, and mechanics. In CAD, problems of interest include the design of curves and surfaces with an associated set of attached frames (see, for example, [32, 61, 62] and references therein). In manufacturing, the geometry of swept volumes (see, for example, [3, 4, 5]) is closely related to the problem of path generation for machine tools. In Chapters 12, 17, and 18 we will examine problems in robotics, polymer science, and mechanics where knowledge of rigid-body kinematics is useful.

References

[1] Ball, R.S., *A Treatise on the Theory of Screws*, Cambridge University Press, Cambridge, 1900.

[2] Bishop, R., There is more than one way to frame a curve, *Am. Math. Mon.*, 82, 246–251, 1975.

[3] Blackmore, D., Leu, M.C., and Shih, F., Analysis and modelling of deformed swept volumes, *Comput.-Aided Design*, 26(4), 315–326, 1994.

[4] Blackmore, D., Leu, M.C., and Wang, L.P., The sweep-envelope differential equation algorithm and its application to NC machining verification, *Comput.-Aided Design*, 29(9), 629–637, 1997.

[5] Blackmore, D., Leu, M.C., Wang, L.P., and Jiang, H., Swept volume: a retrospective and prospective view, *Neural, Parallel and Scientific Computations*, 5, 81–102, 1997.

[6] Blaschke, W., Euklidische kinematik und nichteuklidische geometrie I,II, *Zeitschrift für Mathematik und Physik*, 60, 61–91, 203–204, 1911.

[7] Blaschke, W. and Müller, H.R., *Ebene Kinematik*, Verlag von R. Oldenbourg, München, Germany, 1956.

[8] Bottema, O. and Roth, B., *Theoretical Kinematics*, Dover Publications, New York, 1979.

[9] Brockett, R.W., Robotic manipulators and the product of exponentials formula, in *Mathematical Theory of Networks and Systems*, Fuhrman, A., Ed., 120–129, Springer-Verlag, New York, 1984.

[10] Călugăreanu, G., L'integrale de Gauss et l'analyse des noeuds tridimensionnels, *Revue de Mathématiques Pures et Appliquées*, 4, 5–20, 1959.

[11] Chasles, M., Note sur les propriétés générales du systéme de deux corps semblables entr'eux et placés d'une manière quelconque dans l'espace; et sur le désplacement fini ou infiniment petit d'un corps solids libre, *Férussac, Bulletin des Sciences Mathématiques*, 14, 321–326, 1830.

[12] Chirikjian, G.S., Theory and Applications of Hyper-Redundant Robotic Manipulators, Ph.D. dissertation, School of Engineering and Applied Science, California Institute of Technology, Pasadena, CA, 1992.

[13] Chirikjian, G.S. and Zhou, S., Metrics on motion and deformation of solid models, *J. of Mech. Design*, 120, 252–261, 1998.

[14] Clifford, W.K., Preliminary sketch of biquaternions, in *Mathematical Papers*, Tucker, R., Ed., MacMillan, London, 1882.

[15] Craig, J.J., *Introduction to Robotics, Mechanics and Control*, Addison-Wesley, Reading, MA, 1986.

[16] Denavit, J. and Hartenberg, R.S., A kinematic notation for lower-pair mechanisms based on matrices, *J. of Appl. Mech.*, 215–221, 1955.

[17] Donelan, P.S. and Gibson, C.G., First-order invariants of euclidean motions, *Acta Applicandae Mathematicae*, 24, 233–251, 1991.

[18] Donelan, P.S. and Gibson, C.G., On the hierarchy of screw systems, *Acta Applicandae Mathematicae*, 32, 267–296, 1993.

[19] Etzel, K.R., *Biquaternion Theory and Applications to Spatial Kinematics*, M.S. thesis, University of California, Irvine, CA, 1996.

[20] Etzel, K.R. and McCarthy, J.M., Spatial motion interpolation in an image space of $SO(4)$, Proc. of 1996 ASME Design Engineering Technical Conference and Computers in Engineering Conference, Irvine, CA, August 18-22, 1996.

[21] Eyring, H., The resultant electric moment of complex molecules, *Phys. Rev.*, 39, 746–748, 1932.

[22] Fanghella, P. and Galletti, C., Metric relations and displacement groups in mechanism and robot kinematic, *J. of Mech. Design, Trans. of the ASME*, 117, 470–478, 1995.

[23] Fary, I., Sur la courbure totale d'une courbe gauche faisant un noeud, *Bulletin de la Société Mathématique de France*, 77, 128–138, 1949.

[24] Fischer, I.S., *Dual-Number Methods in Kinematics, Statics, and Dynamics*, CRC Press, Boca Raton, FL, 1999.

[25] Flory, P.J., *Statistical Mechanics of Chain Molecules*, Wiley InterScience, New York, 1969.

[26] Frenchel, W., Uber Krümmung und Windung geschlossenen Raumkurven, *Mathematische Annalen*, 101, 238–252, 1929.

[27] Fuller, F.B., The writhing number of a space curve, *Proc. Nat. Acad. Sci. USA*, 68(4), 815–819, 1971.

[28] Ge, Q.J., On the matrix algebra realization of the theory of biquaternions, Proc. of the 1994 ASME Mechanisms Conference, DE 70, 425–432, 1994.

[29] Grünwald, J., Ein Abbildungsprinzip welches die ebene geometrie und kinematik mit der räumlichen geometrie verknüpft, *Sitzunsberichte Akademie Wissenschaflichen Wien*, 120, 677–741, 1911.

[30] Hervè J.M., Analyse structurelle des mècanismes par groupe des dèplacements, *Mechanisms and Mach. Theor.*, 13, 437–450, 1978.

[31] Inonu, E. and Wigner, E.P., On the contraction of groups and their representatations, *Proc. of the Nat. Acad. of Sci.*, 39(6), 510–524, June 15, 1953.

[32] Kallay, M. and Ravani, B., Optimal twist vectors as a tool for interpolating a network of curves with a minimum energy surface, *Comput. Aided Geometric Design*, 7, 1990.

[33] Karger, A. and Novák, J., *Space Kinematics and Lie Groups*, Gordon and Breach Science, New York, 1985.

[34] Kazerounian, K. and Rastegar, J., Object norms: a class of coordinate and metric independent norms for displacement, *Flexible Mechanisms, Dynamics, and Analysis*, ASME DE 47, 271–275, 1992.

[35] Klok, F., Two moving coordinate frames for sweeping along a 3D trajectory, *Comput. Aided Geometric Design*, 3, 217–229, 1986.

[36] Kyatkin, A.B. and Chirikjian, G.S., Regularization of a nonlinear convolution equation on the Euclidean group, *Acta. Appl. Math.*, 53, 89–123, 1998.

[37] Martinez, J.M.R., Representation of the Euclidean group and kinematic mappings, *Proc. 9th World Congress on the Theor. of Mach. and Mechanisms*, 2, 1594–1600, 1996.

[38] Martinez, J.M.R. and Duffy, J., On the metrics of rigid body displacement for infinite and finite bodies, *ASME J. of Mech. Design*, 117, 41–47, 1995.

[39] Mattice, W.L. and Suter, U.W., *Conformational Theory of Large Molecules: The Rotational Isomeric State Model in Macromolecular Systems*, John Wiley & Sons, New York, 1994.

[40] McCarthy, J.M., Planar and spatial rigid motion as special cases of spherical and 3-spherical motion, *J. of Mech., Transmissions, and Automation in Design*, 105, 569–575, 1983.

[41] McCarthy, J.M., *An Introduction to Theoretical Kinematics*, MIT Press, Cambridge, MA, 1990.

[42] Millman, R.S. and Parker, G.D., *Elements of Differential Geometry*, Prentice-Hall, Englewood Cliffs, NJ, 1977.

[43] Milnor, J., On the total curvature of knots, *Ann. of Math.*, 52, 248–257, 1950.

[44] Murray, R.M., Li, Z., and Sastry, S.S., *A Mathematical Introduction to Robotic Manipulation*, CRC Press, Boca Raton, FL, 1994.

[45] Oka, S., Zur theorie der statistischen Molekülgestalt hochpolymerer Kettenmoleküle unter Berücksichtigung der Behinderung der freien Drehbarkeit, *Proc. Physico-Math. Soc. of Japan*, 24, 657–672, 1942.

[46] Park, F.C., Distance metrics on the rigid-body motions with applications to mechanism design, *J. of Mech. Design, Trans. of the ASME*, 117, 48–54, 1995.

[47] Park, F.C. and Brockett, R.W., Kinematic dexterity of robotic mechanisms, *The Int. J. of Robotics Res.*, 13(1), 1–15, 1994.

[48] Park, F.C. and Ravani, B., Bézier curves on riemannian manifolds and lie groups with kinematics applications, *ASME J. of Mech. Design*, 117, 36–40, 1995.

[49] Pohl, W.F., The self-linking number of a closed space curve, *J. of Math. and Mech.*, 17(10), 975–985, 1968.

[50] Pohl, W.F., Some integral formulas for space curves and their generalizations, *Am. J. of Math.*, 90, 1321–1345, 1968.

[51] Ravani, B. and Roth, B., Motion synthesis using kinematic mapping, *ASME J. of Mech., Transmissions and Autom. in Design*, 105, 460–467, 1983.

[52] Rooney, J., A comparison of representations of general spatial screw displacements, *Environ. and Plann.*, B5, 45–88, 1978.

[53] Segre, B., Sulla torsione integrale delle curve chiuse sghembe, *Atti della Accademia Nazionale del Lincei, Rendiconti*, 3, 422–426, 1947.

[54] Segre, B., Una nuova caratterizzazione della sfera, *Atti della Accademia Nazionale del Lincei, Rendiconti*, 3, 420–422, 1947.

[55] Selig, J.M., *Geometrical Methods in Robotics*, Springer, New York, 1996.

[56] Stéphanos, M.C., Mémoire sur la représentation des homographies binaires par des points de l'space avec application à l'étude des rotations sphériques, *Mathematische Annalen*, 22, 299–367.

[57] Study, E., *Geometrie der Dynamen*, Teubner Verlag, Leipzig, Germany, 1903.

[58] Study, E., Grundlagen und ziele der analytischen kinematik, *Sitzungsberichte der Berliner Matematischen Gesselschaft*, 104, 36–60, 1912.

[59] Tsai, L.-W., *Robot Analysis: The Mechanics of Serial and Parallel Manipulators*, John Wiley & Sons, New York, 1999.

[60] Vologodskii, A., *Topology and Physics of Circular DNA*, CRC Press, Boca Raton, FL, 1992.

[61] Wang, W. and Joe, B., Robust computation of rotation minimizating frame for sweep surface modeling, preprint, 1995.

[62] Wesselink, W. and Veltkamp, R.C., Interactive design of constrained variational curves, *Comput. Aided Geometric Design*, 12(5), 533–546, 1995.

[63] White, J.H., Self-linking and the Gauss integral in higher dimensions, *Am. J. of Math.*, 91, 693–728, 1969.

[64] Wigner, E., On unitary representations of the inhomogeneous Lorentz group, *Ann. of Math.*, 40(1), 149–204, 1939.

[65] Žefran, M., Kumar, V., and Croke, C., Metrics and connections for rigid-body kinematics, *The Int. J. of Robotics Res.*, 18(2), 243–258, 1999.

Chapter 7

Group Theory

In this chapter we illustrate how some ideas introduced previously in the context of rotations and rigid-body motions generalize to other kinds of transformations. We also learn some new things about rotations and rigid-body motions.

7.1 Introduction

Group theory is a mathematical generalization of the study of symmetry. Before delving into the numerous definitions and fundamental results of group theory, we begin this chapter with a discussion of symmetries, or more precisely, operations that preserve symmetries in geometrical objects and equations. For in-depth treatments of group theory and its applications see [1, 2, 3, 5, 7, 8, 12, 13, 14, 18, 19, 20, 21, 22, 23, 24, 27, 29, 32, 36, 40, 41] and the other references cited throughout this chapter. Our discussion of group theory follows these in-depth treatments.

7.1.1 Motivational Examples

Consider the set of numbers $\{0, 1, 2\}$ with addition "modulo 3." This means that when two numbers are added together and the result is greater than 3, 3 is subtracted from the result until a number in $\{0, 1, 2\}$ is obtained. For instance, $1 + 1 \equiv 2 \, (\text{mod} \, 3)$, whereas $1 + 2 \equiv 0 \, (\text{mod} \, 3)$. Modular arithmetic is common to our everyday experiences when dealing with time. For instance if one works with a 12-hour clock, telling time is essentially an exercise in modulo 12 arithmetic [e.g., if it is 7 o'clock now, then 7 hours later it is $7 + 7 \equiv 2 \, (\text{mod} \, 12)$]. The only difference between telling time and modulo 12 arithmetic is that we say, "12 o'clock," instead of, "0 o'clock."

We can make the following addition table for $\{0, 1, 2\}$ with the operation $+_{[3]}$ (addition modulo 3)[1] that reads

$+_{[3]}$	0	1	2
0	0	1	2
1	1	2	0
2	2	0	1

If $k \in \{0, 1, 2\}$, a similar table results for the multiplication of the numbers $W_k = e^{2\pi i k/3}$ which reads

---

[1] We use both $+_{[3]}$ and $+ \, (\text{mod} \, 3)$ to mean addition modulo 3.

$$
\begin{array}{c|ccc}
\cdot & W_0 & W_1 & W_2 \\
\hline
W_0 & W_0 & W_1 & W_2 \\
W_1 & W_1 & W_2 & W_0 \\
W_2 & W_2 & W_0 & W_1
\end{array}
\quad .
$$

Hence $(\{0, 1, 2\}, + \,(\mathrm{mod}\,3))$, and $(\{W_0, W_1, W_2\}, \cdot)$ are in some sense equivalent. This sense is made more precise in Subsection 7.2.4.

Now consider the equilateral triangle shown in Fig. 7.1. We attach labels a, b, c to the vertices and $1, 2, 3$ to the positions in the plane to which the vertices correspond at the beginning of our thought experiment. The relative position of a with respect to the center point 0 is denoted with the vector $\mathbf{x}_{0a}$. Next consider all spatial rotations that move the triangle from its initial orientation to one in which the alphabetically labeled vertices are again matched (though possibly in a different way) to the numbered spaces. In fact, there are six kinds of motion. We can: (1) do nothing and leave the correspondence $a \to 1, b \to 2, c \to 3$; (2) rotate counterclockwise by $2\pi/3$ radians around $\mathbf{x}_{0a} \times \mathbf{x}_{0b}$ (the cross product of vectors $\mathbf{x}_{0a}$ and $\mathbf{x}_{0b}$) resulting in the new matching $a \to 2, b \to 3, c \to 1$; (3) rotate counterclockwise by $4\pi/3$ (or equivalently clockwise by $-2\pi/3$) around $\mathbf{x}_{0a} \times \mathbf{x}_{0b}$ resulting in $a \to 3, b \to 1, c \to 2$; (4) rotate the original triangle by an angle π around $\mathbf{x}_{0a}$ resulting in the correspondence $a \to 1, b \to 3, c \to 2$; (5) rotate the original triangle by an angle π around $\mathbf{x}_{0b}$, resulting in the correspondence $a \to 3, b \to 2, c \to 1$; or (6) rotate the original triangle by an angle π around $\mathbf{x}_{0c}$, resulting in the correspondence $a \to 2, b \to 1, c \to 3$. We label the preceding set of operations as $g_0, \ldots, g_5$ where $g_0 = e$ is called the identity (or "do nothing") operation. We note that while there are an infinite number of motions that will take a triangle back into itself in this way, they are all of the "same kind" as one of the six mentioned previously, provided we look only at where the motion starts and ends.

The natural issue to consider is what happens if one of these operations is followed by another. We denote this composition of relative motion as $g_i \circ g_j$ if first we perform g_j then g_i.[2] For instance, $g_2 \circ g_1$ is first a clockwise rotation by $2\pi/3$ followed by a clockwise rotation by $4\pi/3$. The result is a rotation by 2π which is the same as if no motion had been performed. Thus $g_2 \circ g_1 = e$. Likewise we can do this for all combinations of elements. If these elements truly represent all possible operations that map the vertices to the underlying positions in the plane, the composition of such operations must result in another one of these operations. In general, a *Latin square* is a table indicating how the composition of operations results in new ones. In the present example it can be constructed as

$$
\begin{array}{c|cccccc}
\circ & e & g_1 & g_2 & g_3 & g_4 & g_5 \\
\hline
e & e & g_1 & g_2 & g_3 & g_4 & g_5 \\
g_1 & g_1 & g_2 & e & g_4 & g_5 & g_3 \\
g_2 & g_2 & e & g_1 & g_5 & g_3 & g_4 \\
g_3 & g_3 & g_5 & g_4 & e & g_2 & g_1 \\
g_4 & g_4 & g_3 & g_5 & g_1 & e & g_2 \\
g_5 & g_5 & g_4 & g_3 & g_2 & g_1 & e
\end{array}
\quad . \tag{7.1}
$$

This table is constructed by calculating $g_i \circ g_j$ where i indexes the row and j indexes the column. If we consider only the block

$$
\begin{array}{c|ccc}
\circ & e & g_1 & g_2 \\
\hline
e & e & g_1 & g_2 \\
g_1 & g_1 & g_2 & e \\
g_2 & g_2 & e & g_1
\end{array}
\tag{7.2}
$$

---

[2] If we were using rotation matrices to describe these relative motions, they would be multiplied in the opposite order.

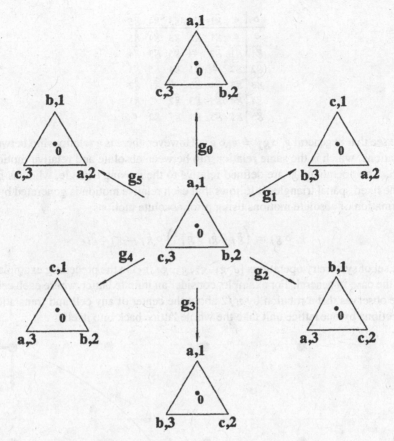

FIGURE 7.1
Symmetry operations of the equilateral triangle.

corresponding to rotations in the plane, the set of transformations called C_3 results. More generally, C_n is the set of planar rotational symmetry operations of a regular n-gon.

Any finite set of transformations for which a Latin square results (i.e., with each transformation appearing once and only once in each row and column) is called a *finite group* when, in addition, the associative law $g_i \circ (g_j \circ g_k) = (g_i \circ g_j) \circ g_k$ holds. The Latin square for a finite group is called a *Cayley table*. The fact that each element appears once and only once implies that for each operation one can find a unique operation, which when composed, will result in the identity. Thus each element in a group has an inverse.

We now consider a variant of the example shown in Fig. 7.1. Consider the same basic set of transformations, except now everything is defined with respect to the labels fixed in space 1, 2, and 3. Whereas g_1 was a rotation about $x_{0a} \times x_{0b}$, we will now consider the transformation $\hat{g}_1$ that performs the rotation by the same angle as g_1, only now around the vector $x_{01} \times x_{02}$. Likewise, $\hat{g}_i$ is the version of g_i with space-fixed labels replacing the labels fixed in the moving triangle. The two operations $\hat{g}_i$ and g_i are the same when the moving and fixed triangles initially coincide, but differ after the triangle has been displaced.

In the center figure in Fig. 7.1, a is coincident with 1 and b is coincident with 2 so it appears that g_1 and $\hat{g}_1$ are the same. However, this is not so when one applies these transformations to the other triangles in the figure. Working through all the products, one finds

$\circ$	e	$\hat{g}_1$	$\hat{g}_2$	$\hat{g}_3$	$\hat{g}_4$	$\hat{g}_5$
e	e	$\hat{g}_1$	$\hat{g}_2$	$\hat{g}_3$	$\hat{g}_4$	$\hat{g}_5$
$\hat{g}_1$	$\hat{g}_1$	$\hat{g}_2$	e	$\hat{g}_5$	$\hat{g}_3$	$\hat{g}_4$
$\hat{g}_2$	$\hat{g}_2$	e	$\hat{g}_1$	$\hat{g}_4$	$\hat{g}_5$	$\hat{g}_3$
$\hat{g}_3$	$\hat{g}_3$	$\hat{g}_4$	$\hat{g}_5$	e	$\hat{g}_1$	$\hat{g}_2$
$\hat{g}_4$	$\hat{g}_4$	$\hat{g}_5$	$\hat{g}_3$	$\hat{g}_2$	e	$\hat{g}_1$
$\hat{g}_5$	$\hat{g}_5$	$\hat{g}_3$	$\hat{g}_4$	$\hat{g}_1$	$\hat{g}_2$	e

$$\text{(7.3)}$$

Hence we see that in general $g_i \circ g_j \neq \hat{g}_i \circ \hat{g}_j$. However, there is a relationship between these two sets of operations, which is the same relationship between absolute and relative motions discussed in Chapter 6. The operations g_i are defined relative to the moving triangle, whereas $\hat{g}_i$ are defined relative to the fixed spatial triangle. It follows that each relative motion is generated by a similarity-like transformation of absolute motions using prior absolute motions

$$g_i \circ g_j = \left(\hat{g}_j \circ \hat{g}_i \circ \hat{g}_j^{-1} \right) \circ \hat{g}_j = \hat{g}_j \circ \hat{g}_i. \tag{7.4}$$

While the set of symmetry operations $\{e, g_1, g_2, g_3, g_4, g_5\}$ in the preceding example is finite, this need not be the case in general. For example, consider an infinite lattice where each cell is an $L \times L$ square. One observes that a rotation by $\pi/2$ about the center of any cell and translations along the x and y directions by one lattice unit take the whole lattice back into itself.

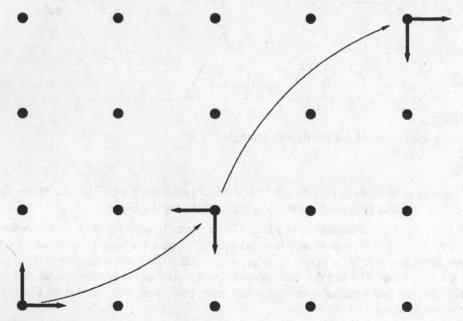

FIGURE 7.2
Symmetry operations of the square lattice.

We can express any such lattice transformation as

$$x'(j, k, l) = x \cos(l\pi/2) - y \sin(l\pi/2) + L \cdot j$$
$$y'(j, k, l) = x \sin(l\pi/2) + y \cos(l\pi/2) + L \cdot k$$

where $l \in [0, 3]$, $j, k \in \mathbb{Z}$ and (x, y) and (x', y') are the coordinates of cell centers. Figure 7.2 illustrates the composition of this kind of transformation. Since $\mathbb{Z}$ is infinite, so too is this set of transformations, which is used in the field of crystallography. It is easy to show that the composition of two such transformations results in another one of the same kind, and that each can be inverted. The associative law holds as well.

Thus far we have seen two sets of operations (one finite and one infinite) which reflect discrete symmetries. The concept of a group is not, however, limited to discrete symmetry operations. Consider, for example, the set of all rigid-body transformations that leave a sphere unchanged in shape or location. This is nothing more than the set of rotations, $SO(3)$, studied in Chapter 5. It has an infinite number of elements in the same way that a solid ball in $\mathbb{R}^3$ consists of an infinite number of points. That is, it is a continuous set of transformations. The full set of rigid-body motions, $SE(3)$, is also a continuous set of transformations, but unlike $SO(3)$, it is not bounded. $SO(3)$ is an example of something called a compact Lie group, whereas $SE(3)$ is a noncompact Lie group (see Appendix G).

As a fifth motivational example of transformations which leave something unchanged, consider Newton's first and second laws expressed in the form of the equation

$$\mathbf{F} = m\frac{d^2\mathbf{x}}{dt^2}. \tag{7.5}$$

This is *the* fundamental equation of all of non-relativistic mechanics, and it forms the foundation for most of mechanical, civil, and aerospace engineering. It describes the position, $\mathbf{x}$, of a particle of mass m subjected to a force $\mathbf{F}$, when all quantities are viewed in an inertial (non accelerating) frame of reference. It has been known for centuries that this equation works independent of what inertial frame of reference is used. This can be observed by changing spatial coordinates

$$\mathbf{x}^1 = R_1\mathbf{x}^0 + \mathbf{b}_1 + \mathbf{v}_1 t_0,$$

where R_1 is a constant (time-independent) rotation matrix relating the orientation of the two frames, t_0 is the time as measured in frame 0, and $\mathbf{b}_1 + \mathbf{v}_1 t_0$ is the vector relating the position of the origin of reference frame 1 relative to reference frame 0 ($\mathbf{b}_1$ and $\mathbf{v}_1$ are constant vectors). By observing that forces transform from one frame to another as $\mathbf{F}^1 = R_1\mathbf{F}^0$, it is easy to see that

$$\mathbf{F}^0 = m\frac{d^2\mathbf{x}^0}{dt_0^2} \qquad \Leftrightarrow \qquad \mathbf{F}^1 = m\frac{d^2\mathbf{x}^1}{dt_1^2}$$

since rotations can be inverted. Furthermore, in classical mechanics, the choice of when to begin measuring time is arbitrary, and so the transformation $t_1 = t_0 + a_1$ leaves Newton's laws unchanged as well. All together this can be written as

$$\begin{pmatrix} \mathbf{x}^1 \\ t^1 \\ 1 \end{pmatrix} = \begin{pmatrix} R_1 & \mathbf{v}_1 & \mathbf{b}_1 \\ \mathbf{0}^T & 1 & a_1 \\ \mathbf{0}^T & 0 & 1 \end{pmatrix} \begin{pmatrix} \mathbf{x}^0 \\ t^0 \\ 1 \end{pmatrix}. \tag{7.6}$$

A second change of reference frames relating $\mathbf{x}^2$ to $\mathbf{x}^1$ can be performed in exactly the same way. The composition of these changes results in the relationship between reference frames 0 and 2. It can be verified by multiplication of matrices of the form in Eq. (7.6) that this composition of transformations results in a transformation of the same kind. These invertible transformations reflect the so-called *Galilean invariance* of mechanics. The set of all such *Galilean transformations* contains both the set of rotations and rigid body motions.

Finally, we note that all of the examples considered thus far can be represented as invertible matrices with special structure. The set of all $N \times N$-dimensional matrices with real (or complex) entries is denoted as $GL(N, \mathbb{R})$ [or $GL(N, \mathbb{C})$].

With these examples we hope the reader has gained some intuition into sets of transformations and why such transformations can be important in the physical sciences and engineering. In the subsequent sections of this chapter we formalize the concept and properties of groups in general. It is not critical that everything in these sections be crystal clear after the first reading. However, for readers unfamiliar with the basic concepts of group theory, it is important to have at least seen this material before proceeding to the following chapters.

7.1.2 General Terminology

A review of terminology, notation, and basic results from the mathematics literature is presented in this section. (See, for example, [6, 15].) We begin with a few standard definitions, which are abbreviated so that the reader gets a feeling for the underlying concepts without the overhead required for complete rigor. It is assumed that the reader is familiar with the concepts of a set and function.

DEFINITION 7.1   *A* (closed) *binary operation is a law of composition which takes any two elements of a set and returns an element of the same set (i.e., $g_1 \circ g_2 \in G$ whenever $g_1, g_2 \in G$).*

DEFINITION 7.2   *A* group *is a set G together with a (closed) binary operation $\circ$ such that for any elements $g, g_1, g_2, g_3 \in G$ the following properties hold*

- *$g_1 \circ (g_2 \circ g_3) = (g_1 \circ g_2) \circ g_3$.*

- *There exists an element $e \in G$ such that $e \circ g = g \circ e = g$.*

- *For every element $g \in G$ there is an element $g^{-1} \in G$ such that $g^{-1} \circ g = g \circ g^{-1} = e$.*

The first of the preceding properties is called associativity; the element e is called the identity of G; and g^{-1} is called the inverse of $g \in G$. In order to distinguish between the group and the set, the former is denoted $(G, \circ)$, unless the operation is understood from the context, in which case G refers to both the set and the group. In the special case when for every two elements $g_1, g_2 \in G$, it is true that $g_1 \circ g_2 = g_2 \circ g_1$, the group is called *commutative* (or *Abelian*[3]); otherwise it is called *noncommutative* (or *non-Abelian*).

We have already seen five examples of noncommutative groups (all the sets of transformations described in Subsection 7.1.1 of this chapter), as well as four commutative ones that are central to standard Fourier analysis: $(\mathbb{R}, +)$ (real numbers with addition); $(T, + \,(\mathrm{mod}\, 2\pi))$ (a circle with circular addition); $(\mathbb{Z}, +)$ (the integers with addition); and $C_N = (Z_N, + \,(\mathrm{mod}\, N))$ (the integers $0, \ldots, N-1$ with addition modulo N, or equivalently, planar rotations that preserve a regular N−gon centered at the origin).

We define

$$h_g = g \circ h \circ g^{-1} \qquad \text{and} \qquad h^g = g^{-1} \circ h \circ g. \qquad (7.7)$$

In various texts both h_g and h^g are called *conjugation* of the element h by g. When we refer to conjugation, we will specify which of the previous definitions are being used. Of course they are related as $h^g = h_{g^{-1}}$. Note however that

$$\left(h^{g_1}\right)^{g_2} = g_2^{-1} \circ \left(g_1^{-1} \circ h \circ g_1\right) \circ g_2 = (g_1 \circ g_2)^{-1} \circ h \circ (g_1 \circ g_2) = h^{g_1 \circ g_2}$$

whereas $(h_{g_1})_{g_2} = h_{g_2 \circ g_1}$.

---

[3] This was named after Niels Henrick Abel (1802–1829).

For groups that are Abelian, conjugation leaves an element unchanged: $h_g = h^g = h$. For noncommutative groups (which is more often the case in group theory), conjugation usually does not result in the same element again. However, it is always true that conjugation of the product of elements is the same as the product of the conjugations

$$h_1^g \circ h_2^g = (h_1 \circ h_2)^g \qquad \text{and} \qquad (h_1)_g \circ (h_2)_g = (h_1 \circ h_2)_g.$$

A *subgroup* is a subset of a group ($H \subseteq G$) which is itself a group such that it is closed under the group operation of G, $e \in H$, and $h^{-1} \in H$ whenever $h \in H$. The notation for subgroup is $H \leq G$. One kind of subgroup, called a *conjugate subgroup,* is generated by conjugating all of the elements of an arbitrary subgroup with a fixed element g of the group. This is denoted as $gHg^{-1} = \{g \circ h \circ g^{-1} | h \in H\}$ for a single $g \in G$. If $H_1, H_2 \leq G$ and $gH_1g^{-1} = H_2$ for a $g \in G$, then H_1 and H_2 are said to be conjugate to each other. A subgroup $N \leq G$ which is conjugate to itself so that $gNg^{-1} = N$ for *all* $g \in G$ is called a *normal* subgroup of G. In this case, we use the notation $N \trianglelefteq G$. For sets H and G such that $H \subseteq G$ and $H \neq G$ we write $H \subset G$. H is then called a proper subset of G. Likewise, for groups if $H \leq G$ and $H \neq G$, we write $H < G$ (H is a proper subgroup of G), and for a normal subgroup we write $N \triangleleft G$ when $N \neq G$ (N is a proper normal subgroup of G).

Example 7.1
The group C_3 is both an Abelian and normal proper subgroup of the group of symmetry operations of the equilateral triangle. (We will prove this shortly.)   ☐

A *transformation group* $(G, \circ)$ is said to act on a set S if $g \cdot x \in S$ is defined for all $x \in S$ and $g \in G$ and has the properties

$$e \cdot x = x \qquad \text{and} \qquad (g_1 \circ g_2) \cdot x = g_1 \cdot (g_2 \cdot x) \in S$$

for all $x \in S$ and $e, g_1, g_2 \in G$. The operation $\cdot$ defines the *action* of G on S.

If any two elements of $x_1, x_2 \in S$ can be related as $x_2 = g \cdot x_1$ for some $g \in G$, then G is said to act *transitively* on S.

Example 7.2
The group of rotations is a transformation group that acts transitively on the sphere, because any point on the sphere can be moved to any other point using an appropriate rotation. The group of rotations $SO(N)$ also acts on $\mathbb{R}^N$, though not transitively because a point on any sphere centered at the origin remains on that sphere after a rotation.   ☐

Example 7.3
The group of rigid-body motions $SE(N)$ is a transformation group that acts transitively on $\mathbb{R}^N$.

A group $(G, \circ)$ is called a *topological* group if the mapping $(g_1, g_2) \rightarrow g_1 \circ g_2^{-1}$ is continuous. Continuity is defined in the context of an appropriate topology (as defined in Appendix B).   ☐

DEFINITION 7.3   *A Hausdorff space*[4] *X is called* locally compact *if for each $x \in X$ and every open set U containing x there exists an open set W such that $Cl(W)$ is compact and $x \in$*

---

[4]See Appendix B.

$W \subseteq Cl(W) \subseteq U$. A locally compact group *is a group for which the underlying set is locally compact [17, 28]*.

DEFINITION 7.4   *The* Cartesian product *of two sets G and H is the set $G \times H$ which consists of all pairs of the form (g, h) for all $g \in G$ and $h \in H$. The* direct product *of two groups $(G, \circ)$ and $(H, \hat{\circ})$ is the group $(P, \odot) = (G, \circ) \times (H, \hat{\circ})$ such that $P = G \times H$, and for any two elements $p, q \in P$, e.g., $p = (g_i, h_j)$, $q = (g_k, h_l)$, the group operation is defined as $p \odot q = (g_i \circ g_k, h_j \hat{\circ} h_l)$.*

Let the group $(G, \circ)$ be a transformation group that acts on the set H where $(H, +)$ is an Abelian group. Then the *semi-direct product* of $(G, \circ)$ and $(H, +)$ is the group $(P, \hat{\circ}) = (H, +) \rtimes_\varphi (G, \circ)$ such that $P = H \times G$, and for any two elements $p, q \in P$ the group operation is defined as $p \hat{\circ} q = (g_i \circ h_l + h_j, g_i \circ g_k)$.

The symbol φ is used here to distinguish further between the symbols $\rtimes_\varphi$ and $\rtimes$. In more general mathematical contexts our definition of a semi-direct product is a special case of a more general definition, in which case the subscript φ takes on an important meaning.

For semi-direct products, it is always the case that H is a normal subgroup of P

$$H \triangleleft P.$$

As an example of a semi-direct product, we observe that

$$SE(N) = \mathbb{R}^N \rtimes_\varphi SO(N).$$

We now examine a more exotic (though geometrically intuitive) example of a group.

Example 7.4 Transformations of a Ribbon

Consider the ribbon defined in Chapter 6. Recently, a group of deformations that acts on ribbons to produce new ribbons has been defined [25]. This group consists of a set of ordered triplets of the form $g = (\mathbf{a}, \alpha(s), M(s))$ where $\mathbf{a} \in \mathbb{R}^3$ is a position vector, and for $s \in [0, 1]$, the smooth bounded functions $M(s) \in SO(3)$ and $\alpha(s) \in \mathbb{R}$ are defined. The group law is written as

$$(\mathbf{a}_1, \alpha_1(s), M_1(s)) \circ (\mathbf{a}_2, \alpha_2(s), M_2(s)) = (\mathbf{a}_1 + \mathbf{a}_2, \alpha_1(s) + \alpha_2(s), M_1(s)M_2(s)) \qquad (7.8)$$

for all $s \in [0, 1]$.   ☐

A ribbon is specified by the ordered pair $p = (\mathbf{x}(s), \mathbf{n}(s))$, where $\mathbf{x}(s)$ is the backbone curve and $\mathbf{n}(s)$ is the unit normal to the backbone curve for each value of s. The action of the group defined in Eq. (7.8) on a ribbon is defined as

$$\begin{aligned} g \cdot p &= (\mathbf{a}, \alpha(s), M(s)) \cdot (\mathbf{x}(s), \mathbf{n}(s)) \\ &= \left(\mathbf{x}(0) + \mathbf{a} + \int_0^s e^{\alpha(\sigma)} M(\sigma) \mathbf{x}'(\sigma) d\sigma, M(s)\mathbf{n}(s) \right) \end{aligned}$$

where $\mathbf{x}'(\sigma)$ is the derivative of $\mathbf{x}(\sigma)$. It is easy to confirm by direct substitution that

$$g_1 \cdot (g_2 \cdot p) = (g_1 \circ g_2) \cdot p.$$

7.2 Finite Groups

In this section, all of the groups are finite. We begin by enumerating all of the possible groups with 2, 3, and 4 elements and then introduce the concepts of a permutation and matrix representation.

7.2.1 Multiplication Tables

Since the group composition operation (also called group multiplication) assigns to each ordered pair of group elements a new group element, one can always describe a finite group in the form of a multiplication table. Entries of the form $g_j \circ g_k$ are inserted in this table, where g_k determines in which column and g_j determines in which row the result of the product is placed. This convention is illustrated as

$$
\begin{array}{c|ccccc}
\circ & e & \cdots & g_k & \cdots & g_n \\
\hline
e & e & \cdots & g_k & \cdots & g_n \\
\vdots & \vdots & \cdots & \vdots & \cdots & \vdots \\
g_j & g_j & \cdots & g_j \circ g_k & \cdots & g_j \circ g_n \\
\vdots & \vdots & \cdots & \vdots & \ddots & \vdots \\
g_n & g_n & \cdots & g_n \circ g_k & \cdots & g_n \circ g_n.
\end{array}
$$

The number of elements in a finite group G is called the *order* of G and is denoted as $|G|$.

We consider in the following the simplest abstract groups, which are those of order 2, 3 and 4. If $|G| = 1$, things are not interesting because the only element must be the identity element: $G = (\{e\}, \circ)$. If $|G| = 2$, things are not much more interesting. Here we have

$$
\begin{array}{c|cc}
\circ & e & a \\
\hline
e & e & a \\
a & a & e
\end{array}.
$$

Note that the form of this table is exactly the same as the form of those for the groups $(\{1, -1\}, \cdot)$ and $\mathbb{Z}_2 = (\{0, 1\}, + \ (\mathrm{mod}\ 2))$. Groups with tables that have the same structure are abstractly the same group, even though they may arise in completely different situations and the group elements may be labeled differently. Later this concept will be formalized with the definition of *isomorphism*.

Considering the case when $|G| = 3$ we see that the table must look like

$$
\begin{array}{c|ccc}
\circ & e & a & b \\
\hline
e & e & a & b \\
a & a & b & e \\
b & b & e & a
\end{array}.
$$

By definition of the identity, the first row and column must be of the preceding form. Less obvious, but also required, is that $a \circ a = b$ must hold for this group. If this product had resulted in e or a it would not be possible to fill in the rest of the table while observing the constraint that every element appear in each row and column only once.

Moving on to the case when $|G| = 4$, we see that this is the first example where the table is not completely determined by the number of elements. The following three Latin squares can be formed

$$
\begin{array}{c|cccc}
\circ & e & a & b & c \\
\hline
e & e & a & b & c \\
a & a & e & c & b \\
b & b & c & e & a \\
c & c & b & a & e
\end{array}
\tag{7.9}
$$

$$
\begin{array}{c|cccc}
\circ & e & a & b & c \\
\hline
e & e & a & b & c \\
a & a & e & c & b \\
b & b & c & a & e \\
c & c & b & e & a
\end{array}
\tag{7.10}
$$

$$
\begin{array}{c|cccc}
\circ & e & a & b & c \\
\hline
e & e & a & b & c \\
a & a & b & c & e \\
b & b & c & e & a \\
c & c & e & a & b
\end{array}
\tag{7.11}
$$

Right away we observe by looking at the tables in Eqs. (7.9) and (7.10) that the corresponding sets of transformations cannot be equivalent under renaming of elements because the number of times the identity element, e, appears on the diagonal is different. On the other hand it may not be clear that these two exhaust the list of all possible *nonisomorphic* tables. For the tables in Eqs. (7.9) and (7.10), one can show that the choice of element in the $(2, 2)$ entry in the table, together with how the 2×2 block in the lower right corner of the table is arranged, completely determines the rest of the table. An exhaustive check for associativity confirms that these are the Cayley tables of two groups of order 4.

Regarding the third preceding table, if the b in the $(2, 2)$ entry is replaced with c, the table is completely determined as

$$
\begin{array}{c|cccc}
\circ & e & a & b & c \\
\hline
e & e & a & b & c \\
a & a & c & e & b \\
b & b & e & c & a \\
c & c & b & a & e
\end{array}
$$

However, this table is "the same as" the table in Eq. (7.11) because it just represents a reordering (or renaming) of elements. In particular, renaming $e \to e, a \to a, b \to c$, and $c \to b$ shows that these two tables are essentially the same, and so it is redundant to consider them both. Likewise Eqs. (7.10) and (7.11) are isomorphic under the mapping $e \to e, a \to b, b \to a, c \to c$.

At this point it may be tempting to believe that every Latin square is the Cayley table for a group. However, this is not true. For instance,

$$
\begin{array}{c|ccccc}
\circ & e & a & b & c & d \\
\hline
e & e & a & b & c & d \\
a & a & e & d & b & c \\
b & b & c & a & d & e \\
c & c & d & e & a & b \\
d & d & b & c & e & a
\end{array}
$$

is a valid Latin square, but $(a \circ b) \circ c \neq a \circ (b \circ c)$, and so the associative property fails to hold.

We observe from the symmetry about the major diagonal of all of the tables in Eqs. (7.9)–(7.11) that $g_j \circ g_k = g_k \circ g_j$, and therefore all of the operations defined by these tables are commutative.

It may be tempting to think that all finite groups are Abelian, but this is not true, as can be observed from the Cayley table for the group of symmetry operations of the equilateral triangle.

While it is true that one can always construct a symmetric group table of any dimension corresponding to an Abelian group and certain theorems of group theory provide results when a group of a given order must be Abelian (such as if $|G|$ is a prime number or the square of a prime number), the groups of most interest in applications are often those that are not commutative.

For a finite group G, one can define the *order of an element* $g \in G$ as the smallest positive number r for which $g^r = e$, where $g^r = g \circ \cdots \circ g$ (r times), and by definition $g^0 = e$. The cyclic group C_n has the property that $|C_n| = n$, and every element of C_n can be written as g^k for some $k \in \{0, \dots, n-1\}$ and some $g \in C_n$ where $g^n = e$.

7.2.2 Permutations and Matrices

The group of permutations of n letters (also called the *symmetric group*) is denoted as S_n. It is a finite group containing $n!$ elements. The elements of S_n can be arranged in any order, and for any fixed arrangement we label the elements of S_n as σ_{i-1} for $i = 1, \dots, n!$. We denote an arbitrary element $\sigma \in S_n$ as

$$\sigma = \begin{pmatrix} 1 & 2 & \dots & n \\ \sigma(1) & \sigma(2) & \dots & \sigma(n) \end{pmatrix}.$$

Changing the order of the columns in the previous element does not change the element. So in addition to the above expression,

$$\sigma = \begin{pmatrix} 2 & 1 & \dots & n \\ \sigma(2) & \sigma(1) & \dots & \sigma(n) \end{pmatrix} = \begin{pmatrix} n & 2 & \dots & 1 \\ \sigma(n) & \sigma(2) & \dots & \sigma(1) \end{pmatrix}$$

where stacked dots denote those columns not explicitly listed.

As an example of a permutation group, the elements of S_3 are

$$\sigma_0 = \begin{pmatrix} 1 & 2 & 3 \\ 1 & 2 & 3 \end{pmatrix}; \quad \sigma_1 = \begin{pmatrix} 1 & 2 & 3 \\ 2 & 3 & 1 \end{pmatrix}; \quad \sigma_2 = \begin{pmatrix} 1 & 2 & 3 \\ 3 & 1 & 2 \end{pmatrix};$$

$$\sigma_3 = \begin{pmatrix} 1 & 2 & 3 \\ 2 & 1 & 3 \end{pmatrix}; \quad \sigma_4 = \begin{pmatrix} 1 & 2 & 3 \\ 3 & 2 & 1 \end{pmatrix}; \quad \sigma_5 = \begin{pmatrix} 1 & 2 & 3 \\ 1 & 3 & 2 \end{pmatrix}.$$

There is more than one convention for what these numbers mean and how the product of two permutations is defined. This is much like the way the product of symmetry operations of the equilateral triangle can be generated as $g_i \circ g_j$ or $\hat{g}_i \circ \hat{g}_j$. The convention we use can be interpreted in the following way. Reading the expression $\sigma_i \circ \sigma_j$ from right to left we say "each k is sent to $\sigma_i(\sigma_j(k))$."

Using this convention we calculate

$$\sigma_2 \circ \sigma_3 = \begin{pmatrix} 1 & 2 & 3 \\ 3 & 1 & 2 \end{pmatrix} \circ \begin{pmatrix} 1 & 2 & 3 \\ 2 & 1 & 3 \end{pmatrix} = \begin{pmatrix} 1 & 2 & 3 \\ 1 & 3 & 2 \end{pmatrix} = \sigma_5.$$

The computation explicitly is $\sigma_2(\sigma_3(1)) = \sigma_2(2) = 1$, $\sigma_2(\sigma_3(2)) = \sigma_2(1) = 3$, $\sigma_2(\sigma_3(3)) = \sigma_2(3) = 2$. However, in one's mind it is convenient to simply read from upper right to lower left yielding the sequence $1 \to 2 \to 1, 2 \to 1 \to 3, 3 \to 3 \to 2$, with the first and last numbers of each composition kept as columns of the new permutation.

With this composition rule, the group table is

$\circ$	σ_0	σ_1	σ_2	σ_3	σ_4	σ_5
σ_0	σ_0	σ_1	σ_2	σ_3	σ_4	σ_5
σ_1	σ_1	σ_2	σ_0	σ_4	σ_5	σ_3
σ_2	σ_2	σ_0	σ_1	σ_5	σ_3	σ_4
σ_3	σ_3	σ_5	σ_4	σ_0	σ_2	σ_1
σ_4	σ_4	σ_3	σ_5	σ_1	σ_0	σ_2
σ_5	σ_5	σ_4	σ_3	σ_2	σ_1	σ_0

We note that the table for the symmetry operations on the equilateral triangle is the same as the table for the group of permutations of three letters. Permutations are an important example of finite groups. An important theorem of group theory (Cayley's theorem [14]) states that *every* finite group is isomorphic (i.e., the same to within an arbitrary relabling) to a subgroup of a permutation group.

We can assign to each permutation of n letters, σ, an invertible $n \times n$ matrix, $D(\sigma)$, that has the property

$$D\left(\sigma_i \circ \sigma_j\right) = D\left(\sigma_i\right) D\left(\sigma_j\right). \tag{7.12}$$

The mapping $\sigma \to D(\sigma)$ is called a *matrix representation* of S_n. The matrices $D(\sigma)$ are constructed in a straightforward way: if the permutation assigns $i \to j$, put a 1 in the j, i entry of the matrix. Otherwise insert a zero.

For example, under the mapping D, the six elements of S_3 have matrix representations [1]

$$D(\sigma_0) = \begin{pmatrix} 1 & 0 & 0 \\ 0 & 1 & 0 \\ 0 & 0 & 1 \end{pmatrix}$$

$$D(\sigma_1) = \begin{pmatrix} 0 & 0 & 1 \\ 1 & 0 & 0 \\ 0 & 1 & 0 \end{pmatrix}$$

$$D(\sigma_2) = \begin{pmatrix} 0 & 1 & 0 \\ 0 & 0 & 1 \\ 1 & 0 & 0 \end{pmatrix}$$

$$D(\sigma_3) = \begin{pmatrix} 0 & 1 & 0 \\ 1 & 0 & 0 \\ 0 & 0 & 1 \end{pmatrix}$$

$$D(\sigma_4) = \begin{pmatrix} 0 & 0 & 1 \\ 0 & 1 & 0 \\ 1 & 0 & 0 \end{pmatrix}$$

$$D(\sigma_5) = \begin{pmatrix} 1 & 0 & 0 \\ 0 & 0 & 1 \\ 0 & 1 & 0 \end{pmatrix}.$$

One observes by direct calculation that Eq. (7.12) holds.

Given a vector $\mathbf{x} = [x_1, \ldots, x_n]^T \in \mathbb{R}^n$ and a permutation $\sigma \in S_n$, we find that

$$D(\sigma)\mathbf{x} = \begin{pmatrix} x_{\sigma^{-1}(1)} \\ \vdots \\ x_{\sigma^{-1}(n)} \end{pmatrix} \quad \text{and} \quad \mathbf{x}^T D(\sigma) = \begin{pmatrix} x_{\sigma(1)} \\ \vdots \\ x_{\sigma(n)} \end{pmatrix}$$

both define actions of S_n on $\mathbb{R}^n$. For example, if $n = 3$, then

$$D(\sigma_1)\mathbf{x} = \begin{pmatrix} 0 & 0 & 1 \\ 1 & 0 & 0 \\ 0 & 1 & 0 \end{pmatrix} \begin{pmatrix} x_1 \\ x_2 \\ x_3 \end{pmatrix} = \begin{pmatrix} x_3 \\ x_1 \\ x_2 \end{pmatrix}.$$

Under this action we see that the component x_i is *sent to place* $\sigma_1(i)$, and the component in the ith place after the permutation is $x_{\sigma_1^{-1}(i)}$.

Since these matrix representations can be generated for any permutation, and every finite group is a subgroup of a permutation group, it follows that every finite group can be represented with square matrices composed of ones and zeros with exactly a single one in each row and column. These are called *permutation matrices*.

We now examine a somewhat more exotic example of a finite group constructed from simpler groups.

Example 7.5 The Wreath Product and Mathematical Biology

Given a finite group $(G, \circ)$, the Cartesian product of the set G with itself n times is denoted as G^n. The *wreath product* of G and S_n is the group $G \wr S_n = (G^n \times S_n, \diamond)$ where the product of two elements in $G^n \times S_n$ is defined as

$$(h_1, \ldots, h_n; \sigma) \diamond (g_1, \ldots, g_n; \pi) = \left(h_1 \circ g_{\sigma^{-1}(1)}, \ldots, h_n \circ g_{\sigma^{-1}(n)}; \sigma\pi \right).$$

As an example of a wreath product, consider the *hyper-octahedral group* $\mathbb{Z}_2 \wr S_n$. This group has been used recently to model the statistics of changes in circular DNA caused by random removal and reattachment of segments [33]. See Fig. 7.3 for an illustration of this process on circular DNA consisting of n base pairs. All possible operations defined by two cuts, a flip, and reattachment into a single loop may be viewed as elements of $\mathbb{Z}_2 \wr S_n$, as are the compositions of two such operations. The number of base pairs in the circle is preserved under these operations, but the relative proximity of base pairs gets mixed up. Mathematically, this is an action of $\mathbb{Z}_2 \wr S_n$ on the circle divided into n segments. □

7.2.3 Cosets and Orbits

In this section we review basic concepts concerning the decomposition of a group into cosets and the decomposition of a set on which a group acts into orbits. Since many terms from set theory are used, we recommend that readers unfamiliar with this topic consult Appendix B.

Coset Spaces and Quotient Groups

Given a subgroup $H \leq G$, and any element $g \in G$, the *left coset* gH is defined as

$$gH = \{g \circ h | h \in H\}.$$

Similarly, the right coset Hg is defined as

$$Hg = \{h \circ g | h \in H\}.$$

In the special case when $g \in H$, the corresponding left and right cosets are equal to H. More generally for all $g \in G$, $g \in gH$, and $g_1 H = g_2 H$ if and only if $g_2^{-1} \circ g_1 \in H$. Likewise for right cosets $Hg_1 = Hg_2$ if and only if $g_1 \circ g_2^{-1} \in H$.

Any group is divided into disjoint left (right) cosets, and the statement, "g_1 and g_2 are in the same left (right) coset" is an equivalence relation. This may be written explicitly for the case of left cosets as

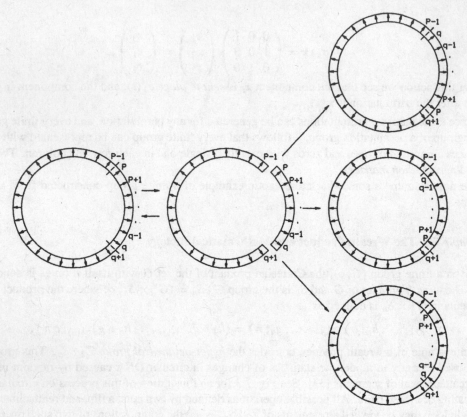

FIGURE 7.3
The action of $\mathbb{Z}_2 \wr S_n$ on closed-loop DNA.

$$g_1 \sim g_2 \quad \Leftrightarrow \quad g_1^{-1} \circ g_2 \in H. \tag{7.13}$$

Since H is a subgroup (and hence is itself a group), it is easy to verify that

$$g^{-1} \circ g = e \in H, \tag{7.14}$$

$$g_1^{-1} \circ g_2 \in H \Rightarrow \left(g_1^{-1} \circ g_2\right)^{-1} \in H, \tag{7.15}$$

and

$$g_1^{-1} \circ g_2 \in H, \quad g_2^{-1} \circ g_3 \in H \quad \Rightarrow \quad \left(g_1^{-1} \circ g_2\right) \circ \left(g_2^{-1} \circ g_3\right) = g_1^{-1} \circ g_3 \in H. \tag{7.16}$$

Hence $\sim$ as defined in Eq. (7.13) is an equivalence relation since Eq. (7.14) says $g \sim g$, Eq. (7.15) says $g_1 \sim g_2$ implies $g_2 \sim g_1$, and Eq. (7.16) says that $g_1 \sim g_2$ and $g_2 \sim g_3$ implies $g_1 \sim g_3$. An analogous argument holds for right cosets with the equivalence relation defined as

$$g_1 \sim g_2 \Leftrightarrow g_1 \circ g_2^{-1} \in H$$

instead of Eq. (7.13). Sometimes instead of $\sim$, one uses the notation

$$g_1 \equiv g_2 \quad \mathrm{mod}\, H$$

to denote either of the equivalence relations previously mentioned when left or right has been specified in advance.

An important property of gH and Hg is that they have the same number of elements as H. Since the group is divided into disjoint cosets, each with the same number of elements, it follows that the number of cosets must divide without remainder the number of elements in the group. The set of all left (or right) cosets is called the left (or right) *coset space,* and is denoted as G/H (or $H\backslash G$). One then writes

$$|G/H| = |H\backslash G| = |G|/|H|.$$

This result is called *Lagrange's theorem* [6].

Analogous to the way a coset is defined, the conjugate of a subgroup H for a given $g \in G$ is defined as

$$gHg^{-1} = \left\{ g \circ h \circ g^{-1} | h \in H \right\}.$$

Recall that a subgroup $N \leq G$ is called *normal* if and only if $gNg^{-1} \subseteq N$ for all $g \in G$. This is equivalent to the conditions $g^{-1}Ng \subseteq N$, and so we also write $gNg^{-1} = N$ and $gN = Ng$ for all $g \in G$.

When N is a normal subgroup of G, we use the notation $N \unlhd G$, and when it is both a proper and normal subgroup we write $N \lhd G$.

THEOREM 7.1
If $N \unlhd G$, then the coset space G/N together with the binary operation $(g_1N)(g_2N) = (g_1 \circ g_2)N$ is a group (called the quotient group).

PROOF   Note that for the operation defined in this way to make sense, the result should not depend on the particular choices of g_1 and g_2. Instead, it should only depend on which coset in G/N the elements g_1 and g_2 belong to. That is, if $g_1N = h_1N$ and $g_2N = h_2N$, with $g_i \neq h_i$, it should nonetheless be the case that the products are the same: $(g_1 \circ g_2)N = (h_1 \circ h_2)N$. This is equivalent to showing that $(g_1 \circ g_2)^{-1} \circ (h_1 \circ h_2) \in N$ whenever $g_1^{-1} \circ h_1 \in N$ and $g_2^{-1} \circ h_2 \in N$. Expanding out the product we see that

$$
\begin{aligned}
(g_1 \circ g_2)^{-1} \circ (h_1 \circ h_2) &= g_2^{-1} \circ g_1^{-1} \circ h_1 \circ h_2 \\
&= g_2^{-1} \circ \left(g_1^{-1} \circ h_1 \right) \circ h_2 \\
&= \left(g_2^{-1} \circ h_2 \right) \circ h_2^{-1} \circ \left(g_1^{-1} \circ h_1 \right) \circ h_2.
\end{aligned}
\tag{7.17}
$$

Since $g_1^{-1} \circ h_1 \in N$ and N is normal, $h_2^{-1} \circ (g_1^{-1} \circ h_1) \circ h_2 \in N$ also. Since $g_2^{-1} \circ h_2 \in N$ and N is closed under the group operation, it follows that the whole product in Eq. (7.17) is in N.

With this definition of product, it follows that the identity of G/N is $eN = N$, and the inverse of $gN \in G/N$ is $g^{-1}N \in G/N$. The associative law follows from the way the product is defined and the fact that G is a group. ∎

Orbits and Stabilizers

Recall that a set X on which a group G acts is called a G-set. When the group G acts transitively on the set X, every two elements of the set $x_1, x_2 \in X$ are related as $x_2 = g \cdot x_1$ for some $g \in G$. When G acts on X but *not* transitively, X is divided into multiple equivalence classes by G. The equivalence class containing $x \in X$ is called the *orbit* of x and is formally defined as

$$Orb(x) = \{g \cdot x | g \in G\}.$$

The set of all orbits is denoted X/G, and it follows that

$$X = \bigcup_{\sigma \in X/G} \sigma \quad \text{and} \quad |X| = \sum_{i=1}^{|X/G|} |Orb(x_i)|,$$

where x_i is a representative of the ith orbit. The hierarchy established by these definitions is $x_i \in Orb(x_i) \in X/G$. When G acts transitively on X, there is only one orbit, and this is the whole of X. In this special case we write $X/G = X = Orb(x)$ for any x.

The subset of all elements of G that leaves a particular element $x \in X$ fixed is called a *stabilizer, stability subgroup, little group,* or *isotropy subgroup* of x. It is formally defined as

$$G_x = \{g \in G | g \cdot x = x\}.$$

The fact that G_x is a subgroup of G follows easily since $e \cdot x = x$ by definition, $x = g \cdot x$ implies

$$g^{-1} \cdot x = g^{-1} \cdot (g \cdot x) = \left(g^{-1} \circ g\right) \cdot x = e \cdot x = x,$$

and for any $g_1, g_2 \in G_x$

$$(g_1 \circ g_2) \cdot x = g_1 \cdot (g_2 \cdot x) = g_1 \cdot x = x.$$

Hence G_x contains the identity, $g^{-1} \in G_x$ for every $g \in G_x$, and G_x is closed under the group operation. We, therefore, write $G_x \leq G$.

We have seen thus far that G divides X into disjoint orbits. Isotropy subgroups of G are defined by the property that they leave elements of X fixed. The next theorem relates these two phenomena.

THEOREM 7.2

If G is a finite group and X is a G-set, then for each $x \in X$

$$|Orb(x)| = |G/G_x|.$$

PROOF   Define the mapping $m : Orb(x) \to G/G_x$ as $m(g \cdot x) = gG_x$. This is a mapping that takes in elements of the orbit $g \cdot x \in Orb(x)$ and returns the coset $gG_x \in G/G_x$. This mapping is well-defined since $g_1 \cdot x = g_2 \cdot x$ implies $g_2^{-1} \circ g_1 \in G_x$, and hence $g_1 G_x = g_2 G_x$. If we can show that this mapping is bijective, the number of elements in $Orb(x)$ and G/G_x must be the same. Surjectivity follows immediately from the definition of m. To see that m is injective, one observes that $g_1 G_x = g_2 G_x$ implies $g_2^{-1} \circ g_1 \in G_x$, and thus $x = (g_2^{-1} \circ g_1) \cdot x$, or equivalently $g_2 \cdot x = g_1 \cdot x$.

∎

Another interesting property of isotropy groups is expressed below.

THEOREM 7.3

Given a group G acting on a set X, the conjugate of any isotropy group G_x by $g \in G$ for any $x \in X$ is an isotropy group, and in particular

$$gG_x g^{-1} = G_{g \cdot x}. \tag{7.18}$$

PROOF $h \in G_x$ means $h \cdot x = x$. Likewise, $k \in G_{g \cdot x}$ means $k \cdot (g \cdot x) = g \cdot x$. Let $k = g \circ h \circ g^{-1}$. Then

$$k \cdot (g \cdot x) = (k \circ g) \cdot x = (g \circ h) \cdot x = g \cdot (h \cdot x) = g \cdot x.$$

Hence $k = g \circ h \circ g^{-1} \in G_{g \cdot x}$ whenever $h \in G_x$, and so we write $gG_xg^{-1} \subseteq G_{g \cdot x}$. On the other hand, *any* $k \in G_{g \cdot x}$ satisfies $k \cdot (g \cdot x) = g \cdot x$. We can write this as

$$g^{-1} \cdot (k \cdot (g \cdot x)) = g^{-1} \cdot (g \cdot x)$$

or

$$\left(g^{-1} \circ k \circ g\right) \cdot x = e \cdot x = x.$$

In other words, $g^{-1} \circ k \circ g = h \in G_x$. From this it follows that all $k \in G_{g \cdot x}$ must be of the form $k = g \circ h \circ g^{-1}$ for some $h \in G_x$, and so we write $gG_xg^{-1} \supseteq G_{g \cdot x}$. ∎

7.2.4 Mappings

DEFINITION 7.5 *A homomorphism is a mapping from one group into another* $h : (G, \circ) \to (H, \hat{\circ})$ *such that*[5]

$$h\,(g_1 \circ g_2) = h\,(g_1)\,\hat{\circ}h\,(g_2).$$

The word "into" refers to the fact that the values $h(g)$ for *all* $g \in G$ must be contained in H, but it is possible that elements of H exist for which there are no counterparts in G. This is illustrated in Fig. 7.4.

It follows immediately from this definition that $h(g) = h(g \circ e) = h(g)\hat{\circ}h(e)$, and so $h(e)$ is the identity in H. Likewise, $h(e) = h(g \circ g^{-1}) = h(g)\hat{\circ}h(g^{-1})$ and so $(h(g))^{-1} = h(g^{-1})$. Thus a homomorphism $h : G \to H$ maps inverses of elements in G to the inverses of their counterparts in H, and the identity of G is mapped to the identity in H.

In general, a homomorphism will map more elements of G to the identity of H than just the identity of G. The set of all $g \in G$ for which $h(g) = h(e)$ is called the *kernel* of the homomorphism and is denoted as $\mathrm{Ker}(h)$. It is easy to see from the definition of homomorphism that if $g_1, g_2 \in \mathrm{Ker}(h)$, then so are their inverses and products. Thus $\mathrm{Ker}(h)$ is a subgroup of G, and moreover it is a normal subgroup because given any $g \in \mathrm{Ker}(h)$ and $g_1 \in G$,

$$h\left(g_1^{-1} \circ g \circ g_1\right) = (h\,(g_1))^{-1}\,\hat{\circ}h(g)\hat{\circ}h\,(g_1) = (h\,(g_1))^{-1}\,\hat{\circ}h\,(g_1) = h(e).$$

That is, conjugation of $g \in \mathrm{Ker}(h)$ by any $g_1 \in G$ results in another element in $\mathrm{Ker}(h)$, and so $\mathrm{Ker}(h)$ is a normal subgroup of G and so we write $\mathrm{Ker}(h) \trianglelefteq G$.

In general, a homomorphism $h : G \to H$ will map all the elements of G to some subset of H. This subset is called the *image* of the homomorphism, and we write $\mathrm{Im}(h) \subseteq H$. More specifically, since a homomorphism maps the identity of G to the identity of H, and inverses in G map to inverses in H, and since for any $g_1, g_2 \in G$ we have $h(g_1)\hat{\circ}h(g_2) = h(g_1 \circ g_2) \in \mathrm{Im}(h)$, it follows that $\mathrm{Im}(h)$ is a subgroup of H, and we write $\mathrm{Im}(h) \le H$. (We note that even more generally if $K \le G$, then the image of K in H under the homomorphism h is also a subgroup of H.) A homomorphism that is surjective[6] is called an *epimorphism*. A homomorphism that is injective is called a *monomorphism*. A one-to-one homomorphism of G onto H is called an *isomorphism*, and when such an isomorphism exists between groups, the groups are called *isomorphic* to each other. If H and G are isomorphic, we write $H \cong G$. An isomorphism of a group G onto itself is called an *automorphism*. Conjugation

---

[5]Our notational choice of using $h(\cdot)$ as a function and $h(g)$ as an element of H is to emphasize that some part of H is, in a sense, parameterized by G.

[6]See Appendix B for a review of definitions from set theory.

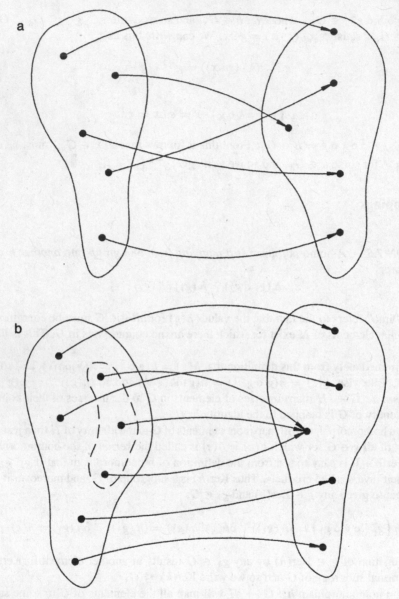

FIGURE 7.4
Illustration of homomorphisms (a) isomorphism; (b) epimorphism (dashed lines denote a fiber)
(continued).

of all elements in a group by one fixed element is an example of an automorphism. For a finite group, an automorphism is essentially a relabeling of elements. Isomorphic groups are fundamentally the same group expressed in different ways.

Example 7.6 **(five examples):**

The two groups $(\{0, 1, 2\}, + \pmod 3)$ and $(\{W_0, W_1, W_2\}, \cdot)$ from Section 7.1.1 are isomorphic; the

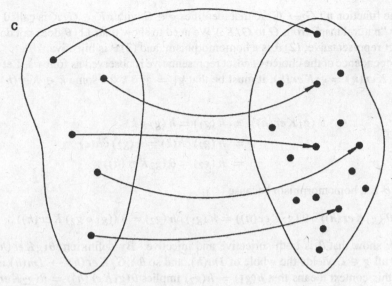

FIGURE 7.4

(Cont.) **Illustration of homomorphisms (c) monomorphism.**

group $(\{e^{i\theta}\}, \cdot)$ and the set of all rotation matrices

$$\begin{pmatrix} \cos\theta & -\sin\theta \\ \sin\theta & \cos\theta \end{pmatrix}$$

under matrix multiplication for $0 \le \theta < 2\pi$ are isomorphic; given a fixed matrix $A \in GL(N, \mathbb{R})$, $x \to Ax$ for all $x \in \mathbb{R}^N$ defines an automorphism of $(\mathbb{R}^N, +)$ onto itself; the group of real numbers under addition, and the set of upper triangular 2×2 matrices of the form

$$\begin{pmatrix} 1 & x \\ 0 & 1 \end{pmatrix}$$

with operation of matrix multiplication are isomorphic; the mapping $\sigma \to D(\sigma)$ defining matrix representations of permutations in Section 7.2.2 is an isomorphism. ☐

The Homomorphism Theorem

There are a number of interesting relationships between the concepts of homomorphism and quotient groups. For our purposes the following will be of the greatest importance.

THEOREM 7.4

Let G and H be groups and $h : G \to H$ be a homomorphism. Then $G/Ker(h) \cong Im(h)$.

PROOF   Define $\theta : G/Ker(h) \to H$ as

$$\theta(\eta(g)) = h(g)$$

for all $g \in G$ where $\eta : G \to G/Ker(h)$ is defined as

$$\eta(g) = gKer(h). \tag{7.19}$$

(In general, the function $\eta : G \to G/K$ that identifies $g \in G$ with $gK \in G/K$ is called the "natural projection" or "natural map" from G to G/K.) We need to show that: (1) θ does not depend on the choice of coset representative; (2) θ is a homomorphism; and (3) θ is bijective.

(1) The independence of the choice of coset representative is observed as follows. Let $g_1, g_2 \in G$. Then given $g_1 Ker(h) = g_2 Ker(h)$, it must be that $g_1 = g_2 \circ k$ for some $k \in Ker(h)$, and so we have

$$
\begin{aligned}
\theta\left(g_1 Ker(h)\right) &= h\left(g_1\right) = h\left(g_2 \circ k\right) \\
&= h\left(g_2\right) \hat{\circ} h(k) = h\left(g_2\right) \hat{\circ} h(e) \\
&= h\left(g_2\right) = \theta\left(g_2 Ker(h)\right).
\end{aligned}
$$

(2) Clearly θ is a homomorphism because

$$
\theta\left(g_1 Ker(h)\right) \hat{\circ} \theta\left(g_2 Ker(h)\right) = h\left(g_1\right) \hat{\circ} h\left(g_2\right) = \theta\left(\left(g_1 \circ g_2\right) Ker(h)\right).
$$

(3) We now show that θ is both surjective and injective. By definition, $\theta(gKer(h)) = h(g)$. Running over all $g \in G$ yields the whole of $Im(h)$, and so $\theta : G/Ker(h) \to Im(h)$ is surjective. Injectivity in this context means that $h(g_1) = h(g_2)$ implies $\theta(g_1 Ker(h)) = \theta(g_2 Ker(h))$ for all $g_1, g_2 \in G$. When $h(g_1) = h(g_2)$, we can write

$$
h(e) = h\left(g_2\right)_0^{-1} h\left(g_1\right) = h\left(g_2^{-1} \circ g_1\right)
$$

which means that $g_2^{-1} \circ g_1 \in Ker(h)$. Therefore it must follow that $g_1 Ker(h) = g_2 Ker(h)$. ∎

The preceding theorem is sometimes called the "fundamental theorem of homomorphisms" or the "first isomorphism theorem." There are, in fact, a number of other isomorphism theorems which will not be discussed here.

G-Morphisms

Let X and Y be two sets on which a group G acts. To distinguish between the two different kinds of actions, denote $g \cdot x \in X$ and $g \bullet y \in Y$ for all $x \in X$ and $y \in Y$.

A *G-morphism* is a mapping $f : X \to Y$ with the property

$$
f(g \cdot x) = g \bullet f(x). \tag{7.20}
$$

This property is often called *G-equivariance*.

We now review three important examples of G-morphisms found in [35].

Example 7.7

Suppose that $g \bullet y = y$ for all $y \in Y$ and $g \in G$. Then $f : X \to Y$ with the property

$$
f(g \cdot x) = f(x)
$$

for all $x \in X$ is a G-morphism that is constant on each orbit $\sigma \in X/G$. ⬜

Example 7.8

Let Y be the set of all subgroups of G. That is, $H \in Y \Leftrightarrow H \leq G$. Define the action $g \bullet H = gHg^{-1}$. Then the function $f : X \to Y$ defined as $f(x) = G_x$ (the stabilizer of x) satisfies the equivariance property because

$$
f(g \cdot x) = G_{g \cdot x} = gG_x g^{-1} = g \bullet f(x). \quad ⬜
$$

Example 7.9

Let M be a set and $g \odot m \in M$ for all $m \in M$ and $g \in G$ define an action. Let X be the Cartesian product $X = G \times M$. Let G act on X as

$$g \cdot (h, m) = \left(g \circ h \circ g^{-1}, g \odot m \right)$$

for $g \in G$ and $(h, m) \in X$. Following [35], let $Z \subset G \times M$ be defined as

$$Z = \{(h, m) | h \odot m = m\}.$$

It may be verified that $g \cdot (h, m) \in Z$ and $g^{-1} \cdot (h, m) \in Z$ and

$$gZ = \{g \cdot (h, m) | (h, m) \in Z\} = Z.$$

Since $|gZ| = |Z|$, the mapping $f : Z \to Z$ defined as $f(h, m) = g \cdot (h, m)$ must be bijective.

It may then be shown that both $\theta : Z \to M$ defined as $\theta(h, m) = m$, and $\tau : Z \to G$ defined as $\tau(h, m) = h$ are G-morphisms.

The inverse images (or *fibers*) of θ and τ are the subsets of Z that map to any given $m \in M$ and $h \in G$, respectively. Recall that for any mapping $f : X \to Y$, the inverse image $f^{-1}(y)$ is the set of all $x \in X$ such that $f(x) = y \in Y$. It may be shown that the inverse images of the preceding G-morphisms are

$$\theta^{-1}(m) = \{(h, m) | h \in G_m\}$$

and

$$\tau^{-1}(h) = \{(h, m) | m \in F(h)\}$$

where

$$F(h) = \{m | h \odot m = m\} \subseteq M. \quad \square$$

Counting Formulas

The G-morphisms τ and θ of the previous section provide a tool for counting the number of elements in the set Z. We can evaluate $|Z|$ by first counting how many elements are in $\tau^{-1}(h)$ for a given h, then sum over all h

$$|Z| = \sum_{m \in F(h)} \left| \tau^{-1}(h) \right|,$$

or we can count the elements of Z as

$$|Z| = \sum_{h \in G_m} \left| \theta^{-1}(m) \right|.$$

Both of these equations can be viewed as special cases of Eq. (B.1) in Appendix B with the function taking the value 1 on every element of the set Z, and where membership in a fiber is the equivalence relation.

We exclude the subset $\{e\} \times M$ from Z, and define

$$P = \bigcup_{h \in G - \{e\}} F(h).$$

We can then perform the counting computations [35]

$$|Z - \{e\} \times M| = \sum_{h \in G - \{e\}} |F(h)|$$

and

$$|Z - \{e\} \times M| = \sum_{m \in P} (|G_m| - 1) = \sum_{Orb(m) \in P/G} \frac{|G|}{|G_m|} (|G_m| - 1)$$

where in the sum on the right side it is understood that m is an arbitrary representative of $Orb(m)$. The last equality holds because $|G_{h \odot m}| = |G_m|$ is constant on each orbit $Orb(m)$, and from the orbit-stabilizer theorem and Lagrange's theorem

$$|Orb(m)| = |G/G_m| = |G|/|G_m|.$$

In Section 7.3.4 the formula

$$\sum_{h \in G - \{e\}} |F(h)| = \sum_{Orb(m) \in P/G} \frac{|G|}{|G_m|} (|G_m| - 1). \tag{7.21}$$

is used to classify the finite subgroups of $SO(3)$.

7.2.5   Conjugacy Classes, Class Functions, and Class Products

Conjugacy Classes and Conjugate Subgroups

Two elements $a, b \in G$ are said to be *conjugate* to each other if $a = g^{-1} \circ b \circ g$ for some $g \in G$. Conjugacy of two elements is an equivalence relation because the following properties hold: $a = e^{-1} \circ a \circ e$ (reflexivity); $a = g^{-1} \circ b \circ g$ implies $b = g \circ a \circ g^{-1}$ (symmetry); and $a = g_1^{-1} \circ b \circ g_1$ and $b = g_2^{-1} \circ c \circ g_2$ implies $a = (g_2 \circ g_1)^{-1} \circ c \circ (g_2 \circ g_1)$ (transitivity).

If we conjugate all the elements of a subgroup $H \leq G$ with respect to a particular $g \in G$, the result is also a subgroup, called a *conjugate subgroup*, denoted $g^{-1} H g$. It follows that $g^{-1} H g$ is a subgroup because: $e \in H$ and hence $e = g^{-1} \circ e \circ g = e \in g^{-1} H g$; for any $h, h^{-1} \in H$ we have $(g^{-1} \circ h \circ g)^{-1} = g^{-1} \circ h^{-1} \circ g \in g^{-1} H g$; and for any $h_1, h_2 \in H$ we have $(g^{-1} \circ h_1 \circ g) \circ (g^{-1} \circ h_1 \circ g) = g^{-1} \circ (h_1 \circ h_2) \circ g \in g^{-1} H g$.

If instead of fixing an element g and conjugating a whole subgroup by g, we choose a fixed element $b \in G$ and calculate $g^{-1} \circ b \circ g$ for *all* $g \in G$, the result is a set of elements in G called the *conjugacy class* (or simply *class* for short) containing b. All elements in a class are conjugate to every other element in the same class. An element of the group can only belong to one class. That is, the group is divided (or partitioned) into disjoint classes. This follows from the more general fact that an equivalence relation (of which conjugacy is one example) partitions a set into disjoint subsets. If C_i denotes the ith conjugacy class and $|C_i|$ denotes the number of group elements in this class, then

$$\sum_{i=1}^{\alpha} |C_i| = |G| \tag{7.22}$$

where α is the number of classes. Equation (7.22) is called the *class equation*. Clearly, $\alpha \leq |G|$ since $|C_i|$ are positive integers. Only in the case when G is Abelian is $|C_i| = 1$ for all i, and hence this is the only case where $\alpha = |G|$. By denoting C_1 as the class containing the identity element, we see that $|C_1| = 1$, i.e., it contains *only* the identity element, since $g^{-1} \circ e \circ g = e$ for all $g \in G$.

Class Functions and Sums

An important concept we will see time and time again throughout this book is the *class function*. This is a function $C : G \to \mathbb{C}$ (recall that $\mathbb{C}$ denotes the complex numbers) which is constant on conjugacy classes. That is, its value is the same for all elements of each conjugacy class. This means that

$$C(g) = C\left(h^{-1} \circ g \circ h\right) \quad \text{or} \quad C(h \circ g) = C(g \circ h)$$

for all values of $g, h \in G$. The previous two equations are completely equivalent because the change of variables $g \to h \circ g$ is invertible.

A very special kind of class function on a discrete (and not necessarily finite) group is the delta function

$$\delta(h) = \begin{cases} 1 & \text{for } h = e \\ 0 & \text{for } h \neq e \end{cases}.$$

The "shifted" version of this function: $\delta_{g_1}(g) = \delta(g_1^{-1} \circ g)$ for any $g_1 \in G$, while not a class function, has some useful properties that will be used shortly. By defining the operation $(L(g_1)f)(g) = f(g_1^{-1} \circ g) = f_{g_1}(g)$ we see that

$$\left(L\left(g_1\right) L\left(g_2\right) f\right)(g) = \left(L\left(g_1\right) f_{g_2}\right)(g) = f_{g_2}\left(g_1^{-1} \circ g\right) = \left(L\left(g_1 \circ g_2\right) f\right)(g). \tag{7.23}$$

The preceding equalities can be a great source of confusion when seen for the first time. The temptation is to switch the order of g_1^{-1} and g_2^{-1}.

Using shifted δ-functions, we can define the *class sum function* as

$$C_i(g) = \sum_{a \in C_i} \delta\left(a^{-1} \circ g\right). \tag{7.24}$$

This is a class function with the property

$$C_i(g) = \begin{cases} 1 & \text{for } g \in C_i \\ 0 & \text{for } g \notin C_i \end{cases}.$$

It is therefore the characteristic function (see Appendix B) of the set C_i. Clearly, any class function can be written as

$$C = \sum_{i=1}^{\alpha} \gamma_i C_i.$$

Central to the extension of Fourier analysis to the case of finite groups in Chapter 8 is that the definition of convolution generalizes in a natural way

$$(f_1 * f_2)(g) = \sum_{h \in G} f_1(h) f_2\left(h^{-1} \circ g\right) = \sum_{h \in G} f_1\left(g \circ h^{-1}\right) f_2(h).$$

The second preceding equality results from the invertible change of variables $h \to g \circ h^{-1}$.

Here f_1 and f_2 are complex-valued functions with group-valued argument. In general, $f_1 * f_2 \neq f_2 * f_1$. However, if either f_1 or f_2 is a class function, we have

$$\begin{aligned} (f * C)(g) &= \sum_{h \in G} f(h) C\left(h^{-1} \circ g\right) \\ &= \sum_{h \in G} f(h) C\left(g \circ h^{-1}\right) \\ &= \sum_{h \in G} C\left(g \circ h^{-1}\right) f(h) \\ &= (C * f)(g). \end{aligned}$$

Furthermore, since $\delta_{g_1}(g)$ is zero unless $g = g_1$, it is easy to see that

$$
\begin{aligned}
\left(\delta_{g_1} * \delta_{g_2}\right)(g) &= \sum_{h \in G} \delta\left(g_1^{-1} \circ h\right) \delta\left(g_2^{-1} \circ h^{-1} \circ g\right) \\
&= \delta\left(g_2^{-1} \circ g_1^{-1} \circ g\right) = \delta\left((g_1 \circ g_2)^{-1} \circ g\right) \\
&= \delta_{g_1 \circ g_2}(g).
\end{aligned}
$$

Hence,

$$
\delta_{g_1} * \delta_{g_1^{-1}} = \delta.
$$

This means that for a class function, C,

$$
\delta_{g_1} * C * \delta_{g_1^{-1}} = \delta_{g_1} * \left(C * \delta_{g_1^{-1}}\right) = \delta_{g_1} * \left(\delta_{g_1^{-1}} * C\right) = \left(\delta_{g_1} * \delta_{g_1^{-1}}\right) * C = C.
$$

Likewise, one finds that for an arbitrary function on the group, $f(g)$,

$$
\left(\delta_{g_1} * f * \delta_{g_1^{-1}}\right)(g) = f\left(g_1^{-1} \circ g \circ g_1\right),
$$

and so the *only* functions for which $\delta_{g_1} * f * \delta_{g_1^{-1}} = f$ are class functions.

Products of Class Sum Functions for Finite Groups

Given two class sum functions C_i and C_j (which, being class functions, must satisfy $\delta_h * C_i * \delta_{h^{-1}} = C_i$ and $\delta_h * C_j * \delta_{h^{-1}} = C_j$), we may write

$$
\begin{aligned}
C_i * C_j &= \left(\delta_h * C_i * \delta_{h^{-1}}\right) * \left(\delta_h * C_j * \delta_{h^{-1}}\right) \\
&= \delta_h * C_i * \left(\delta_{h^{-1}} * \delta_h\right) * C_j * \delta_{h^{-1}} \\
&= \delta_h * C_i * C_j * \delta_{h^{-1}}.
\end{aligned}
$$

Therefore, the convolution product of two class sum functions must be a class function. Since all class functions are expandable as a linear combination of class sum functions we have

$$
C_i * C_j = \sum_{k=1}^{\alpha} c_{ij}^k C_k, \tag{7.25}
$$

where c_{ij}^k are called *class constants*. Recall that α is the number of conjugacy classes in the group.

Since the convolution of class functions (and hence class sum functions) commutes, it follows that $c_{ij}^k = c_{ji}^k$.

Analogous to the Cayley table for a group we can construct a class product table that illustrates the convolution of class sum functions. Note that this table must be symmetric about the diagonal.

$*$	C_1	$\cdots$	C_k	$\cdots$	C_n
C_1	C_1	$\cdots$	C_k	$\cdots$	C_n
$\vdots$	$\vdots$	$\cdots$	$\vdots$	$\cdots$	$\vdots$
C_j	C_j	$\cdots$	$C_j * C_k$	$\cdots$	$C_j * C_n$
$\vdots$	$\vdots$	$\cdots$	$\vdots$	$\ddots$	$\vdots$
C_n	C_n	$\cdots$	$C_n * C_k$	$\cdots$	$C_n * C_n$

The $(1, 1)$ entry in the table is $C_1 * C_1 = C_1$ because C_1 is the conjugacy class $\{e\}$, and hence $C_1 = \delta$, which yields itself when convolved with itself. Likewise, $C_1 * C_i = C_i * C_1 = C_i$, and so $c_{i,1}^1 = \delta_{i,1}$. More generally, the coefficient c_{ij}^1 is found as follows. First we expand out

$$
\left(C_i * C_j\right)(g) = \sum_{h \in G} \left(\sum_{a \in C_i} \delta\left(a^{-1} \circ h\right)\right) \left(\sum_{b \in C_j} \delta\left(b^{-1} \circ h^{-1} \circ g\right)\right)
$$

$$
= \sum_{a \in C_i} \sum_{b \in C_j} \left(\sum_{h \in G} \delta\left(a^{-1} \circ h\right) \delta\left(b^{-1} \circ h^{-1} \circ g\right)\right).
$$

The last term in parenthesis is only nonzero when $h = a$, and so the inner summation vanishes and

$$
\left(C_i * C_j\right)(g) = \sum_{a \in C_i} \sum_{b \in C_j} \delta\left((a \circ b)^{-1} \circ g\right).
$$

The coefficient c_{ij}^1 may be viewed as the number of times the preceding summand reduces to $\delta(g)$. $c_{ij}^1 \neq 0$ only when it is possible to write $b = a^{-1}$. If C is the class containing c, we denote C^{-1} to be the class containing c^{-1}. (We have not defined a product of conjugacy classes, so C^{-1} should not be interpreted as the inverse of C, but rather only as the set containing the inverses of all the elements of C.) Usually $C \neq C^{-1}$ but it is always true that $|C| = |C^{-1}|$. Using this notation, $c_{ij}^1 \neq 0$ only when $C_i = C_j^{-1}$. In this case, $c_{ij}^1 = |C_i|$ since we can go through all $a \in C_i$, and for each of them we can find $b = a^{-1} \in C_i^{-1}$. Therefore,

$$
c_{ij}^1 = |C_i| \delta_{C_i, C_j^{-1}} = \begin{cases} |C_i| & \text{for } C_i = C_j^{-1} \\ 0 & \text{for } C_i \neq C_j^{-1} \end{cases}.
$$

Double Cosets and their Relationship to Conjugacy Classes

We now review the concept of double cosets. This concept is closely related to that of cosets as described in Section 7.2.3. At the end of this subsection, we examine a relationship between double cosets and conjugacy classes.

Let $H < G$ and $K < G$. Then for any $g \in G$, the set

$$
HgK = \{h \circ g \circ k | h \in H, k \in K\} \tag{7.26}
$$

is called the *double coset* of H and K, and any $g' \in HgK$ (including $g' = g$) is called a *representative* of the double coset. Though a double coset representative can often be written with two or more different pairs (h_1, k_1) and (h_2, k_2) so that $g' = h_1 \circ g \circ k_1 = h_2 \circ g \circ k_2$, we only count g' once in HgK. Hence $|HgK| \leq |G|$, and in general $|HgK| \neq |H| \cdot |K|$. If redundancies were not excluded in Eq. (7.26), we might expect the last inequality to be an equality. Having said this, we note that there are special cases when $|HgK| = |H| \cdot |K| = |G|$. For example, if every $g \in G$ can be uniquely decomposed as $g = h \circ k$ for $h \in H$ and $k \in K$, then

$$
H(h \circ k)K = H(h \circ e \circ k)K = (Hh)e(kK) = HeK = G.
$$

In general, the set of all double cosets of H and K is denoted $H \backslash G / K$. Hence we have the hierarchy $g \in HgK \in H \backslash G / K$.

THEOREM 7.5
Membership in a double coset is an equivalence relation on G. That is, G is partitioned into disjoint double cosets, and for $H < G$ and $K < G$ either $Hg_1 K \cap Hg_2 K = \emptyset$ or $Hg_1 K = Hg_2 K$.

PROOF   Suppose Hg_1K and Hg_2K have an element in common. If

$$h_1 \circ g_1 \circ k_1 = h_2 \circ g_2 \circ k_2$$

then

$$g_1 = \left(h_1^{-1} \circ h_2\right) \circ g_2 \circ \left(k_2 \circ k_1^{-1}\right).$$

The quantities in parenthesis are in H and K, respectively. Then

$$Hg_1K = H\left(h \circ g_2 \circ k\right)K = (Hh)g_2(kK) = Hg_2K. \quad \blacksquare$$

From Theorem 7.5 it follows that for any finite group G, we can write

$$G = \bigcup_{i=1}^{|H\backslash G/K|} H\sigma_i K \quad \text{and} \quad |G| = \sum_{i=1}^{|H\backslash G/K|} |H\sigma_i K|$$

where σ_i is an arbitrary representative of the ith double coset. However, unlike single cosets where $|gH| = |H|$, we cannot assume that all the double cosets have the same size. In the special case mentioned above where we can make the unique decomposition $g = h \circ k$, the fact that double cosets must either be disjoint or identical means $H\backslash G/K = \{G\}$ and $|H\backslash G/K| = 1$.

THEOREM 7.6

Given $H < G$ and $K < G$ and $g_1, g_2 \in G$ such that $g_1K \cap Hg_2K \neq \emptyset$, then $g_1K \subseteq Hg_2K$.

PROOF   Let $g_1 \circ k \in g_1K$. If $g_1 \circ k \in Hg_2K$ then $g_1 \circ k = h_1 \circ g_2 \circ k_1$ for some $h_1 \in H$ and $k_1 \in K$. Then we can write $g_1 = h_1 \circ g_2 \circ k_1 \circ k^{-1}$. Hence for all $k_2 \in K$, we have $g_1 \circ k_2 = h_1 \circ g_2 \circ k_1 \circ k^{-1} \circ k_2 = h_1 \circ g_2 \circ k_3 \in HgK$ where $k_3 = k_1 \circ k^{-1} \circ k_2 \in K$, and so $g_1K \subseteq Hg_2K. \quad \blacksquare$

We note without proof the important result of Frobenius that the number of single left cosets gK that appear in HgK is $\gamma_g = |H/(H \cap K_g)|$ where $K_g = gKg^{-1} < G$ is the subgroup K conjugated by g. Note that $H \cap K_g \leq H$ and $H \cap K_g \leq K_g$. This means that

$$HgK = \bigcup_{i=1}^{\gamma_g} \sigma_i K \tag{7.27}$$

where σ_i is a coset representative of $\sigma_i K \subseteq HgK$. An analogous result follows for right cosets. In the special case when H is a normal subgroup of G, we find $\gamma_g = 1$ for all $g \in G$, and so $HgH = gH$.

We now examine the relationship between double cosets and conjugacy classes. Recall that $G \times G$ denotes the direct product of G and G consisting of all pairs of the form (g, h) for $g, h \in G$ and with the product $(g_1, h_1)(g_2, h_2) \overset{\triangle}{=} (g_1 \circ g_2, h_1 \circ h_2)$. Let $G \cdot G$ denote the subset of $G \times G$ consisting of all pairs of the form $(g, g) \in G \times G$. Clearly $G \cdot G$ is a subgroup of $G \times G$.

There are some interesting features of the double coset space $(G \cdot G)\backslash(G \times G)/(G \cdot G)$.[7] For instance, without loss of generality, double coset representatives of $(G \cdot G)(g', g'')(G \cdot G)$ can always

---

[7] The parentheses around $G \times G$ and $G \cdot G$ are simply to avoid incorrectly reading of $G \cdot G\backslash G \times G/G \cdot G$ as $G \cdot (G\backslash G) \times (G/G) \cdot G$.

be written as (e, g). This follows because

$$(g_1, g_1)\left(g', g''\right)(g_2, g_2) = \left(g_1 \circ g' \circ g_2, g_1 \circ g'' \circ g_2\right)$$
$$= (g_1, g_1)\left(e, g'' \circ \left(g'\right)^{-1}\right)\left(g' \circ g_2, g' \circ g_2\right).$$

Since $(g' \circ g_2, g' \circ g_2) \in G \cdot G$ for all $g' \in G$, we can always define $(e, g) = (e, g'' \circ (g')^{-1})$ to be the coset representative. Furthermore, for any $h \in G$, the representatives (e, g) and $(e, h \circ g \circ h^{-1})$ belong to the same double coset since

$$(g_1, g_1)(e, g)(g_2, g_2) = \left(g_1 \circ h^{-1}, g_1 \circ h^{-1}\right)\left(e, h \circ g \circ h^{-1}\right)(h \circ g_2, h \circ g_2).$$

This means that elements of the same conjugacy class of G yield representatives of the same double coset under the mapping $g \to (e, g)$. Conversely, if

$$(G \cdot G)(e, g)(G \cdot G) = (G \cdot G)\left(e, g'\right)(G \cdot G),$$

it must be possible to find $g, g', g_1, g_2, g_3, g_4 \in G$ such that

$$(g_1, g_1)(e, g)(g_2, g_2) = (g_3, g_3)\left(e, g'\right)(g_4, g_4).$$

Expanding out the calculation on both sides and equating terms means $g_1 \circ g_2 = g_3 \circ g_4$ and $g_1 \circ g \circ g_2 = g_3 \circ g' \circ g_4$. Rewriting the first of these as $g_2 = g_1^{-1} \circ g_3 \circ g_4$ and substituting into $g' = g_3^{-1} \circ g_1 \circ g \circ g_2 \circ g_4^{-1}$ yields

$$g' = g_3^{-1} \circ g_1 \circ g \circ \left(g_1^{-1} \circ g_3 \circ g_4\right) \circ g_4^{-1} = \left(g_3^{-1} \circ g_1\right) \circ g \circ \left(g_3^{-1} \circ g_1\right)^{-1}.$$

This means that representatives (e, g) and (e, g') of the same double coset $(G \cdot G)(g_i, g_j)(G \cdot G)$ always map to elements of the same conjugacy class of G.

Written in a different way, we have the following result:

THEOREM 7.7

Given a finite group G, there is a bijective mapping between the set of double cosets $(G \cdot G)\backslash(G \times G)/(G \cdot G)$ and the set of conjugacy classes of G, and hence

$$\alpha = |(G \cdot G)\backslash(G \times G)/(G \cdot G)|,$$

where α is the number of conjugacy classes of G.

PROOF Let $C(g)$ denote the conjugacy class containing g and $\mathcal{C}$ denote the set of all conjugacy classes [i.e., $C(g) \in \mathcal{C}$ for all $g \in G$]. Let the function $f : \mathcal{C} \to (G \cdot G)\backslash(G \times G)/(G \cdot G)$ be defined as

$$f(C(g)) = (G \cdot G)(e, g)(G \cdot G).$$

Then for any $h \in G$ the previous discussion indicates that

$$f(C(g)) = f\left(C\left(h \circ g \circ h^{-1}\right)\right) = (G \cdot G)\left(e, h \circ g \circ h^{-1}\right)(G \cdot G) = (G \cdot G)(e, g)(G \cdot G),$$

and so f is defined consistently in the sense that all elements of the same class are mapped to the same double coset.

Since a double coset representative of $(G \cdot G)(g', g'')(G \cdot G)$ can always be chosen of the form (e, g) for some $g \in G$, and since

$$(G \cdot G)(e, g_1)(G \cdot G) = (G \cdot G)(e, g_2)(G \cdot G) \quad \Leftrightarrow \quad g_1 \in \mathcal{C}(g_2),$$

it follows that f is a bijection between $\mathcal{C}$ and $(G \cdot G)\backslash(G \times G)/(G \cdot G)$, and therefore $|\mathcal{C}| = |(G \cdot G)\backslash(G \times G)/(G \cdot G)|$. ∎

7.2.6 Examples of Definitions: Symmetry Operations on the Equilateral Triangle (Revisited)

Note that any element of the group of symmetry operations on the equilateral triangle can be written as products of powers of the elements g_1 and g_3 defined in Fig. 7.1. In such cases, we say that these elements are *generators* of the group. In the case when a whole group is generated by a single element, it is called a *cyclic* group. A cyclic group is always Abelian because all elements are of the form a^n, and $a^n \circ a^m = a^{n+m} = a^m \circ a^n$.

The proper subgroups of this group can be observed from the table in Eq. (7.1). They are $\{e\}$, $H_1 = \{e, g_1, g_2\}$, $H_2 = \{e, g_3\}$, $H_3 = \{e, g_4\}$, and $H_4 = \{e, g_5\}$. The conjugacy classes for this group are

$$C_1 = \{e\}; \qquad C_2 = \{g_1, g_2\}; \qquad C_3 = \{g_3, g_4, g_5\}.$$

The class multiplication constants are

$$c_{11}^1 = c_{21}^2 = c_{31}^3 = c_{12}^2 = c_{22}^2 = c_{13}^3 = 1;$$
$$c_{22}^1 = c_{32}^3 = c_{23}^3 = 2;$$
$$c_{33}^1 = c_{33}^2 = 3.$$

The subgroups H_2, H_3, and H_4 are conjugate to each other, e.g.,

$$g_1 H_2 g_1^{-1} = H_4; \qquad g_1 H_3 g_1^{-1} = H_2; \qquad g_1 H_4 g_1^{-1} = H_3.$$

H_1 is conjugate to itself

$$g H_1 g^{-1} = H_1$$

for all $g \in G$. Hence, H_1 is a normal subgroup.

The left cosets are

$$H_1 = eH_1 = g_1 H_1 = g_2 H_1; \qquad \{g_3, g_4, g_5\} = g_3 H_1 = g_4 H_1 = g_5 H_1;$$
$$H_2 = eH_2 = g_3 H_2; \quad \{g_1, g_4\} = g_1 H_2 = g_4 H_2; \quad \{g_2, g_5\} = g_2 H_2 = g_5 H_2;$$
$$H_3 = eH_3 = g_4 H_3; \quad \{g_2, g_3\} = g_2 H_3 = g_3 H_3; \quad \{g_1, g_5\} = g_1 H_3 = g_5 H_3;$$
$$H_4 = eH_4 = g_5 H_4; \quad \{g_1, g_3\} = g_1 H_4 = g_3 H_4; \quad \{g_2, g_4\} = g_2 H_4 = g_4 H_4.$$

In the present example we have

$$G/G = \{e\}$$
$$G/H_1 = \{\{e, g_1, g_2\}, \{g_3, g_4, g_5\}\}$$
$$G/H_2 = \{\{e, g_3\}, \{g_1, g_4\}, \{g_2, g_5\}\}$$
$$G/H_3 = \{\{e, g_4\}, \{g_2, g_3\}, \{g_1, g_5\}\}$$
$$G/H_4 = \{\{e, g_5\}, \{g_1, g_3\}, \{g_2, g_4\}\}.$$

The right cosets are

$$H_1 = H_1 e = H_1 g_1 = H_1 g_2; \qquad \{g_3, g_4, g_5\} = H_1 g_3 = H_1 g_4 = H_1 g_5;$$
$$H_2 = H_2 e = H_2 g_3; \qquad \{g_1, g_5\} = H_2 g_1 = H_2 g_5; \qquad \{g_2, g_4\} = H_2 g_2 = H_2 g_4;$$
$$H_3 = H_3 e = H_3 g_4; \qquad \{g_1, g_3\} = H_3 g_1 = H_3 g_3; \qquad \{g_2, g_5\} = H_3 g_2 = H_3 g_5;$$
$$H_4 = H_4 e = H_4 g_5; \qquad \{g_1, g_4\} = H_4 g_1 = H_4 g_4; \qquad \{g_2, g_3\} = H_4 g_2 = H_4 g_3.$$

The corresponding coset spaces $H \backslash G$ are

$$G \backslash G = \{e\}$$
$$H_1 \backslash G = \{\{e, g_1, g_2\}, \{g_3, g_4, g_5\}\}$$
$$H_2 \backslash G = \{\{e, g_3\}, \{g_1, g_5\}, \{g_2, g_4\}\}$$
$$H_3 \backslash G = \{\{e, g_4\}, \{g_1, g_3\}, \{g_2, g_5\}\}$$
$$H_4 \backslash G = \{\{e, g_5\}, \{g_1, g_4\}, \{g_2, g_3\}\}.$$

Note that $G/H_1 = H_1 \backslash G$, which follows from H_1 being a normal subgroup. The coset space G/H_1 (or $H_1 \backslash G$) is therefore a group.

The Cayley table for G/H_1 is

$\circ$	H_1	gH_1
H_1	H_1	gH_1
gH_1	gH_1	H_1

Since all groups with two elements are isomorphic to each other, we can say immediately that $G/H_1 \cong H_2 \cong H_3 \cong H_4 \cong \mathbb{Z}_2$.

Examples of double cosets and double coset spaces associated with G are

Example 7.10

$$H_1 g H_1 = g H_1$$

for all $g \in G$ (and hence $H_1 \backslash G / H_1 = G/H_1$).   ☐

Example 7.11

$$H_2 g_3 H_2 = H_2 e H_2 = e H_2 = H_2$$

and

$$H_2 g_1 H_2 = H_2 g_2 H_2 = H_2 g_4 H_2 = H_2 g_5 H_2 = g_2 H_2 \cup g_4 H_2$$

(and hence $H_2 \backslash G / H_2 = \{H_2, g_2 H_2 \cup g_4 H_2\}$ and $|H_2 \backslash G / H_2| = 2$.). If $g \in H_2$ then $H_2 g H_2 = g H_2 = H_2$ and $\gamma_g = 1$ since $H_2 / (H_2 \cap H_2) = H_2 / H_2 = \{e\}$. If $g \in g_4 H_2 \cup g_2 H_2$ then $H_2 g H_2 = g_4 H_2 \cup g_2 H_2$ and $H_2/(H_2 \cap g H_2 g^{-1}) = H_2/(H_2 \cap H_4) = H_2/\{e\} = H_2$. Hence in this case $\gamma_g = 2$. Then for any $\sigma_2 \in H_2$ and $\sigma_1 \in G - H_2$, we have $G = H_2 \sigma_1 H_2 \cup H_2 \sigma_2 H_2$.   ☐

Example 7.12

$$H_1 g H_2 = G$$

for all $g \in G$ and so $H_1 \backslash G / H_2 = \{G\}$ and $|H_1 \backslash G / H_2| = 1$.   ☐

Example 7.13

$$H_3 e H_2 = H_3 g_1 H_2 = H_3 g_3 H_2 = H_3 g_4 H_2 = \{e, g_1, g_3, g_4\} = g_1 H_2 \cup e H_2.$$

We therefore write

$$H_3 g_2 H_2 = H_3 g_5 H_2 = \{g_2, g_5\} = g_2 H_2$$

and so $H_3 \backslash G / H_2 = \{g_1 H_2 \cup e H_2, g_2 H_2\}$ and $|H_3 \backslash G / H_2| = 2$.   ☐

7.3   Lie Groups

In this section we examine a very important kind of group. These are the Lie groups.[8] We begin the first subsection of this section with some motivation and an example. In the second subsection the rigorous definitions are provided. In previous chapters we already examined in great detail two very important Lie groups for engineering applications: the rotation group and the Euclidean motion group. We revisit rotation and rigid-body motion with the tools of group theory.

7.3.1   An Intuitive Introduction to Lie Groups

From elementary calculus we know that for a constant $x \in \mathbb{C}$ and a complex-valued function $y(t)$ that

$$\frac{dy}{dt} = xy; \qquad y(0) = y_0 \qquad \Rightarrow \qquad y(t) = e^{xt} y_0.$$

Likewise if $Y(t)$ and X take values in $\mathbb{C}^{N \times N}$, we have

$$\frac{dY}{dt} = XY; \qquad Y(0) = Y_0 \qquad \Rightarrow \qquad Y(t) = e^{tX} Y_0.$$

Here the matrix exponential is defined via the same power series that defines the scalar exponential function

$$e^{tX} = \mathbb{I}_{N \times N} + \sum_{n=1}^{\infty} \frac{t^n X^n}{n!}.$$

If we specify that $Y_0 = \mathbb{I}_{N \times N}$, then $e^{0X} = \mathbb{I}_{N \times N}$. In general, the exponential of a matrix with finite entries will be invertible because (see Appendix D)

$$\det(e^X) = e^{trace(X)} \neq 0.$$

By multiplying out the power series for $e^{t_1 X}$ and $e^{t_2 X}$, one also sees that $e^{t_1 X} e^{t_2 X} = e^{(t_1 + t_2) X}$. Choosing $t_1 = t$ and $t_2 = -t$, the result of this product of exponentials is the identity matrix. Since matrix multiplication is associative, it follows that all the properties of an Abelian group are satisfied by the matrix exponential e^{tX} under the operation of matrix multiplication. Since $t \in \mathbb{R}$, this is an example of a one-dimensional group. This is one of the simplest examples of a Lie group. For an intuitive understanding at this point, it is sufficient to consider a Lie group to be one which is continuous in the same sense that $\mathbb{R}^N$ is. The group operation of a Lie group is also continuous in the sense that the product of slightly perturbed group elements are sent to an element "near" the product

---

[8] These are named after the Norwegian mathematician Marius Sophus Lie (1842–1899).

of the unperturbed elements. For a Lie group, the inverse of a perturbed element is also "close to" the inverse of the unperturbed one. These conditions are referred to as the continuity of the product and inversion of group elements.

If $X_1, X_2 \in \mathbb{C}^{N \times N}$, then in general $e^{X_1} e^{X_2} \neq e^{X_2} e^{X_1}$. A sufficient (though not necessary) condition for equality to hold is $X_1 X_2 = X_2 X_1$ (see Appendix D).

Now, if there are a set of linearly independent matrices $\{X_1, \ldots, X_n\}$, we calculate $e^{t_i X_i}$, and it happens to be the case that the product of the resulting exponentials is closed under all possible multiplications and all values of $t_1, \ldots, t_n \in \mathbb{R}$, the result is an n-dimensional continuous group.[9] When working with matrices $A_i \in \mathbb{C}^{N \times N}$, norms of differences provide a measure of distance $(d(A_i, A_j) = \|A_i - A_j\|)$ that can be used to verify continuity of the product and inversion of group elements. A Lie group that is also a set of matrices is called a *matrix Lie group*.

For example, given

$$X_1 = \begin{pmatrix} 0 & 1 & 0 \\ 0 & 0 & 0 \\ 0 & 0 & 0 \end{pmatrix} \qquad X_2 = \begin{pmatrix} 0 & 0 & 1 \\ 0 & 0 & 0 \\ 0 & 0 & 0 \end{pmatrix} \qquad X_3 = \begin{pmatrix} 0 & 0 & 0 \\ 0 & 0 & 1 \\ 0 & 0 & 0 \end{pmatrix}$$

the corresponding matrix exponentials are

$$e^{t_1 X_1} = \begin{pmatrix} 1 & t_1 & 0 \\ 0 & 1 & 0 \\ 0 & 0 & 1 \end{pmatrix} \qquad e^{t_2 X_2} = \begin{pmatrix} 1 & 0 & t_2 \\ 0 & 1 & 0 \\ 0 & 0 & 1 \end{pmatrix} \qquad e^{t_3 X_3} = \begin{pmatrix} 1 & 0 & 0 \\ 0 & 1 & t_3 \\ 0 & 0 & 1 \end{pmatrix}.$$

This choice of the X_i is special because they are *nilpotent* matrices, i.e., they satisfy $(X_i)^K = 0_{N \times N}$ for some finite $K \in \mathbb{Z}^+$ (where in this case $K = 2$ and $N = 3$). It is also true that any weighted sum of the matrices X_i is nilpotent: $(\alpha X_1 + \beta X_2 + \gamma X_3)^3 = 0_{3 \times 3}$. Hence, the matrix exponentials in this case have a very simple form. It is clear that no matter in what order we multiply the above matrix exponentials, the result will be one of the form

$$g = \begin{pmatrix} 1 & \alpha & \beta \\ 0 & 1 & \gamma \\ 0 & 0 & 1 \end{pmatrix}$$

for some value of $\alpha, \beta, \gamma \in \mathbb{R}$. In particular, one can make the Euler-angle-like factorization

$$g = \begin{pmatrix} 1 & 0 & \beta \\ 0 & 1 & 0 \\ 0 & 0 & 1 \end{pmatrix} \begin{pmatrix} 1 & 0 & 0 \\ 0 & 1 & \gamma \\ 0 & 0 & 1 \end{pmatrix} \begin{pmatrix} 1 & \alpha & 0 \\ 0 & 1 & 0 \\ 0 & 0 & 1 \end{pmatrix}.$$

Furthermore, we see that

$$g_1 \circ g_2 = \begin{pmatrix} 1 & \alpha_1 & \beta_1 \\ 0 & 1 & \gamma_1 \\ 0 & 0 & 1 \end{pmatrix} \begin{pmatrix} 1 & \alpha_2 & \beta_2 \\ 0 & 1 & \gamma_2 \\ 0 & 0 & 1 \end{pmatrix} = \begin{pmatrix} 1 & \alpha_1 + \alpha_2 & \beta_1 + \beta_2 + \alpha_1 \gamma_2 \\ 0 & 1 & \gamma_1 + \gamma_2 \\ 0 & 0 & 1 \end{pmatrix},$$

and so there is closure in the sense that the product of matrices of this form results in a matrix of the same form. It is also clear that the choice $\alpha_2 = -\alpha_1$, $\gamma_2 = -\gamma_1$, and $\beta_2 = -\beta_1 + \alpha_1 \gamma_1$ means $g_2 = g_1^{-1}$ exists for every $g_1 \in G$. Hence this set is a group under matrix multiplication. This group is isomorphic to the 3-D *Heisenberg group* and 1-D *Galilei group*. It is an example of a more general class of groups called nilpotent Lie groups.

---

[9] The number n (the dimension of the group) is generally different from N (the dimension of the matrix).

Every element of the Heisenberg group can be generated by the exponential

$$\exp \begin{pmatrix} 0 & \alpha' & \beta' \\ 0 & 0 & \gamma' \\ 0 & 0 & 0 \end{pmatrix} = \begin{pmatrix} 1 & \alpha' & \beta' + \alpha'\gamma'/2 \\ 0 & 1 & \gamma' \\ 0 & 0 & 1 \end{pmatrix}.$$

7.3.2   Rigorous Definitions

A *Lie group* $(G, \circ)$ is a group for which the set G is an analytic manifold, with the operations $(g_1, g_2) \to g_1 \circ g_2$ and $g \to g^{-1}$ being analytic (see Appendix G for definitions) [38]. The dimension of a Lie group is the dimension of the associated manifold G.

Since a metric (distance) function $d(\cdot, \cdot)$ exists for G, we have a way to measure "nearness," and hence continuity is defined in the usual way.

A *matrix Lie group* $(G, \circ)$ is a Lie group where G is a set of square matrices and the group operation is matrix multiplication. Another way to say this is that a matrix Lie group is a (closed) Lie subgroup of $GL(N, \mathbb{R})$ or $GL(N, \mathbb{C})$ for some $N \in \{1, 2, \dots\}$. We will deal exclusively with matrix Lie groups.

Given a matrix Lie group, elements sufficiently close to the identity are written as $g(t) = e^{tX}$ for some $X \in \mathcal{G}$ (the Lie algebra of G) and t near 0. (See Appendix C for definition of Lie algebra.) For matrix Lie groups, the corresponding Lie algebra is denoted with small letters. For example, the Lie algebras of the groups $GL(N, \mathbb{R})$, $SO(N)$, and $SE(N)$ are respectively denoted as $gl(N, \mathbb{R})$, $so(N)$, and $se(N)$.

In the case of the rotation and motion groups, the exponential mapping is surjective (see Chapters 5 and 6), so all group elements can be parameterized as matrix exponentials of Lie algebra elements.

The matrices $\frac{dg}{dt}g^{-1}$ and $g^{-1}\frac{dg}{dt}$ are respectively called the *tangent vectors* or *tangent elements* at g. Evaluated at $t = 0$, these vectors reduce to X.

The *adjoint* operator is defined as

$$Ad\,(g_1)\,X = \frac{d}{dt}\left(g_1 e^{tX} g_1^{-1}\right)|_{t=0} = g_1 X g_1^{-1}. \tag{7.28}$$

This gives a homomorphism $Ad : G \to GL(\mathcal{G})$ from the group into the set of all invertible linear transformations of $\mathcal{G}$ onto itself. It is a homomorphism because

$$Ad\,(g_1)\,Ad\,(g_2)\,X = g_1 \left(g_2 X g_2^{-1}\right) g_1^{-1} = (g_1 g_2)\,X\,(g_1 g_2)^{-1} = Ad\,(g_1 g_2)\,X.$$

It is linear because

$$Ad(g)\,(c_1 X_1 + c_2 X_2) = g\,(c_1 X_1 + c_2 X_2)\,g^{-1} = c_1 g X_1 g^{-1} + c_2 g X_2 g^{-1}$$
$$= c_1 Ad(g)X_1 + c_2 Ad(g)X_2.$$

In the special case of a 1-parameter subgroup when $g = g(t)$ is an element close to the identity,[10] we can approximate $g(t) \approx \mathbb{I}_{N \times N} + tX$ for small t. Then we get $Ad(\mathbb{I}_{N \times N} + tX)Y = Y + t(XY - YX)$. The quantity

$$XY - YX = [X, Y] = \frac{d}{dt}\,(Ad(g(t))Y)\,|_{t=0} \overset{\triangle}{=} ad(X)Y \tag{7.29}$$

---

[10]In the context of matrix Lie groups, distance is defined naturally as a matrix norm of the difference of two group elements.

is called the *Lie bracket* of the elements $X, Y \in \mathcal{G}$. The last equality in Eq. (7.29) defines $ad(X)$. One observes from this definition that

$$Ad(\exp tX) = \exp(t \cdot ad(X)). \tag{7.30}$$

It is clear from the definition in Eq. (7.29) that the Lie bracket is linear in each entry

$$[c_1 X_1 + c_2 X_2, Y] = c_1 [X_1, Y] + c_2 [X_2, Y]$$

and

$$[X, c_1 Y_1 + c_2 Y_2] = c_1 [X, Y_1] + c_2 [X, Y_2].$$

Furthermore, the Lie bracket is anti-symmetric

$$[X, Y] = -[Y, X], \tag{7.31}$$

and hence $[X, X] = 0$. Given a basis $\{X_1, \ldots, X_n\}$ for the Lie algebra $\mathcal{G}$, the Lie bracket of any two elements will result in a linear combination of all basis elements. This is written as

$$[X_i, X_j] = \sum_{k=1}^{n} C_{ij}^k X_k.$$

The constants C_{ij}^k are called the *structure constants* of the Lie algebra $\mathcal{G}$. Note that in contrast to the class multiplication constants c_{ij}^k that we saw in Section 7.2.5, the structure constants are antisymmetric: $C_{ij}^k = -C_{ji}^k$.

It can be checked that for any three elements of the Lie algebra, the *Jacobi identity* is satisfied

$$[X_1, [X_2, X_3]] + [X_2, [X_3, X_1]] + [X_3, [X_1, X_2]] = 0. \tag{7.32}$$

As a result of the Jacobi identity, $ad(X)$ satisfies

$$ad([X, Y]) = ad(X)ad(Y) - ad(Y)ad(X).$$

An important relationship called the *Baker-Campbell-Hausdorff formula* exists between the Lie bracket and matrix exponential (see [4, 9, 16]). Namely, the product of two Lie group elements written as exponentials of Lie algebra elements can be expressed as

$$e^X e^Y = e^{Z(X,Y)}$$

where

$$Z(X, Y) = X + Y + \frac{1}{2}[X, Y] + \frac{1}{12}([X, [X, Y]] + [Y, [Y, X]]) + \cdots.$$

This expression is verified by expanding e^X, e^Y, and $e^{Z(X,Y)}$ in Taylor series and comparing terms.

In summary, the matrix exponential provides a tool for parameterizing connected Lie groups. We already saw two ways of parameterizing the Heisenberg group using the matrix exponential. We can exponentiate each basis element of the Lie algebra and multiply the results (in any order), or we can exponentiate a weighted sum of all the basis elements. Both parameterization methods are suitable in more general contexts. We have already seen this in great detail in the context of the rotation and Euclidean motion groups.

7.3.3 Examples

Examples of Lie groups which act on $\mathbb{R}^N$ (N-dimensional Euclidean space) include

1. The rotation group $SO(N)$, which consists of all $N \times N$ special orthogonal matrices R (i.e., $R R^T = I$, $\det R = +1$). As with all matrix groups, the group operation is matrix multiplication, and elements of $SO(N)$ act on $x \in \mathbb{R}^N$ as $x' = Rx \in \mathbb{R}^N$. $SO(N)$ is an $N(N-1)/2$-dimensional Lie group. The corresponding Lie algebra is denoted $so(N)$.

2. The group $SO(p, q)$ consists of all $N \times N$ matrices ($N = p + q$) with unit determinant that preserve the matrix

$$J(p, q) = \begin{pmatrix} -\mathbb{I}_{p \times p} & 0_{p \times q} \\ 0_{q \times p} & \mathbb{I}_{q \times q} \end{pmatrix}.$$

That is, $Q \in SO(p, q)$ satisfies

$$Q^T J(p, q) Q = J(p, q).$$

$SO(p, q)$ is an $N(N-1)/2$-dimensional Lie group. The corresponding Lie algebra is denoted $so(p, q)$.

3. The special Euclidean group (also called the Euclidean motion group, Euclidean group, or simply the motion group) $SE(N) = \mathbb{R}^N \rtimes_\varphi SO(N)$, which consists of all pairs $g = (R, \mathbf{b})$, where $R \in SO(N)$ and $\mathbf{b} \in \mathbb{R}^N$, has the group law $g_1 \circ g_2 = (R_1 R_2, R_1 b_2 + b_1)$ and acts on $\mathbb{R}^N$ as $x' = Rx + \mathbf{b}$. $SE(N)$ is the group of rigid body motions of N-dimensional Euclidean space and is itself an $N(N+1)/2$-dimensional Lie group. The corresponding Lie algebra is denoted $se(N)$. Discrete subgroups of $SE(N)$ include the crystallographic space groups, i.e., the symmetry groups of regular lattices.

4. The scale-Euclidean (or similitude) group, $SIM(N)$, which consists of all pairs $(e^a R, b)$, has the group law $g_1 \circ g_2 = (e^{a_1 + a_2} R_1 R_2, R_1 b_2 + b_1)$ and acts on $\mathbb{R}^N$ by translation, rotation, and dilation as $x' = e^a Rx + b$. It is a $(1 + N(N+1)/2)$-dimensional Lie group, with corresponding Lie algebra $sim(N)$.

5. The real special linear group $SL(N, \mathbb{R})$, which consists of all $N \times N$ matrices L with $\det L = +1$ acts on $\mathbb{R}^N$ ($x' = Lx$) by rotation, stretch, and shear in such a way that volume is preserved, and parallel lines remain parallel. It is an $(N^2 - 1)$-dimensional Lie group with corresponding Lie algebra $sl(N, \mathbb{R})$.

6. The special affine group of $\mathbb{R}^N$ [denoted $\mathbb{R}^N \rtimes_\varphi GL^+(N, \mathbb{R})$], which is the set of all pairs $g = (e^a L, b)$, acts on objects in $\mathbb{R}^N$ by translation, rotation, shear, stretch, and dilation.[11] In short, this is the most general type of deformation of Euclidean space which transforms parallel lines into parallel lines. The special affine group is $N(N+1)$-dimensional.

7.3.4 Demonstration of Theorems with $SO(3)$

In this section we demonstrate many of the definitions and results of theorems using the rotation group $SO(3)$ and related groups, and sets on which these groups act.

---

[11]The notation $GL^+(N, \mathbb{R})$ stands for the subgroup of the *general linear group* consisting of $N \times N$ matrices with real entries and positive determinants.

The Baker-Campbell-Hausdorff Formula

For $SO(3)$ the Baker-Campbell-Hausdorff formula

$$e^X e^Y = e^{Z(X,Y)}$$

can be interpreted as a statement of what the axis and angle of rotation are for the composition of two rotations. That is, in the case of $SO(3)$

$$\exp(\theta_1 N_1)\exp(\theta_2 N_2) = \exp(\theta_3 N_3),$$

and we can find θ_3 and N_3 for given (θ_i, N_i) for $i = 1, 2$.

By expanding out

$$\exp(\theta_i N_i) = \mathbb{I} + \sin\theta_i N_i + (1 - \cos\theta_i) N_i^2$$

and matching terms, one finds [34]

$$\cos\frac{\theta_3}{2} = \cos\frac{\theta_1}{2}\cos\frac{\theta_2}{2} - (\mathbf{n}_1 \cdot \mathbf{n}_2)\sin\frac{\theta_1}{2}\sin\frac{\theta_2}{2}$$

and

$$N_3 = aN_1 + bN_2 + 2c[N_1, N_2]$$

where

$$a\sin\frac{\theta_3}{2} = \sin\frac{\theta_1}{2}\cos\frac{\theta_2}{2}$$

$$b\sin\frac{\theta_3}{2} = \cos\frac{\theta_1}{2}\sin\frac{\theta_2}{2}$$

$$c\sin\frac{\theta_3}{2} = \sin\frac{\theta_1}{2}\sin\frac{\theta_2}{2}.$$

Isotropy Groups, Homogeneous Manifolds, and Orbits

$R \in SO(3)$ acts on $\mathbf{x} \in \mathbb{R}^3$ in a natural way with matrix multiplication. If we consider the set of all rotations that keep a particular $\mathbf{x} \in \mathbb{R}^3$ fixed, one finds this set to be all rotation matrices of the form

$$\text{ROT}(\mathbf{x}/\|\mathbf{x}\|, \theta) = S(\mathbf{x})R_3(\theta)S^{-1}(\mathbf{x})$$

for $\theta \in (0, 2\pi)$ where S is any rotation matrix with third column $\mathbf{x}/\|\mathbf{x}\|$. The set of all matrices of this form is the little group (stability subgroup or isotropy subgroup) of $\mathbf{x}$. It is isomorphic to $SO(2)$ since each element in the little group is related to a planar rotation by conjugation. Since the little group of $\mathbf{x}$ and $\mathbf{x}/\|\mathbf{x}\|$ are the same, it suffices to consider the little groups of points on the unit sphere. Each point on S^2 defines the axis of a little group, and every little group can be identified with a point in S^2. Since every little group is isomorphic with $SO(2)$, one writes

$$SO(3)/SO(2) \cong S^2.$$

Hence the homogeneous space $SO(3)/SO(2)$ [set of all left cosets of $SO(2)$ in $SO(3)$] is identified with the unit sphere S^2. Similarly for right cosets we write

$$SO(2)\backslash SO(3) \cong S^2.$$

The unit vector specifying the axis of rotation of each little group serves as a name tag by which the little group is identified. We also note that

$$SU(2)/U(1) \cong S^2/\mathbb{Z}_2.$$

$SO(3)$ divides $\mathbb{R}^3$ into concentric spherical orbits centered at the origin. Each orbit is specified by its radius, and so the set of all orbits is equated with the set of all radii

$$\mathbb{R}^3/SO(3) = \mathbb{R}^+ \cup \{0\}.$$

Here we use the notation $/$ in the sense of X/G where G is a group acting on X. Likewise, the orbits of S^2 under the action of $SO(2)$ are parallel circles in the planes normal to the axis of rotation. This set of circles can be identified with either the set of points $[-1, 1]$ along the axis of rotation, or with the Euler angle $0 \leq \beta \leq \pi$. In the latter case we write

$$S^2/SO(2) \cong [0, \pi].$$

Since $S^2 \cong SO(2)\backslash SO(3)$, it follows that we can write the *double coset decomposition*

$$SO(2)\backslash SO(3)/SO(2) \cong [0, \pi].$$

Since there is a two-to-one surjective homomorphism from $SU(2)$ onto $SO(3)$, the homomorphism theorem says that

$$SU(2)/[\mathbb{I}_{2\times 2}, -\mathbb{I}_{2\times 2}] \cong SO(3).$$

This is an example of the quotient of a Lie group with respect to a finite normal subgroup being isomorphic to a Lie group of the same dimension as the original group. Another example of this is

$$SO(2) \cong \mathbb{R}/2\pi\mathbb{Z}.$$

Other homogeneous spaces that arise are

$$SO(N)/SO(N-1) \cong S^{N-1}$$

and

$$SO(N, 1)/SO(N) \cong H^N$$

(where H^N is a hyperboloid in $\mathbb{R}^{N+1}$).

Quotient groups that are of general interest are

$$GL(N, \mathbb{R})/SL(N, \mathbb{R}) \cong \mathbb{R} - \{0\}$$

and

$$GL^+(N, \mathbb{R})/SL(N, \mathbb{R}) \cong \mathbb{R}^+.$$

These result from the homomorphism theorem with the homomorphisms det $: GL(N, \mathbb{R}) \to \mathbb{R} - \{0\}$ and det $: GL^+(N, \mathbb{R}) \to \mathbb{R}^+$, respectively.

Conjugacy Classes and Class Functions

Since every rotation can be parameterized with the exponential map $\exp(\theta X)$, which is the same as a conjugated matrix $R_3(\theta)$, it follows that functions of the form

$$\chi(\theta) = \text{trace}(\exp(\theta X)) = 2\cos\theta + 1$$

are class functions for $SO(3)$ where θ is the angle of rotation and X is normalized. One can reason further that any function $\tilde{f}$ of a class function is also a class function since $\tilde{f}(\chi(\theta))$ is also invariant under conjugation. Choosing $f = \tilde{f} \circ \chi$ (here $\circ$ means composition of functions) results in the class function $f(\theta)$. Since $\chi(\theta)$ is even, it follows that $f(\theta) = f(-\theta)$.

From the preceding discussion, it is clear that every even function of only the angle of rotation is a class function of $SO(3)$. The following theorem makes a stronger statement.

THEOREM 7.8
A function in $\mathcal{L}^2(SO(3))$ with nonzero measure is a class function if and only if it is an even function of the angle (and not the axis) of rotation.

PROOF   By definition, matrix invariants are functions of matrix-valued argument that are invariant under similarity transformation: $I_i(A) = I_i(SAS^{-1})$ for all invertible S of compatible dimension. The square-integrable class functions for $SO(3)$ with nonzero measure must be functions of the three matrix invariants of $A \in SO(3)$. That is, $f(A) = f_1(I_1(A), I_2(A), I_3(A))$, where $\det(\lambda 1 - A) = \lambda^3 - I_1(A)\lambda^2 + I_2(A)\lambda - I_3(A) = 0$. Note that $I_3(A) = \det(A) = +1$, and any element of $SO(3)$ can be written as $A(\theta, \omega) = R\, Z(\theta)\, R^T$ where θ and ω are the angle and axis of rotation, $Z(\theta)$ is rotation around z-axis, and R is any rotation matrix R with third column ω. Any element of $SO(3)$ can be parameterized as

$$A(\theta, \omega) = e^{\theta[\omega]} = \mathbb{I}_{3\times 3} + \sin\theta[\omega] + (1 - \cos\theta)[\omega]^2$$

where θ is the angle of rotation, and ω is the unit vector specifying the axis of rotation. The matrix $[\omega]$ is the skew-symmetric matrix such that $[\omega]x = \omega \times x$ for any vector $x \in \mathbb{R}^3$. From this, it is clear that $I_1(A) = \text{trace}(A) = 2\cos\theta + 1$ and $I_2(A) = \frac{1}{2}[(\text{trace}(A))^2 - \text{trace}(A^2)]$ are even functions of the rotation angle only, and so well-behaved class functions on $SO(3)$ must be of the form $f(A) = f_2(\theta) = f_2(-\theta)$. On the other hand, any square-integrable even function $f_2(\theta)$ can be written in the form $f_1(I_1(A), I_2(A), I_3(A))$ for an appropriate choice of f_1 since a power series in $I_1(A)$ and $I_2(A)$ [or just in $I_1(A)$] is equivalent to a Fourier series of an even function of θ, and a Fourier series converges for any $f(\theta) \in \mathcal{L}^2(SO(3))$.   ∎

Classification of Finite Subgroups

In this section we use Eq. (7.21) to determine conditions that any finite subgroup $G < SO(3)$ must satisfy. Other approaches that yield the same result can be found in [1, 11]. To be consistent with the literature, we simplify the form of the equations and write $n = |G|$, $r = |P/G|$, and $n_i = |G_m|$, where m is in the ith orbit. Since every rotation other than the identity keeps two (antipodal) points fixed, it follows in the case of a finite rotation that

$$\sum_{h \in G - \{e\}} |F(h)| = 2(|G| - 1).$$

Then Eq. (7.21) is rewritten as

$$2(n - 1) = \sum_{i=1}^{r} \frac{n}{n_i}(n_i - 1),$$

or equivalently

$$2 - \frac{2}{n} = r - \sum_{i=1}^{r} \frac{1}{n_i}. \tag{7.33}$$

Since n, r, and n_i must all be positive integers, the set of possible solutions is very small. In fact, it may be shown that Eq. (7.33) can only hold when $r = 2$ or $r = 3$. When $r = 2$, the condition

$n_1 = n_2 = n$ is the only way in which Eq. (7.33) is satisfied. This is nothing more than the finite subgroup of rotations in a plane by angles $2\pi/n$ (called the *cyclic group*) denoted as C_n. When $r = 3$, there are three possibilities for the values $(n_1, n_2, n_3; n)$ (where the numbers are arranged so that $n_1 \leq n_2 \leq n_3 < n$). These are:

the *dihedral* groups for any integer $k \geq 2$

$$D_k : \quad (2, 2, k; 2k);$$

the *tetrahedral* group

$$T_{12} : \quad (2, 3, 3; 12);$$

the *cubo-octahedral* group

$$O_{24} : \quad (2, 3, 4; 24);$$

and the *icosa-dodecahedral* group

$$I_{60} : \quad (2, 3, 5; 60).$$

D_k is the group of discrete spatial rotations that bring a regular k-gon back into itself. It includes C_k rotations about the normal to the face of the polygon through its center, as well as rotations by π around the axes through the center and each vertex of the polygon. When k is even, rotations by π around the axes passing through the edges and center of the polygon are also in D_k.

T_{12}, O_{24}, and I_{60} are the groups of rotational symmetry operations that preserve the platonic solids for which they are named. An alternate derivation of these results can be found in [11].

The platonic solids arise in nature in several contexts. For instance, the simplest viruses (i.e., those with the smallest genetic code) have evolved so that they only have one kind of protein in their protective shell. In order to best approximate a sphere (which is the surface that has the maximum volume to surface area), the protein molecules that these viruses produce fold into sections of the viral shell that self-assembles into an icosahedron [10, 26, 31].

7.3.5   Calculating Jacobians

Given a finite-dimensional matrix Lie group, an orthonormal basis for the Lie algebra can always be found when an appropriate inner product is defined. Such a basis can be constructed by the Gram-Schmidt orthogonalization procedure starting with any Lie algebra basis (see Appendix C).

One can define an inner product between elements of the Lie algebra as

$$(X, Y) = \frac{1}{2}\mathrm{Re}\left[\mathrm{trace}\left(XWY^\dagger\right)\right] \tag{7.34}$$

where W is a Hermitian weighting matrix with positive eigenvalues, and $\mathrm{Re}[z]$ is the real part of the complex number z. In the case of a real matrix Lie group, the Lie algebra will consist of real matrices, and so

$$(X, Y) = \frac{1}{2}\mathrm{trace}\left(XWY^T\right),$$

where W is a real symmetric positive definite matrix. Usually, but not always, W will be a scalar multiple of the identity.

It is easy to see that Eq. (7.34) follows the general definition of an inner product on a *real* vector space given in the Appendix C (in the mathematician's sense rather than the physicist's) because even though the Lie algebra basis elements X_i may be complex, all the scalars that multiply these basis elements will be real.

Given an orthonormal basis $X_1, \ldots, X_n$ for the Lie algebra, we can project the left and right tangent operators onto this basis to yield elements of the right- and left-Jacobian matrices

$$(J_R)_{ij} = \left(g^{-1} \frac{\partial g}{\partial x_i}, X_j \right) \quad \text{and} \quad (J_L)_{ij} = \left(\frac{\partial g}{\partial x_i} g^{-1}, X_j \right) \tag{7.35}$$

where $g = g(x_1, \ldots, x_n)$. Note that $J_R(h \circ g) = J_R(g)$ and $J_L(g \circ h) = J_L(g)$. Here $\{x_1, \ldots, x_n\}$ is the set of local coordinates used to parameterize a neighborhood of the group around the identity. For the groups considered here, these parameterizations extend over the whole group with singularities of measure zero.

The above right and left Jacobians are natural extensions of the Jacobians presented in Chapter 4, with an appropriate concept of inner product of vectors in the tangent space to the group defined at every group element. For an n-dimensional matrix Lie group, we denote two special kinds of vector fields as

$$V_L(g) = \sum_{i=1}^n v_i X_i g \quad \text{and} \quad V_R(g) = \sum_{i=1}^n v_i g X_i.$$

The subscripts of V denote on which side the Lie algebra basis element X_i appears. Here $X_i g$ and $g X_i$ are simply matrix products, and $\{v_i\}$ are real numbers.[12] For a vector field $V(g)$ on a matrix Lie group (which need not be left- or right-invariant), the left- and right-shift operations are defined as

$$L(h)V(g) = h V(g) \quad \text{and} \quad R(h)V(g) = V(g) h$$

where $h \in G$. Then it is clear that V_R is *left-invariant* and V_L is *right-invariant* in the sense that

$$L(h)V_R(g) = V_R(h \circ g) \quad \text{and} \quad R(h)V_L(g) = V_L(g \circ h).$$

This means that there are left- and right-invariant ways to extend the inner product $(\cdot, \cdot)$ on the Lie algebra over the whole group. Namely, for all $Y, Z \in \mathcal{G}$, we can define right and left inner products respectively as

$$(gY, gZ)_g^R \triangleq (Y, Z) \quad \text{and} \quad (Yg, Zg)_g^L \triangleq (Y, Z)$$

for any $g \in G$. In this way, the inner product of two invariant vector fields $Y_R(g)$ and $Z_R(g)$ [or $Y_L(g)$ and $Z_L(g)$] yields

$$\left((Y_R(g), Z_R(g))_g^R \right) = \left((Y_L(g), Z_L(g))_g^L \right) = (Y, Z).$$

Let G be a matrix Lie group with $\{X_i\}$ being an orthonormal basis for $\mathcal{G}$. Then when $Y = \sum_{i=1}^n y_i X_i$ and $Z = \sum_{i=1}^n z_i X_i$, it follows that

$$(Y, Z) = \sum_{i=1}^n y_i z_i.$$

In this notation, the (i, j) entry of the left Jacobian is $\partial g / \partial x_i$ projected on the jth basis element for the tangent space to G at g as

$$(J_L)_{ij} = \left(\frac{\partial g}{\partial x_i}, X_j g \right)_g^L = \left(\frac{\partial g}{\partial x_i} g^{-1}, X_j \right).$$

Likewise,

$$(J_R)_{ij} = \left(\frac{\partial g}{\partial x_i}, g X_j \right)_g^R = \left(g^{-1} \frac{\partial g}{\partial x_i}, X_j \right).$$

---

[12] We restrict the discussion to real vector fields.

$SE(2)$ Revisited

According to the formulation of Chapter 5, $SE(2)$ can be parameterized as

$$g(x_1, x_2, \theta) = \exp(x_1 X_1 + x_2 X_2) \exp(\theta X_3) = \begin{pmatrix} \cos\theta & -\sin\theta & x_1 \\ \sin\theta & \cos\theta & x_2 \\ 0 & 0 & 1 \end{pmatrix}$$

where

$$X_1 = \begin{pmatrix} 0 & 0 & 1 \\ 0 & 0 & 0 \\ 0 & 0 & 0 \end{pmatrix}; \qquad X_2 = \begin{pmatrix} 0 & 0 & 0 \\ 0 & 0 & 1 \\ 0 & 0 & 0 \end{pmatrix}; \qquad X_3 = \begin{pmatrix} 0 & -1 & 0 \\ 1 & 0 & 0 \\ 0 & 0 & 0 \end{pmatrix}.$$

Using the weighting matrix

$$W = \begin{pmatrix} 1 & 0 & 0 \\ 0 & 1 & 0 \\ 0 & 0 & 2 \end{pmatrix},$$

one observes that $(X_i, X_j) = \delta_{ij}$.

The Jacobians for this parameterization, basis, and weighting matrix are then of the form

$$J_R = \begin{pmatrix} \cos\theta & -\sin\theta & 0 \\ \sin\theta & \cos\theta & 0 \\ 0 & 0 & 1 \end{pmatrix}$$

and

$$J_L = \begin{pmatrix} 1 & 0 & 0 \\ 0 & 1 & 0 \\ x_2 & -x_1 & 1 \end{pmatrix}.$$

The J_L and J_R calculated in this way are identical to those calculated using the technique of Chapter 6. Note that

$$\det(J_L) = \det(J_R) = 1.$$

We now examine the Jacobians for the exponential parameterization

$$g(v_1, v_2, \alpha) = \exp(v_1 X_1 + v_2 X_2 + \alpha X_3) = \exp\begin{pmatrix} 0 & -\alpha & v_1 \\ \alpha & 0 & v_2 \\ 0 & 0 & 0 \end{pmatrix}.$$

The closed-form expression for this exponential was given previously (see Section 6.1.1). We find

$$J_R = \begin{pmatrix} \frac{\sin\alpha}{\alpha} & \frac{\cos\alpha - 1}{\alpha} & 0 \\ \frac{1-\cos\alpha}{\alpha} & \frac{\sin\alpha}{\alpha} & 0 \\ \frac{\alpha v_1 - v_2 + v_2\cos\alpha - v_1\sin\alpha}{\alpha^2} & \frac{v_1 + \alpha v_2 - v_1\cos\alpha - v_2\sin\alpha}{\alpha^2} & 1 \end{pmatrix}.$$

$$J_L = \begin{pmatrix} \frac{\sin\alpha}{\alpha} & \frac{1-\cos\alpha}{\alpha} & 0 \\ \frac{\cos\alpha - 1}{\alpha} & \frac{\sin\alpha}{\alpha} & 0 \\ \frac{\alpha v_1 + v_2 - v_2\cos\alpha - v_1\sin\alpha}{\alpha^2} & \frac{-v_1 + \alpha v_2 + v_1\cos\alpha - v_2\sin\alpha}{\alpha^2} & 1 \end{pmatrix}.$$

It follows that

$$\det(J_L) = \det(J_R) = \frac{-2(\cos\alpha - 1)}{\alpha^2}.$$

Jacobians for the 'ax + b' Group

After having examined rotations and general rigid-body motions in some detail, one may be tempted to believe that bi-invariant integration measures always exist for Lie groups. We now provide an example that demonstrates otherwise.

The so-called "$ax + b$ group," or affine group of the line, may be viewed as the set of all matrices of the form

$$g(a, b) = \begin{pmatrix} a & b \\ 0 & 1 \end{pmatrix}.$$

It acts on the real line as

$$\begin{pmatrix} x' \\ 1 \end{pmatrix} = g(a, b) \begin{pmatrix} x \\ 1 \end{pmatrix}.$$

That is, $x' = ax + b$ (hence the name).

An orthonormal basis for the Lie algebra of this group is

$$X_1 = \begin{pmatrix} 1 & 0 \\ 0 & 0 \end{pmatrix}; \qquad X_2 = \begin{pmatrix} 0 & 1 \\ 0 & 0 \end{pmatrix}.$$

The weighting matrix W for which this basis is orthonormal is

$$W = \begin{pmatrix} 2 & 0 \\ 0 & 2 \end{pmatrix}.$$

A straightforward calculation shows that

$$J_R = \begin{pmatrix} 1/a & 0 \\ 0 & 1/a \end{pmatrix}; \qquad J_L = \begin{pmatrix} 1/a & 0 \\ -b/a & 1 \end{pmatrix}^T.$$

Hence,

$$\det(J_R) = \frac{1}{a^2} \neq \det(J_L) = \frac{1}{a}.$$

Jacobians for $GL(2, \mathbb{R})$

Elements of $GL(2, \mathbb{R})$ are invertible 2×2 real matrices

$$g(x_1, x_2, x_3, x_4) = \begin{pmatrix} x_1 & x_2 \\ x_3 & x_4 \end{pmatrix}.$$

An orthonormal basis for the Lie algebra $gl(2, \mathbb{R})$ is

$$X_1 = \begin{pmatrix} 1 & 0 \\ 0 & 0 \end{pmatrix}; \qquad X_2 = \begin{pmatrix} 0 & 1 \\ 0 & 0 \end{pmatrix}; \qquad X_3 = \begin{pmatrix} 0 & 0 \\ 1 & 0 \end{pmatrix}; \qquad \text{and} \qquad X_4 = \begin{pmatrix} 0 & 0 \\ 0 & 1 \end{pmatrix}.$$

The weighting matrix for which these basis elements are orthonormal is $W = 2\mathbb{I}_{2 \times 2}$.

The Jacobians in this parameterization, basis, and weighting are

$$J_R = \frac{1}{\det g} \begin{pmatrix} x_4 & 0 & -x_2 & 0 \\ 0 & x_4 & 0 & -x_2 \\ -x_3 & 0 & x_1 & 0 \\ 0 & -x_3 & 0 & x_1 \end{pmatrix}$$

and

$$J_L = \frac{1}{\det g} \begin{pmatrix} x_4 & -x_3 & 0 & 0 \\ -x_2 & x_1 & 0 & 0 \\ 0 & 0 & x_4 & -x_3 \\ 0 & 0 & -x_2 & x_1 \end{pmatrix}.$$

The determinants are

$$\det(J_L) = \det(J_R) = \frac{1}{|\det g|^2}.$$

Jacobians for $SL(2, \mathbb{R})$ and $\mathbb{R}^2 \rtimes_\varphi SL(2, \mathbb{R})$

A basis for the Lie algebra $sl(2, \mathbb{R})$ is

$$X_1 = \begin{pmatrix} 0 & -1 \\ 1 & 0 \end{pmatrix}; \qquad X_2 = \begin{pmatrix} 1 & 0 \\ 0 & -1 \end{pmatrix}; \qquad \text{and} \qquad X_3 = \begin{pmatrix} 0 & 1 \\ 1 & 0 \end{pmatrix}.$$

This basis is orthonormal with respect to the weighting matrix $W = \mathbb{I}_{2 \times 2}$. The *Iwasawa decomposition* allows one to write an arbitrary $g \in SL(2, \mathbb{R})$ in the form [37]

$$g = u_1(\theta) u_2(t) u_3(\xi)$$

where

$$u_1(\theta) = \exp(\theta X_1) = \begin{pmatrix} \cos\theta & -\sin\theta \\ \sin\theta & \cos\theta \end{pmatrix};$$

$$u_2(t) = \exp(t X_2) = \begin{pmatrix} e^t & 0 \\ 0 & e^{-t} \end{pmatrix};$$

$$u_3(\xi) = \exp\left(\frac{\xi}{2}(X_3 - X_1)\right) = \begin{pmatrix} 1 & \xi \\ 0 & 1 \end{pmatrix}.$$

The right Jacobian is

$$J_R(\theta, t, \xi) = \frac{1}{2} \begin{pmatrix} e^{-2t} + e^{2t}(1 + \xi^2) & -2e^{2t}\xi & e^{2t} - e^{-2t}(1 + e^{4t}\xi^2) \\ -2\xi & 2 & 2\xi \\ -1 & 0 & 1 \end{pmatrix}.$$

The left Jacobian is

$$J_L(\theta, t, \xi) = \frac{1}{2} \begin{pmatrix} 2 & 0 & 0 \\ 0 & 2\cos 2\theta & 2\sin 2\theta \\ -e^{2t} & -e^{2t}\sin 2\theta & e^{2t}\cos 2\theta \end{pmatrix}.$$

It is easy to verify that

$$\det(J_R(\theta, t, \xi)) = \det(J_L(\theta, t, \xi)) = \frac{1}{2} e^{2t}.$$

Hence, $SL(2, \mathbb{R})$ is *unimodular* (which means the determinants of the left and right Jacobians are the same).

In the matrix exponential parameterization

$$g(a, b, c) = \exp \begin{pmatrix} c & -a+b \\ a+b & -c \end{pmatrix} = \begin{pmatrix} \cosh x + \frac{c}{x}\sinh x & \frac{(a-b)e^{-x}(-1+e^{2x})}{2x} \\ \frac{(a+b)e^{-x}(-1+e^{2x})}{2x} & \cosh x - \frac{c}{x}\sinh x \end{pmatrix}$$

(where $x = \sqrt{-a^2 + b^2 + c^2}$), one may verify that

$$det\,(J_L) = det\,(J_R) = -\frac{\sinh^2 x}{x^2}.$$

Jacobians for the Scale-Euclidean Group

A basis for the Lie algebra of the scale-Euclidean group $SIM(2)$ is

$$X_1 = \begin{pmatrix} 0 & 0 & 1 \\ 0 & 0 & 0 \\ 0 & 0 & 0 \end{pmatrix}; \qquad X_2 = \begin{pmatrix} 0 & 0 & 0 \\ 0 & 0 & 1 \\ 0 & 0 & 0 \end{pmatrix};$$

$$X_3 = \begin{pmatrix} 0 & -1 & 0 \\ 1 & 0 & 0 \\ 0 & 0 & 0 \end{pmatrix}; \qquad \text{and} \qquad X_4 = \begin{pmatrix} 1 & 0 & 0 \\ 0 & 1 & 0 \\ 0 & 0 & 0 \end{pmatrix}.$$

This basis is orthonormal with respect to the weighting matrix

$$W = \begin{pmatrix} 1 & 0 & 0 \\ 0 & 1 & 0 \\ 0 & 0 & 2 \end{pmatrix}.$$

The parameterization

$$\begin{aligned} g\,(x_1, x_2, \theta, a) &= \exp\,(x_1 X_1 + x_2 X_2) \exp\,(\theta X_3 + a X_4) \\ &= \begin{pmatrix} e^a \cos\theta & -e^a \sin\theta & x_1 \\ e^a \sin\theta & e^a \cos\theta & x_2 \\ 0 & 0 & 1 \end{pmatrix} \end{aligned}$$

extends over the whole group.

The Jacobians are

$$J_R = \begin{pmatrix} e^{-a}\cos\theta & -e^{-a}\sin\theta & 0 & 0 \\ e^{-a}\sin\theta & e^{-a}\cos\theta & 0 & 0 \\ 0 & 0 & 1 & 0 \\ 0 & 0 & 0 & 1 \end{pmatrix}$$

and

$$J_L = \begin{pmatrix} 1 & 0 & 0 & 0 \\ 0 & 1 & 0 & 0 \\ x_2 & -x_1 & 1 & 0 \\ -x_1 & -x_2 & 0 & 1 \end{pmatrix}.$$

Note that

$$det\,(J_R) = e^{-2a} \neq det\,(J_L) = 1.$$

As a general rule, subgroups of the affine group with elements of the form

$$g = \begin{pmatrix} A & \mathbf{b} \\ \mathbf{0}^T & 1 \end{pmatrix}$$

will have left and right Jacobians whose determinants are different unless $det(A) = 1$.

Parameterization and Jacobians for $SL(2, \mathbb{C})$ and $SO(3, 1)$

We now illustrate the construction of the Jacobians for a group of complex matrices.

$SL(N, \mathbb{C})$ is the group of all $n \times n$ complex matrices with unit determinant. It follows that the matrix exponential of any traceless complex matrix results in an element of $SL(N, \mathbb{C})$.

The case of $n = 2$ plays a very important role in physics, and it may play a role in engineering applications as well.

A general element of the Lie algebra $sl(2, \mathbb{C})$ is of the form

$$X = \begin{pmatrix} x_1 + ix_2 & x_3 + ix_4 \\ x_5 + ix_6 & -x_1 - ix_2 \end{pmatrix},$$

and exponentiating matrices of this form result in an element of $SL(2, \mathbb{C})$

$$\exp X = \begin{pmatrix} \cosh x + (x_1 + ix_2)(\sinh x)/x & (x_3 + ix_4)(\sinh x)/x \\ (x_5 + ix_6)(\sinh x)/x & \cosh x - (x_1 + ix_2)(\sinh x)/x \end{pmatrix}$$

$$= \mathbb{I}_{2\times 2} \cosh x + X \frac{\sinh x}{x}$$

where

$$x = i\sqrt{\det(X)} = \sqrt{(x_1 + ix_2)^2 + (x_3 + ix_4)(x_5 + ix_6)}.$$

The basis elements for the Lie algebra $sl(2, \mathbb{C})$ may be taken as

$$X_1 = \begin{pmatrix} 0 & i \\ i & 0 \end{pmatrix}; \qquad X_2 = \begin{pmatrix} 0 & -1 \\ 1 & 0 \end{pmatrix}; \qquad X_3 = \begin{pmatrix} i & 0 \\ 0 & -i \end{pmatrix};$$

$$X_4 = \begin{pmatrix} 0 & 1 \\ 1 & 0 \end{pmatrix}; \qquad X_5 = \begin{pmatrix} 0 & i \\ -i & 0 \end{pmatrix}; \qquad \text{and} \qquad X_6 = \begin{pmatrix} 1 & 0 \\ 0 & -1 \end{pmatrix}.$$

Three of the preceding basis elements are exactly the same as those for $su(2)$, indicating that $SU(2)$ is a subgroup of $SL(2, \mathbb{C})$. This is an example where the inner product

$$(X, Y) = \frac{1}{2}\mathrm{Re}\left(\mathrm{trace}\left(XY^\dagger\right)\right)$$

for the Lie algebra makes $\{X_i\}$ an orthonormal basis.

We note that for $SL(2, \mathbb{C})$ the parameterization

$$\exp(v_1 X_1 + v_2 X_2)\exp(v_3 X_3 + v_4 X_4)\exp(v_5 X_5 + v_6 X_6) = \begin{pmatrix} a & b \\ c & d \end{pmatrix} \qquad (7.36)$$

where

$$a = e^{v_1}\cos v_2 + ie^{v_1}\sin v_2 + \left(e^{v_1}\cos v_2 + ie^{v_1}\sin v_2\right)(iv_4 + v_3)(iv_6 + v_5)$$
$$b = \left(e^{v_1}\cos v_2 + ie^{v_1}\sin v_2\right)(iv_4 + v_3)$$
$$c = \left(e^{-v_1}\cos v_2 - ie^{-v_1}\sin v_2\right)(iv_6 + v_5)$$
$$d = e^{-v_1}\cos v_2 - ie^{-v_1}\sin v_2$$

can be more convenient than $\exp(\sum_{i=1}^{6} v_i X_i)$.

$SL(2, \mathbb{C})$ is intimately related to the *proper Lorentz group* $SO(3, 1)$. The Lorentz group is the group of linear transformations $A \in \mathbb{R}^{4\times 4}$ that preserves the quadratic form

$$Q(\mathbf{x}) = \pm\left(x_1^2 + x_2^2 + x_3^2 - x_4^2\right).$$

That is,

$$Q(A\mathbf{x}) = Q(\mathbf{x}).$$

The *proper Lorentz group* satisfies the further constraint

$$\det(A) = +1.$$

The first three dimensions have the physical meaning of spatial directions, and $x_4 = ct$ is temporal.

More generally, the matrix group consisting of $(p + q) \times (p + q)$ dimensional matrices that preserve the quadratic form

$$Q(\mathbf{x}) = \sum_{i=1}^{p} x_i^2 - \sum_{j=p+1}^{p+q} x_j^2$$

and have determinant $+1$ is denoted as $SO(p, q)$. The two most important examples are $SO(N) = SO(N, 0)$ and the proper Lorentz group, $SO(3, 1)$.

The basis elements for the Lie algebra of the proper Lorentz group are

$$X_1 = \begin{pmatrix} 0 & 0 & 0 & 0 \\ 0 & 0 & -1 & 0 \\ 0 & 1 & 0 & 0 \\ 0 & 0 & 0 & 0 \end{pmatrix}; \quad X_2 = \begin{pmatrix} 0 & 0 & 1 & 0 \\ 0 & 0 & 0 & 0 \\ -1 & 0 & 0 & 0 \\ 0 & 0 & 0 & 0 \end{pmatrix}; \quad X_3 = \begin{pmatrix} 0 & -1 & 0 & 0 \\ 1 & 0 & 0 & 0 \\ 0 & 0 & 0 & 0 \\ 0 & 0 & 0 & 0 \end{pmatrix};$$

$$X_4 = \begin{pmatrix} 0 & 0 & 0 & 1 \\ 0 & 0 & 0 & 0 \\ 0 & 0 & 0 & 0 \\ 1 & 0 & 0 & 0 \end{pmatrix}; \quad X_5 = \begin{pmatrix} 0 & 0 & 0 & 0 \\ 0 & 0 & 0 & 1 \\ 0 & 0 & 0 & 0 \\ 0 & 1 & 0 & 0 \end{pmatrix}; \quad \text{and} \quad X_6 = \begin{pmatrix} 0 & 0 & 0 & 0 \\ 0 & 0 & 0 & 0 \\ 0 & 0 & 0 & 1 \\ 0 & 0 & 1 & 0 \end{pmatrix}.$$

These basis elements are orthonormal with respect to the inner product $(X, Y) = \frac{1}{2}\text{trace}(XY^T)$. The first three basis elements are isomorphic to those for $so(3)$, indicating that spatial rotations form a subgroup of proper Lorentz transformations.

Just as $SU(2)$ is a double cover of $SO(3)$, $SL(2, \mathbb{C})$ is a double cover for the Lorentz group, $SO(3, 1)$. Using group-theoretic notation, one states this as

$$SU(2)/\mathbb{Z}_2 \cong SO(3) \quad \text{and} \quad SL(2, \mathbb{C})/\mathbb{Z}_2 \cong SO(3, 1).$$

In other words, there are homomorphisms $SU(2) \to SO(3)$ and $SL(2, \mathbb{C}) \to SO(3, 1)$ with kernel $\mathbb{Z}_2$, meaning that these are two-to-one homomorphisms.

There is a one-to-one correspondence between the basis elements of the Lie algebras $so(3)$ and $su(2)$, and similarly for the Lie algebras $so(3, 1)$ and $sl(2, \mathbb{C})$. These correspondences preserve the Lie bracket. It follows that the Jacobians and adjoint matrices are the same when such correspondences exist.

In the parameterization Eq. (7.36), the right Jacobian for $SL(2, \mathbb{C})$ is

$$J_R = \begin{pmatrix} a & b & v_3 & v_4 & c & d \\ e & g & -v_4 & v_3 & f & c \\ v_5 & v_6 & \frac{1}{2} & 0 & h & k \\ -v_6 & v_5 & 0 & \frac{1}{2} & -k & h \\ 0 & 0 & 0 & 0 & \frac{1}{2} & 0 \\ 0 & 0 & 0 & 0 & 0 & \frac{1}{2} \end{pmatrix}$$

where

$$
\begin{aligned}
a &= 1 + 2v_3v_5 - 2v_4v_6 \\
b &= 2v_5v_4 + 2v_6v_3 \\
c &= -v_5^2v_3 + v_6^2v_3 + 2v_5v_4v_6 - v_5 \\
d &= v_6^2v_4 - v_6 - v_5^2v_4 - 2v_6v_3v_5 \\
e &= -2v_5v_4 - 2v_6v_3 \\
f &= v_6 - v_6^2v_4 + v_5^2v_4 + 2v_6v_3v_5 \\
g &= 1 + 2v_3v_5 - 2v_4v_6 \\
h &= v_6^2/2 - v_5^2/2 \\
k &= -v_6v_5.
\end{aligned}
$$

A similar exercise gives J_L, and it may be shown that

$$
\det(J_R) = \det(J_L) = \frac{1}{16}.
$$

7.3.6   The Killing Form

A *bilinear form*, $B(X, Y)$ for $X, Y \in \mathcal{G}$ is said to be Ad-invariant if

$$
B(X, Y) = B(Ad(g)X, Ad(g)Y)
$$

for any $g \in G$. In the case of real matrix Lie groups (and the corresponding Lie algebras), which are the ones of most interest in engineering applications, a symmetric ($B(X, Y) = B(Y, X)$) and Ad-invariant bilinear form, called the *Killing form*[13] is defined as

$$
B(X, Y) = \text{trace}(ad(X)ad(Y)). \tag{7.37}
$$

It can be shown that this form is written as [39]

$$
B(X, Y) = \lambda\text{trace}(XY) + \mu(\text{trace}X)(\text{trace}Y)
$$

for some constant real numbers μ and λ that are group dependent.

The Killing form is important in the context of harmonic analysis because the Fourier transform and inversion formula can be explained for large classes of groups which are defined by the behavior of their Killing form. The classification of Lie groups according to the properties of the Killing form is based on *Cartan's Criteria*[14] [38]. For example, a Lie group is called *nilpotent* if $B(X, Y) = 0$ for all $X, Y \in \mathcal{G}$. A Lie group is called *solvable* if and only if for all $X, Y, Z \in \mathcal{G}$ the equality $B(X, [Y, Z]) = 0$ holds. *Semi-simple* Lie groups are those for which $B(X, Y)$ is nondegenerate [i.e., the determinant of the $n \times n$ matrix with elements $B(X_i, X_j)$ is nonzero where $\{X_1, \ldots, X_n\}$ ($n \geq 2$) is a basis for $\mathcal{G}$]. For example, the Heisenberg groups are nilpotent, the rotation groups are semi-simple, and groups of rigid-body motions are solvable.

---

[13]This was named after Wilhelm Karl Joseph Killing (1847–1923).

[14]This was named after the mathematician Elie Cartan (1869–1951) who made major contributions to the theory of Lie groups.

7.3.7 The Matrices of $Ad(G)$, $ad(X)$, and $B(X, Y)$

The formal coordinate-independent definitions of the adjoints $Ad(g)$ and $ad(X)$ and the Killing form $B(X, Y)$ are central to the theory of Lie groups. And while such definitions are sufficient for mathematicians to prove many fundamental properties, it is useful for the engineer interested in group theory to have such concepts illustrated with matrices.

As with all linear operators, $Ad(g)$ and $ad(X)$ are expressed as matrices using an appropriate inner product and concrete basis for the Lie algebra. In particular, with the inner product defined earlier for Lie algebras we have

$$[Ad(g)]_{ij} = \left(X_i, Ad(g)X_j\right) = \left(X_i, gX_jg^{-1}\right) \tag{7.38}$$

and

$$[ad(X)]_{ij} = \left(X_i, ad(X)X_j\right) = \left(X_i, [X, X_j]\right). \tag{7.39}$$

Another way to view these is that if we define the linear operation $\vee$ such that it converts Lie algebra basis elements X_i to elements of the natural basis element $\mathbf{e}_i \in \mathbb{R}^n$,

$$(X_i)^\vee = \mathbf{e}_i,$$

then the matrix with elements given in Eq. (7.38) will be

$$[Ad(g)] = \left[\left(gX_1g^{-1}\right)^\vee, \ldots, \left(gX_ng^{-1}\right)^\vee\right].$$

It is this matrix that relates left and right Jacobians. Using the $\vee$ notation, we may write

$$J_L = \left[\left(\frac{\partial g}{\partial x_1}g^{-1}\right)^\vee, \ldots, \left(\frac{\partial g}{\partial x_n}g^{-1}\right)^\vee\right]$$

and

$$J_R = \left[\left(g^{-1}\frac{\partial g}{\partial x_1}\right)^\vee, \ldots, \left(g^{-1}\frac{\partial g}{\partial x_n}\right)^\vee\right].$$

Since

$$\left(g\left(g^{-1}\frac{\partial g}{\partial x_1}\right)g^{-1}\right)^\vee = \left(\frac{\partial g}{\partial x_1}g^{-1}\right)^\vee$$

it follows that

$$J_L = [Ad(g)]J_R.$$

Hence, if the Jacobians are known, we can write

$$[Ad(g)] = J_L J_R^{-1}.$$

The matrix with elements given in Eq. (7.39) will be

$$[ad(X)] = \left[([X, X_1])^\vee, \ldots, ([X, X_n])^\vee\right].$$

This then gives a concrete tool with which to calculate the $n \times n$ matrix with entries

$$[B]_{ij} = B\left(X_i, X_j\right) = \operatorname{tr}\left([ad(X_i)][ad(X_j)]\right).$$

B is then degenerate if and only if

$$\det([B]) = 0.$$

If $\det([B]) \neq 0$ the Lie algebra is called *semi-simple*. If $[B]_{ij} = 0$ for all i, j, the Lie algebra is called *nilpotent*.

Relationship between $ad(X)$, $B(X, Y)$, and the Structure Constants

Recall that the structure constants of a real Lie algebra are defined by

$$[X_i, X_j] = \sum_{k=1}^{N} C_{ij}^k X_k.$$

From the anti-symmetry of the Lie bracket Eq. (7.31) and the Jacobi identity Eq. (7.32), respectively, we see that

$$C_{ij}^k = -C_{ji}^k$$

and

$$\sum_{j=1}^{N} \left(C_{ij}^l C_{km}^j + C_{mj}^l C_{ik}^j + C_{kj}^l C_{mi}^j \right) = 0.$$

The matrix entries of $[ad(X_k)]_{ij}$ are related to the structure constants as

$$[ad\,(X_k)]_{ij} = \left(X_i, [X_k, X_j]\right) = \left(X_i, \sum_{m=1}^{N} C_{kj}^m X_m\right).$$

For a real Lie algebra, the inner product

$$(X, Y) = \frac{1}{2}\text{trace}\left(X W Y^T\right)$$

is linear in the second argument, and so

$$[ad\,(X_k)]_{ij} = \sum_{m=1}^{N} (X_i, X_m)\, C_{kj}^m = C_{kj}^i.$$

Then

$$\begin{aligned}
B\left(X_i, X_j\right) &= \text{trace}\left(ad\,(X_i)\,ad\,(X_j)\right) \\
&= \sum_{m=1}^{N}\sum_{n=1}^{N} [ad\,(X_i)]_{mn}\,\big[ad\,(X_j)\big]_{nm} \\
&= \sum_{m=1}^{N}\sum_{n=1}^{N} C_{in}^m C_{jm}^n.
\end{aligned}$$

We now illustrate these concepts with some concrete examples.

The $ax + b$ Group

$$gX_1 g^{-1} = \begin{pmatrix} a & b \\ 0 & 1 \end{pmatrix}\begin{pmatrix} 1 & 0 \\ 0 & 0 \end{pmatrix}\begin{pmatrix} 1/a & -b/a \\ 0 & 1 \end{pmatrix} = \begin{pmatrix} 1 & -b \\ 0 & 0 \end{pmatrix} = X_1 - bX_2.$$

$$gX_2 g^{-1} = \begin{pmatrix} a & b \\ 0 & 1 \end{pmatrix}\begin{pmatrix} 0 & 1 \\ 0 & 0 \end{pmatrix}\begin{pmatrix} 1/a & -b/a \\ 0 & 1 \end{pmatrix} = \begin{pmatrix} 0 & a \\ 0 & 0 \end{pmatrix} = aX_2.$$

Therefore,

$$[Ad(g)] = \begin{pmatrix} 1 & 0 \\ -b & a \end{pmatrix}.$$

Similarly, the calculation

$$[X, X_1] = \begin{pmatrix} x & y \\ 0 & 0 \end{pmatrix} \begin{pmatrix} 1 & 0 \\ 0 & 0 \end{pmatrix} - \begin{pmatrix} 1 & 0 \\ 0 & 0 \end{pmatrix} \begin{pmatrix} x & y \\ 0 & 0 \end{pmatrix} = \begin{pmatrix} 0 & -y \\ 0 & 0 \end{pmatrix}$$

and

$$[X, X_2] = \begin{pmatrix} x & y \\ 0 & 0 \end{pmatrix} \begin{pmatrix} 0 & 1 \\ 0 & 0 \end{pmatrix} - \begin{pmatrix} 0 & 1 \\ 0 & 0 \end{pmatrix} \begin{pmatrix} x & y \\ 0 & 0 \end{pmatrix} = \begin{pmatrix} 0 & x \\ 0 & 0 \end{pmatrix},$$

and so

$$[ad(X)] = \begin{pmatrix} 0 & 0 \\ -y & x \end{pmatrix}.$$

Substitution into the definition shows that

$$[B] = \begin{pmatrix} 1 & 0 \\ 0 & 0 \end{pmatrix}.$$

Clearly this is degenerate, and the $ax + b$ group is not semi-simple.

$SO(3)$

When the inner product is normalized so that $(X_i, X_j) = \delta_{ij}$,

$$[Ad(g)] = J_L J_R^{-1},$$

where J_R and J_L are given in Chapter 4 in a variety of parameterizations.

A straightforward calculation shows

$$[ad(X)] = X$$

and

$$[B] = -2\mathbb{I}_{3 \times 3}.$$

Hence $SO(3)$ is semi-simple.

$SE(2)$

Substitution into the definitions yields

$$[Ad(g)] = \begin{pmatrix} \cos\theta & -\sin\theta & x_2 \\ \sin\theta & \cos\theta & -x_1 \\ 0 & 0 & 1 \end{pmatrix}.$$

And when

$$X = \begin{pmatrix} 0 & -\alpha & v_1 \\ \alpha & 0 & v_2 \\ 0 & 0 & 0 \end{pmatrix},$$

$$[ad(X)] = \begin{pmatrix} 0 & -\alpha & v_2 \\ \alpha & 0 & -v_1 \\ 0 & 0 & 0 \end{pmatrix},$$

and

$$[B] = \begin{pmatrix} 0 & 0 & 0 \\ 0 & 0 & 0 \\ 0 & 0 & -2 \end{pmatrix}.$$

This is clearly degenerate, and $SE(2)$ is therefore not semi-simple. [Neither is $SE(3)$.]

These examples are enough to illustrate the computation for real Lie groups. We note that $GL(N, \mathbb{R})$ and $SL(N, \mathbb{R})$ are both semi-simple.

7.4   Summary

In this chapter we presented a comprehensive introduction to group theory. We considered both finite and Lie groups. Many good books already exist on group theory (for example, see the reference list below). Our goal here was to present group theory in a manner that builds on the mathematical knowledge of engineers. This was done particularly in the context of Lie groups. There, our treatment of matrix Lie groups used basic linear algebra. Concepts such as the adjoint and Killing form were presented in such a way that concrete matrices could be computed (rather than the usual treatment of these concepts as abstract linear operators in mathematics books).

The techniques and terminology of this chapter provide the foundations for Chapter 8 and the remainder of the book.

References

[1] Artin, M., *Algebra*, Prentice Hall, Upper Saddle River, NJ, 1991.

[2] Alperin, J.L. and Bell, R.B., *Groups and Representations*, Springer-Verlag, New York, 1995.

[3] Baglivo, J. and Graver, J.E., *Incidence and Symmetry in Design and Architecture*, Cambridge University Press, New York, 1983.

[4] Baker, H.F., Alternants and continuous groups, *Proc. London Math. Soc.*, (2nd series), 3, 24–47, 1904.

[5] Barut, A.O. and Raçzka, R., *Theory of Group Representations and Applications*, World Scientific, Singapore, 1986.

[6] Birkhoff, G. and MacLane, S., *A Survey of Modern Algebra*, 4th ed., Macmillan, New York, 1977.

[7] Bishop, D.M., *Group Theory and Chemistry*, Dover, New York, 1973.

[8] Bradley, C.J. and Cracknell, A.P., *The Mathematical Theory of Symmetry in Solids*, Clarendon Press, Oxford, 1972.

[9] Campbell, J.E., On a law of combination of operators, *Proc. London Mathematical Society*, 29, 14–32, 1897.

[10] Casper, D.L.D. and Klug, A., Physical Principles in the construction of regular viruses, *Cold Spring Harbor Symposium on Quantitative Biology*, 27, 1–24, 1962.

[11] Coxeter, H.S.M., *Regular Polytopes*, 2nd ed., MacMillan, New York, 1963.

[12] Eisenhart, L.P., *Continuous Groups of Transformations*, Dover, New York, 1961.

[13] Gel'fand, I.M., Minlos, R.A., and Shapiro, Z.Ya., *Representations of the Rotation and Lorentz Groups and their Applications*, Macmillan, New York, 1963.

[14] Humphreys, J.F., *A Course in Group Theory*, Oxford University Press, New York, 1996.

[15] Herstein, I.N., *Topics in Algebra*, 2nd ed., John Wiley & Sons, New York, 1975.

[16] Hausdorff, F., Die symbolische exponentialformel in der gruppentheorie, *Berichte der Sachsichen Akademie der Wissenschaften*, 58, 19–48, Leipzig, Germany, 1906.

[17] Husain, T., *Introduction to Topological Groups*, W.B. Sanders, Philadelphia, 1966.

[18] Inui, T., Tanabe, Y., and Onodera, Y., *Group Theory and Its Applications in Physics*, 2nd ed., Springer-Verlag, Berlin, 1996.

[19] Isaacs, I.M., *Character Theory of Finite Groups*, Dover, NY, 1976.

[20] James, G. and Liebeck, M., *Representations and Characters of Groups*, Cambridge University Press, Cambridge, 1993.

[21] Janssen, T., *Crystallographic Groups*, North-Holland/American Elsevier, New York, 1973.

[22] Johnston, B.L. and Richman, F.R., *Numbers and Symmetry: An Introduction to Algebra*, CRC Press, Boca Raton, FL, 1997.

[23] Kettle, S.F.A., *Symmetry and Structure: Readable Group Theory for Chemists*, John Wiley & Sons, New York, 1995.

[24] Klein, F., *Lectures on the Icosahedron*, Kegan Paul, London, 1913.

[25] Lawton, W., Raghavan, R., and Viswanathan, R., Ribbons and groups: a thin rod theory for catheters and proteins, *J. of Physics A: Mathematical and General*, 32(9), 1709–1735, 1999.

[26] Liljas, L., The structure of spherical viruses, *Prog. Biophys. Molec. Biol.*, 48, 1–36, 1986.

[27] Miller, W., *Symmetry Groups and their Applications*, Academic Press, New York, 1972.

[28] Montgomery, D. and Zippin, L., *Topological Transformation Groups*, Interscience, New York, 1955.

[29] Naimark, M.A., *Linear Representations of the Lorentz Group*, Macmillan, New York, 1964.

[30] Neuenschwander, D., *Probabilities on the Heisenberg Group: Limit Theorems and Brownian Motion*, Lecture Notes in Mathematics, No. 1630, Springer-Verlag, Berlin, 1996.

[31] Rossmann, M.G. and Johnson, J.E., Icosahedral RNA virus structure, *Annu. Rev. Biochem.*, 58, 533–73, 1989.

[32] Sattinger, D.H. and Weaver, O.L., *Lie Groups and Algebras with Applications to Physics, Geometry, and Mechanics*, Springer-Verlag, New York, 1986.

[33] Schoolfield, C.H., *Random Walks on Wreath Products of Groups and Markov Chains on Related Homogeneous Spaces*, Ph.D. dissertation, Dept. of Mathematical Sciences, Johns Hopkins University, Baltimore, 1998.

[34] Selig, J.M., Geometrical methods in robotics, *Springer Monogr. in Comput. Sci.*, New York, 1996.

[35] Sternberg, S., *Group Theory and Physics*, Cambridge University Press, New York, 1994.

[36] Sudarshan, E.C.G. and Mukunda, N., *Classical Dynamics: A Modern Perspective*, Wiley-InterScience, New York, 1974.

[37] Sugiura, M., *Unitary Representations and Harmonic Analysis*, 2nd ed., North-Holland, Amsterdam, 1990.

[38] Varadarajan, V.S., *Lie Groups, Lie Algebras, and their Representations*, Springer-Verlag, New York, 1984.

[39] Vilenkin, N.J. and Klimyk, A.U., *Representation of Lie Groups and Special Functions*, Vols. 1–3, Kluwer Academic, Dordrecht, Holland, 1991.

[40] Weyl, H., *The Classical Groups*, Princeton University Press, Princeton, NJ, 1946.

[41] Želobenko, D.P., Compact lie groups and their representations, *Transl. of Math. Monogr., Am. Math. Soc.*, Providence, RI, 1973.

Chapter 8

Harmonic Analysis on Groups

In this chapter we review Fourier analysis of functions on certain kinds of noncommutative groups. We begin with the simplest case of finite groups and then consider compact Lie groups. Finally, methods to handle noncompact noncommutative unimodular groups are discussed. To this end, we review in some depth what it means to differentiate and integrate functions on Lie groups. The rotation and motion groups are discussed in great detail in Chapters 9 and 10.

In their most general form, the results of this chapter are as follows: Given functions $f_i(g)$ for $i = 1, 2$ which are square integrable with respect to an invariant integration measure μ on a unimodular group $(G, \circ)$, i.e.,

$$\mu\left(|f_i(g)|^2\right) = \int_G |f_i(g)|^2 \, d(g) < \infty$$

and

$$\mu\left(f_i(h \circ g)\right) = \mu\left(f_i(g \circ h)\right) = \mu\left(f_i(g)\right) \qquad \forall \qquad h \in G,$$

we can define the convolution product

$$(f_1 * f_2)(g) = \int_G f_1(h) f_2\left(h^{-1} \circ g\right) d(h),$$

and the Fourier transform

$$\mathcal{F}(f) = \hat{f}(p) = \int_G f(g) U\left(g^{-1}, p\right) d(g)$$

where $U(\cdot, p)$ is a unitary matrix function [called an irreducible unitary representation (IUR)] for each value of the parameter p. The Fourier transform defined in this way has corresponding inversion, convolution, and Parseval theorems

$$f(g) = \int_{\hat{G}} \operatorname{trace}\left[\hat{f}(p) U(g, p)\right] d\nu(p),$$

$$\mathcal{F}(f_1 * f_2) = \hat{f}_2(p) \hat{f}_1(p)$$

and

$$\int_G |f(g)|^2 d(g) = \int_{\hat{G}} \left\|\hat{f}(p)\right\|^2 d\nu(p).$$

Here $\|\cdot\|$ is the Hilbert-Schmidt norm, $\hat{G}$ (which is defined on a case-by-case basis, and is the space of all p values) is called the *dual* of the group G, and ν is an appropriately chosen integration measure (in a generalized sense) on $\hat{G}$.

In addition to the condition of square integrability, we will assume functions to be as well-behaved as required for a given context. For instance, when discussing differentiation of functions on Lie groups with elements parameterized as $g(x_1, \ldots, x_n)$, we will assume all the partial derivatives of $f(g(x_1, \ldots, x_n))$ in all the parameters x_i exist. When considering noncompact groups, we will assume that functions $f(g)$ are in $\mathcal{L}^p(G)$ for all $p \in \mathbb{Z}^+$ and rapidly decreasing in the sense that

$$\int_G |f(g(x_1, \ldots, x_n))| x_1^{p_1} \cdots x_n^{p_n} d(g(x_1, \ldots, x_n)) < \infty$$

for all $p_i \in \mathbb{Z}^+$.

In the case of finite groups, all measures are counting measures, and these integrals are interpreted (to within a constant) as sums. In the case of unimodular Lie groups, integration by parts provides a tool to derive operational properties that convert differential operators acting on functions to algebraic operations on the Fourier transforms of functions, much like classical Fourier analysis.

For historical perspective and abstract aspects of noncommutative harmonic analysis see [3, 19, 21, 24, 31, 44, 51, 65, 71, 72, 80].

8.1   Fourier Transforms for Finite Groups

Let $(G, \circ)$ be a finite group with $|G| = n$ elements, $g_1, \ldots, g_n$. Let $f(g_i)$ be a function which assigns a complex value to each group element. Recall that this is written as $f : G \to \mathbb{C}$. Analogous to the standard (Abelian) convolution product and discrete Fourier transform we saw in Chapter 2, the convolution product of two complex-valued functions on G is written as

$$(f_1 * f_2)(g) = \sum_{h \in G} f_1(h) f_2\left(h^{-1} \circ g\right) = \sum_{i=1}^{|G|} f_1(g_i) f_2\left(g_i^{-1} \circ g\right). \tag{8.1}$$

Since we are summing over h, we can make the change of variables $k = h^{-1} \circ g$ and write the convolution in the alternate form

$$(f_1 * f_2)(g) = \sum_{k \in G} f_1\left(g \circ k^{-1}\right) f_2(k).$$

However, unlike the case of Abelian (in particular, additive) groups where $\circ = +$, in the more general noncommutative context we can not switch the order of elements on either side of the group operation. This means that in general, the convolution of two arbitrary functions on G does not commute: $(f_1 * f_2)(g) \neq (f_2 * f_1)(g)$. There are, of course, exceptions to this. For example, if $f_1(g)$ is a scalar multiple of $f_2(g)$ or if one or both functions are class functions,[1] the convolution will commute.

A number of problems can be formulated as convolutions on finite groups. These include generating the statistics of random walks, as well as coding theory. (See [62] for an overview.) In some applications, instead of computing the convolution, one is interested in the inverse problem of solving the convolution equation

$$(f_1 * f_2)(g) = f_3(g)$$

where f_1 and f_3 are given and f_2 must be found. This problem is often referred to as *deconvolution*.

For a finite group with $|G| = n$ elements, the convolution Eq. (8.1) is performed in $\mathcal{O}(n^2)$ arithmetic operations when the preceding definition is used. The deconvolution can be performed as

---

[1] Recall that a class function has the property $f(g \circ h) = f(h \circ g)$.

a matrix inversion in the following way. Assume f_1 and f_3 are given, and define $A_{ij} = f_1(g_i \circ g_j^{-1})$, $x_j = f_2(g_j)$ and $b_i = f_3(g_i)$. Then inversion of the matrix equation

$$Ax = b$$

will solve the deconvolution problem. However, it requires $\mathcal{O}(n^\gamma)$ computations where $2 \leq \gamma \leq 3$ is the exponent required for matrix inversion and multiplication. (See Appendix D for a discussion.) If Gaussian elimination is used, $\gamma = 3$. Thus it is desirable from a computational perspective to generalize the concept of the Fourier transform, and particularly the FFT, as a tool for efficiently computing convolutions and solving deconvolution problems. In order to do this, a generalization of the unitary function $u(x, \omega) = \exp(i\omega x)$ is required. The two essential properties of this function are

$$\exp(i\omega(x_1 + x_2)) = \exp(i\omega x_1) \cdot \exp(i\omega x_2)$$

and

$$\exp(i\omega x)\overline{\exp(i\omega x)} = 1.$$

The concept of a *unitary group representation* discussed in the following subsection is precisely what is required to define the Fourier transform of a function on a finite group.

In essence, if we are able to find a particular set of homomorphisms from G into a set of unitary matrices $\{U^\lambda(g)\}$ indexed by λ [i.e., for each λ, $U^\lambda(g)$ is a unitary matrix for all $g \in G$], then for functions $f : G \to \mathbb{C}$ we can define the matrix function

$$\mathcal{T}^\lambda(f) = \tilde{f}^\lambda = \sum_{g \in G} f(g)U^\lambda(g). \tag{8.2}$$

Since $U^\lambda(g \circ h) = U^\lambda(g)U^\lambda(h)$, it may be shown (and will be shown later) that this transformation has the following properties under left and right shifts and convolutions

$$\mathcal{T}^\lambda(L(h)f) = U^\lambda(h)\tilde{f}^\lambda \quad \text{and} \quad \mathcal{T}^\lambda(R(h)f) = \tilde{f}^\lambda U^\lambda\left(h^{-1}\right) \tag{8.3}$$

and

$$\mathcal{T}^\lambda(f_1 * f_2) = \tilde{f}_1^\lambda \tilde{f}_2^\lambda. \tag{8.4}$$

Recall from Chapter 7 that $(L(h)f)(g) = f(h^{-1} \circ g)$ and $(R(h)f)(g) = f(g \circ h)$.

8.1.1 Representations of Finite Groups

An N−dimensional matrix representation of a group G is a homomorphism from the group into a set of invertible matrices, $D : G \to GL(N, \mathbb{C})$. Following from the definition of a homomorphism, $D(e) = \mathbb{I}_{N \times N}$, $D(g_1 \circ g_2) = D(g_1)D(g_2)$, and $D(g^{-1}) = D^{-1}(g)$. Here $\mathbb{I}_{N \times N}$ is the $N \times N$ dimensional identity matrix, and the product of two representations is ordinary matrix multiplication.

If the mapping $D : G \to D(G)$ is bijective, then the representation D is called *faithful*. Two representations D and D' are said to be *equivalent* if they are similar as matrices, i.e., if $SD(g) = D(g)S$ for some $S \in GL(N, \mathbb{C})$ and all $g \in G$ where S is independent of g. In this case, the notation $D \cong D'$ is used. A representation is called *reducible* if it is equivalent to the direct sum of other representations. In other words, if a representation can be block-diagonalized, it is reducible. Thus by an appropriate similarity transformation, any finite dimensional representation matrix can be decomposed by successive similarity transformations until the resulting blocks can no longer be reduced into smaller ones. These smallest blocks are called *irreducible* representations and form the foundations upon which all other representation matrices can be built. It is common to use the notation

$$B = A_1 \oplus A_2 = \begin{pmatrix} A_1 & 0 \\ 0 & A_2 \end{pmatrix}$$

for block diagonal matrices. B is said to be the *direct sum* of A_1 and A_2. If instead of two blocks, B consists of M blocks on the diagonal, the following notation is used

$$B = \sum_{m=1}^{M} \bigoplus A_m = \begin{pmatrix} A_1 & 0 & 0 & \cdots & 0 \\ 0 & A_2 & 0 & \cdots & \vdots \\ 0 & 0 & \ddots & 0 & \vdots \\ \vdots & \vdots & 0 & \ddots & 0 \\ 0 & \cdots & \cdots & 0 & A_M \end{pmatrix}.$$

In the case when $A_1 = A_2 = \cdots = A_M = A$, we say that $B = \bigoplus MA$. If not for the $\bigoplus$ this would be easily confused with scalar multiplication MA, but this will not be an issue here.

Representations can be considered as linear operators which act on functions on a group. Depending on what basis is used, these operators are expressed as different representation matrices which are equivalent under similarity transformation. However, from a computational perspective the choice of basis can have a significant impact.

The two standard representations which act on functions on a group are called the *left-regular* and *right-regular* representations. They are defined respectively as

$$(L(g)f)(h) = f\left(g^{-1} \circ h\right) \overset{\Delta}{=} f_h^L(g) \qquad (R(g)f)(h) = f(h \circ g) \overset{\Delta}{=} f_h^R(g) \qquad (8.5)$$

for all $g, h \in G$.

The fact that these are representations may be observed as follows

$$(L(g_1)(L(g_2)f))(h) = \left(L(g_1)f_{g_2}^L\right)(h) = f_{g_2}^L\left(g_1^{-1} \circ h\right) =$$
$$f\left(g_2^{-1} \circ \left(g_1^{-1} \circ h\right)\right) = f\left((g_1 \circ g_2)^{-1} \circ h\right) = (L(g_1 \circ g_2)f)(h),$$
$$(R(g_1)(R(g_2)f))(h) = \left(R(g_1)f_{g_2}^R\right)(h) = f((h \circ g_1) \circ g_2) = (R(g_1 \circ g_2)f)(h).$$

Note also that left and right shifts commute with each other, but left (right) shifts do not generally commute with other left (right) shifts.

The matrix elements of these representations are expressed using the inner product

$$(f_1, f_2) = \sum_{g \in G} f_1(g)\overline{f_2(g)}.$$

Shifted versions of the delta function

$$\delta(h) = \begin{cases} 1 & \text{for } h = e \\ 0 & \text{for } h \neq e \end{cases}$$

for all $h \in G$ may be used as a basis for all functions on the finite group G. That is, any function $f(g)$ can be expressed as $f(g) = \sum_{h \in G} f(h)\delta_g(h)$ where $\delta_g(h) = \delta(g^{-1} \circ h)$ is a left-shifted version of $\delta(h)$.[2] An $N \times N$ matrix representation of G is then defined element by element by computing

$$T_{ij}(g) = \left(\delta_{g_i}, T(g)\delta_{g_j}\right)$$

---

[2] Actually, $\delta(h)$ is a special function in the sense that $\delta(g^{-1} \circ h) = \delta(h \circ g^{-1})$ and $\delta(g^{-1}) = \delta(g)$, and thus it doesn't matter if the shifting is from the left or right, and also $\delta_g(h) = \delta_h(g)$.

where the operator T can be either L or R. Because of the properties of δ functions, this matrix representation has the property that $T(e) = \mathbb{I}_{N \times N}$, and the diagonal elements of $T(g)$ for $g \neq e$ are all zero. In the case when $T = L$, this can be seen by carrying out the computation

$$L_{ij}(g) = \sum_{h \in G} \delta\left(g_i^{-1} \circ h\right) \delta\left(g_j^{-1} \circ g^{-1} \circ h\right) = \delta\left(g_j^{-1} \circ g^{-1} \circ g_i\right) = \delta\left(g \circ \left(g_j \circ g_i^{-1}\right)\right),$$

which reduces to $L_{ii}(g) = \delta(g)$ for diagonal elements. A similar computation follows for $R(g)$.

In either case, taking the trace of $T(g)$ defines the *character* of T

$$\chi(g) = \text{trace}(T(g)) = \sum_{i=1}^{|G|} \delta(g) = \begin{cases} |G| & \text{for } g = e \\ 0 & \text{for } g \neq e. \end{cases} \tag{8.6}$$

For all $g_i \in G$, let $\pi_j(g_i)$ be a $d_j \times d_j$ irreducible unitary representation of G (in contrast to T, which may be reducible). Thus

$$\pi_j(g_k \circ g_l) = \pi_j(g_k)\,\pi_j(g_l),$$

and

$$\pi_j(e) = \mathbb{I}_{d_j \times d_j} \qquad \pi_j^{\dagger}(g) = \pi_j\left(g^{-1}\right).$$

(Recall that $\dagger$ denotes complex conjugate transpose.)

Let n_j be the number of times $\pi_j(g)$ appears in the decomposition of $T(g)$. Then, since the trace is invariant under similarity transformation, it follows from Eq. (8.6) that

$$\chi(g) = \sum_{k=1}^{N} n_k \chi_k(g) = \begin{cases} |G| & \text{for } g = e \\ 0 & \text{for } g \neq e \end{cases} \tag{8.7}$$

where $\chi_k(g) = \text{trace}(\pi_k(g))$, and N is the number of inequivalent irreducible unitary representations (IURs).

THEOREM 8.1

If G is a finite group, $g \in G$, and $D(g)$ is a matrix representation, then a unitary representation equivalent to $D(g)$ results from the similarity transformation

$$U(g) = M^{-1} D(g) M \qquad \text{where} \qquad M = \left(\sum_{g \in G} D^{\dagger}(g) D(g) \right)^{\frac{1}{2}}.$$

PROOF The matrix square root above is well defined and results in a Hermitian matrix ($M^{\dagger} = M$) since M^2 is a positive definite Hermitian matrix (see Appendix D). Furthermore, by the definition of a representation and the general property $(X_1 X_2)^{\dagger} = X_2^{\dagger} X_1^{\dagger}$, we have for an arbitrary $h \in G$

$$D^{\dagger}(h) M^2 D(h) = \sum_{g \in G} D^{\dagger}(h) D^{\dagger}(g) D(g) D(h) = \sum_{g \in G} D^{\dagger}(g \circ h) D(g \circ h) = M^2.$$

Multiplication on the left and right by M^{-1} shows that the product of $M D(h) M^{-1} = U(h)$ with its Hermitian conjugate is the identity matrix. ∎

THEOREM 8.2 (Schur's lemma) [69]

If $\pi_{\lambda_1}(g)$ and $\pi_{\lambda_2}(g)$ are irreducible matrix representations of a group G (in complex vector spaces V_1 and V_2 respectively) and

$$\pi_{\lambda_2}(g)X = X\pi_{\lambda_1}(g)$$

for all $g \in G$, then X must either be the zero matrix or an invertible matrix. Furthermore, $X = 0$ when $\pi_{\lambda_1}(g) \not\cong \pi_{\lambda_2}(g)$, X is invertible when $\pi_{\lambda_1}(g) \cong \pi_{\lambda_2}(g)$, and $X = c\mathbb{I}$ for some $c \in \mathbb{C}$ when $\pi_{\lambda_1}(g) = \pi_{\lambda_2}(g)$.

PROOF

Case 1: $\pi_{\lambda_1}(g) = \pi_{\lambda_2}(g)$. Let $(\mathbf{x}, \lambda)$ be an eigenvalue/eigenvector pair for the matrix X, i.e., a solution to the equation $X\mathbf{x} = \lambda\mathbf{x}$. If $\pi_{\lambda_1}(g)X = X\pi_{\lambda_1}(g)$, it follows that

$$X\pi_{\lambda_1}(g)\mathbf{x} = \pi_{\lambda_1}(g)X\mathbf{x} = \lambda\pi_{\lambda_1}(g)\mathbf{x},$$

and thus $\mathbf{x}'(g) = \pi_{\lambda_1}(g)\mathbf{x}$ is also an eigenvector of X. Thus the space of all eigenvectors (which is a subspace of V_1) is invariant under the action of representations $\pi_{\lambda_1}(g)$. But by definition $\pi_{\lambda_1}(g)$ is irreducible, and so the vectors $\mathbf{x} \neq \mathbf{0}$ must span the *whole* vector space V_1. Hence, the only way $X\mathbf{x} = \lambda\mathbf{x}$ can hold *for all* $\mathbf{x} \in V_1$ is if $X = \lambda\mathbb{I}$.

Case 2: $\pi_{\lambda_1}(g) \not\cong \pi_{\lambda_2}(g)$. If V_1 is d_1-dimensional and V_2 is d_2-dimensional, then X must have dimensions $d_2 \times d_1$. Since $d_1 \neq d_2$, without loss of generality let us choose $d_2 < d_1$. Then we can always find a nontrivial d_1-dimensional vector $\mathbf{n}$ such that $X\mathbf{n} = \mathbf{0}$ and the set of all such vectors (called the nullspace of X) is a vector subspace of V_1. It follows that

$$X\pi_{\lambda_1}(g)\mathbf{n} = \pi_{\lambda_2}(g)X\mathbf{n} = \mathbf{0}$$

and so $\pi_{\lambda_1}(g)\mathbf{n}$ is also in the nullspace. But by definition, no vector subspace of V_1 other than $\{\mathbf{0}\} \subset V_1$ and V_1 itself can be invariant under $\pi_{\lambda_1}(g)$ since it is irreducible. Thus since $\mathbf{n} \neq \mathbf{0}$ the nullspace of X must equal V_1. The only way this can be so is if $X = 0$. ∎

The next subsection reviews the very important and somewhat surprising fact that each π_j appears in the decomposition of the (left or right) regular representation T exactly d_j times and that the number of inequivalent irreducible representations of G is the same as the number of conjugacy classes in G. We therefore write

$$T(g) \cong \sum_{j=1}^{\alpha} \bigoplus d_j\pi_j(g) \tag{8.8}$$

where α is the number of conjugacy classes of G.

On first reading, Eq. (8.8) can be taken as fact, and the following subsection can be skipped. It can be read at a later time to gain full understanding.

We now present a theorem which illustrates the orthogonality of matrix elements of IURs.

THEOREM 8.3 (Orthogonality of IUR Matrix Elements) [20, 32]

Given irreducible representatation matrices π_{λ_1} and π_{λ_2}, their elements satisfy the orthogonality relations

$$\sum_{g \in G} \pi_{ij}^{\lambda_1}(g)\overline{\pi_{kl}^{\lambda_2}(g)} = \frac{|G|}{d_{\lambda_1}}\delta_{\lambda_1,\lambda_2}\delta_{i,k}\delta_{j,l}. \tag{8.9}$$

PROOF   For any two irreducible representations π_{λ_1} and π_{λ_2}, we can define the matrix

$$A = \sum_{g \in G} \pi_{\lambda_1}(g) E \, \pi_{\lambda_2}(g^{-1}) \tag{8.10}$$

for any matrix E of compatible dimensions.   Due to the invariance of summation over the group under shifts and homomorphism property of representations, we observe that

$$\pi_{\lambda_1}(h) A \, \pi_{\lambda_2}\left(h^{-1}\right) = \sum_{g \in G} \pi_{\lambda_1}(h)\pi_{\lambda_1}(g) E \, \pi_{\lambda_2}(g^{-1})\pi_{\lambda_2}\left(h^{-1}\right)$$

$$= \sum_{g \in G} \pi_{\lambda_1}(h \circ g) E \, \pi_{\lambda_2}\left((h \circ g)^{-1}\right) = A,$$

or equivalently,

$$\pi_{\lambda_1}(h) A = A \, \pi_{\lambda_2}(h).$$

Thus, from Schur's lemma,

$$A = a\delta_{\lambda_1,\lambda_2} \mathbb{I}_{d_{\lambda_1} \times d_{\lambda_2}},$$

where $d_\lambda = \dim(\pi_\lambda(g))$ and $\mathbb{I}_{d_{\lambda_1} \times d_{\lambda_2}}$ when $\pi_{\lambda_1}(g) \not\cong \pi_{\lambda_2}(g)$ is the rectangular matrix with (i, j)th element $\delta_{i,j}$ and all others zero. When $\lambda_1 = \lambda_2$ it becomes an identity matrix.

Following [20, 32], we choose $E = E^{j,l}$ to be the matrix with elements $E_{p,q}^{j,l} = \delta_{p,j}\delta_{l,q}$. Making this substitution into Eq. (8.10) and evaluating the (i, j)th element of the result yields

$$a_{j,l}\delta_{i,k}\delta_{\lambda_1,\lambda_2} = \sum_{p,q} \sum_{g \in G} \pi_{i,p}^{\lambda_1}(g)\delta_{p,j}\delta_{l,q}\pi_{q,k}^{\lambda_2}\left(g^{-1}\right) = \sum_{g \in G} \pi_{i,j}^{\lambda_1}(g)\pi_{l,k}^{\lambda_2}\left(g^{-1}\right). \tag{8.11}$$

The constant $a_{j,l}$ is determined by taking the trace of both sides when $\lambda_1 = \lambda_2 = \lambda$. The result is $a_{j,l} = (|G|/d_\lambda)\delta_{j,l}$. Finally, using the unitarity of the representations we see that Eq. (8.9) holds.
∎

The result of this theorem will be very useful in the proofs that follow.

8.1.2   Characters of Finite Groups

In this section we prove two fundamental results used in the inversion formula and Plancherel equality for finite groups: (1) the number of irreducible representations in the decomposition of the regular representation is the same as the number of conjugacy classes in the group; and (2) the number of times an irreducible representation appears in the decomposition of the regular representation is equal to the dimension of the irreducible representation.

The proofs of these statements follow from orthogonality properties of characters, which themselves need to be proved. Our proofs follow those given in [20, 32].

THEOREM 8.4   (First Orthogonality of Characters)

$$\sum_{g \in G} \chi_{\lambda_1}(g)\overline{\chi_{\lambda_2}(g)} = |G|\delta_{\lambda_1,\lambda_2} \tag{8.12}$$

for finite groups.

This may be written in the equivalent form

$$\sum_{k=1}^{\alpha} |C_k| \chi_{\lambda_1}(g_{C_k}) \overline{\chi_{\lambda_2}(g_{C_k})} = |G| \delta_{\lambda_1, \lambda_2} \tag{8.13}$$

where g_{C_k} is any representative element of C_k (the kth conjugacy class), i.e., $g_{C_k} \in C_k \subset G$, and $|C_k|$ is the number of elements in C_k. It follows that $\sum_{k=1}^{\alpha} |C_k| = |G|$.

PROOF   Equation (8.12) results from evaluating Eq. (8.9) at $i = j$ and $k = l$ and summing over all i and k. It can be written in the alternate form

$$\sum_{k=1}^{\alpha} \sum_{g \in C_k} \chi_{\lambda_1}(g) \overline{\chi_{\lambda_2}(g)} = |G| \delta_{\lambda_1, \lambda_2},$$

where the summation over the group is decomposed into a summation over classes and summation within each class. Because characters are class functions, and hence constant on conjugacy classes, we have

$$\sum_{g \in C_k} \chi_{\lambda_1}(g) \overline{\chi_{\lambda_2}(g)} = |C_k| \chi_{\lambda_1}(g_{C_k}) \overline{\chi_{\lambda_2}(g_{C_k})}. \quad \blacksquare$$

THEOREM 8.5 (Second Orthogonality of Characters)

$$\sum_{\lambda=1}^{N} \chi_\lambda(g_{C_i}) \overline{\chi_\lambda(g_{C_j})} = \frac{|G|}{|C_i|} \delta_{i,j} \tag{8.14}$$

where $g_{C_i} \in C_i$ and $g_{C_j} \in C_j$ are class representatives, and N is the number of inequivalent irreducible representations of G.

PROOF   Recall that the class sum function C_i is defined as $C_i(g) = \sum_{h \in C_i} \delta(h^{-1} \circ g)$. It then follows using the definition Eq. (8.2) that

$$\tilde{C}_i^\lambda = \sum_{g \in G} C_i(g) U^\lambda(g) = \sum_{g \in G} \left(\sum_{h \in C_i} \delta\left(h^{-1} \circ g\right) \right) U^\lambda(g) = \sum_{h \in C_i} U^\lambda(h).$$

Since by definition every conjugacy class is closed under conjugation, we know that

$$C(h \circ g) = C(g \circ h). \tag{8.15}$$

Application of Eqs. (8.2)–(8.15) results in

$$U^\lambda(h) \tilde{C}_i^\lambda = \tilde{C}_i^\lambda U^\lambda(h)$$

for all $h \in G$. Taking $U^\lambda = \pi_\lambda$ to be irreducible, it follows from Schur's lemma that

$$\tilde{C}_i^\lambda = \kappa_{i,\lambda} \mathbb{I}_{d_\lambda \times d_\lambda} \tag{8.16}$$

for some scalar $\kappa_{i,\lambda}$. We find the value of this constant by taking the trace of both sides of Eq. (8.16) and observing that

$$\kappa_{i,\lambda} \cdot d_\lambda = \text{trace} \left(\sum_{h \in C_i} \pi_\lambda(h) \right) = |C_i| \text{trace}(\pi_\lambda(h)) = |C_i| \chi_\lambda(g_{C_i})$$

for any $g_{C_i} \in C_i$. Hence

$$\tilde{C}_i^\lambda = \frac{|C_i|}{d_\lambda} \chi_\lambda \left(g_{C_i} \right) \mathbb{I}_{d_\lambda \times d_\lambda}. \tag{8.17}$$

The equality

$$\tilde{C}_i^\lambda \tilde{C}_j^\lambda = \sum_{k=1}^\alpha c_{ij}^k \tilde{C}_k^\lambda$$

follows from Eqs. (7.25) and (8.4). Using Eq. (8.17), this is rewritten as

$$|C_i||C_j| \chi_\lambda \left(g_{C_i} \right) \chi_\lambda \left(g_{C_j} \right) = d_\lambda \sum_{k=1}^\alpha c_{ij}^k \chi_\lambda \left(g_{C_k} \right).$$

Dividing by $|C_i||C_j|$, summing over λ, and changing the order of summations yields

$$\sum_{\lambda=1}^N \chi_\lambda \left(g_{C_i} \right) \chi_\lambda \left(g_{C_j} \right) = \frac{1}{|C_i||C_j|} \sum_{k=1}^\alpha c_{ij}^k \sum_{\lambda=1}^N d_\lambda \chi_\lambda \left(g_{C_k} \right) = \frac{1}{|C_i||C_j|} \sum_{k=1}^\alpha c_{ij}^k |G| \delta \left(g_{C_k} \right).$$

The last equality follows from Eq. (8.7). Since $k = 1$ is the class containing the identity, and we know from Section 7.2.5 that $c_{ij}^1 = |C_i| \delta_{C_i, C_j^{-1}}$, it follows that

$$\frac{1}{|C_i||C_j|} \sum_{k=1}^\alpha c_{ij}^k |G| \delta \left(g_{C_k} \right) = \frac{|G|}{|C_i||C_j|} c_{ij}^1 = \frac{|G|}{|C_j|} \delta_{C_i, C_j^{-1}}.$$

This is written as Eq. (8.14) by observing from the unitarity of π_λ that

$$\chi_\lambda \left(g_{C_j^{-1}} \right) = \overline{\chi_\lambda \left(g_{C_j} \right)},$$

and so switching the roles of C_j and C_j^{-1} in the preceding equations yields

$$\sum_{\lambda=1}^N \chi_\lambda \left(g_{C_i} \right) \overline{\chi_\lambda \left(g_{C_j} \right)} = \frac{|G|}{|C_i|} \delta_{C_i, C_j}.$$

Recognizing that $\delta_{C_i, C_j} = \delta_{i,j}$ completes the proof. ∎

THEOREM 8.6
The number of distinct irreducible representations of a finite group is equal to the number of its conjugacy classes.

PROOF   Following [32], we can view $v_i^\lambda = \sqrt{|C_i|} \chi_\lambda(g_{C_i})$ for $i = 1, \ldots, \alpha$ as the components of an α-dimensional vector, $\mathbf{v}^\lambda \in \mathbb{C}^\alpha$. In this interpretation, the first orthogonality of characters then says that all such vectors are orthogonal to each other (when using the standard inner product on $\mathbb{C}^\alpha$). But by definition λ can have N values, each corresponding to a different irreducible representation. Thus it must be the case that $N \leq \alpha$ because N orthogonal vectors cannot exist in an α-dimensional vector space if $N > \alpha$.

Now let $w_i^C = \chi_i(g_C)$ be the components of an N-dimensional vector $\mathbf{w}^C \in \mathbb{C}^N$. (Note that for each component of the vector $\mathbf{w}^C$, the class is the same and the character function varies, whereas for the components of $\mathbf{v}^\lambda$, the character function is the same but the class varies.) The second orthogonality of characters says that this set of vectors is orthogonal in $\mathbb{C}^N$. The number of such vectors, α, must then obey $\alpha \leq N$.

The two results $N \leq \alpha$ and $\alpha \leq N$ can only hold if $\alpha = N$. ∎

THEOREM 8.7
The number of times each irreducible representation of a finite group appears in the decomposition of the regular representation is equal to its dimension.

PROOF   The regular representation is decomposed into the direct sum of irreducible ones as

$$T(g) = S^{-1} \left(\sum_{\lambda=1}^{N} \bigoplus n_\lambda \pi_\lambda(g) \right) S$$

for some invertible matrix S. Taking the trace of both sides yields

$$\chi(g) = \sum_{\lambda=1}^{N} n_\lambda \chi_\lambda(g).$$

Multiplying both sides by $\overline{\chi_{\lambda'}(g)}$ and summing over G we get

$$\sum_{g \in G} \chi(g) \overline{\chi_{\lambda'}(g)} = \sum_{\lambda=1}^{N} n_\lambda \sum_{g \in G} \chi_\lambda(g) \overline{\chi_{\lambda'}(g)} = n_{\lambda'} |G|.$$

For finite groups, we already saw in Eq. (8.7) that $\chi(g) = |G|\delta(g)$, and so

$$\sum_{g \in G} \chi(g) \overline{\chi_{\lambda'}(g)} = |G| \overline{\chi_{\lambda'}(e)} = |G| \text{trace} \left(\mathbb{I}_{d_{\lambda'} \times d_{\lambda'}} \right) = |G| d_{\lambda'}.$$

Thus $n_{\lambda'} = d_{\lambda'}$. ∎

It follows from this that every Abelian group has exactly $|G|$ characters, and each of them appears only once.

8.1.3   Fourier Transform, Inversion, and Convolution Theorem

Since the trace of a matrix is invariant under similarity transformation, the trace of both sides of Eq. (8.8) yields the equality

$$\text{trace}(T(g)) = \sum_{j=1}^{\alpha} d_j \chi_j(g), \tag{8.18}$$

where $\chi_j(g) = \text{trace}[\pi_j(g)]$ is called the *character* of the irreducible representation $\pi_j(g)$. The characters are all examples of *class* functions [i.e., $\chi_j(h^{-1} \circ g \circ h) = \chi_j(g)$], since they have the property that they are constant on conjugacy classes of the group. This property results directly from the fact that π_j is a representation (and therefore a homomorphism) and the invariance of the trace under similarity transformation. In fact, it may be shown that any class function can be expanded as a weighted sum of characters.

Evaluating both sides of Eq. (8.18) at $g = e$ (where the representations are $d_j \times d_j$ identity matrices), we get *Burnside's formula*

$$|G| = \sum_{j=1}^{\alpha} d_j^2. \tag{8.19}$$

(Recall that $|G|$ is the number of group elements.) Evaluating Eq. (8.18) at $g \neq e$ yields

$$0 = \sum_{j=1}^{\alpha} d_j \chi_j(g), \qquad \forall \quad g \neq e. \tag{8.20}$$

With these facts, we are now ready to define the Fourier transform of a function on a finite group and the corresponding inversion formula.

DEFINITION 8.1 *The Fourier transform of a complex-valued function $f : G \to \mathbb{C}$ is the matrix-valued function with representation-valued argument defined as*

$$\mathcal{F}(f) = \hat{f}(\pi_j) = \sum_{g \in G} f(g) \pi_j(g^{-1}). \tag{8.21}$$

One refers to $\hat{f}(\pi_j)$ as the Fourier transform at the representation π_j, whereas the whole collection of matrices $\{\hat{f}(\pi_j)\}$ for $j = 1, \ldots, \alpha$ is called the spectrum of the function f.

Note that unlike $\mathcal{T}^{\lambda}(f)$ defined in Eq. (8.2), the representations are now irreducible, and the Fourier transform is defined with $\pi_j(g^{-1})$ instead of $\pi_j(g)$.

THEOREM 8.8
The following inversion formula reproduces f from its spectrum

$$f(g) = \frac{1}{|G|} \sum_{j=1}^{\alpha} d_j \mathrm{trace}\left[\hat{f}(\pi_j) \pi_j(g) \right], \tag{8.22}$$

where the sum is taken over all inequivalent irreducible representations.

PROOF By definition of the Fourier transform and properties of the representations π_j,

$$\hat{f}(\pi_j)\pi_j(g) = \sum_{h \in G} f(h)\pi_j(h^{-1})\pi_j(g)$$

$$= \sum_{h \in G} f(h)\pi_j(h^{-1} \circ g)$$

$$= f(g)\mathbb{I}_{d_j \times d_j} + \sum_{h \neq g} f(h)\pi_j(h^{-1} \circ g).$$

Taking the trace of both sides,

$$\mathrm{trace}\left[\hat{f}(\pi_j)\,\pi_j(g) \right] = d_j f(g) + \sum_{h \neq g} f(h)\chi_j\left(h^{-1} \circ g \right).$$

Multiplying both sides by d_j and summing over all j yields

$$\sum_{j=1}^{\alpha} d_j \text{trace} \left[\hat{f}\left(\pi_j\right) \pi_j(g) \right] = \left(\sum_{j=1}^{\alpha} d_j^2 \right) f(g) + \sum_{h \neq g} f(g) \sum_{j=1}^{\alpha} d_j \chi_j \left(h^{-1} \circ g \right).$$

Using Burnside's formula [Eq. (8.19)] and Eq. (8.20), we see that

$$\sum_{j=1}^{\alpha} d_j \text{trace} \left[\hat{f}\left(\pi_j\right) \pi_j(g) \right] = |G| f(g)$$

which, after division of both sides by $|G|$, results in the inversion formula.   ∎

While not directly stated in the preceding proof, the inversion formula works because of the orthogonality Eq. (8.9) and the completeness of the set of IURs.

REMARK 8.1   Note that while we use irreducible representations that are unitary, because of the properties of the trace, the formulation would still work even if the representations were not unitary. ∎

THEOREM 8.9
The Fourier transform of the convolution of two functions on G is the matrix product of the Fourier transform matrices

$$\mathcal{F}\left((f_1 * f_2)(g)\right)\left(\pi_j\right) = \hat{f}_2\left(\pi_j\right) \hat{f}_1\left(\pi_j\right),$$

where the order of the products matters.

PROOF   By definition we have

$$\mathcal{F}\left((f_1 * f_2)(g)\right)\left(\pi_j\right) = \sum_{g \in G} \left(\sum_{h \in G} f_1(h) f_2\left(h^{-1} \circ g\right) \right) \pi_j\left(g^{-1}\right).$$

Making the change of variables $k = h^{-1} \circ g$ and substituting $g = h \circ k$, the previous expression becomes

$$\sum_{k \in G} \sum_{h \in G} f_1(h) f_2(k) \pi_j\left(k^{-1} \circ h^{-1}\right).$$

Using the homomorphism property of π_j and the commutativity of scalar-matrix multiplication and summation, this can be split into

$$\left(\sum_{k \in G} f_2(k) \pi_j\left(k^{-1}\right) \right) \left(\sum_{h \in G} f_1(h) \pi_j\left(h^{-1}\right) \right) = \hat{f}_2\left(\pi_j\right) \hat{f}_1\left(\pi_j\right).   ∎$$

REMARK 8.2   The reversal of order of the Fourier transform matrices relative to the order of convolution of functions is a result of the $\pi_j(g^{-1})$ in the definition of the Fourier transform. Often, in the literature the Fourier transform is defined using $\pi_j(g)$ in the definition instead of $\pi_j(g^{-1})$. This keeps the order of the product of the Fourier transform matrices the same as the order of convolution of functions in the convolution theorem. However, for the inversion formula to work, $\pi_j(g)$ must then be replaced by $\pi_j(g^{-1})$. As a matter of personal preference, and consistency with the way in which the standard Fourier series and transform were defined in Chapter 2, we view the inversion formula as an expansion of a function $f(g)$ in harmonics (matrix elements of the IURs) instead of their complex conjugates. ∎

LEMMA 8.1

The generalization of the Parseval equality (called the Plancherel theorem in the context of group theory) is written as

$$\sum_{g \in G} |f(g)|^2 = \frac{1}{|G|} \sum_{j=1}^{\alpha} d_j \left\| \hat{f}(\pi_j) \right\|_2^2,$$

where $\| \cdot \|_2$ denotes the Hilbert-Schmidt norm of a matrix. A more general form of this equality is

$$\sum_{g \in G} f_1\left(g^{-1}\right) f_2(g) = \frac{1}{|G|} \sum_{j=1}^{\alpha} d_j \operatorname{trace}\left[\hat{f}_1(\pi_j) \, \hat{f}_2(\pi_j) \right].$$

PROOF   Applying the inversion formula to the result of the convolution theorem one finds

$$(f_1 * f_2)(g) = \frac{1}{|G|} \sum_{j=1}^{\alpha} d_j \operatorname{trace}\left[\hat{f}_2(\pi_j) \, \hat{f}_1(\pi_j) \, \pi_j(g) \right].$$

Evaluating this result at $g = e$ results in the preceding second statement. In the particular case when $f_1(g) = \overline{f(g^{-1})}$ and $f_2(g) = f(g)$, this reduces to the first statement. ∎

In the special case when f is a class function, the general formulation is simplified. Any class function can be expanded as a finite linear combination of characters (which forms a complete orthogonal basis for the space of all square integrable class functions)

$$f(g) = \frac{1}{|G|} \sum_{i=1}^{\alpha} f_i \chi_i(g)$$

where

$$f_i = \sum_{g \in G} f(g) \overline{\chi_i(g)}.$$

Furthermore, the Plancherel equality for class functions reduces to

$$\sum_{g \in G} |f(g)|^2 = \frac{1}{|G|} \sum_{i=1}^{\alpha} |f_i|^2.$$

The constants f_i are related to the Fourier transforms as: $\hat{f}(\pi_i) = \frac{f_i}{d_i} \mathbb{I} d_i \times d_i$.

Analogous to the way class functions are constant on classes, one can also define functions constant on cosets. For instance, a function satisfying $f(g) = f(g \circ h)$ for all $h \in H < G$ is constant on all left cosets, and is hence viewed as a function on the coset space G/H. Using the property of such functions, the Fourier transform is written in this special case as

$$\hat{f}(\pi_j) = \sum_{g \in G} f(g \circ h) \left(\pi_j(g)\right)^\dagger = \pi_j(h) \sum_{k \in G} f(k) \left(\pi_j(k)\right)^\dagger,$$

where the change of variables $k = g \circ h$ has been made. Summing both sides over H, and dividing by $|H|$ gives

$$\hat{f}(\pi_j) = \Delta(\pi_j)\,\hat{f}(\pi_j)$$

where

$$\Delta(\pi_j) = \frac{1}{|H|} \sum_{h \in H} \pi_j(h)$$

is a matrix which constrains the structure of $\hat{f}(\pi_j)$. For some values of j, $\Delta(\pi_j) = 0$, hence forcing $\hat{f}(\pi_j)$ to be zero. For others it will have a structure which causes only parts of $\hat{f}(\pi_j)$ to be zero. A similar construction follows for functions $f(g) = f(h \circ g)$ for all $h \in H < G$ which are constant on right cosets, and hence are functions on $H\backslash G$.

The material presented in this section is well known to mathematicians who concentrate on finite groups. References which explain this material from different perspectives include [20, 23, 33, 56]. Concrete representations of crystallographic space groups can be found in [41, 42].

We note also that fast Fourier transform and inversion algorithms for broad classes of finite groups have been developed in the recent literature [12, 16, 62, 52]. This is discussed in the next section.

8.1.4   Fast Fourier Transforms for Finite Groups

At the cores of the fast Fourier transforms reviewed at the end of Chapter 2 and fast polynomial transforms reviewed in Chapter 3 were recurrence formulae used to decompose the sums defining discrete transforms. This has been extended to the context of finite groups in a series of papers by Diaconis and Rockmore [16], Rockmore [62], Maslen [52], and Maslen and Rockmore [53]. We shall not go into all of the technical detail of fast Fourier transforms for groups but will rather only examine the key enabling concepts.

Given a function $F : S \to V$, where S is a finite set and V is a vector space, one can make the decomposition

$$\sum_{s \in S} F(s) = \sum_{[s] \in S/\sim} \left(\sum_{x \in [s]} F(x) \right) \tag{8.23}$$

for any equivalence relation $\sim$. (See Appendix B for an explanation of terms.)

In the context of arbitrary groups, two natural equivalence classes are immediately apparent. These are conjugacy classes and cosets. In the case of conjugacy classes, Eq. (8.23) may be written as

$$\sum_{g \in G} F(g) = \sum_{i=1}^{\alpha} \sum_{g \in C_i} F(g).$$

The decomposition when the equivalence classes are chosen to be cosets is

$$\sum_{g \in G} F(g) = \sum_{\sigma \in G/H} \left(\sum_{s \in \sigma} F(s) \right). \tag{8.24}$$

Since each left coset is of the form $\sigma = gH$, it follows that $|\sigma| = |H|$ and so $|G| = |G/H| \cdot |H|$, and the number of left cosets is $|G/H| = |G|/|H|$.

Let $s_\sigma \in \sigma$ for each $\sigma \in G/H$. s_σ is called a coset representative. With the observation that

$$\sum_{s \in \sigma} F(s) = \sum_{h \in H} F(s_\sigma \circ h),$$

one rewrites Eq. (8.24) as

$$\sum_{g \in G} F(g) = \sum_{\sigma \in G/H} \left(\sum_{h \in H} F(s_\sigma \circ h) \right).$$ (8.25)

This is a key step in establishing a recursive computation of the Fourier transform for groups. However, this alone does not yield a fast evaluation of an arbitrary sum of the form $\sum_{g \in G} F(g)$. The second essential feature is that the Fourier transform of a function on a finite group is special because

$$F(g) = f(g)\pi\left(g^{-1}\right)$$

where $\pi(g)$ is an irreducible matrix representation of the group, and therefore $\pi(g_1 \circ g_2) = \pi(g_1)\pi(g_2)$. This allows one to write Eq. (8.21) as

$$\sum_{g \in G} f(g)\pi\left(g^{-1}\right) = \sum_{\sigma \in G/H} \left(\sum_{h \in H} f_\sigma(h)\pi\left(h^{-1}\right) \right) \pi\left(s_\sigma^{-1}\right)$$ (8.26)

where $f_\sigma(h) = f(s_\sigma \circ h)$. In general, since the restriction of an irreducible representation to a subgroup is reducible, one writes it as a direct sum

$$\pi(h) \cong \sum_k \bigoplus c_k \pi_k^H(h).$$ (8.27)

When using a special kind of basis (called "H-adapted"), the equivalence sign in Eq. (8.27) becomes an equality. The representations $\pi_k^H(h)$ are irreducible, and c_k is the number of times $\pi_k^H(h)$ appears in the decomposition of $\pi(h)$. Since $\sum_k c_k \dim(\pi_k^H) = \dim(\pi)$, it follows that it is more efficient to compute the Fourier transforms of all the functions $f_\sigma(h)$ in the subgroup H using the irreducible representations as $\sum_{h \in H} f_\sigma(h)\pi_k^H(h^{-1})$ for all $\sigma \in G/H$ and all values of k, and recompose the Fourier transform on the whole group using Eqs. (8.26) and (8.27).

Given a "tower of subgroups"

$$G_n \subset \cdots \subset G_2 \subset G_1 \subset G,$$ (8.28)

one can write Eq. (8.25) as

$$\sum_{g \in G} F(g) = \sum_{\sigma_1 \in G/G_1} \sum_{\sigma_2 \in G_1/G_2} \cdots \sum_{\sigma_n \in G_{n-1}/G_n} \sum_{h \in G_n} F\left(s_{\sigma_1} \circ \cdots \circ s_{\sigma_n} \circ h\right)$$ (8.29)

and recursively decompose the representations $\pi(g)$ into smaller and smaller ones as they are restricted to smaller and smaller subgroups.

The computational performance of this procedure is generally better than the $\mathcal{O}(|G|^2)$ computations required for direct evaluation of the whole spectrum of a function on a finite group. For some classes of groups $\mathcal{O}(|G| \log |G|)$ or $\mathcal{O}(|G|(\log |G|)^2)$ performance can be achieved. In general, at least $\mathcal{O}(|G|^{\frac{3}{2}})$ performance is possible.[3] The computational complexity depends on the structure of the group, and in particular, on how many subgroups are cascaded in the tower in Eq. (8.28). The following is a specific case from the literature.

---

[3]For the $\mathcal{O}(\cdot)$ symbol to make sense in this context, the groups under consideration must be members of infinite classes so that $|G|$ can become arbitrarily large and asymptotic complexity estimates have meaning.

THEOREM 8.10 [4]

Let $G = G_1 \times G_2 \times \cdots \times G_k$ be the direct product of finite groups $G_1, \ldots, G_k$. Then the Fourier transform and its inverse can be computed in $\mathcal{O}(|G| \sum_{i=1}^{k} |G_i|)$ arithmetic operations.

A number of works have addressed various aspects of the complexity of computation of the Fourier transform and convolution of functions on finite groups. These include [5, 10] in the commutative case, and [4, 7, 34, 64] in the noncommutative context. An efficient Fourier transform for wreath product groups is derived in [63]. Applications of fast Fourier transforms for finite groups are discussed in [35, 36, 37, 38, 76, 78].

8.2   Differentiation and Integration of Functions on Lie Groups

8.2.1   Derivatives, Gradients, and Laplacians of Functions on Lie Groups

Given a function $f : G \to \mathbb{C}$ with $g \in G$ parameterized as $g(x_1, \ldots, x_n)$ and $f(g(x_1, \ldots, x_n))$ differentiable in all the parameters x_i, we define for any element of the Lie algebra, X

$$X^L f(g) = \frac{d}{dt} (f((\mathbb{I} + tX)g))|_{t=0} = \lim_{t \to 0} \frac{f((\exp tX)g) - f(g)}{t}. \tag{8.30}$$

Similarly, let

$$X^R f(g) = \frac{d}{dt} (f(g(\mathbb{I} + tX)))|_{t=0} = \lim_{t \to 0} \frac{f(g(\exp tX)) - f(g)}{t}. \tag{8.31}$$

The superscripts L and R denote on which side of the argument the infinitesimal operation is applied. It follows from the fact that left and right shifts do not interfere with each other that

$$\left(L(g_0) X^R f \right)(g) = X^R f \left(g_0^{-1} \circ g \right) = X^R (L(g_0) f)(g)$$

and

$$\left(R(g_0) X^L f \right)(g) = X^L f(g \circ g_0) = X^L (R(g_0) f)(g).$$

We note in passing that the (right) Taylor series about $g \in G$ of an an analytic function on a Lie group G is expanded as [79]

$$f(g \circ \exp tX) = \sum_{k=0}^{\infty} \left(\frac{d^k}{dt^k} f(g \circ \exp tX) \right)|_{t=0} \frac{t^k}{k!}. \tag{8.32}$$

The left Taylor series follows analogously.

Particular examples of the operators X^L and X^R are X_i^R and X_i^L where X_i is a basis element for the Lie algebra.

These operators inherit the Lie bracket properties from $\mathcal{G}$

$$[X, Y]^L = -\left[X^L, Y^L \right] \quad \text{and} \quad [X, Y]^R = -\left[X^R, Y^R \right].$$

More sophisticated differential operators can be constructed from these basic ones in a straightforward way. For instance, the left and right gradient vectors at the identity can be defined as

$$\text{grad}_e^R f = \sum_{i=1}^{N} X_i X_i^R f(g)|_{g=e} \quad \text{and} \quad \text{grad}_e^L f = \sum_{i=1}^{N} X_i X_i^L f(g)|_{g=e}.$$

These gradients are two ways to describe the same element of the Lie algebra $\mathcal{G}$, i.e., $\text{grad}_e^R f = \text{grad}_e^L f$. We may likewise generate images of these gradients at any tangent to the group by defining

$$\text{grad}_g^R f = \sum_{i=1}^N g X_i X_i{}^R f(g) \qquad \text{and} \qquad \text{grad}_g^L f = \sum_{i=1}^N X_i g X_i{}^L f(g).$$

The products $g X_i$ and $X_i g$ are matrix products, or more generally can be written as $d/dt(g \circ \exp t X_i)|_{t=0}$ and $d/dt(\exp t X_i \circ g)|_{t=0}$.

The collection of all grad_g^R (resp. grad_g^L) can be viewed as a vector field.

We may also define the left and right divergence of a vector field, $Y(g) = \sum_{i=1}^N Y_i(g) X_i(g)$, on the group as

$$\text{div}_L(Y) = \left(\text{grad}_g^L, Y(g) \right)_g^L \qquad \text{div}_R(Y) = \left(\text{grad}_g^R, Y(g) \right)_g^R,$$

where appropriate left or right inner products on the tangent space of G at g are used.

Given a right-invariant vector field $Y^L = \sum_{i=1}^N y_i X_i g$ where y_i are constants, we can also define differential operators of the form

$$Y^L f = \text{div}_L(Y f) = \sum_{i=1}^N y_i X_i{}^L f,$$

and analogously for the left-invariant version. These are analogous to the directional derviatives.

Analogous to the Abelian case, the Laplacian operator is defined as the divergence of gradient vectors.

Using the orthogonality of Lie algebra basis elements, we have

$$\text{div}_L \left(\text{grad}_L(f) \right) = \sum_{i=1}^N \left(X_i{}^L \right)^2 f \tag{8.33}$$

$$\text{div}_R \left(\text{grad}_R(f) \right) = \sum_{i=1}^N \left(X_i{}^R \right)^2 f. \tag{8.34}$$

Hence, whenever $\sum_{i=1}^N (X_i{}^L)^2 = \sum_{i=1}^N (X_i{}^R)^2$, the Laplacian for the group is written without subscripts as $\text{div}(\text{grad}(f))$.

8.2.2 Integration Measures on Lie Groups and their Homogeneous Spaces

In this section we review integration on Lie groups and on their left and right coset spaces. Obviously, one could not go very far in extending concepts from classical Fourier analysis to the noncommutative case without some concept of integration. Thus it is important to get the basic idea of the material in this section before proceeding.

Haar Measures and Shifted Functions

Let G be a locally compact group (and in particular, a Lie group). On G, two natural integration measures exist. These are called the *left* and *right Haar* measures [27, 58].[4]

---

[4] The Haar measure, named after Alfréd Haar (1885–1933), is used for abstract locally compact groups. In the particular case of Lie groups, the Haar measure is sometimes called a *Hurwitz measure*, named after Adolf Hurwitz (1859–1919).

The measures μ_L and μ_R respectively have the properties

$$\mu_L(f) = \int_G f(g)d_L(g) = \int_G (L(a)f)(g)d_L(g) = \int_G f\left(a^{-1} \circ g\right)d_L(g)$$

and

$$\mu_R(f) = \int_G f(g)d_R(g) = \int_G (R(a)f)(g)d_R(g) = \int_G f(g \circ a)d_R(g)$$

for any $a \in G$, where we will assume $f(g)$ to be a "nice" function. $f(g)$ is assumed to be "nice" in the sense that $f(g) \in \mathcal{L}^p(G, d_{L,R})$ for all $p \in \mathbb{Z}^p$ (where $d_{L,R}$ stands for either d_L or d_R), and in addition is infinitely differentiable in each coordinate parameter.

The measures μ_L and μ_R are invariant under left, $L(a)$, and right, $R(a)$, shifts respectively. However, if the left Haar measure is evaluated under right shifts, and the right Haar measure is evaluated under left shifts, one finds

$$\mu_L(f) = \Delta_L(a) \int_G f(g \circ a)d_L(g)$$

and

$$\mu_R(f) = \Delta_R(a) \int_G f\left(a^{-1} \circ g\right)d_R(g),$$

where $\Delta_L(g)$ and $\Delta_R(g)$ are respectively called the left and right *modular* functions. In general, these functions are related to each other as $\Delta_L(g) = \Delta_R(g^{-1})$, and these functions are both homomorphisms from G into the multiplicative group of positive real numbers under scalar multiplication.

A group for which $\Delta_L(g) = \Delta_R(g) = 1$ is called *unimodular*. For such groups, we can set $d(g) = d_L(g) = d_R(g)$. Integration of "nice" functions on such groups is, by definition, shift invariant. In addition, integration on unimodular groups is invariant under inversions of the argument

$$\int_G f\left(g^{-1}\right)d(g) = \int_G f(g)d(g).$$

All but two of the groups on which we will integrate in this book [including $(\mathbb{R}, +)$, all compact groups, the motion groups, and all direct products of the aforementioned] are unimodular. The two exceptions that we consider are the affine and scale-Euclidean groups, which are of importance in image analysis and computer vision.

Integration on G/H and $H\backslash G$

Let G be a Lie group and H be a Lie subgroup of G. Recall that G/H and $H\backslash G$, respectively, denote the spaces of all left and right cosets. Coset spaces are also refered to as quotient, factor, or homogeneous spaces. Integration and orthogonal expansions on such spaces is a field of study in its own right (see, for example, [11, 75, 82, 84, 86]). In this section we review the fundamental concepts of invariant and quasi-invariant integration measures on G/H. The formulation follows in an exactly analogous manner for $H\backslash G$.

The phrase "invariant integration of functions on homogeneous spaces of groups" is by no means a standard one in the vocabulary of most engineers. However, the reader has, perhaps without knowing it, already seen in Chapters 2, 4, and 5 how to integrate functions on some of the simplest examples of homogeneous spaces: $\mathbb{R}/2\pi\mathbb{Z} \cong S^1$ (the unit circle), $SO(3)/SO(2) \cong S^2$ (the unit sphere), and $SE(3)/\mathbb{R}^3 \cong SO(3)$. Knowledge of the more general theory is important in the context of noncommutative harmonic analysis, because construction of the irreducible unitary representations of many groups (utilizing the method of induced representations) is achieved using these ideas. In Chapter 10 we use this method to generate the IURs of the Euclidean motion group.

Let $g_H \in gH$ be an arbitrary coset representative. Then $\nu_{G/H}(f)$ is called a G−invariant measure of $f \in \mathcal{L}^2(G/H) \cap C(G/H)$ if

$$\nu_{G/H}(f) = \int_{G/H} f(a \circ g_H) d_{G/H}(g_H) = \int_{G/H} f(g_H) d_{G/H}(g_H) \qquad (8.35)$$

for all $a \in G$. Here $d_{G/H}(\cdot)$ returns the same value for all values of the argument from the same coset, and the integral can intuitively be thought of as adding function values once over each coset. Another way to write Eq. (8.35) is

$$\nu_{G/H}(f) = \int_{G/H} f(a \circ (g \circ h)) d(gH) = \int_{G/H} f(g \circ h) d(gH) \qquad (8.36)$$

for any $h \in H$, where $d(gH) \overset{\Delta}{=} d_{G/H}(g_H)$ for all $g_H \in gH$. $d(gH)$ takes the coset as its argument, while $d_{G/H}(g_H)$ takes a coset representative (which is a group element) as its argument. Similarly, instead of writing f as a function on the group that is constant on left cosets [i.e., $f(g) = f(g \circ h)$ for all $h \in H$], we can write all of the preceding expressions in terms of $\tilde{f}(gH) \overset{\Delta}{=} f(g_H) = f(g)$.

For example, we can identify any unit vector with elements of $S^2 \cong SO(3)/SO(2)$. In this case Eq. (8.36) is written as

$$\int_{S^2} \tilde{f}\left(R^T \mathbf{u}\right) d\mathbf{u} = \int_{S^2} \tilde{f}(\mathbf{u}) d\mathbf{u}$$

where $d\mathbf{u}$ is the usual integration measure on S^2 and $R \in SO(3)$ is an arbitrary rotation.

Invariant measures of suitable functions on cosets exist whenever the modular function of G restricted to H is the same as the modular function of H [86]. Since we are most interested in unimodular groups (and their unimodular subgroups) this condition is automatically satisfied.

The existence of a G−invariant measure on G/H allows one to write [8, 29, 86]

$$\int_G f(g) d_G(g) = \int_{G/H} \left(\int_H f(g_H \circ h) d_H(h)\right) d_{G/H}(g_H) \qquad (8.37)$$

for any well-behaved $f \in \mathcal{L}^2(G)$ where d_G and d_H are, respectively, the invariant integration measures on G and H. Since the arguments of the volume elements are descriptive enough, one can drop the subscripts G, H, and G/H. Doing this in combination with the notation in Eq. (8.36), we may rewrite Eq. (8.37) as

$$\int_G f(g) d(g) = \int_{G/H} \left(\int_H f(g \circ h) d(h)\right) d(gH) \qquad (8.38)$$

where $g \in gH$ is taken to be the coset representative. In the special case when $f(g)$ is a left-coset function, Eq. (8.38) reduces to

$$\int_G f(g) d(g) = \int_{G/H} \tilde{f}(gH) d(gH)$$

where it is assumed that $d(h)$ is normalized so that $\text{Vol}(H) = \int_H d(h) = 1$.

For example, let $SO(3)$ be parameterized with ZYZ Euler angles, $g = R_3(\alpha) R_2(\beta) R_3(\gamma)$. Taking $H \cong SO(2)$ to be the subgroup of all $R_3(\gamma)$, and identifying all matrices of the form $R_3(\alpha) R_2(\beta)$ with points on the unit sphere [with (α, β) serving as spherical coordinates], one writes

$$\int_{SO(3)} f(g) d(g) = \int_{S^2} \left(\int_{SO(2)} f\left((R_3(\alpha) R_2(\beta)) R_3(\gamma)\right) d\left(R_3(\gamma)\right)\right) d\mathbf{u}(\alpha, \beta),$$

where $\mathbf{u}(\alpha, \beta) = R_3(\alpha)R_2(\beta)\mathbf{e}_3$. Even more explicitly, this is

$$\frac{1}{8\pi^2} \int_0^{2\pi} \int_0^{\pi} \int_0^{2\pi} f\left(R_3(\alpha)R_2(\beta)R_3(\gamma)\right) \sin\beta \, d\alpha \, d\beta \, d\gamma$$

$$= \frac{1}{4\pi} \int_0^{\pi} \int_0^{2\pi} \left(\frac{1}{2\pi} \int_0^{2\pi} f\left((R_3(\alpha)R_2(\beta))\, R_3(\gamma)\right) d\gamma\right) \sin\beta \, d\alpha \, d\beta.$$

If we are given a unimodular group G with unimodular subgroups K and H such that $K \leq H$, we may use the facts that

$$\int_G f(g) d_G(g) = \int_{G/H} \int_H f(g_H \circ h)\, d_H(h) d_{G/H}(g_H)$$

and

$$\int_H f(h) d_H(h) = \int_{H/K} \int_K f(h_K \circ k)\, d_K(k) d_{H/K}(h_K)$$

to decompose the integral of any nice function $f(g)$ as

$$\int_G f(g) d_G(g) = \int_{G/H} \int_{H/K} \int_K f(g_H \circ h_K \circ k)\, d_K(k) d_{H/K}(h_K)\, d_{G/H}(g_H).$$

In the other notation this is

$$\int_G f(g) d(g) = \int_{G/H} \int_{H/K} \int_K f(g \circ h \circ k) d(k) d(hK) d(gH). \tag{8.39}$$

For example, using the fact that any element of $SE(3)$ can be decomposed as the product of translation and rotation subgroups, $g = T(\mathbf{a}) \circ A(\alpha, \beta, \gamma)$, the integral of a function over $SE(3)$

$$\int_{SE(3)} f(g) d(g) = \int_{\mathbb{R}^3} \int_{SO(3)} f(\mathbf{a}, A) dA d\mathbf{a}$$

can be written as

$$\int_{SE(3)/SO(3)} \int_{SO(3)/SO(2)} \int_{SO(2)} f\left(T(\mathbf{a}) \circ (R_3(\alpha)R_2(\beta)) \circ R_3(\gamma)\right) d\left(R_3(\gamma)\right) d\mathbf{u}(\alpha, \beta) d\mathbf{a}.$$

We can also decompose the integral using the subgroups of $(\mathbb{R}^3, +)$, but this is less interesting since the decomposition of a function on $\mathbb{R}^3$ is just

$$\int_{\mathbb{R}^3} F(\mathbf{a}) d\mathbf{a} = \int_{\mathbb{R}} \int_{\mathbb{R}} \int_{\mathbb{R}} F\left(a_1\mathbf{e}_1 + a_2\mathbf{e}_2 + a_3\mathbf{e}_3\right) da_1 da_2 da_3.$$

Another interesting thing to note (when certain conditions are met) is the decomposition of the integral of a function on a group in terms of two subgroups and a double coset space

$$\int_G f(g) d(g) = \int_K \int_{K \backslash G/H} \int_H f(k \circ g \circ h) d(h) d(KgH) d(k).$$

A particular example of this is the integral over $SO(3)$, which can be written as

$$\int_{SO(3)} = \int_{SO(2)} \int_{SO(2) \backslash SO(3)/SO(2)} \int_{SO(2)}.$$

8.2.3 Constructing Invariant Integration Measures

In this subsection we explicitly construct integration measures on several Lie groups, and give a short explanation of how this can be done in general.

According to Lie theory there always *exists* both unique left-invariant and right-invariant integration measures (up to an arbitrary scaling) on arbitrary Lie groups. Usually the left- and right-invariant integration measures are different. For the case of compact Lie groups, a unique bi-invariant (left and right invariant) measure exists, and the most natural scaling in this case is the one for which $\int_G d(g) = 1$. For the case of $SO(3)$ we constructed this bi-invariant measure based on the physically intuitive relationship between angular velocity and the rate of change of rotation parameters. The fact that $SE(3)$ has a bi-invariant integration measure which is the product of those for $SO(3)$ and $\mathbb{R}^3$ was also demonstrated earlier.

When it is possible to view the manifold of a Lie group as a hyper-surface in $\mathbb{R}^N$, volume elements based on the Riemannian metric can be used as integration measures. Invariance under left or right shifts must then be established by an appropriate choice of scaling matrix [W_0 in Eq. (4.12)]. The theory of differential forms can also be used to generate left (right)-invariant integration measures.

Left- and right-invariant integration measures are then found as

$$d_L(g) = |\det(J_R)| \, dx_1 \cdots dx_n \qquad \text{and} \qquad d_R(g) = |\det(J_L)| \, dx_1 \cdots dx_n.$$

Generally, $d_L(g) \neq d_R(g)$. When equality holds, the group is called *unimodular*. The reason for this will be explained in Section 8.2.4.

We already had a preview of this general theory in action for $SO(3)$ and $SE(3)$. In those cases, as with the following examples, it is possible to identify each orthonormal basis element X_i of the Lie algebra with a basis element $\mathbf{e}_i \in \mathbb{R}^n$ through the $\vee$ operation. Having introduced the concept of a Lie algebra, and an inner product on the Lie algebras of interest here, we need not use the $\vee$ opcration in the context of the examples that follow.

In any coordinate patch, elements of an n-dimensional matrix Lie group can be parameterized as $g(x_1, \ldots, x_n)$. In a patch near the identity, two natural parameterizations are

$$g(x_1, \ldots, x_n) = \prod_{i=1}^n \exp(x_i X_i) \qquad \text{and} \qquad \hat{g}(x_1, \ldots, x_n) = \exp\left(\sum_{i=1}^n x_i X_i\right). \tag{8.40}$$

In practice these are often the only parameterizations we need. Usually, for noncommutative groups, these parameterizations do not yield the same result, but by some nonlinear transformation of coordinates they can be related as $\hat{g}(\mathbf{x}) = g(\mathbf{f}(\mathbf{x}))$. We have already seen this for the case of the Heisenberg group, the rotation group, and the Euclidean motion group.

For a matrix Lie group, the operation of partial differentiation with respect to each of the parameters x_i is well defined. Thus we can compute the left and right "tangent operators"

$$T_i^L(g) = \frac{\partial g}{\partial x_i} g^{-1} \qquad \text{and} \qquad T_i^R(g) = g^{-1} \frac{\partial g}{\partial x_i}.$$

They have the property that they convert the matrix group elements g into their derivatives

$$T_i^L(g)g = \frac{\partial g}{\partial x_i} = g T_i^R(g).$$

As a result of the product rule for differentiation of matrices, we have $\partial(g \circ g_0)/\partial x_i = (\partial g/\partial x_i)g_0$, and likewise for left shifts. This means that $T_i^L(g \circ g_0) = T_i^L(g)$ and $T_i^R(g_0 \circ g) = T_i^R(g)$.

If either of these operators is evaluated at $g = e$, and if either of the "natural" parameterizations in Eq. (8.40) is used, one sees that the tangent operators reduce to Lie algebra basis vectors

$$T_i^L(g)\Big|_{g=e} = T_i^R(g)\Big|_{g=e} = X_i.$$

We may use the tangent operators T_i^L and T_i^R defined previously to construct left- or right-invariant integration measures on the group. For the moment, let us restrict the discussion to the left-invariant case. First, associate with each tangent operator $T_i^L(g)$ an $n \times 1$ array, the components of which are the projection of $T_i^L(g)$ onto the basis vectors $\{X_j\}$. This projection is performed using one of the inner products on the Lie algebra in Section 7.3.5. The array corresponding to $T_i^L(g)$ is denoted as $(T_i^L(g))^\vee$. A Jacobian matrix is generated by forming an $n \times n$ matrix whose columns are these arrays

$$J_L(g) = \left[\left(T_1^L(g)\right)^\vee, \ldots, \left(T_n^L(g)\right)^\vee \right], \tag{8.41}$$

and similarly for J_R. When the basis elements X_i are chosen to be orthonormal with respect to the inner product $(X_i, X_j) = \text{Re}[\text{trace}(X_i X_j^\dagger)]$, it is clear that $J_L(e) = \mathbb{I}_{n \times n}$.

An infinitesimal volume element located at the identity element of the group can be viewed as an n-dimensional box in the Lie algebra $\mathcal{G} = \mathcal{G}(e)$. It is then parameterized by $\sum_{i=1}^n y_i X_i$ where y_i ranges from 0 to dx_i, and its volume is $dx_1 \ldots dx_n$. We may view the volume at this element as $d_L(e) = |\det(J_R(e))| dx_1 \ldots dx_n$. The volume at $g \neq e$ is computed in a similar way. Using the first parameterization in Eq. (8.40), we have a volume element parameterized by $0 \leq y_i \leq dx_i$ for $i = 1, \ldots, n$

$$g(x_1 + y_1, \ldots, x_n + y_n) = \prod_{i=1}^n \exp\left((x_i + y_i) X_i\right).$$

Using the smallness of each dx_i, we write

$$g(x_1 + dx_1, \ldots, x_n + dx_n) = g(x_1, \ldots, x_n) \left(\mathbb{I}_{N \times N} + \sum_{i=1}^n T_i^R(g) dx_i\right)$$

$$= \left(\mathbb{I}_{N \times N} + \sum_{i=1}^n T_i^L(g) dx_i\right) g(x_1, \ldots, x_n).$$

Recall that g is an $N \times N$ matrix, whereas the dimension of the Lie group consisiting of such matrices has dimension n.

Each of the tangent vectors provides a direction in the tangent space $\mathcal{G}(g)$ defining an edge of the differential box, which is now deformed. The volume of this deformed box is

$$d_L(g(x_1, \ldots, x_n)) = |\det(J_R(g))| dx_1 \ldots dx_n \tag{8.42}$$

because the Jacobian matrix $J_R(g)$ reflects how the coordinate system in $\mathcal{G}(g)$ is distorted relative to the one in the Lie algebra $\mathcal{G}(e)$. Due to the left-invariance of $T_i^R(g)$, and hence $J_R(g)$, we have that $d_L(g_0 \circ g) = d_L(g)$. By a simple change of variables, this means integration of well-behaved functions with respect to this volume element is also left invariant

$$\int_G f(g) d_L(g) = \int_G f(g) d_L(g_0 \circ g) = \int_G f\left(g_0^{-1} \circ g\right) d_L(g).$$

A completely analogous derivation results in a right-invariant integration measure.

Finally, it is worth noting that from a computational point of view, the first parameterization in Eq. (8.40) is often the most convenient to use in analytical and numerical computations of $T_i^L(g)$ and $T_i^R(g)$, and hence the corresponding Jacobians. This is because

$$\frac{\partial g}{\partial x_i} = g(x_1, 0, \ldots, 0) g(0, x_2, 0, \ldots, 0) \cdots \frac{\partial}{\partial x_i} g(0, \ldots, x_i, 0, \ldots, 0) \cdots g(0, \ldots, 0, x_n),$$

and

$$g^{-1}(x_1, \ldots, x_n) = g^{-1}(0, \ldots, 0, x_n) \cdots g^{-1}(0, \ldots, x_i, 0, \ldots, 0) \cdots g^{-1}(x_1, 0, \ldots, 0)$$

and so there is considerable cancellation in the products of these matrices in the definition of $T_i^R(g)$ and $T_i^L(g)$.

8.2.4   The Relationship between Modular Functions and the Adjoint

We saw in the previous section that $d_L(g)$ and $d_R(g)$ are defined and invariant under left and right shifts, respectively. But what happens if $d_L(g)$ is shifted from the right, or $d_R(g)$ is shifted from the left?

We see $T_i^R(g \circ g_0) = g_0^{-1} T_i^R(g) g_0$ and $T_i^L(g_0 \circ g) = g_0 T_i^L(g) g_0^{-1}$. This means that $(T_i^R(g \circ g_0))^\vee = [Ad(g_0^{-1})](T_i^R(g))^\vee$ and $(T_i^L(g_0 \circ g))^\vee = [Ad(g_0)](T_i^L(g))^\vee$. These transformations of tangent vectors propagate through to the Jacobian, and we see that

$$d_L(g \circ g_0) = \Delta_L(g_0)\, d(g) \qquad \text{and} \qquad d_R(g_0 \circ g) = \Delta_R(g_0)\, d(g)$$

where

$$\Delta_L\left(g_0^{-1}\right) = \|[\det(Ad(g_0))]\| = \Delta_R(g_0).$$

We will compute the adjoint for each of the groups of interest in this book, and hence the modular function.

Since

$$\Delta_L(g) \cdot \Delta_R(g) = 1,$$

it is common to define

$$\Delta_L(g) = \Delta(g) \qquad \text{and} \qquad \Delta_R(g) = \frac{1}{\Delta(g)}.$$

Then the following statements hold

$$\int_G f\left(g^{-1}\right) d(g) = \int_G f(g)\Delta\left(g^{-1}\right) d(g)$$

$$\int_G f(g_0 \circ g)\, d_R(g) = \Delta(g_0) \int_G f(g)\, d_R(g)$$

$$\int_G f(g \circ g_0)\, d_L(g) = \Delta\left(g_0^{-1}\right) \int_G f(g)\, d_L(g).$$

The modular function is a continuous homomorphism $\Delta : G \to (\mathbb{R}^+, \cdot)$. The homomorphism property follows from the fact that shifting twice on the right (first by g_2, then by g_1) gives

$$\Delta(g_1 \circ g_2) = \Delta(g_1) \cdot \Delta(g_2).$$

Continuity of the modular function follows from the fact that matrix Lie groups are analytic. On a compact Lie group, all continuous functions are bounded, i.e., it is not possible for a function with compact support to shoot to infinity anywhere and still maintain continuity. But by the homomorphism property, products of modular functions result after each right shift. The only way this can remain finite and positive after an arbitrary number of right shifts is if $\Delta(g) = 1$. Hence, all compact Lie groups are unimodular. As it turns out, other kinds of Lie groups of interest in engineering are also unimodular, and this will prove to be of great importance in later chapters.

8.2.5   Examples of Volume Elements

For $SE(2)$ in the matrix exponential parameterization,

$$d(g) = c_1 dx_1 dx_2 d\theta = c_2 \frac{\sin^2 \alpha/2}{\alpha^2} dv_1 dv_2 d\alpha,$$

where c_1 is an arbitrary scaling constant.

The bi-invariant integration measure for $GL(2, \mathbb{R})$ is

$$d(g) = \frac{1}{|\det g|^2} dx_1 dx_2 dx_3 dx_4.$$

We state without proof that more generally, $GL(N, \mathbb{R})$ is unimodular with

$$d(g) = \frac{1}{|\det g|^N} dx_1 \cdots dx_{N^2} \qquad (8.43)$$

where

$$g = \begin{pmatrix} x_1 & x_2 & \cdots & x_N \\ x_{N+1} & x_{N+2} & \cdots & \vdots \\ \vdots & \vdots & \ddots & \vdots \\ x_{N^2-N+1} & x_{N^2-N+2} & \cdots & x_{N^2} \end{pmatrix}.$$

The dimension of $GL(N, \mathbb{R})$ as a Lie group is $n = N^2$.

One possible normalization for the volume element of $SL(2, \mathbb{R})$ [with integration over the $SO(2)$ subgroup normalized to unity and unit normalization for the two noncompact subgroups] is then

$$d(g(\theta, t, \xi)) = \frac{1}{2\pi} e^{2t} d\theta dt d\xi.$$

Using exactly the same argument as for $SE(2)$ presented in Chapter 6, one can show that the product $d_L(g(\theta, t, \xi)) \cdot d(\mathbf{x})$ for $\mathbf{x} \in \mathbb{R}^2$ is the bi-invariant volume element for $\mathbb{R}^2 \rtimes_\varphi SL(2, \mathbb{R})$.

8.3   Harmonic Analysis on Lie Groups

The results presented for finite groups carry over in a straightforward way to many kinds of groups for which the invariant integration measures exist. In this section we discuss representations of Lie groups, and the decomposition of functions on certian Lie groups into a weighted sum of the matrix elements of the IURs.

8.3.1   Representations of Lie Groups

The representation theory of Lie groups has many physical applications. This theory evolved with the development of quantum mechanics, and today serves as a natural language in which to articulate the quantum theory of angular momentum. There is also a very close relationship between Lie group representations and the theory of special functions [17, 57, 81, 83]. It is this connection that allows for our concrete treatment of harmonic analysis on the rotation and motion groups in Chapters 9 and 10.

Let K denote either $\mathbb{R}^N$ or $\mathbb{C}^N$. Consider a transformation group $(G, \circ)$ that acts on vectors $\mathbf{x} \in K$ as $g \circ \mathbf{x} \in K$ for all $g \in G$. The *left quasi-regular representation* of G is the group $GL(\mathcal{L}^2(K))$ [the set of all linear transformations of $\mathcal{L}^2(K)$ with operation of composition]. That is, it has elements $L(g)$ and group operation $L(g_1)L(g_2)$, such that the linear operators $L(g)$ act on scalar-valued functions $f(\mathbf{x}) \in \mathcal{L}^2(K)$ (the set of all square-integrable complex-valued functions on K) in the following way

$$L(g)f(\mathbf{x}) = f\left(g^{-1} \circ \mathbf{x}\right).$$

Since $GL(\mathcal{L}^2(K))$ is a group of linear transformations, all that needs to be shown to prove that L is a representation of the group G is to show that $L : G \to GL(\mathcal{L}^2(K))$ is a homomorphism

$$(L(g_1)L(g_2)f)(\mathbf{x}) = (L(g_1)(L(g_2)f))(\mathbf{x}) = (L(g_1)f_{g_2})(\mathbf{x})$$
$$= f\left(g_2^{-1} \circ g_1^{-1} \circ \mathbf{x}\right) = f\left((g_1 \circ g_2)^{-1} \circ \mathbf{x}\right) = L(g_1 \circ g_2)f(\mathbf{x}).$$

Here we have used the notation $f_{g_2}(\mathbf{x}) = f(g_2^{-1} \circ \mathbf{x})$. In other words, $L(g_1 \circ g_2) = L(g_1)L(g_2)$. When G is a matrix group, g is interpreted as an $N \times N$ matrix, and we write $g \circ \mathbf{x} = g\mathbf{x}$.

The *right quasi-regular representation* of a matrix Lie group G defined by

$$R(g)f(\mathbf{x}) = f\left(\mathbf{x}^T g\right)$$

is also a representation.

Matrices corresponding to these operators are generated by selecting an appropriate basis for the invariant subspaces of $\mathcal{L}^2(K)$. This leads to the irreducible matrix representations of the group. In the case of noncompact noncommutative groups like the Euclidean motion group, the invariant subspaces will have an infinite number of basis elements, and so the representation matrices will be infinite dimensional.

8.3.2 Compact Lie Groups

The formulation developed in the previous section for finite groups can almost be copied word-for-word to formulate the Fourier pair for functions on compact Lie groups with invariant integration replacing summation over the group. Instead of doing this, we present the representation theory and Fourier expansions on compact groups in a different way. This formulation is also applicable to the finite case. We leave it to the reader to decide which is preferable as a tool for learning the subject.

As we saw in Section 8.2.2, every compact Lie group has a natural integration measure which is invariant to left and right shifts. We denote $U(g, \lambda)$ to be the λth irreducible unitary representation matrix of the compact Lie group G and $U_{i,j}(g, \lambda)$ is its (i, j)th matrix elements.

Thus, analogous to the Fourier transform on the circle and line, it makes sense to define

$$\hat{f}(\lambda) = \int_G f(g)U\left(g^{-1}, \lambda\right)d(g), \tag{8.44}$$

or in component form

$$\hat{f}_{i,j}(\lambda) = \int_G f(g)U_{i,j}\left(g^{-1}, \lambda\right)d(g).$$

The collection of all λ values is denoted as $\hat{G}$ and is called the dual of the group G. Unlike the case of finite groups where $\hat{G} = \{1, \ldots, \alpha\}$, for the case of compact groups, $\hat{G}$ contains a countably infinite number of elements. The collection of Fourier transforms $\{\hat{f}(\lambda)\}$ for all $\lambda \in \hat{G}$ is called the spectrum of the function f.

The convolution of two square-integrable functions on a compact Lie group is defined as

$$(f_1 * f_2)(g) = \int_G f_1(h) f_2\left(h^{-1} \circ g\right) d(h).$$

Since

$$U(g_1 \circ g_2, \lambda) = U(g_1, \lambda) U(g_2, \lambda)$$

and $d(g_1 \circ g) = d(g \circ g_1) = d(g)$ for any fixed $g_1 \in G$, the convolution theorem

$$\mathcal{F}((f_1 * f_2)(g)) = \hat{f}_2(\lambda) \hat{f}_1(\lambda),$$

follows from

$$\mathcal{F}((f_1 * f_2)(g)) = \int_G \left(\int_G f_1(h) f_2\left(h^{-1} \circ g\right) d(h) \right) U\left(g^{-1}, \lambda\right) d(g)$$

by making the change of variables $k = h^{-1} \circ g$ (and replacing all integrations over g by ones over k) and changing the order of integration, which results in

$$\int_G \int_G f_1(h) f_2(k) U\left((h \circ k)^{-1}, \lambda\right) d(k) d(h) =$$

$$\left(\int_G f_2(k) U\left(k^{-1}, \lambda\right) d(k) \right) \left(\int_G f_1(h) U\left(h^{-1}, \lambda\right) d(h) \right) = \hat{f}_2(\lambda) \hat{f}_1(\lambda).$$

The fundamental fact that allows for the reconstruction of a function from its spectrum is that every irreducible representation of a compact group is equivalent to $U(g, \lambda)$ for some value of λ, and $U(g, \lambda_1)$ and $U(g, \lambda_2)$ are not equivalent if $\lambda_1 \neq \lambda_2$. This result is Schur's lemma for compact groups.

Analogous to the case of finite groups, if $U(g)$ and $V(g)$ are irreducible matrix representations of a compact group G with dimensions $d_1 \times d_1$ and $d_2 \times d_2$, respectively, and

$$AU(g) = V_{\cdot}(g) A,$$

where the dimensions of the matrix A are $d_2 \times d_1$, and if $U(g)$ is not equivalent to $V(g)$, all the entries in the matrix A must be zero. If $U(g) \cong V(g)$ then A is uniquely determined up to an arbitrary scaling, and if $U(g) = V(g)$ then $A = a\mathbb{I}_{d_1 \times d_1}$ for some constant a.

In particular, we can define the matrix

$$A = \int_G U(g, \lambda_1) E U\left(g^{-1}, \lambda_2\right) d(g) \tag{8.45}$$

for any matrix E of compatible dimensions, and due to the invariance of the integration measure and homomorphism property of representations, we observe that

$$U(h, \lambda_1) A U\left(h^{-1}, \lambda_2\right) = \int_G U(h, \lambda_1) U(g, \lambda_1) E U\left(g^{-1}, \lambda_2\right) U\left(h^{-1}, \lambda_2\right) d(g)$$

$$= \int_G U(h \circ g, \lambda_1) E U\left((h \circ g)^{-1}, \lambda_2\right) d(g) = A,$$

or equivalently,

$$U(h, \lambda_1) A = AU(h, \lambda_2).$$

Thus from Schur's lemma,

$$A = a\delta_{\lambda_1, \lambda_2} \mathbb{I}_{d(\lambda_1) \times d(\lambda_2)},$$

where $d(\lambda) = \dim(U(g, \lambda))$ and $\mathbb{I}_{d(\lambda_1) \times d(\lambda_2)}$ when $U(h, \lambda_1) \not\cong U(h, \lambda_2)$ is the rectangular matrix with (i, j)th element $\delta_{i,j}$ and all others zero. When $\lambda_1 = \lambda_2$ it becomes an identity matrix.

Following Sugiura's notation [73], we choose $E = E^{j,l}$ to be the matrix with elements $E_{p,q}^{j,l} = \delta_{p,j}\delta_{l,q}$. Making this substitution into Eq. (8.45) and evaluating the (i, j)th element of the result yields

$$a_{j,l}\delta_{i,k}\delta_{\lambda_1,\lambda_2} = \sum_{p,q} \int_G U_{i,p}(g, \lambda_1)\,\delta_{p,j}\delta_{l,q}U_{q,k}\left(g^{-1}, \lambda_2\right)d(g). \qquad (8.46)$$

The constant $a_{j,l}$ is determined by taking the trace of both sides when $\lambda_1 = \lambda_2 = \lambda$. The result is $a_{j,l} = \delta_{j,l}/d(\lambda)$. Finally, using the unitarity of the representations $U(g, \lambda)$, we see that the matrix elements of irreducible unitary representations of a compact group G satisfy

$$\delta_{j,l}\delta_{i,k}\delta_{\lambda_1,\lambda_2} = d(\lambda) \cdot \int_G U_{i,j}(g, \lambda_1)\,\overline{U_{k,l}(g, \lambda_2)}dg \qquad (8.47)$$

where

$$\int_G dg = 1.$$

This means that if

$$f(g) = \sum_{\lambda \in \hat{G}} \sum_{i,j=1}^{d(\lambda)} c_{i,j}(\lambda)U_{i,j}(g, \lambda),$$

then the orthogonality dictates that

$$c_{i,j}(\lambda) = d(\lambda)\hat{f}(\lambda)_{i,j}.$$

In other words, the inversion formula

$$f(g) = \sum_{\lambda \in \hat{G}} d(\lambda)\text{trace}\left(\hat{f}(\lambda)U(g, \lambda)\right) \qquad (8.48)$$

holds, provided that the collection of matrix elements $U_{i,j}(g, \lambda)$ is *complete* in the set of square integrable functions on G. Fortunately, this is the case, as stated in the following Peter-Weyl theorem. As a result, we also have the Plancherel formulae

$$\int_G f_1(g)\overline{f_2(g)}d(g) = \sum_{\lambda \in \hat{G}} d(\lambda)\text{trace}\left(\hat{f}_1(\lambda)\hat{f}_2^\dagger(\lambda)\right)$$

and

$$\int_G |f(g)|^2 d(g) = \sum_{\lambda \in \hat{G}} d(\lambda)\|\hat{f}(\lambda)\|^2.$$

THEOREM 8.11 (Peter-Weyl)[5] [60]
The collection of functions $\{\sqrt{d(\lambda)}U_{i,j}(g, \lambda)\}$ for all $\lambda \in \hat{G}$ and $1 \le i, j \le d(\lambda)$ form a complete orthonormal basis for $\mathcal{L}^2(G)$. The Hilbert space $\mathcal{L}^2(G)$ can be decomposed into orthogonal subspaces

$$\mathcal{L}^2(g) = \sum_{\lambda \in \hat{G}} \bigoplus V_\lambda. \qquad (8.49)$$

---

[5]Hermann Weyl (1885–1955) was a Swiss mathematician who contributed to group representation theory and its applications in physics, and F. Peter was his student.

For each fixed value of λ, *the functions* $U_{i,j}(g, \lambda)$ *form a basis for the subspace* V_λ.

Analogous to the finite-group case, the left and right regular representations of the group are defined on the space of square integrable functions on the group $\mathcal{L}^2(G)$ *as*

$$(L(g)k)(h) = k\left(g^{-1} \circ h\right) \quad and \quad (R(g)k)(h) = k(h \circ g),$$

where $k \in \mathcal{L}^2(G)$. *Both of these representations can be decomposed as*

$$T(g) \cong \sum_{\lambda \in \hat{G}} \bigoplus d(\lambda) U(g, \lambda),$$

where $d(\lambda)$ *is the dimension of* $U(g, \lambda)$. *Here* $T(g)$ *stands for either* $L(g)$ *or* $R(g)$. *Each of the irreducible representations* $U(g, \lambda)$ *acts only on the corresponding subspaces* V_λ.

Finally, we note that analogous to the case of finite groups, the Fourier series on a compact group reduces to a simpler form when the function is constant on cosets or conjugacy classes. These formulae are the same as the finite case with integration replacing summation over the group.

For further reading, see the classic work on compact Lie groups by Želobenko [87].

8.3.3   Noncommutative Unimodular Groups in General

The striking similarity between the Fourier transform, inversion, and Plancherel formulae for finite and compact groups results in large part from the existence of invariant integration measures. A number of works have been devoted to abstract harmonic analysis on locally compact unimodular groups (e.g., [47, 54, 55, 67, 70]). See [61] for an introduction.

Bi-invariant integration measures do not exist for arbitrary Lie groups. However, they do exist for a large enough set of noncompact Lie groups for harmonic analysis on many such groups to have been addressed. Harmonic analysis for general classes of unimodular Lie groups was undertaken by Harish-Chandra[6] [28]. The classification of irreducible representations of a wide variety of groups was pioneered by Langlands [45]. See [40, 43] for more on the representation theory of Lie groups.

Expansions of functions in terms of IURs have been made in the mathematics and physics literature for many groups including the Galilei group [46], the Lorentz group [6, 18, 59, 66, 85], and the Heisenberg group [15, 68, 77]. We shall not review expansions on these groups here.

The noncompact noncommutative groups of greatest relevance to engineering problems is the Euclidean motion group. Harmonic analysis on this group is reviewed in great detail in Chapter 10.

We now review some general operational properties for unimodular Lie groups. See [74] for further reading.

Operational Properties for Unimodular Lie Groups

Recall that a unimodular group G is one which has a left and right invariant integration measure, and hence $\Delta(g) = 1$.

Something that is the same in both the commutative and noncommutative contexts is that a differential operator acting on the convolution of two functions is the convolution of the derivative of one of the functions with the other. Only in the noncommutative case the kind of operator matters:

$$X_i^L(f_1 * f_2) = \left(X_i^L f_1\right) * f_2 \quad and \quad X_i^R(f_1 * f_2) = f_1 * \left(X_i^R f_2\right).$$

---

[6]Harish-Chandra (1923–1983) was a mathematics professor at Princeton University.

There are many other similarities between the "operational properties" of the Abelian and non-commutative cases. In order to see the similarities, another definition is required.

Given an $N_\lambda \times N_\lambda$ matrix representation $U(g, \lambda)$ of G we can define a representation, $u(X, \lambda)$, of the Lie algebra $\mathcal{G}$ as

$$u(X, \lambda) = \frac{d}{dt} \left(U(\exp tX), \lambda \right)|_{t=0} .$$

This is called a representation of $\mathcal{G}$ because it inherits the Lie bracket from the differential operators X^L and X^R

$$u([X, Y]) = [u(X), u(Y)].$$

By substitution we see that

$$\mathcal{F}\left(X^L f\right) = \frac{d}{dt} \int_G f((\exp tX) \circ g) U\left(g^{-1}, \lambda\right) d(g) = \frac{d}{dt} \int_G f(g) U\left(g^{-1} \circ \exp(-tX), \lambda\right) d(g)$$

$$= \left(\int_G f(g) U\left(g^{-1}, \lambda\right) d(g) \right) \left(\frac{d}{dt} U(\exp(-tX), \lambda) \right) = -\hat{f}(\lambda) u(X, \lambda).$$

Similarly, for the right derivative,

$$\mathcal{F}(X^R f) = -u(X, \lambda) \hat{f}(\lambda).$$

Using the Lie bracket property of the Lie algebra representations, expanding the bracket, and using the previous operational properties, we see that

$$\mathcal{F}\left(\left[X^L, Y^L \right] f \right) = \int_G \left(\left[X^L, Y^L \right] f(g) \right) U\left(g^{-1}, \lambda\right) d(g) = [u(X, \lambda), u(Y, \lambda)] \hat{f}(\lambda)$$

and

$$\mathcal{F}\left(\left[X^R, Y^R \right] f \right) = \int_G \left(\left[X^R, Y^R \right] f(g) \right) U\left(g^{-1}, \lambda\right) d(g) = \hat{f}(\lambda) [u(X, \lambda), u(Y, \lambda)].$$

Note that this is an operational property with no analog in classical (Abelian) Fourier analysis.

The Fourier transform of the Laplacian of a function is transformed as

$$\mathcal{F}\left(\text{div}_R \left(\text{grad}_R(f) \right) \right) = \sum_{i=1}^{N} \left(u\left(X_i, \lambda\right) \right)^2 \hat{f}(\lambda)$$

and

$$\mathcal{F}\left(\text{div}_L \left(\text{grad}_L(f) \right) \right) = \sum_{i=1}^{N} \hat{f}(\lambda) \left(u\left(X_i, \lambda\right) \right)^2 .$$

For the examples we will examine [such as $SO(3)$ and $SU(2)$], the matrices $\sum_{i=1}^{N} (u(X_i, \lambda))^2$ are multiples of the identity and the subscripts L and R can be dropped so that $\mathcal{F}(\text{div}(\text{grad}(f))) = -\tilde{\alpha}(\lambda) \hat{f}(\lambda)$ where $\tilde{\alpha}(\lambda)$ is a positive real-valued function. When this happens, the heat equation

$$\frac{\partial^2 f}{\partial t^2} = K^2 \, \text{div}(\text{grad}((f))$$

may be solved in closed form for $f(g, t)$ subject to initial conditions $f(g, 0) = \delta(g)$. From this fundamental solution, the solution for any initial conditions $f(g, 0) = \tilde{f}(g)$ can be generated by convolution.

The Adjoint of Differential Operators

Consider a unimodular Lie group with the inner product of complex-valued square-integrable functions defined as

$$(f_1, f_2) = \int_G \overline{f_1(g)} f_2(g) d(g).$$

The properties resulting from invariant integration on G means that it is easy to calculate the *adjoint*, $(X_i^R)^*$, of the differential operators

$$X_i^R f = \frac{d}{dt} \left(f \left(g \circ \exp[t X_i] \right) \right) \big|_{t=0},$$

as defined by the equality

$$\left(\left(X_i^R \right)^* f_1, f_2 \right) = \left(f_1, X_i^R f_2 \right).$$

To explicitly calculate the adjoint, observe that

$$\left(f_1, X_i^R f_2 \right) = \int_G \overline{f_1(g)} \frac{d}{dt} \left(f_2 \left(g \circ \exp[t X_i] \right) \right)_{t=0} d(g)$$

$$= \frac{d}{dt} \left(\int_G \overline{f_1(g)} f_2 \left(g \circ \exp[t X_i] \right) d(g) \right) \bigg|_{t=0}.$$

Letting $h = g \circ \exp[t X_i]$, substituting for g, and using the invariance of integration under shifts results in

$$\frac{d}{dt} \left(\int_G \overline{f_1 \left(h \circ \exp[-t X_i] \right)} f_2(h) d(h) \right)_{t=0} = \int_G \frac{d}{dt} \overline{\left(f_1 \left(h \circ \exp[-t X_i] \right) \right)_{t=0}} f_2(h) d(h)$$

$$= - \int_G \overline{X_i^R f_1(h)} f_2(h) d(h) = \left(-X_i^R f_1, f_2 \right).$$

Hence, we conclude that since X_i^R is real,

$$\left(X_i^R \right)^* = -X_i^R.$$

It is often convenient to consider the self-adjoint form of an operator, and in this context it is easy to see that the operator defined as

$$Y_i^R = -i X_i^R$$

satisfies the self-adjointness condition

$$\left(Y_i^R \right)^* = Y_i^R.$$

Before moving on to the construction of explicit representation matrices for $SO(3)$ and $SE(N)$ in Chapters 9 and 10, a very general technique for generating group representations from representations of subgroups is examined in the next section. This method is applicable to both discrete and continuous groups.

8.4   Induced Representations and Tests for Irreducibility

In this chapter we have assumed that a complete set of IURs for a group have either been given in advance or have been obtained by decomposing the left or right regular representations. In practice, other methods for generating IURs are more convenient than decomposition of the regular representation. One such method for generating representations of a subgroup H of a group G for which representations are already known is called *subduction*. This is nothing more than the restriction of the representation $D(g)$ of G to H. This is often denoted $D(h) = D \downarrow H$. If $D(h)$ is irreducible for $h \in H$, then so too is $D(g)$ for $g \in G$. But in general, irreducibility of $D(g)$ does not guarantee irreducibility of $D(h)$.

A very general method for creating representations of groups from representations of subgroups exists. It is called the method of *induced representations*. Given a representation $D(h)$ for $h \in H < G$, the corresponding induced representation is denoted in the literature in a variety of ways including $D \uparrow G$, $D(H) \uparrow G$, $U = \text{ind}_H^G(D)$, or $U^D = \text{ind}_{H,D}^G$. It is this method that will be of importance to us when constructing representations of $SE(2)$ and $SE(3)$.

The method of induced representations for finite groups was developed by Frobenius [22], Weyl, and Clifford [13]. The extension and generalization of these ideas to locally compact groups (and in particular to Lie groups) was performed by Mackey [47]–[50]. The collection of methods for generating induced representations of a locally compact group from a normal subgroup $N \lhd G$ and subgroups of G/N is often called the "Mackey machine."

In general, these representations will be reducible. It is therefore useful to have a general test to determine when a representation is reducible. We discuss here both the method of induced representations and general tests for irreducibility. These methods are demonstrated in subsequent chapters with the rotation and motion groups.

8.4.1 Finite Groups

Induced Representations of Finite Groups

Let $\{D^\lambda\}$ for $\lambda = 1, \ldots, \alpha$ be a complete set of IUR matrices for a subgroup $H < G$ and $dim(D^\lambda) = d_\lambda$. Then the matrix elements of an induced representation of G with dimension $d_\lambda \cdot |G/H|$ are defined by the expression [2, 14, 32]

$$U_{i\mu, j\nu}^\lambda(g) = \delta_{ij}(g) D_{\mu\nu}^\lambda \left(g_i^{-1} \circ g \circ g_j \right) \tag{8.50}$$

where $g, g_i, g_j \in G$, and

$$\delta_{ij}(g) = \begin{cases} 1 & \text{for } g \circ g_j \in g_i H \\ 0 & \text{for } g \circ g_j \notin g_i H. \end{cases}$$

This guarantees that $U_{i\mu, j\nu}^\lambda(g) = 0$ unless $g_i^{-1} \circ g \circ g_j \in H$.

From the previous definitions we verify that

$$\sum_{j=1}^{|G/H|} \sum_{\nu=1}^{d_\lambda} U_{i\mu, j\nu}^\lambda(g_1) U_{j\nu, k\eta}^\lambda(g_2) = \sum_{j=1}^{|G/H|} \sum_{\nu=1}^{d_\lambda} \delta_{ij}(g_1) D_{\mu\nu}^\lambda \left(g_i^{-1} \circ g \circ g_j \right) \delta_{jk}(g_2) D_{\nu\eta}^\lambda \left(g_j^{-1} \circ g \circ g_k \right)$$

$$= \sum_{j=1}^{|G/H|} \delta_{ij}(g_1) \delta_{jk}(g_2) \sum_{\nu=1}^{d_\lambda} D_{\mu\nu}^\lambda \left(g_i^{-1} \circ g \circ g_j \right) D_{\nu\eta}^\lambda \left(g_j^{-1} \circ g \circ g_k \right).$$

Since D^λ is a representation,

$$\sum_{\nu=1}^{d_\lambda} D_{\mu\nu}^\lambda \left(g_i^{-1} \circ g \circ g_j \right) D_{\nu\eta}^\lambda \left(g_j^{-1} \circ g \circ g_k \right) = D_{\mu\eta}^\lambda \left(g_i^{-1} \circ (g_1 \circ g_2) \circ g_k \right).$$

Likewise,

$$\sum_{j=1}^{|G/H|} \delta_{ij}(g_1)\,\delta_{jk}(g_2) = \delta_{ik}(g_1 \circ g_2),$$

since if $g_i^{-1} \circ g_1 \circ g_j \in H$ and $g_j^{-1} \circ g_2 \circ g_k \in H$, then the product

$$g_i^{-1} \circ g_1 \circ g_j \circ g_j^{-1} \circ g_2 \circ g_k = g_i^{-1} \circ g_1 \circ g_2 \circ g_k \in H.$$

Therefore,

$$\sum_{j=1}^{|G/H|} \sum_{\nu=1}^{d_\lambda} U_{i\mu,j\nu}^\lambda(g_1)\, U_{j\nu,k\eta}^\lambda(g_2) = U_{i\mu,k\eta}^\lambda(g_1 \circ g_2).$$

We note that the four indicies j, ν, k, η may be contracted to two while preserving the homomorphism property. There is more than one way to do this. We may denote

$$\tilde{U}_{IJ}^\lambda(g) = U_{i\mu,j\nu}^\lambda(g)$$

where

$$I = i + |G/H| \cdot (\mu - 1), \qquad J = j + |G/H| \cdot (\nu - 1)$$

or

$$I = d_\lambda \cdot (i - 1) + \mu, \qquad J = d_\lambda \cdot (j - 1) + \nu.$$

Either way,

$$\tilde{U}_{IK}^\lambda(g_1 \circ g_2) = \sum_{J=1}^{|G/H| \cdot d_\lambda} \tilde{U}_{IJ}^\lambda(g_1)\, \tilde{U}_{JK}^\lambda(g_2).$$

In this way standard two-index matrix operations can be performed.

For applications of induced representations of finite groups in chemistry and solid-state physics see [2, 32].

Checking Irreducibility of Representations of Finite Groups

THEOREM 8.12

A representation $U(g)$ of a finite group G is irreducible if and only if the character $\chi(g) = \text{trace}(U(g))$ satisfies

$$\frac{1}{|G|} \sum_{g \in G} \chi(g)\overline{\chi(g)} = 1. \tag{8.51}$$

PROOF   For an arbitrary representation of G we write

$$U(g) = S^{-1}(c_1\pi_1 \oplus \cdots \oplus c_n\pi_n)\, S,$$

where S is an invertible matrix and c_i is the number of times π_i appears in the decomposition of U. Since the trace is invariant under similarity transformation,

$$\chi(g) = \sum_{i=1}^{n} c_i \chi_i(g),$$

where $\chi_i(g) = \text{trace}(\pi_i(g))$. Substituting into Eq. (8.51) and using the orthogonality of characters yields the condition

$$\sum_{i=1}^{n} c_i^2 = 1.$$

Since c_i is a nonnegative integer, this equality holds only if one of the c_i's is equal to one and all the others are equal to zero. On the other hand, if U is irreducible $c_i = \delta_{ij}$ for some $j \in \{1, \ldots, n\}$. Therefore, when U satisfies Eq. (8.51) it must be the case that $U \cong \pi_j$ for some j, and U is therefore irreducible. ∎

8.4.2 Lie Groups

Induced Representations of Lie Groups

A formal definition of the induced representations (see, for example, [9, 14, 26, 49]) $D(H) \uparrow G$, where $D(H)$ are representations of subgroup H of group G, is the following.

DEFINITION 8.2   *Let V be the space of complex values of functions $\phi(\sigma)$, $\sigma \in G/H$. The action of operators $U(g) \in D \uparrow G$ ($g \in G$) (where D, which is a representation of subgroup H of group G, acts in V) is such that*

$$(U(g)\phi)(\sigma) = D\left(s_\sigma^{-1} g\, s_{g^{-1}\sigma}\right) \phi\left(g^{-1}\sigma\right), \tag{8.52}$$

where s_σ is an arbitrary representative of the coset $\sigma \in G/H$.

We observe the homomorphism property by first defining $\phi_g(\sigma) = (U(g)\phi)(\sigma)$. Then

$$
\begin{aligned}
(U(g_1) U(g_2)\phi)(\sigma) &= \left(U(g_1)\phi_{g_2}\right)(\sigma) = D\left(s_\sigma^{-1} g_1\, s_{g_1^{-1}\sigma}\right) \phi_{g_2}\left(g_1^{-1}\sigma\right) \\
&= D\left(s_\sigma^{-1} g_1\, s_{g_1^{-1}\sigma}\right) D\left(s_{g_1^{-1}\sigma}^{-1} g_2\, s_{g_2^{-1} g_1^{-1}\sigma}\right) \phi\left(g_1^{-1} g_2^{-1}\sigma\right) \\
&= D\left(s_\sigma^{-1} g_1 g_2\, s_{(g_1 g_2)^{-1}\sigma}\right) \phi\left((g_1 g_2)^{-1}\sigma\right) = (U(g_1 g_2)\phi)(\sigma).
\end{aligned}
$$

Since in general, the homogeneous space G/H will be a manifold of dimension $\dim(G/H) = \dim(G) - \dim(H)$, it follows that matrix elements of $U(g)$ are calculated as

$$U_{i,j}(g) = \left(e_i, U(g)e_j\right),$$

where $\{e_k\}$ for $k = 1, 2, \ldots$ is a complete set of orthonormal eigenfunctions for $L^2(G/H)$, and the inner product is defined as

$$(\phi_1, \phi_2) = \int_{G/H} \overline{\phi_1(\sigma)}\phi_2(\sigma)d(\sigma).$$

Recall that in this context $d(\sigma)$ denotes an integration measure for G/H.

Since this method will generally produce infinite-dimensional representation matrices, it is clear that in the case of compact Lie groups, these representations will generally be reducible. However, as we shall see, the infinite-dimensional representations generated in this way for certain noncompact noncommutative Lie groups will be irreducible.

The next section provides tests for determining when a representation is reducible.

Checking Irreducibility of Representations of Compact Lie Groups

Several means are available to quickly determine if a given representation of a compact Lie group is irreducible or not. Since in many ways the representation theory and harmonic analysis for compact Lie groups is essentially the same as for finite groups (with summation over the group replaced with invariant integration) the following theorem should not be surprising.

THEOREM 8.13

A representation $U(g)$ of a compact Lie group G with invariant integration measure $d(g)$ normalized so that $\int_G d(g) = 1$ is irreducible if and only if the character $\chi(g) = \text{trace}(U(g))$ satisfies

$$\int_G \chi(g)\overline{\chi(g)}d(g) = 1. \tag{8.53}$$

PROOF   Same as for Theorem 8.12 with integration replacing summation.   ∎

Another test for irreducibility of compact Lie group representations is based solely on their continuous nature and is not simply a direct extension of concepts originating from the representation theory of finite groups. For a Lie group, we can evaluate any given representation $U(g)$ at a one-parameter subgroup generated by exponentiating an element of the Lie algebra as $U(\exp(tX_i))$. Expanding this matrix function of t in a Taylor series, the linear term in t will have the constant coefficient matrix

$$u(X_i) = \left. \frac{dU(\exp(tX_i))}{dt} \right|_{t=0}. \tag{8.54}$$

If U is reducible, then all of the $u(X_i)$ must simultaneously be block-diagonalizable by the same similarity transformation. If no such similarity transformation can be found, then we can conclude that U is irreducible.

THEOREM 8.14 ([25] [59] [81])

Given a Lie group G, associated Lie algebra $\mathcal{G}$, and finite-dimensional $u(X_i)$ as defined in Eq. (8.54), then

$$U(\exp(tX_i)) = \exp[tu(X_i)], \tag{8.55}$$

where $X_i \in \mathcal{G}$. Furthermore, if the matrix exponential parameterization

$$g(x_1, \ldots, x_n) = \exp\left(\sum_{i=1}^{n} x_i X_i\right) \tag{8.56}$$

is surjective, then when the $u(X_i)$'s are not simultaneously block-diagonalizable,

$$U(g) = \exp\left(\sum_{i=1}^{n} x_i u(X_i)\right) \tag{8.57}$$

is an irreducible representation for all $g \in G$.

PROOF   For the exponential parameterization (8.56), one observes

$$g(tx_1, \ldots, tx_n) \circ g(\tau x_1, \ldots, \tau x_n) = g((t + \tau)x_1, \ldots, (t + \tau)x_n)$$

for all $t, \tau \in \mathbb{R}$, i.e., the set of all $g(tx_1, \ldots, tx_n)$ forms a one-dimensional (Abelian) subgroup of G for fixed values of x_i. From the definition of a representation it follows that

$$U\left(g\left(((t + \tau)x_1, \ldots, (t + \tau)x_n)\right)\right) = U\left(g\left(tx_1, \ldots, tx_n\right)\right) U\left(g\left(\tau x_1, \ldots, \tau x_n\right)\right)$$

$$= U\left(g\left(\tau x_1, \ldots, \tau x_n\right)\right) U\left(g\left(tx_1, \ldots, tx_n\right)\right). \quad (8.58)$$

Define

$$\tilde{U}(x_1, \ldots, x_n) = U\left(g\left(x_1, \ldots, x_n\right)\right).$$

Then differentiating Eq. (8.58) with respect to τ and setting $\tau = 0$,

$$\frac{d}{dt}\tilde{U}(tx_1, \ldots, tx_n) = \frac{d}{d\tau}\tilde{U}(\tau x_1, \ldots, \tau x_n)|_{\tau=0} \tilde{U}(tx_1, \ldots, tx_n).$$

But since infinitesimal operations commute, it follows from Eq. (8.54) that

$$\frac{d}{d\tau}\tilde{U}(\tau x_1, \ldots, \tau x_n)|_{\tau=0} = \sum_{i=1}^{n} x_i\, u\,(X_i).$$

We therefore have the matrix differential equation

$$\frac{d}{dt}\tilde{U}(tx_1, \ldots, tx_n) = \left(\sum_{i=1}^{n} x_i\, u\,(X_i)\right) \tilde{U}(tx_1, \ldots, tx_n)$$

subject to the initial conditions

$$\tilde{U}(0, \ldots, 0) = \mathbb{I}_{dim(U) \times dim(U)}.$$

The solution is therefore

$$\tilde{U}(tx_1, \ldots, tx_n) = \exp\left(t \sum_{i=1}^{n} x_i\, u\,(X_i)\right).$$

Evaluating at $t = 1$, we find Eq. (8.57) and setting all $x_j = 0$ except x_i, we find Eq. (8.55). The irreducibility of these representations follows from the assumed properties of $u(X_i)$. ∎

Essentially the same argument can be posed in different terminology by considering what happens to subspaces of function spaces on which representation operators (not necessarily matrices) act. This technique, which we will demonstrate for $SU(2)$ in the next chapter, also can be used for noncompact noncommutative Lie groups (which generally have infinite-dimensional represention matrices). The concrete example of $SE(2)$ is presented in Chapter 10 to demonstrate this.

8.5 Wavelets on Groups

The concept of wavelet transforms discussed in Chapter 3 extends to the group-theoretical setting with appropriate concepts of translation, dilation, and modulation. We discuss the modulation-translation case in Section 8.5.1 and the dilation-translation case in Section 8.5.2.

8.5.1   The Gabor Transform on a Unimodular Group

Let G be a unimodular group with invariant integration measure dg and $\gamma \in L^2(G)$ be normalized so that $\|\gamma\| = (\gamma, \gamma)^{1/2} = 1$. The matrix representations of G in a suitable basis are denoted as $U(g, \lambda)$ for $\lambda \in \hat{G}$ and $g \in G$.

In direct analogy with the Gabor transform on the real line and unit circle, we can define the transform

$$\tilde{f}_\gamma(h, \lambda) = \int_G f(g) \overline{\gamma (h^{-1} \circ g)} U\left(g^{-1}, \lambda\right) dg. \tag{8.59}$$

Here $h \in G$ is the shift, and $U(g^{-1}, \lambda)$ is the modulation.

This group-theoretical Gabor transform is inverted by following the same steps as in cases of the line and circle. We first multiply $\tilde{f}_\gamma(h, \lambda)$ by $\gamma(h^{-1} \circ g)$ and integrate over all shifts:

$$\int_G \tilde{f}_\gamma(h, \lambda) \gamma \left(h^{-1} \circ g\right) dh = \int_G \left(\int_G \overline{\gamma (h^{-1} \circ g)} \gamma \left(h^{-1} \circ g\right) dh\right) f(g) U\left(g^{-1}, \lambda\right) dg.$$

Using the invariance of integration under shifts and inversions, this simplifies to

$$\int_G \tilde{f}_\gamma(h, \lambda) \gamma \left(h^{-1} \circ g\right) dh = \|\gamma\|^2 \int_G f(g) U\left(g^{-1}, \lambda\right) dg.$$

Since $\|\gamma\|^2 = 1$, this is nothing more than the Fourier transform of f and the inverse Fourier transform is used to recover f. Hence, we write:

$$f(g) = \int_{\hat{G}} d\nu(\lambda) \text{trace} \left(U(g, \lambda) \int_G \tilde{f}_\gamma(h, \lambda) \gamma \left(h^{-1} \circ g\right) dh\right). \tag{8.60}$$

In the case of a compact or finite group this integral becomes a weighted sum.

8.5.2   Wavelet Transforms on Groups based on Dilation and Translation

Let $\varphi(g; s) \in L^2(G \times \mathbb{R}^+)$, and call $\varphi(g; 1)$ the mother wavelet. Then define

$$\varphi^p_{s,h}(g) = |s|^{-p} \varphi \left(h^{-1} \circ g; s\right).$$

At this point, the structure of $\varphi(g; s)$ will be left undetermined, as long as the Hermitian matrices

$$K_\lambda = \int_0^\infty \hat{\varphi}(\lambda, s) \hat{\varphi}^\dagger(\lambda, s) |s|^{-2} ds \tag{8.61}$$

have nonzero eigenvalues for all $\lambda \in \hat{G}$.

We define the continuous wavelet transform on G as

$$\tilde{f}_p(s, h) = \int_G \overline{\varphi^p_{s,h}(g)} f(g) dg. \tag{8.62}$$

Let $\varphi_{-1}(g; s) \triangleq \varphi(g^{-1}; s)$. Then Eq. (8.62) can be written as

$$\tilde{f}_p(s, h) = |s|^{-p} \int_G f(g) \overline{\varphi_{-1} \left(g^{-1} \circ h, s\right)} dg = |s|^{-p} f * \overline{\varphi_{-1}}(h; s).$$

Applying the Fourier transform to $\overline{\varphi_{-1}}$ we see that

$$
\begin{aligned}
\mathcal{F}\left(\overline{\varphi_{-1}}\right) &= \int_G \overline{\varphi_{-1}(g; s)} U\left(g^{-1}, \lambda\right) dg \\
&= \int_G \overline{\varphi\left(g^{-1}; s\right)} U\left(g^{-1}, \lambda\right) dg \\
&= \int_G \overline{\varphi(g; s)} U(g, \lambda) dg \\
&= (\mathcal{F}(\varphi))^\dagger,
\end{aligned}
$$

and hence by the convolution theorem

$$
\mathcal{F}\left(\tilde{f}_p(s, h)\right) = |s|^{-p} \hat{\varphi}^\dagger(\lambda, s) \hat{f}(\lambda).
$$

Multiplying on the left by $s^{p-2}\hat{\varphi}$, integrating over all values of s, and inverting the resulting matrix K_λ, we see that

$$
\mathcal{F}(f) = K_\lambda^{-1} \int_0^\infty \mathcal{F}\left(\tilde{f}_p(s, h)\right) |s|^{p-2} ds.
$$

Finally, the Fourier inversion theorem applied to the above equation reproduces f. Hence we conclude that when a suitable mother wavelet exists, the inverse of the continuous wavelet transform on a unimodular group is of the form

$$
f(g) = \int_{\hat{G}} d(\lambda) \text{trace} \left(K_\lambda^{-1} \int_0^\infty \int_G \tilde{f}_p(s, h) U\left(h^{-1}, \lambda\right) dh |s|^{p-2} ds \right) U(g, \lambda). \tag{8.63}
$$

A natural question to ask at this point is how does one construct suitable functions $\varphi(g; s)$? While the dilation operation is not natural in the context of functions on groups, the diffusion operator is very natural (at least in the context of the unimodular Lie groups of interest in this book). Let $\kappa(g; s)$ be a function with the property that

$$
\kappa(g; s_1) * \kappa(g; s_2) = \kappa(g; s_1 + s_2).
$$

Then define

$$
\varphi(g; s) = \phi(g) * \kappa(g; s).
$$

We investigate conditions on ϕ in Chapter 9 when $G = SO(3)$. See [1, 39] for background on the concept of coherent states and their relationship to group-theoretic aspects of wavelet transforms.

8.6 Summary

This chapter provided a broad introduction to harmonic analysis on finite, compact, and noncompact unimodular groups. In order to gain a fuller understanding of this subject, the classic books in the bibliography for this chapter should be consulted. However, our treatment is complete enough for the engineer or scientist to use these techniques in applications. The somewhat abstract material in this chapter lays the foundations for concrete harmonic analysis on $SO(3)$ and $SE(3)$ in the two chapters to follow. Chapters 11–18 then examine computational issues and their impact on applications.

References

[1] Ali, S.T., Antoine, J.-P., and Gazeau, J.-P., *Coherent States, Wavelets and their Generalizations,* Graduate Texts in Contemporary Physics, Springer, New York, 2000.

[2] Altmann, S.L., *Induced Representations in Crystals and Molecules,* Academic Press, New York, 1977.

[3] Arthur, J., Harmonic analysis and group representations, *Notices of the Am. Math. Soc.,* 47(1), 26–34, 2000.

[4] Atkinson, M.D., The complexity of group algebra computations, *Theor. Comput. Sci.,* 5, 205–209, 1977.

[5] Apple, G. and Wintz, P., Calculation of Fourier transforms on finite Abelian groups, *IEEE Trans. on Inf. Theor.,* IT-16, 233–236, 1970.

[6] Bargmann, V., Irreducible unitary representations of the Lorentz group, *Ann. of Math.,* 48(3), 568–640, 1947.

[7] Beth, T., On the computational complexity of the general discrete Fourier transform, *Theor. Comput. Sci.,* 51, 331–339, 1987.

[8] Bröcker, T. and tom Dieck, T., *Representations of Compact Lie Groups,* Springer-Verlag, New York, 1985.

[9] Bruhat, F., Sur les représentations induites de groupes de Lie, *Bull. Soc. Math. France,* 84, 97–205, 1956.

[10] Cairns, T.W., On the fast Fourier transform on finite Abelian groups, *IEEE Trans. on Comput.,* 20, 569–571, 1971.

[11] Cartan, É., Sur la détermination d'un système orthogonal complet dans un espace de Riemann symmetrique clos, *Rend. Circ. Mat. Polermo,* 53, 217–252, 1929.

[12] Clausen, M., Fast generalized Fourier transforms, *Theor. Comput. Sci.,* 67, 55–63, 1989.

[13] Clifford, A.H., Representations induced in an invariant subgroup, *Ann. Math.,* 38, 533–550, 1937.

[14] Coleman, A.J., *Induced Representations with Applications to S_n and $GL(n)$,* Queen's University, Kingston, Ontario, 1966.

[15] Corwin, L. and Greenleaf, F.P., *Representations of Nilpotent Lie Groups and their Applications Part 1: Basic Theory and Examples,* Cambridge Studies in Advanced Mathematics 18, Cambridge University Press, New York, 1990.

[16] Diaconis, P. and Rockmore, D., Efficient computation of the Fourier transform on finite groups, *J. of the Am. Math. Soc.,* 3(2), 297–332, 1990.

[17] Dieudonné, J., *Special Functions and Linear Representations of Lie Groups,* American Mathematical Society, Providence, RI, 1980.

[18] Dobrev, V.K., *Harmonic Analysis on the n-Dimensional Lorentz Group and its Application to Conformal Quantum Field Theory,* Springer-Verlag, New York, 1977.

[19] Dunkl, C.F. and Ramirez, D.E., *Topics in Harmonic Analysis*, Appleton-Century-Crofts, Meredith Corporation, New York, 1971.

[20] Fässler, A. and Stiefel, E., *Group Theoretical Methods and their Applications*, Birkhäuser, Boston, 1992.

[21] Folland, G.B., *A Course in Abstract Harmonic Analysis*, CRC Press, Boca Raton, FL, 1995.

[22] Frobenius, F.G., Über relationen zwischen den charakteren einer gruppe und denen ihrer untergruppen, *Sitz. Preuss. Akad. Wiss.*, 501–515, 1898.

[23] Fulton, W. and Harris, J., *Representation Theory: A First Course*, Springer-Verlag, 1991.

[24] Gross, K.I., Evolution of noncommutative harmonic analysis, *Am. Math. Mon.*, 85(7), 525–548, 1978.

[25] Gelfand, I.M., Minlos, R.A., and Shapiro, Z.Ya., *Representations of the Rotation and Lorentz Groups and their Applications*, Macmillan, New York, 1963.

[26] Gurarie, D., *Symmetry and Laplacians. Introduction to Harmonic Analysis, Group Representations and Applications*, Elsevier Science, The Netherlands, 1992.

[27] Haar, A., Der maßbegriff in der theorie der kontinuierlichen gruppen, *Ann. of Math.*, 34, 147–169, 1933.

[28] Harish-Chandra, *Collected Papers*, Vol. I-IV, Springer-Verlag, 1984.

[29] Helgason, S., *Differential Geometry and Symmetric Spaces*, Academic Press, New York, 1962.

[30] Hermann, R., *Fourier Analysis on Groups and Partial Wave Analysis*, W.A. Benjamin, New York, 1969.

[31] Hewitt, E. and Ross, K.A., *Abstract Harmonic Analysis I, II*, Springer-Verlag, Berlin, 1963, 1970.

[32] Inui, T., Tanabe, Y., and Onodera, Y., *Group Theory and its Applications in Physics*, 2nd ed., Springer-Verlag, 1996.

[33] Janssen, T., *Crystallographic Groups*, North Holland/American Elsevier, New York, 1972.

[34] Karpovsky, M.G., Fast Fourier transforms on finite non-Abelian groups, *IEEE Trans. on Comput.*, C-26(10), 1028–1030, 1977.

[35] Karpovsky, M.G. and Trachtenberg, E.A., Fourier transform over finite groups for error detection and error correction in computation channels, *Inf. and Control*, 40, 335–358, 1979.

[36] Karpovsky, M.G. and Trachtenberg, E.A., Some optimization problems for convolution systems over finite groups, *Inf. and Control*, 34, 227–247, 1977.

[37] Karpovsky, M.G., Error detection in digital devices and computer programs with the aid of linear recurrent equations over finite commutative groups, *IEEE Trans. on Comput.*, C-26(3), 208–218, 1977.

[38] Karpovsky, M.G. and Trachtenberg, E.A., Statistical and computational performance of a class of generalized Wiener filters, *IEEE Trans. on Inf. Theor.*, IT-32(2), 303–307, 1986.

[39] Klauder, J.R. and Skagerstam, B.S., *Coherent States: Application in Physics and Mathematical Physics*, World Scientific, Singapore, 1985.

[40] Knapp, A.W., *Representation Theory of Semisimple Groups, an Overview Based on Examples*, Princeton University Press, Princeton, NJ, 1986.

[41] Koster, G.F., *Space Groups and their Representations*, Academic Press, New York, 1957.

[42] Kovalev, O.V., *Representations of the Crystallographic Space Groups: Irreducible Representations, Induced Representations, and Corepresentations*, Stokes, H.T., and Hatch, D.M., Eds., (transl.) Worthey, G.C., Gordon and Breach Scientific, Yverdon, Switzerland; Langhome, PA, 1993.

[43] Kirillov, A.A., Ed., *Representation Theory and Noncommutative Harmonic Analysis I, II*, Springer-Verlag, Berlin, 95, 1988.

[44] Kunze, R., L_p Fourier transforms on locally compact unimodular groups, *Trans. of the Am. Math. Soc.*, 89, 519–540, 1958.

[45] Langlands, R.P., On the classification of irreducible representations of real algebraic groups, Notes, Institute for Advanced Study, Princeton, NJ, 1973.

[46] Loebl, E.M., Ed., *Group Theory and its Applications*, Vols. I, II, III, Academic Press, New York, 1968.

[47] Mackey, G.W., Induced representations of locally compact groups I, *Ann. of Math.*, 55, 101–139, 1952.

[48] Mackey, G.W., Unitary representations of group extensions, *Acta Math.*, 99, 265–311, 1958.

[49] Mackey, G.W., *Induced Representations of Groups and Quantum Mechanics*, W.A. Benjamin, New York, Amsterdam, 1968.

[50] Mackey, G.W., *The Theory of Unitary Group Representations*, The University of Chicago Press, Chicago, 1976.

[51] Mackey, G.W., *The Scope and History of Commutative and Noncommutative Harmonic Analysis*, American Mathematical Society, 1992.

[52] Maslen, D.K., *Fast Transforms and Sampling for Compact Groups*, Ph.D. dissertation, Dept. of Mathematics, Harvard University, Cambridge, MA, 1993.

[53] Maslen, D.K. and Rockmore, D.N., Generalized FFTs — a survey of some recent results, *DIMACS Ser. in Discrete Math. and Theor. Comput. Sci.*, 28, 183–237, 1997.

[54] Mautner, F.I., Unitary representations of locally compact groups II, *Ann. Math.*, 52, 1950.

[55] Mautner, F.I., Note on the Fourier inversion formula on groups, *Trans. Am. Math. Soc.*, 78, 1955.

[56] Miller, W., *Symmetry Groups and their Applications*, Academic Press, New York, 1972.

[57] Miller, W., *Lie Theory and Special Functions*, Academic Press, New York, 1968.

[58] Nachbin, L., *The Haar Integral*, Van Nostrand, Princeton, NJ, 1965.

[59] Naimark, M.A., *Linear Representations of the Lorentz Group*, Macmillan, New York, 1964.

[60] Peter, F. and Weyl, H., Die vollständigkeit der primitiven darstellungen einer geschlossenen kontinuierlichen gruppe, *Math. Ann.*, 97, 735–755, 1927.

[61] Robert, A., *Introduction to the Representation Theory of Compact and Locally Compact Groups*, London Mathematical Society Lecture Note Series 80, Cambridge University Press, Cambridge, 1983.

[62] Rockmore, D.N., Efficient computation of Fourier inversion for finite groups, *J. of the Assoc. for Comput. Mach.*, 41(1), 31–66, 1994.

[63] Rockmore, D.N., Fast Fourier transforms for wreath products, *Appl. and Computational Harmonic Anal.*, 4, 34–55, 1995.

[64] Roziner, T.D., Karpovsky, M.G., and Trachtenberg, E.A., Fast Fourier transforms over finite groups by multiprocessor systems, *IEEE Trans., Acoust., Speech, and Signal Process.*, 38(2), 226–240, 1990.

[65] Rudin, W., *Fourier Analysis on Groups*, InterScience, New York, 1962.

[66] Rühl, W., *The Lorentz Group and Harmonic Analysis*, W.A. Benjamin, New York, 1970.

[67] Saito, K., On a duality for locally compact groups, *Tohoku Math. J.*, 20, 355–367, 1968.

[68] Schempp, W., *Harmonic Analysis on the Heisenberg Nilpotent Lie Group, with Applications to Signal Theory*, Pitman Research Notes in Mathematics Series 147, Longman Scientific & Technical, London, 1986.

[69] Schur, I., Über die darstellung der endlichen gruppen durch gebrochene lineare substitutionen, *J. für Math.*, 127, 20–50, 1904.

[70] Segal, I.E., An Extension of Plancherel's Formula to Separable Unimodular Groups, *Ann. of Math.*, 52(2), 272–292, 1950.

[71] Serre, J.-P., *Linear Representations of Finite Groups*, Springer-Verlag, New York, 1977.

[72] Simon, B., *Representations of Finite and Compact Groups*, Graduate Studies in Mathematics, Vol. 10, American Mathematical Society, Providence, RI, 1996.

[73] Sugiura, M., *Unitary Representations and Harmonic Analysis*, 2nd ed., North-Holland, Amsterdam, 1990.

[74] Taylor, M.E., *Noncommutative Harmonic Analysis*, Math. Surv. and Monogr., American Mathematical Society, Providence, RI, 1986.

[75] Terras, A., *Harmonic Analysis on Symmetric Spaces and Applications I*, Springer-Verlag, New York, 1985.

[76] Terras, A., *Fourier Analysis on Finite Groups and Applications*, London Mathematical Society Student Texts 43, Cambridge University Press, Cambridge, 1999.

[77] Thangavelu, S., *Harmonic Analysis on the Heisenberg Group*, Birkhäuser, Boston, 1998.

[78] Trachtenberg, E.A. and Karpovsky, M.G., Filtering in a communication channel by Fourier transforms over finite groups, in *Spectral Techniques and Fault Detection*, Karpovsky, M.G., Ed., Academic Press, New York, 1985.

[79] Varadarajan, V.S., *Lie Groups, Lie Algebras, and their Representations*, Springer-Verlag, New York, 1984.

[80] Varadarajan, V.S., *An Introduction to Harmonic Analysis on Semisimple Lie Groups*, Cambridge University Press, New York, 1989.

[81] Vilenkin, N.J. and Klimyk, A.U., *Representation of Lie Groups and Special Functions*, Vols. 1–3, Kluwer Academic, Dordrecht, Holland 1991.

[82] Wallach, N.R., *Harmonic Analysis on Homogeneous Spaces*, Marcel Dekker, New York, 1973.

[83] Wawrzyńczyk, A., *Group Representations and Special Functions, Mathematics and its Applications* (East European series), Reidel, D., Dordrecht, Holland, 1984.

[84] Weyl, H., Harmonics on homogeneous manifolds, *Ann. of Math.*, 35, 486–499, 1934.

[85] Wigner, E., On unitary representations of the inhomogeneous Lorentz group, *Ann. of Math.*, 40(1), 149–204, 1939.

[86] Williams, F.L., *Lectures on the Sprectum of $L^2(\Gamma \backslash G)$*, Pitman Research Notes in Mathematics Series 242, Longman Scientific & Technical, London, 1991.

[87] Želobenko, D.P., *Compact Lie Groups and their Representations, Translations of Mathematical Monographs*, American Mathematical Society, Providence, RI, 1973.

Chapter 9

Representation Theory and Operational Calculus for $SU(2)$ and $SO(3)$

In many of the chapters that follow, $SO(3)$ representations will be essential in applications. While the range of applications we will explore is broad (covering topics from solid mechanics to rotational Brownian motion and satellite attitude estimation), we do not cover every possible application. For instance, group representations are well known as a descriptive tool in theoretical physics. In this chapter we briefly explain why this is so, but since the intent of this work is to inform an engineering audience, we do not pursue this in depth. We mention also in passing that the representation theory of $SO(3)$ has been used in application engineering areas not covered in this book. One notable area is in the design of stereo sound systems [8].

The current chapter presents a very coordinate-dependent introduction to the representation theory of the groups $SU(2)$ and $SO(3)$. The emphasis is on explicit calculation of matrix elements of IURs and how these quantities transform under certain differential operators.

9.1 Representations of $SU(2)$ and $SO(3)$

9.1.1 Irreducible Representations from Homogeneous Polynomials

Several methods are available for generating a complete set of irreducible unitary representations of $SO(3)$. See, for example, the classic references [7, 18, 29].

Perhaps the most straightforward computational technique is based on the recognition that the space of *analytic functions* (i.e., functions that can be expanded and approximated to arbitrary precision with a Taylor series) of the form $f(\mathbf{z}) = \tilde{f}(z_1, z_2)$ for all $\mathbf{z} \in \mathbb{C}^2$ can be decomposed into subspaces of *homogeneous polynomials* on which each of the IURs of $SU(2)$ act [6, 18, 28, 29]. Since there is a two-to-one homomorphism from $SU(2)$ onto $SO(3)$, representations of $SO(3)$ are a subset of those for $SU(2)$.

A homogeneous polynomial of degree l is one of the form

$$P_l(z_1, z_2) = \sum_{k=0}^{l} c_k z_1^k z_2^{l-k}.$$

Here z_i^k is z_i to the power k. The set of all $P_l(z_1, z_2)$ is a subspace of the space of all analytic functions of the form $\tilde{f}(z_1, z_2)$. An inner product on this complex vector space may be defined as

$$\left(\sum_{k=0}^{l} a_k z_1^k z_2^{l-k}, \sum_{k=0}^{l} b_k z_1^k z_2^{l-k} \right) \triangleq \sum_{k=0}^{l} k!(l-k)! a_k \overline{b_k}. \tag{9.1}$$

With respect to this inner product, the following are elements of an orthonormal basis for the set of all $P_l(z_1, z_2)$

$$e_j^m(\mathbf{z}) = \tilde{e}_j^m(z_1, z_2) = \frac{z_1^{j+m} z_2^{j-m}}{[(j+m)!(j-m)!]^{1/2}}. \tag{9.2}$$

Here the range of indices is $j = 0, 1/2, 1, 3/2, 2, \ldots$ and the superscripts $m \in \{-j, -j+1, \ldots, j-1, j\}$ are half-integer values with unit increment. Substituting Eq. (9.2) into (9.1), one verifies that

$$\left(e_j^m, e_j^n \right) = \delta_{m,n}.$$

The matrix elements of IURs for $SU(2)$ are found explicitly as

$$e_j^n\left(A^{-1} \mathbf{z} \right) = \sum_{m=-j}^{j} U_{mn}^j(A) e_j^m(\mathbf{z}). \tag{9.3}$$

Using the inner product Eq. (9.1) one writes

$$U_{kn}^j(A) = \left(e_j^k(\mathbf{z}), e_j^n\left(A^{-1} \mathbf{z} \right) \right).$$

Here we use the notation that the matrices $U^j(A)$ have elements $U_{kn}^j(A)$. In the notation of Chapter 8, the corresponding quantities would be respectively labeled $U(g, j)$ and $U_{kn}(g, j)$, or $U^j(g)$ and $U_{kn}^j(g)$.

Discussions of the irreducibility and unitarity of these representations can be found in [29, 32]. Taking this for granted, one can use Eq. (9.3) as a way to enumerate the IURs of $SU(2)$.

We shall not be concerned with proving the irreducibility or unitarity of $U^j(A)$ at this point (which can be done using the tools of Chapter 8), but rather their explicit calculation. This is demonstrated in the following for $j = 0, 1/2, 1, 3/2, 2$. Taking

$$A = \begin{pmatrix} a & b \\ -\overline{b} & \overline{a} \end{pmatrix} \in SU(2),$$

(and so $a\overline{a} + b\overline{b} = 1$) we have

$$A^{-1} = \begin{pmatrix} \overline{a} & -b \\ \overline{b} & a \end{pmatrix}.$$

Then $\mathbf{z}' = A^{-1} \mathbf{z}$, so that $e_j^m(A^{-1} \mathbf{z}) = \tilde{e}_j^m(z_1', z_2')$. Consider the case when $j = 1/2$. Then

$$e_{1/2}^{-1/2}(\mathbf{z}) = z_2 \quad \text{and} \quad e_{1/2}^{1/2}(\mathbf{z}) = z_1.$$

Therefore we write

$$e_{1/2}^{-1/2}\left(A^{-1} \mathbf{z} \right) = \overline{b} z_1 + a z_2 = a e_{1/2}^{-1/2}(\mathbf{z}) + \overline{b} e_{1/2}^{1/2}(\mathbf{z})$$

and

$$e_{1/2}^{1/2}\left(A^{-1} \mathbf{z} \right) = \overline{a} z_1 - b z_2 = -b e_{1/2}^{-1/2}(\mathbf{z}) + \overline{a} e_{1/2}^{1/2}(\mathbf{z}).$$

From Eq. (9.3) we see that the IUR matrix $U^{1/2}$ is found from the coefficients in the previous equation as

$$
\begin{pmatrix} e_{1/2}^{-1/2}(A^{-1}z) \\ e_{1/2}^{1/2}(A^{-1}z) \end{pmatrix} = \begin{pmatrix} U_{-1/2,-1/2}^{1/2} & U_{-1/2,1/2}^{1/2} \\ U_{1/2,-1/2}^{1/2} & U_{1/2,1/2}^{1/2} \end{pmatrix}^T \begin{pmatrix} e_{1/2}^{-1/2}(z) \\ e_{1/2}^{1/2}(z) \end{pmatrix}.
$$

Hence we write

$$
U^{1/2}(A) = \begin{pmatrix} a & -b \\ b & \bar{a} \end{pmatrix}.
$$

This is a 2×2 unitary matrix related to A by similarity as

$$
U^{1/2}(A) = T_{1/2}^{-1} A T_{1/2}
$$

where

$$
T_{1/2} = T_{1/2}^{-1} = \begin{pmatrix} 1 & 0 \\ 0 & -1 \end{pmatrix}.
$$

For $j = 1$ we have

$$
e_1^{-1}(z) = z_2^2/\sqrt{2} \qquad e_1^0(z) = z_1 z_2 \qquad e_1^1(z) = z_1^2/\sqrt{2}.
$$

Then

$$
e_1^{-1}(A^{-1}z) = \frac{1}{\sqrt{2}} \left(\bar{b} z_1 + a z_2 \right)^2 = a^2 e_1^{-1}(z) + \sqrt{2} b a e_1^0(z) + \bar{b}^2 e_1^1(z);
$$

$$
e_1^0 \left(A^{-1}z \right) = -\sqrt{2} a b e_1^{-1}(z) + \left(\bar{a} a - b \bar{b} \right) e_1^0(z) + \sqrt{2} \bar{a} \bar{b} e_1^1(z);
$$

$$
e_1^1 \left(A^{-1}z \right) = \frac{1}{\sqrt{2}} \left(\bar{a} z_1 - b z_2 \right)^2 = b^2 e_1^{-1}(z) - \sqrt{2} \bar{a} b e_1^0(z) + \bar{a}^2 e_1^1(z).
$$

The coefficients are collected together as

$$
U^1 = \begin{pmatrix} a^2 & -\sqrt{2} ab & b^2 \\ \sqrt{2} \bar{b} a & \bar{a} a - b \bar{b} & -\sqrt{2} \bar{a} b \\ \bar{b}^2 & \sqrt{2} \bar{a} \bar{b} & \bar{a}^2 \end{pmatrix}.
$$

We note that this matrix can be related by similarity transformation to the rotation matrix $R(A) \in SO(3)$ defined by the Cayley-Klein parameters

$$
U^1(A) = T_1^{-1} R(A) T_1 \tag{9.4}
$$

where

$$
T_1 = \frac{-1}{\sqrt{2}} \begin{pmatrix} -1 & 0 & 1 \\ i & 0 & i \\ 0 & -\sqrt{2} & 0 \end{pmatrix}
$$

is unitary.

The analogous calculations for $j = 3/2$ and $j = 2$ respectively yield

$$
U^{3/2} = \begin{pmatrix} a^3 & -\sqrt{3} b a^2 & \sqrt{3} a b^2 & -b^3 \\ \sqrt{3} \bar{b} a^2 & a^2 \bar{a} - 2 b \bar{b} a & b^2 \bar{b} - 2 \bar{a} a b & \sqrt{3} \bar{a} b^2 \\ \sqrt{3} \bar{b}^2 a & 2 \bar{b} a \bar{a} - b \bar{b}^2 & \bar{a}^2 a - 2 \bar{a} b \bar{b} & -\sqrt{3} \bar{a}^2 b \\ \bar{b}^3 & \sqrt{3} \bar{a} \bar{b}^2 & \sqrt{3} \bar{a}^2 \bar{b} & \bar{a}^3 \end{pmatrix}
$$

and

$$
U^2 = \begin{pmatrix}
a^4 & -2a^3b & \sqrt{6}a^2b^2 & -2b^3a & b^4 \\
2\bar{b}a^3 & \bar{a}a^3 - 3b\bar{b}a^2 & \sqrt{6}(-\bar{a}a^2b + b^2\bar{b}a) & 3\bar{a}ab^2 - b^3\bar{b} & -2\bar{a}b^3 \\
\sqrt{6}\bar{b}^2a^2 & \sqrt{6}(\bar{b}\bar{a}a^2 - \bar{b}^2ba) & \bar{a}^2a^2 - 4\bar{a}a\bar{b}b + b^2\bar{b}^2 & \sqrt{6}(-\bar{a}^2ab + \bar{a}b^2\bar{b}) & \sqrt{6}\bar{a}^2b^2 \\
2\bar{b}^3a & 3\bar{b}^2\bar{a}a - b\bar{b}^3 & \sqrt{6}(\bar{a}^2ab - \bar{a}b\bar{b}^2) & \bar{a}^3a - 3\bar{a}^2b\bar{b} & -2\bar{a}^3b \\
\bar{b}^4 & 2\bar{b}^3\bar{a} & \sqrt{6}\bar{a}^2\bar{b}^2 & 2\bar{a}^3\bar{b} & \bar{a}^4
\end{pmatrix}.
$$

For values of $j \geq 3/2$, it follows from the definition of irreducibility that U^j is not equivalent to any direct sum of 1, A, and $R(A)$ since U^j cannot be block diagonalized.

The IURs presented previously are one of an infinite number of choices which are all equivalent under similarity transformation. For instance, one might choose to enumerate basis elements from positive to negative superscripts instead of the other way around. Writing,

$$
\begin{pmatrix}
e_{1/2}^{1/2}(A^{-1}\mathbf{z}) \\
e_{1/2}^{-1/2}(A^{-1}\mathbf{z})
\end{pmatrix}
=
\begin{pmatrix}
U_{1/2,1/2}^{1/2} & U_{1/2,-1/2}^{1/2} \\
U_{-1/2,1/2}^{1/2} & U_{-1/2,-1/2}^{1/2}
\end{pmatrix}^T
\begin{pmatrix}
e_{1/2}^{1/2}(\mathbf{z}) \\
e_{1/2}^{-1/2}(\mathbf{z})
\end{pmatrix}
$$

results in an IUR equivalent to $U^{1/2}$

$$
S_{1/2}\, U^{1/2}\, S_{1/2}^{-1} = \begin{pmatrix} \bar{a} & \bar{b} \\ -b & a \end{pmatrix}.
$$

Analogously for $j = 1$ we can write

$$
S_1 U^1 S_1^{-1} = \begin{pmatrix}
\bar{a}^2 & \sqrt{2}\bar{a}\bar{b} & \bar{b}^2 \\
-\sqrt{2}\bar{a}b & \bar{a}a - b\bar{b} & \sqrt{2}\bar{b}a \\
b^2 & -\sqrt{2}ab & a^2
\end{pmatrix},
$$

and likewise for all U^j where S_j is the matrix which has the effect of reflecting both rows and columns of U^j about their centers.

Finally we note that the matrix elements U_{mn}^l can be calculated in this parameterization as [32]

$$
U_{mn}^l(A) = \frac{1}{2\pi}\left[\frac{(l-m)!(l+m)!}{(l-n)!(l+n)!}\right]^{\frac{1}{2}} \int_0^{2\pi} \left(ae^{i\phi} + \bar{b}\right)^{l-n} \left(-be^{i\phi} + \bar{a}\right)^{l+n} e^{i(m-l)\phi}d\phi. \quad (9.5)
$$

9.1.2   The Adjoint and Tensor Representations of $SO(3)$

Let $X \in so(3)$ be an arbitrary 3×3 skew-symmetric matrix. Then for $R \in SO(3)$,

$$
Ad(R)X = RXR^T
$$

is the adjoint representation of $SO(3)$ with the property

$$
Ad(R_1R_2) = Ad(R_1)\,Ad(R_2). \quad (9.6)
$$

In Chapter 5 we saw the rule

$$
\text{vect}\left(RXR^T\right) = R\text{vect}(X).
$$

It follows that the matrix of the adjoint representation is

$$[Ad(R)] = R.$$

From Eq. (9.4) we therefore observe that $Ad(R)$ is equivalent (under similarity transformation) to $U^1(R)$.

The homomorphism property Eq. (9.6) does not depend on X being in $SO(3)$. We could have chosen $X \in \mathbb{R}^{3\times3}$ arbitrarily, and the homomorphism property would still hold. In this more general context,

$$T(R)X = RXR^T$$

is called a *tensor representation* of $SO(3)$. Let $(X)^\vee \in \mathbb{R}^9$ be the vector of independent numbers formed by sequentially stacking the columns of the matrix X. Then we may define a matrix $[T(R)]$ of the tensor representation $T(R)$ as

$$[T(R)](X)^\vee = \left(RXR^T\right)^\vee.$$

$[T(R)]$ is then a 9×9 matrix representation of $SO(3)$. As we shall see shortly, $[T(R)]$ [and hence $T(R)$] is a reducible representation.

It is well known that any matrix can be decomposed into a skew-symmetric part and a symmetric part. The symmetric part can further be decomposed into a part that is a constant multiple of the identity and a part that is traceless. Hence for arbitrary $X \in \mathbb{R}^{3\times3}$, we write

$$X = X_{skew} + X_{sym1} + X_{sym2}$$

where

$$X_{sym1} = \frac{1}{3}\text{tr}(X)\mathbb{I}; \quad X_{skew} = \frac{1}{2}(X - X^T); \quad X_{sym2} = \frac{1}{2}(X + X^T) - \frac{1}{3}\text{tr}(X)\mathbb{I}.$$

Let the set of all matrices of the form X_{sym1} be called V_1, the set of all X_{skew} be called V_2 [which is the same as $so(3)$], and the set of all X_{sym2} be called V_3. Each of these spaces is invariant under similarity transformation by a rotation matrix. That is, if $Y \in V_i$ then $RYR^T \in V_i$. This is the very definition of reducibility of the representation $T(R)$ on the space $\mathbb{R}^{3\times3}$. We therefore write

$$\mathbb{R}^{3\times3} = V_1 \oplus V_2 \oplus V_3.$$

We may then define representations on each of these subspaces. Since $T_1(R)X_{sym1} = RX_{sym1}R^T = X_{sym1}$, it is clear that this is equivalent to the representation $U^0(R) = 1$. Since $X_{skew} \in SO(3)$, it follows that the representation $T_2(R)$ acting on elements of V_2 is the same as the adjoint representation, which we already know is equivalent to $U^1(R)$. Finally, we note that $T_3(R)X_{sym2} = RX_{sym2}R^T$ is a five-dimensional representation. This may be shown to be irreducible and hence equivalent to $U^2(R)$. We therefore write

$$T \cong U^0 \oplus U^1 \oplus U^2.$$

Another way to say this is that there exists a matrix $S \in GL(9, \mathbb{C})$ such that the 9×9 matrix $[T(R)]$ is block-diagonalized as

$$[T(R)] = S^{-1}\begin{pmatrix} U^0(R) & 0_{1\times3} & 0_{1\times5} \\ 0_{3\times1} & U^1(R) & 0_{3\times5} \\ 0_{5\times1} & 0_{5\times3} & U^2(R) \end{pmatrix} S.$$

Explicitly, the matrix $[T(R)]$ is the *tensor product* of R with itself. That is,

$$[T(R)] = R \otimes R \tag{9.7}$$

where $A \otimes B$ is defined for any two matrices

$$A = \begin{pmatrix} a_{11} & a_{12} & \cdots & a_{1q} \\ a_{21} & a_{22} & \cdots & \vdots \\ \vdots & \vdots & \ddots & \vdots \\ a_{p1} & a_{p2} & \cdots & a_{pq} \end{pmatrix} \in \mathbb{R}^{p \times q}$$

and

$$B = \begin{pmatrix} b_{11} & b_{12} & \cdots & b_{1s} \\ b_{21} & b_{22} & \cdots & \vdots \\ \vdots & \vdots & \ddots & \vdots \\ b_{r1} & b_{r2} & \cdots & b_{rs} \end{pmatrix} \in \mathbb{R}^{r \times s}$$

as

$$A \otimes B \triangleq \begin{pmatrix} a_{11}B & a_{12}B & \cdots & a_{1q}B \\ a_{21}B & a_{22}B & \cdots & \vdots \\ \vdots & \vdots & \ddots & \vdots \\ a_{p1}B & a_{p2}B & \cdots & a_{pq}B \end{pmatrix} \in \mathbb{R}^{pr \times qs}.$$

From this definition, it is clear that if $C \in \mathbb{R}^{q \times l}$ and $D \in \mathbb{R}^{s \times t}$ for some l and t, then

$$(A \otimes B)(C \otimes D) = \begin{pmatrix} a_{11}B & a_{12}B & \cdots & a_{1q}B \\ a_{21}B & a_{22}B & \cdots & \vdots \\ \vdots & \vdots & \ddots & \vdots \\ a_{p1}B & a_{p2}B & \cdots & a_{pq}B \end{pmatrix} \begin{pmatrix} c_{11}D & c_{12}D & \cdots & c_{1l}D \\ c_{21}D & c_{22}D & \cdots & \vdots \\ \vdots & \vdots & \ddots & \vdots \\ c_{q1}D & c_{q2}D & \cdots & c_{ql}D \end{pmatrix}$$

$$= (AC) \otimes (BD). \tag{9.8}$$

If A and B are square, then

$$\text{trace}(A \otimes B) = \text{trace}(A) \cdot \text{trace}(B)$$

and

$$\det(A \otimes B) = (\det A)^{dim B} \cdot (\det B)^{dim A} = \det(B \otimes A).$$

Note that this is in contrast to

$$\text{trace}(A \oplus B) = \text{trace}(A) + \text{trace}(B)$$

and

$$\det(A \oplus B) = (\det A) \cdot (\det B)$$

for the direct sum.

When A, B, and X are square and of the same dimension, the tensor product plays an important role because

$$Y = AXB^T$$

is converted as

$$Y^\vee = (B \otimes A)X^\vee, \tag{9.9}$$

or equivalently

$$\left(Y^T\right)^\vee = (A \otimes B)\left(X^T\right)^\vee.$$

Hence, Eq. (9.7) is a special case of Eq. (9.9).

One can generalize the concept of a tensor representation further. We postpone discussion of this until Chapter 18 where applications of the representation theory of $SO(3)$ in mechanics are discussed.

In the remainder of this chapter we examine the form of the representation matrices in various parameterizations and review harmonic analysis on $SU(2)$ and $SO(3)$ as concrete examples of the general theory presented in the previous chapter.

9.1.3 Irreducibility of the Representations $U_i(g)$ of $SU(2)$

We now demonstrate the irreducibility of the matrix representations $U_i(g)$ using three techniques: (1) by checking that the characters satisfy Eq. (8.53); (2) by observing that the infinitesimal matrices

$$u(X_i, l) = \frac{d}{dt} U^l (\exp t X_i)|_{t=0}$$

are not all block-diagonalizable by the same matrix under similarity transformation; and (3) by observing how the operator corresponding to each $U_i(g)$ acts on the space of homogeneous polynomials.

Method 1:

When $SU(2)$ is parameterized using Euler angles, the character functions $\chi_l(g) = \text{trace}(U^l(g))$ take the form (see Eq. (9.20) or [32, p. 359])

$$\chi_l(g) = \sum_{m=-l}^{l} e^{-im(\alpha+\gamma)} P_{mm}^l(\cos\beta) = \frac{\sin(l + \frac{1}{2})\theta}{\sin\frac{\theta}{2}},$$

where θ is the angle of rotation. Using the orthogonalities of $e^{im\phi}$ and the functions $P_{mn}^l(\cos\theta)$ [see Eq. (9.25)], it may be shown that

$$\int_{SU(2)} \chi_{l_1}(g)\overline{\chi_{l_2}(g)}dg = \delta_{l_1, l_2}.$$

When $l_1 = l_2$, the condition of Theorem 8.13 is met, and so the representations U^l are all irreducible.

Method 2:

A basis for the Lie algebra $SU(2)$ is

$$X_1 = \frac{i}{2}\begin{pmatrix} 0 & 1 \\ 1 & 0 \end{pmatrix}; \qquad X_2 = \frac{1}{2}\begin{pmatrix} 0 & -1 \\ 1 & 0 \end{pmatrix}; \qquad X_3 = \frac{i}{2}\begin{pmatrix} 1 & 0 \\ 0 & -1 \end{pmatrix}.$$

The corresponding one-parameter subgroups are

$$g_1(t) = \exp(t X_1) = \begin{pmatrix} \cos\frac{t}{2} & i\sin\frac{t}{2} \\ i\sin\frac{t}{2} & \cos\frac{t}{2} \end{pmatrix}; \tag{9.10}$$

$$g_2(t) = \exp(t X_2) = \begin{pmatrix} \cos\frac{t}{2} & -\sin\frac{t}{2} \\ \sin\frac{t}{2} & \cos\frac{t}{2} \end{pmatrix}; \tag{9.11}$$

$$g_3(t) = \exp(tX_3) = \begin{pmatrix} e^{i\frac{t}{2}} & 0 \\ 0 & e^{-i\frac{t}{2}} \end{pmatrix}. \tag{9.12}$$

The matrices

$$u(X_i, l) = \left.\frac{dU^l(\exp(tX_i))}{dt}\right|_{t=0}$$

corresponding to infinitesimal motion for this case for $i = 1$ look like

$$u\left(X_1, \frac{1}{2}\right) = \begin{pmatrix} 0 & -\frac{i}{2} \\ -\frac{i}{2} & -\frac{i}{2} \end{pmatrix};$$

$$u(X_1, 1) = \begin{pmatrix} 0 & -\frac{i}{\sqrt{2}} & 0 \\ -\frac{i}{\sqrt{2}} & 0 & -\frac{i}{\sqrt{2}} \\ 0 & -\frac{i}{\sqrt{2}} & 0 \end{pmatrix};$$

$$u\left(X_1, \frac{3}{2}\right) = \begin{pmatrix} 0 & -i & 0 & 0 \\ -\frac{i\sqrt{3}}{2} & 0 & -i & 0 \\ 0 & -i & 0 & -\frac{i\sqrt{3}}{2} \\ 0 & 0 & -\frac{i\sqrt{3}}{2} & 0 \end{pmatrix}.$$

For $i = 2$,

$$u\left(X_2, \frac{1}{2}\right) = \begin{pmatrix} 0 & \frac{1}{2} \\ -\frac{1}{2} & -\frac{1}{2} \end{pmatrix};$$

$$u(X_2, 1) = \begin{pmatrix} 0 & \frac{1}{\sqrt{2}} & 0 \\ -\frac{1}{\sqrt{2}} & 0 & -\frac{1}{\sqrt{2}} \\ 0 & -\frac{1}{\sqrt{2}} & 0 \end{pmatrix};$$

$$u\left(X_2, \frac{3}{2}\right) = \begin{pmatrix} 0 & \frac{\sqrt{3}}{2} & 0 & 0 \\ -\frac{\sqrt{3}}{2} & 0 & 1 & 0 \\ 0 & -1 & 0 & \frac{\sqrt{3}}{2} \\ 0 & 0 & -\frac{\sqrt{3}}{2} & 0 \end{pmatrix}.$$

For $i = 3$,

$$u\left(X_3, \frac{1}{2}\right) = \begin{pmatrix} \frac{i}{2} & 0 \\ 0 & 0 \end{pmatrix};$$

$$u(X_3, 1) = \begin{pmatrix} i & 0 & 0 \\ 0 & 0 & 0 \\ 0 & 0 & -i \end{pmatrix};$$

$$u\left(X_3, \frac{3}{2}\right) = \begin{pmatrix} \frac{3i}{2} & 0 & 0 & 0 \\ 0 & \frac{i}{2} & 0 & 0 \\ 0 & 0 & -\frac{i}{2} & 0 \\ 0 & 0 & 0 & -\frac{3i}{2} \end{pmatrix}.$$

It may be verified that for all $i = 1, 2, 3$, the preceding matrices (as well as those for all values of l) are not simultaneously block-diagonalizable [7].

Method 3:

For homogeneous polynomials of the form

$$f(z_1, z_2) = \sum_{n=-l}^{l} a_n z_1^{l-n} z_2^{l+n},$$

one observes that

$$f(z_1, z_2) = z_2^{2l} f(z_1/z_2, 1)$$

for $z_2 \neq 0$. It is convenient to define $\varphi(x) = f(x, 1)$. Hence, the operator $U^l(g)$ acting on the space of all homogeneous polynomials of degree l whose matrix elements are defined in Eq. (9.3) may be rewritten in the form [32]

$$\hat{U}^l(g)\varphi(z) = (\beta z + \overline{\alpha})^{2l} \, \varphi\left(\frac{\alpha z - \overline{\beta}}{\beta z + \overline{\alpha}}\right)$$

(with l taking the place of j, $\alpha = \overline{a}$, and $\beta = \overline{b}$). Evaluated at the one-parameter subgroups, this becomes

$$\hat{U}^l(\exp(tX_1))\,\varphi(x) = \left(ix \sin\frac{t}{2} + \cos\frac{t}{2}\right)^{2l} \varphi\left(\frac{x\cos\frac{t}{2} + i\sin\frac{t}{2}}{ix\sin\frac{t}{2} + \cos\frac{t}{2}}\right);$$

$$\hat{U}^l(\exp(tX_2))\,\varphi(x) = \left(x \sin\frac{t}{2} + \cos\frac{t}{2}\right)^{2l} \varphi\left(\frac{x\cos\frac{t}{2} - \sin\frac{t}{2}}{x\sin\frac{t}{2} + \cos\frac{t}{2}}\right);$$

$$\hat{U}^l(\exp(tX_3))\,\varphi(x) = e^{-itl}\,\varphi\left(xe^{it}\right).$$

One defines differential operators based on these representation operators as

$$\hat{u}_i^l = \left.\frac{d\hat{U}^l(\exp(tX_i))}{dt}\right|_{t=0}.$$

Note that $\hat{u}_i^l$ are differential operators corresponding to the matrices $u(X_i, l)$. Explicitly one finds

$$\hat{u}_1^l = ilx + \frac{i}{2}\left(1 - x^2\right)\frac{d}{dx};$$

$$\hat{u}_2^l = -lx + \frac{1}{2}\left(1 + x^2\right)\frac{d}{dx};$$

$$\hat{u}_3^l = i\left(x\frac{d}{dx} - l\right).$$

These operators act on the space of all polynomials of the form

$$\varphi(x) = \sum_{k=-l}^{l} a_k x^{l-k}$$

which may be equipped with an inner product $(\cdot, \cdot)$ defined such that

$$\left(\varphi, x^{l-k}\right) = (l - k)!(l + k)! a_k.$$

Orthonormal basis elements for this space of polynomials are taken as

$$e_k^l(x) = \frac{x^{l-k}}{\sqrt{(l-k)!(l+k)!}}$$

for $-l \leq k \leq l$.

The irreducibility of the representation operators $U^l(g)$ may then be observed from the way the differential operators act on the basis elements. To this end, it is often convenient to use the operators [32]

$$\hat{H}_+^l = i\left(\hat{u}_1^l + i\hat{u}_2^l\right) = -\frac{d}{dx};$$

$$\hat{H}_-^l = i\left(\hat{u}_1^l - i\hat{u}_2^l\right) = -2lx + x^2\frac{d}{dx};$$

$$\hat{H}_3^l = i\hat{u}_3^l = l - x\frac{d}{dx}.$$

One then observes that

$$\hat{H}_+^l e_n^l(x) = -\sqrt{(l-n)(l+n+1)}e_{n+1}^l(x);$$

$$\hat{H}_-^l e_n^l(x) = -\sqrt{(l+n)(l-n+1)}e_{n-1}^l(x);$$

$$\hat{H}_3^l e_n^l(x) = ne_n^l(x).$$

While subspaces of functions that consist of scalar multiples of $e_n^l(x)$ are stable under $\hat{H}_3^l$, it is clear that no subspaces (other than zero and the whole space) can be left invariant under $\hat{H}_+^l$, $\hat{H}_-^l$, and $\hat{H}_3^l$ simultaneously since $\hat{H}_+^l$ and $\hat{H}_-^l$ always "push" basis elements to "adjacent" subspaces. Hence the representation operators $\hat{U}(g)$ are irreducible, as must be their corresponding matrices.

9.2   Some Differential Geometry of S^3 and $SU(2)$

The relationship between $SO(3)$, $SU(2)$, and S^3 was established in Chapter 5. Here we calculate some geometric quantities using several different methods and show how they reduce to the same thing.

9.2.1   The Metric Tensor

The metric tensor for the group $SU(2)$ may be found using the parameterization of the rotation matrix in terms of Euler angles and coordinates of the unit sphere in $\mathbb{R}^4$ (see, for example, [7, 32]). For the range of the Euler angles $0 \leq \alpha < 2\pi$, $0 \leq \beta < \pi$, $0 \leq \gamma < 2\pi$ this tensor corresponds to the metric tensor on $SO(3)$, whereas for $SU(2) \gamma \in [-2\pi, 2\pi]$.

The covariant metric tensor is

$$G = \begin{bmatrix} 1 & 0 & 0 \\ 0 & 1 & \cos\beta \\ 0 & \cos\beta & 1 \end{bmatrix} \tag{9.13}$$

where the element G_{11} corresponds to the $G_{\beta\beta}$ element. From a group-theoretic perspective, we recognize that

$$J_L^T J_L = J_R^T J_R = G.$$

From a purely geometric perspective, the vector

$$\mathbf{x}(\alpha, \beta, \gamma) = \begin{pmatrix} \cos\frac{\beta}{2}\cos\left(\frac{\alpha+\gamma}{2}\right) \\ \cos\frac{\beta}{2}\sin\left(\frac{\alpha+\gamma}{2}\right) \\ -\sin\frac{\beta}{2}\sin\left(\frac{\alpha-\gamma}{2}\right) \\ \sin\frac{\beta}{2}\cos\left(\frac{\alpha-\gamma}{2}\right) \end{pmatrix}$$

defines points on the unit sphere in $\mathbb{R}^4$. The Jacobian

$$J = \left[\frac{\partial\mathbf{x}}{\partial\beta}; \frac{\partial\mathbf{x}}{\partial\alpha}; \frac{\partial\mathbf{x}}{\partial\gamma} \right]$$

may be used to define the same metric tensor as

$$G = J^T W J$$

where W is a scalar multiple of $\mathbb{I}_{4\times 4}$.

Regardless of how G is computed we see that

$$G^{-1} = \begin{bmatrix} 1 & 0 & 0 \\ 0 & \frac{1}{\sin^2\beta} & -\frac{\cos\beta}{\sin^2\beta} \\ 0 & -\frac{\cos\beta}{\sin^2\beta} & \frac{1}{\sin^2\beta} \end{bmatrix}, \tag{9.14}$$

and for each differentiable $f(A)$ on $SO(3)$ we define

$$\nabla_A f = \left(\frac{\partial f}{\partial\beta}, \frac{\partial f}{\partial\alpha}, \frac{\partial f}{\partial\gamma} \right)^T.$$

This is *not* the gradient defined in Chapter 4, which in this notation would be

$$\operatorname{grad} f = \sum_{j=1}^{3} g^{ij} \left(\nabla_A f \right)_j \mathbf{v}_j$$

where

$$\mathbf{v}_j = \frac{\partial\mathbf{x}}{\partial\alpha_i}$$

where $\alpha_1 = \beta$, $\alpha_2 = \alpha$, and $\alpha_3 = \gamma$. The corresponding unit vectors are $\mathbf{e}_{\alpha_j}$.

It is easy to see that [for real function $f(A)$]

$$\int_{SO(3)} \left((\nabla_A f)^T G^{-1} \nabla_A f \right) d(A)$$
$$= \frac{1}{8\pi^2} \int_{\beta=0}^{\pi} \int_{\alpha=0}^{2\pi} \int_{\gamma=0}^{2\pi} \left(\frac{\partial f}{\partial\beta}\frac{\partial f}{\partial\beta} + \frac{1}{\sin^2\beta}\frac{\partial f}{\partial\alpha}\frac{\partial f}{\partial\alpha} + \frac{1}{\sin^2\beta}\frac{\partial f}{\partial\gamma}\frac{\partial f}{\partial\gamma} \right.$$
$$\left. - \frac{2\cos\beta}{\sin^2\beta}\frac{\partial f}{\partial\alpha}\frac{\partial f}{\partial\gamma} \right) \sin\beta \, d\beta \, d\alpha \, d\gamma.$$

It may be checked by integration by parts that

$$\int_{SO(3)} \left((\nabla_A f)^T G^{-1} \nabla_A f \right) d(A) = \int_{SO(3)} \left(f, -\nabla_A^2 f \right) d(A) \tag{9.15}$$

where the Laplacian $\nabla_A^2 = \operatorname{div}(\operatorname{grad} f)$ is given in Eq. (9.19). We note that we may write a gradient vector in the unit vector notations as

$$\operatorname{grad}(f) = \frac{\partial f}{\partial \beta} \mathbf{e}_\beta + \frac{1}{\sin \beta} \frac{\partial f}{\partial \alpha} \mathbf{e}_\alpha + \frac{1}{\sin \beta} \frac{\partial f}{\partial \gamma} \mathbf{e}_\gamma ,$$

where the unit vectors $\mathbf{e}_\beta$, $\mathbf{e}_\alpha$, $\mathbf{e}_\gamma$ satisfy the inner product relations

$$(\mathbf{e}_\beta, \mathbf{e}_\alpha) = (\mathbf{e}_\beta, \mathbf{e}_\gamma) = 0$$
$$(\mathbf{e}_\alpha, \mathbf{e}_\gamma) = (\mathbf{e}_\gamma, \mathbf{e}_\alpha) = \cos \beta.$$

In this notation, the relation Eq. (9.15) is written as

$$\int_{SO(3)} \|\operatorname{grad}(f)\|^2 \, d(A) = \int_{SO(3)} \left(f, -\nabla_A^2 f \right) d(A).$$

We mention also a definition of the δ-function on the rotation group

$$\delta(A, A') = \begin{cases} \frac{8\pi^2}{\sin \beta} \delta(\alpha - \alpha') \delta(\beta - \beta') \delta(\gamma - \gamma') & \text{for } \sin \beta' \neq 0 \\ \frac{2\delta(\beta)}{\sin \beta} & \text{for } \sin \beta' = 0 \end{cases}.$$

The δ-function has the properties that

$$\int_{SO(3)} f(A) \, \delta(A, A') \, d(A) = f(A') \tag{9.16}$$

and

$$\int_{SO(3)} \delta(A, A') \, d(A) = 1.$$

The preceding definition is valid for either $SO(3)$ or a hemisphere in $\mathbb{R}^4$. Using the group operation of $SO(3)$, the Dirac delta function of two arguments is replaced with one of a single argument

$$\delta(A, A') = \delta\left(\left(A' \right)^T A \right).$$

9.2.2 Differential Operators and Laplacian for $SO(3)$ in Spherical Coordinates

Spherical coordinates in $\mathbb{R}^4$ can be chosen as

$$\mathbf{r}(\theta_1, \theta_2, \theta_3) = \begin{bmatrix} r_1 \\ r_2 \\ r_3 \\ r_4 \end{bmatrix} = \begin{bmatrix} \cos \theta_1 \\ \sin \theta_1 \cos \theta_2 \\ \sin \theta_1 \sin \theta_2 \cos \theta_3 \\ \sin \theta_1 \sin \theta_2 \sin \theta_3 \end{bmatrix}$$

for $\theta_1 \in [0, \pi]$, $\theta_2 \in [0, \pi]$, $\theta_3 \in [0, 2\pi]$.

Using the four-dimensional matrix representation of rotation corresponding to $\mathbf{r}$ as in Chapter 5, i.e., by defining $R_{4 \times 4}(r)$ such that

$$R_{4 \times 4}(\mathbf{r}) q = r q$$

(where r is the unit quaternion corresponding to $\mathbf{r} \in \mathbb{R}^4$ and rq is quaternion multiplication of r and q), one has

$$R_{4 \times 4}(\mathbf{r}) = \begin{bmatrix} r_1 & -r_2 & -r_3 & -r_4 \\ r_2 & r_1 & -r_4 & r_3 \\ r_3 & r_4 & r_1 & -r_2 \\ r_4 & -r_3 & r_2 & r_1 \end{bmatrix}.$$

Hence, $R_{4 \times 4}(\mathbf{e}_1)$ is the identity matrix and corresponds to no rotation. If one considers infinitesimal motions on the sphere S^3 in each of the remaining coordinate directions in the vicinity of the point $\mathbf{r} = \mathbf{e}_1$, the corresponding rotations will be of the form

$$R_{4 \times 4}(\mathbf{e}_1 + \epsilon \mathbf{e}_i) = \mathbb{I}_{4 \times 4} + \epsilon X_{i-1}$$

for $i = 2, 3, 4$ and $|\epsilon| << 1$. It is easy to see that $X_{i-1} = R_{4 \times 4}(\mathbf{e}_{i-1})$ and explicitly have the form

$$X_1 = \begin{bmatrix} 0 & -1 & 0 & 0 \\ 1 & 0 & 0 & 0 \\ 0 & 0 & 0 & -1 \\ 0 & 0 & 1 & 0 \end{bmatrix}; \quad X_2 = \begin{bmatrix} 0 & 0 & -1 & 0 \\ 0 & 0 & 0 & 1 \\ 1 & 0 & 0 & 0 \\ 0 & -1 & 0 & 0 \end{bmatrix}; \quad X_3 = \begin{bmatrix} 0 & 0 & 0 & -1 \\ 0 & 0 & -1 & 0 \\ 0 & 1 & 0 & 0 \\ 1 & 0 & 0 & 0 \end{bmatrix}. \quad (9.17)$$

Analogous to the way the cross product in $\mathbb{R}^3$ takes unit basis vectors into unit basis vectors as $\mathbf{e}_1 \times \mathbf{e}_2 = \mathbf{e}_3$, $\mathbf{e}_2 \times \mathbf{e}_3 = \mathbf{e}_1$, $\mathbf{e}_3 \times \mathbf{e}_1 = \mathbf{e}_2$, one finds that under the Lie bracket operation (which in this case we take to be half of matrix commutator),

$$[X_i, X_j] = \frac{1}{2}(X_i X_j - X_j X_i),$$

that

$$[X_1, X_2] = X_3; \qquad [X_2, X_3] = X_1; \qquad [X_3, X_1] = X_2.$$

This indicates that the matrices X_i are representations of the basis elements of the Lie algebra $so(3)$. These elements are not unique, as can be seen by the fact that

$$R_{4 \times 4}(\mathbf{e}_1 + \epsilon R(\mathbf{r})\mathbf{e}_i) = \mathbb{I}_{4 \times 4} + \epsilon R_{4 \times 4} X_{i-1} R_{4 \times 4}^T.$$

Hence any set of basis elements of the form $\tilde{X}_i = R_{4 \times 4} X_i R_{4 \times 4}^T$ is equally acceptable.

Differential motions on the surface of the sphere S^3 can be used to define differential operators which act on real valued-functions of the form $f(\mathbf{r}(\theta_1, \theta_2, \theta_3))$ as

$$D_i f = \frac{d}{dt} f\left(\exp(tX_i)\,\mathbf{r}(\theta_1, \theta_2, \theta_3)\right)|_{t=0}$$

for $i = 1, 2, 3$. Here

$$\exp(A) = \mathbb{I}_{4 \times 4} + \sum_{n=1}^{\infty} \frac{A^n}{n!}$$

is the matrix exponential for any $A \in \mathbb{R}^{4 \times 4}$.

After a little work, it can be shown that the explicit form of these operators (in a basis, $\{\tilde{X}_i\}$ different than ours) is [35]

$$\tilde{D}_1 = -\cos\theta_2 \frac{\partial}{\partial\theta_1} + \sin\theta_2 \cot\theta_1 \frac{\partial}{\partial\theta_2} - \frac{\partial}{\partial\theta_3};$$

$$\tilde{D}_2 = -\sin\theta_2 \cos\theta_3 \frac{\partial}{\partial\theta_1} + (\sin\theta_3 - \cot\theta_1 \cos\theta_2 \cos\theta_3) \frac{\partial}{\partial\theta_2}$$
$$+ \left(\cot\theta_1 \frac{\sin\theta_3}{\sin\theta_2} + \cot\theta_2 \cos\theta_3\right) \frac{\partial}{\partial\theta_3};$$

$$\tilde{D}_3 = -\sin\theta_2 \sin\theta_3 \frac{\partial}{\partial\theta_1} - (\cos\theta_3 + \cot\theta_1 \cos\theta_2 \sin\theta_3) \frac{\partial}{\partial\theta_2}$$
$$+ \left(-\cot\theta_1 \frac{\cos\theta_3}{\sin\theta_2} + \cot\theta_2 \sin\theta_3\right) \frac{\partial}{\partial\theta_3}.$$

The Laplacian, which is invariant under the choice of $SO(3)$ basis, is defined as

$$\nabla^2 = (D_1)^2 + (D_2)^2 + (D_3)^2 = \left(\tilde{D}_1\right)^2 + \left(\tilde{D}_2\right)^2 + \left(\tilde{D}_3\right)^2$$

$$= \frac{\partial^2}{\partial\theta_1^2} + 2\cot\theta_1 \frac{\partial}{\partial\theta_1} + \frac{\cot\theta_2}{\sin^2\theta_1} \frac{\partial}{\partial\theta_2} + \frac{1}{\sin^2\theta_1} \frac{\partial^2}{\partial\theta_2^2} + \frac{1}{\sin^2\theta_1 \sin^2\theta_2} \frac{\partial^2}{\partial\theta_3^2}. \qquad (9.18)$$

9.3   Matrix Elements of $SU(2)$ Representations as Eigenfunctions of the Laplacian

Using the matrices g_k in Eqs. (9.10)–(9.12), elements of $SU(2)$ are parameterized with ZXZ Euler angles as

$$g(\alpha, \beta, \gamma) = g_3(\alpha)g_1(\beta)g_3(\gamma) = \begin{pmatrix} \cos\frac{\beta}{2} e^{i(\alpha+\gamma)/2} & i\sin\frac{\beta}{2} e^{i(\alpha-\gamma)/2} \\ -i\sin\frac{\beta}{2} e^{-i(\alpha-\gamma)/2} & \cos\frac{\beta}{2} e^{-i(\alpha+\gamma)/2} \end{pmatrix}.$$

Here $0 \le \alpha < 2\pi$, $-2\pi \le \gamma < 2\pi$, and $0 \le \beta \le \pi$. The invariant integration measure is

$$d(g) = C\sin\beta d\alpha d\beta d\gamma,$$

and the constant C is usually set so that

$$\int_{SU(2)} d(g) = 1.$$

That is, $C = 1/16\pi^2$.

Identifying elements of $SU(2)$ with points on the sphere S^3, the differential-geometric Laplacian operator is written in the Euler angles as

$$\nabla^2 = \frac{\partial^2}{\partial\beta^2} + \cot\beta \frac{\partial}{\partial\beta} + \frac{1}{\sin^2\beta} \left(\frac{\partial^2}{\partial\alpha^2} - 2\cos\beta \frac{\partial^2}{\partial\alpha\partial\gamma} + \frac{\partial^2}{\partial\gamma^2}\right). \qquad (9.19)$$

This is the same as the "group-theoretic" Laplacian

$$\nabla^2 = \left(X_1^R\right)^2 + \left(X_2^R\right)^2 + \left(X_3^R\right)^2 = \left(X_1^L\right)^2 + \left(X_2^L\right)^2 + \left(X_3^L\right)^2.$$

The preceding second equality follows from the $Ad(g)$-invariance of the inner product on the Lie algebra $su(2)$. The explicit form of the operators X_i^L and X_i^R are given in Section 9.10.

The eigenfunctions $u(\alpha, \beta, \gamma)$ which satisfy

$$\nabla^2 u = \lambda u$$

can be found using the separation of variables $u(\alpha, \beta, \gamma) = u_1(\alpha)u_2(\gamma)u_3(\beta)$ which reduces this PDE to the three Sturm-Liouville problems

$$u_1'' + m^2 u_1 = 0 \qquad u_1(0) = u_1(2\pi) \qquad u_1'(0) = u_1'(2\pi)$$
$$u_2'' + n^2 u_2 = 0 \qquad u_2(-2\pi) = u_2(2\pi) \qquad u_2'(-2\pi) = u_2'(2\pi)$$
$$\left(\sin\beta u_3'\right)' + \left[l(l+1)\sin\beta - \frac{n^2 - 2mn\cos\beta + m^2}{\sin\beta}\right] u_3 = 0.$$

The previous equation in $u_3(\beta)$ is transformed to a singular Sturm-Liouville problem in the variable $x = \cos\beta$, whose solution is written as $P_{mn}^l(x)$. That is,

$$\left[\frac{d}{dx}\left((1-x^2)\frac{d}{dx}\right) - \frac{m^2 + n^2 - 2mnx}{1-x^2} + l(l+1)\right] P_{mn}^l = 0.$$

The composite solution $u(\alpha, \beta, \gamma)$ for each set of l, m, n can be written as $U_{mn}^l(g(\alpha, \beta, \gamma))$. These are the matrix elements of the irreducible unitary representations of $SU(2)$ given by

$$U_{mn}^l(g(\alpha, \beta, \gamma)) = i^{m-n} e^{-i(m\alpha + n\gamma)} P_{mn}^l(\cos\beta). \tag{9.20}$$

The functions $P_{mn}^l(\cos\beta)$ are generalizations of the associated Legendre functions and are given by the Rodrigues formula [32]

$$P_{mn}^l(x) = \frac{(-1)^{l-m}}{2^l} \left[\frac{(l+m)!}{(l-n)!(l+n)!(l-m)!}\right]^{\frac{1}{2}} \times$$
$$(1+x)^{-(m+n)/2}(1-x)^{(n-m)/2} \frac{d^{l-m}}{dx^{l-m}}\left[(1-x)^{l-n}(1+x)^{l+n}\right]. \tag{9.21}$$

They can also be calculated by the integral

$$P_{mn}^l(\cos\beta) = \frac{i^{n-m}}{2\pi} \left[\frac{(l-m)!(l+m)!}{(l-n)!(l+n)!}\right]^{\frac{1}{2}} \int_0^{2\pi} \left(\cos\frac{\beta}{2}e^{i\phi/2} + i\sin\frac{\beta}{2}e^{-i\phi/2}\right)^{l-n}$$
$$\times \left(\cos\frac{\beta}{2}e^{-i\phi/2} + i\sin\frac{\beta}{2}e^{i\phi/2}\right)^{l+n} e^{im\phi} d\phi, \tag{9.22}$$

or relative to the Jacobi polynomials as

$$P_{mn}^l(\cos\beta) = \left[\frac{(l-m)!(l+m)!}{(l-n)!(l+n)!}\right]^{\frac{1}{2}} \sin^{m-n}\frac{\beta}{2}\cos^{m+n}\frac{\beta}{2} P_{l-m}^{(m-n,m+n)}(\cos\beta).$$

These functions satisfy certain symmetry relations including

$$P_{mn}^l(x) = (-1)^{m+n} P_{nm}^l(x) \qquad P_{mn}^l(x) = (-1)^{m-n} P_{-m,-n}^l(x) \tag{9.23}$$
$$P_{mn}^l(x) = P_{-n,-m}^l(x) \qquad P_{mn}^l(x) = (-1)^{l+n} P_{-m,n}^l(-x). \tag{9.24}$$

The functions $P^l_{mn}(z)$ are also given as in [32]

$$P^l_{mn}(z) = \left[\frac{(l-n)!(l+m)!}{(l-m)!(l+n)!}\right]^{1/2} \frac{(1-z)^{\frac{m-n}{2}}(1+z)^{\frac{m+n}{2}}}{2^m(m-n)!}$$
$$_2F_1\left(l+m+1, -l+m; m-n+1; \frac{1-z}{2}\right)$$

where $_2F_1$ is the hypergeometric function. This expression is valid for $m-n \geq 0$; the corresponding expressions for other possible values of m, n may be found using the properties of the hypergeometric functions and the symmetry properties of $P^l_{mn}(z)$. Other expressions for $P^l_{mn}(z)$ in terms of the Krawtchouk polynomials can be found in [32], and classical Fourier series expansions of $P^l_{mn}(\cos\theta)$ can be found in [2, 3, 23].

They also satisfy certain recursion relations including [7, 32, 31][1]

$$\cos\beta\, P^l_{mn} = \frac{[(l^2-m^2)(l^2-n^2)]^{\frac{1}{2}}}{l(2l+1)} P^{l-1}_{mn} + \frac{mn}{l(l+1)} P^l_{mn}$$
$$+ \frac{[(l+1)^2-m^2]^{\frac{1}{2}}[(l+1)^2-n^2]^{\frac{1}{2}}}{(l+1)(2l+1)} P^{l+1}_{mn};$$

$$\frac{1}{2}c^l_n P^l_{m,n+1} + \frac{1}{2}c^l_{-n} P^l_{m,n-1} = \frac{m-n\cos\beta}{\sin\beta} P^l_{mn};$$

$$-\frac{1}{2}c^l_m P^l_{m+1,n} - \frac{1}{2}c^l_{-m} P^l_{m-1,n} = \frac{n-m\cos\beta}{\sin\beta} P^l_{mn};$$

where $c^l_n = \sqrt{(l-n)(l+n+1)}$. Note that $c^l_{-n} = c^l_{n-1}$.

Furthermore, they can be related to functions of classical physics such as the Legendre polynomials,

$$P_l(x) = P^l_{00}(x)$$

and the associated Legendre polynomials

$$P^n_l(x) = C_n\left[\frac{(l+n)!}{(l-n)!}\right]^{\frac{1}{2}} P^l_{-n,0}(x)$$

where $C_n = 1$ for $n > 0$ and $C_n = (-1)^n$ for $n < 0$. It follows from the fact that they are solutions of a Sturm-Liouville problem that the following orthogonality relations hold

$$\int_0^\pi P^l_{mn}(\cos\beta) P^s_{mn}(\cos\beta)\sin\beta\,d\beta = \frac{2}{2l+1}\delta_{ls}. \tag{9.25}$$

These and other properties of the functions $P^l_{mn}(x)$ and relationships with other special functions are described in great detail in the three-volume set by Vilenkin and Klimyk [32].

It is easy to see that given the orthogonality properties of the functions $P^l_{mn}(\cos\beta)$ that the matrix elements U^l_{mn} form an orthogonal set of functions on $SU(2)$

$$\int_{SU(2)} U^l_{mn}(g)\overline{U^s_{pq}(g)}d(g) = \frac{1}{2l+1}\delta_{ls}\delta_{mp}\delta_{nq}.$$

---

[1]Our notation is consistent with Vilenkin and Klimyk [32]. The functions that Gel'fand, Minlos, and Shapiro [7] call P^l_{mn} differ from ours by a factor of i^{m-n}, which results in slightly different looking recurrence relations. Varshalovich, Moskalev, and Khersonskii [31] use ZYZ Euler angles. Their relations are different from ours by a factor of $(-1)^{m-n}$ for reasons which are explained in Section 9.4.

· Hence, we expand functions on $SU(2)$ in a Fourier series as (see, for example, [12, 13, 14])

$$f(g) = \sum_{l=0,\frac{1}{2},1,\frac{3}{2},\ldots} (2l+1) \sum_{m=-l}^{l} \sum_{n=-l}^{l} \hat{f}_{mn}^{l} U_{nm}^{l}(g) \tag{9.26}$$

where

$$\hat{f}_{mn}^{l} = \int_{SU(2)} f(g) U_{mn}^{l}(g^{-1}) d(g). \tag{9.27}$$

The homomorphism property $U^{l}(g_1 \circ g_2) = U^{l}(g_1)U^{l}(g_2)$ holds because of the derivation in Section 9.1.1, and $\{U^{l}\}$ for $l = 0, 1/2, 1, 3/2, \ldots$ is a complete set of irreducible unitary representations for $SU(2)$. Taking these facts for granted, the Plancherel equality and convolution theorem hold as special cases of the general theory.

For $SO(3)$ the formulas look almost exactly the same. Only now

$$\hat{f}_{mn}^{l} = \int_{SO(3)} f(A) U_{mn}^{l}(A^{-1}) dA, \tag{9.28}$$

and

$$f(A) = \sum_{l=0}^{\infty} (2l+1) \sum_{m=-l}^{l} \sum_{n=-l}^{l} \hat{f}_{mn}^{l} U_{nm}^{l}(A), \tag{9.29}$$

which results from the completeness relation

$$\sum_{l=0}^{\infty} (2l+1) \sum_{m=-l}^{l} \sum_{n=-l}^{l} U_{mn}^{l}\left(R^{-1}\right) U_{nm}^{l}(A) = \delta\left(R^{-1}A\right). \tag{9.30}$$

Explicitly, the first few matrices $P^{l}(\cos\beta)$ with elements $P_{mn}^{l}(\cos\beta)$ are

$$P^0 = P_{00}^0 = 1;$$

$$P^{\frac{1}{2}} = \begin{pmatrix} P_{-\frac{1}{2},-\frac{1}{2}}^{\frac{1}{2}} & P_{-\frac{1}{2},\frac{1}{2}}^{\frac{1}{2}} \\ P_{\frac{1}{2},-\frac{1}{2}}^{\frac{1}{2}} & P_{\frac{1}{2},\frac{1}{2}}^{\frac{1}{2}} \end{pmatrix} = \begin{pmatrix} \cos\frac{\beta}{2} & -\sin\frac{\beta}{2} \\ \sin\frac{\beta}{2} & \cos\frac{\beta}{2} \end{pmatrix};$$

$$P^1 = \begin{pmatrix} P_{-1,-1}^1 & P_{-1,0}^1 & P_{-1,1}^1 \\ P_{0,-1}^1 & P_{0,0}^1 & P_{0,1}^1 \\ P_{1,-1}^1 & P_{1,0}^1 & P_{1,1}^1 \end{pmatrix} = \begin{pmatrix} \frac{1+\cos\beta}{2} & -\frac{\sin\beta}{\sqrt{2}} & \frac{1-\cos\beta}{2} \\ \frac{\sin\beta}{\sqrt{2}} & \cos\beta & -\frac{\sin\beta}{\sqrt{2}} \\ \frac{1-\cos\beta}{2} & \frac{\sin\beta}{\sqrt{2}} & \frac{1+\cos\beta}{2} \end{pmatrix}.$$

$SU(2)$ representation matrices U^{l} are easily constructed as the matrix product of these matrices with diagonal ones which depend on α and γ.

Many more of the functions $P_{mn}^{l}(\cos\beta)$ can be found in [31], and all of them can be found up to any finite value of l by using the Rodrigues formula or recursion relations given earlier in this section. For instance, it is easy to verify that

$$P^2 = \begin{pmatrix} P_{-2,-2}^2 & P_{-2,-1}^2 & P_{-2,0}^2 & P_{-2,1}^2 & P_{-2,2}^2 \\ P_{-1,-2}^2 & P_{-1,-1}^2 & P_{-1,0}^2 & P_{-1,1}^2 & P_{-1,2}^2 \\ P_{0,-2}^2 & P_{0,-1}^2 & P_{0,0}^2 & P_{0,1}^2 & P_{0,2}^2 \\ P_{1,-2}^2 & P_{1,-1}^2 & P_{1,0}^2 & P_{1,1}^2 & P_{1,2}^2 \\ P_{2,-2}^2 & P_{2,-1}^2 & P_{2,0}^2 & P_{2,1}^2 & P_{2,2}^2 \end{pmatrix} = \begin{pmatrix} a & -b & c & -g & h \\ b & d & -e & k & -g \\ c & e & f & -e & c \\ g & k & e & d & -b \\ h & g & c & b & a \end{pmatrix}$$

where

$$a = \frac{(1 + \cos \beta)^2}{4}$$

$$b = \frac{\sin \beta (1 + \cos \beta)}{2}$$

$$c = \frac{1}{2} \sqrt{\frac{3}{2}} \sin^2 \beta$$

$$d = \frac{2 \cos^2 \beta + \cos \beta - 1}{2}$$

$$e = \sqrt{\frac{3}{2}} \sin \beta \cos \beta$$

$$f = \frac{3 \cos^2 \beta - 1}{2}$$

$$g = \frac{(1 - \cos \beta) \sin \beta}{2}$$

$$h = \frac{(1 - \cos \beta)^2}{4}$$

$$k = \frac{-2 \cos^2 \beta + \cos \beta + 1}{2}.$$

9.4 $SO(3)$ Matrix Representations in Various Parameterizations

In this section we consider the representations of $SO(3)$ when using several different parameterizations.

9.4.1   Matrix Elements Parameterized with Euler Angles

The Fourier series for $SO(3)$ using ZXZ Euler angles is almost identical to that for $SU(2)$, with the following two exceptions: (1) only integer values of l appear in the Fourier series, and therefore only matrix representations with odd dimensions such as U^0, U^1, U^2, etc., need to be calculated; and (2) the integration in the Fourier transform is over half the range, with invariant integration measure $\frac{1}{8\pi^2} \sin \beta d\beta d\alpha d\gamma$. For $SO(3)$ we have $0 \leq \gamma < 2\pi$, whereas for $SU(2)$ we have $-2\pi \leq \gamma < 2\pi$.

In the literature, a range of equivalent (though frustratingly varied) formulae for matrix elements of the IURs of $SO(3)$ are provided. For instance, it is often convenient to consider the representations of $SO(3)$ parameterized in ways other than ZXZ Euler angles. When ZYZ Euler angles are used,

$$R_3(\alpha) R_2(\beta) R_3(\gamma) = R_3(\alpha) \left(R_3(\pi/2) R_1(\beta) R_3(-\pi/2) \right) R_3(\gamma)$$
$$= R_3(\alpha + \pi/2) R_1(\beta) R_3(\gamma - \pi/2),$$

and so

$$R_{ZYZ}(\alpha, \beta, \gamma) = R_{ZXZ}(\alpha + \pi/2, \beta, \gamma - \pi/2).$$

Hence, evaluating the matrix elements Eq. (9.20), we find

$$U_{mn}^l \left(R_{ZYZ}(\alpha, \beta, \gamma) \right) = e^{i(n-m)\pi/2} U_{mn}^l \left(R_{ZXZ}(\alpha, \beta, \gamma) \right) = e^{-im\alpha} P_{mn}^l (\cos \beta) e^{-in\gamma} \tag{9.31}$$

since the factor $e^{i(n-m)\pi/2} = i^{n-m}$ cancels with the i^{m-n} in Eq. (9.20). In the physics literature one often finds the similar expression

$$D_{mn}^l (R_{ZYZ}(\alpha, \beta, \gamma)) = e^{-im\alpha} d_{mn}^l (\cos \beta) e^{-in\gamma},$$

which are called the *Wigner D-functions* [24, 31],[2] and $d_{mn}^l(\cos \beta) = (-1)^{m-n} P_{mn}^l (\cos \beta)$.

Since $U^l(R_{ZYZ}(\alpha, \beta, \gamma))$ and $U^l(R_{ZXZ}(\alpha, \beta, \gamma))$ are equivalent under similarity transformation by diagonal matrices with elements of the form $v_{mn} = i^m \delta_{m,n}$ (no sum over repeated indices) on their diagonals (which are unitary matrices), the distinction between representations parameterized using ZXZ and ZYZ is really unimportant, since a change of basis in the definition of the representations has the same effect as the change of parameterization. Likewise, two similarity transformations of this form relate $U^l(R_{ZYZ})$ to $D^l(R_{ZYZ})$. Hence, independent of whether ZXZ or ZYZ Euler angles are used, or a factor of 1, i^{m-n}, i^{n-m}, $(-1)^{m-n}$ appears in the definition of matrix elements, they are all unitarily equivalent (i.e., equal under similarity transformation by a unitary matrix). Finally, another variant is that some texts use $e^{im\alpha}$ and $e^{in\gamma}$ in the definition of matrix elements instead of $e^{-im\alpha}$ and $e^{-in\gamma}$. This too is acceptable, since if U^l is a unitary representation then so is $\overline{U^l}$. While any of these choices is valid, for the sake of concreteness, we will stick with the ZXZ convention when using Euler angles and the factor i^{m-n} unless otherwise stated. Other choices will have slightly different-looking recurrence relations than those given in Section 9.3.

When ZYZ Euler angles are used, the $SO(3)$ matrix elements are related to the spherical harmonics as

$$U_{m0}^l (R_{ZYZ}(\alpha, \beta, \gamma)) = e^{-im\alpha} P_{m0}^l (\cos \beta) = (-1)^m \sqrt{\frac{4\pi}{2l+1}} \overline{Y_l^m}(\beta, \alpha)$$

and

$$U_{0n}^l (R_{ZYZ}(\alpha, \beta, \gamma)) = P_{0n}^l (\cos \beta) e^{-in\gamma} = \sqrt{\frac{4\pi}{2l+1}} \overline{Y_l^n}(\beta, \gamma).$$

It follows from the homomorphism property that

$$U_{0n}^l (R_1 R_2) = \sum_{m=-l}^{l} U_{0m}^l (R_1) U_{mn}^l (R_2).$$

This means

$$\overline{Y_l^n} \left(\theta', \phi' \right) = \sum_{m=-l}^{l} \overline{Y_l^m}(\theta, \phi) U_{mn}^l (\alpha, \beta, \gamma) \tag{9.32}$$

where θ' and ϕ' are the transformed spherical coordinates such that

$$\mathbf{u}' = \mathbf{u} \left(\theta', \phi' \right) = R_{ZYZ}(\alpha, \beta, \gamma) \mathbf{u}(\theta, \phi) = R\mathbf{u}.$$

Using the notation $Y_l^m(\mathbf{u}(\theta, \phi)) = Y_l^m(\theta, \phi)$ and conjugating both sides of Eq. (9.32), one writes

$$Y_l^n(R\mathbf{u}) = \sum_{m=-l}^{l} \overline{U_{mn}^l(R)} Y_l^m(\mathbf{u}).$$

Substitution of R^T for R and using the unitarity of the representations U^l gives

$$Y_l^n(R^T \mathbf{u}) = \sum_{m=-l}^{l} U_{nm}^l(R) Y_l^m(\mathbf{u}). \tag{9.33}$$

---

[2]These are named after Eugene Wigner (1902–1995), a pioneer in the application of group theory in quantum mechanics. He shared the 1963 Nobel prize for physics.

9.4.2   Axis-Angle and Cayley Parameterizations

We now review the form which the matrix elements U_{mn}^l take when rotations are parameterized as $R(\theta, \lambda, \nu) = \exp(\theta N(\lambda, \nu))$ where $N\mathbf{x} = \mathbf{n} \times \mathbf{x}$, and $\mathbf{n}$ is a unit vector defining the axis of rotation. The ZXZ Euler angles are related to the angle of rotation, θ, and polar and azimuthal angles (λ, ν) of $\mathbf{n}$ as in Eq. (5.4.4). Making appropriate substitutions and using trigonometric rules, one finds [31]

$$U_{mn}^l(R(\theta, \lambda, \nu)) = i^{m-n} e^{-i(m-n)\nu} \left(\frac{1 - i \tan \theta/2 \cos \lambda}{\sqrt{1 + \tan^2 \theta/2 \cos^2 \lambda}} \right)^{m+n} P_{mn}^l(x)$$

where x satisfies

$$\sin x/2 = \sin \theta/2 \sin \lambda.$$

In particular,

$$U^1 = \begin{pmatrix} U_{1,1}^1 & U_{1,0}^1 & U_{1,-1}^1 \\ U_{0,1}^1 & U_{0,0}^1 & U_{0,-1}^1 \\ U_{-1,1}^1 & U_{-1,0}^1 & U_{-1,-1}^1 \end{pmatrix} = \begin{pmatrix} a^2 & ab & -c^2 \\ -a\overline{b} & 1 - 2c\overline{c} & \overline{a}b \\ -(\overline{c})^2 & -\overline{a}b & (\overline{a})^2 \end{pmatrix}$$

where

$$\begin{aligned} a &= \cos(\theta/2) - i \sin(\theta/2) \cos \lambda \\ b &= -i\sqrt{2} \sin(\theta/2) \sin \lambda\, e^{-i\nu} \\ c &= \sin(\theta/2) \sin \lambda\, e^{-i\nu}. \end{aligned}$$

In general,

$$\chi_l(\theta) = \text{trace}\left(U^l(R(\theta, \lambda, \nu)) \right) = \frac{\sin(l + \frac{1}{2})\theta}{\sin \theta/2},$$

and it is easy to verify from the preceding matrix U^1 that

$$\chi_1(\theta) = 1 + 2\cos\theta = \frac{\sin 3\theta/2}{\sin \theta/2}.$$

The differential operators considered in the case of the Euler angles can be written explicitly in this parameterization as well. For instance, the Laplacian is written as

$$\nabla^2 = \frac{\partial^2}{\partial \theta^2} + \cot \theta/2 \frac{\partial}{\partial \theta} + \frac{1}{4\sin^2 \theta/2} \left(\frac{\partial^2}{\partial \lambda^2} + \cos \lambda \frac{\partial}{\partial \lambda} + \frac{1}{\sin^2 \nu} \frac{\partial^2}{\partial \nu^2} \right).$$

Regardless of the parameterization, the effect on the matrix elements U_{mn}^l is the same.

Moses [15, 16, 17] derived the matrix elements for the parameterization

$$R(\theta_1, \theta_2, \theta_3) = \exp \begin{pmatrix} 0 & -\theta_3 & \theta_2 \\ \theta_3 & 0 & -\theta_1 \\ -\theta_2 & \theta_1 & 0 \end{pmatrix}.$$

These matrix elements are of the form

$$\begin{aligned} U_{mn}^l(R(\theta_1, \theta_2, \theta_3)) &= (-1)^{2l+m+n} \left[\frac{(l-m)!}{(l+m)!(l-n)!(l+n)!} \right]^{\frac{1}{2}} \\ &\times (\sin \theta/2)^{m-n} \left(\frac{-\theta_1 + i\theta_2}{\theta} \right)^{m-n} \\ &\times \left(\cos \theta/2 - i\frac{\theta_3}{\theta} \sin \theta/2 \right)^{m+n} P_{l-m}^{(m-n, m+n)}\left((1 - \theta_3^2/\theta^2)\cos\theta + \theta_3^2/\theta^2 \right) \end{aligned}$$

where $\theta = \sqrt{\theta_1^2 + \theta_2^2 + \theta_3^2}$ and $P_n^{(\alpha,\beta)}(x)$ are the Jacobi polynomials.

In particular,

$$U^1 = \begin{pmatrix} U_{1,1}^1 & U_{1,0}^1 & U_{1,-1}^1 \\ U_{0,1}^1 & U_{0,0}^1 & U_{0,-1}^1 \\ U_{-1,1}^1 & U_{-1,0}^1 & U_{-1,-1}^1 \end{pmatrix} = \begin{pmatrix} c^2 & -\sqrt{2}a\bar{b}c & a^2(\bar{b})^2 \\ \sqrt{2}abc & -a^2|b|^2 + |c|^2 & -\sqrt{2}a\bar{b}c \\ a^2b^2 & \sqrt{2}ab\bar{c} & (\bar{c})^2 \end{pmatrix}$$

where now

$$a = \sin\theta/2; \qquad b = \frac{-\theta_1 - i\theta_2}{\theta}; \qquad c = \cos\theta/2 - i\frac{\theta_3}{\theta}\sin\theta/2.$$

9.5   Sampling and FFT for $SO(3)$ and $SU(2)$

Recently, sampling and fast Fourier transform techniques for the rotation group have been developed [10, 11]. Essentially, the double coset decomposition of $SO(3)$ [or $SU(2)$] corresponding to ZXZ Euler angles yields matrix elements of the IUR matrices which lend themselves to fast transforms in each coordinate (Euler angle).

Using the notational simplification

$$f(g(\alpha, \beta, \gamma)) = f(\alpha, \beta, \gamma),$$

we rewrite Eqs. (9.26) and (9.27) in band-limited form explicitly as

$$f(\alpha, \beta, \gamma) = \sum_{l=0,\frac{1}{2},1,\frac{3}{2},\dots<B} (2l+1) \sum_{m=-l}^{l} \sum_{n=-l}^{l} \hat{f}_{nm}^l i^{m-n} e^{-i(m\alpha+n\gamma)} P_{mn}^l(\cos\beta) \qquad (9.34)$$

where

$$\hat{f}_{mn}^l = \frac{1}{16\pi^2} \int_{\beta=0}^{\pi} \int_{\gamma=-2\pi}^{2\pi} \int_{\alpha=0}^{2\pi} f(\alpha, \beta, \gamma) i^{m-n} e^{i(n\alpha+m\gamma)} P_{nm}^l(\cos\beta) \sin\beta \, d\alpha \, d\beta \, d\gamma. \qquad (9.35)$$

We have written this for the $SU(2)$ case, and the $SO(3)$ case follows easily.

The whole spectrum [collection of Fourier transform matrix elements in Eq. (9.34)] can be calculated fast in principle by using the classical FFT over α and γ and fast transforms over β. By using a quadrature rule, Eq. (9.34) can be sampled in each coordinate at $\mathcal{O}(B)$ values to exactly compute the integral. The whole of $SU(2)$ is then sampled at $N = \mathcal{O}(B^3)$ points. Explicitly, we first calculate

$$\tilde{f}_n(\beta, \gamma) = \int_0^{2\pi} f(\alpha, \beta, \gamma) e^{in\alpha} d\alpha$$

for all $n \in \{-B, \dots, B\}$ and all sample values of β and γ. This requires $\mathcal{O}(B^2 \cdot B \log B)$ operations. Then we calculate

$$\tilde{\tilde{f}}_{mn}(\beta) = \int_{-2\pi}^{2\pi} \tilde{f}_n(\beta, \gamma) e^{im\gamma} d\gamma$$

in $\mathcal{O}(B^2 \cdot B \log B)$ operations [$\mathcal{O}(B \log B)$ for each value of n and β]. Finally,

$$\hat{f}_{mn}^l = \frac{1}{16\pi^2} i^{m-n} \int_0^{\pi} \tilde{\tilde{f}}_{mn}(\beta) P_{nm}^l(\cos\beta) \sin\beta \, d\beta$$

is calculated in $\mathcal{O}(B^2 \cdot B(\log B)^2)$ operations $[\mathcal{O}(B(\log B)^2)$ for each value of m and $n]$.

Since the limiting calculation is the $\mathcal{O}(B(\log B)^2)$ required for the fast transform in the variable β as described in [11], the whole procedure is $\mathcal{O}(N(\log N)^2)$ for all values of m, n, l up to the band-limit. For an alternative (but not as fast) numerical approach see [23]. If the expansion in β is done directly [i.e., using $\mathcal{O}(B^2)$ operations instead of using an $\mathcal{O}(B(\log B)^2)$ fast polynomial transform], then the $SU(2)$ Fourier transforms for all m, m, l up to the band-limit can be performed in $\mathcal{O}(B^4) = \mathcal{O}(N^{4/3})$ arithmetic operations when the FFT or DFT is used in α and γ.

The cost of reconstructing a function on $SU(2)$ from its spectrum is on the same order as computing the whole spectrum.

Since convolution of functions on $SU(2)$ with band-limit B requires the multiplication of matrices of dimensions $(2l + 1) \times (2l + 1)$ for $l = 0, \ldots, B$, the cost of convolution will be

$$\mathcal{O}\left(\sum_{l=0}^{B}(2l + 1)^\gamma\right) = \mathcal{O}(B^{\gamma+1}).$$

Hence, when Gaussian elimination is used ($\gamma = 3$), it is clear that the order of computation of the Fourier transforms will be no greater than the cost of convolution [even when the $\mathcal{O}(B^4)$ version is used].

The only difference between the computation for $SU(2)$ and $SO(3)$ is the range of integration, values of l used, and the normalization of the volume element.

One can imagine FFTs for $SU(2)$ and $SO(3)$ in parameterizations other than Euler angles. To our knowledge, this is an open problem.

9.6   Wavelets on the Sphere and Rotation Group

9.6.1   Inversion Formulas for Wavelets on Spheres

In this section we examine an inversion formula for one of the spherical wavelet transforms given in Chapter 4. This inversion formula uses properties of the rotation group such as inversion of arguments in integrands under shifts and the completeness of the IUR matrix elements of $SO(3)$. We then define a diffusion-based wavelet transform for the sphere and derive its inversion formula.

Inverting Gabor Wavelet Transforms for the Sphere

The concepts of modulation and translation are generalized to the sphere as multiplication by a spherical harmonic and rotation, respectively.

Recall from Chapter 4 that the spherical Gabor transform is defined as Eq. (4.40)

$$\tilde{f}_l^m(R) = \int_{S^2} f(\mathbf{u})\overline{\psi_l^m(\mathbf{u}, R)}d\mathbf{u}.$$

Here the mother wavelet is the function $\psi \in \mathcal{L}^2(S^2)$, and $\psi_l^m(\mathbf{u}, R)$ is defined as

$$\psi_l^m(\mathbf{u}, R) = Y_l^m(\mathbf{u})\psi(R^{-1}\mathbf{u})$$

where $R \in SO(3)$.

Analogous to the case of the circle, we calculate

$$\sum_{l=0}^{\infty}\sum_{m=-l}^{l}\int_{SO(3)}\tilde{f}_l^m(R)\psi_l^m(\mathbf{u}, R)dR =$$

$$\int_{SO(3)} \int_{S^2} f(\mathbf{v}) \left(\sum_{l=0}^{\infty} \sum_{m=-l}^{l} Y_l^m(\mathbf{u})\overline{Y_l^m(\mathbf{v})} \right) \psi(R^{-1}\mathbf{u})\overline{\psi(R^{-1}\mathbf{v})}d\mathbf{v}dR.$$

The term in parenthesis previously is rewritten as

$$\sum_{l=0}^{\infty} \sum_{m=-l}^{l} Y_l^m(\mathbf{u})\overline{Y_l^m(\mathbf{v})} = \delta_{S^2}(\mathbf{u}, \mathbf{v})$$

where $\delta_{S^2}(\cdot, \cdot)$ is the Dirac delta function for the sphere. This means that

$$\sum_{l=0}^{\infty} \sum_{m=-l}^{l} \int_{SO(3)} \tilde{f}_l^m(R)\psi_l^m(\mathbf{u}, R)dR = f(\mathbf{u}) \int_{SO(3)} \overline{\psi(A^{-1}\mathbf{u})}\psi(A^{-1}\mathbf{u})dA.$$

All we have done is use the properties of the Dirac delta and changed the name of the variable of integration for the $SO(3)$ integral. The above integral over $SO(3)$ can be simplified by observing that if we write $\mathbf{u} = R(\mathbf{e}_3, \mathbf{u})\mathbf{e}_3$ and use the invariance of integration on $SO(3)$ under shifts,

$$\int_{SO(3)} \overline{\psi(A^{-1}\mathbf{u})}\psi(A^{-1}\mathbf{u})dA$$

$$= \int_{SO(3)} \overline{\psi\left(\left(R(\mathbf{e}_3, \mathbf{u})^{-1} A \right)^{-1} \mathbf{e}_3 \right)}\psi\left(\left(R(\mathbf{e}_3, \mathbf{u})^{-1} A \right)^{-1} \mathbf{e}_3 \right) dA$$

$$= \int_{SO(3)} \overline{\psi(A^{-1}\mathbf{e}_3)}\psi\left(A^{-1}\mathbf{e}_3 \right) dA.$$

But $\psi(A^{-1}\mathbf{e}_3)$ is a function of only two of the Euler angles, and constant with respect to the third. This is because $\text{ROT}[\mathbf{e}_3, \theta]\mathbf{e}_3 = \mathbf{e}_3$. Therefore

$$\int_{SO(3)} \overline{\psi(A^{-1}\mathbf{u})}\psi\left(A^{-1}\mathbf{u} \right) dA = \int_{S^2} \overline{\psi(\mathbf{u}')}\psi(\mathbf{u}') d\mathbf{u}' = \|\psi\|_2^2.$$

Hence we write the inversion formula

$$f(\mathbf{u}) = \frac{1}{\|\psi\|_2^2} \sum_{l=0}^{\infty} \sum_{m=-l}^{l} \int_{SO(3)} \tilde{f}_l^m(R)\psi_l^m(\mathbf{u}, R)dR. \qquad (9.36)$$

Diffusion-Based Wavelets for the Sphere

Let $h(\mathbf{u}, t)$ be the solution to the heat equation on the sphere

$$\frac{\partial h}{\partial t} = \nabla_{\mathbf{u}}^2 h$$

(where $\nabla_{\mathbf{u}}^2$ is the Laplacian for the sphere) subject to the initial conditions $h(\mathbf{u}, 0) = \delta(\mathbf{u}, \mathbf{e}_3)$.

We define a diffusion-based wavelet transform of a real-valued function $f \in \mathcal{L}^2(S^2)$ as

$$\tilde{f}(R, t) = \int_{S^2} f(\mathbf{u})\overline{h(R^{-1}\mathbf{u}, t)}d\mathbf{u}$$

where $R \in SO(3)$ and $t \in \mathbb{R}^+$. In the preceding equation we note that

$$\overline{h(R^{-1}\mathbf{u}, t)} = h\left(R^{-1}\mathbf{u}, t \right)$$

since $h(\mathbf{u}, t)$ is a real-valued function. We use the complex conjugate notation to be consistent with transforms presented earlier.

In order to invert, we write explicitly

$$h(\mathbf{u}, t) = \sum_{l=0}^{\infty} e^{-l(l+1)t} Y_l^0(\mathbf{u})$$

and using Eq. (9.33) we have

$$h\left(R^{-1}\mathbf{u}, t\right) = \sum_{l=0}^{\infty} e^{-l(l+1)t} \sum_{m=-l}^{l} U_{0m}^l(R) Y_l^m(\mathbf{u}).$$

We seek an inversion formula of the form

$$f(\mathbf{u}) = \int_0^{\infty} w(t) \int_{SO(3)} \tilde{f}(R, t) h\left(R^{-1}\mathbf{u}, t\right) dR\, dt$$

where $w(t)$ is a function to be determined.

First performing the $SO(3)$ integral, we see that

$$\int_{SO(3)} \tilde{f}(R, t) h\left(R^{-1}\mathbf{u}, t\right) dR = \int_{S^2} f(\mathbf{v}) \int_{SO(3)} \overline{h\left(R^{-1}\mathbf{v}, t\right)} h\left(R^{-1}\mathbf{u}, t\right) dR\, d\mathbf{v} =$$

$$\int_{S^2} f(\mathbf{v}) \int_{SO(3)} \sum_{l=0}^{\infty} e^{-l(l+1)t} \sum_{m=-l}^{l} U_{0m}^l(R) Y_l^m(\mathbf{u}) \sum_{l'=0}^{\infty} e^{-l'(l'+1)t} \sum_{m'=-l'}^{l'} \overline{U_{0m'}^{l'}(R) Y_{l'}^{m'}(\mathbf{v})}\, dR\, d\mathbf{v} =$$

$$\int_{S^2} f(\mathbf{v}) \sum_{l=0}^{\infty} \sum_{l'=0}^{\infty} e^{-l(l+1)t} e^{-l'(l'+1)t} \sum_{m=-l}^{l} \sum_{m'=-l'}^{l'} Y_l^m(\mathbf{u}) \overline{Y_{l'}^{m'}(\mathbf{v})} \left(\int_{SO(3)} U_{0m}^l(R) \overline{U_{0m'}^{l'}(R)} dR \right) d\mathbf{v}.$$

The term in parenthesis simplifies using the orthogonality of IUR matrix elements as

$$\int_{SO(3)} U_{0m}^l(R) \overline{U_{0m'}^{l'}(R)} dR = \frac{1}{2l+1} \delta_{ll'} \delta_{mm'}.$$

We therefore write

$$\int_{SO(3)} \tilde{f}(R, t) h\left(R^{-1}\mathbf{u}, t\right) dR =$$

$$\int_{S^2} f(\mathbf{v}) \sum_{l=0}^{\infty} \sum_{l'=0}^{\infty} e^{-l(l+1)t} e^{-l'(l'+1)t} \sum_{m=-l}^{l} \sum_{m'=-l'}^{l'} Y_l^m(\mathbf{u}) \overline{Y_{l'}^{m'}(\mathbf{v})} \frac{1}{2l+1} \delta_{ll'} \delta_{mm'} d\mathbf{v} =$$

$$\int_{S^2} f(\mathbf{v}) \sum_{l=0}^{\infty} \frac{e^{-2l(l+1)t}}{2l+1} \sum_{m=-l}^{l} Y_l^m(\mathbf{u}) \overline{Y_l^m(\mathbf{v})} d\mathbf{v}.$$

At this point we can consider the integral over t and write

$$\int_0^{\infty} w(t) \int_{SO(3)} \tilde{f}(R, t) h\left(R^{-1}\mathbf{u}, t\right) dR\, dt =$$

$$\int_{S^2} f(\mathbf{v}) \sum_{l=0}^{\infty} \left[\int_0^{\infty} \frac{e^{-2l(l+1)t}}{2l+1} w(t) dt \right] \sum_{m=-l}^{l} Y_l^m(\mathbf{u}) \overline{Y_l^m(\mathbf{v})} d\mathbf{v}.$$

From the completeness relation for spherical harmonics,

$$\sum_{l=0}^{\infty} \sum_{m=-l}^{l} Y_l^m(\mathbf{u})\overline{Y_l^m(\mathbf{v})} = \delta(\mathbf{u}, \mathbf{v}),$$

we see that $f(\mathbf{u})$ can be recovered if $w(t)$ can be found such that

$$\int_0^{\infty} \frac{e^{-2l(l+1)t}}{2l+1} w(t)dt = 1$$

for all $l \in \{0, 1, 2, \dots\}$. What remains to be seen is if a function $w(t)$ can be found that satisfies this condition.

9.6.2   Diffusion-Based Wavelets for $SO(3)$

Let $h(A, t)$ for $A \in SO(3)$ and $t \in \mathbb{R}^+$ be the solution to the heat equation on $SO(3)$

$$\frac{\partial h}{\partial t} = \nabla_A^2 h$$

subject to the initial conditions $h(A, 0) = \delta(A)$. In this subsection we seek the inversion formula for a wavelet transform of the form

$$\tilde{f}(R, t) = \int_{SO(3)} f(A)\overline{h\left(R^{-1}A, t\right)}dA.$$

The inversion formula we seek is of the form

$$f(A) = \int_0^{\infty} w(t) \int_{SO(3)} \tilde{f}(R, t)h\left(R^{-1}A, t\right) dR dt.$$

The explicit solution to the heat equation on $SO(3)$ as a Fourier series is of the form

$$h(A, t) = \sum_{l=0}^{\infty} (2l+1)e^{-l(l+1)t} \sum_{m=-l}^{l} U_{mm}^l(A).$$

The shifted version is

$$
\begin{aligned}
h\left(R^{-1}A, t\right) &= \sum_{l=0}^{\infty} (2l+1)e^{-l(l+1)t} \sum_{m=-l}^{l} U_{mm}^l\left(R^{-1}A\right) \\
&= \sum_{l=0}^{\infty} (2l+1)e^{-l(l+1)t} \sum_{m=-l}^{l} \sum_{n=-l}^{l} U_{mn}^l\left(R^{-1}\right) U_{nm}^l(A) \\
&= \sum_{l=0}^{\infty} (2l+1)e^{-l(l+1)t} \sum_{m=-l}^{l} \sum_{n=-l}^{l} \overline{U_{nm}^l(R)} U_{nm}^l(A).
\end{aligned}
$$

We seek an inversion formula by calculating

$$f(A) = \int_0^{\infty} w(t) \left[\int_{SO(3)} \int_{SO(3)} f(Q)\overline{h\left(R^{-1}Q, t\right)}h\left(R^{-1}A, t\right) dQ dR\right] dt.$$

We begin by substituting the explicit form of the shifted solutions to the heat equation in the preceding term in brackets as

$$\int_{SO(3)} \int_{SO(3)} f(Q) \overline{h\left(R^{-1}Q, t\right)} h\left(R^{-1}A, t\right) dQ dR =$$

$$\int_{Q \in SO(3)} f(Q) \left[\int_{R \in SO(3)} \left(\sum_{l=0}^{\infty} (2l+1) e^{-l(l+1)t} \sum_{m=-l}^{l} \sum_{n=-l}^{l} U_{nm}^l(R) \overline{U_{nm}^l(Q)} \right) \right.$$

$$\left. \left(\sum_{l'=0}^{\infty} (2l'+1) e^{-l'(l'+1)t} \sum_{m'=-l'}^{l'} \sum_{n'=-l'}^{l'} \overline{U_{n'm'}^{l'}(R)} U_{n'm'}^{l'}(A) \right) dR \right] dQ.$$

Using the orthogonality of the matrix elements U_{mn}^l, we find that this reduces to

$$\int_{Q \in SO(3)} f(Q) \left[\sum_{l=0}^{\infty} (2l+1) e^{-2l(l+1)t} \sum_{m=-l}^{l} \sum_{n=-l}^{l} \overline{U_{nm}^l(Q)} U_{nm}^l(A) \right] dQ. \qquad (9.37)$$

But

$$\hat{f}_{mn}^l = \int_{SO(3)} f(Q) \overline{U_{nm}^l(Q)} dQ$$

is just the $SO(3)$ Fourier transform of $f(A)$, and the inverse transform is also present in Eq. (9.37), which allows us to write

$$\int_0^{\infty} w(t) \int_{SO(3)} \int_{SO(3)} f(Q) \overline{h\left(R^{-1}Q, t\right)} h\left(R^{-1}A, t\right) dQ dR dt = f(A)$$

when

$$\int_0^{\infty} e^{-2l(l+1)t} w(t) dt = 1$$

for all values of l. However this condition clearly cannot hold, and so this concept of diffusion-based wavelets for $SO(3)$ must be augmented in some way in order to obtain an inversion formula.

What has been gained by this exercise (and the verifications of the inversion formulae in the previous sections) is a demonstration of how $SO(3)$ IURs can be used as an analytical tool.

9.7   Helicity Representations

In this section we construct the representations of the little group $H_{\hat{u}} \subset SO(3)$. Recall that $H_{\hat{u}}$ is defined as the group which leaves the point $\hat{u} \in S^2$ fixed. Representations of $H_{\hat{u}}$ are important in the constructions of Chapter 10.

To calculate the representations of $H_{\hat{u}}$ explicitly, we first choose a particular coset representative $\hat{u} = e_3 \in S^2 \cong SO(3)/SO(2)$. The vector $\hat{u}$ is invariant with respect to rotations from the $SO(2)$ subgroup of $SO(3)$, and for this particular choice of $\hat{u}$ we not only have $H_{\hat{u}} \cong SO(2)$, but rather $H_{\hat{u}} = SO(2)$. That is, the general statement

$$\text{ROT}[\mathbf{u}, \theta] \mathbf{u} = \mathbf{u} \qquad (9.38)$$

reduces to $R_3(\theta) \hat{u} = \hat{u}$ for this choice of coset representative.

For each $\mathbf{u} \in S^2$ we may define $R_\mathbf{u} \in SO(3)/SO(2)$ such that

$$R_\mathbf{u} \hat{\mathbf{u}} = \mathbf{u}.$$

Explicitly, this rotation matrix is the one which has an axis pointing in the direction defined by $\hat{\mathbf{u}} \times \mathbf{u}$ and has a rotation angle whose sin is $\|\hat{\mathbf{u}} \times \mathbf{u}\|$. Hence, using the notation of Chapter 5 (Section 2.3),

$$R_\mathbf{u} = R(\hat{\mathbf{u}}, \mathbf{u}) = e^{\text{matr}[\hat{\mathbf{u}} \times \mathbf{u}]}$$

where matr[$\mathbf{c}$] is the skew-symmetric matrix such that $(\text{matr}[\mathbf{c}])\mathbf{x} = \mathbf{c} \times \mathbf{x}$.

For any $A \in SO(3)$ it follows from the definition of $R_\mathbf{u}$ that

$$R_{A^{-1}\mathbf{u}} \hat{\mathbf{u}} = A^{-1}\mathbf{u}.$$

Multiplying both sides by A, making the replacement $\mathbf{u} = R_\mathbf{u} \hat{\mathbf{u}}$ on the right side, and multiplying both sides by $R_\mathbf{u}^{-1}$ means

$$\left(R_\mathbf{u}^{-1} A R_{A^{-1}\mathbf{u}}\right) \hat{\mathbf{u}} = \hat{\mathbf{u}}.$$

Therefore, $Q(\mathbf{u}, A) \triangleq (R_\mathbf{u}^{-1} A R_{A^{-1}\mathbf{u}}) \in H_{\hat{\mathbf{u}}}$. It follows that

$$Q(\mathbf{u}, A) Q\left(A^{-1}\mathbf{u}, A^{-1}B\right) = Q(\mathbf{u}, B). \tag{9.39}$$

This expression has significance in quantum mechanics [34].

The representations of $H_{\hat{\mathbf{u}}}$ may be taken to be of the form

$$\Delta_s : \phi \rightarrow e^{is\phi} ; \ 0 \leq \phi \leq 2\pi;$$

and $s = 0, \pm1, \pm2, \ldots$. Here $\phi = \theta(Q(\mathbf{u}, A))$ is the angle of rotation of the matrix $Q(\mathbf{u}, A)$. The representations Δ_s form the usual Fourier basis for $S^1 \cong SO(2) \cong H_{\hat{\mathbf{u}}}$.

We now derive the form of $Q(\mathbf{u}, A)$ explicitly. At first sight this would appear to be a complicated function of $\mathbf{u}$ and A. We show that this is in fact not the case.

We begin by observing that

$$R_{A^{-1}\mathbf{u}} = R(\hat{\mathbf{u}}, A^{-1}\mathbf{u}) = e^{\text{matr}[\hat{\mathbf{u}} \times (A^{-1}\mathbf{u})]}.$$

Using the general cross-product rules in Chapter 5 (Section 2.2.2), one finds that

$$\hat{\mathbf{u}} \times \left(A^{-1}\mathbf{u}\right) = A^{-1}[(A\hat{\mathbf{u}}) \times \mathbf{u})]$$

and

$$\text{matr}\left[A^{-1}\{(A\hat{\mathbf{u}}) \times \mathbf{u}\}\right] = A^{-1}\text{matr}[(A\hat{\mathbf{u}}) \times \mathbf{u}]A.$$

Since conjugation commutes with the matrix exponential, it follows that

$$R_{A^{-1}\mathbf{u}} = A^{-1} R(A\hat{\mathbf{u}}, \mathbf{u})A = A^{-1} e^{\text{matr}[(A\hat{\mathbf{u}}) \times \mathbf{u}]}A.$$

Substitution of this into the definition of $Q(\mathbf{u}, A)$, and using the fact that

$$R_\mathbf{u}^{-1} = \exp(-\text{matr}[\hat{\mathbf{u}} \times \mathbf{u}]) = \exp(\text{matr}[\mathbf{u} \times \hat{\mathbf{u}}]),$$

one finds

$$Q(\mathbf{u}, A) = e^{\text{matr}[\mathbf{u} \times \hat{\mathbf{u}}]} e^{\text{matr}[(A\hat{\mathbf{u}}) \times \mathbf{u}]} A. \tag{9.40}$$

This is a considerable simplification.

9.8   Induced Representations

One can seek representations of $SO(3)$ by using the method of induced representations together with the knowledge of $SO(2)$ representations. The method of induction yields representations of the form

$$\left(\mathcal{U}^s(A)\varphi\right)(\mathbf{u}) = \Delta_s\left(R_{\mathbf{u}}^{-1} A R_{A^{-1}\mathbf{u}}\right)\varphi\left(A^{-1}\mathbf{u}\right), \qquad (9.41)$$

where $A \in SO(3)$, Δ_s are the helicity representations of $H_{\hat{\mathbf{u}}} \cong SO(2)$ and $s = 0, \pm 1, \pm 2, \ldots$ discussed in Section 9.7.

It may be shown that the $\Delta_s(Q(\mathbf{u}, A))$ factor [in formula Eq. (9.41)] may be expressed as

$$\Delta_s(Q(\hat{\mathbf{u}}, A)) = e^{is(\alpha+\gamma)}, \qquad (9.42)$$

where α and γ are Euler angles of rotation around the z-axis, for the vector $\hat{\mathbf{u}} = (0, 0, 1)$ and arbitrary rotation $A \in SO(3)$ (see [29]).

Using this fact, the following basis functions for S^2 may be defined:

$$h_{ms}^l(\beta, \alpha) = \Delta_s(Q(\hat{\mathbf{u}}, P))(-1)^{(l-s)} \sqrt{\frac{2l+1}{4\pi}} \, \overline{\tilde{U}_{m,-s}^l(P)} \qquad (9.43)$$

where $\tilde{U}_{nm}^l$ are defined relative to Eq. (9.20) as

$$\tilde{U}_{mn}^l = i^{m-n} U_{mn}^l.$$

This then differs by a factor of $(-1)^{n-m}$ from the Wigner functions (which are defined in ZYZ Euler angles). Again α, β, γ are ZXZ Euler angles of $P \in SO(3)$. We note that Eq. (9.43) does not depend on γ.

For $s = 0$, h_{ms}^l is the same as Y_l^m to within a constant factor. From the orthogonalities of the IURs of $SO(3)$ we observe for fixed s that

$$\left(h_{m's}^{l'}, h_{ms}^l\right) = \delta_{mm'}\delta_{ll'}$$

where $(\cdot, \cdot)$ is the usual inner product for $L^2(S^2)$.

Under the rotation A, these functions are transformed as

$$\left(\mathcal{U}^s(A) h_{ms}^l\right)(\mathbf{u}) = \Delta_s(Q(\mathbf{u}, A)) h_{ms}^l\left(A^{-1}\mathbf{u}\right) \qquad (9.44)$$

where $\mathbf{u} = P\,\hat{\mathbf{u}}$ is found by rotation P (for arbitrary γ) of $\hat{\mathbf{u}}$. This transformation law may be written as (see [29])

$$\left(\mathcal{U}^s(A) h_{ms}^l\right)(P\,\hat{\mathbf{u}}) = \Delta_s(Q(P\hat{\mathbf{u}}, A))\,(-1)^{(l-s)} \sqrt{\frac{2l+1}{4\pi}}\,\cdot$$

$$\Delta_s\left(Q\left(\hat{\mathbf{u}}, A^{-1}P\right)\right)\overline{\tilde{U}_{m,-s}^l\left(A^{-1}P\right)}. \qquad (9.45)$$

Using the multiplication law Eq. (9.39) and the property

$$\Delta_s(Q(\hat{\mathbf{u}}, A))^{-1} = \Delta_s\left(Q\left(\hat{\mathbf{u}}, A^{-1}\right)\right)$$

we may see that

$$\Delta_s(Q(P\hat{u}, A))\Delta_s\left(Q\left(\hat{u}, A^{-1}P\right)\right) = \Delta_s(Q(\hat{u}, P))\Delta_s\left(Q\left(\hat{u}, P^{-1}A\right)\right)$$

$$\Delta_s\left(Q\left(\hat{u}, A^{-1}P\right)\right) = \Delta_s(Q(\hat{u}, P)). \tag{9.46}$$

The group property, the unitarity, and Eq. (9.46) allow us to write a transformation law as

$$\left(\mathcal{U}^s(A)h^l_{ms}\right)(\mathbf{u}) = \Delta_s(Q(\hat{u}, P))(-1)^{(l-s)}\sqrt{\frac{2l+1}{4\pi}}\sum_{k=-l}^{l}\overline{\tilde{U}^l_{mk}(A^{-1})}\,\overline{\tilde{U}^l_{k,-s}(P)}$$

$$= \sum_{k=-l}^{l}\tilde{U}^l_{km}(A)\Delta_s(Q(\hat{u}, P))(-1)^{(l-s)}\sqrt{\frac{2l+1}{4\pi}}\,\overline{\tilde{U}^l_{k,-s}(P)}$$

$$= \sum_{k=-l}^{l}\tilde{U}^l_{km}(A)h^l_{ks}(\mathbf{u}). \tag{9.47}$$

The matrix elements of this representation are

$$\left(h^{l'}_{m's}, \mathcal{U}^s(A)h^l_{ms}\right) = \tilde{U}^l_{m'm}\delta_{ll'}.$$

9.9 The Clebsch-Gordan Coefficients and Wigner 3jm Symbols

In the case of the group $SO(2)$, the representations are all one dimensional, and they are of the form $u^l(\theta) = e^{il\theta}$. It is easy to see that the product of two $SO(2)$ representations is

$$u^{l_1}(\theta)u^{l_2}(\theta) = u^{l_1+l_2}(\theta) = \sum_{l=-\infty}^{\infty}\delta_{l,l_1+l_2}u^l(\theta).$$

Since the product of two matrix elements of IURs of a group is a function on the group, we can ask the more general question, "if the product of two matrix elements is expressed in a Fourier series on the group, what do the Fourier coefficients (or transforms) of this product look like?"

In fact, it can be shown [31][3] that in the case when $SU(2)$ and $SO(3)$ are parameterized using ZYZ Euler angles and the matrix elements in Eq. (9.20) are used, then

$$U^{l_1}_{m_1,n_1}(\alpha, \beta, \gamma)U^{l_2}_{m_2,n_2}(\alpha, \beta, \gamma) = \sum_{l=|l_1-l_2|}^{l_1+l_2}\sum_{m,n=-l}^{l}C(l_1, l_2; m_1, m_2; n_1, n_2; l, n, m)U^l_{m,n}(\alpha, \beta, \gamma). \tag{9.48}$$

It can also be shown that the decomposition

$$C(l_1, l_2; m_1, m_2; n_1, n_2; l, n, m) = C(l_1, m_1; l_2, m_2|l, m)\,C(l_1, n_1; l_2, n_2|l, n)$$

holds. The coefficients $C(l_1, m_1; l_2, m_2|l, m)$ are called the *Clebsch-Gordan coefficients* (CGCs).

---

[3]Our choice of matrix elements U^l_{mn} is different from these by a factor of $(-1)^{m-n}$.

It is easy to see from this definition that by taking $n_1 = m_1$, $n_2 = m_2$, and $n = m$, the magnitude of these coefficients is calculated as

$$|C(l_1, m_1; l_2, m_2|l, m)| = \sqrt{(2j+1)\left(\mathcal{F}\left(U^{l_1}_{m_1, m_1} U^{l_2}_{m_2, m_2}\right)\right)_{mm}}$$

where $\mathcal{F}(\cdot)_{mm}$ denotes the $(m, m)^{th}$ element of the $SO(3)$ Fourier transform. However, this only defines the CGCs to within a sign, which is fixed by the convention [31]

$$C(l_1, m_1; l_2, m_2|l, m)\, C(l_1, l_1; l_2, -l_2|l, l_1 - l_2) = (2j+1)\left(\mathcal{F}\left(U^{l_1}_{m_1, l_1} U^{l_2}_{m_2, -l_2}\right)\right)_{m, l_1 - l_2}$$

and

$$C(l_1, l_1; l_2, -l_2|l, l_1 - l_2) > 0.$$

The CGCs can be explicitly calculated using the integral

$$C(l_1, m_1; l_2, m_2 \mid l_3, m_3) = \frac{(-1)^{l_1 - l + m_2}}{2^{l_1 + l_2 + l_3 + 1}}$$

$$\times \left[\frac{(l_3 + m_3)!(l_1 + l_2 - l_3)!(l_1 + l_2 + l_3 + 1)!(2l_3 + 1)}{(l_1 - m_1)!(l_1 + m_1)!(l_2 - m_2)!(l_2 + m_2)!(l_3 - m_3)!(l_2 + l_3 - l_1)!(l_1 + l_3 - l_2)!}\right]^{\frac{1}{2}}$$

$$\times \int_{-1}^{1} (1-x)^{l_1 - m_1}(1+x)^{l_2 - m_2}\frac{d^{l_3 - m_3}}{dx^{l_3 - m_3}}\left[(1-x)^{l_2 + l_3 - l_1}(1+x)^{l_1 + l_3 - l_2}\right]dx. \qquad (9.49)$$

They can be generated recursively and posses a number of symmetries including

$$C(l_1, m_1; l_2, m_2 \mid l_3, m_3) = (-1)^{(m_2 - m_3)}(-1)^{(l_1 + l_2 + l_3)}\sqrt{\frac{2l_3 + 1}{2l_2 + 1}}$$

$$\times C(l_1, m_1; l_3, -m_3 \mid l_2, -m_2). \qquad (9.50)$$

The CGCs are related to the *Wigner 3jm symbols* often used in the physics literature as [31]

$$\begin{pmatrix} j_1 & j_2 & j \\ m_1 & m_2 & m \end{pmatrix} = \frac{(-1)^{m+j+2j_1}}{\sqrt{2j+1}} C(j_1, -m_1; j_2, -m_2|j, m) \qquad (9.51)$$

and

$$C(j_1, m_1; j_2, m_2|j, m) = (-1)^{m+j_1-j_2}\sqrt{2j+1}\begin{pmatrix} j_1 & j_2 & j \\ m_1 & m_2 & -m \end{pmatrix}. \qquad (9.52)$$

In physics, the Racah coefficients [20] and related coefficients called the Wigner 6jm and 9jm coefficients arise. A number of works have considered computational aspects of computing these coefficients using recurrence relations. See, for example, [1, 5, 9, 21, 22, 25, 26, 27, 33, 36].

9.10   Differential Operators for $SO(3)$

Let $A \in SO(3)$ be an arbitrary rotation, and $f(A)$ be a function which assigns a complex number to each value of A. Analogous to the definition of the partial derivative (or directional derivative) of a complex-valued function of an $\mathbb{R}^N$-valued argument, we can define differential operators which act on functions of rotation-valued argument. Only now, there are two choices corresponding to whether the operation is applied to the right or left. These take the form

$$X_{\mathbf{n}}^L f(A) = \lim_{\epsilon \to 0} \frac{1}{\epsilon} [f(\mathrm{ROT}[\mathbf{n}, -\epsilon] \cdot A) - f(A)] = \frac{df(\mathrm{ROT}[\mathbf{n}, -t] \cdot A)}{dt}\Big|_{t=0} \qquad (9.53)$$

and

$$X_{\mathbf{n}}^R f(A) = \lim_{\epsilon \to 0} \frac{1}{\epsilon} [f(A \cdot \mathrm{ROT}[\mathbf{n}, \epsilon]) - f(A)] = \frac{df(A \cdot \mathrm{ROT}[\mathbf{n}, t])}{dt}\Big|_{t=0}. \qquad (9.54)$$

In the case of the left operator, $-\epsilon$ is used above to be consistent with notations that will follow in subsequent chapters. Note that for small motions,

$$\mathrm{ROT}[\mathbf{n}, \theta] \approx \mathbb{I} + \theta N = \mathbb{I} + \theta (n_1 X_1 + n_2 X_2 + n_3 X_3)$$

where

$$X_1 = \begin{pmatrix} 0 & 0 & 0 \\ 0 & 0 & -1 \\ 0 & 1 & 0 \end{pmatrix}; \qquad X_2 = \begin{pmatrix} 0 & 0 & 1 \\ 0 & 0 & 0 \\ -1 & 0 & 0 \end{pmatrix}; \qquad X_3 = \begin{pmatrix} 0 & -1 & 0 \\ 1 & 0 & 0 \\ 0 & 0 & 0 \end{pmatrix}.$$

We now find the explicit forms of the operators $X_{\mathbf{n}}^L$ and $X_{\mathbf{n}}^R$ in any 3-parameter description of rotation $A = A(q_1, q_2, q_3)$. Expanding in a Taylor series, one writes

$$X_{\mathbf{n}}^R f = \sum_{i=1}^{3} \frac{\partial f}{\partial q_i} \frac{\partial q_i^R}{\partial \epsilon}\Big|_{\epsilon=0}$$

where $\{q_i^R\}$ are the parameters such that $A(q_1, q_2, q_3)\mathrm{ROT}[\mathbf{n}, \epsilon] = A(q_1^R, q_2^R, q_3^R)$.

The coefficients $\frac{\partial q_i^R}{\partial \epsilon}|_{\epsilon=0}$ are determined by observing two different looking, though equivalent, ways of writing $A \cdot \mathrm{ROT}[\mathbf{n}, \epsilon]$ for small ϵ

$$A + \epsilon AN \approx A \cdot \mathrm{ROT}[\mathbf{n}, \epsilon] \approx A + \epsilon \sum_{i=1}^{3} \frac{\partial A}{\partial q_i} \frac{\partial q_i^R}{\partial \epsilon}\Big|_{\epsilon=0}.$$

We then have that

$$N = \sum_{i=1}^{3} A^T \frac{\partial A}{\partial q_i} \frac{\partial q_i^R}{\partial \epsilon}\Big|_{\epsilon=0},$$

or

$$\mathbf{n} = \mathrm{vect}(N) = \sum_{i=1}^{3} \mathrm{vect}\left(A^T \frac{\partial A}{\partial q_i}\right) \frac{\partial q_i^R}{\partial \epsilon}\Big|_{\epsilon=0},$$

which is written as $\mathbf{n} = J_R \frac{d\mathbf{q}^R}{d\epsilon}|_{\epsilon=0}$. This allows us to solve for

$$\frac{d\mathbf{q}^R}{d\epsilon}\Big|_{\epsilon=0} = J_R^{-1}\mathbf{n}.$$

J_R is the "body" Jacobian calculated in Eq. (5.43) for the ZXZ Euler angles. Its inverse is

$$J_R^{-1} = \begin{pmatrix} \sin\gamma/\sin\beta & \cos\gamma/\sin\beta & 0 \\ \cos\gamma & -\sin\gamma & 0 \\ -\cot\beta\sin\gamma & -\cot\beta\cos\gamma & 1 \end{pmatrix}.$$

Making the shorthand notation $X_{e_i}^R = X_i^R$, we then write for the ZXZ Euler angles

$$X_1^R = -\cot\beta\sin\gamma\frac{\partial}{\partial\gamma} + \frac{\sin\gamma}{\sin\beta}\frac{\partial}{\partial\alpha} + \cos\gamma\frac{\partial}{\partial\beta};$$

$$X_2^R = -\cot\beta\cos\gamma\frac{\partial}{\partial\gamma} + \frac{\cos\gamma}{\sin\beta}\frac{\partial}{\partial\alpha} - \sin\gamma\frac{\partial}{\partial\beta};$$

$$X_3^R = \frac{\partial}{\partial\gamma}.$$

For the ZYZ Euler-angles, these same operators take the form

$$X_1^R = \cot\beta\cos\gamma\frac{\partial}{\partial\gamma} - \frac{\cos\gamma}{\sin\beta}\frac{\partial}{\partial\alpha} + \sin\gamma\frac{\partial}{\partial\beta};$$

$$X_2^R = -\cot\beta\sin\gamma\frac{\partial}{\partial\gamma} + \frac{\sin\gamma}{\sin\beta}\frac{\partial}{\partial\alpha} + \cos\gamma\frac{\partial}{\partial\beta};$$

$$X_3^R = \frac{\partial}{\partial\gamma}.$$

Analogous to the directional derivative in $\mathbb{R}^3$, we have $X_{\mathbf{n}}^R = \sum_{i=1}^3 n_i X_i^R$.

The operators $X_{e_i}^L = X_i^L$ can be derived in a completely analogous way using the spatial Jacobian instead of the body Jacobian. Or they can be derived from $X_{e_i}^R$ by observing that

$$\text{ROT}[\mathbf{n}, \epsilon]A = A\left(A^T\text{ROT}[\mathbf{n}, \epsilon]A\right) = A\,\text{ROT}\left[A^T\mathbf{n}, \epsilon\right].$$

This means that

$$X_{e_i}^L = X_{-A^T e_i}^R = -(\nabla_\mathbf{q}f)\cdot\left(J_L^{-1}e_i\right)$$

Explicitly in terms of ZXZ Euler angles,

$$J_L^{-1} = J_R^{-1}A^T = \begin{pmatrix} -\sin\alpha\cot\beta & \cos\alpha\cot\beta & 1 \\ \cos\alpha & \sin\alpha & 0 \\ \sin\alpha/\sin\beta & -\cos\alpha/\sin\beta & 0 \end{pmatrix},$$

and so

$$X_1^L = \sin\alpha\cot\beta\frac{\partial}{\partial\alpha} - \cos\alpha\frac{\partial}{\partial\beta} - \sin\alpha/\sin\beta\frac{\partial}{\partial\gamma}$$

$$X_2^L = -\cos\alpha\cot\beta\frac{\partial}{\partial\alpha} - \sin\alpha\frac{\partial}{\partial\beta} + \cos\alpha/\sin\beta\frac{\partial}{\partial\gamma}$$

$$X_3^L = -\frac{\partial}{\partial\alpha}.$$

When ZYZ Euler angles are used, this takes the form

$$X_1^L = \cos\alpha\cot\beta\frac{\partial}{\partial\alpha} + \sin\alpha\frac{\partial}{\partial\beta} - \cos\alpha/\sin\beta\frac{\partial}{\partial\gamma}$$

$$X_2^L = \sin\alpha\cot\beta\frac{\partial}{\partial\alpha} - \cos\alpha\frac{\partial}{\partial\beta} - \sin\alpha/\sin\beta\frac{\partial}{\partial\gamma}$$

$$X_3^L = -\frac{\partial}{\partial\alpha}.$$

9.11   Operational Properties

We now examine the effect of the operators X_i^R and X_i^L on the matrix elements U_{mn}^l. Due to the homomorphism property of representations, we write

$$X_i^R U_{mn}^l(A) = \lim_{\epsilon \to 0} \frac{1}{\epsilon} \left[\sum_{k=-l}^{l} U_{mk}^l(A) U_{kn}^l \left(\text{ROT}[e_i, \epsilon] \right) - U_{mn}^l(A) \right] \tag{9.55}$$

and

$$X_i^L U_{mn}^l(A) = \lim_{\epsilon \to 0} \frac{1}{\epsilon} \left[\sum_{k=-l}^{l} U_{mk}^l \left(\text{ROT}[e_i, -\epsilon] \right) U_{kn}^l(A) - U_{mn}^l(A) \right]. \tag{9.56}$$

Expanding the integral in Eq. (9.22) for $\beta = \epsilon$, we find that

$$U_{mn}^l \left(\text{ROT}(e_1, \epsilon) \right) = \delta_{mn} + \frac{i}{2} c_{-n}^l \epsilon \delta_{m+1,n} + \frac{i}{2} c_n^l \epsilon \delta_{m-1,n} + \mathcal{O}\left(\epsilon^2\right).$$

Again we use the definition $c_n^l = \sqrt{(l-n)(l+n+1)}$. Using the fact that

$$U_{mn}^l \left(\text{ROT}(e_2, \epsilon) \right) = i^{n-m} U_{mn}^l \left(\text{ROT}(e_1, \epsilon) \right),$$

this allows us to write

$$U_{mn}^l \left(\text{ROT}(e_2, \epsilon) \right) = \delta_{mn} - \frac{1}{2} c_{-n}^l \epsilon \delta_{m+1,n} + \frac{1}{2} c_n^l \epsilon \delta_{m-1,n} + \mathcal{O}\left(\epsilon^2\right).$$

Substituting these into Eqs. (9.55) and (9.56), we find that

$$X_1^R U_{mn}^l = \frac{1}{2} i c_{-n}^l U_{m,n-1}^l + \frac{1}{2} i c_n^l U_{m,n+1}^l;$$

$$X_2^R U_{mn}^l = -\frac{1}{2} c_{-n}^l U_{m,n-1}^l + \frac{1}{2} c_n^l U_{m,n+1}^l;$$

$$X_1^L U_{mn}^l = -\frac{1}{2} i c_{-m-1}^l U_{m+1,n}^l - \frac{1}{2} i c_{m-1}^l U_{m-1,n}^l;$$

$$X_2^L U_{mn}^l = \frac{1}{2} c_{-m-1}^l U_{m+1,n}^l - \frac{1}{2} c_{m-1}^l U_{m-1,n}^l.$$

The operators X_3^R and X_3^L can be applied directly to the closed-form formula Eq. (9.20) to result in

$$X_3^R U_{mn}^l = -in U_{mn}^l;$$

$$X_3^L U_{mn}^l = im U_{mn}^l.$$

By repeated application of these rules one finds

$$\left(X_1^R\right)^2 U_{mn}^l = -\frac{1}{4}c_{-n}^l c_{-n+1}^l U_{m,n-2}^l - \frac{1}{4}\left(c_{-n}^l c_{n-1}^l + c_n^l c_{-n-1}^l\right) U_{mn}^l - \frac{1}{4}c_n^l c_{n+1}^l U_{m,n+2}^l;$$

$$\left(X_2^R\right)^2 U_{mn}^l = \frac{1}{4}c_{-n}^l c_{-n+1}^l U_{m,n-2}^l - \frac{1}{4}\left(c_{-n}^l c_{n-1}^l + c_n^l c_{-n-1}^l\right) U_{mn}^l + \frac{1}{4}c_n^l c_{n+1}^l U_{m,n+2}^l;$$

$$\left(X_3^R\right)^2 U_{mn}^l = -n^2 U_{mn}^l;$$

$$X_1^R X_2^R U_{m,n}^l = -\frac{i}{4}c_{-n}^l c_{-n+1}^l U_{m,n-2}^l + \frac{i}{4}\left(-c_{-n}^l c_{n-1}^l + c_n^l c_{-n-1}^l\right) U_{m,n}^l + \frac{i}{4}c_n^l c_{n+1}^l U_{m,n+2}^l;$$

$$X_2^R X_1^R U_{m,n}^l = -\frac{i}{4}c_{-n}^l c_{-n+1}^l U_{m,n-2}^l + \frac{i}{4}\left(c_{-n}^l c_{n-1}^l - c_n^l c_{-n-1}^l\right) U_{m,n}^l + \frac{i}{4}c_n^l c_{n+1}^l U_{m,n+2}^l;$$

$$X_1^R X_3^R U_{m,n}^l = \frac{n}{2}\left(c_{-n}^l U_{m,n-1}^l + c_n^l U_{m,n+1}^l\right);$$

$$X_3^R X_1^R U_{m,n}^l = \frac{n-1}{2}c_{-n}^l U_{m,n-1}^l + \frac{n+1}{2}c_n^l U_{m,n+1}^l;$$

$$X_3^R X_2^R U_{m,n}^l = \frac{i(n-1)}{2}c_{-n}^l U_{m,n-1}^l - \frac{i(n+1)}{2}c_n^l U_{m,n+1}^l;$$

$$X_2^R X_3^R U_{m,n}^l = \frac{in}{2}\left(c_{-n}^l U_{m,n-1}^l - c_n^l U_{m,n+1}^l\right).$$

Since $c_{-n}^l c_{n-1}^l + c_n^l c_{-n-1}^l = 2(l^2 - n^2) + 2l$, one finds that for arbitrary constant scalars D_1 and D_2,

$$\left[D_1\left(\left(X_1^R\right)^2 + \left(X_2^R\right)^2\right) + D_2\left(X_3^R\right)^2\right] = -D_1\left[l(l+1) + n^2\left(D_2/D_1 - 1\right)\right]U_{mn}^l. \tag{9.57}$$

The Laplacian[4] is given by

$$\nabla^2 = \left(X_1^R\right)^2 + \left(X_2^R\right)^2 + \left(X_3^R\right)^2.$$

This is exactly the same result from the differential-geometric definition of the Laplacian given in Eq. (9.18). The Laplacian commutes with all the operators mentioned previously, as well as with left and right shifts, and can also be written as

$$\nabla^2 = \left(X_1^L\right)^2 + \left(X_2^L\right)^2 + \left(X_3^L\right)^2.$$

When $D_1 = D_2 = 1$ in Eq. (9.57) the Laplacian results, and

$$\nabla^2 U_{mn}^l = -l(l+1)U_{mn}^l$$

as it must from its differential-geometric definition.

Other differential and algebraic operators also transform matrix elements into matrix elements. One of the many such operators found in [31] is

$$\sin\beta\frac{\partial}{\partial\beta}\left(U_{mn}^l\right) = -\frac{(l+1)[(l^2-m^2)(l^2-n^2)]^{\frac{1}{2}}}{l(2l+1)}U_{mn}^{l-1} - \frac{mn}{l(l+1)}U_{mn}^l$$

$$+ \frac{l[(l+1)^2-m^2]^{\frac{1}{2}}[(l+1)^2-n^2]^{\frac{1}{2}}}{(l+1)(2l+1)}U_{mn}^{l+1}. \tag{9.58}$$

Note that unlike the operators X_i^R and X_i^L, the right side of these equalities have shifts in the value of l but leave m and n unchanged.

---

[4] The negative of this Laplacian is often denoted in physics as J^2, and in mathematics it is called the Casimir operator.

9.12   Classification of Quantum States According to Representations of $SU(2)$

Conservation laws are laws of nature observed in physical experiments. Each conservation law states the fact that some physical quantities are left invariant under the action of some group of transformations. Important groups of transformations include translations, rotations, and reflections. These groups lead to the conservation of momentum, angular momentum, and parity, respectively.

The invariance of equations of motion under a group of transformations leads to the classification of quantum states as eigenvalues of representations of a group of transformations. Let us write the Schrödinger equation

$$i\frac{\partial \psi}{\partial t} = \hat{H}\psi$$

where $\hat{H}$ is an operator representing the Hamiltonian of the system and ψ is a wave function. Let us assume that the equation is invariant under some transformation $\hat{A}$. $\hat{A}$ is an operator and acts on wave functions as $(\hat{A}\psi)(x) = \psi(A^{-1} \cdot x)$, where A is an element of the group of transformations G. We note that the transformation law assumes that the operators $\hat{A}$ are representations of the group G which act on the space of wave functions $\psi(x)$. Invariance of Schrödinger's equation means that the transformed wave function $\psi' = \hat{A}\psi$ must still satisfy the Schrödinger equation

$$i\frac{\partial \psi'}{\partial t} = \hat{H}\psi'.$$

If we assume that the transformation is independent of time, then we may write

$$i\hat{A}\frac{\partial \psi}{\partial t} = \hat{H}\hat{A}\psi.$$

We observe that the Schrödinger equation is invariant if the Hamiltonian commutes with the transformation operator A

$$\hat{H}\hat{A} = \hat{A}\hat{H}.$$

According to Schur's lemma, if the operator $\hat{H}$ commutes with all the irreducible representation operators of the group G, then $\hat{H}$ must be proportional to the identity operator.

Thus for the direct sum of different irreducible representations of G, the Hamiltonian may be represented in diagonal form. Therefore, the Hamiltonian has the simplest form in the basis where the wave functions are the basis functions of representations of the group of transformations G. If we assume that the quantum system is invariant with respect to the rotation group $SO(3)$, this leads to the classification of quantum states according to quantum numbers l, m, which enumerate different basis functions of representations of $SO(3)$. Wave functions (quantum states) may be considered as a sum over different irreducible representation of $SO(3)$, enumerated by $l = 0, 1, 2, \ldots$. Thus, for spherically symmetric systems (such as the hydrogen atom), the wave functions may be represented in the form

$$\psi(r, \theta, \phi) = \sum_{l=0}^{\infty} \sum_{m=-l}^{l} \sum_{n=l+1}^{\infty} c_{lm} R_l(r, n) Y_l^m(\theta, \phi)$$

where $Y_l^m(\theta, \phi)$ are spherical harmonics, c_{lm} are constant coefficients, and $R_l(r, n)$ represents a radial dependence (n is an additional radial quantum number). The number l which enumerates irreducible representation of $SO(3)$ has the meaning of angular momentum of the system, and $m = -l, \ldots, l$ has the meaning of the projection of angular momentum on a fixed axis. It may be shown from the Schrödinger equation for the radial function $R_l(r, n)$ that $n = l + 1, l + 2, \ldots$.

We note that in quantum mechanics the state with $l = 0$ is called the s state, $l = 1$ is the p state, $l = 2$ is the d state and so on.

Because $l = 0, 1, 2, \ldots$, angular momentum can have only discrete values. This property is called *quantization* of angular momentum. If X_3 is the generator of $SO(3)$ for rotations around the z-axis, it acts on quantum spherical states as

$$X_3 \psi_{l,m} = m \psi_{l,m}$$

because of the corresponding property of spherical harmonics. The operator $X^2 = X_1^2 + X_2^2 + X_3^2$ acts as

$$X^2 \psi_{l,m} = -l(l+1) \psi_{l,m}.$$

If we assume that the system has several non-interacting particles, then the Hamiltonian should be invariant under the combined transformation which is a direct product of rotations of $SO(3)$ for each of the individual particles [and the corresponding spherical dependence on $\theta_1, \phi_1, \theta_2, \phi_2$ may be written as a product of spherical harmonics $Y_{l_1}^{m_1}(\theta_1, \phi_1)$ and $Y_{l_2}^{m_2}(\theta_2, \phi_2)$]. The direct product of irreducible representations of $SO(3)$ may be written as a sum over irreducible representations of $SO(3)$. This leads to the summation rules for angular momentum. The direct product of irreducible representations A_{l_1} and A_{l_2} of $SO(3)$, enumerated by l_1 and l_2 may be written as a sum over irreducible representations A_l, where $l = |l_1 - l_2|, \ldots, l_1 + l_2$.

Because of this, the sum of two angular momenta l_1 and l_2 may have values from $l = |l_1 - l_2|$ to $l = l_1 + l_2$. The projections of momenta m_1 and m_2 are added as ordinary numbers $m = m_1 + m_2$ because the Clebsch-Gordan coefficients $C(l_1, m_1; l_2, m_2 | l, m)$ are not equal to zero only for $m_1 + m_1 = m$.

So far we have discussed physical systems which are scalar under transformation, i.e., the ψ functions are scalar-valued functions. Many important physical systems may be described as vector-, spinor-, or even tensor-valued functions. Vector-valued functions may be described as functions which transform under a representation operator $\hat{A}$ of $SO(3)$ as

$$\left(\hat{A} \psi_i\right)(x) = \sum_j A_{ij} \psi_j \left(A^{-1} \cdot x\right)$$

where $A \in SO(3)$. Because $A \in SO(3)$ is equivalent to an $l = 1$ representation of $SO(3)$, these functions describe particles with $l = 1$ internal momentum. We mention that transformations of ψ functions in the internal space of ψ describe the *internal* angular momentum of particles. Since particles in non-relativistic quantum mechanics are electrons (which have non-integer internal angular momentum), vector functions are rarely used in quantum mechanics (more general vector functions are used in relativistic quantum mechanics to describe photons).

Particles with non-integer internal angular momentum (called spin) are described by spinor-valued functions. A spinor-valued function transforms under the action of $SU(2)$ (corresponding to the rotation matrix A) in the space of ψ_i (i is an index describing internal degrees of freedom). The transformation law for spinor functions is

$$\left(\hat{A} \psi_i\right)(x) = \sum_j A'_{ij} \psi_j \left(A^{-1} \cdot x\right)$$

where $A \in SO(3)$ and $A' \in SU(2)$. The matrix $A' = \cos(\theta/2) + i(\sigma \cdot \mathbf{n}) \sin(\theta/2)$ corresponds to the rotation matrix A parameterized with $(\mathbf{n}, \theta)$ where the unit vector $\mathbf{n}$ defines the axis of rotation, and θ is the angle of rotation. Matrices σ_i (Pauli matrices) are proportional to the basis elements of the Lie algebra $SU(2)$

$$\sigma_1 = \begin{pmatrix} 0 & 1 \\ 1 & 0 \end{pmatrix}; \qquad \sigma_2 = \begin{pmatrix} 0 & -i \\ i & 0 \end{pmatrix}; \qquad \sigma_3 = \begin{pmatrix} 1 & 0 \\ 0 & -1 \end{pmatrix}.$$

Because the $SU(2)$ matrix in the fundamental representation corresponds to the $l = 1/2$ representation of $SU(2)$, spinor functions describe particles with internal half-integer spin.

See [4, 19, 24, 31] for more in-depth treatments of the relationship between the representation theory of $SU(2)$ and quantum mechanics.

9.13 Representations of $SO(4)$ and $SU(2) \times SU(2)$

We have already seen that rotations in $\mathbb{R}^4$ can be described using two independent special unitary matrices as

$$Y = AXB^\dagger$$

where

$$X = \begin{pmatrix} x_4 - ix_2 & -x_1 - ix_3 \\ x_1 - ix_3 & x_4 + ix_2 \end{pmatrix}.$$

Furthermore, the pair $(A, B) \in SU(2) \times SU(2)$ is redundant in the sense that $(-A, -B)$ describes the same rotation as (A, B). In group-theoretic language, $SU(2) \times SU(2)$ is the double covering group of $SO(4)$ in the same way that $SU(2)$ is the double cover of $SO(3)$. One writes

$$(SU(2) \times SU(2))/[-\mathbb{I}, \mathbb{I}] \cong SO(4).$$

This is useful to know, since it means we need not construct the IURs of $SO(4)$ from scratch. The property Eq. (9.8) of the tensor product indicates that if $U^j(A)$ and $U^{j'}(B)$ are representations of $SU(2)$, then

$$\mathcal{U}^{j,j'}(A, B) = U^j(A) \otimes U^{j'}(B)$$

is a representation of $SU(2) \times SU(2)$. It may be shown that the set of all representations of the form $\mathcal{U}^{j,j'}(A, B)$ is complete, and each representation is unitary [29]. Furthermore, the complete set of representations for $SO(4)$ is just the subset of those for $SU(2) \times SU(2)$ for which $j + j'$ is an integer.

We note that due to the close relationship between representations of $SO(4)$ and those for $SU(2)$, the sampling and FFT techniques discussed earlier may be applied in this case as well. This and the relationship between $SO(4)$ and $SE(3)$ would seem to lead to a promising way to perform fast approximate FFTs for $SE(3)$. The difficulty is that the stereographic projection relating $SO(4)$ and $SE(3)$ is only accurate for small rotations. This means that probability density functions on $SO(4)$ used to approximate PDFs on $SE(3)$ would require an extremely large band limit in order to achieve any reasonable accuracy. That is, the uncertainty principle will not allow functions which simultaneously have a small band limit and small compact support. Therefore, we must consider the representation theory of $SE(N)$ for $N = 2$ and $N = 3$ and seek fast transforms in order that some of the problems posed in subsequent chapters can be solved using efficient numerical techniques.

9.14 Summary

In this chapter we presented a concrete overview of the representation theory of $SO(3)$ and $SU(2)$ together with harmonic analysis on these groups. Irreducible unitary representation matrices were given in terms of several parameterizations. It was shown how differential operators acting on functions of these groups can be defined in a concrete way, and how the matrix elements of IURs behave under these operators. The results of this chapter are used in Chapter 10 to define the Fourier transform and operational calculus for $SE(3)$, and in Chapters 16 and 17 in the context of applications.

References

[1] Bretz, V., Improved method for calculation of angular-momentum coupling coefficients, *Acta Physica Academiae Scientiarum Hungarucae*, 40(4), 255–259, 1976.

[2] Bunge, H.J., Über eine Fourier-Entwicklung verallgemeinerter kugelfunktionen, *Mber. Dt. Acad. Wiss.*, 9, 652–658, 1967.

[3] Bunge, H.J., Calculation of the Fourier coefficients of the generalized spherical functions, *Kristall Techn.*, 9, 939–963, 1974.

[4] Edmonds, A.R., *Angular momentum in quantum mechanics*, 2nd ed., Princeton University Press, Princeton, NJ, 1963.

[5] Fack, V., Vanderjeugt, J., and Rao, K.S., Parellel computation of recoupling coefficients using transputers, *Comput. Phys. Commun.*, 71(3), 285–304, 1992.

[6] Fässler, A. and Stiefel, E., *Group Theoretical Methods and Their Applications*, (transl.,) Wong, B.D., Birkhäuser, Boston, 1992.

[7] Gel'fand, I.M., Minlos, R.A., and Shapiro, Z.Ya., *Representations of the Rotation and Lorentz Groups and their Applications*, Macmillan, New York, 1963.

[8] Hannabuss, K., Sound and symmetry, *The Math. Intelligencer*, 19(4), 16–20, 1997.

[9] Larson, E.G., Li, M.S., and Larson, G.C., Some comments on the electrostatic potential of a molecule, *Int. J. of Quantum Chem.*, Suppl. 26, 181–205, 1992.

[10] Maslen, D.K., *Fast Transforms and Sampling for Compact Groups*, Ph.D. dissertation, Dept. of Mathematics, Harvard University, Cambridge, MA, 1993.

[11] Maslen, D.K. and Rockmore, D.N., Generalized FFTs — a survey of some recent results, *DIMACS Ser. in Discrete Math. and Theor. Comput. Sci.*, 28, 183–237, 1997.

[12] Mayer, R.A., Fourier series of differentiable functions on SU(2), *Duke Math. J.*, 34, 549–554, 1967.

[13] Mayer, R.A., Summation of Fourier series on compact groups, *Am. J. of Math.*, 89, 661–692, 1967.

[14] Mayer, R.A., Localization for Fourier series on $SU(2)$, *Trans. of the Am. Math. Soc.*, 130, 414–424, 1968.

[15] Moses, H.E., Irreducible Representations of the Rotation Group in Terms of the Axis and Angle of Rotation, *Ann. of Phys.*, 42, 343–346, 1967.

[16] Moses, H.E., Irreducible representations of the rotation group in terms of the axis and angle of rotation, *Ann. of Phys.*, 37, 224–226, 1966.

[17] Moses, H.E., Irreducible representations of the rotation group in terms of Euler's theorem, *Il Nuovo Cimento*, 40(4), 1120–1138, 1965.

[18] Naimark, M.A., *Linear Representations of the Lorentz Group*, Macmillan, New York, 1964.

[19] Normand, J.-M., *A Lie Group: Rotations in Quantum Mechanics*, North-Holland, Amsterdam, 1980.

[20] Racah, G., Theory of complex spectra. I, II and III, *Phys. Rev.*, 61, 186–197, 1942; 62, 438–462, 1942; 63, 367–382, 1943.

[21] Rao, K.S. and Venkatesh, K., New fortran programs for angular-momentum coefficients, *Comput. Phys. Commun.*, 15(3-4), 227–235, 1978.

[22] Rao, K.S., Rajeswari, V., and Chiu, C.B., A new fortran program for the 9-J angular-momentum coefficient, *Comput. Phys. Commun.*, 56(2), 231–248, 1989.

[23] Risbo, T., Fourier transform summation of Legendre series and D-functions, *J. of Geodesy*, 70(7), 383–396, 1996.

[24] Rose, M.E., *Elementary Theory of Angular Momentum*, John Wiley & Sons, 1957 (Dover Ed., 1995).

[25] Schulten, K. and Gordon, R.G., Recursive evaluation of 3J and 6J coefficients, *Comput. Phys. Commun.*, 11(2), 269–278, 1976.

[26] Scott, N.S., Milligan, P., and Riley, H.W.C., The parallel computation of racah coefficients using transputers, *Comput. Phys. Commun.*, 46(1), 83–98, 1987.

[27] Sherborne, B.S. and Stedman, G.E., Recursive generation of Cartesian angular-momentum coupling trees for *SO*(3), *Comput. Phys. Commun.*, 59(2), 417–428, 1990.

[28] Sugiura, M., *Unitary Representations and Harmonic Analysis*, 2nd ed., North-Holland, Amsterdam, 1990.

[29] Talman, J., *Special Functions*, W.A. Benjamin, Amsterdam, 1968.

[30] Torresani, B., Position-frequency analysis for signals defined on spheres, *Signal Process.*, 43, 341–346, 1995.

[31] Varshalovich, D.A., Moskalev, A.N., and Khersonskii, V.K., *Quantum Theory of Angular Momentum*, World Scientific, Singapore, 1988.

[32] Vilenkin, N.J. and Klimyk, A.U., *Representation of Lie Groups and Special Functions*, Vols. 1–3, Kluwer Academic, Dordrecht, Holland, 1991.

[33] Wei, L.Q., Unified approach for exact calculation of angular momentum coupling and recoupling coefficients, *Comput. Phys. Commun.*, 120(2-3), 222–230, 1999.

[34] Wightman, A.S., On the localizability of quantum mechanical systems, *Rev. Mod. Phys.*, 34, 845–872, 1962.

[35] Willsky, A.S. Dynamical systems defined on groups: structural properties and estimation, Ph.D. dissertation, MIT, Boston, 1973.

[36] Zhao, D.Q. and Zare, R.N., Numerical computation of 9-J symbols, *Mol. Phys.*, 65(5), 1263–1268, 1988.

Chapter 10

Harmonic Analysis on the Euclidean Motion Groups

10.1 Introduction

The Euclidean motion group, $SE(N)$,[1] is the semidirect product of $\mathbb{R}^N$ with the special orthogonal group, $SO(N)$. That is, $SE(3) = \mathbb{R}^3 \rtimes_\varphi SO(3)$. We denote elements of $SE(N)$ as $g = (\mathbf{a}, A) \in SE(N)$ where $A \in SO(N)$ and $\mathbf{a} \in \mathbb{R}^N$. For any $g = (\mathbf{a}, A)$ and $h = (\mathbf{r}, R) \in SE(N)$, the group law is written as $g \circ h = (\mathbf{a} + A\mathbf{r}, AR)$, and $g^{-1} = (-A^T\mathbf{a}, A^T)$. Alternately, one may represent any element of $SE(N)$ as an $(N + 1) \times (N + 1)$ homogeneous transformation matrix of the form

$$H(g) = \begin{pmatrix} A & \mathbf{a} \\ \mathbf{0}^T & 1 \end{pmatrix}.$$

Clearly, $H(g)H(h) = H(g \circ h)$ and $H(g^{-1}) = H^{-1}(g)$, and the mapping $g \to H(g)$ is an isomorphism between $SE(N)$ and the set of homogeneous transformation matrices.

$SE(N)$ is a solvable Lie group, and general methods for constructing unitary representations of solvable Lie groups have been known for some time (see, for example, [2, 18, 23]). In the past 40 years, the representation theory and harmonic analysis for the Euclidean groups have been developed in the pure mathematics and mathematical physics literature. The study of matrix elements of irreducible unitary representation of $SE(3)$ was initiated by Vilenkin [41] in 1957 (some particular matrix elements are also given in [42]). The most complete study of $\widetilde{SE}(3)$ [the universal covering group of $SE(3)$] with application to the harmonic analysis was given by Miller in [19]. The representations of $SE(3)$ were also studied in [10, 16, 24, 37].

However, despite the considerable progress in mathematical developments of the representation theory of $SE(3)$, these achievements have not yet been widely incorporated in engineering and applied fields. In subsequent chapters we try to fill this gap. In this chapter, we review the representation theory of $SE(3)$, derive the matrix elements of the irreducible unitary representations, and define the Fourier transform for $SE(3)$. We derive symmetry and operational properties of the Fourier transform and give explicit examples of Fourier transforms of functions on the motion group.

---

[1] Recall from Chapter 6 that the notation $SE(N)$ comes from the terminology, "Special Euclidean group of N dimensional space."

321

10.2   Matrix Elements of IURs of $SE(2)$

Each element of $SE(2)$ is parameterized in either rectangular or polar coordinates as

$$g(a_1, a_2, \theta) = \begin{pmatrix} \cos\theta & -\sin\theta & a_1 \\ \sin\theta & \cos\theta & a_2 \\ 0 & 0 & 1 \end{pmatrix}$$

or

$$g(a, \phi, \theta) = \begin{pmatrix} \cos\theta & -\sin\theta & a\cos\phi \\ \sin\theta & \cos\theta & a\sin\phi \\ 0 & 0 & 1 \end{pmatrix},$$

where $a = \|\mathbf{a}\|$.

A unitary representation of $SE(2)$ (see [22, 35, 37, 40, 42] for general definition) is defined by the unitary operator

$$U(g, p)\tilde{\varphi}(\mathbf{x}) \overset{\Delta}{=} e^{-ip(\mathbf{a}\cdot\mathbf{x})} \tilde{\varphi}\left(A^T \mathbf{x}\right) \overset{\Delta}{=} \tilde{\varphi}_g(\mathbf{x}), \tag{10.1}$$

for each $g = (\mathbf{a}, A) = g(a_1, a_2, \theta) \in SE(2)$. Here $p \in \mathbb{R}^+$, and $\mathbf{x} \cdot \mathbf{y} = x_1 y_1 + x_2 y_2$. The vector $\mathbf{x}$ is a unit vector ($\mathbf{x} \cdot \mathbf{x} = 1$).

By definition, group representations observe the homomorphism property, which in this case is seen as follows

$$\begin{aligned} (U(g, p)U(h, p)\tilde{\varphi})(\mathbf{x}) &= (U(g, p)(U(h, p)\tilde{\varphi}))(\mathbf{x}) \\ &= (U(g, p)\tilde{\varphi}_h)(\mathbf{x}) = e^{-ip(\mathbf{a}\cdot\mathbf{x})}\tilde{\varphi}_h\left(A^T\mathbf{x}\right) \\ &= e^{-ip(\mathbf{a}\cdot\mathbf{x})}e^{-ip(\mathbf{r}\cdot(A^T\mathbf{x}))}\tilde{\varphi}\left(R^T A^T \mathbf{x}\right) \\ &= e^{-ip(\mathbf{a}+A\mathbf{r})\cdot\mathbf{x}}\tilde{\varphi}\left((AR)^T\mathbf{x}\right) \\ &= (U(g \circ h, p)\tilde{\varphi})(\mathbf{x}). \end{aligned}$$

Since $\mathbf{x}$ is a unit vector, the function $\tilde{\varphi}(\mathbf{x}) \overset{\Delta}{=} \tilde{\varphi}(\cos\psi, \sin\psi) \equiv \varphi(\psi)$ is a function on the unit circle. Henceforth we will not distinguish between $\tilde{\varphi}$ and φ.

Any function $\varphi(\psi) \in \mathcal{L}^2(S^1)$ can be expressed as a weighted sum (Fourier series) of orthonormal basis functions as $\varphi(\psi) = \sum_{n\in\mathbb{Z}} c_n e^{in\psi}$. Likewise, the matrix elements of the operator $U(g, p)$ are expressed in this basis as

$$u_{mn}(g, p) = \left(e^{im\psi}, U(g, p)e^{in\psi}\right) = \frac{1}{2\pi}\int_0^{2\pi} e^{-im\psi} e^{-i(a_1 p\cos\psi + a_2 p\sin\psi)} e^{in(\psi-\theta)} d\psi \tag{10.2}$$

$\forall$ $m, n \in \mathbb{Z}$, where the inner product $(\cdot, \cdot)$ is defined as

$$(\varphi_1, \varphi_2) = \frac{1}{2\pi}\int_0^{2\pi} \overline{\varphi_1(\psi)}\varphi_2(\psi)d\psi.$$

It is easy to see that $(U(g, p)\varphi_1, U(g, p)\varphi_2) = (\varphi_1, \varphi_2)$ and that $U(g, p)$ is therefore unitary with respect to this inner product.

A number of works including [22], [37] and [40], have shown that the matrix elements of this representation are expressed as

$$u_{mn}(g(a, \phi, \theta), p) = i^{n-m} e^{-i[n\theta+(m-n)\phi]} J_{n-m}(pa) \tag{10.3}$$

where $J_\nu(x)$ is the νth order Bessel function.

From this expression, and the fact that $U(g, p)$ is a unitary representation, we have that

$$u_{mn}\left(g^{-1}(a, \phi, \theta), p\right) = u_{mn}^{-1}(g(a, \phi, \theta), p)$$

$$= \overline{u_{nm}(g(a, \phi, \theta), p)} = i^{n-m} e^{i[m\theta + (n-m)\phi]} J_{m-n}(pa). \quad (10.4)$$

Henceforth no distinction will be made between the operator $U(g, p)$ and the corresponding infinite dimensional matrix with elements $u_{mn}(g, p)$.

Symmetry Properties

The matrix elements are related by the symmetries

$$\overline{u_{mn}(g, p)} = (-1)^{m-n} u_{-m, -n}(g, p), \quad (10.5)$$

$$u_{mn}(g(-a, \phi, \theta), p) \overset{\Delta}{=} u_{mn}(g(a, \phi \pm \pi, \theta), p) = (-1)^{m-n} u_{m,n}(g(a, \phi, \theta), p) \quad (10.6)$$

and

$$(-1)^{m-n} u_{m,n}(g(a, \phi - \theta, -\theta), p) = \overline{u_{nm}(g(a, \phi, \theta), p)}. \quad (10.7)$$

The equality in Eq. (10.7) follows from Eqs. (10.4) and (10.6).

10.2.1 Irreducibility of the Representations $U(g, p)$ of $SE(2)$

Since $SE(2)$ is neither compact nor commutative, the representation matrices will be infinite dimensional. Therefore, it will be more convenient to show the irreducibility of the operator $U(g, p)$ rather than the corresponding matrix. To this end, we examine one parameter subgroups generated by exponentiating linearly independent basis elements of the Lie algebra $SE(2)$. Using the basis

$$X_1 = \begin{pmatrix} 0 & 0 & 1 \\ 0 & 0 & 0 \\ 0 & 0 & 0 \end{pmatrix}; \qquad X_2 = \begin{pmatrix} 0 & 0 & 0 \\ 0 & 0 & 1 \\ 0 & 0 & 0 \end{pmatrix}; \qquad X_3 = \begin{pmatrix} 0 & -1 & 0 \\ 1 & 0 & 0 \\ 0 & 0 & 0 \end{pmatrix};$$

one finds

$$g_1(t) = \exp(tX_1) = \begin{pmatrix} 1 & 0 & t \\ 0 & 1 & 0 \\ 0 & 0 & 1 \end{pmatrix};$$

$$g_2(t) = \exp(tX_2) = \begin{pmatrix} 1 & 0 & 0 \\ 0 & 1 & t \\ 0 & 0 & 1 \end{pmatrix};$$

$$g_3(t) = \exp(tX_3) = \begin{pmatrix} \cos t & -\sin t & 0 \\ \sin t & \cos t & 0 \\ 0 & 0 & 1 \end{pmatrix}.$$

The corresponding differential operators $\tilde{X}_i^R$ (in polar coordinates) are

$$\tilde{X}_1^R = \cos(\theta - \phi)\frac{\partial}{\partial a} + \frac{\sin(\theta - \phi)}{a}\frac{\partial}{\partial \phi}$$

$$\tilde{X}_2^R = -\sin(\theta - \phi)\frac{\partial}{\partial a} + \frac{\cos(\theta - \phi)}{a}\frac{\partial}{\partial \phi}$$

$$\tilde{X}_3^R = \frac{\partial}{\partial \theta}.$$

The operators $\tilde{X}_i^L$ are

$$\tilde{X}_1^L = \cos\phi \frac{\partial}{\partial a} - \frac{\sin\phi}{a} \frac{\partial}{\partial \phi}$$

$$\tilde{X}_2^L = \sin\phi \frac{\partial}{\partial a} + \frac{\cos\phi}{a} \frac{\partial}{\partial \phi}$$

$$\tilde{X}_3^L = \frac{\partial}{\partial \theta} + \frac{\partial}{\partial \phi}.$$

These are found in a completely analogous way to the operators X_i^R and X_i^L for $SO(3)$. The tilde is to distinguish the $SE(2)$ case from the $SO(3)$ case. These operators act on functions on the group.

It follows from Eq. (10.1) that

$$U(g_1(t), p)\varphi(\psi) = e^{-ipt\cos\psi}\varphi(\psi);$$

$$U(g_2(t), p)\varphi(\psi) = e^{-ipt\sin\psi}\varphi(\psi);$$

$$U(g_3(t), p)\varphi(\psi) = \varphi(\psi - t).$$

Differentiating with respect to t and setting $t = 0$, we define the operators

$$\hat{X}_i(p)\varphi(\psi) \triangleq \left.\frac{dU(\exp(tX_i), p)\varphi(\psi)}{dt}\right|_{t=0}.$$

Explicitly,

$$\hat{X}_1(p)\varphi(\psi) = -ip\cos\psi\,\varphi(\psi);$$

$$\hat{X}_2(p)\varphi(\psi) = -ip\sin\psi\,\varphi(\psi);$$

$$\hat{X}_3(p)\varphi(\psi) = -\frac{d\varphi}{d\psi}.$$

Analogous to $SU(2)$, we define the operators

$$\hat{Y}_+(p) = \hat{X}_1(p) + i\hat{X}_2(p); \quad \hat{Y}_-(p) = \hat{X}_1(p) - i\hat{X}_2(p); \quad \hat{Y}_3(p) = \hat{X}_3(p).$$

Since $\varphi \in \mathcal{L}^2(S^1)$, basis elements are of the form $e^{ik\psi}$. These basis elements are transformed by the operators $\hat{Y}_\pm$ and $\hat{Y}_3$ as [40]

$$\hat{Y}_+(p)e^{ik\psi} = -ipe^{i(k+1)\psi}; \qquad \hat{Y}_-(p)e^{ik\psi} = -ipe^{i(k-1)\psi}; \qquad \hat{Y}_3(p)e^{ik\psi} = -ike^{ik\psi}.$$

As in the case of $SU(2)$, $\hat{Y}_+$ and $\hat{Y}_+$ always "push" basis elements to "adjacent" subspaces. Since no subspaces are left invariant by $\hat{Y}_\pm$, the representation operators $U(g, p)$ must be irreducible.

10.3   The Fourier Transform for $SE(2)$

Given the representations from the previous section, we are ready for the following definition.

DEFINITION 10.1   *[35] The Fourier transform of a rapidly decreasing function[2] $f \in \mathcal{L}^2(G)$ [where $G = SE(2)$] and its inverse transform are defined as*

$$\mathcal{F}(f) = \hat{f}(p) = \int_G f(g) U\left(g^{-1}, p\right) d(g)$$

and

$$\mathcal{F}^{-1}\left(\hat{f}\right) = f(g) = \int_0^\infty \text{trace}\left(\hat{f}(p) U(g, p)\right) p dp.$$

As with the Fourier transform of functions on $\mathbb{R}^N$,

$$\mathcal{F}\mathcal{F}^{-1}\left(\hat{f}\right) = \hat{f} \quad \mathcal{F}^{-1}\mathcal{F}(f) = f,$$

and so we write symbolically that

$$\mathcal{F}\mathcal{F}^{-1} = \mathcal{F}^{-1}\mathcal{F} = id$$

where id is the identity operator. A proof that these identities hold is given in [35]. The fact that the inverse transform works depends on $\{U(g, p)\}$ being a complete set of irreducible representations, and the fact that it is unitary allows us to write $U(g^{-1}, p) = U^\dagger(g, p)$ instead of computing the inverse of an infinite dimensional matrix. In Section 10.4.2 we show that the inversion formula works.

The matrix elements of the transform can be calculated using the matrix elements of $U(g, p)$ defined in Eq. (10.3) as

$$\hat{f}_{mn}(p) = \left(e^{im\psi}, \hat{f}(p) e^{in\psi}\right) = \int_G f(g) u_{mn}\left(g^{-1}, p\right) d(g).$$

Likewise, the inverse transform can be written in terms of elements as

$$f(g) = \sum_{n, m \in \mathbb{Z}} \int_0^\infty \hat{f}_{mn}(p) u_{nm}(g, p) p dp.$$

10.4 Properties of Convolution and Fourier Transforms of Functions on $SE(2)$

In Section 10.4.1, it is shown that the Fourier transform defined in Section 10.3 possesses the convolution property in an analogous way to the usual Fourier transform. In Section 10.4.2, the inversion formula is proved. In Section 10.4.3, Parseval's inequality is proved. In Section 10.4.4, some of the operational properties of the Fourier transform pair for functions on $SE(2)$ are derived.

10.4.1 The Convolution Theorem

Let us assume that there are real scalar-valued functions $f_1(\cdot)$, $f_2(\cdot) \in \mathcal{L}^2(G)$ where $G = SE(2)$. Recall that one of the most powerful properties of the Fourier transform of functions on $\mathbb{R}^N$ is that

---

[2] $f(\cdot)$ is rapidly decreasing if $\lim_{r \to \infty} r^n f(g(r, \phi, \theta)) = 0$ for all $n \in \mathbb{Z}^+$. Examples include functions $f(\cdot)$ which are zero outside of a compact subset of $SE(2)$ [i.e., if $f(\cdot)$ has compact support], and $f(g) = e^{-a^2 r^2} f(\phi, \theta)$ where $f(\cdot)$ is bounded.

the Fourier transform of the convolution of two functions is the product of the Fourier transforms of the functions. This property extends to the concept of a Fourier transform for functions on $SE(2)$. The proof of this fact presented here follows that of Sugiura [35].

Given that

$$(f_1 * f_2)(g) = \int_G f_1(h) f_2 \left(h^{-1} \circ g \right) d(h),$$

one gets

$$
\begin{aligned}
\mathcal{F}(f_1 * f_2) &= \int_G (f_1 * f_2)(g) U \left(g^{-1}, p \right) d(g) \\
&= \int_G \left(\int_G f_1(h) f_2 \left(h^{-1} \circ g \right) d(h) \right) U \left(g^{-1}, p \right) d(g).
\end{aligned}
\tag{10.8}
$$

Switching the order of integration, one gets

$$\mathcal{F}(f_1 * f_2) = \int_G \left(\int_G f_2 \left(h^{-1} \circ g \right) U \left(g^{-1}, p \right) d(g) \right) f_1(h) d(h).
\tag{10.9}$$

Since $d(\cdot)$ is left and right invariant

$$\int_G f(h \circ g) d(g) = \int_G f(g \circ h) d(g) = \int_G f \left(g^{-1} \right) d(g) = \int_G f(g) d(g)$$

for any function $F \in \mathcal{L}^2(SE(2))$. These facts allow us to write the inner integral in Eq. (10.9) as

$$\int_G f_2 \left(h^{-1} \circ (h \circ g) \right) U \left((h \circ g)^{-1}, p \right) d(g) = \int_G f_2(g) U \left(g^{-1} \circ h^{-1}, p \right) d(g).$$

Since $U(g, p)$ is a representation of $SE(2)$,

$$U \left(g^{-1} \circ h^{-1}, p \right) = U \left(g^{-1}, p \right) U \left(h^{-1}, p \right),$$

and so

$$
\begin{aligned}
\mathcal{F}(f_1 * f_2) &= \int_G \left(\int_G f_2(g) U \left(g^{-1}, p \right) U \left(h^{-1}, p \right) d(g) \right) f_1(h) d(h) \\
&= \left(\int_G f_2(g) U \left(g^{-1}, p \right) d(g) \right) \left(\int_G f_1(h) U \left(h^{-1}, p \right) d(h) \right) \\
&= \mathcal{F}(f_2) \mathcal{F}(f_1) = \hat{f}_2(p) \hat{f}_1(p).
\end{aligned}
$$

The order in which $U(g^{-1}, p)$ and $U(h^{-1}, p)$ appear is important because they are representations of $SE(2)$, which is not a commutative group. As seen previously, the fact that $U(g, p)$ is a representation is critical for the separation required for the convolution theorem to hold. The noncommutative nature of $SE(2)$ expressed in $U(g, p)$ is responsible for the reversed order of the product of the Fourier transforms. Some authors define the Fourier transform of functions on groups differently so that the order of the product of the transform is the same as the order of the convolved functions. We use the definition of Fourier transform given in Section 10.3 because it is the most analogous to the standard Fourier transform.

10.4.2 Proof of the Inversion Formula

We now present a proof that the Fourier inversion formula for functions on $SE(2)$ actually works. This proof is coordinate-dependent to avoid the introduction of additional mathematical machinery. Summation notation is used in certain places for simplicity, i.e., repeated indices indicate summation from $-\infty$ to ∞. A very elegant coordinate-independent proof can be found in [35]. The proof presented here is a variant of one found in [40].

Rapidly decreasing functions $f(g) \in \mathcal{L}^2(SE(2))$ where $g = g(a, \phi, \theta)$ can be expressed in a series of the form

$$f(g) = f(a, \phi, \theta) = \sum_{j,k \in \mathbb{Z}} F_{jk}(a) e^{-ij\phi} e^{-ik\theta}.$$

The matrix elements of the Fourier transform of this function (as defined in Section 10.3) are

$$\hat{f}_{mn} = \int_G f(g) u_{mn}(g^{-1}, p) \, d(g)$$

$$= \frac{1}{(2\pi)^2} \int_{-\pi}^{\pi} \int_{-\pi}^{\pi} \int_0^\infty \left(F_{jk}(a) e^{-ij\phi} e^{-ik\theta} \right) \left(i^{n-m} e^{i[m\theta + (n-m)\phi]} J_{m-n}(pa) \right) a \, da \, d\phi \, d\theta.$$

Rearranging the integrals, this is rewritten as

$$\left(i^{n-m} \int_0^\infty F_{jk}(a) J_{m-n}(pa) a \, da \right) \left(\frac{1}{2\pi} \int_{-\pi}^{\pi} e^{-ij\phi} e^{i(n-m)\phi} d\phi \right) \left(\frac{1}{2\pi} \int_{-\pi}^{\pi} e^{-ik\theta} e^{im\theta} d\theta \right)$$

$$= \delta_{k,m} \delta_{j,n-m} \left(i^{n-m} \int_0^\infty F_{jk}(a) J_{m-n}(pa) a \, da \right) = i^{n-m} \int_0^\infty F_{n-m,m}(a) J_{m-n}(pa) a \, da.$$

$\delta_{p,q}$ is equal to 1 if $p = q$ and zero otherwise, and there is no summation over indices in the last preceding term.

The fact that this Fourier transform matrix with elements $\hat{f}_{mn}$ is inverted using the inversion formula to reconstruct $f(g)$ is seen as follows:

$$f(g) = \int_0^\infty \text{trace}(\hat{f}(p) U(g, p)) p \, dp = \sum_{m,n \in \mathbb{Z}} \int_0^\infty \hat{f}_{mn} u_{nm}(g, p) p \, dp$$

$$= \sum_{m,n \in \mathbb{Z}} \int_0^\infty \left(i^{n-m} \int_{a=0}^\infty F_{n-m,m}(a) J_{m-n}(pa) a \, da \right) \left(i^{m-n} e^{-i[m\theta + (n-m)\phi]} J_{m-n}(pa) \right) p \, dp$$

$$= \sum_{m,n \in \mathbb{Z}} e^{-i[m\theta + (n-m)\phi]} \int_0^\infty \left(\int_{a=0}^\infty F_{n-m,m}(a) J_{m-n}(pa) a \, da \right) J_{m-n}(pa) p \, dp.$$

From here, the inversion is exactly the same as the Hankel transform pair Eqs. (3.36)–(3.37)

$$\hat{\phi}(p) = \int_0^\infty \phi(a) J_\nu(pa) a \, da,$$

then

$$\phi(a) = \int_0^\infty \hat{\phi}(p) J_\nu(pa) p \, dp.$$

In our case, $\hat{\phi}(p) = \hat{F}_{n-m,m}(p)$, $\phi(a) = F_{n-m,m}(a)$, and $\nu = m - n$. This means that

$$\sum_{m,n \in \mathbb{Z}} \int_0^\infty \hat{f}_{mn} u_{nm}(g, p) p \, dp = \sum_{m,n \in \mathbb{Z}} e^{-i[m\theta + (n-m)\phi]} F_{n-m,m}(a)$$

$$= \sum_{j,k \in \mathbb{Z}} F_{jk}(a) e^{-ij\phi} e^{-ik\theta}$$

$$= f(g).$$

This last step was simply a renaming of variables: $k = m$ and $n - m = j$. Since the sums over n and m are over $\mathbb{Z}$, these shifts do not change the range of summation.

10.4.3 Parseval's Equality

We now present a proof of Parseval's equality (also called the Plancherel formula) so that in later chapters we have a mechanism for regularizing integral equations on $SE(2)$ using the Fourier transform. The proof presented here is similar to the one found in [35].

We begin by defining

$$f^*(g) = \overline{f(g^{-1})} \in \mathcal{L}^2(SE(2)), \qquad \forall \quad f \in \mathcal{L}^2(SE(2)).$$

Then

$$\begin{aligned} h(g) &= f * f^*(g) \\ &= \int_G f(h) f^*(h^{-1} \circ g) d(h) = \int_G f(h) \overline{f(g^{-1} \circ h)} d(h). \end{aligned}$$

Evaluating this at the identity element $g = e$,

$$h(e) = \int_G f(h) \overline{f(h)} d(h) = \int_G |f(g)|^2 \, d(g).$$

The function $h(g)$ can also be expressed via the inversion formula as

$$h(g) = \int_0^\infty \operatorname{trace}(\hat{h}(p) U(g, p)) p \, dp.$$

Evaluated at the identity element, $U(e, p)$ is the identity operator, and so

$$h(e) = \int_0^\infty \operatorname{trace}(\hat{h}(p)) p \, dp = \int_0^\infty \operatorname{trace}\left(\hat{f}^*(p) \hat{f}(p)\right) p \, dp.$$

But

$$\mathcal{F}_{mn}(f^*) = \int_G \overline{f\left(g^{-1}\right)} u_{mn}\left(g^{-1}, p\right) d(g) = \int_G \overline{f(g)} u_{mn}(g, p) \, d(g).$$

This follows from the fact that $G = SE(2)$ is a unimodular group (possessing a volume element which is left and right invariant), and thus for any function $f \in \mathcal{L}^2(G)$

$$\int_G f\left(g^{-1}\right) d(g) = \int_G f(g) \, d(g).$$

From Eq. (10.4) we can then write

$$\mathcal{F}_{mn}(f^*) = \int_G \overline{f(g) u_{nm}\left(g^{-1}, p\right)} d(g) = \overline{\hat{f}}_{nm} = \overline{\hat{f}}_{mn}^T = \left(\hat{f}_{mn}\right)^\dagger,$$

i.e., $\mathcal{F}(f^*) = (\hat{f})^\dagger$ is the complex conjugate transpose of the Fourier transform matrix $\hat{f}$. This means that

$$h(e) = \int_0^\infty \operatorname{trace}\left(\hat{f}^\dagger(p) \hat{f}(p)\right) p \, dp = \int_0^\infty \left\|\hat{f}(p)\right\|_2^2 p \, dp.$$

Equating the two expressions for $h(e)$, we get Parseval's equality for $SE(2)$

$$\int_G |f(g)|^2 \, d(g) = \int_0^\infty \left\|\hat{f}(p)\right\|_2^2 p \, dp.$$

10.4.4 Operational Properties

Given a differentiable function f (with derivative f') such that $f, f' \in \mathcal{L}^2(\mathbb{R})$, recall that the Fourier transform of f' is defined using the inversion formula in Eq. (2.9) by differentiating under the integral

$$\frac{df}{dx} = \frac{1}{\sqrt{2\pi}} \frac{d}{dx} \int_{-\infty}^{\infty} \hat{f}(\omega) u(x, \omega) d\omega = \frac{1}{\sqrt{2\pi}} \int_{-\infty}^{\infty} \hat{f}(\omega) \frac{\partial u}{\partial x} d\omega. \tag{10.10}$$

Since $\frac{\partial u}{\partial x} = i\omega u(x, \omega)$, we get that the Fourier transform of df/dx is $i\omega \hat{f}(\omega)$ where $\hat{f}(\omega)$ is the Fourier transform of $f(x)$.

The same argument can be used to generate the Fourier transform matrices of derivatives of functions $f(g)$ where $g \in SE(2)$. For example, using the inverse transform given in Section 10.3 we get that

$$a \frac{\partial f}{\partial a} = a \int_0^\infty \hat{f}_{nm} \frac{\partial u_{mn}}{\partial a} p \, dp. \tag{10.11}$$

Since $u_{mn}(g(a, \phi, \theta), p) = i^{n-m} e^{-i[n\theta + (m-n)\phi]} J_{n-m}(pa)$, one gets that $\partial u_{mn}/\partial a = i^{n-m} e^{-i[n\theta + (m-n)\phi]} J'_{n-m}(pa)p$. Integrating the part of the expression in Eq. (10.11) which depends on p by parts

$$a \int_0^\infty \hat{f}_{nm}(p) J'_{n-m}(pa) p^2 dp = -\int_0^\infty \frac{d}{dp}\left(p^2 \hat{f}_{nm}(p)\right) J_{n-m}(pa) dp$$

$$= -\int_0^\infty \left(2\hat{f}_{nm}(p) + p\hat{f}'_{nm}(p)\right) J_{n-m}(pa) p \, dp.$$

Here we have used the fact that

$$a J'_{n-m}(pa) = \frac{\partial}{\partial p} J_{n-m}(pa).$$

This means that

$$a \frac{\partial f}{\partial a} = -\int_0^\infty \left(2\hat{f}_{nm} + p\hat{f}'_{nm}\right) u_{mn} p \, dp,$$

and so

$$\mathcal{F}\left(a \frac{\partial f}{\partial a}\right) = -\left(2\hat{f} + p\frac{d\hat{f}}{dp}\right). \tag{10.12}$$

We may then use Parseval's equality together with integration by parts to show that

$$\int_G \left| a \frac{\partial f}{\partial a} \right|^2 d(g) = \int_0^\infty \left\| p\frac{d\hat{f}}{dp} \right\|_2^2 p \, dp.$$

We find that the transforms of the derivatives of $f(g)$ with respect to ϕ and θ are

$$\mathcal{F}\left(\frac{\partial f}{\partial \theta}\right) = D\hat{f} \tag{10.13}$$

and

$$\mathcal{F}\left(\frac{\partial f}{\partial \phi}\right) = \hat{f}D - D\hat{f} \tag{10.14}$$

where $D_{mn} = -im\delta_{mn}$ (no sum) is a diagonal matrix. This comes from the fact that $\partial u_{mn}/\partial \theta = (-in)i^{n-m} e^{-i[n\theta+(m-n)\phi]} J_{n-m}(pa) = -in u_{mn}$ (no sum), and $\partial u_{mn}/\partial \phi = -i(m-n)i^{n-m} e^{-i[n\theta+(m-n)\phi]} J_{n-m}(pa) = -i(m-n)u_{mn}$ (no sum).

The transforms of other derivatives follow from these by composition.

10.5   Differential Operators for $SE(3)$

The left and right differential operators $\tilde{X}_i^L$ and $\tilde{X}_i^R$ for $i = 1, \ldots, 6$ acting on functions on $SE(3)$ are calculated much like they were for the cases of $SO(3)$ and $SE(2)$.

For small translational (rotational) displacements from the identity along (about) the ith coordinate axis, the homogeneous transforms representing infinitesimal motions look like

$$H_i(\epsilon) \overset{\Delta}{=} \exp\left(\epsilon \tilde{X}_i\right) \approx \mathbb{I}_{4\times4} + \epsilon \tilde{X}_i$$

where

$$\tilde{X}_1 = \begin{pmatrix} 0 & 0 & 0 & 0 \\ 0 & 0 & -1 & 0 \\ 0 & 1 & 0 & 0 \\ 0 & 0 & 0 & 0 \end{pmatrix}; \quad \tilde{X}_2 = \begin{pmatrix} 0 & 0 & 1 & 0 \\ 0 & 0 & 0 & 0 \\ -1 & 0 & 0 & 0 \\ 0 & 0 & 0 & 0 \end{pmatrix}; \quad \tilde{X}_3 = \begin{pmatrix} 0 & -1 & 0 & 0 \\ 1 & 0 & 0 & 0 \\ 0 & 0 & 0 & 0 \\ 0 & 0 & 0 & 0 \end{pmatrix};$$

$$\tilde{X}_4 = \begin{pmatrix} 0 & 0 & 0 & 1 \\ 0 & 0 & 0 & 0 \\ 0 & 0 & 0 & 0 \\ 0 & 0 & 0 & 0 \end{pmatrix}; \quad \tilde{X}_5 = \begin{pmatrix} 0 & 0 & 0 & 0 \\ 0 & 0 & 0 & 1 \\ 0 & 0 & 0 & 0 \\ 0 & 0 & 0 & 0 \end{pmatrix}; \quad \tilde{X}_6 = \begin{pmatrix} 0 & 0 & 0 & 0 \\ 0 & 0 & 0 & 0 \\ 0 & 0 & 0 & 1 \\ 0 & 0 & 0 & 0 \end{pmatrix}.$$

Sometimes in the literature $\tilde{X}_1, \tilde{X}_2, \tilde{X}_3$ are denoted J_1, J_2, J_3 and $\tilde{X}_4, \tilde{X}_5, \tilde{X}_6$ are denoted P_1, P_2, P_3 because they correspond respectively to infinitesimal rotations and translations about the 1, 2, and 3 axes. We use $\tilde{X}_i$ instead of X_i to avoid confusion with the $SO(3)$ case.

The commutation relations

$$\left[J_i, J_j\right] = \epsilon_{ijk} J_k \qquad \left[J_i, P_j\right] = \epsilon_{ijk} P_k \qquad \left[P_i, P_j\right] = 0$$

describe the Lie algebra $SE(3)$.

It is often convenient to write these in vector form as

$$\left(\tilde{X}_1\right)^\vee = \begin{pmatrix} 1 \\ 0 \\ 0 \\ 0 \\ 0 \\ 0 \end{pmatrix}; \quad \left(\tilde{X}_2\right)^\vee = \begin{pmatrix} 0 \\ 1 \\ 0 \\ 0 \\ 0 \\ 0 \end{pmatrix}; \quad \left(\tilde{X}_3\right)^\vee = \begin{pmatrix} 0 \\ 0 \\ 1 \\ 0 \\ 0 \\ 0 \end{pmatrix};$$

$$\left(\tilde{X}_4\right)^\vee = \begin{pmatrix} 0 \\ 0 \\ 0 \\ 1 \\ 0 \\ 0 \end{pmatrix}; \quad \left(\tilde{X}_5\right)^\vee = \begin{pmatrix} 0 \\ 0 \\ 0 \\ 0 \\ 1 \\ 0 \end{pmatrix}; \quad \left(\tilde{X}_6\right)^\vee = \begin{pmatrix} 0 \\ 0 \\ 0 \\ 0 \\ 0 \\ 1 \end{pmatrix}.$$

Given that elements of $SE(3)$ (viewed as homogeneous transforms) are parameterized as $H = H(\mathbf{q})$, the differential operators take the form

$$\tilde{X}_i^R f(H) = \lim_{\epsilon \to 0} \frac{1}{\epsilon} \left[f(H \circ H_i(\epsilon)) - f(H)\right] = \frac{df(H \circ (\mathbb{I} + t\tilde{X}_i))}{dt}\bigg|_{t=0} \qquad (10.15)$$

$$\tilde{X}_i^L f(H) = \lim_{\epsilon \to 0} \frac{1}{\epsilon} \left[f(H_i^{-1}(\epsilon) \circ H) - f(H) \right] = \frac{df((\mathbb{I} - t\tilde{X}_i) \circ H)}{dt} \Bigg|_{t=0}. \tag{10.16}$$

Since H and $H_i(\epsilon)$ are 4×4 matrices, we henceforth drop the "$\circ$" notation since it is understood as matrix multiplication.

Analogous to the $SO(3)$ case, we observe for the case of $\tilde{X}_i^R$ that

$$H + \epsilon H \tilde{X}_i = H H_i(\epsilon) = H + \epsilon \sum_{j=1}^{6} \frac{\partial H}{\partial q_j} \frac{\partial q_j^{R,i}}{\partial \epsilon} \Bigg|_{\epsilon=0}$$

We then have that

$$\tilde{X}_i = \sum_{j=1}^{6} H^{-1} \frac{\partial H}{\partial q_j} \frac{\partial q_j^{R,i}}{\partial \epsilon} \Bigg|_{\epsilon=0},$$

or

$$\left(\tilde{X}_i \right)^{\vee} = \sum_{j=1}^{6} \left(H^{-1} \frac{\partial H}{\partial q_j} \right)^{\vee} \frac{\partial q_j^{R,i}}{\partial \epsilon} \Bigg|_{\epsilon=0},$$

which is written as $(\tilde{X}_i)^{\vee} = \mathcal{J}_R(\mathbf{q}) \frac{d\mathbf{q}^{R,i}}{d\epsilon} |_{\epsilon=0}$ where $\mathcal{J}_R$ is the $SE(3)$ right Jacobian defined in Chapter 6. This allows us to solve for

$$\frac{d\mathbf{q}^{R,i}}{d\epsilon} \Bigg|_{\epsilon=0} = \mathcal{J}_R^{-1}(\tilde{X}_i)^{\vee},$$

which is used to calculate

$$\tilde{X}_i^R f = \sum_{j=1}^{6} \frac{\partial f}{\partial q_j} \frac{\partial q_j^{R,i}}{\partial \epsilon} \Bigg|_{\epsilon=0}$$

We use the tilde to distinguish between the full motion and rotation operators. For the case when the rotations are parameterized with ZXZ Euler angles α, β, γ, and translations are parameterized in Cartesian coordinates a_1, a_2, a_3, one finds

$$\tilde{X}_i^R = \begin{cases} X_i^R & \text{for } i = 1, 2, 3 \\ (R^T \nabla_{\mathbf{a}})_{i-3} & \text{for } i = 4, 5, 6 \end{cases} \tag{10.17}$$

where X_i^R is defined in Section 9.10, and $(\nabla_{\mathbf{a}})_i = \partial / \partial a_i$.

The operators $\tilde{X}_i^L$ are calculated from $\tilde{X}_i^R$ by observing that $H J H^{-1} = Ad(H)J$, and so

$$\tilde{X}_i^L = \begin{cases} X_i^L + \sum_{k=1}^{3} (\mathbf{a} \times \mathbf{e}_i) \cdot \mathbf{e}_k \partial / \partial a_k & \text{for } i = 1, 2, 3 \\ -\partial / \partial a_{i-3} & \text{for } i = 4, 5, 6 \end{cases} \tag{10.18}$$

For a more abstract treatment of differential operators on $SE(N)$ (including combinations of the preceding operators that commute with left and right shifts) see [8, 38].

10.6 Irreducible Unitary Representations of $SE(3)$

We define unitary representations of $SE(3)$ (see [19, 37, 24, 10, 42] for discussions and definitions) in the following way.

We start to construct the representation of the motion group in the space of functions $\varphi(\mathbf{p}) \in \mathcal{L}^2(\hat{\mathbf{T}})$, where $\hat{T}$ is a dual (frequency) space of the $\mathbb{R}^3$ subgroup. Functions $\varphi(\mathbf{p})$ corresponding to the Fourier transforms of the functions $\varphi(\mathbf{r}) \in \mathcal{L}^2(T)$, where $T = \mathbb{R}^3$, are defined as

$$\varphi(\mathbf{p}) = \frac{1}{(2\pi)^{3/2}} \int_T e^{-i\mathbf{p}\cdot\mathbf{r}} \varphi(\mathbf{r}) \, d\mathbf{r}. \tag{10.19}$$

In the current context, we use the same notation $\varphi(\mathbf{p})$ for the Fourier transform of the function as for the function $\varphi(\mathbf{r})$ itself in order to avoid cumbersome notations. The argument $\mathbf{p}$ or $\mathbf{r}$ is sufficient to specify which we are considering.

The rotation subgroup $SO(3)$ of the motion group acts on $\hat{T}$ by rotations, so $\hat{T}$ is divided into orbits S_p, where S_p are S^2 spheres of radius $p = |\mathbf{p}|$. The translation operator acts on $\varphi(\mathbf{p})$ as

$$(U(\mathbf{a}, \mathbb{I})\varphi)(\mathbf{p}) = e^{-i\mathbf{p}\cdot\mathbf{a}} \varphi(\mathbf{p}). \tag{10.20}$$

Therefore, the irreducible representations of the motion group may be built on spaces $\varphi(\mathbf{p}) \in \mathcal{L}^2(S_p)$, with the inner product defined as

$$(\varphi_1, \varphi_2) = \int_{\Theta=0}^{\pi} \int_{\Phi=0}^{2\pi} \overline{\varphi_1(\mathbf{p})} \, \varphi_2(\mathbf{p}) \sin\Theta \, d\Theta \, d\Phi, \tag{10.21}$$

where $\mathbf{p} = (p \sin\Theta \cos\Phi, \, p \sin\Theta \sin\Phi, \, p \cos\Theta)$, and $p > 0, 0 \leq \Theta \leq \pi, 0 \leq \Phi \leq 2\pi$.

The inner product (φ_1, φ_2) is invariant with respect to transformations

$$\varphi(\mathbf{p}) \rightarrow e^{i\alpha} \varphi\left(A^{-1}\mathbf{p}\right), \tag{10.22}$$

where $A \in SO(3)$ and $0 \leq \alpha \leq 2\pi$.

The parameter α in Eq. (10.22) may, in general, depend on p and group element $A \in SO(3)$. In this case, different functions $\alpha_s(p, A)$ [where s enumerates the irreducible representations of $SO(2)$], which are nonlinear functions of group element A, correspond to different irreducible representations of the motion group. Functions $\varphi(\mathbf{p})$, thus, may have different *internal* properties with respect to rotations.

With the help of functions $\alpha_s(p, A)$ we may construct the representations of $G = SE(3) \simeq \hat{T} \rtimes_{\varphi} SO(3)$ from representations of its subgroup $G' = \hat{T} \rtimes_{\varphi} SO(2)$ using the method of induced representations. In our case (we disregard for the moment the translation group $\hat{T}$), $G = SO(3)$, $H = SO(2)$, and $\sigma = \mathbf{p} \in S_p \simeq SO(3)/SO(2)$.

To construct the representations of the motion group explicitly, we choose a particular vector $\hat{\mathbf{p}} = (0, 0, p)$ on each orbit S_p. The vector $\hat{\mathbf{p}}$ is invariant with respect to rotations from the $SO(2)$ subgroup of $SO(3)$

$$\Lambda \hat{\mathbf{p}} = \hat{\mathbf{p}}; \quad \Lambda \in H_{\hat{\mathbf{p}}} = SO(2), \tag{10.23}$$

where $H_{\hat{\mathbf{p}}}$ is a little group of $\hat{\mathbf{p}}$. For each $\mathbf{p} \in S_p$ we may find $R_\mathbf{p} \in SO(3)/SO(2)$, such that

$$R_\mathbf{p} \hat{\mathbf{p}} = \mathbf{p}.$$

Then for any $A \in SO(3)$, one may check that

$$\left(R_\mathbf{p}^{-1} A R_{A^{-1}\mathbf{p}}\right) \hat{\mathbf{p}} = \hat{\mathbf{p}}.$$

Therefore, $Q(\mathbf{p}, A) \stackrel{\Delta}{=} (R_\mathbf{p}^{-1} A R_{A^{-1}\mathbf{p}}) \in H_{\hat{\mathbf{p}}}$. The representations of $H_{\hat{\mathbf{p}}}$ may be taken to be of the form

$$\Delta_s: \quad \phi \rightarrow e^{is\phi}, \quad 0 \leq \phi \leq 2\pi$$

for $s = 0, \pm 1, \pm 2, \ldots$.

Thus we may construct the induced representation $(\hat{T} \rtimes_\varphi \Delta_s(H_{\hat{\mathbf{p}}})) \uparrow SE(3)$ of the motion group from the representations of its subgroup $\hat{T} \rtimes_\varphi H_{\hat{\mathbf{p}}}$.

DEFINITION 10.2 *The unitary representations $U^s(\mathbf{a}, A)$ of $SE(3)$, which act on the space of functions $\varphi(\mathbf{p})$ with the inner product Eq. (10.21), are defined by*

$$\left(U^s(\mathbf{a}, A)\varphi\right)(\mathbf{p}) = e^{-i\mathbf{p}\cdot\mathbf{a}} \Delta_s \left(R_\mathbf{p}^{-1} A R_{A^{-1}\mathbf{p}}\right) \varphi\left(A^{-1}\mathbf{p}\right), \tag{10.24}$$

where $A \in SO(3)$, Δ_s are representations of $H_{\hat{\mathbf{p}}}$, and $s = 0, \pm 1, \pm 2, \ldots$.

Explicitly, if A is parameterized as in Eq. (5.60), and $\mathbf{p}$ has polar and azimuthal angles Θ, Φ, then [24]

$$\Delta_s \left(R_\mathbf{p}^{-1} A R_{A^{-1}\mathbf{p}}\right) = \left(\frac{(1 - 2b\bar{b})\sin\Theta - abe^{i\Phi}(1 + \cos\Theta) + \bar{a}be^{i\Phi}(1 - \cos\Theta)}{(1 - 2b\bar{b})\sin\Theta + abe^{i\Phi}(1 - \cos\Theta) - \bar{a}be^{-i\Phi}(1 + \cos\Theta)}\right)^s.$$

Each representation characterized by $p = \|\mathbf{p}\|$ and s is irreducible [they, however, become reducible if we restrict $SE(3)$ to $SO(3)$, i.e., when $a = \|\mathbf{a}\| = 0$]. They are unitary because $(U^s(\mathbf{a}, A)\varphi_1, U^s(\mathbf{a}, A)\varphi_2) = (\varphi_1, \varphi_2)$.

We note that Eq. (10.24) may be written in the form

$$\left(U^s(\mathbf{a}, A)\varphi\right)(\mathbf{u}) = e^{-ip\mathbf{u}\cdot\mathbf{a}} \Delta_s \left(R_\mathbf{u}^{-1} A R_{A^{-1}\mathbf{u}}\right) \varphi\left(A^{-1}\mathbf{u}\right), \tag{10.25}$$

where $\mathbf{p} = p\mathbf{u}$ and $\mathbf{u}$ is a unit vector. Here $\varphi(\cdot)$ is defined on the unit sphere.

Representations Eq. (10.24), which we denote below by $U^s(g, p)$, satisfy the homomorphism properties

$$U^s(g_1 \circ g_2, p) = U^s(g_1, p) \cdot U^s(g_2, p),$$

where $\circ$ is the group operation. The corresponding multiplication law for the $Q(\mathbf{p}, A)$ factors is [43]

$$Q(\mathbf{p}, A) Q\left(A^{-1}\mathbf{p}, A^{-1}B\right) = Q(\mathbf{p}, B). \tag{10.26}$$

Analogous to the planar case, we define $g_i(t) = \exp(t\tilde{X}_i)$ for $i = 1, \ldots, 6$ and evaluate the representation Eq. (10.25). The result for $k = 1, 2, 3$ is

$$U(g_k(t), p)\varphi(\psi) = \Delta_s \left(R_\mathbf{u}^{-1} \text{ROT}[\mathbf{e}_k, t] R_{\text{ROT}[\mathbf{e}_k, -t]\mathbf{u}}\right) \varphi(\text{ROT}[\mathbf{e}_k, -t]\mathbf{u});$$

$$U^s\left(g_{k+3}(t), p\right)\varphi(\psi) = e^{-ipt\mathbf{e}_k\cdot\mathbf{u}}\varphi(\mathbf{u}).$$

Differentiating with respect to t and setting $t = 0$, we define the operators

$$\hat{X}_i(p, s)\varphi(\mathbf{u}) = \left.\frac{dU^s(\exp(t\tilde{X}_i), p)\varphi(\mathbf{u})}{dt}\right|_{t=0}$$

for $i = 1, \ldots, 6$. Explicitly, with $\mathbf{u} = \mathbf{u}(\Theta, \Phi)$,

$$\hat{J}_1\varphi(\mathbf{u}) \stackrel{\triangle}{=} \hat{X}_1(p, s)\varphi(\mathbf{u}) = \left(\sin\Phi\frac{\partial}{\partial\Theta} - is\frac{\cos\Phi}{\sin\Theta} + \cos\Phi\cot\Theta\frac{\partial}{\partial\Phi}\right)\varphi(\mathbf{u})$$

$$\hat{J}_2\varphi(\mathbf{u}) \stackrel{\triangle}{=} \hat{X}_2(p, s)\varphi(\mathbf{u}) = \left(-\cos\Phi\frac{\partial}{\partial\Theta} - is\frac{\sin\Phi}{\sin\Theta} + \sin\Phi\cot\Theta\frac{\partial}{\partial\Phi}\right)\varphi(\mathbf{u})$$

$$\hat{J}_3\varphi(\mathbf{u}) \stackrel{\triangle}{=} \hat{X}_3(p, s)\varphi(\mathbf{u}) = -\frac{\partial}{\partial\Phi}\varphi(\mathbf{u})$$

$$\hat{P}_k\varphi(\mathbf{u}) \stackrel{\triangle}{=} \hat{X}_{k+3}(p, s)\varphi(\mathbf{u}) = -ip\,(\mathbf{e}_k \cdot \mathbf{u})\,\varphi(\mathbf{u}) \tag{10.27}$$

for $k = 1, 2, 3$.

10.7   Matrix Elements

To obtain the matrix elements of the unitary representations, we use the group property

$$U^s(\mathbf{a}, A) = U^s(\mathbf{a}, \mathbb{I}) \cdot U^s(0, A). \tag{10.28}$$

The basis eigenfunctions of the irreducible representations Eq. (10.24) of $SE(3)$ may be enumerated by the integer numbers l, m (for each s and p). We note that the values $l(l+1)$, m, ps, and $-p^2$ correspond to the eigenvalues of the operators $\hat{\mathbf{J}}^2$, $\hat{J}^3$, $\hat{\mathbf{P}} \cdot \hat{\mathbf{J}} = \sum_{i=1}^3 \hat{P}_i \hat{J}_i$ and $\hat{\mathbf{P}} \cdot \hat{\mathbf{P}} = \sum_{i=1}^3 \hat{P}_i \hat{P}_i$ (where $\hat{J}_i$, $\hat{P}_i$, $i = 1, 2, 3$ are rotation and translation operators defined in Eq. (10.27) and $\hat{J}^3 = i\hat{J}_3$, $\hat{P}^3 = i\hat{P}_3$) which may be diagonalized simultaneously (i.e., they commute). The restrictions for the l, m, s numbers are $l \geq |s|$; $l \geq |m|$.

The basis functions may be expressed in the form [19]

$$h^l_{ms}(\mathbf{u}(\Theta, \Phi)) = Q^l_{s,m}(\cos\Theta)\, e^{i(m+s)\Phi}, \tag{10.29}$$

where

$$Q^l_{-s,m}(\cos\Theta) = (-1)^{l-s}\sqrt{\frac{2l+1}{4\pi}}\, P^l_{sm}(\cos\Theta), \tag{10.30}$$

and generalized Legendre functions $P^l_{ms}(\cos\Theta)$ are given as in Vilenkin and Klimyk [42].

It may be shown that these basis functions are transformed under the rotations $h^l_{ms}(\mathbf{u}) \rightarrow \Delta_s(Q(\mathbf{u}, A))\, h^l_{ms}(A^{-1}\mathbf{u})$ as (see Chapter 9 for the proof)

$$\left(U^s(0, A)\, h^l_{ms}\right)(\mathbf{u}) = \sum_{n=-l}^{l} \tilde{U}^l_{nm}(A) h^l_{ns}(\mathbf{u}), \tag{10.31}$$

where the matrix elements $\tilde{U}^l_{nm}(A)$ are

$$\tilde{U}^l_{mn}(A) = e^{-im\alpha}\, (-1)^{n-m}\, P^l_{mn}(\cos\beta)\, e^{-in\gamma}, \tag{10.32}$$

where α, β, γ are ZXZ Euler angles of the rotation. We note that the rotation matrix elements do not depend on s.

The translation matrix elements are given by the integral [19]

$$\left(h^{l'}_{m's}, U^s(\mathbf{a}, \mathbb{I}) h^l_{ms}\right) = [l', m' \mid p, s \mid l, m](\mathbf{a})$$

$$= \int_{\Theta=0}^{\pi} \int_{\Phi=0}^{2\pi} Q^{l'}_{s,m'}(\cos\Theta) e^{-i(m'+s)\Phi}\, e^{-i\mathbf{p}\cdot\mathbf{a}}$$

$$Q^l_{s,m}(\cos\Theta) e^{i(m+s)\Phi}\, \sin\Theta\, d\Theta\, d\Phi. \tag{10.33}$$

These are written in closed form as [19]

$$[l', m' \mid p, s \mid l, m](\mathbf{a}) = (4\pi)^{1/2} \sum_{k=|l'-l|}^{l'+l} i^k \sqrt{\frac{(2l'+1)(2k+1)}{(2l+1)}}\, j_k(p\,a)\, C(k, 0; l', s \mid l, s)$$

$$C(k, m-m'; l', m' \mid l, m)\, Y_k^{m-m'}(\theta, \phi), \tag{10.34}$$

where θ, ϕ are polar and azimuthal angles of $\mathbf{a}$, and $C(k, m - m'; l', m' \mid l, m)$ are Clebsch-Gordan coefficients (see, for example, [13]).

Finally, using the group property Eq. (10.28), the matrix elements of the unitary representation $U^s(g, p)$ Eq. (10.24) (for $s = 0, \pm 1, \pm 2, \ldots$) are expressed as

$$U^s_{l',m';l,m}(\mathbf{a}, A; p) = \sum_{j=-l}^{l} [l', m' \mid p, s \mid l, j] (\mathbf{a}) \tilde{U}^l_{jm}(A). \qquad (10.35)$$

Because Eq. (10.34) contains only half-integer Bessel functions, all matrix elements may be expressed in terms of elementary functions. Below we have shown several matrix elements in explicit form (using the notation $a = \|\mathbf{a}\|$ and $p = \|\mathbf{p}\|$)

$$U^0_{0,0;0,0}(\mathbf{a}, A; p) = \frac{\sin(a\,p)}{a\,p};$$

$$U^0_{1,0;0,0}(\mathbf{a}, A; p) = \frac{i\sqrt{3}\cos(\theta)\,(-\cos(a\,p) + \sin(a\,p)/(a\,p))}{a\,p};$$

$$U^0_{1,-1;0,0}(\mathbf{a}, A; p) = \frac{-i\sqrt{\frac{3}{2}}\,e^{i\phi}\,(a\,p\cos(a\,p) - \sin(a\,p))\sin(\theta)}{a^2\,p^2};$$

$$U^0_{1,1;0,0}(\mathbf{a}, A; p) = \frac{i\sqrt{\frac{3}{2}}\,e^{-i\phi}\,(a\,p\cos(a\,p) - \sin(a\,p))\sin(\theta)}{a^2\,p^2};$$

$$U^0_{2,1;0,0}(\mathbf{a}, A; p) = -\frac{1}{a^3\,p^3}\sqrt{\frac{15}{8}}\,e^{-i\phi}\,(3a\,p\cos(a\,p) - 3\sin(a\,p) + a^2\,p^2\sin(a\,p))\sin(2\theta);$$

$$U^0_{1,-1;1,1}(\mathbf{a}, A; p) = \frac{3}{4a^3\,p^3}\,e^{-i\alpha-i\gamma}(2a\,e^{2i\alpha}\,p\cos(a\,p)\left(\sin\frac{\beta}{2}\right)^2$$

$$-6ae^{2i\alpha}\,p\cos(a\,p))(\cos\theta)^2\left(\sin\frac{\beta}{2}\right)^2 - 2e^{2i\alpha}\left(\sin\frac{\beta}{2}\right)^2\sin(a\,p)$$

$$+2a^2\,e^{2i\alpha}\,p^2\left(\sin\frac{\beta}{2}\right)^2\sin(a\,p)(\sin\theta)^2 + 6e^{2i\alpha}(\cos\theta)^2\left(\sin\frac{\beta}{2}\right)^2\sin(a\,p)$$

$$-6a\,e^{2i\phi}\,p\left(\cos\frac{\beta}{2}\right)^2\cos(a\,p)(\sin\theta)^2 + 6e^{2i\phi}\left(\cos\frac{\beta}{2}\right)^2\sin(a\,p)(\sin\theta)^2$$

$$-2a^2\,e^{2i\phi}\,p^2\left(\cos\frac{\beta}{2}\right)^2\sin(a\,p)(\sin\theta)^2 + 3a\,e^{i\phi+i\alpha}\,p\cos(a\,p)\sin(\beta)\sin(2\theta)$$

$$-3e^{i\phi+i\alpha}\sin(\beta)\sin(a\,p)\sin(2\theta) + a^2\,e^{i\phi+i\alpha}\,p^2\sin(\beta)\sin(a\,p)\sin(2\theta), \qquad (10.36)$$

where $a = \mid \mathbf{a} \mid, \theta, \phi$ are polar and azimuthal angles of $\mathbf{a}$, and α, β, γ are Euler angles of the $SO(3)$ rotation.

Symmetry Properties

We note symmetry properties of the matrix elements. Using the property of the Clebsch-Gordan coefficients [13]

$$C(l_1, m_1; l_2, m_2 \mid l, m) = (-1)^{(l_1+l_2+l)} C(l_1, -m_1; l_2, -m_2 \mid l, -m), \qquad (10.37)$$

(for integer l_1, l_2, l), the property

$$Y_l^{-m}(\theta, \phi) = (-1)^m \overline{Y}_l^m(\theta, \phi)$$

and the symmetry relation [42]

$$P_{m\,n}^l(x) = (-1)^{(m-n)} P_{-m,-n}^l(x),$$

it may be shown that

$$\overline{U_{l',m';l,m}^s}(\mathbf{a}, A; p) = (-1)^{(l'-l)}(-1)^{(m'-m)} U_{l',-m';l,-m}^s(\mathbf{a}, A; p). \tag{10.38}$$

Also, it may be observed from the transformation law Eq. (10.24) that the complex conjugate transformation is related with the $(-s)$ transformation evaluated at $(-\mathbf{a}, A)$. Explicitly,

$$\overline{U_{l',m';l,m}^{-s}}(-\mathbf{a}, A; p) = (-1)^{(m'-m)} U_{l',-m';l,-m}^s(\mathbf{a}, A; p). \tag{10.39}$$

We also list here a unitarity relation

$$U_{l',m';l,m}^s\left(-A^{-1}\mathbf{a}, A^{-1}; p\right) = \overline{U_{l,m;l',m'}^s(\mathbf{a}, A; p)}. \tag{10.40}$$

Orthogonality

Using the orthogonality of the rotation matrix elements $\tilde{U}_{mn}^l$, the integral expression Eq. (10.33) for the translation matrix elements, and the integral representation for the δ-function

$$\int_{\mathbb{R}^3} e^{i(\mathbf{p}-\mathbf{p}')\cdot\mathbf{r}} d^3\mathbf{r} = (2\pi)^3 \delta\left(\mathbf{p} - \mathbf{p}'\right) \tag{10.41}$$

it may be shown [19] that the $SE(3)$ matrix elements satisfy the orthogonality relation

$$\int_{\mathbb{R}^3} d\mathbf{a} \int_{SO(3)} dA \, \overline{U_{l_1,m_1;j_1,k_1}^{s_1}}(\mathbf{a}, A; p_1) U_{l,m;j,k}^s(\mathbf{a}, A; p)$$

$$= 2\pi^2 \delta_{l_1 l} \, \delta_{j_1 j} \, \delta_{m_1 m} \, \delta_{k_1 k} \, \delta_{s_1 s} \, \frac{\delta(p_1 - p)}{p^2} \tag{10.42}$$

where $d\mathbf{a} = a^2 \, da \, \sin\theta \, d\theta \, d\phi$.

10.8   The Fourier Transform for $SE(3)$

Here we define the Fourier transform of functions $f(\mathbf{a}, A) \in \mathcal{L}^2(SE(3))$. The inner product of two such functions is given by

$$(f_1, f_2) = \int_{\mathbb{R}^3} \int_{SO(3)} \overline{f_1(\mathbf{a}, A)} \, f_2(\mathbf{a}, A) \, dA \, d\mathbf{a}. \tag{10.43}$$

To define the Fourier transform for functions on $SE(3)$, we have to use a complete orthogonal basis for functions on this group. The completeness of matrix elements Eq. (10.35) depends in part on the completeness of the rotation matrix elements $\tilde{U}_{mn}^l(A)$ on $SO(3)$ [19]. The orthogonality relation for the matrix elements is given in Eq. (10.42). So, using the unitary representations $U(g, p)$ Eq. (10.24) (for $s = 0, \pm 1, \pm 2, \ldots$), we may define the Fourier transform of functions on the motion group.

DEFINITION 10.3 *For any absolutely- and square-integrable complex-valued function $f(\mathbf{a}, A)$ on $SE(3)$ we define the Fourier transform as*

$$\mathcal{F}(f) = \hat{f}(p) = \int_{SE(3)} f(g) U\left(g^{-1}, p\right) d(g)$$

where $g = (\mathbf{a}, A) \in SE(3)$ and $d(g) = dA\, d\mathbf{a}$.

The matrix elements of the transform are given in terms of matrix elements Eq. (10.35) as

$$\hat{f}^s_{l',m';l,m}(p) = \int_{SE(3)} f(\mathbf{a}, A) \overline{U^s_{l,m;l',m'}(\mathbf{a}, A; p)}\, dA\, d\mathbf{a} \tag{10.44}$$

where we have used the unitarity property.

The inverse Fourier transform is defined by

$$f(g) = \mathcal{F}^{-1}(\hat{f}) = \frac{1}{2\pi^2} \int_{SE(3)} \text{trace}(\hat{f}(p) U(g, p))\, p^2\, dp. \tag{10.45}$$

Explicitly

$$f(\mathbf{a}, A) = \frac{1}{2\pi^2} \sum_{s=-\infty}^{\infty} \sum_{l'=|s|}^{\infty} \sum_{l=|s|}^{\infty} \sum_{m'=-l'}^{l'} \sum_{m=-l}^{l} \int_0^{\infty} p^2\, dp\, \hat{f}^s_{l,m;l',m'}(p) U^s_{l',m';l,m}(\mathbf{a}, A; p). \tag{10.46}$$

We note that we may use any unitary equivalent representation $T^{\dagger} U(g, p) T$ (where T is a unitary transformation, which does not depend on g) to define the Fourier transform.

Convolution of Functions

Recall that the convolution integral of functions $f_1, f_2 \in \mathcal{L}^2(SE(3))$ may be defined as

$$(f_1 * f_2)(g) = \int_{SE(3)} f_1(h)\, f_2\left(h^{-1} \circ g\right) d(h). \tag{10.47}$$

One of the most powerful properties of the Fourier transform of functions on $\mathbf{R}^N$ is that the Fourier transform of the convolution of two functions is the product of the Fourier transform of the functions. This property persists also for the convolution of functions on the group, namely

$$\mathcal{F}(f_1 * f_2) = \mathcal{F}(f_2)\, \mathcal{F}(f_1) \tag{10.48}$$

or, in the matrix form

$$(\mathcal{F}(f_1 * f_2))^s_{l',m';l,m}(p) = \sum_{j=|s|}^{\infty} \sum_{k=-j}^{j} \left(\hat{f}_2\right)^s_{l',m';j,k}(p) \left(\hat{f}_1\right)^s_{j,k;l,m}(p). \tag{10.49}$$

Parseval/Plancherel Equality

This form of Parseval equality is valid

$$\int_{SE(3)} |f(\mathbf{a}, A)|^2 \, dA da$$

$$= \frac{1}{2\pi^2} \sum_{s=-\infty}^{\infty} \sum_{l'=|s|}^{\infty} \sum_{l=|s|}^{\infty} \sum_{m'=-l'}^{l'} \sum_{m=-l}^{l} \int_0^{\infty} \left| \hat{f}_{l',m';l,m}^s(p) \right|^2 p^2 \, dp$$

$$= \frac{1}{2\pi^2} \int_0^{\infty} \left\| \hat{f}(p) \right\|_2^2 p^2 \, dp, \tag{10.50}$$

where the Hilbert-Schmidt norm of $\hat{f}(p)$ is given by

$$\left\| \hat{f}(p) \right\|_2^2 = \sum_{s=-\infty}^{\infty} \sum_{l'=|s|}^{\infty} \sum_{l=|s|}^{\infty} \sum_{m'=-l'}^{l'} \sum_{m=-l}^{l} \left| \hat{f}_{l',m';l,m}^s(p) \right|^2.$$

A similar relation holds for the inner product

$$\int_{SE(3)} (f(\mathbf{a}, A), g(\mathbf{a}, A)) \, dA da$$

$$= \frac{1}{2\pi^2} \sum_{s=-\infty}^{\infty} \sum_{l'=|s|}^{\infty} \sum_{l=|s|}^{\infty} \sum_{m'=-l'}^{l'} \sum_{m=-l}^{l} \int_0^{\infty} \overline{\hat{f}_{l',m';l,m}^s(p)} \hat{g}_{l',m';l,m}^s(p) \, p^2 \, dp$$

$$= \frac{1}{2\pi^2} \int_0^{\infty} \text{trace} \left(\hat{f}^{\dagger}(p) \hat{g}(p) \right) p^2 \, dp, \tag{10.51}$$

where $\hat{f}_{l,m;l',m'}^{\dagger} = \overline{\hat{f}_{l',m';l,m}}$ is the Hermitian conjugate.

Symmetries

For the real function $f(\mathbf{a}, A)$, we note a symmetry property of the Fourier transform, which follows from symmetry Eq. (10.38) of the matrix elements

$$\overline{\hat{f}_{l',m';l,m}^s(p)} = (-1)^{(l'-l)} (-1)^{(m'-m)} \hat{f}_{l',-m';l,-m}^s(p). \tag{10.52}$$

Others analogous to those for $SE(2)$ also exist.

Contraction of Indices

It is convenient to rewrite the 4-index Fourier transform matrix element $\hat{f}_{l',m';l,m}^s(p)$ as a 2-index matrix $\hat{f}_{ij}^s(p)$. To satisfy the matrix product definition in Eq. (10.49) we arrange l, m indices in a row (we show an example for $s = 0$), 0 (for $l = 0$); $-1, 0, 1$ (for $l = 1$); $-2, -1, 0, 1, 2$ (for $l = 2$); $\ldots$, which corresponds to $j = 1, 2, 3, 4, \ldots$.
Explicitly

$$\hat{f}_{l',m';l,m}^s(p) = \hat{f}_{ij}^s(p), \tag{10.53}$$

where $i = l'(l' + 1) + m' - s^2 + 1$; $j = l(l + 1) + m - s^2 + 1$.

In particular, any 4-index matrix such as $\hat{f}^0_{l',m';l,m}$ where $0 \le l, l' \le L$, $|m| \le l$ and $|m'| \le l'$ can be written as an equivalent $(L+1)^2 \times (L+1)^2$ matrix with two indices using the rule

$$l'(l'+1) + m' + 1 \to i; \qquad l(l+1) + m + 1 \to j$$

where

$$\hat{f}^0_{i,j} \triangleq \hat{f}^0_{l',m';l,m}.$$

In this way the product is preserved. That is,

$$\sum_{j=1}^{(L+1)^2} \hat{f}^0_{i,j}\hat{g}^0_{j,k} = \sum_{l=0}^{L} \sum_{m=-l}^{l} \hat{f}^0_{l',m';l,m}\hat{g}^0_{l,m;l'',m''}$$

where $k = l''(l''+1) + m'' + 1$. For instance, if $L = 1$, we write

$$\hat{f}^0 = \begin{pmatrix} \hat{f}^0_{0,0;0,0} & \hat{f}^0_{0,0;1,-1} & \hat{f}^0_{0,0;1,0} & \hat{f}^0_{0,0;1,1} \\ \hat{f}^0_{1,-1;0,0} & \hat{f}^0_{1,-1;1,-1} & \hat{f}^0_{1,-1;1,0} & \hat{f}^0_{1,-1;1,1} \\ \hat{f}^0_{1,0;0,0} & \hat{f}^0_{1,0;1,-1} & \hat{f}^0_{1,0;1,0} & \hat{f}^0_{1,0;1,1} \\ \hat{f}^0_{1,1;0,0} & \hat{f}^0_{1,1;1,-1} & \hat{f}^0_{1,1;1,0} & \hat{f}^0_{1,1;1,1} \end{pmatrix}. \tag{10.54}$$

For other approaches to the representation theory and harmonic analysis on the Euclidean motion groups see [1, 4, 6, 11, 12, 25, 26, 27, 28, 29, 30, 31, 32, 33, 34, 36, 39]. For applications of representations of motion groups with scale changes, see [20, 21].

10.9 Operational Properties

The general formulation of operational properties for Lie group Fourier transforms described in Chapter 8 is illustrated here concretely. In particular, when $G = SE(3)$,

$$u\left(\tilde{X}_i, p\right) \triangleq \frac{d}{dt}\left(U\left(\exp t\tilde{X}_i\right), p\right)\Big|_{t=0}, \tag{10.55}$$

and by substitution we see that

$$\begin{aligned} \mathcal{F}\left(\tilde{X}^L f\right) &= \frac{d}{dt}\int_G f\left(\left(\exp t\tilde{X}_i\right) \circ g\right)\Big|_{t=0} U\left(g^{-1}, p\right) d(g) \\ &= \frac{d}{dt}\int_G f(g) U\left(g^{-1} \circ \exp\left(-t\tilde{X}_i\right), p\right)\Big|_{t=0} d(g) \\ &= \left(\int_G f(g) U\left(g^{-1}, p\right) d(g)\right)\left(\frac{d}{dt}U\left(\exp\left(-t\tilde{X}_i\right), p\right)\right)\Big|_{t=0} \\ &= -\hat{f}(p)u\left(\tilde{X}_i, p\right). \end{aligned}$$

Similarly, for the right derivative,

$$\mathcal{F}\left(\tilde{X}^R_i f\right) = u\left(\tilde{X}_i, p\right)\hat{f}(p).$$

Hence, evaluating Eq. (10.55) for $i = 1, \ldots, 6$ results in a number of operational properties. However, these are not the only operational properties. Certain others result from the particular parameterization used. For instance, there are operational properties associated with Euler angles and derivatives with respect to $a = \|\mathbf{a}\|$. In the following subsections we explore these in detail.

10.9.1 Properties of Translation Differential Operators

It may be observed from the integral representation Eqs. (10.33) and (10.35) that the matrix elements of the motion group satisfy the relation

$$\nabla_{\mathbf{a}}^2 U_{l',m';l,m}^s(\mathbf{a}, A; p) = \left(-p^2\right) U_{l',m';l,m}^s(\mathbf{a}, A; p) \tag{10.56}$$

where $\nabla_{\mathbf{a}}^2$ is the Laplacian with respect to $\mathbf{a}$. This equation leads to the Fourier transform property

$$\mathcal{F}\left(\nabla_{\mathbf{a}}^2 f(\mathbf{a}, A)\right) = \left(-p^2\right) \hat{f}(p). \tag{10.57}$$

For a differentiable function $f(\mathbf{a}, A)$, which is rapidly decreasing as $a = \|\mathbf{a}\| \to \infty$ and less singular than $\frac{1}{a}$ at $a \to 0$, the following relation is valid

$$\mathcal{F}\left(a\frac{\partial}{\partial a} f(\mathbf{a}, A)\right) = -\left(p\frac{d\hat{f}(p)}{dp} + 3\hat{f}(p)\right). \tag{10.58}$$

This may be proven using the equation

$$\frac{\partial}{\partial a} f(\mathbf{a}, A) = \frac{1}{2\pi^2} \int_0^\infty \text{trace}\left(\hat{f}(p)\frac{\partial}{\partial a}U(g, p)\right) p^2\, dp, \tag{10.59}$$

the fact that matrix elements of $U(g, p)$ are functions of the product (pa) [which may be seen from Eq. (10.34)], and integrating Eq. (10.59) by parts.

The more general property

$$\mathcal{F}\left(\frac{1}{a^{n-1}}\frac{\partial}{\partial a}\left(a^n f(\mathbf{a}, A)\right)\right) = (n-3)\hat{f}(p) - p\frac{d\hat{f}(p)}{dp}$$

follows from Eq. (10.58).

We mention also the related property

$$\int_{SE(3)} |a\frac{\partial}{\partial a} f(\mathbf{a}, A)|^2 \, dA\, da = \frac{1}{2\pi^2} \int_0^\infty \left\|p\frac{d\hat{f}(p)}{dp}\right\|_2^2 p^2\, dp,$$

which follows from the Parseval equality in Eqs. (10.50) and (10.58).

The equality $i\frac{\partial}{\partial a_3} e^{-i\mathbf{p}\cdot\mathbf{a}} = p\cos\theta\, e^{-i\mathbf{p}\cdot\mathbf{a}} = -p\sqrt{\frac{4\pi}{3}} Q_{00}^1(\cos\theta) e^{-i\mathbf{p}\cdot\mathbf{a}}$ [where θ is the polar angle of $\mathbf{p}$ and $Q_{-s,m}^l(\cos\theta)$ is defined in Eq. (10.30)] and the explicit expressions for the Clebsch-Gordan coefficients lead to the relation

$$i\frac{\partial}{\partial a_3} U_{l',m';l,m}^s(\mathbf{a}, A; p) = -p\left(\frac{(l'^2 - m'^2)(l'^2 - s^2)}{(2l'+1)(2l'-1)l'^2}\right)^{1/2} U_{l'-1,m';l,m}^s(\mathbf{a}, A; p)$$

$$- p\frac{m's}{l'(l'+1)} U_{l',m';l,m}^s(\mathbf{a}, A; p)$$

$$- p\left(\frac{((l'+1)^2 - m'^2)((l'+1)^2 - s^2)}{(2l'+1)(l'+1)^2(2l'+3)}\right)^{1/2} U_{l'+1,m';l,m}^s(\mathbf{a}, A; p).$$

We may get from this relation

$$\mathcal{F}\left(i\frac{\partial}{\partial a_3}f(\mathbf{a}, A)\right) = -p\left(\frac{((l+1)^2 - m^2)((l+1)^2 - s^2)}{(2l+1)(l+1)^2(2l+3)}\right)^{1/2} \hat{f}^s_{l',m';l+1,m}(p)$$

$$- p\frac{m\,s}{l(l+1)} \hat{f}^s_{l',m';l,m}(p)$$

$$- p\left(\frac{(l^2 - m^2)(l^2 - s^2)}{(2l+1)(2l-1)l^2}\right)^{1/2} \hat{f}^s_{l',m';l-1,m}(p).$$

For operators $P^\pm = i\frac{\partial}{\partial a_1} \pm \frac{\partial}{\partial a_2}$, which act on the exponential as

$$P^\pm e^{-i\mathbf{p}\cdot\mathbf{a}} = \pm p\sqrt{\frac{8\pi}{3}}\, Q^1_{0,\pm 1}(\cos\theta)\, e^{\mp i\phi} e^{-\mathbf{p}\cdot\mathbf{a}}$$

(θ, ϕ are polar and azimuthal angles of $\mathbf{p}$), we have the relations

$$P^+ U^s_{l',m';l,m}(\mathbf{a}, A; p) = -p\left(\frac{(l'-1-m')(l'-m')(l'^2 - s^2)}{(2l'+1)(2l'-1)l'^2}\right)^{1/2} U^s_{l'-1,m'+1;l,m}(\mathbf{a}, A; p)$$

$$- p\frac{\sqrt{(l'-m')(l'+1+m')}\,s}{l'(l'+1)} U^s_{l',m'+1;l,m}(\mathbf{a}, A; p)$$

$$+ p\left(\frac{(l'+1+m')(l'+2+m')((l'+1)^2 - s^2)}{(2l'+1)(l'+1)^2(2l'+3)}\right)^{1/2}$$
$$U^s_{l'+1,m'+1;l,m}(\mathbf{a}, A; p)$$

and

$$P^- U^s_{l',m';l,m}(\mathbf{a}, A; p) = -p\left(\frac{(l'-1+m')(l'+m')(l'^2 - s^2)}{(2l'+1)(2l'-1)l'^2}\right)^{1/2} U^s_{l'-1,m'-1;l,m}(\mathbf{a}, A; p)$$

$$- p\frac{\sqrt{(l'+m')(l'+1-m')}\,s}{l'(l'+1)} U^s_{l',m'-1;l,m}(\mathbf{a}, A; p)$$

$$- p\left(\frac{(l'+1-m')(l'+2-m')((l'+1)^2 - s^2)}{(2l'+1)(l'+1)^2(2l'+3)}\right)^{1/2}$$
$$U^s_{l'+1,m'-1;l,m}(\mathbf{a}, A; p).$$

These relations yield the operational properties

$$\mathcal{F}(P^+ f(\mathbf{a}, A)) = -p\left(\frac{(l+1-m)(l+2-m)((l+1)^2 - s^2)}{(2l+1)(l+1)^2(2l+3)}\right)^{1/2} \hat{f}^s_{l',m';l+1,m-1}(p)$$

$$- p\frac{\sqrt{(l+m)(l+1-m)}\,s}{l(l+1)} \hat{f}^s_{l',m';l,m-1}(p)$$

$$+ p\left(\frac{(l-1+m)(l+m)(l^2 - s^2)}{(2l+1)(2l-1)l^2}\right)^{1/2} \hat{f}^s_{l',m';l-1,m-1}(p)$$

and

$$\mathcal{F}(P^- f(\mathbf{a}, A)) = p \left(\frac{(l+1+m)(l+2+m)((l+1)^2 - s^2)}{(2l+1)(l+1)^2(2l+3)} \right)^{1/2} \hat{f}^s_{l',m';l+1,m+1}(p)$$

$$- p \frac{\sqrt{(l-m)(l+1+m)}\, s}{l(l+1)} \hat{f}^s_{l',m';l,m+1}(p)$$

$$- p \left(\frac{(l-1-m)(l-m)(l^2 - s^2)}{(2l+1)(2l-1)l^2} \right)^{1/2} \hat{f}^s_{l',m';l-1,m+1}(p).$$

We note that $\hat{f}^s_{l',m';l,m}(p) = 0$ if $| m'(m) | > l'(l)$ because of the definition of $\tilde{U}^l_{mn}$. Using integration by parts together with the property of Eq. (10.57), it may be shown that

$$\int_{SE(3)} | \nabla_{\mathbf{a}} f(g) |^2 \, d(g) = \int_{SE(3)} \left(f(g), -\nabla_{\mathbf{a}}^2 f(g) \right) d(g)$$

$$= \frac{1}{2\pi^2} \int_0^\infty \text{trace} \left(p^2 \hat{f}^\dagger(p) \hat{f}(p) \right) p^2 \, dp. \tag{10.60}$$

10.9.2   Properties of Rotational Differential Operators

From the fact that [37, 42]

$$\nabla_A^2 \tilde{U}^l_{mn}(A) = (-l(l+1)) \tilde{U}^l_{mn}(A), \tag{10.61}$$

where the Laplacian operator on $SO(3)$ is given by

$$\nabla_A^2 = \frac{1}{\sin\beta} \frac{\partial}{\partial\beta} \left(\sin\beta \frac{\partial}{\partial\beta} \right) + \frac{1}{\sin^2\beta} \left(\frac{\partial^2}{\partial\alpha^2} - 2\cos\beta \frac{\partial^2}{\partial\alpha\partial\gamma} + \frac{\partial^2}{\partial\gamma^2} \right) \tag{10.62}$$

(recall that α, β, γ are ZXZ Euler angles of A), it follows that

$$\mathcal{F} \left(\nabla_A^2 f(\mathbf{a}, A) \right)^s_{l',m';l,m} = (-l'(l'+1)) \hat{f}^s_{l',m';l,m}(p). \tag{10.63}$$

The straightforward relation

$$\mathcal{F} \left(i \frac{\partial}{\partial\gamma} f(\mathbf{a}, A) \right)^s_{l',m';l,m} = m' \hat{f}^s_{l',m';l,m}(p)$$

follows from Eqs. (10.32) and (10.35) and Fourier transform definition (10.45). The equations [7]

$$J_- \tilde{U}^l_{mn}(A) = i\sqrt{(l+n)(l-n+1)}\, \tilde{U}^l_{m,n-1}(A)$$

$$J_+ \tilde{U}^l_{mn}(A) = -i\sqrt{(l-n)(l+n+1)}\, \tilde{U}^l_{m,n+1}(A),$$

where

$$J_- = e^{i\gamma} \left(-\cot\beta \frac{\partial}{\partial\gamma} + \frac{1}{\sin\beta} \frac{\partial}{\partial\alpha} + i\frac{\partial}{\partial\beta} \right)$$

and

$$J_+ = e^{-i\gamma} \left(\cot\beta \frac{\partial}{\partial\gamma} - \frac{1}{\sin\beta} \frac{\partial}{\partial\alpha} + i\frac{\partial}{\partial\beta} \right)$$

lead to the relations

$$\mathcal{F}(J_- f(\mathbf{a}, A))^s_{l',m';l,m} = i\sqrt{(l' + m' + 1)(l' - m')} \, \hat{f}^s_{l',m'+1;l,m}(p)$$

and

$$\mathcal{F}(J_+ f(\mathbf{a}, A))^s_{l',m';l,m} = -i\sqrt{(l' - m' + 1)(l' + m')} \, \hat{f}^s_{l',m'-1;l,m}(p).$$

10.9.3 Other Operational Properties

Scaling Property

For $\tilde{f}(\mathbf{a}, A) = f(\frac{\mathbf{a}}{k}, A)$, where $k \neq 0$ is any real number, the following scaling property is valid

$$\mathcal{F}(\tilde{f}(\mathbf{a}, A)) = k^3 \hat{f}(k\, p).$$

Shift Property

We note also the shift property

$$\mathcal{F}\left(f(g_1 \circ g \circ g_2^{-1})\right) = U\left(g_2^{-1}, p\right) \cdot \hat{f}(p) \cdot U(g_1, p) = U^\dagger(g_2, p) \cdot \hat{f}(p) \cdot U(g_1, p),$$

where $g_1, g_2 \in SE(3)$, $U^\dagger(g_2, p)$ is the Hermitian conjugate of $U(g_2, p)$ and $\hat{f}(p)$ is the Fourier transform of $f(g) = f(\mathbf{a}, A)$.

Properties of the δ-function

The δ-function $\delta(g, g')$ is defined in Appendix E. Using the Fourier transform definition and the property of Eq. (9.16) of δ-function, it is easy to see that

$$\mathcal{F}(\delta(g, e)) = \mathbb{I},$$

where $\mathbb{I}$ is a unit operator [an identity matrix after contraction of indices according to Eq. (10.53) for each s]. Using the convolution property in Eq. (10.48), the Fourier transform of the convolution

$$(f * \delta)(g) = \int_{SE(3)} f(h) \, \delta\left(h^{-1} \circ g\right) d(h)$$

is $\mathcal{F}(f)$ [where we use the notation $\delta(g)$ for $\delta(g, e)$]. It means that

$$(f * \delta)(g) = f(g). \tag{10.64}$$

Analogously,

$$(\delta * f)(g) = f(g). \tag{10.65}$$

We note that the convolution in Eq. (10.65) may be written as

$$(\delta * f)(g) = \int_{SE(3)} f(h) \, \delta\left(g \circ h^{-1}\right) d(h).$$

Comparing the $SE(3)$ equivalent of Eq. (9.16) and the convolution equations we have

$$\delta(g, g_1) = \delta\left(g_1^{-1} \circ g\right) = \delta\left(g \circ g_1^{-1}\right).$$

10.10   Analytical Examples

Here we give some examples which illustrate the Fourier transform and inverse Fourier transform and allow us to perform analytical calculations.

First, let us consider rapidly decreasing spherically symmetric functions

$$f(\mathbf{a}, A) = f(a).$$

Because this function does not depend on the Euler angles of rotation, only $U^s_{l',m';0,0}(\mathbf{a}, A; p)$ elements contribute to the Fourier transform. Moreover, $s = 0$ because $\mid s \mid \le l, l'$. Finally, examining the expression in Eq. (10.34) and using the fact that

$$\int_{\theta=0}^{\pi} \int_{\phi=0}^{2\pi} Y_l^m \sin\theta \, d\theta \, d\phi = \sqrt{4\pi} \, \delta_{l0} \, \delta_{m0},$$

we see that only the element $U^0_{0,0;0,0}(\mathbf{a}, A; p)$ contributes to the Fourier transform.

For the function $f(a) = e^{-a}$ the Fourier transform gives

$$\hat{f}^0_{0,0;0,0}(p) = \int_{SE(3)} f(a)\overline{U^0_{0,0;0,0}(\mathbf{a}, A; p)} \, d^3\mathbf{a} \, dA$$

$$= 4\pi \int_0^{\infty} e^{-a} \frac{\sin(pa)}{p} a \, da = \frac{8\pi}{(1+p^2)^2}, \tag{10.66}$$

where we have used the fact that $\int_{SO(3)} dA = 1$.

The inverse Fourier transform reproduces the original function as

$$f(a) = \frac{1}{2\pi^2} \int_0^{\infty} \hat{f}^0_{0,0;0,0}(p) U^0_{0,0;0,0}(\mathbf{a}, A; p) \, p^2 dp$$

$$= \frac{1}{2\pi^2} \int \frac{8\pi}{(1+p^2)^2} \frac{\sin(pa)}{a} \, p \, dp = e^{-a}.$$

For the function $f(a) = e^{-a^2}$ we have for the Fourier transform $\hat{f}(p) = (\pi)^{3/2} e^{-p^2/4}$. The inverse Fourier transform gives the original function.

Another example is the function

$$f(\mathbf{a}, A) = e^{-a} \cos\theta \cos\beta$$

where θ is the polar angle of $\mathbf{a}$ and β is Euler angle (around the x-axis) of rotation A. Because $U^1_{00}(A) = \cos\beta$, it may be shown from Eqs. (10.34) and (10.35) that only $\hat{f}^0_{1,0;0,0}(p); \hat{f}^0_{1,0;1,0}(p);$ $\hat{f}^0_{1,0;2,0}(p); \hat{f}^1_{1,0;1,0}(p); \hat{f}^1_{1,0;2,0}(p); \hat{f}^{-1}_{1,0;1,0}(p); \hat{f}^{-1}_{1,0;2,0}(p)$ can give nonzero contributions. Direct computations show that the Fourier transform elements are (we show only nonzero matrix elements)

$$\hat{f}^0_{1,0;0,0}(p) = -\frac{8i\pi}{3\sqrt{3}} \frac{p}{(1+p^2)^2};$$

$$\hat{f}^0_{1,0;2,0}(p) = -\frac{16i\pi}{\sqrt{135}} \frac{p}{(1+p^2)^2};$$

$$\hat{f}^1_{1,0;2,0}(p) = -\frac{8i\pi}{3\sqrt{5}} \frac{p}{(1+p^2)^2};$$

$$\hat{f}_{1,0;2,0}^{-1}(p) = -\frac{8i\pi}{3\sqrt{5}} \frac{p}{(1+p^2)^2}. \tag{10.67}$$

For the inverse Fourier transform we obtain the following expression for the trace in Eq. (10.45)

$$\text{trace}\left(\hat{f}(p)U(g,p)\right) = 8\pi \cos\theta \, \cos\beta \, \frac{(\sin(pa) - pa\cos(pa))}{(1+p^2)^2 \, pa^2}.$$

The p integration in Eq. (10.45) reproduces the original function.
We computed also the Fourier transform of the function

$$f(\mathbf{a}, A) = a^2 e^{-a} \cos\theta \, \cos\beta.$$

The nonzero matrix elements are

$$\hat{f}_{1,0;0,0}^{0}(p) = \frac{32i\pi}{3\sqrt{3}} \frac{p(p^2 - 5)}{(1+p^2)^4};$$

$$\hat{f}_{1,0;2,0}^{0}(p) = \frac{64i\pi}{\sqrt{135}} \frac{p(p^2 - 5)}{(1+p^2)^4};$$

$$\hat{f}_{1,0;2,0}^{1}(p) = \frac{32i\pi}{3\sqrt{5}} \frac{p(p^2 - 5)}{(1+p^2)^4};$$

$$\hat{f}_{1,0;2,0}^{-1}(p) = \frac{32i\pi}{3\sqrt{5}} \frac{p(p^2 - 5)}{(1+p^2)^4}. \tag{10.68}$$

The inverse Fourier transform gives the original function.
We chop all zero elements for $l(l') > 2$. Therefore, after contraction of four indices to two using Eq. (10.53), the Fourier transform of the examples in Eqs. (10.67) and (10.68) may be written as a block-diagonal matrix

$$\hat{F} = \begin{bmatrix} \hat{F}_{-1} & & \\ & \hat{F}_0 & \\ & & \hat{F}_1 \end{bmatrix} \tag{10.69}$$

The nonzero blocks are the 9×9 matrix $\hat{F}_0$ and two 8×8 matrices $\hat{F}_{-1}$, $\hat{F}_1$ (lower indices correspond to s index). Using Eq. (10.53), these matrices may be depicted as

$$\hat{F}_0 = \begin{bmatrix} 0 & \cdots & 0 & & 0 \\ \vdots & & & & \vdots \\ \hat{f}_{31}^0 & \cdots & \hat{f}_{37}^0 & & 0 \\ \vdots & & & & \vdots \\ 0 & \cdots & 0 & & 0 \end{bmatrix} \tag{10.70}$$

where $\hat{f}_{31}^0 = \hat{f}_{1,0;0,0}^0(p)$, $\hat{f}_{37}^0 = \hat{f}_{1,0;2,0}^0(p)$. The other matrices are

$$
\hat{F}_{\pm 1} = \begin{bmatrix} 0 & \cdots & 0 & & 0 \\ 0 & \cdots & \hat{f}_{26}^{\pm 1} & & 0 \\ \vdots & & & & \vdots \\ 0 & \cdots & 0 & & 0 \end{bmatrix}
\tag{10.71}
$$

where $\hat{f}_{26}^{\pm 1} = \hat{f}_{1,0;2,0}^{\pm 1}(p)$.

10.11   Linear-Algebraic Properties of Fourier Transform Matrices for $SE(2)$ and $SE(3)$

In this section we state several theorems that characterize the structure of $SE(N)$ Fourier transform matrices for $N = 2$ and $N = 3$.

THEOREM 10.1 [17]
Complex eigenvalues of matrices with symmetries of the form in Eqs. (10.5) and (10.52) appear in conjugate pairs.

PROOF
Case 1: $SE(2)$.

The eigenvalue problem is written in component form as

$$
\lambda x_m = \sum_n \hat{f}_{mn} x_n = (-1)^m \sum_n (-1)^n \overline{\hat{f}_{-m,-n}} x_n.
$$

Multiplying both sides by $(-1)^m$ and regrouping terms we get

$$
\sum_n (-1)^n \overline{\hat{f}_{-m,-n}} x_n = \sum_n \overline{\hat{f}_{-m,-n}} \left((-1)^n x_n \right) = \lambda \left((-1)^m x_m \right).
$$

Changing the variables (m, n) to $(-m, -n)$, and taking the complex conjugate of both sides, one finds

$$
\sum_n \hat{f}_{m,n} \left((-1)^n \overline{x_{-n}} \right) = \overline{\lambda} \left((-1)^m \overline{x_{-m}} \right),
$$

indicating that if (λ, x_m) is an eigenvalue/vector pair, then so is $(\overline{\lambda}, (-1)^m \overline{x_{-m}})$.

Case 2: $SE(3)$.

The eigenvalue problem is written for each s block as

$$
\sum_{l,m} \hat{f}_{l',m';l,m}^s x_{l,m}^s = \lambda x_{l',m'}^s,
$$

where there is summation over l and m.

Substitution using symmetry in Eq. (10.52) yields

$$
\sum_{l,m} (-1)^{(l'-l)} (-1)^{(m'-m)} \overline{\hat{f}_{l',-m';l,-m}^s} x_{l,m}^s
$$

$$
= (-1)^{(l'+m')} \sum_{l,m} \overline{\hat{f}_{l',-m';l,-m}^s} \left((-1)^{(l+m)} x_{l,m}^s \right) = \lambda x_{l',m'}^s.
$$

Multiplication on both sides by $(-1)^{(l'+m')}$ and complex conjugation yield

$$\sum_{l,m} \hat{f}^s_{l',-m';l,-m} \left((-1)^{(l+m)} \overline{x^s_{l,m}} \right) = \bar{\lambda} \left((-1)^{(l'+m')} \overline{x^s_{l',m'}} \right).$$

Finally, changing the dummy variables (m, m') to $(-m, -m')$, we get

$$\sum_{l,m} \hat{f}^s_{l',m';l,m} \left((-1)^{(l-m)} \overline{x^s_{l,-m}} \right) = \bar{\lambda} \left((-1)^{(l'-m')} \overline{x^s_{l',-m'}} \right),$$

indicating that for every eigenvalue/vector pair $(\lambda, x^s_{l',m'})$, there is also a pair $(\bar{\lambda}, (-1)^{(l'-m')} \overline{x^s_{l',-m'}})$.

∎

While the Fourier transform is generally used for harmonic analysis of square-integrable functions in this book, there are a few notable exceptions. These are the generalized (singularity) functions (or distributions). These are discussed here and used throughout the book.

The Dirac delta function for $\mathbb{R}^N$ is defined to have the following properties:

$$\int_{\mathbb{R}^N} \delta(\mathbf{r}) d^N r = 1, \qquad \int_{\mathbb{R}^N} f(\mathbf{r}) \delta(\mathbf{x} - \mathbf{r}) d\mathbf{r} = (f * \delta)(\mathbf{x}) = f(\mathbf{x}),$$

where $d\mathbf{r} = dr_1 dr_2 \dots dr_N$.

The Dirac delta function on $SO(N)$ has the analogous properties

$$\int_{SO(N)} \delta(\mathcal{R}) d\mathcal{R} = 1, \qquad \int_{SO(N)} f(\mathcal{R}) \delta\left(\mathcal{R}^T R \right) d\mathcal{R} = (f * \delta)(R) = f(R).$$

It follows directly from the invariance of integration under shifts and inversion of the arguments of functions on $\mathbb{R}^N$ and $SO(N)$ that

$$\delta(\mathbf{x} - \mathbf{r}) = \delta(\mathbf{r} - \mathbf{x}) \quad \text{and} \quad \delta\left(\mathcal{R}^T R \right) = \delta\left(R^T \mathcal{R} \right).$$

The delta function for $SE(N)$ is the product of these

$$\delta(g) = \delta(\mathbf{r}) \delta(A) \quad \text{for} \quad g = (\mathbf{r}, A) \in SE(N).$$

The integration over $SE(N)$ in the convolution integral may be rewritten as integration over position and orientation separately

$$(f_1 * f_2)(\mathbf{x}, R) = \int_{SO(N)} \int_{\mathbb{R}^N} f_1(\xi, \mathcal{R}) f_2 \left(\mathcal{R}^T(\mathbf{x} - \xi), \mathcal{R}^T R \right) d\xi d\mathcal{R} \qquad (10.72)$$

where $g = (\mathbf{x}, R)$ and $h = (\xi, \mathcal{R})$ are elements of $SE(N)$.

Using this notation, it is easy to see that the Fourier transforms of functions such as $f(R)\delta(\mathbf{x})$ and $f(\mathbf{x})\delta(R)$ reduce to

$$\int_{SO(N)} f(R) U \left(0, R^T; p \right) dR$$

and

$$\int_{\mathbb{R}^N} f(\mathbf{x}) U(-\mathbf{x}, \mathbb{I}; p) d\mathbf{x}$$

respectively.

THEOREM 10.2
Functions $f(g) \in \mathcal{L}^2(SE(N))$ with Fourier transforms satisfying the normality condition

$$\hat{f}^\dagger(p)\hat{f}(p) = \hat{f}(p)\hat{f}^\dagger(p)$$

include
a) *$f(g)$ which satisfy the condition*

$$\overline{f(g)} = \pm f\left(g^{-1}\right).$$

We call these functions symmetric (antisymmetric).
b) *$f(g)$ which is a class function, i.e.,*

$$f(g) = f\left(h^{-1} \circ g \circ h\right)$$

for any g, $h \in G$.
c) *$f(g)$ with a Fourier transform matrix which is proportional to a unitary matrix.*

PROOF
   Cases (a) and (b) may be shown easily using the Fourier transform definition, the unitarity of $U(g, p)$, and the invariance of integration with $d(g)$ under shifts and inversion of the argument. In particular, it is easy to see that the Fourier transform of a symmetric (antisymmetric) function is a Hermitian (skew-Hermitian) matrix

$$\hat{F}^\dagger(p) = \pm \hat{F}(p);$$

Case (c) follows directly from the definition of unitary matrices.   ∎

   We may also find normal matrices which are not in these categories. For example, a function with the Fourier transform matrix $\hat{f}^\dagger(p) = e^{ic(p)} \hat{f}(p)$ [where $c(p)$ is some function] is also normal. The next two theorems examine how broad the sets of class and symmetric/antisymmetric functions are for $SE(N)$.

THEOREM 10.3
There are no class functions in $\mathcal{L}^2(SE(N))$ for which $\int_{SE(N)} |f(g)| \, d(g) > 0$.

PROOF
   For a function to be a class function on $SE(N)$, necessary conditions are that $f(g_1^{-1} \circ g \circ g_1) = f(g)$ for $g_1 = (\mathbf{0}, A_1)$ and $f(g_2^{-1} \circ g \circ g_2) = f(g)$ for $g_2 = (\mathbf{r}_1, I)$. That is, the definition must hold for general automorphisms, and so it must hold for pure rotations and translations individually. Using the notation $f(g) = f(\mathbf{r}, A)$, these conditions are written as

$$f(\mathbf{r}, A) = f\left(A_1^T \mathbf{r}, A_1^T A A_1\right) \tag{10.73}$$

and

$$f(\mathbf{r}, A) = f\left(\mathbf{r} + (A - \mathbb{I})\mathbf{r}_1, A\right). \tag{10.74}$$

Equation (10.74) can only be true for arbitrary $A \neq \mathbb{I}$ if it has no dependence on $\mathbf{r}$. This leads to

Case 1: $f(\mathbf{r}, A) = f_1(A)$.

If, however, $f(\mathbf{r}, A) = 0$ for all $A \neq \mathbb{I}$, then Eq. (10.74) is also satisfied. The only way this can be true and $\int_{SE(N)} |f(g)| \, d(g) > 0$ is when

Case 2: $f(\mathbf{r}, A) = f_2(\mathbf{r}) \delta(A)$.

Neither of these functions are square integrable on $SE(N)$. While this completes the proof, it is interesting to note that in order to satisfy Eq. (10.73), $f_1(A)$ must be a class function on $SO(N)$, and $f_2(\mathbf{r}) = f_2(|\mathbf{r}|) = f_2(r)$.

From Chapter 6 we know that all class functions for $SO(3)$ are functions of the rotation angle, and so $f_1(A) = f_1(\theta)$. ∎

THEOREM 10.4

There exist nontrivial square-integrable symmetric and antisymmetric functions on $SE(N)$.

PROOF

The proof is by construction. Let $f_i : \mathbb{R}^{N \times N} \times \mathbb{R}^N \to \mathbb{C}(or\ \mathbb{R})$ for $i = 1, 2, 3, 4$ be square integrable functions on $\mathbb{R}^{N \times N \times N}$ such that

$$f_1\left((-1)^n B, (-1)^m \mathbf{y}\right) = f_1(B, \mathbf{y}),$$
$$f_2\left((-1)^n B, (-1)^m \mathbf{y}\right) = (-1)^m f_2(B, \mathbf{y}),$$
$$f_3\left((-1)^n B, (-1)^m \mathbf{y}\right) = (-1)^n f_3(B, \mathbf{y}),$$
$$f_4\left((-1)^n B, (-1)^m \mathbf{y}\right) = (-1)^{(n+m)} f_4(B, \mathbf{y}),$$

for all $B \in \mathbb{R}^{N \times N}$ and $\mathbf{y} \in \mathbb{R}^N$ and $m, n \in \{0, 1\}$. Then it is easy to confirm by direct substitution of $g^{-1} = (A^T, -A^T \mathbf{r})$ for $g = (A, \mathbf{r}) \in SE(N)$ in the above functions that

$$f_1\left(A + A^T, A^{-1/2} \mathbf{r}\right)$$

$$f_3\left(A + A^T, A^{-1/2} \mathbf{r}\right)$$

$$f_1\left(A - A^T, A^{-1/2} \mathbf{r}\right)$$

$$f_4\left(A - A^T, A^{-1/2} \mathbf{r}\right)$$

are symmetric, and

$$f_2\left(A + A^T, A^{-1/2} \mathbf{r}\right)$$

$$f_4\left(A + A^T, A^{-1/2} \mathbf{r}\right)$$

$$f_3\left(A - A^T, A^{-1/2} \mathbf{r}\right)$$

$$f_2\left(A - A^T, A^{-1/2} \mathbf{r}\right)$$

are antisymmetric, and they are all square integrable by definition (the $A^{-1/2}$ is defined as rotation around the same axis as in A by half of the angle in opposite direction). For instance, if $f(g) = f(\mathbf{r}, A) = f_1(A + A^T, A^{-1/2}\mathbf{r})$, then $f(g^{-1}) = f(-A^T\mathbf{r}, A^T) = f_1(A^T + A, (A^T)^{-1/2}(-A^T\mathbf{r}))$. But since any matrix commutes with powers of itself, and $A^T = A^{-1}$, $(A^T)^{-1/2}(-A^T\mathbf{r}) = (A^{-1})^{-1/2}(-A^{-1}\mathbf{r}) = A^{1/2}(-A^{-1}\mathbf{r}) = -A^{-1/2}\mathbf{r}$. Clearly then, $f(g^{-1}) = f_1(A^T + A, -A^{-1/2}\mathbf{r}) = f(g)$ in this case. The other cases follow in the same way. ∎

In the case of $SE(3)$, if the rotation matrix is parameterized using the axis and angle of rotation, $A = A(\theta, \omega)$, and the direction of the axis of rotation is parameterized by polar and azimuthal angles, $\omega = \omega(\alpha_1, \alpha_2)$, then functions of the form $f(\theta, \alpha_1, \alpha_2; r)$ where $f(\pm\theta, \alpha_1, \alpha_2; r) = \pm f(\theta, \alpha_1, \alpha_2; r)$ are symmetric/antisymmetric.

10.12   Summary

In this chapter we described one possible definition of the irreducible unitary representations and Fourier transform for $SE(2)$ and $SE(3)$. Operational properties of the Fourier transform under differential operators acting on functions on these groups were investigated. These operational properties are important in the context of applications in Chapter 17 (see also [5]). Chapter 11 addresses computational issues of the Fourier transforms for $SE(N)$ which are used in Chapters 12 and 13.

References

[1] Arnal, D. and Cortet, J.C., Star representation of E(2), *Letters in Mathematical Physics*, 20, 141–149, 1990.

[2] Auslander, L. and Moore, C.C., *Unitary Representations of Solvable Lie Groups*, Mem. Amer. Math. Soc., No. 62, 1966.

[3] Ballesteros, A., Celeghini, E., Giachetti, R., and Tarlini, M., An R-matrix approach to the quantization of the Euclidean group E(2), *J. of Phys. and Math.*, 26, 7495–7501, 1993.

[4] Bohm, M. and Junker, G., Path integration over the n-dimensional Euclidean group, *J. of Math. Phys.*, 30, 1195–1197, 1989.

[5] Chirikjian, G.S. and Kyatkin, A.B., An operational calculus for the Euclidean motion group with applications in robotics and polymer science, *J. Fourier Anal. and Appl.*, to appear.

[6] Feinsilver, P., Lie algebras and recurrence relations IV: representations of the Euclidean group and Bessel functions, *Acta Applicandae Math.*, 43, 289–316, 1996.

[7] Gel'fand, I.M., Minlos, R.A., and Shapiro, Z. Ya., *Representations of the Rotation and Lorentz Groups and Their Applications*, Pergamon Press, New York, 1963.

[8] Gonzalez, F.B., Bi-invariant differential operators on the Euclidean motion group and applications to generalized Radon transforms, *Achiv for Matematik*, 26(2), 191–204, 1988.

[9] Gross, K.I. and Kunze, R.A., Bessel functions and representation theory, *J. of Functional Anal.*, 22(2), 73–105, 1976.

[10] Gurarie, D., *Symmetry and Laplacians. Introduction to Harmonic Analysis, Group Representations and Applications*, Elsevier Science, The Netherlands, 1992.

[11] Humi, M., Representations and invariant equations of E(3), *J. of Math. Phys.*, 28, 2807–2811, 1987.

[12] Isham, C.J. and Klauder, J.R., Coherent states for n-dimensional Euclidean groups E(n) and their application, *J. of Math. Phys.*, 32, 607–620, 1991.

[13] Jones, M.N., *Spherical Harmonics and Tensors for Classical Field Theory*, Research Studies Press Ltd., England, 1985.

[14] Kopský, V., Translation Normalizers of Euclidean Groups, *J. of Math. and Phys.*, 34, 1548–1556, 1993.

[15] Kopský, V., Translational normalizers of Euclidean groups (II), *J. of Math. and Phys.*, 34, 1557–1576, 1993.

[16] Kumahara, K. and Okamoto, K., An analogue of the Paley-Wiener theorem for the Euclidean motion group, *Osaka J. Math.*, 10, 77–92, 1973.

[17] Kyatkin, A.B. and Chirikjian, G.S., Regularization of a nonlinear convolution equation on the Euclidean group, *Acta. Appl. Math*, 53, 89–123, 1998.

[18] Mackey, G.W., *Induced Representations of Groups and Quantum Mechanics*, W.A. Benjamin, New York, 1968.

[19] Miller, W., Jr., *Lie Theory and Special Functions*, Academic Press, New York, 1968 (see also Miller, W., Jr., Some applications of the representation theory of the Euclidean group in three-space, *Commun. Pure App. Math.*, 17, 527–540, 1964).

[20] Moses, H.E. and Quesada, A.F., The expansion of physical quantities in terms of the irreducible representations of the scale-Euclidean group and applications to the construction of scale-invariant correlation functions part I, *Arch. of Ration. Mech. and Anal.*, 44(3), 217–248, 1972.

[21] Moses, H.E. and Quesada, A.F., The expansion of physical quantities in terms of the irreducible representations of the scale-Euclidean group and applications to the construction of scale-invariant correlation functions part II, *Arch. of Ration. Mech. and Anal.*, 50, 194–236, 1973.

[22] Orihara, A., Bessel functions and the Euclidean motion group, *Tohoku Math. J.*, 13, 66–71, 1961.

[23] Pukanszky, L., Unitary representations of solvable Lie groups, *Ann. Sci. Ecol. Norm. Sup.*, 4(4), 457–608, 1971.

[24] Rno, J.S., *Clebsch-Gordan Coefficients and Special Functions Related to the Euclidean Group in Three-Space*, Ph.D., thesis, University of Minnesota, 1973.

[25] Rno, J.S., Clebsch-Gordan coefficients and special functions related to the Euclidean group in three-space, *J. of Math. Phys.*, 15(12), 2042–2047, 1974.

[26] Rno, J.S., Harmonic analysis on the Euclidean group in three-space, I, II, *J. of Math. Phys.*, 26, 675–677, 1985; 26, 2186–2188, 1985.

[27] Rozenblyum, A.V. and Rozenblyum, L.V., Orthogonal polynomials of several variables related to representations of groups of Euclidean motion, *Differentsial'nye Uravneniya*, 22(11), 1972–1977, 1986.

[28] Rozenblyum, A.V., Representations of lie groups and multidimensional special functions, *Acta Applicandae Mathematicae*, 29, 171–240, 1992.

[29] Rubin, R.L., Harmonic analysis on the group of rigid motions of the Euclidean plane, *Studia Math.*, 62, 125–141, 1978.

[30] Sakai, K., Irreducible unitary representations of the group of motions in 3-dimensional Euclidean space and the spherical Bessel functions, *Rev. of the Marine Tech. Coll.*, 8, 93–115, 1964 (in Japanese).

[31] Sakai, K., On the representations of the motion group of n-dimensional Euclidean space I, *Sci. Rep. of the Kagoshima Univers.*, 16, 25–33, 1967.

[32] Sakai, K., Some remarks on unitary representations of the Euclidean motion group in Π_m-spaces, *Sci. Rep. of the Kagoshima Univers.*, 29, 13–26, 1980.

[33] Sakai, K., On indecomposable unitary representations of the 2-dimensional Euclidean motion group in finite dimensional indefinite inner product spaces I, *Sci. Rep. of the Kagoshima Univers.*, 29, 27–51, 1980.

[34] Sakai, K., On indecomposable unitary representations of the 2-dimensional Euclidean motion group in finite dimensional indefinite inner product spaces II, *Sci. Rep. of the Kagoshima Univers.*, 30, 1–21, 1981.

[35] Sugiura, M., *Unitary Representations and Harmonic Analysis,* 2nd ed., Elsevier Science, The Netherlands, 1990.

[36] Symons, J., Irreducible representations of the group of movements of the Euclidean plane, *J. of the Aust. Math. Soc.*, 18, 78–96, 1974.

[37] Talman, J., *Special Functions*, W.A. Benjamin, Amsterdam, 1968.

[38] Takiff, S., Invariant polynomials on Lie algebras of inhomogeneous unitary and special orthogonal groups, *Trans. Amer. Math. Soc.*, 170, 221–230, 1972.

[39] Torres del Castillo, G.F., Spin-weighted cylindrical harmonics and the Euclidean group of the plane, *J. of Math. Phys.*, 34, 3856–3862, 1993.

[40] Vilenkin, N.J., Bessel functions and representations of the group of Euclidean motions, *Uspehi Mat. Nauk.*, 11(3), 69–112, 1956 (in Russian).

[41] Vilenkin, N.J., Akim, E.L., and Levin, A.A., The matrix elements of irreducible unitary representations of the group of Euclidean three-dimensional space motions and their properties, *Dokl. Akad. Nauk SSSR*, 112, 987–989, 1957 (in Russian).

[42] Vilenkin, N.J. and Klimyk, A.U., *Representation of Lie Groups and Special Functions,* Vol. 1–3, Kluwer Academic, The Netherlands, 1991.

[43] Wightman, A.S., On the Localizability of Quantum Mechanical Systems, *Rev. Mod. Phys.*, 34, 845–872, 1962.

Chapter 11

Fast Fourier Transforms for Motion Groups

In this chapter we apply techniques from noncommutative harmonic analysis to the development of fast numerical algorithms for the computation of convolution integrals on motion groups. In particular, we focus on the group of rigid-body motions in the plane and 3-space. Using IURs in operator form, we write the Fourier transform of functions on the motion group as an integral over the product space $SE(3) \times S^2$. The integral form of the Fourier transform matrix elements allows us to apply Fast Fourier Transform (FFT) methods developed previously for $\mathbb{R}^3$, S^2, and $SO(3)$ to considerably speed up the numerical computation of convolutions of functions on $SE(3)$. Such convolutions play an important role in a number of engineering disciplines. Numerical algorithms for the fast computation of the Fourier transform are given and the complexity of the numerical implementations are discussed. The Fourier transform on the 3D "discrete motion group" (semi-direct product of the icosahedral group with the translation group) is also developed and the results of the numerical implementation are discussed. Finally, we examine alternative approaches to computing fast motion-group convolutions. These include the FFT for the direct product group $SO(3) \times \mathbb{R}^3$, and the use of reducible representations of $SE(3)$.

11.1 Preliminaries: Direct Convolution and the Cost of Interpolation

If we let $g = (\mathbf{r}, R)$ and $h = (\mathbf{a}, A)$ be elements of $SE(n)$, then the convolution

$$(f_1 * f_2)(g) = \int_{SE(n)} f_1(h) f_2\left(h^{-1} \circ g\right) d(h)$$

is written as

$$(f_1 * f_2)(\mathbf{r}, R) = \int_{SO(n)} \int_{\mathbb{R}^n} f_1(\mathbf{a}, A) f_2\left(A^T(\mathbf{r} - \mathbf{a}), A^T R\right) d\mathbf{a} \, dA. \tag{11.1}$$

If we compute this convolution integral by direct evaluation of each of N values of g and sum over N values of h, then a total of $\mathcal{O}(N^2)$ computations are required. Here we assume there are N_r samples in the $\mathbb{R}^n$ part, and N_R samples in the $SO(n)$ part. If there are $\mathcal{O}(S)$ samples in each coordinate direction, this means that in the case of $SE(3)$ we get $N = \mathcal{O}(S^6)$, and $\mathcal{O}(S^{12})$ computations are required to perform a direct convolution. Hence, even $S \approx 10$ is prohibitive (at the time of the writing of this book).

If the functions f_i are given numerically with the $\mathbb{R}^n$ part given on a Cartesian lattice, the direct evaluation of Eq. (11.1) will require the interpolation of function values off grid. We, therefore, devote the first section of this chapter to issues related to the complexity of interpolation.

11.1.1 Fast Exact Fourier Interpolation

We can, without loss of any information, interpolate finite values of a periodic band-limited function on the plane (or, equivalently, a band-limited function on T^2) sampled at Cartesian grid points to points on a polar grid. This is shown in the left side of Fig. 11.1, where function values are specified at each intersection of the straight solid lines. The first step in this procedure is

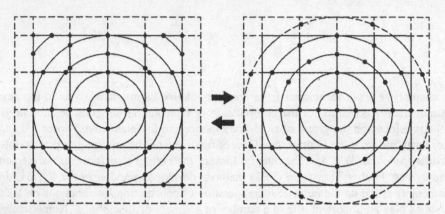

FIGURE 11.1
Interpolation from Cartesian to polar grid.

the interpolation of the given values to each of the points marked with circular dots. Since each straight grid line (which can be viewed as an unwrapped circle) defines a one-dimensional Fourier series, this step amounts to the evaluation of a one-dimensional Fourier series at nonequally spaced points. Since band-limited Fourier series can be viewed as polynomials of the form $\sum_{n=0}^{N_r^{1/2}-1} a_n Z^n$ where Z is a complex exponential, each vertical and horizontal evaluation can be performed in $\mathcal{O}(N_r^{\frac{1}{2}}(\log N_r^{\frac{1}{2}})^2)$ [1, 2]. Since there are $\mathcal{O}(N_r^{\frac{1}{2}})$ such lines, the total procedure of interpolating values from the Cartesian to polar grid requires $\mathcal{O}(N_r^{\frac{1}{2}} \cdot N_r^{\frac{1}{2}}(\log N_r^{\frac{1}{2}})^2) = \mathcal{O}(N_r(\log N_r)^2)$ computations. This procedure is an exact one in exact arithmetic. And since we are careful to choose a polar grid for which there are a sufficient number of intersections of circles with each Cartesian grid line, no information is lost.

Once the values have been interpolated to each of the concentric circles, they define a band-limited Fourier series on each circle that can be evaluated at equally spaced angles within each circle. This is again an $\mathcal{O}(N_r(\log N_r)^2)$ computation since there are $\mathcal{O}(N_r^{\frac{1}{2}})$ circles. It too is exact in the sense that it conserves the original information defined at the N_r grid points.

In order for the information of the function values to be recoverable to the Cartesian grid from the polar one, a sufficient number of circles extending outside of the original Cartesian grid must have values defined on them. In the example shown, there is only one circle that extends beyond the Cartesian grid. The straight dashed lines in Fig. 11.1 indicate where opposing sides of the grid are glued.

To see that this procedure is reversible, first evaluate the Fourier series on each circle at the points on the circle that intersect the Cartesian grid. Then fit the band-limited Fourier series on each straight line that passes through the values at the points of intersection. This fitting of a Fourier series to irregularly spaced data on each line is performed with the same order of computations as the evaluation of Fourier series at irregular points [1, 2]. In order for there to be enough information to reconstruct the original band-limited Fourier series on the outer edges of the Cartesian grid, the information contained in the partially outlying circle is required. For finer grids, multiple such circles are required. In the current example, the corner values could not be determined from the polar grid unless the information in the circle extending outside of the polar grid is included. The values marked with an asterisk are found using a two-step process whereby function values on surrounding lines must be evaluated at Cartesian grid points before there is enough data available to exactly recover the values.

From the previous discussion, we observe that interpolation back and forth between polar and Cartesian grids can be performed in $\mathcal{O}(N_r(\log N_r)^2)$ computations. This procedure of Fourier interpolation between Cartesian and polar grids is "exact" in the sense that in exact arithmetic it reversibly takes values from one kind of grid into another.

The number of function values on each circle enclosed in the Cartesian grid increases linearly with the radius. This ensures that the information in a band-limited function on T^2 is exactly transferred to the collection of band-limited functions on the circles of the polar grid using the procedure described previously. Another way to view this is that a band-limited periodic function on the plane is set to zero outside of a finite window with dimensions equal to the period of the function. The version of this function sampled in polar coordinates defines a new function that, in principle, can be nonzero outside of this finite window. Given the standard assumptions in numerical Fourier analysis, the values on all circles not fully contained within the finite square grid will be made arbitrarily close to zero by appropriate choice of the band-limit and zero padding.

The circles partially enclosed in the Cartesian grid diminish in significance with radius. In the context of the grid shown in Fig. 11.1, we see that on circles from the center outward there are 1, 4, 4, 12, 12, 8 values. This means that the band-limit for functions of the polar angle for each of these circles will be the numbers given in the previous sentence. At radius p, the band limit will be $\mathcal{O}(p)$, with $p \leq \mathcal{O}(N_r^{\frac{1}{2}})$.

11.1.2 Approximate Spline Interpolation

In spline interpolation, only the N_s nearest grid points to the point at which the interpolated value is desired enter the calculation. This necessarily means that some information about the function is destroyed in the spline interpolation process. However, since the functions in question are all assumed to be band-limited Fourier series, they do not oscillate on the length scale of the distance between sample points. Hence, a polynomial spline of sufficiently high order will approximate well the local neighborhood of the point at which the interpolated value is to be determined. While this technique is an approximate one, it has been used with great success in compute tomography [5, 7, 25], image processing [6, 12, 18, 24, 31], and other fields [3, 35].

The benefit of this approach is that only $\mathcal{O}(N_s N_r) = \mathcal{O}(N_r)$ computations are required. The value of N_s is chosen for a given finite error threshold. The drawback is that unlike Fourier interpolation, it is not mathematically exact in exact arithmetic. For more discussion of splines see [28].

In the sections that follow, we use the notation $\epsilon(N_r)$ to denote the relative complexity of Fourier interpolation to spline interpolation. That is, for Fourier interpolation

$$\epsilon(N_r) = (\log N_r)^2,$$

whereas for spline interpolation

$$\epsilon(N_r) = N_s = \mathcal{O}(1).$$

11.2   A Fast Algorithm for the Fourier Transform on the 2D Motion Group

Numerical algorithms for Fourier transforms on the 2D motion group were developed in [17] (continuous motion group) and [16] (discrete motion group). For completeness we describe in the following a new algorithm for the continuous motion group Fourier transform using FFT methods. This algorithm is similar to the discrete motion group algorithm used in [16] but uses $\exp(i\, m\, \Phi)$ as basis functions instead of pulse functions. We begin with the representation operators for the 2D motion group

$$U(g, p)\tilde{\varphi}(\mathbf{x}) = e^{-ip(\mathbf{r}\cdot\mathbf{x})}\tilde{\varphi}\left(R^T\mathbf{x}\right), \tag{11.2}$$

which are defined for each $g = (\mathbf{r}, R(\theta)) \in SE(2)$. Here $p \in \mathbb{R}^+$, and $\mathbf{x}\cdot\mathbf{y} = x_1y_1 + x_2y_2$. The vector $\mathbf{x}$ is a unit vector ($\mathbf{x}\cdot\mathbf{x} = 1$), so $\tilde{\varphi}(\mathbf{x}) = \tilde{\varphi}(\cos\psi, \sin\psi) \equiv \varphi(\psi)$ is a function on the unit circle. Henceforth we will not distinguish between $\tilde{\varphi}$ and φ.

Any function $\varphi(\psi) \in \mathcal{L}^2([0, 2\pi))$ can be expressed as a weighted sum of orthonormal basis functions as $\varphi(\psi) = \sum_n a_n e^{in\psi}$. Likewise, the matrix elements of the operator $U(g, p)$ are expressed in this basis as [32]

$$U_{mn}(g, p) = \left(e^{im\psi}, U(g, p)e^{in\psi}\right) = \frac{1}{2\pi}\int_0^{2\pi} e^{-im\psi}e^{-i(r_1 p\cos\psi + r_2 p\sin\psi)}e^{in(\psi-\theta)}\, d\psi \tag{11.3}$$

$\forall\; m, n \in \mathbf{Z}$. The inner product $(\cdot, \cdot)$ is defined as

$$(\varphi_1, \varphi_2) = \frac{1}{2\pi}\int_0^{2\pi} \overline{\varphi_1(\psi)}\varphi_2(\psi)d\psi.$$

It is easy to see that $(U(g, p)\varphi_1, U(g, p)\varphi_2) = (\varphi_1, \varphi_2)$, and that $U(g, p)$ is, therefore, unitary with respect to this inner product.

We may express the Fourier matrix elements as an integral[1]

$$\hat{f}_{mn}(p) = \int_{\mathbf{r}\in\mathbb{R}^2}\int_{\theta=0}^{2\pi}\int_{\psi=0}^{2\pi} f(\mathbf{r}, \theta)\, e^{in\psi}e^{i(\mathbf{p}\cdot\mathbf{r})}e^{-im(\psi-\theta)}\, d^2r\, d\theta\, d\psi.$$

We compute the band-limited approximation of the Fourier transform for $|m|, |n| \leq S$ harmonics. Furthermore, we assume that it is computed at $N_p = \mathcal{O}(S)$ points along the p-interval, and we assume that the order of sampling points in an $\mathbb{R}^2$ region is of the order of S^2 ($N_r = \mathcal{O}(S^2)$) and the number of sampling points of orientation angle θ is $N_R = \mathcal{O}(S)$. In this way the total number of points sampled in $SE(2)$ is $N = \mathcal{O}(S^3)$, which is on the same order as the total number of sample points in the Fourier domain.

For estimates of complexity of numerical algorithms we introduce the following notations:

N_r	Number of samples on $\mathbb{R}^2$
N_R	Number of samples on $SO(2)$
N_p	Number of samples of p interval
N_u	Number of samples on $[0, 2\pi)$
N_F	Total number of harmonics .

---

[1]We use the notation $\mathbf{dr}$ and $d^n r$ to both mean $dr_1, \ldots, dr_n$ to within a constant factor.

We assume that only a finite number of harmonics are required for an accurate approximation of the function $f(g)$. Hence only matrix elements in the range $|s| < S$ and $l, l' < L$ are computed. We assume also that $L = \mathcal{O}(S)$. We have made the following assumptions about the number of samples in terms of S:

N_r	$\mathcal{O}(S^2)$
N_R	$\mathcal{O}(S)$
N_p	$\mathcal{O}(S)$
N_u	$\mathcal{O}(S)$
N_F	$\mathcal{O}(S^2)$

From these definitions, $N = N_r \cdot N_R = \mathcal{O}(S^3)$ and $N_p \cdot N_F = \mathcal{O}(N)$.

We may perform first $\mathbb{R}^2$ integration using the usual FFT

$$f_1(\mathbf{p}, \theta) = \int_{\mathbb{R}^2} f(\mathbf{r}, \theta)\, e^{i(\mathbf{p} \cdot \mathbf{r})} d^2 r$$

in $\mathcal{O}(N_R N_r \log N_r)$ computations.

Then we perform interpolation to the polar coordinate mesh. For N_s-point spline interpolation this may be performed in $\mathcal{O}(N_s N_r N_R)$ computations. If Fourier interpolation is used, this becomes $\mathcal{O}((\log N_r)^2 N_r N_R)$ computations.

The next step is to perform integration on $SO(2)$

$$f_2^{(m)}(p, \psi) = \int_{SO(2)} f_1(p, \psi, \theta)\, e^{im\theta} d\theta.$$

This may be computed in $\mathcal{O}(N_r N_R \log N_R)$ computations. Then, ψ integration

$$\hat{f}_{mn}(p) = \int_0^{2\pi} \left[f_2^{(m)}(p, \psi) e^{-im\psi} \right] e^{in\psi} d\psi$$

may be computed in $\mathcal{O}(S N_p S \log S) = \mathcal{O}(N_r N_R \log N_R)$ computations.

We denote the total number of samples on $SE(2)$ as $N = N_r N_R = \mathcal{O}(S^3)$. Thus the total number of arithmetic operations used to compute the direct Fourier transform is $\mathcal{O}(N \log N) + \mathcal{O}(N\epsilon(N^{\frac{1}{2}}))$. When spline interpolation is used, the first term in this complexity estimate dominates for large values of N, whereas the second term dominates when Fourier interpolation is used.

The Fourier inversion formula for $SE(2)$ with matrix elements $U_{mn}(g, p)$ written in integral form is

$$f(g) = \frac{1}{2\pi} \int_0^\infty p\, dp \sum_{m,n=-\infty}^{\infty} \hat{f}_{nm}(p) \int_0^{2\pi} e^{i(n-m)\psi} e^{-i\mathbf{r} \cdot \mathbf{p}} e^{-in\theta} d\psi.$$

This may be rewritten as

$$f(g) = \frac{1}{2\pi} \sum_{n=-\infty}^{\infty} e^{-in\theta} \int_{\mathbb{R}^2} \tilde{f}_n(\mathbf{p}) e^{-i\mathbf{r} \cdot \mathbf{p}} d^2 p$$

where

$$\tilde{f}_n(\mathbf{p}) = \sum_{m=-\infty}^{\infty} \hat{f}_{nm}(p) e^{i(n-m)\psi}.$$

If we assume that the above sums over m and n are truncated at $\pm S$ where $S = \mathcal{O}(N_R)$ and $\tilde{f}_n(\mathbf{p})$ is a band-limited function of $\mathbf{p}$ for each value of n, then it may be shown that the Fourier inversion can be performed in the same order of computations as the set of Fourier transforms. Note that this is faster than if the ψ-integration is performed first, even though that integration results in closed-form solutions for the matrix elements $U_{mn}(g, p)$.

The matrix product in the convolution may be performed in $\mathcal{O}(N^{(\gamma+1)/3})$ computations, where $2 \leq \gamma \leq 3$ is the cost of matrix multiplication.

11.3   Algorithms for Fast $SE(3)$ Convolutions Using FFTs

Here we describe an algorithm for computing 3D continuous motion group convolutions using FFTs. We use irreducible unitary representations of $SE(3)$ (as described in Chapter 10) to calculate Fourier matrix elements in integral form.

To write the Fourier transform in matrix form we calculate the matrix elements of $U(\mathbf{r}, R; p)$ as

$$U^s_{l',m';l,m}(\mathbf{r}, R; p) = \int_{S^2} \overline{h^s_{l'm'}(\mathbf{u})} \left(U(\mathbf{r}, R; p)h^s_{lm} \right)(\mathbf{u})\, d\mathbf{u}$$

where $d\mathbf{u} = \sin \Theta d\Theta\, d\Phi$ and $h^s_{l'm'}(\mathbf{u}) = h^s_{l'm'}(\mathbf{u}(\Theta, \Phi))$ are generalized spherical harmonics defined in Chapter 9 and used in Chapter 10. Here $g = (\mathbf{r}, R) \in SE(3)$.

11.3.1   Direct Fourier Transform

We use basis eigenfunctions $h^s_{lm}(\mathbf{u})$ to write Fourier matrix elements in the integral form

$$\hat{f}^s_{l',m';l,m}(p)$$
$$= \int_{\mathbf{u} \in S^2} \int_{\mathbf{r} \in \mathbb{R}^3} \int_{R \in SO(3)} f(\mathbf{r}, R)\, h^s_{lm}(\mathbf{u})\, e^{ip\,\mathbf{u}\cdot\mathbf{r}}\, \overline{\Delta_s(Q(\mathbf{u}, R))h^s_{l'm'}(R^{-1}\mathbf{u})}\, d\mathbf{u}\, d^3r\, dR$$

where dR is the normalized invariant integration measure on $SO(3)$.

Recall from Chapter 10 that the basis functions may be expressed in the form [21, 22]

$$h^s_{lm}(\mathbf{u}(\Theta, \Phi)) = Q^l_{s,m}(\cos \Theta)\, e^{i(m+s)\Phi} \tag{11.4}$$

where

$$Q^l_{-s,m}(\cos \Theta) = (-1)^{l-s} \sqrt{\frac{2l+1}{4\pi}}\, P^l_{s\,m}(\cos \Theta),$$

and generalized Legendre functions $P^l_{m\,s}(\cos \Theta)$ are given as in Vilenkin and Klimyk [33].

Under the rotation R these functions are transformed as

$$\left(U^s(\mathbf{0}, R; p) h^s_{lm} \right)(\mathbf{u}) = \Delta_s(Q(\mathbf{u}, R)) h^s_{lm}\left(R^{-1}\mathbf{u} \right) = \sum_{n=-l}^{l} U^l_{nm}(R)h^s_{ln}(\mathbf{u}).$$

$U^l_{nm}(R)$ are matrix elements of $SO(3)$ representations

$$U^l_{mn}(A) = e^{-im\alpha}\, (-1)^{n-m}\, P^l_{mn}(\cos \beta)\, e^{-in\gamma}, \tag{11.5}$$

where α, β, γ are ZXZ Euler angles of the rotation and $P^l_{mn}(\cos \beta)$ is a generalization of the associated Legendre functions [32]. ($U^l_{m,n}$ were called $\tilde{U}^l_{m,n}$ in Chapter 10).

Thus the Fourier transform matrix elements may be written in the form

$$\hat{f}_{l',m';l,m}^{s}(p)$$

$$= \int_{\mathbf{u}\in S^2} \int_{\mathbf{r}\in\mathbb{R}^3} \int_{R\in SO(3)} f(\mathbf{r}, R)\, h_{lm}^{s}(\mathbf{u})\, e^{i p \mathbf{u}\cdot\mathbf{r}} \sum_{n=-l'}^{l'} \overline{U_{nm'}^{l'}(R)}\overline{h_{l'n}^{s}(\mathbf{u})}\, d\mathbf{u}\, d^3 r\, dR.$$

For estimates of complexity of numerical algorithms we introduce the following notations:

N_r	Number of samples on $\mathbb{R}^3$
N_R	Number of samples on $SO(3)$
N_p	Number of samples of p interval
N_u	Number of samples on S^2
N_F	Total number of harmonics.

We assume that only a finite number of harmonics are required for an accurate approximation of the function $f(g)$. Hence only matrix elements in the range $|s| < S$ and $l, l' < L$ are computed. We assume also that $L = \mathcal{O}(S)$. We have made the following assumptions about the number of samples in terms of S:

N_r	$\mathcal{O}(S^3)$
N_R	$\mathcal{O}(S^3)$
N_p	$\mathcal{O}(S)$
N_u	$\mathcal{O}(S^2)$
N_F	$\mathcal{O}(S^5)$.

From these definitions, $N = N_r \cdot N_R = \mathcal{O}(S^6)$ and $N_p \cdot N_F = \mathcal{O}(N)$.

Our algorithm for the numerical computation of the direct Fourier transform is as follows:

a) First, we compute

$$f_1(R, \mathbf{p}) = \int_{\mathbb{R}^3} f(\mathbf{r}, R)\, e^{i\mathbf{p}\cdot\mathbf{r}}\, d^3 r$$

using FFTs for a Cartesian lattice in $\mathbb{R}^3$. This integral may be computed in $\mathcal{O}(N_r \log(N_r) N_R)$ computations. The resulting Fourier transform is computed on a rectangular grid. We need to perform interpolation to the spherical coordinate grid in order to compute values

$$f_1(R; p, \mathbf{u}).$$

The complexity of this interpolation is discussed in Section 11.1. In general, $\mathcal{O}(N_r \epsilon(N_r))$ computations will be required for each different sampled rotation. For high precision numerical approximations, 3D spline interpolation techniques may be used (see, for example, [5, 28]). For N_s-point spline interpolation for all values of rotation, the complexity of interpolation is of $\mathcal{O}(N_s N_r N_R)$ computations, i.e., $\epsilon(N_r) = N_s$. Since spline interpolation only uses a small subset of the sample points, we assume $N_s = \mathcal{O}(1)$. For exactly reversible interpolation (in exact arithmetic), Fourier interpolation can be used. In this case $\epsilon(N_r) = \mathcal{O}((\log N_r)^2)$, for reasons that are explained in Section 11.1.

b) Then we perform integration on $SO(3)$

$$(f_2)_{nm'}^{l'}(p, \mathbf{u}) = \int_{SO(3)} f_1(R, p, \mathbf{u})\overline{U_{nm'}^{l'}(R)}\, dR.$$

This is the Fourier transform on the rotation group, computed for different values of p and $\mathbf{u}$. A fast Fourier transform technique for $SO(3)$ has been developed by Maslen and Rockmore which calculates the forward and inverse Fourier transform of band-limited functions on $SO(3)$ in $\mathcal{O}(N_R(\log N_R)^2)$ arithmetic operations for N_R sample points [20]. This may be applied to compute f_2 for all values of p and $\mathbf{u}$ and all indices in $\mathcal{O}(N_p N_u N_R (\log N_R)^2)$. We assume that $N_p N_u \approx N_r$, thus the order of computations is $\mathcal{O}(N_r N_R (\log N_R)^2)$.

c) Then we may perform integrations on the unit sphere

$$\hat{f}^s_{l',m';l,m}(p) = \sum_{n=-l'}^{l'} \int_{S^2} (f_2)^{l'}_{nm'}(p,\mathbf{u}) h^s_{lm}(\mathbf{u}) \overline{h^s_{l'n}(\mathbf{u})}\, d\mathbf{u}. \tag{11.6}$$

Using expression Eq. (10.29) for the basis functions we write Eq. (11.6) in the form

$$\hat{f}^s_{l',m';l,m}(p)$$
$$= \sum_{n=-l'}^{l'} \int_{S^2} (f_2)^{l'}_{nm'}(p,\mathbf{u}) Q^l_{s,m}(\cos\Theta) Q^{l'}_{s,n}(\cos\Theta) \exp(i(m-n)\Phi)\, d\Phi \sin\Theta d\Theta.$$

We may perform integration with respect to Φ using the FFT on S^1

$$(f_3)_{l',m',n;m}(p,\Theta) = \int_0^{2\pi} \left[(f_2)^{l'}_{nm'}(p,\Phi,\Theta) \exp(-in\Phi) \right] \exp(im\Phi) d\Phi.$$

Integrations may be performed in $\mathcal{O}(N_p N_\Theta S^3 N_\Phi \log N_\Phi)$, where $N_\Phi = \mathcal{O}(S)$ is the number of samplings on the Φ interval. Thus the order of computations is $\mathcal{O}(N_r N_R \log N_R)$.

Then we perform integration with respect to Θ

$$(f_4)^s_{l',m';l,m,n}(p) = \int_0^\pi \left[(f_3)_{l',m',n;m}(p,\Theta) Q^{l'}_{s,n}(\cos\Theta) \right] Q^l_{s,m}(\cos\Theta) \sin\Theta d\Theta. \tag{11.7}$$

Using the fact that $Q^l_{s,m}(\cos\Theta)$ is $P^l_{-s,m}(\cos\Theta)$ (up to a constant coefficient), the integration (summation) for all but l fixed indices may be performed using the Driscoll and Healy technique [10, 11, 20] in $\mathcal{O}(S^5 N_p N_\Theta (\log N_\Theta)^2)$, where $N_\Theta = \mathcal{O}(S)$ is the number of samples on the Θ interval. Thus the order of computations is $\mathcal{O}(N_R^{5/3} N_r^{2/3} (\log N_R)^2)$. The Θ integration may be performed in $\mathcal{O}(S^6 N_p N_\Theta) = \mathcal{O}(N_R^2 N_r^{2/3})$ operations using plain integration.

Then the matrix elements of the $SE(3)$-Fourier transform may be found by the summation

$$\hat{f}^s_{l',m';l,m}(p) = \sum_{n=-l'}^{l'} (f_4)^s_{l',m';l,m,n}(p)$$

which is of the order $\mathcal{O}(S^6 N_p) = \mathcal{O}(N_R^2 N_p)$.

Thus the total order of computations of the direct Fourier transform is $\mathcal{O}(N_r N_R (\log(N_r) + (\log N_R)^2 + \epsilon(N_r)) + N_R^{5/3} N_r^{2/3} (\log N_R)^2)$. Under the assumption that $N_r = \mathcal{O}(N_R)$ and using the notation $N_r N_R = N$ [N is the total number of samples on $SE(3)$] we write the leading order terms as $\mathcal{O}(N^{7/6} (\log N)^2 + N \epsilon(N^{\frac{1}{2}}))$. For plain Θ integration [i.e., if the fast generalized Legendre transform is not used to evaluate Eq. (11.7)] the estimate becomes $\mathcal{O}(N^{4/3} + N \epsilon(N^{\frac{1}{2}}))$.

11.3.2 Inverse Fourier Transform

The inverse Fourier transform integral may be written as

$$f(\mathbf{r}, R) = \frac{1}{2\pi^2} \int_{p=0}^{\infty} \int_{\mathbf{u}\in S^2} \sum_{s,l,m,l',m'} \hat{f}_{l',m';l,m}^s(p)\, \overline{h_{lm}^s(\mathbf{u})} \exp(-ip\mathbf{u}\cdot\mathbf{r})$$

$$\sum_{n=-l'}^{l'} U_{nm'}^{l'}(R) h_{l'n}^s(\mathbf{u})\, p^2\, dp\, d\mathbf{u}.$$

where

$$\sum_{s,l,m,l',m'} = \sum_{s=-\infty}^{\infty} \sum_{l=|s|}^{\infty} \sum_{m=-l}^{l} \sum_{l'=|s|}^{\infty} \sum_{m'=-l'}^{l'}.$$

A band-limited approximation results when the restrictions $s \leq S$ and $l, l' \leq L = \mathcal{O}(S)$ are imposed. We note that $p^2\,dp\,d\mathbf{u} = d^3p$. Our algorithm for the inverse Fourier transform is as follows.

a) We first compute

$$(g_1)_{l',m';n}^s (p, \mathbf{u}) = \sum_{l=|s|}^{L} \left[\sum_{m=-l}^{l} \hat{f}_{l',m';l,m}^s(p)\, \overline{h_{lm}^s(\mathbf{u})} \right] h_{l'n}^s(\mathbf{u}) \qquad (11.8)$$

for fixed values of l', m', n, s. The summation may be performed in the following way. We first perform the summation in square brackets. Using the expression for basis functions in Eq. (10.29) this summation may be written as

$$(g_{11})_{l',m';l}^s (p, \Theta, \Phi) = \exp(-is\Phi) \sum_{m=-l}^{l} \hat{f}_{l',m';l,m}^s(p)\, Q_{s,m}^l(\cos\Theta)\, \exp(-im\Phi).$$

Replacing the summation limits $|m| \leq l$ by $|m| \leq L = \mathcal{O}(S)$ [and assuming that the corresponding elements $\hat{f}_{l',m';l,m}^s(p)$ are zero for $|m| > l$ for given l] we may compute this sum using the one-dimensional FFT for fixed values of other indices. This may be computed in $\mathcal{O}(S^4 N_p N_\Theta S \log S) = \mathcal{O}(N_R^{5/3} N_r^{2/3} \log N_R)$. We note that the term $\exp(-is\Phi)$ is canceled with the corresponding term from $h_{l'n}^s(\mathbf{u})$ in Eq. (11.8).

Then we compute the summation

$$(g_{12})_{l',m'}^s (p, \Theta, \Phi) = \sum_{l=|s|}^{L} (g_{11})_{l',m';l}^s (p, \Theta, \Phi)$$

which may be performed in $\mathcal{O}(N_p N_\Theta N_\Phi S^4) = \mathcal{O}(N_r N_R^{4/3})$ computations. The product

$$(g_1)_{l',m';n}^s (p, \mathbf{u}) = (g_{12})_{l',m'}^s (p, \Theta, \Phi)\, Q_{s,n}^{l'}(\cos\Theta)\, \exp(in\Phi) \qquad (11.9)$$

may be computed in the same amount of computations.

Thus, the sum in Eq. (11.8) may be performed in $\mathcal{O}(N^{7/6} \log N)$ computations. Then we interpolate from a spherical coordinate grid to a rectangular grid. This requires $\mathcal{O}(N_r N_R^{4/3} \epsilon(N_r)) = \mathcal{O}(N^{7/6} \epsilon(N^{\frac{1}{2}}))$ operations.

Application of the Driscoll-Healy fast transform technique gives an additional savings in computation of sum in Eq. (11.8). We define formally the $\hat{f}_{l',m';l,m}^s(p)$ matrix elements to be zero for

$|m|, |s| > l$ and extend the limits of summation in Eq. (11.8) from $l = 0$ and from $m = -L$ to $m = L$. Then the summation with respect to l may be performed first as

$$(g_{11})_{l',m';m}^s (p, \Theta) = \sum_{l=0}^{L} \hat{f}_{l',m';l,m}^s (p) \, Q_{s,m}^l (\cos \Theta)$$

using the fast tansform technique in $\mathcal{O}(S^4 N_p \, S (\log S)^2) = \mathcal{O}(N_R^{4/3} N_r^{2/3} (\log N_R)^2)$ computations. Then the summation

$$(g_{12})_{l',m'}^s (p, \Theta, \Phi) = \sum_{m=-L}^{L} (g_{11})_{l',m';m}^s (p, \Theta) \exp(-im\Phi)$$

may be performed using the one-dimensional FFT in $\mathcal{O}(S^3 N_p N_\Theta S \log S) = \mathcal{O}(N_R^{4/3} N_r^{2/3} \log N_R)$.
  b) Next we compute integrals of the form

$$(g_2)_{l',m';n}^s (\mathbf{r}) = \int_{\mathbb{R}^3} (g_1)_{l',m';n}^s (\mathbf{p}) \exp(-i\mathbf{p} \cdot \mathbf{r}) \, d^3 p$$

using 3D FFTs. This requires $\mathcal{O}(S^4 \, N_r \, \log(N_r)) = \mathcal{O}(N^{7/6} \log N)$ operations to compute.
  c) The function $f(\mathbf{r}, R)$ may be recovered by the summation

$$f(\mathbf{r}, R) = \sum_{s=-S}^{S} \left[\sum_{l'=|s|}^{L} \sum_{m'=-l'}^{l'} \sum_{n=-l'}^{l'} U_{nm'}^{l'} (R) \, (g_2)_{l',m';n}^s (\mathbf{r}) \right].$$

The expression in square brackets is a set of inverse Fourier transforms on $SO(3)$ for fixed values of $\mathbf{r}$ and s. It takes $\mathcal{O}(N_r \, S \, N_R \, (\log N_R)^2) = \mathcal{O}(N^{7/6} (\log N)^2)$ operations to compute.
  Thus the total order of computation for the inverse Fourier transform is $\mathcal{O}(N^{7/6} ((\log N)^2 + N \, \epsilon(N^{\frac{1}{2}}))$.

Convolution of Functions

  The convolution integral

$$(f_1 * f_2)(g) = \int_{SE(3)} f_1(h) \, f_2(h^{-1} \circ g) \, d(h) \tag{11.10}$$

may be written as a matrix product in Fourier space

$$(\mathcal{F}(f_1 * f_2))_{l',m';l,m}^s (p) = \sum_{j=|s|}^{\infty} \sum_{k=-j}^{j} \left(\hat{f}_2 \right)_{l',m';j,k}^s (p) \left(\hat{f}_1 \right)_{j,k;l,m}^s (p).$$

When $f_1(g)$ and $f_2(g)$ are band-limited in the sense defined earlier, the matrix product may be computed directly in $\mathcal{O}(N_p S^7) = \mathcal{O}(N_r^{1/3} N_R^{7/3}) = \mathcal{O}(N^{4/3})$ operations, which may be the largest time-consuming computation. We note that a fast matrix multiplication algorithm may be applied for $n \times n = 2^m \times 2^m$ matrices, which is of the order of $n^{\log_2 7}$ instead of n^3 [26, 27, 29, 34]. Using this algorithm, the matrix product may be computed in $\mathcal{O}(N^{(\log_2 7 + 1)/3})$ computations. Since fast matrix multiplications is an active research field in its own right, we characterize the order of computations for the convolution product as $\mathcal{O}(N^{(\gamma+1)/3})$, where $2 \leq \gamma \leq 3$ indicates the cost of matrix multiplication.
  Therefore, the total order of computations of convolution is, at most, $\mathcal{O}(N^{(\gamma+1)/3}) + \mathcal{O}(N^{7/6} (\log N)^2) + \mathcal{O}(N^{7/6} \epsilon(N^{\frac{1}{2}}))$. Thus when $\epsilon(N_r) \leq \mathcal{O}((\log N_r)^2)$, the preceding algorithm provides very considerable savings compared to the direct integration in Eq. (11.10), which is of the order of $\mathcal{O}(N_r^2 N_R^2) = \mathcal{O}(N^2)$.

11.4 Fourier Transform on the Discrete Motion Group of the Plane

The subgroup of $SE(2)$ where $\theta = 2\pi i/N_R$ for $i = 0, \ldots, N_R - 1$ is called the *discrete motion group* of the plane. While $SE(n)$ and crystallographic space groups have been studied extensively in the literature, the discrete motion groups have received very little attention.

11.4.1 Irreducible Unitary Representations of the Discrete Motion Group

We define in this section matrix elements of the irreducible unitary representations (IURs) of the discrete motion group $G = \mathbb{R}^2 \rtimes_\varphi C_{N_R}$, where C_{N_R} is the N_R-element finite subgroup of $SO(2)$ (i.e., the group of rotational symmetries of a regular planar N_R-gon), and the notation $\rtimes_\varphi$ means semi-direct product (which in the present context means nothing more than that group elements are expressed as homogeneous transforms).

The IURs $U(\mathbf{a}, A)$ of $SE(2)$ act on functions $f(\mathbf{u}) \in \mathcal{L}^2(S)$ [S is a unit circle, $\mathbf{u} = (\cos\Theta, \sin\Theta)^T$ is a vector to a point on the unit circle] with the inner product

$$(\varphi_1 \cdot \varphi_2) = \frac{1}{2\pi} \int_{S^1} \overline{\varphi_1(\Theta)}\, \varphi_2(\Theta)\, d\Theta$$

where Θ is an angle on the unit circle. These operators are defined by the expression

$$\left(U_p(\mathbf{a}, A; p)\varphi\right)(\mathbf{u}) = e^{-ip\mathbf{u}\cdot\mathbf{a}}\, \varphi\left(A^{-1}\mathbf{u}\right) = e^{-i\mathbf{p}\cdot\mathbf{a}}\, \varphi\left(A^{-1} \cdot (\mathbf{p}/p)\right), \tag{11.11}$$

where $A \in SO(2)$, $p \in \mathbb{R}^+$, $\mathbf{p} = p\mathbf{u}$ is the vector to arbitrary points in the dual (frequency) space of $\mathbb{R}^2$ (p is its magnitude and $\mathbf{u}$ is its direction).

We choose a pulse orthonormal basis $\varphi_{N_R,n}(\mathbf{u})$ on S^1, i.e., we subdivide the circle into identical segments F_n and choose the φ-functions to satisfy the orthonormality relations

$$\frac{1}{2\pi} \int_{S^1} \varphi_{N_R,n}(\mathbf{u})\, \varphi_{N_R,m}(\mathbf{u})\, d\Theta = \delta_{nm}.$$

We may choose the orthonormal functions as

$$\varphi_{N_R,n}(\mathbf{u}) = \begin{cases} (N_R)^{1/2} & \text{if } \mathbf{u} \in F_n \\ 0 & \text{otherwise.} \end{cases}$$

$n = 0, \ldots, N_R - 1$ enumerates different segments. We denote these pulse functions as δ-like functions $\varphi_{N_R,n}(\mathbf{u}) = (1/N_R)^{1/2}\, \delta_{N_R}(\mathbf{u}, \mathbf{u_n})$, where $\mathbf{u_n}$ is the vector to the center of the F_n segment. The matrix elements in this basis are

$$U_{mn}(A, \mathbf{r}; p) = \frac{1}{2\pi} \int_S \varphi_{N_R,m}(\mathbf{u})\, e^{-ip\,\mathbf{u}\cdot\mathbf{r}}\, \varphi_{N_R,n}\left(A^{-1}\mathbf{u}\right) d\Theta. \tag{11.12}$$

Using the "delta-function" notation, this may be written as

$$U_{mn}(A, \mathbf{r}; p) = \frac{1}{2\pi N_R} \int_S \delta_{N_R}(\mathbf{u}, \mathbf{u_m})\, e^{-ip\,\mathbf{u}\cdot\mathbf{r}}\, \delta_{N_R}\left(A^{-1}\mathbf{u}, \mathbf{u_n}\right) d\Theta.$$

This integral may be approximated as

$$U_{mn}(A, \mathbf{r}; p) \approx 1/N_R\, e^{-ip\,\mathbf{u_m}\cdot\mathbf{r}}\, \delta_{N_R}\left(A^{-1}\mathbf{u_m}, \mathbf{u_n}\right). \tag{11.13}$$

We approximate delta-functions as

$$1/N_R \, \delta_{N_R} \left(A^{-1} \mathbf{u}_m, \mathbf{u}_n \right) = \delta_{A^{-1} \mathbf{u}_m, \mathbf{u}_n} = \begin{cases} 1 & \text{if } A^{-1} \mathbf{u}_m = \mathbf{u}_n \\ 0 & \text{otherwise} \end{cases}$$

which means that we restrict rotations to the rotations A_j from the finite subgroup C_{N_R} of $SO(2)$, and $A_j^{-1} \mathbf{u}_m = \mathbf{u}_{m-j} = \mathbf{u}_n$.

Thus the matrix elements of the irreducible unitary representations of the "discrete" motion subgroup $G = \mathbb{R}^2 \rtimes_\varphi C_{N_R}$ are given as

$$U_{mn} \left(A_j, \mathbf{r}; p \right) = e^{-i \, p \, \mathbf{u}_m \cdot \mathbf{r}} \, \delta_{A_j^{-1} \mathbf{u}_m, \mathbf{u}_n}, \tag{11.14}$$

where $\delta_{A_j^{-1} \mathbf{u}_m, \mathbf{u}_n} = \delta_{m-j, n}$ in this case.

11.4.2   Fourier Transforms on the Discrete Motion Group

Though motivated as an approximation, the matrix elements in Eq. (11.14) are exact expressions for the matrix elements of the unitary representations of the discrete motion group. However, this set of matrix elements in Eq. (11.14) is incomplete. This means that the direct and inverse Fourier transforms, defined using these matrix elements, would reproduce the original function with $\mathcal{O}(1/N_R)$ error, i.e.,

$$\mathcal{F}^{-1} \left(\mathcal{F}(f(A_i, \mathbf{r})) = f(A_i, \mathbf{r}) \left(1 + \mathcal{O} \left(\frac{1}{N_R} \right) \right).$$

The reason for this is that summing through all possible segments cannot replace integration over all possible angles on the circle. It is also clear that the additional continuous parameter which enumerates possible angles inside each segment on the circle must give the complete set of the matrix elements.

Thus the matrix elements are modified as

$$U_{mn} \left(A_j, \mathbf{r}; p, \Phi \right) = e^{-i \, p \, \mathbf{u}_m^\Phi \cdot \mathbf{r}} \, \delta_{A_j^{-1} \mathbf{u}_m, \mathbf{u}_n} \tag{11.15}$$

where $\mathbf{u}_k^\Phi$ denotes the vector to the angle Θ on the unit circle on the interval

$$F_k = [2\pi \, k/N_R , \, 2\pi \, (k+1)/N_R],$$

where $k = 0, \ldots, N_R - 1$ (Φ measures the angle on this segment).

The completeness relation

$$\sum_{m=0}^{N_R-1} \sum_{n=0}^{N_R-1} \int_0^\infty \int_0^{2\pi/N_R} \overline{U_{mn}(A_i, \mathbf{r}_1; p, \Phi)} \, U_{mn} \left(A_j, \mathbf{r}_2; p, \Phi \right) \, p \, dp \, d\Phi$$

$$= (2\pi)^2 \, \delta^2 \left(\mathbf{r}_1 - \mathbf{r}_2 \right) \, \delta_{A_i, A_j} \tag{11.16}$$

$(d\Phi = d\Theta)$ is exact, because the integration is now over the whole space of $\mathbf{p}$ values. This may be proven using the integral representation of the δ-function

$$\frac{1}{(2\pi)^2} \int_{\mathbb{R}^2} e^{-i \, \mathbf{p} \cdot \mathbf{r}} \, d^2 p = \delta^2(\mathbf{r}).$$

The orthogonality relation is written as

$$\sum_{i=0}^{N_R-1} \int_{\mathbb{R}^2} \overline{U_{mn}(A_i, \mathbf{r}; p, \Phi)}\, U_{m'n'}(A_i, \mathbf{r}; p', \Phi')\, d^2 r$$

$$= (2\pi)^2 \frac{\delta(p-p')}{p}\, \delta_{m,m'}\, \delta_{n,n'}\, \delta(\Phi - \Phi'). \tag{11.17}$$

The direct Fourier transform is defined as

$$\hat{f}_{mn}(p, \Phi) = \sum_{i=0}^{N_R-1} \int_{\mathbb{R}^2} f(A_i, \mathbf{r})\, U_{mn}^{-1}(A_i, \mathbf{r}; p, \Phi)\, d^2 r. \tag{11.18}$$

The vector $\mathbf{u}_m^\Phi$, which is inside the segment F_m, may be found by the rotation A_m (which transforms F_0 to F_m) from $\mathbf{u}_0^\Phi$, $\mathbf{u}_m^\Phi = A_m \mathbf{u}_0^\Phi$. The parameter Φ denotes the position inside the segment F_0.

The inverse Fourier transform is

$$\mathcal{F}^{-1}\left(\hat{f}\right) = \frac{1}{4\pi^2} \sum_{m=0}^{N_R-1} \sum_{n=0}^{N_R-1} \int_0^\infty \int_0^{2\pi/N_R} \hat{f}_{mn}(p, \Phi)\, U_{nm}(A_i, \mathbf{r}; p, \Phi)\, p\, dp\, d\Phi. \tag{11.19}$$

We note that this result is in agreement with [13].

We define a convolution on the discrete motion group as

$$F(\mathbf{r}, A_j) = \frac{1}{2\pi N_R} \sum_{i=0}^{N_R-1} \int_{\mathbb{R}^2} f_1(\mathbf{a}, A_i)\, f_2\left(A_i^{-1}(\mathbf{r} - \mathbf{a}), A_{j-i}\right) d^2 a. \tag{11.20}$$

(The normalization factor corresponds in the limit $N_R \to \infty$ to the normalization of the convolution on the continuous (Euclidean) motion group in [17]).

The Fourier transform of the convolution of functions on the discrete motion group is just a product of the Fourier matrices with the corresponding normalization

$$\hat{F}_{mn}^\Phi(p) = \frac{A}{2\pi N_R} \sum_{k=0}^{N_R-1} \left(\hat{f}_2\right)_{mk}^\Phi (p)\, \left(\hat{f}_1\right)_{kn}^\Phi (p) \tag{11.21}$$

where A is the area of a compact region of $\mathbb{R}^2 \ni \mathbf{r}$ where the FFT is computed. The functions must have support inside the area A, and are considered periodic outside of this region. The area factor arises because the discrete Fourier transform may be computed as an approximation of the continuous case using

$$r_1 \to \frac{L}{N_r} i; \quad p_1 \to \frac{2\pi}{L} i$$

(and analogously for the 2nd component). Here L is the length of the compact region in the x (y) direction. While the factor L is canceled out of equations if we apply direct and inverse discrete Fourier transform to the function, it appears in the convolution defined in Eq. (11.20).

11.4.3 Efficiency of Computation of $SE(2)$ Convolution Integrals Using the Discrete-Motion-Group Fourier Transform

In this section we show that using the Fourier transform on the discrete motion group is a fast method to compute convolution integrals on the discrete motion group (we assume in the following that the finite rotation group has N_R elements). Convolution on the discrete motion group is an approximation to $SE(2)$ convolution. Particularly, we show that the convolution of a function $f(\mathbf{r}, A_i)$ sampled at $N = N_R \cdot N_r$ points (N_r is the number of samples in an $\mathbb{R}^2$ region) may be performed in $\mathcal{O}(N \log N_r) + \mathcal{O}(N N_R)$ operations instead of $\mathcal{O}(N_g^2)$ computations in the direct coordinate space integration using the "plain" integration in Eq. (11.20). The structure of the matrix elements in Eq. (11.15) allows one to apply fast Fourier methods and reduce the amount of computations [without the application of FFT the amount of computations using the Fourier transform method is $\mathcal{O}(N^2/N_R)$].

First, we estimate the amount of computations to perform the direct and inverse Fourier transforms of $f(g)$. We assume that we restrict p values to a finite interval and sample it at N_p points, and sample the Φ values at N_Φ points. We also assume that the total number of harmonics $N_p N_\Phi N_R^2 = N = N_R N_r$.

Let us consider the direct Fourier transform in Eq. (11.18). Each term i (for fixed A_i) gives one nonzero term in each row and column of the Fourier matrix $\hat{f}_{mn}^\Phi(p)$ [because only one element in each row and column of $U_{mn}^{-1}(g; p, \Phi)$ is nonzero]. For each fixed i we may compute the usual FFT of $f(\mathbf{r}; A_i)$, which may be computed in $\mathcal{O}(N_r \log(N_r))$ operations. The Fourier transform elements found by FFT are computed on a square grid of $\mathbf{p}$ values. We may, however, interpolate the Fourier elements computed on the grid to the Fourier elements computed at polar coordinates. The radial part p is determined by the length of $\mathbf{p}$ and the angular part determines the indices m, Φ (the other index n is determined uniquely for given A_i). We interpolate values on a square grid to values on a polar grid. Such an interpolation may be performed in $\mathcal{O}(N_r \in (N_r))$ computations. Each term i in Eq. (11.18) may be computed in $\mathcal{O}(N_r \log(N_r))$ computations. The whole Fourier matrix may be calculated in $\mathcal{O}(N_R N_r \log(N_r))$ computations.

Again, one element from each row and column is used in the computation of the inverse Fourier transform for each rotation element A_i. After inverse interpolation to Cartesian coordinates [which may be done in $\mathcal{O}(N_r \in (N_r))$ computations], the inverse Fourier integration may be performed in $\mathcal{O}(N_r \log(N_r))$ for each of the N_R nonzero matrix elements of U using the FFT. Thus in $\mathcal{O}(N_R N_r \log(N_r))$ computations we reproduce the function for all A_i.

The matrix product of $\hat{f}_{mn}^\Phi(p)$ may be computed directly in $\mathcal{O}(N_R^3)$ computations[2] for each value of p and parameters Φ. This means that the convolution (which is a matrix product of Fourier matrices) may be performed in $\mathcal{O}(N_R^3 N_p N_\Phi) = \mathcal{O}(N_R N)$ computations.

Therefore, the convolution of functions on the discrete motion group may be performed in $\mathcal{O}(N \log N_r) + \mathcal{O}(N N_R) + O(N \in (N_r))$ using Fourier methods on the discrete motion group and the usual FFT, without assuming any special matrix multiplication technique.

It may be shown that without the application of the FFT the convolution may be performed using the Fourier transform in $\mathcal{O}(N^2/N_R)$, which is still faster than evaluating the direct integration.

11.5   Fourier Transform for the 3D "Discrete" Motion Groups

In this section we develop fast approximate algorithms for computing the Fourier transform of functions on the discrete motion groups, G_{N_R}, which are the semi-direct product of the continuous translation group $\mathbf{T} = (\mathbb{R}^3, +)$ and a finite subgroup $I_{N_R} \subset SO(3)$ with N_R elements. I_{N_R} can be the icosahedral, cubo-octahedral, or tetrahedral rotational symmetry groups. Our formulation is

---

[2]Estimates as fast as $\mathcal{O}(N_R^{2.38})$ have been reported (see [8] and references therein) which, depending on how they are implemented, have the potential to increase speed further.

completely general, though in discussions of numerical implementations we focus on the icosahedral discrete motion group, where the number of elements is $N_R = 60$. In Section 11.5.1 the mathematical formulation is presented and in Section 11.5.2 the computational complexity of implementing Fourier transforms and convolution of functions on G_{N_R} is discussed.

11.5.1 Mathematical Formulation

Instead of using spherical harmonics in Eq. (10.29) as basis functions as was done when calculating matrix elements of the IURs of the continuous motion group, we now choose pulse functions $\varphi_{N_R,n}(\mathbf{u})$ on S^2, i.e., we subdivide the sphere into spherical regions and choose the φ-functions to satisfy the orthonormality relations

$$\int_{S^2} \varphi_{N_R,n}(\mathbf{u}) \, \varphi_{N_R,m}(\mathbf{u}) \, d\mathbf{u} = \delta_{nm}$$

where $d\mathbf{u} = \sin \Theta d\Theta \, d\Phi$, and (Θ, Φ) are spherical coordinates.[3] For example, in the case of I_{60} we may subdivide the sphere into 20 equilateral triangles or 12 regular pentagons. These figures can be used as the support for pulse functions, but as we will see shortly, it is convenient to subdivide these regular figures so that 60 congruent (but irregular) regions, F_n, result. We then choose the orthonormal functions as

$$\varphi_{N_R,n}(\mathbf{u}) = \begin{cases} (\frac{N_R}{4\pi})^{1/2} & \text{if } \mathbf{u} \in F_n \\ 0 & \text{otherwise} \end{cases}$$

Here $n = 1, \ldots, N_R$ enumerates the different congruent polygonal regions on the sphere. We denote these δ-like functions as $\varphi_{N_R,n}(\mathbf{u}) = (4\pi/N_R)^{1/2} \delta_{N_R}(\mathbf{u}, \mathbf{u_n})$, where $\mathbf{u_n}$ is a vector from the center of the sphere to a point in F_n.

The matrix elements may be found in this basis as

$$U_{mn}^s(A, \mathbf{r}; p) = \int_{S^2} \varphi_{N_R,m}(\mathbf{u}) \, e^{-i \, p \, \mathbf{u} \cdot \mathbf{r}} \Delta_s \left(R_{\mathbf{u}}^{-1} A R_{A^{-1}\mathbf{u}} \right) \varphi_{N_R,n} \left(A^{-1}\mathbf{u} \right) d\mathbf{u}. \qquad (11.22)$$

Using the delta-function notations, this integral may be written as

$$U_{mn}^s(A, \mathbf{r}; p) = \frac{4\pi}{N_R} \int_{S^2} \delta_{N_R}(\mathbf{u}, \mathbf{u_m}) \, e^{-i \, p \, \mathbf{u} \cdot \mathbf{r}} \Delta_s \left(R_{\mathbf{u}}^{-1} A R_{A^{-1}\mathbf{u}} \right) \delta_{N_R} \left(A^{-1}\mathbf{u}, \mathbf{u_n} \right) d\mathbf{u}.$$

This integral may be approximated as

$$U_{mn}^s(A, \mathbf{r}; p) \approx 4\pi/N_R \, e^{-i \, p \, \mathbf{u_m} \cdot \mathbf{r}} \Delta_s \left(R_{\mathbf{u_m}}^{-1} A R_{A^{-1}\mathbf{u_m}} \right) \delta_{N_R} \left(A^{-1}\mathbf{u_m}, \mathbf{u_n} \right). \qquad (11.23)$$

We again approximate delta-functions as

$$4\pi/N_R \, \delta_{N_R} \left(A^{-1}\mathbf{u_m}, \mathbf{u_n} \right) = \delta_{A^{-1}\mathbf{u_m}, \mathbf{u_n}} = \begin{cases} 1 & \text{if } A^{-1}\mathbf{u_m} = \mathbf{u_n} \\ 0 & \text{otherwise} \end{cases}$$

which means that we discretize the rotation group, i.e., we restrict rotations to the rotations A_j from the finite subgroup I_{N_R} of $SO(3)$ and $A_j^{-1}\mathbf{u_m} = \mathbf{u_n}$.

Thus matrix elements of irreducible unitary representations of the "discrete" motion subgroup G_{N_R} are given as

$$U_{mn}^s \left(A_j, \mathbf{r}; p \right) = e^{-i \, p \, \mathbf{u_m} \cdot \mathbf{r}} \Delta_s \left(R_{\mathbf{u_m}}^{-1} A_j R_{\mathbf{u_n}} \right) \delta_{A_j^{-1}\mathbf{u_m}, \mathbf{u_n}}. \qquad (11.24)$$

---

[3] The set of functions $\{\varphi_{N_R,n}(\mathbf{u})\}$ is of course not complete in $\mathcal{L}^2(S^2)$ and therefore is not a basis in which to expand matrix elements of the IURs, but we take this into account shortly.

Here s enumerates the representations of the little group, which is a finite subgroup $C_n \subset SO(2)$ in this case, and $s = 0, 1, \ldots, n - 1$ for C_n. Although the expression in Eq. (11.24) has been derived as an approximation of the continuous expression in Eq. (11.22), the matrix elements in Eq. (11.24) are exact expressions for the matrix elements of the irreducible unitary representations of G_{N_R}, i.e., the relation $U(g_1, p) \cdot U(g_2, p) = U(g_1 \circ g_2, p)$ holds.

We note $R_{\mathbf{u}_m} \mathbf{u}_0 = A_m \mathbf{u}_0 = \mathbf{u}_m$, where $\mathbf{u}_0$ is the center of the spherical polygon F_0 which is chosen arbitrarily from the set $\{F_i\}$ containing N_R faces.

However, due to the incompleteness of the set of functions $\{\varphi_{N_R,n}\}$, the Fourier transform matrix elements defined by Eq. (11.24) form an incomplete set of matrix elements. A complete set of elements are defined in the following way.

Let us allow the vector $\mathbf{u}_m^w$ to point to an arbitrary position, w, inside the spherical shape which forms the support for the pulse basis function $\varphi_{N_R,m}$.

It is then possible to verify that the following matrix elements form a complete set of matrix elements for G_{N_R}

$$U_{mn}^s \left(A_j, \mathbf{r}; p, w \right) = e^{-i\, p\, \mathbf{u}_m^w \cdot \mathbf{r}} \, \Delta_s \left(R_{\mathbf{u}_m}^{-1} A_j R_{\mathbf{u}_n} \right) \delta_{A_j^{-1} \mathbf{u}_m, \mathbf{u}_n}. \qquad (11.25)$$

In particular, it is easy to check that $U(g_1; p, w) \cdot U(g_2; p, w) = U(g_1 \circ g_2; p, w)$.

As an example we consider the case when I_{N_R} is the icosahedral subgroup of the $SO(3)$, which has $N_R = 60$ elements. This is the largest finite subgroup of $SO(3)$ [14]. If we subdivide the sphere into 20 equilateral spherical triangles (see Fig. 11.2), this group has 6 axes of rotation of order 5 (i.e, rotation $2\pi/5 \cdot n, n = 0, 1, 2, 3, 4$ around each axis) located at the triangle corners, 10 axes of order 3 located at the triangle centers, and 15 axes of order 2 located in the middle point of each triangle side.

Different representations of this "discrete" motion group may be classified according to different choices of little group C_n. This corresponds to choosing orthogonal pulse functions with differently shaped support. To illustrate the possible cases, consider the tessellations of the sphere in Fig. 11.2. In order to have a complete set of matrix elements for non-trivial little group, C_n, we have to consider all possible $s = 0, \ldots, n-1$. The representations for different s can be viewed as blocks in a 60×60 representation matrix.

The possible choices for little group corresponding to the shapes illustrated in Fig. 11.2 are

1) Little group C_5: For 12 regular spherical pentagons, such as ECGJH, chosen as the support for spherical pulse functions we have 12×12 representations. The little group of $\mathbf{u}_0$ is C_5, thus $s = 0, 1, 2, 3, 4$. These representation matrices, each corresponding to an element of the little group enumerated by a value of s, may be viewed as blocks in 60×60 representation matrices of G.

2) Little group C_3: Twenty equilateral spherical triangles are chosen as pulse functions on S^2, such as triangle ABO. We have 20 vectors $\mathbf{u}_m$ pointing to the centers of triangles, i.e., the representations and Fourier matrices consist of 20×20 nonzero blocks (for each fixed p and s). The little group of the arbitrary, chosen vector $\mathbf{u}_0$ is C_3, thus we have three inequivalent representations of the little group for each value of p enumerated by $s = 0, 1, 2$. These representations may be combined as blocks to form 60×60 representation matrices of G.

3) Little group C_2: Thirty spherical parallelograms (such as OEAC) are chosen as the support of pulse functions and we have 30×30 representations matrices, the little group is C_2, and $s = 0, 1$.

4) Trivial little group: The 4-sided figure ODCF of Fig. 11.2 may be chosen as the support for pulse functions. We have 60 such figures; the little group is trivial in this case since they possess no rotational symmetry. The representation matrices are then 60×60. We note that some other divisions of the sphere into 60 equal spherical figures lead to equivalent representations (for example the choice of the triangles ACO or ECO). These representations are irreducible.

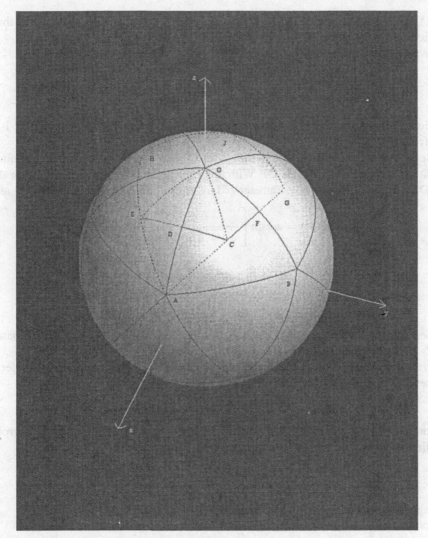

FIGURE 11.2
Illustration for basis function of discrete motion group.

We note that each row and column of the 60×60 representation matrices corresponding to the trivial little group contain only one non-zero element. The matrix elements of the representations are written in this case as

$$U_{mn}\left(A_j, \mathbf{r}; p, w\right) = e^{-i\, p\, \mathbf{u}_m^w \cdot \mathbf{r}}\, \delta_{A_j^{-1}\mathbf{u}_m, \mathbf{u}_n}. \tag{11.26}$$

Henceforth we restrict the discussion to this case.

The direct Fourier transform is defined as

$$\hat{f}_{mn}(p, w) = \sum_{i=0}^{N_R-1} \int_{\mathbb{R}^3} f\left(A_i, \mathbf{r}\right) U_{mn}^{-1}\left(A_i, \mathbf{r}; p, w\right) d^3r \tag{11.27}$$

where it depends now on w. The vector $\mathbf{u}_m^w$, which is inside the figure F_m may be found by the rotation A_m (which transforms F_0 to F_m) from $\mathbf{u}_0^w$, as $\mathbf{u}_m^w = A_m \mathbf{u}_0^w$. The parameter w, thus denotes the position inside the figure F_0.

The inverse Fourier transform is

$$\mathcal{F}^{-1}\left(\hat{f}\right) = \frac{1}{8\pi^3} \sum_{m=0}^{N_R-1} \sum_{n=0}^{N_R-1} \int_0^\infty \int_{F_0} \hat{f}_{mn}(p, w) \, U_{nm}\left(A_i, \mathbf{r}; p, w\right) \, p^2 \, dp \, d^2w \qquad (11.28)$$

(integration with respect to w is over the area of the basis figure F_0). The vector $\mathbf{u}_m^w$ is found by rotation from $\mathbf{u}_0^w$.

We choose the z-axis passing through the vertex O. The $\Phi=0$ circular arc then contains side OD and the $\Phi = 2\pi/5$ arc contains side OF. With this choice, the border of the area F_0 for the 4-sided figure ODCF of Fig. 11.2 may be parameterized in terms of the spherical angles Θ, Φ as

$$\Theta(\Phi) = \arcsin\left[\frac{0.525732}{\sqrt{1 - 0.723605 \sin^2(\Phi')}}\right],$$

where

$$\Phi' = \Phi, \quad \text{if } 0 \le \Phi \le \pi/5,$$

and

$$\Phi' = 2\pi/5 - \Phi, \quad \text{for } \pi/5 \le \Phi \le 2\pi/5.$$

This dependence is derived using the relationships between the angles of spherical triangles (e.g., the "sin"- and "cos"-theorems from spherical trigonometry).

Due to the completeness of matrix elements the application of direct and inverse Fourier transform reproduces function on the "discrete" motion group G

$$\mathcal{F}^{-1}\mathcal{F}(f(g)) = f(g).$$

The completeness of the set of IURs we have developed for the 3D discrete motion group may be seen by first observing the integral representation for the δ-function in $\mathbb{R}^3$

$$\int_{\mathbb{R}^3} e^{-i\,\mathbf{p}\cdot\mathbf{r}} \, d^3p = (2\pi)^3 \delta(\mathbf{r}).$$

Here integration is through the Fourier space, which is parameterized by $\mathbf{p}$.

The completeness relation

$$\sum_{m=0}^{N_R-1} \sum_{n=0}^{N_R-1} \int_0^\infty \int_{F_0} \overline{U_{mn}\left(A_i, \mathbf{r}_1; p, w\right)} \, U_{mn}\left(A_j, \mathbf{r}_2; p, w\right) \, p^2 \, dp \, d^2w$$

$$= (2\pi)^3 \delta^3\left(\mathbf{r}_1 - \mathbf{r}_2\right) \delta_{A_i, A_j}. \qquad (11.29)$$

($d^2w = \sin\Theta d\Theta \, d\Phi$) then follows because the integration is over the whole space of $\mathbf{p} = p\mathbf{u}$ values. Repeated integration along the boundary of the basis figures F_i gives zero contribution, because the functions are nonsingular and the integration measure along the boundaries is zero.

The orthogonality relation is written as

$$\sum_{i=0}^{N_R-1} \int_{\mathbb{R}^3} \overline{U_{mn}(A_i, \mathbf{r}; p, w)} \, U_{m'n'}\left(A_i, \mathbf{r}; p', w'\right) \, d^3r$$

$$= (2\pi)^3 \frac{\delta(p - p')}{p^2} \delta_{m,m'} \delta_{n,n'} \delta^2\left(w - w'\right). \qquad (11.30)$$

Using the orthogonality relations one may check that the convolution properties and Plancherel (Parseval) identity are exact for square integrable functions on the discrete motion group.

11.5.2 Computational Complexity

We now analyze the computational complexity of an algorithm for the numerical implementation of convolution of functions on the discrete motion group (with the icosahedral group as the rotation subgroup). In particular, we show that the convolution of functions $f_1(\mathbf{r}, A_i)$ and $f_2(\mathbf{r}, A_i)$ sampled at $N = N_R \cdot N_r$ points (N_r is the number of samples in a region in $\mathbb{R}^3$ and N_R is the order of a finite rotation subgroup, which is 60 for the icosahedral group) may be performed in $\mathcal{O}(N \, (\epsilon(N_r) + \log N_r)) + \mathcal{O}(N \, N_R^{\gamma-2})$ operations (where again $2 \leq \gamma \leq 3$ is the exponent for matrix multiplication) instead of the $\mathcal{O}(N^2)$ computations required for the direct computation of convolution by discretization of the convolution integral and evaluation at each discrete value of translation and rotation. The structure of matrix elements of Eq. (11.26) allows one to apply Fast Fourier methods and reduce the amount of computations. However, even without the application of the FFT, the amount of computations required to compute convolutions using the group-theoretical Fourier transform method is $\mathcal{O}(N^2/N_R)$, which is a savings over brute-force discretization of the convolution integral.

First, we estimate the amount of computations to perform the direct and inverse Fourier transforms of $f(g)$. We assume that we restrict p values to a finite interval and sample it at N_p points and the w region at N_w points. We also assume that the total number of harmonics $N_p N_w N_R^2 = N = N_R N_r$.

Let us consider the direct Fourier transform in Eq. (11.27). Each term i (for fixed A_i) gives one nonzero term in each row and column of the Fourier matrix $\hat{f}_{mn}^w(p)$ [because only one element in each row and column of $U_{mn}^{-1}(g; p, w)$ is nonzero]. For each fixed i we may compute the FFT of $f(\mathbf{r}; A_i)$, which may be computed in $\mathcal{O}(N_r \log(N_r))$ operations. The Fourier transform elements found by FFT are computed on a square grid of $\mathbf{p}$ values. However, we need to interpolate the Fourier elements computed on the grid to the Fourier elements computed in polar (spherical) coordinates. The radial part p is determined by the length of $\mathbf{p}$ and the angular part is determined by the indices m and w (the other index n is determined uniquely for given A_i). If we interpolate the values from the square grid to the values on the polar (spherical) grid, $\mathcal{O}(N_r \epsilon(N_r) N_R)$ computations are required. Each term i in Eq. (11.27) may be computed in $\mathcal{O}(N_r \log(N_r))$ computations. The whole Fourier matrix is found in $\mathcal{O}(N_R N_r \log(N_r))$ computations.

Again, one element from each row and column is used in the computation of the inverse Fourier transform for each rotation element A_i. After inverse interpolation to Cartesian coordinates [which may be done in $\mathcal{O}(N_r \epsilon(N_r))$ computations], the inverse Fourier integration may be performed in $\mathcal{O}(N_r \log(N_r))$ computation using the FFT. Thus in $\mathcal{O}(N_R N_r \log(N_r))$ computations we reproduce the function for each A_i.

The matrix product of $\hat{f}_{mn}(p, w)$ may be computed in $\mathcal{O}(N^\gamma)$ computations for each value of p and parameters w. This means that the convolution (which is a matrix product of Fourier matrices) may be performed in $\mathcal{O}(N_\gamma^3 N_p N_w) = \mathcal{O}(N_R^{\gamma-2} N)$ computations.

Therefore, the convolution of functions on the discrete motion group may be performed in $\mathcal{O}(N \cdot (\log N_r + \epsilon(N_r))) + \mathcal{O}(N \, N_R^{\gamma-2})$ using the Fourier methods on the motion group and FFT. When spline interpolation is used, the term with $\epsilon(N_r)$ can be ignored, whereas this term is more important when Fourier interpolation is used.

11.6 Alternative Methods for Computing $SE(D)$ Convolutions

In this section we examine alternatives to calculating convolutions on $SE(D)$ using Fourier transforms based on representations of $SE(D)$. First we consider a Fourier transform based on reducible unitary representations. Then we consider the Fourier transform for the direct product $\mathbb{R}^D \times SO(D)$ as a tool to compute $SE(D)$ convolutions. Finally, we consider contraction of $SE(D)$ to $SO(D+1)$.

11.6.1   An Alternative Motion-Group Fourier Transform Based on Reducible Representations

Let $g = (\mathbf{a}, A) \in SE(3)$, and consider the following operator that acts on functions $\varphi \in \mathcal{L}^2(SO(3))$

$$\mathcal{U}(g, \mathbf{p})\varphi(R) = e^{i((R\mathbf{p})\cdot\mathbf{a})}\varphi\left(A^T R\right). \tag{11.31}$$

These representations were studied in [15, 23]. We observe that

$$
\begin{aligned}
\mathcal{U}(g_1, \mathbf{p})\left[\mathcal{U}(g_2, \mathbf{p})\varphi(R)\right] &= \mathcal{U}(g_1, \mathbf{p})\left[e^{i((R\mathbf{p})\cdot\mathbf{a}_2)}\varphi\left(A_2^T R\right)\right] \\
&= e^{i((R\mathbf{p})\cdot\mathbf{a}_1)} \cdot e^{i((A_1^T R\mathbf{p})\cdot\mathbf{a}_2)}\varphi\left(A_2^T A_1^T R\right) \\
&= e^{i((R\mathbf{p})\cdot(\mathbf{a}_1+A_1\mathbf{a}_2))}\varphi\left((A_1 A_2)^T R\right) \\
&= \mathcal{U}(g_1 \circ g_2, \mathbf{p})\varphi(R).
\end{aligned}
$$

In this context, the matrix elements corresponding to $\mathcal{U}(g, \mathbf{p})$ are calculated as

$$\mathcal{U}^{l,l'}_{m,n;m',n'} = \sqrt{d_l d_{l'}} \int_{SO(3)} \overline{U^l_{m,n}(R)} e^{i(R\mathbf{p})\cdot\mathbf{a}} U^{l'}_{m',n'}\left(A^T R\right) dR \tag{11.32}$$

where $U^l_{m,n}$ are $SO(3)$ matrix elements, $d_l = 2l+1$ and $\{\sqrt{d_l}\, U^l_{m,n}\}$ is an orthonormal basis for $\mathcal{L}^2(SO(3))$. Using the homomorphism property, we see that

$$U^{l'}_{m',n'}\left(A^T R\right) = \sum_{r=-l'}^{l'} U^{l'}_{m',r}\left(A^T\right) U^{l'}_{r,n'}(R)$$

and

$$U^{l'}_{m',r}\left(A^T\right) = \overline{U^{l'}_{r,m'}(A)}.$$

This allows us to write

$$\mathcal{U}^{l,l'}_{m,n;m',n'}(g, \mathbf{p}) = \sqrt{d_l d_{l'}} \sum_{r=-l'}^{l'} \overline{U^{l'}_{r,m'}(A)} \int_{SO(3)} e^{i(R\mathbf{p})\cdot\mathbf{a}} U^{l'}_{r,n'}(R) \overline{U^l_{m,n}(R)} dR. \tag{11.33}$$

This may be computed in closed form using the expansion of $e^{i(R\mathbf{p})}$ in spherical coordinates, but for computational purposes it is sufficient to keep it as is.

The matrix elements of a Fourier transform based on these matrix elements are computed as

$$\hat{f}^{l,l'}_{m,n;m',n'}(\mathbf{p}) = \int_{SE(3)} f(g)\mathcal{U}^{l,l'}_{m,n;m',n'}\left(g^{-1}, \mathbf{p}\right) dg. \tag{11.34}$$

The reconstruction formula is written as

$$
\begin{aligned}
f(g) &= \int_{\mathbb{R}^3} \text{trace}(\hat{f}(\mathbf{p})\mathcal{U}(g, \mathbf{p}))d\mathbf{p} \tag{11.35} \\
&= \sum_{l=0}^{\infty} \sum_{l'=0}^{\infty} \sum_{m=-l}^{l} \sum_{n=-l}^{l} \sum_{m'=-l'}^{l'} \sum_{n'=-l'}^{l'} \int_{\mathbb{R}^3} \hat{f}^{l,l'}_{m,n;m',n'}(\mathbf{p}) \mathcal{U}^{l,l'}_{m',n';m,n}(g, \mathbf{p}) d\mathbf{p}.
\end{aligned}
$$

The associated completeness relation follows from those for $SO(3)$ and $\mathbb{R}^3$ as

$$\int_{SE(3)} \mathcal{U}^{l,l'}_{m,n;m',n'}(g, \mathbf{p}) \overline{\mathcal{U}^{k,k'}_{r,s;r',s'}(g, \mathbf{p}')} dg = (2\pi)^3 \delta_{kl} \delta_{k'l'} \delta_{mr} \delta_{ns} \delta_{m'r'} \delta_{n's'} \delta\left(\mathbf{p} - \mathbf{p}'\right).$$

Instead of a six-index Fourier transform we can define

$$\hat{f}_{ij} \triangleq \hat{f}^{l,l'}_{m,n;m',n'}$$

where

$$i = \sum_{k=0}^{l} (2k-1)^2 + (2l+1)(l+n) + l + n$$

(and analogously for j).

The multiplication of Fourier transform matrices is usually a computational bottleneck when using the group Fourier transform to calculate convolutions. In the present case, if $N = N_r \cdot N_R$ sample points in $SE(3)$ are taken, and $l, l' \leq \mathcal{O}(N_R^{1/3})$, the two-index Fourier matrices will be $\mathcal{O}(N_R) \times \mathcal{O}(N_R)$. Multiplication of these matrices will require $\mathcal{O}(N_R^\gamma)$ for each value of $\mathbf{p}$. In other words, $\mathcal{O}(N_R^\gamma \cdot N_r) = \mathcal{O}(N^{(1+\gamma)/2})$ computations are required. If matrices are multiplied with complexity $\gamma = 3$, then it is clear that this method will require the same order of computation as brute force numerical integration of the convolution integral. Hence, we conclude that this method is not computationally advantageous.

11.6.2 Computing $SE(D)$ Convolutions Using the Fourier Transform for $\mathbb{R}^D \times SO(D)$

Given the pairs $g = (\mathbf{a}, A)$ and $h = (\mathbf{r}, R)$ where $\mathbf{a}, \mathbf{r} \in \mathbb{R}^D$ and $A, R \in SO(D)$, if no group law is specified, there is no natural definition of the Fourier transform. That is, g and h could be elements of the semi-direct product group $SE(D)$ or the direct product group $\mathbb{R}^D \times SO(D)$. And while the natural choice for the definition of a group Fourier transform is based on the IURs of that group, the similarities between $SE(D)$ and $\mathbb{R}^D \times SO(D)$ are worth exploiting because the IURs of the former are infinite-dimensional, while those of the latter are finite dimensional.

The group law for $\mathbb{R}^D \times SO(D)$ is simply

$$(\mathbf{a}, A) \circ (\mathbf{r}, R) = (\mathbf{a} + \mathbf{r}, AR).$$

The representations are the direct product of representations of $\mathbb{R}^D$ and $SO(D)$

$$V^l(\mathbf{a}, A; \mathbf{p}) = e^{i\mathbf{a} \cdot \mathbf{p}} U^l(A)$$

where $U^l(A)$ are IURs of $SO(D)$. (We are only interested in the cases $D = 2$ and $D = 3$.) The corresponding Fourier transform of a well-behaved function is

$$\hat{f}^l(\mathbf{p}) = \mathcal{F}(f) = \int_{\mathbb{R}^D \times SO(D)} f(g) V^l\left(g^{-1}; \mathbf{p}\right) d(g)$$

$$= \int_{SO(D)} \int_{\mathbb{R}^D} f(\mathbf{a}, A) e^{-i\mathbf{a} \cdot \mathbf{p}} U^l\left(A^{-1}\right) d\mathbf{a} \, dA. \qquad (11.36)$$

The computationally attractive nature of this Fourier transform is emphasized when written in the form

$$\hat{f}^l(\mathbf{p}) = \int_{SO(D)} \left[\int_{\mathbb{R}^D} f(\mathbf{a}, A) e^{-i\mathbf{a} \cdot \mathbf{p}} d\mathbf{a}\right] U^l\left(A^{-1}\right) dA. \qquad (11.37)$$

We compute these integrals by sampling at $N = N_r \cdot N_R$ points [where N_r is the number of samples in $\mathbb{R}^D$ and N_R is the number of samples in $SO(D)$]. For $D = 2$, $N_r = \mathcal{O}(N^{2/3})$, and $N_R = \mathcal{O}(N^{1/3})$, while for $D = 3$, $N_r = \mathcal{O}(N^{1/2})$, and $N_R = \mathcal{O}(N^{1/2})$. The term in parenthesis in Eq. (11.37) is calculated in $\mathcal{O}(N_R \cdot N_r \log N_r)$ operations using the FFT for $\mathbb{R}^D$. We denote

$$\tilde{f}(\mathbf{p}, A) \triangleq \int_{\mathbb{R}^D} f(\mathbf{a}, A) e^{-i\mathbf{a}\cdot\mathbf{p}} d\mathbf{a}.$$

Then Eq. (11.36) becomes

$$\hat{f}^l(\mathbf{p}) = \int_{SO(D)} \tilde{f}(\mathbf{p}, A) U^l \left(A^{-1} \right) dA.$$

This may be computed in $\mathcal{O}(N_r \cdot N_R \, \eta(N_R))$ where $\eta(N_R) = \log N_R$ when $D = 2$, and $\eta(N_R) = (\log N_R)^2$ when $D = 3$. In either case, by adding the complexities of the two steps, Eq. (11.37) is computed for all values of l up to the band limit and all sampled values of $\mathbf{p}$ in $\mathcal{O}(N \, \eta(N))$ computations.

The inverse Fourier transform for $\mathbb{R}^D \times SO(D)$ is computed as

$$f(\mathbf{a}, A) = \frac{1}{(2\pi)^D} \sum_{l=0}^{B-1} d_l \int_{\mathbb{R}^D} \text{trace} \left(\hat{f}^l(\mathbf{p}) e^{i\mathbf{p}\cdot\mathbf{a}} U^l(A) \right) d\mathbf{p}. \tag{11.38}$$

This follows immediately from the Fourier inversion formulas for $\mathbb{R}^D$ and $SO(D)$, and can also be computed in $\mathcal{O}(N \, \eta(N))$ arithmetic operations. When $D = 2$, we have $d_l = 1$ and $B = \mathcal{O}(N_R)$, whereas for $D = 3$ we have $d_l = 2l + 1$ and $B = \mathcal{O}(N_R^{1/3})$.

We now examine the computational complexity of performing $SE(D)$ convolutions using the above fast Fourier transform pair for $\mathbb{R}^D \times SO(D)$. Recall that for $SE(D)$ the convolution of two well-behaved functions is

$$f_3(\mathbf{a}, A) = (f_1 * f_2)(\mathbf{a}, A) = \int_{SO(D)} \int_{\mathbb{R}^D} f_1(\mathbf{r}, R) f_2(R^{-1}(\mathbf{a} - \mathbf{r}), R^{-1}A) d\mathbf{r} \, dR.$$

The $\mathbb{R}^D \times SO(D)$ Fourier transform of this is

$$\hat{f}_3^l(\mathbf{p}) = \int_{SO(D)} \int_{\mathbb{R}^D} e^{-i\mathbf{a}\cdot\mathbf{p}} U^l \left(A^{-1} \right) \times$$
$$\left[\int_{SO(D)} \int_{\mathbb{R}^D} f_1(\mathbf{r}, R) f_2 \left(R^{-1}(\mathbf{a} - \mathbf{r}), R^{-1}A \right) d\mathbf{r} \, dR \right] d\mathbf{a} \, dA. \tag{11.39}$$

Making the change of variables $(R^{-1}(\mathbf{a} - \mathbf{r}), R^{-1}A) = (\mathbf{q}, Q)$, observing that $d\mathbf{a} \, dA = d\mathbf{q} \, dQ$ and performing the outer integration in Eq. (11.39) first, we write

$$\int_{SO(D)} \int_{\mathbb{R}^D} f_2(\mathbf{q}, Q) e^{-i(R\mathbf{q}+\mathbf{r})\cdot\mathbf{p}} U^l \left(Q^{-1} R^{-1} \right) d\mathbf{q} \, dQ = \hat{f}_2^l \left(R^{-1}\mathbf{p} \right) e^{-i\mathbf{p}\cdot\mathbf{r}} U^l \left(R^{-1} \right).$$

This means that

$$\hat{f}_3^l(\mathbf{p}) = \int_{SO(D)} \int_{\mathbb{R}^D} f_1(\mathbf{r}, R) \hat{f}_2^l \left(R^{-1}\mathbf{p} \right) e^{-i\mathbf{p}\cdot\mathbf{r}} U^l \left(R^{-1} \right) d\mathbf{r} \, dR.$$

This may be rewritten as

$$\hat{f}_3^l(\mathbf{p}) = \int_{SO(D)} \hat{f}_2^l \left(R^{-1}\mathbf{p} \right) \tilde{f}_1(\mathbf{p}, R) U^l \left(R^{-1} \right) dR \tag{11.40}$$

where

$$\tilde{f}_1(\mathbf{p}, R) = \int_{\mathbb{R}^D} f_1(\mathbf{r}, R) e^{-i\mathbf{p}\cdot\mathbf{r}} d\mathbf{r}$$

is computed in $\mathcal{O}(N_R N_r \log N_r)$ operations.

Hence, the price of using the Fourier transform for $\mathbb{R}^D \times SO(D)$ instead of that for $SE(D)$ is that pointwise multiplication of Fourier matrices is replaced with the integral in Eq. (11.40). This integral is *not* just the Fourier transform on $SO(D)$ for each fixed value of $\mathbf{p}$ since

$$F^l(R, \mathbf{p}) \overset{\triangle}{=} \hat{f}_2^l\left(R^{-1}\mathbf{p}\right) \tilde{f}_1(\mathbf{p}, R) \qquad (11.41)$$

depends on l.

When $D = 3$, $F^l(R, \mathbf{p})$ is a matrix, each element of which can be calculated in $\mathcal{O}(N_R \cdot N_r \epsilon(N_r))$ operations for all values of $\mathbf{p}$ and R where $\epsilon(\cdot)$ depends on what form of interpolation is used, as described at the beginning of this chapter. This calculation can be done for all $\mathcal{O}(N_R)$ matrix elements using $\mathcal{O}(N^{3/2}\epsilon(N^{1/2}))$ operations. Since

$$\sum_{l=0}^{B-1} d_l^\gamma = \mathcal{O}\left(B^{\gamma+1}\right)$$

operations are required to multiply $F^l(R, \mathbf{p})$ and $U^l(R^{-1})$ for all values of $l < B$ and each value of $\mathbf{p}$ and R, it follows that a total of $\mathcal{O}(B^{\gamma+1} \cdot N_r \cdot N_R)$ arithmetic operations are required to multiply $F^l(R, \mathbf{p})$ and $U^l(R^{-1})$ for all values of l up to the band-limit and all values of $\mathbf{p}$ and R at the sample points. When Gaussian elimination is used, $\gamma = 3$ and the order of computation to perform all the matrix multiplications becomes $\mathcal{O}(B^{(\gamma+1)} \cdot N_r \cdot N_R) = \mathcal{O}(N^{5/3})$ since $B = \mathcal{O}(N_R^{1/3}) = \mathcal{O}(N^{1/6})$.

When $D = 2$, the $SO(D)$ representations are one-dimensional and $\gamma = 0$. In this case, Eq. (11.41) is a scalar function of R and $\mathbf{p}$ that can be calculated for all values of $\mathbf{p}$, R, and l in $\mathcal{O}(N_R^2 \cdot N_r \epsilon(N_r))$ operations. Then Eq. (11.40) can be computed for each value of $\mathbf{p}$ and l in $\mathcal{O}(B \cdot N_r \cdot N_R) = \mathcal{O}(N^{4/3})$ since $B = \mathcal{O}(N_R) = \mathcal{O}(N^{1/3})$.

Hence, it appears that for the $D = 2$ case, this method is on the same order of complexity as when using the $SE(D)$ Fourier transform, while for the $D = 3$ case it is slower. In addition, it must be noted that this technique will always be an approximate one. This is because

$$f(\mathbf{a}, A) = \frac{1}{(2\pi)^D} \sum_{l=0}^{B-1} d_l \int_{\mathbb{R}^D} \operatorname{trace}\left(\hat{f}^l(\mathbf{p}) e^{i\mathbf{p}\cdot\mathbf{a}} U^l(A)\right) d\mathbf{p}$$

is *not* a band-limited function on $SE(D)$ even though it is a band-limited function on $\mathbb{R}^D \times SO(D)$. This can be observed directly by taking the $SE(D)$ Fourier transform. And even if two functions f_1 and f_2 are band-limited on $\mathbb{R}^D \times SO(D)$, if they are convolved on $SE(3)$ the result $f_1 * f_2$ will generally no longer be band-limited on $\mathbb{R}^D \times SO(D)$. This is because $\hat{f}_2^l(R^{-1}\mathbf{p})$ need not have a band-limited expansion in the $SO(D)$ harmonics for each fixed value of $\mathbf{p}$.

11.6.3 Contraction of $SE(D)$ to $SO(D+1)$

The methods for approximating rigid-body motions in D-dimensional space with rotations in $(D+1)$-dimensional space which are discussed in Chapter 6 can be used as a method for performing fast approximate $SE(D)$ convolutions.

Using this idea, functions $f_i : SE(D) \rightarrow \mathbb{C}$ are replaced with functions $f_i' : SO(D+1) \rightarrow \mathbb{C}$ where the conversion from f_i to f_i' is achieved by mapping points from $SE(D)$ onto $SO(D+1)$, with the correspondence $f_i'(A) = f_i(g)$ established when $g \in SE(D)$ is mapped to $A \in SO(D+1)$.

The convolution $f_1' * f_2'$ may then be performed using the fast Fourier transform for $SO(D + 1)$. In Chapter 9, the IURs of $SO(4)$ are shown to be closely related to those for $SU(2)$, and an $\mathcal{O}(N(\log N)^2)$ algorithm exists for $SO(4)$ Fourier transforms [19]. The result is then mapped back to $SE(D)$.

While this method has the best computational speed of any algorithm we have considered for computing motion-group Fourier transforms, the drawback is that the mapping between $SE(D)$ and $SO(D + 1)$ results in a distortion so that

$$\left(f_1' * f_i'\right)(A) \neq (f_1 * f_2)(g).$$

11.7   Summary

Fast numerical algorithms for computing the convolution product of functions on motion groups were derived. These algorithms use the group-theoretic Fourier transform with the irreducible unitary representations written in operator form, and their matrix elements calculated numerically instead of analytically. This, together with interpolation between Cartesian and spherical coordinate grids, made it possible to use well-known FFTs for compactly supported functions on $\mathbb{R}^3$, and more recent FFTs for the sphere and rotation group. Using the Fourier transform based in IURs of $SE(3)$ appears to be more suitable than using alternatives based on other kinds of group Fourier transforms. Other alternatives can also be found in the literature (see, for example, [9]).

References

[1] Aho, A.V., Hopcroft, J.E., and Ullman, J.D., *The Design and Analysis of Computer Algorithms*, Addison-Wesley, Reading, MA, 1974.

[2] Borodin, A. and Munro, I., *The Computational Complexity of Algebraic and Numerical Problems*, Elsevier, New York, 1975.

[3] Bucci, O.M., Gennarelli, C., and Savarese, C., Fast and accurate near-field-far-field transformation by sampling interpolation of plane-polar measurements, *IEEE Trans. on Antennas and Propagation*, 39(1), 48–55, 1991.

[4] Chirikjian, G.S. and Kyatkin, A.B., Algorithms for fast convolutions on motion groups, *Applied and Commutational Harmonic Analysis*, to appear.

[5] Choi, H. and Munson, D.C., Direct-Fourier reconstruction in tomography and synthetic aperture radar, *Int. J. of Imaging Syst. and Technol.*, 9, 1–13, 1998.

[6] Danielsson, P.-E. and Hammerin, M., High-accuracy rotation of images, *CVGIP: Graphical Models and Image Process.*, 54(4), 340–344, 1992.

[7] Deans, S.R., *The Radon Transform and Some of its Applications*, John Wiley & Sons, New York, 1983.

[8] Diaconis, P. and Rockmore, D., Efficient computation of the Fourier transform on finite groups, *J. of the Am. Math. Soci.*, 3(2), 297–332, 1990.

[9] Dooley, A.H., A nonabelian version of the Shannon sampling theorem, *SIAM J. Math. Anal.*, 20(3), 624–633, 1989.

[10] Driscoll, J.R. and Healy, D., Computing Fourier transforms and convolutions on the 2-sphere, *Adv. in Appl. Math.*, 15, 202–250, 1994.

[11] Driscoll, J.R., Healy, D., and Rockmore, D.N., Fast discrete polynomial transform with applications to data analysis for distance transitive graphs, *SIAM J. Computing*, 26, 1066–1099, 1997.

[12] Fraser, D. and Schowengerdt, R.A., Avoidance of additional aliasing in multipass image rotations, *IEEE Trans. on Image Process.*, 3(6), 721–735, 1994.

[13] Gauthier, J.P., Bornard, G., and Sibermann, M., Motion and pattern analysis: harmonic analysis on motion groups and their homogeneous spaces, *IEEE Trans. Syst. Man Cybern.*, 21, 159–172, 1991.

[14] Gurarie, D., *Symmetry and Laplacians. Introduction to Harmonic Analysis, Group Representations and Applications*, Elsevier Science, The Netherlands, 1992.

[15] Kumahara, K. and Okamoto, K., An analogue of the Paley-Wiener theorem for the Euclidean motion group, *Osaka J. Math.*, 10, 77–92, 1973.

[16] Kyatkin, A.B. and Chirikjian, G.S., Template matching as a correlation on the discrete motion group, *Comput. Vision and Image Understanding*, 74(1), 22–35, 1999.

[17] Kyatkin, A.B. and Chirikjian, G.S., Synthesis of binary manipulators using the Fourier transform on the Euclidean group, *ASME J. Mech. Design*, 121, 9–14, 1999.

[18] Larkin, K.G., Oldfield, M.A., and Klemm, H., Fast Fourier method for the accurate rotation of sampled images, *Optics Commun.*, 139, 99–106, 1997.

[19] Maslen, D.K., *Fast Transforms and Sampling for Compact Groups*, Ph.D. dissertation, Dept. of Mathematics, Harvard University, Cambridge, MA, 1993.

[20] Maslen, D.K. and Rockmore, D.N., Generalized FFTs — a survey of some recent results, *DIMACS Ser. in Discrete Math. and Theor. Comput. Sci.*, 28, 183–237, 1997.

[21] Miller, W., *Lie Theory and Special Functions*, Academic Press, New York, 1968.

[22] Miller, W., Some applications of the representation theory of the Euclidean group in three-space, *Commun. Pure App. Math.*, 17, 527–540, 1964.

[23] Orihara, A., Bessel functions and the Euclidean motion group, *Tohoku Math. J.*, 13, 66–71, 1961.

[24] Paeth, A.W., A fast algorithm for general raster rotation, in *Graphics Gems*, Glassner, A.S., Ed., Academic Press, Boston, 179–195, 1990.

[25] Pan, S.X. and Kak, A.C., A computational study of reconstruction algorithms for diffraction tomography: interpolation versus filtered backpropagation, *IEEE Trans. on Acoustics, Speech, and Signal Process.*, ASSP-31(5), 1262–1275, 1983.

[26] Pan, V., How can we speed up matrix multiplication, *SIAM Rev.*, 26, 393–416, 1984.

[27] Pan, V., *How to Multiply Matrices Fast*, Springer-Verlag, Berlin, Heidelberg, 1984.

[28] Speath, H., *Two Dimensional Spline Interpolation Algorithms*, AK Peter, Wellesley, MA, 1995.

[29] Strassen, V., Gaussian elimination is not optimal, *Numerische Math.*, 13, 354–356, 1969.

[30] Talman, J., *Special Functions*, W.A. Benjamin, Amsterdam, 1968.

[31] Unser, M., Thévenaz, P., and Yaroslavsky, L., Convolution-based interpolation for fast, high-quality rotation of images, *IEEE Trans. on Image Process.*, 4(10), 1371–1381, 1995.

[32] Vilenkin, N.J., Bessel functions and representations of the group of Euclidean motions, *Uspehi Mat. Nauk.*, 11, 69–112, 1956 (in Russian).

[33] Vilenkin, N.J. and Klimyk, A.U., *Representation of Lie Group and Special Functions*, Vol. 1–3, Kluwer Academic, The Netherlands, 1991.

[34] Winograd, S., A new algorithm for inner products, *IEEE Trans. Comp.*, C-17, 693–694, 1968.

[35] Yaghjian, A.D. and Woodworth, M.B., Sampling in plane-polar coordinates, *IEEE Trans. on Antennas and Propagation*, 44(5), 696–700, 1996.

Chapter 12

Robotics

12.1 A Brief Introduction to Robotics

Robotics is the study of machines that exhibit some degree of autonomy and flexibility in the tasks that they perform. There are two major kinds of robotic devices: manipulators and mobile robots.

A manipulator is a robot arm and/or hand that is usually fixed to some kind of base. A manipulator can have a *serial chain* topology in which there are no loops formed by the links of the arm. It is also possible to have a *parallel* or *platform* architecture in which one or more loops exist. A third possibility is a *tree-like* topology such as a hand or cooperating serial manipulators. In any of these topologies, the actuators which drive the movement of the arm can be revolute (rotational) such as an electric motor, prismatic (translational) such as a hydraulic cylinder, or some combination of the two (e.g., a screw). The key computational issues in the use of manipulator arms in industrial or service environments all depend on the fast calculation of joint angles[1] which will place the functional end of the arm (called the *end effector*) at the desired position and/or orientation relative to its base. The set of all positions and orientations that an arm can reach is called its *workspace*. The determination of the joint angles that result in the desired end-effector state is called the *inverse kinematics* problem. The *forward kinematics* problem is the problem of finding the position and/or orientation of the end effector when the value of the joint angles is given. As a rule, the forward kinematics problem is very easy to solve for serial manipulators, and the inverse problem is more difficult. However, for parallel manipulators, the opposite is true. And in a sense, hybrid manipulators (in which two or more parallel structures are stacked) inherit the difficult aspects of both serial and parallel manipulators. The *workspace generation* problem is that of determining all positions and/or orientations reachable by a robotic arm (see, for example, [1, 7, 33, 34, 49, 50]).

Excellent introductions to the issues involved in the kinematics, dynamics, and control of manipulator arms can be found in the textbooks [17, 19, 43, 52, 56, 57]. For group-theoretic issues in robotics see [5, 44, 48, 53].

A mobile robot is a machine with wheels, tracks, legs, or other means of propulsion intended to move from one location to another. In the context of mobile robots, we will concentrate on issues in the motion planning of a single rigid-body robot. The most basic problem is that of navigating a mobile robot of known shape through an environment with known obstacles without regard to effects of measurement error, wheel slippage, or the vehicle dynamics. This problem has been addressed extensively in the literature (see, for example, [6, 16, 38]), and the approach presented here is one of a variety of acceptable techniques.

---

[1]The term "joint angles" refers not only to angles but to any generalized actuator displacement such as stroke lengths of prismatic actuators.

Combinations of manipulators and mobile robots, which can be considered the two most basic subsystems, can be arranged in a variety of ways. For example, one or more manipulator arms can be affixed atop a mobile platform which serves as a transport device for the arms. We do not examine combined manipulator and mobile platform systems. The reader interested in this area will find the following references useful: [18, 29].

In this chapter, we examine manipulators and mobile robots using concepts from noncommutative harmonic analysis. In Section 12.2 a particular kind of manipulator (called a *binary* or *discretely actuated* manipulator) is examined in detail, and the concept of a *density function* which describes where such an arm can reach is defined. It is shown in Section 12.3 why this information is useful in solving the inverse kinematics problem for discretely actuated manipulators. Section 12.4 examines the symmetries of the density function inherited from geometrical symmetries of the arm. Section 12.5 then describes the design problem for binary arms, which involves the inversion of convolution equations on the Euclidean motion group. In Section 12.6 we turn to an issue which is of more general interest: the accumulation of error in serial chains (including manipulators). Finally, in Section 12.7 we illustrate how convolution-like integrals arise when considering the set of all positions and orientations a rigid mobile robot can attain without intersecting obstacles.

12.2   The Density Function of a Discretely Actuated Manipulator

The convolution product of real-valued functions on the Euclidean group is used in this section as a computational tool. The primary application is the generation of discretely actuated manipulator workspaces and determination of the density of reachable frames in any portion of the workspace. A discretely actuated manipulator is an arm for which each of the actuators has a finite number of states. This includes manipulators driven by stepper motors and pneumatic cylinders or solenoids. In the case when the discrete-state actuators have only two states, the resulting arm is called a *binary manipulator.* Figure 12.1(a), (b) shows a schematic of a three-bit binary platform and a photo of a manipulator made out of this kind of three-bit unit. This structure is a truss where all the vertices can be thought of as passive hinges, and each of the labeled legs is an actuator which changes length. The labels indicate the two states, with the shorter being labeled "0" and the longer labeled "1." The three-dimensional version of this kind of platform (regardless of whether or not the actuators have discrete states) has six extensible legs and is called a Stewart/Gough platform.

Figure 12.2 shows a three-dimensional binary manipulator which is constructed of a series of Stewart/Gough platforms stacked, or cascaded, on top of each other. Since each actuator (which in this case is a pneumatic cylinder in parallel with a viscous dashpot) has two stable states, each platform has 2^6 states, and the whole arm has $(2^6)^6 = 2^{36}$ states. Binary arms are attractive because they require no feedback control, and they are very inexpensive to construct. However, issues such as inverse kinematics become much more computationally challenging than in the case of continuous actuation.

In general, if a discretely actuated manipulator has P units (where, for example, each unit is a platform) and each unit has K states, then the arm will have K^P states. It will generally be desirable to avoid direct computation of the K^P different configurations of the arm when performing calculations like inverse kinematics (see Section 12.3). In fact, while the concept of discretely actuated manipulators has been in the literature for more than 30 years (see [31, 47, 51]), it seems that the exponential complexity of the problem has been a major stumbling block. See [8, 9, 11, 12, 13, 14, 15, 35, 39, 40, 41] for various approaches to circumventing this complexity.

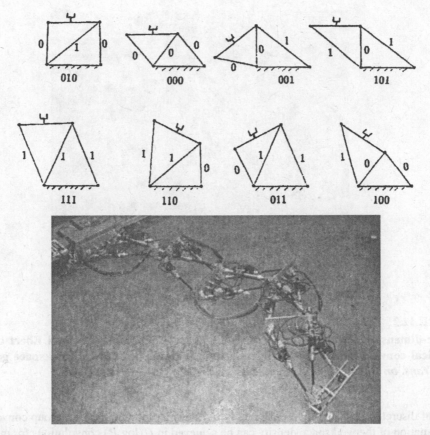

FIGURE 12.1

(a) A 3-bit, 8-state platform manipulator. (b) A concatenation of the units in (a). (Chirikjian, G.S. and Ebert-Uphoff, I., Numerical convolution on the Euclidean group with applications to workspace generation, *IEEE Trans. on Robotics and Autom.*, 14(1), 123–136, 1998. © 1998 IEEE.)

In the context of discrete actuation, the density of frames [number of frames per unit volume in $SE(3)$] in many ways replaces classical measures of dexterity used in robotics (see, for example, [2, 4, 30, 32, 46, 58]) as a scalar function of importance defined over the workspace. This is because density in the neighborhood of a given frame is an indicator of how accurately a discretely actuated manipulator can reach that point/frame.

To compute the workspace density function using brute force enumeration is computationally intractable, e.g., it requires $O(K^P)$ evaluations of the forward kinematic equations for a manipulator with P actuated modules each with K states. In addition, an array storing the density of all volume elements in the workspace must be incremented $O(K^P)$ times if brute force computation is used.

Figure 12.3 shows a schematic of the density of frames reachable by a discretely actuated variable geometry truss manipulator. If there are 30 actuated truss elements (ten modules) and each element has 4 states (and thus the whole manipulator has $4^{30} \approx 10^{18}$ states) the workspace density cannot simply be computed using brute force because this could take years using current computer technology. This combinatorial explosion is a major reason why discrete actuation is not commonly used, despite the fact that the concept is almost three decades old (see, for example, [47, 51]). Having a representation of the density of reachable frames is important for performing inverse kinematics and

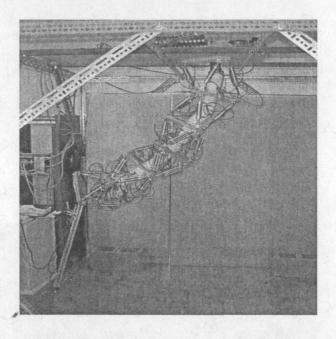

FIGURE 12.2
**A three-dimensional manipulator with 2^{36} states. (Chirikjian, G.S. and Ebert-Uphoff, I.,
Numerical convolution on the Euclidean group with applications to workspace generation,
IEEE Trans. on Robotics and Autom., 14(1), 123–136, 1998. © 1998 IEEE.)**

design of discretely actuated manipulators. Using the concept of Euclidean group convolution, an
approximation of the workspace density can be achieved in $O(\log P)$ convolutions for macroscopically serial (hybrid) manipulators[2] composed of P identical modules. This reduces the computation
time to minutes when convolutions are implemented in an efficient way.

The remainder of this section is organized as follows. In Section 12.2.1 we give the geometrical intuition behind Euclidean group convolution. In Section 12.2.2 we show why this concept
is important for workspace generation of discretely actuated manipulators. In Section 12.2.3 the
computational benefit of this approach is explained. In Section 12.2.4 the two-dimensional case
is considered explicitly. In Section 12.2.5 a direct numerical implementation of convolution is explained in detail for the planar case. Numerical results are generated in Section 12.2.6. Much of the
material presented in this section originally appeared in [11]. The concept of using sweeping as a
tool to generate workspaces was considered in [22].

12.2.1 Geometric Interpretation of Convolution of Functions on $SE(D)$

Suppose there are three frames in space, F_1, F_2, and F_3, as shown in Fig. 12.4. The first frame
can be viewed as fixed, the second frame as moving with respect to the first, and the third frame
as moving with respect to the second. Let the homogeneous transform $\mathcal{H}$ describe the position and
orientation of F_2 w.r.t. F_1, and let H describe the position and orientation of F_3 w.r.t. F_2. Then the
position and orientation of F_3 with respect to F_1 is $H' = \mathcal{H}H$. The position and orientation of F_3
with respect to F_2 can then be written as

$$H = \mathcal{H}^{-1}H'.$$

---

[2]These are a serial cascade of modules where each module may be a serial or parallel kinematic structure.

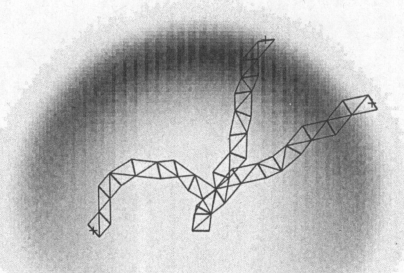

FIGURE 12.3
A discretely actuated manipulator with 4^{30} states with superimposed density function. (Chirikjian, G.S. and Ebert-Uphoff, I., Numerical convolution on the Euclidean group with applications to workspace generation, *IEEE Trans. on Robotics and Autom.*, 14(1), 123–136, 1998. © 1998 IEEE.)

We may divide up $SE(D)$ into volume elements, or "voxels," of finite but small size. The volume of the voxel centered at $H \in SE(D)$ (for $D = 2$ or 3) is denoted $\Delta(H)$, and as the element size is chosen smaller and smaller it becomes closer to the differential volume element $d(H)$.

The motion of F_2 relative to F_1 and the motion of F_3 relative to F_2 can both be considered as elements of $SE(D)$, and no distinction is made between these motions and the transformation matrices $\mathcal{H}$ and H which represent these motions.

Assuming we move $\mathcal{H}$ and H through a finite number of different positions and orientations, let ρ_1 be a function that records how often the $\mathcal{H}$ frames appear in each voxel, divided by the voxel volume $\Delta(\mathcal{H})$. Likewise, let ρ_2 be the function describing how often the H frames appear in each voxel normalized by voxel volume.

To calculate how often the H' frames appear in each voxel in $SE(D)$ for all possible values of $\mathcal{H}$ and H, we may perform the following steps:

- Evaluate $\rho_1 = \rho_1(\mathcal{H})$ (frequency of occurrence of $\mathcal{H}$).

- Evaluate $\rho_2 = \rho_2(H) = \rho_2(\mathcal{H}^{-1}H')$ (frequency of occurrence of $H = \mathcal{H}^{-1}H'$).

- Weight (multiply) the left-shifted density histogram $\rho_2(\mathcal{H}^{-1}H')$ by the number of frames

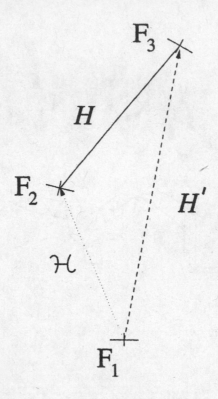

FIGURE 12.4
Concatenation of homogeneous transformations. (Chirikjian, G.S. and Ebert-Uphoff, I., Numerical convolution on the Euclidean group with applications to workspace generation, *IEEE Trans. on Robotics and Autom.*, 14(1), 123–136, 1998. © 1998 IEEE.)

which are doing the shifting. This number is $\rho_1(\mathcal{H})\Delta(\mathcal{H})$ for each $\mathcal{H}$.[3]

- Sum (integrate) over all these contributions

$$(\rho_1 * \rho_2)\left(H'\right) = \int_{SE(D)} \rho_1(\mathcal{H})\rho_2\left(\mathcal{H}^{-1}H'\right) d(\mathcal{H}).$$

As will be seen in subsequent sections, this approach yields an approximation to the density of H' frames which can be computed very efficiently. The number of H' frames in each voxel of $SE(D)$ can be calculated from this density as simply $(\rho_1 * \rho_2)(H')\Delta(H')$.

12.2.2   The Use of Convolution for Workspace Generation

In this section, we show how the concept of the convolution product of functions on $SE(D)$ can be applied to the generation of workspaces.

Let us consider a manipulator that consists of two mechanisms stacked on top of one another. For example, the two mechanisms can be in-parallel platform mechanisms or serial linkages. A frame

---

[3]Note that the product $\rho_2(\mathcal{H}^{-1}H')\rho_1(\mathcal{H})\Delta(\mathcal{H})$ is then an approximation to the sum of histograms which would result by sweeping $\rho_2(H)$ by the homogeneous transforms $\mathcal{H}$ in the voxel with volume $\Delta(\mathcal{H})$. In the limiting case when the volume size becomes small, this approximation becomes better.

F_1 is attached to the bottom of the first mechanism and a frame F_2 is attached to its top, which also defines the bottom of the second mechanism. A third frame, F_3, defines the position and orientation of the top of the second mechanism. This leads back to the situation shown in Fig. 12.4, where now $\mathcal{H}$ describes the homogeneous transformation corresponding to the lower mechanism, H the one for the upper mechanism, and H' the one for the whole manipulator.

If the manipulator is actuated discretely, then each mechanism only has a finite number of different states, which can be described by two finite sets, S_1 and S_2, which contain m_1 and m_2 elements, respectively. The set of all homogeneous transformations that can be attained by the distal end of the manipulator when the base is fixed results from all possible combinations of these two sets,

$$S' = \left\{ H' = \mathcal{H}H : H \in S_1, \mathcal{H} \in S_2 \right\},$$

and hence consists of $m_1 \cdot m_2$ elements.

For a finite set of frames [elements of $SE(D)$] that is very large, it is useful to reduce the amount of data by approximating its information with a density function. This is done by dividing a bounded region in $SE(D)$ into small volume elements, counting how many reachable frames occupy each volume element, and dividing this number by the volume of each element. This density function describes the distribution of frames in the workspace. Figure 12.3 shows the integral of such a distribution over all orientations reachable by a manipulator. The result is a function of position in the plane (gray scale corresponds to density).

If the sets S_1 and S_2 are approximated with density functions $\rho_1(\cdot)$ and $\rho_2(\cdot)$, respectively, then the density function resulting from the convolution of the two is a density function for the whole manipulator

$$\rho\left(H'\right) = (\rho_1 * \rho_2)\left(H'\right).$$

Furthermore, this reasoning can be applied to manipulators consisting of more than two mechanisms (modules) stacked on top of one another. For instance, if four modules are stacked, the density of the lower two is $\rho_1 * \rho_2$ and the density of the upper two is $\rho_3 * \rho_4$ by the reasoning given previously. Treating the lower two modules as one big module, and the upper two as one big module, the density of the collection of all four modules is $(\rho_1 * \rho_2) * (\rho_3 * \rho_4) = \rho_1 * \rho_2 * \rho_3 * \rho_4$.

If the possible configurations of a discretely actuated manipulator consisting of P independent modules are described by sets $S_1, \ldots, S_P$ in the way described previously, then the density of the whole manipulator is derived by multiple convolution as

$$\rho(H') = (\rho_1 * \rho_2 * \ldots * \rho_P)(H').$$

12.2.3 Computational Benefit of this Approach

In this section the computational complexity of the approach described in the preceding section is compared with brute force enumeration of manipulator states.

Suppose a manipulator consisting of P modules is considered and the number of homogeneous transformations in each set S_i is m_i. The explicit (brute force) calculation of all combinations of homogeneous transformations is of the same order as the number of all combinations, $\mathcal{O}(\prod_{i=1}^{P} m_i)$. If $m_1 = \cdots = m_P = K$, then this is an $\mathcal{O}(K^P)$ calculation.

In our approach, density functions are used to describe the frame distribution for each module. The frame distribution of the whole manipulator results by performing P convolutions. These calculations also depend on the dimension of $SE(D)$ for $D = 2, 3$. While we treat D as a constant because it does not change with the number of actuator states, it is worth noting that if a compact subset of $SE(D)$ is divided into $\mathcal{N}_j$ increments in each dimension, then $Q = \prod_{j=1}^{D(D+1)/2} \mathcal{N}_j$ voxels result. Treating Q as constant, the calculation of $\rho_i(\cdot)$ requires $\mathcal{O}(m_i)$ additions to increment the number of frames in each voxel. Since the voxels are uniform in size in the $SE(2)$ case, there is no need to explicitly divide by voxel volume (i.e., this normalization can be performed concurrently with convolution). Thus the calculation of $\rho_i(\cdot)$ is effectively $\mathcal{O}(m_i)$.

Consider the convolution of density functions of any two adjacent modules. The numerical approximation of the convolution integral evaluated at a single point in the support of $(\rho_i * \rho_{i+1})(\cdot)$ becomes a sum over all voxels in the support of $\rho_i(\cdot)$. This calculation must be performed for all voxels in the support of $(\rho_i * \rho_{i+1})(\cdot)$, and so the computations required to perform one convolution are $\mathcal{O}(\text{convolution}) = \mathcal{O}(Q_i \cdot Q_i^*)$, where Q_i and Q_i^* are respectively the number of voxels in the support of $\rho_i(\cdot)$ and $(\rho_i * \rho_{i+1})(\cdot)$. If the voxel size is kept constant after convolution, then $Q_i^* > Q_i$, because the workspace of any two adjacent modules is bigger than either one individually. Treating Q_i and D as constants, the calculation of P convolutions would then be polynomial in P. The order of this polynomial would depend on D. However, if the voxel size is rescaled after convolution, so that $Q_i^* \approx Q_i$, then each convolution is $\mathcal{O}(Q_i^2) = \mathcal{O}(1)$. Hence the total order of this approach is $\mathcal{O}(\sum_{i=1}^{P} m_i) + P \cdot \mathcal{O}(1) = \mathcal{O}(P)$ for this data storage strategy.

In the special case of a manipulator consisting of P identical modules the number of convolutions to be performed can be further reduced by using a different strategy. In this case the frame distribution $\rho(H')$ is calculated from P identical functions as a P-fold convolution

$$\rho = \rho_1^{(P)} = \rho_1 * \rho_1 * \ldots * \rho_1.$$

Note that the repeated convolution of a function with itself generates the following sequence of functions

$$\rho^{(2)} = \rho_1 * \rho_1, \quad \rho^{(4)} = \rho^{(2)} * \rho^{(2)}, \quad \rho^{(8)} = \rho^{(4)} * \rho^{(4)}, \quad \text{etc.},$$

i.e., it is possible to generate $\rho^{(2)}$ by one convolution, $\rho^{(4)}$ by two convolutions, and more generally, $\rho^{(2^n)}$ by n convolutions. Thus, for a manipulator with P identical modules, approximately $\mathcal{O}(\log P)$ convolutions have to be performed, which is an $\mathcal{O}(\log P)$ calculation if the number of voxels is held constant after each convolution (i.e., if we allow voxel size to grow with each convolution).[4]

One can estimate the support of the convolution of two density functions by first making a gross overestimate and doing a very crude (low resolution) convolution. Those voxels which have zero density after convolution can be discarded, and what is left over is a closer overestimation of the support of the convolved functions. This region is smaller than the original estimate, and voxel sizes can be scaled down to get the best resolution for the allowable memory.

12.2.4   Computation of the Convolution Product of Functions on $SE(2)$

In the two dimensional case, the homogeneous transforms H' and $\mathcal{H}$ in the convolution integral can be parameterized as

$$H'(x, y, \theta) = \begin{pmatrix} \cos\theta & -\sin\theta & x \\ \sin\theta & \cos\theta & y \\ 0 & 0 & 1 \end{pmatrix}$$

and

$$\mathcal{H}(\xi, \eta, \alpha) = \begin{pmatrix} \cos\alpha & -\sin\alpha & \xi \\ \sin\alpha & \cos\alpha & \eta \\ 0 & 0 & 1 \end{pmatrix}.$$

We define a parameterized density function $\rho(x, y, \theta)$ by identifying

$$\rho(x, y, \theta) \equiv \rho\left(H'(x, y, \theta)\right),$$

---

[4] In this context the cost of performing a convolution is considered to be a constant, and we are interested in determining the computational cost as a function of the number of manipulator modules P.

which leads to an explicit form of the convolution product on $SE(2)$

$$(\rho_1 * \rho_2)(x, y, \theta) = \int_{SE(2)} \rho_1(\mathcal{H})\rho_2\left(\mathcal{H}^{-1}H'\right) d(\mathcal{H})$$

$$= \int_{-\infty}^{\infty} \int_{-\infty}^{\infty} \int_{-\pi}^{\pi} \rho_1\left(\xi, \eta, \alpha\right) \cdot$$

$$\rho_2\Big((x-\xi)c\alpha + (y-\eta)s\alpha, -(x-\xi)s\alpha + (y-\eta)c\alpha, \theta-\alpha\Big) d\xi d\eta d\alpha,$$

where $c\alpha = \cos\alpha$ and $s\alpha = \sin\alpha$.

In general, if a subset of $SE(D)$ for $D = 2, 3$ is parameterized with $D(D + 1)/2$ variables $q_1, \ldots, q_{D(D+1)/2}$, then within a constant[5]

$$d(\mathcal{H}(\mathbf{q})) = |\det(\mathcal{J})| dq_1 \ldots dq_{D(D+1)/2},$$

where $\mathcal{J}$ (taken to be either $\mathcal{J}_R$ or $\mathcal{J}_L$) is a $D(D + 1)/2 \times D(D + 1)/2$ Jacobian matrix of the parameterization. In the case when $D = 2$ the determinant of the Jacobian matrix is one.

For the following derivation we assume that ρ_1 and ρ_2 are real-valued functions on $SE(D)$, which are nonzero and bounded everywhere, and have "compact support." That is, they vanish outside of a compact (closed and bounded) subset of $SE(2)$, which for simplicity is chosen of the form

$$\left[x_{\min}^{(j)}, x_{\max}^{(j)}\right] \times \left[y_{\min}^{(j)}, y_{\max}^{(j)}\right] \times [-\pi, \pi] \quad \text{for} \quad j = 1, 2.$$

The range of $x - y$ values is chosen to include the support of the workspace density of the concatenated modules.

The convolution product of two such functions on $SE(2)$ can be expressed in the form

$$(f_1 * f_2)(x, y, \theta) = \int_{x_{\min}^{(1)}}^{x_{\max}^{(1)}} \int_{y_{\min}^{(1)}}^{y_{\max}^{(1)}} \int_{-\pi}^{\pi} \rho_1\left(\xi, \eta, \alpha\right)$$

$$\cdot \rho_2\Big((x-\xi)\cos\alpha + (y-\eta)\sin\alpha,$$

$$-(x-\xi)\cos\alpha + (y-\eta)\sin\alpha, (\theta-\alpha)\bmod 2\pi\Big) d\xi d\eta d\alpha,$$

where $\bmod 2\pi$ is used here to mean that the difference $\theta - \alpha$ is taken in the range $[-\pi, \pi)$.[6]

12.2.5 Workspace Generation for Planar Manipulators

In Chapter 8 we showed how the convolution product of functions on Lie groups is defined. In Section 12.2.2 we showed how convolution of functions on $SE(D)$ can be applied to workspace generation. This section describes the details of a numerical implementation which is based on a description of density as a piecewise constant histogram.

We start this section with a summary of the procedure for generating the workspace of a discretely actuated planar manipulator.

---

[5]For compact groups the constant is set so that $\int_G d(g) = 1$, but $SE(D)$ is not compact and so there is no unique way to scale the volume element.

[6]This is different from the standard definition which would put the result in $[0, 2\pi)$.

1. The manipulator is divided into P kinematically independent modules. The modules are numbered from 1 to P, starting at the base with module 1 and increasing up to the most distal module, module P. For each module, one frame is attached to the base of the module and a second one to the top, where the next module is attached. Modules can have a parallel kinematic structure internally, but the modules are all cascaded in a serial way.

2. For each module, $(p = 1, \ldots, P)$ the finite set $S_p = \{H_{i_p}\}$ of all frames the top of module p can attain relative to the bottom is determined. That is, the position and orientation of the upper frame with respect to the lower frame is described by homogeneous transformations, $H_{i_p} \in S_p \subset SE(2)$, while the module undergoes all possible discrete configurations.

3. A compact subset $C_p \subset SE(2)$ is chosen that contains all of the discrete sets S_p. The subset C_p is discretized and a piecewise constant density function/histogram ρ_p is calculated for each S_p to represent this information.

4. Finally, the discrete density functions are convolved in the order

$$\rho_W = \rho_1 * \rho_2 * \cdots * \rho_P$$

to yield an approximation to the density function of the workspace.

The implementation of Steps 1 and 2 of the workspace generation procedure depends on the architecture of the manipulator. Generating these sets is a simple task if the manipulator can be separated into a sufficiently large number of kinematically independent modules of simple structure. We assume that the discrete sets S_p are calculated efficiently either numerically or in closed form for all modules of the manipulator.

For Steps 3 and 4 we have to define discretized density functions used to represent each set S_p. The discretization of the parameter space is described in the following subsection for the case $SE(2)$, and the subsection after that describes the resulting discrete form of the convolution.

Discretization of Parameter Space

For each set S_p arising in Step 3 of the procedure, the support of ρ_P in terms of the parameters (x, y, θ) is of the form $[x_{\min}^{(p)}, x_{\max}^{(p)}] \times [y_{\min}^{(p)}, y_{\max}^{(p)}] \times [-\pi, \pi]$. This set of parameters is divided into elements/voxels of equal size. Note that in the following discussion the superscript "p" is dropped, unless we refer to a particular set S_p. We choose the resolution in the x- and y-directions to be identical, i.e., $\Delta x = \Delta y$, and we explain below how to choose the resolution in the angular direction such that the resulting errors from inaccuracy in position and rotation are of the same order.

We denote by N_1, N_2, and M the number of discretizations in the x-, y-, and θ- directions, respectively, i.e., $\Delta x = \frac{(x_{\max} - x_{\min})}{N_1}$, $\Delta y = \frac{(y_{\max} - y_{\min})}{N_2}$, $\Delta \theta = \frac{(2\pi - 0)}{M}$, and we choose N_1 and N_2 such that $\Delta x = \Delta y$.[7] Each voxel is a volume element of the form $[x_{\min} + i\Delta x, x_{\min} + (i+1)\Delta x] \times [y_{\min} + j\Delta y, y_{\min} + (j+1)\Delta y] \times [-\pi + k\Delta\theta, -\pi + (k+1)\Delta\theta]$ with center coordinates $(x_i, y_j, \theta_k) = (x_{\min} + (i + 0.5)\Delta x, y_{\min} + (j + 0.5)\Delta y, -\pi + (k + 0.5)\Delta\theta)$.

To characterize the error resulting from discretization we consider a homogeneous transform $H \in SE(2)$ corresponding to some exact parameters: $H = H(x, y, \theta)$. H is then compared to the homogeneous transformation $\hat{H}$ corresponding to the rounded coordinates: $\hat{H} = H(\hat{x}, \hat{y}, \hat{\theta})$. By definition (x, y, θ) differs from $(\hat{x}, \hat{y}, \hat{\theta})$ at most by $(\frac{\Delta x}{2}, \frac{\Delta y}{2}, \frac{\Delta \theta}{2})$.

---

[7] To get exact equality it is usually necessary to slightly change one of the workspace boundaries, e.g., to slightly increase $y_{\max}$.

If we apply H and $\hat{H}$ to any vector $\mathbf{v} \in \mathbb{R}^2$ and use $(R, \mathbf{b})$ and $(\hat{R}, \hat{\mathbf{b}})$ to denote the rotation and translation of H and $\hat{H}$, respectively (so that, for instance, $H \cdot \mathbf{v} = R\mathbf{v} + \mathbf{b}$), then the error $\| H \cdot \mathbf{v} - \hat{H} \cdot \mathbf{v} \|$ is bounded as

$$
\begin{aligned}
\left\| H \cdot \mathbf{v} - \hat{H} \cdot \mathbf{v} \right\| &= \left\| (R\mathbf{v} + \mathbf{b}) - \left(\hat{R}\mathbf{v} + \hat{\mathbf{b}} \right) \right\| \\
&= \left\| \left(R\mathbf{v} - \hat{R}\mathbf{v} \right) + (\mathbf{b} - \hat{\mathbf{b}}) \right\|
\end{aligned}
$$

$$
\leq \underbrace{\left\| R\mathbf{v} - \hat{R}\mathbf{v} \right\|}_{\text{rot. part}} + \underbrace{\left\| \mathbf{b} - \hat{\mathbf{b}} \right\|}_{\text{transl. part}}
$$

$$
\leq \frac{\Delta \theta}{2} \| \mathbf{v} \| + \left(\left(\frac{\Delta x}{2} \right)^2 + \left(\frac{\Delta y}{2} \right)^2 \right)^{\frac{1}{2}}.
$$

If V is a set of vectors, then the maximal difference in displacement between transformed versions of a vector $\mathbf{v} \in V$ after transformation by H and $\hat{H}$ is bounded by

$$
\max_{\mathbf{v} \in V} \left\| H \cdot \mathbf{v} - \hat{H} \cdot \mathbf{v} \right\| = \frac{\Delta \theta}{2} \max_{\mathbf{v} \in V} \| \mathbf{v} \| + \left(\left(\frac{\Delta x}{2} \right)^2 + \left(\frac{\Delta y}{2} \right)^2 \right)^{\frac{1}{2}} \tag{12.1}
$$

$$
= \frac{\Delta \theta}{2} \max_{\mathbf{v} \in V} \| \mathbf{v} \| + \frac{\Delta x}{\sqrt{2}}. \tag{12.2}
$$

The last equality holds because $\Delta x = \Delta y$.

In the case of convolution of two workspace densities we can use Eq. (12.1) to balance the error between the angular and translational parts. We denote the smallest simply connected continuous regions containing the sets S_1 and S_2 as C_1 and C_2, respectively. To discretize C_1 we first select a maximal acceptable error e (which results from a trade-off between memory and accuracy). The resolution parameters $\Delta x^{(1)}$, $\Delta y^{(1)}$, $\Delta \theta^{(1)}$ of C_1 are then determined such that the two parts of the error are of the same order and add up to e, i.e.,

$$
\frac{\Delta \theta^{(1)}}{2} \max_{\mathbf{v} \in T(C_2)} \| \mathbf{v} \| \; \overset{!}{=} \; \frac{\Delta x^{(1)}}{\sqrt{2}} \; \overset{!}{=} \; \frac{e}{2},
$$

where $T(C_2) \subset \mathbb{R}^2$ is the union of all projections of constant theta "slices" of C_2 onto the $x - y$ plane.

This results in the choice

$$
\Delta x^{(1)} = \Delta y^{(1)} = \frac{e}{\sqrt{2}}, \qquad \Delta \theta^{(1)} = \frac{e}{\displaystyle\max_{\mathbf{v} \in T(C_2)} \| \mathbf{v} \|}. \tag{12.3}
$$

The step sizes for the discretization of C_2 are chosen analogously. The error bound in Eq. (12.1) was derived in [21].

Numerical Convolution of Histograms on $SE(2)$

Our goal is to store density functions in the form of piecewise constant histograms, i.e., we only want to store average values for each voxel. This section presents convolution in a form applicable to histograms on $SE(2)$. As a first step the integral from the previous section

$$f_3(x, y, \theta) = (f_1 * f_2)(x, y, \theta) = \int_{x_{min}^{(1)}}^{x_{max}^{(1)}} \int_{y_{min}^{(1)}}^{y_{max}^{(1)}} \int_{-\pi}^{\pi} f_1(\xi, \eta, \alpha)$$

$$\cdot f_2\Big((x-\xi)\cos\alpha + (y-\eta)\sin\alpha, -(x-\xi)\sin\alpha + (y-\eta)\cos\alpha,$$

$$(\theta - \alpha)\bmod 2\pi\Big)\ d\xi\, d\eta\, d\alpha$$

is approximated by a Riemann-Stieltjes sum

$$(f_1 * f_2)(x, y, \theta) \approx \Delta\xi\, \Delta\eta\, \Delta\alpha. \sum_{l=0}^{N_1} \sum_{m=0}^{N_2} \sum_{n=0}^{M} f_1(\xi_l, \eta_m, \alpha_n)$$

$$\cdot f_2\Big((x-\xi_l)\cos\alpha_n + (y-\eta_m)\sin\alpha_n,$$

$$- (x-\xi_l)\sin\alpha_n + (y-\eta_m)\cos\alpha_n, (\theta - \alpha_n)\bmod 2\pi\Big).$$

Although the right side of the equation is approximated by a discrete sum, the function f_2 is required to exist for any values of arguments because its arguments generally do not coincide with points on the grid of any discretization. We therefore approximate the functions f_2 for any real-valued arguments by interpolation using function values at neighboring discrete points. Since this problem is frequently encountered in many applications, there exist many different strategies. The simplest strategy is that the function $f_2(x, y, \theta)$ is approximated by the value of $f_2(x_i, y_j, \theta_k)$ (i.e., at the closest point on the grid). Because this can lead to large round-off errors, we use instead linear interpolation. For each coordinate (x, y, θ) we find the indices i, j, k in the grid such that $x_i \le x \le x_{i+1}$, $y_j \le y \le y_{j+1}$, etc., and define the ratios

$$t = \frac{x - x_i}{\Delta x}, \qquad u = \frac{y - y_j}{\Delta y}, \qquad v = \frac{\theta - \theta_k}{\Delta \theta}.$$

The value $f_2(x, y, \theta)$ is then interpolated from the values at eight discrete points (8-point interpolation)

$$\begin{aligned} f_2(x, y, \theta) = \ & (1-t)(1-u)(1-v) f_2(x_i, y_j, \theta_k) \\ & + (t)(1-u)(1-v) f_2(x_{i+1}, y_j, \theta_k) \\ & + (1-t)(u)(1-v) f_2(x_i, y_{j+1}, \theta_k) \\ & + \ldots + (t)(u)(v) f_2(x_{i+1}, y_{j+1}, \theta_{k+1}). \end{aligned}$$

While this constitutes an approximation, as we shall see in the next section, this produces acceptable results.

12.2.6 Numerical Results for Planar Workspace Generation

In this section we present numerical results for the generation of workspaces using the methods presented earlier in this chapter.[8]

The algorithm is implemented on a SUN SPARCstation 5, 110 MHz, in the C programming language. Figures were made using Mathematica version 3.0. The algorithm is applied to a version

---

[8] We thank Dr. Imme Ebert-Uphoff for generating the numerical results which appear here.

of the discretely actuated manipulator shown in Fig. 12.3. The manipulator in Fig. 12.3 consists of 10 modules composed of three legs each, where each leg has two bits (four states). Hence each module has $4^3 = 64$ discrete states. In our example we consider a manipulator consisting of only eight identical modules of this kind, resulting in a manipulator with $(64)^8 \approx 2.8 \cdot 10^{14}$ states. Unless otherwise specified, the width of each platform is chosen as $w = 0.2$ compared to the minimal and maximal actuator lengths of $q_1 = 0.15$ and $q_2 = 0.22$.

For this manipulator we calculate the workspace density corresponding to only the first two modules by brute force ($64^2 = 4096$ states), which we will refer to as W2 in the following. Since all modules are identical, convolution of this density with itself leads to the density of the four-module workspace, W4. Convolving this workspace again with itself leads to the workspace density of the whole manipulator, W8.

For the workspace of four modules it is possible to calculate the results using brute force ($(64)^4 \approx 1.7 \cdot 10^7$ states). In the following we first compare the results of this approach to the results obtained from convolution, and then we quantify the error. Afterward, we show results for the workspace of the 8-module manipulator, which cannot simply be calculated by brute force ($2.8 \cdot 10^{14}$ states).

Figure 12.5(a) is generated by calculating the histogram directly (brute force) and Fig. 12.5(b) is generated by convolution of the density function corresponding to two modules with itself. In the brute force calculation we use linear (8-point) smoothing when incrementing voxels, because this makes the raw data less sensitive to small shifts in the way the grid is superimposed. Eight-point interpolation is also used when evaluating the discrete density functions for the discrete convolution.

In each figure the z-axis corresponds to the angle θ (orientation of end-effector), and the point density per voxel is represented by a gray scale (black representing very high density). The base of the manipulator lies at the origin of the coordinate system. To enhance differences in low density areas, we chose a nonlinear gray scale: each density value is normalized to a value between 0 and 1 (divided by the largest density value in the drawing), and the fourth root of this value is displayed as gray value for W4 (the eighth root for W8).

Convolving the density array in Fig. 12.5(a) with itself results in W8. The result is shown in Fig. 12.6. Figures 12.7(a), (b) show the workspace of the same manipulator as in Fig. 12.6, if the maximal actuator length is decreased to $q_2 = 0.2$ or increased to $q_2 = 0.25$, respectively.

Error Measures

To quantify the error resulting from convolution for W4 (as compared to direct calculation) three different error measures are used. The first two measures compare the density resulting from the brute force approach, ρ, with the density from numerical convolution, $\tilde{\rho}$

$$E_1 = \frac{\sum\limits_{i,j,k} |\rho(x_i, y_j, \theta_k) - \tilde{\rho}(x_i, y_j, \theta_k)|}{\sum\limits_{i,j,k} |\rho(x_i, y_j, \theta_k)|},$$

$$E_2 = \frac{\sum\limits_{i,j,k} |\rho(x_i, y_j, \theta_k) - \tilde{\rho}(x_i, y_j, \theta_k)|^2}{\sum\limits_{i,j,k} |\rho(x_i, y_j, \theta_k)|^2}.$$

The third measure is a function of the shapes of the workspaces (by shape we mean the set of all voxels with nonzero density), by counting the number of voxels which belong to one of the workspaces, but not to the other

$$E_3 = \frac{\# \text{ for which } ((\rho > 0) \& (\tilde{\rho} = 0)) \text{ or } ((\rho = 0) \& (\tilde{\rho} > 0))}{\# \text{ for which } (\rho > 0)}.$$

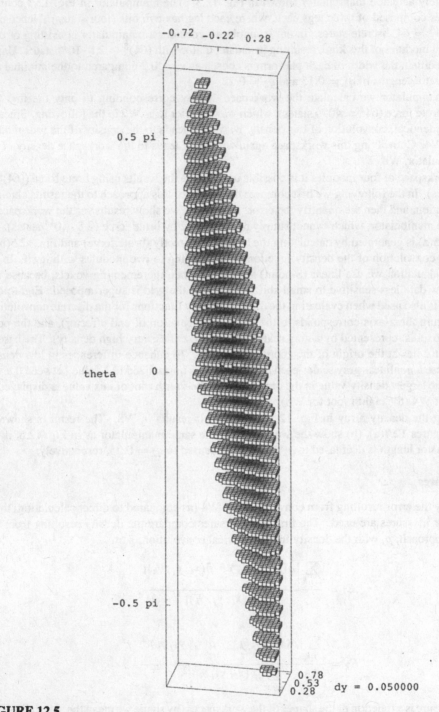

FIGURE 12.5

(a) Workspace density for a 4-module manipulator calculated by brute force. Scale is $\Delta x = \Delta y = 0.05$ and $\Delta\theta = \pi/30$. (Chirikjian, G.S. and Ebert-Uphoff, I., Numerical convolution on the Euclidean group with applications to workspace generation, *IEEE Trans. on Robotics and Autom.*, 14(1), 123–136, 1998. © 1998 IEEE.) *(Continued)*.

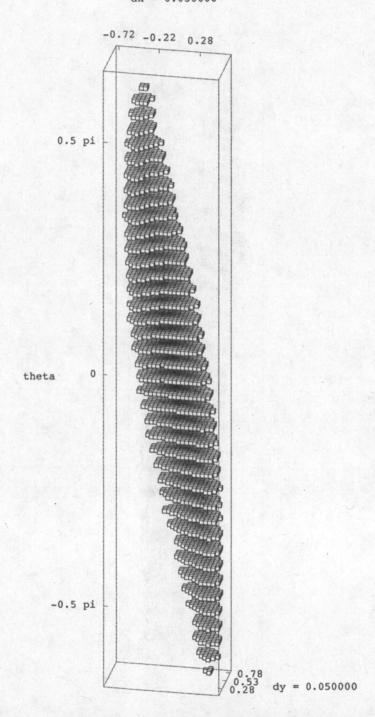

FIGURE 12.5

(Cont.) **(b) Workspace density for a 4-module manipulator using convolution. Scale is $\Delta x =$ $\Delta y = 0.05$ and $\Delta \theta = \pi/30$. (Chirikjian, G.S. and Ebert-Uphoff, I., Numerical convolution on the Euclidean group with applications to workspace generation,** *IEEE Trans. on Robotics and Autom.,* **14(1), 123–136, 1998. © 1998 IEEE.)**

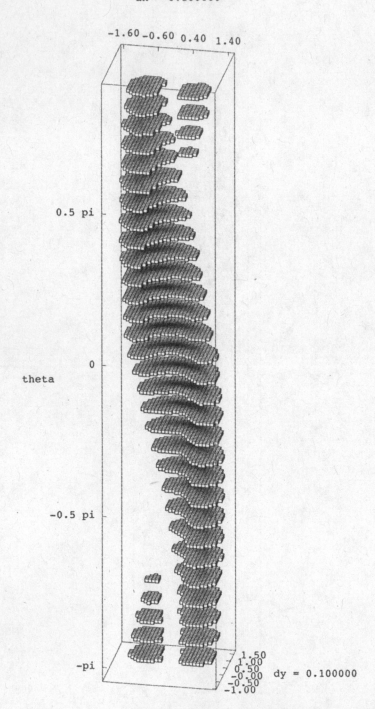

FIGURE 12.6

Workspace density for an 8-module manipulator using convolution. Scale is $\Delta x = \Delta y = 0.1$ and $\Delta \theta = \pi/15$. (Chirikjian, G.S. and Ebert-Uphoff, I., Numerical convolution on the Euclidean group with applications to workspace generation, *IEEE Trans. on Robotics and Autom.,* **14(1), 123–136, 1998. © 1998 IEEE.)**

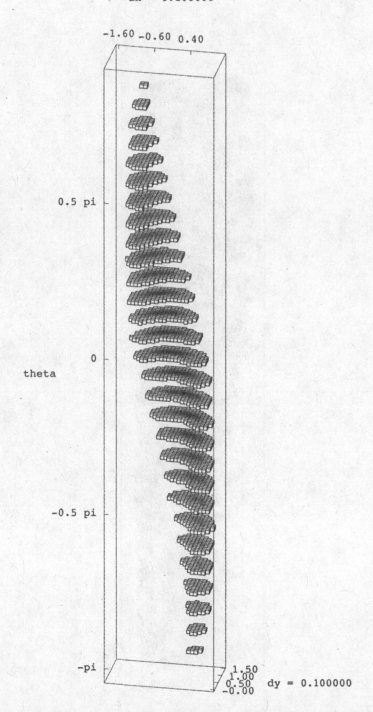

FIGURE 12.7

(a) Workspace density for an 8-module manipulator with kinematic parameters $q_2 = 0.2$. Scale is $\Delta x = \Delta y = 0.1$ and $\Delta \theta = \pi/15$. (Chirikjian, G.S. and Ebert-Uphoff, I., Numerical convolution on the Euclidean group with applications to workspace generation, *IEEE Trans. on Robotics and Autom.*, 14(1), 123–136, 1998. © 1998 IEEE.) *(Continued)*.

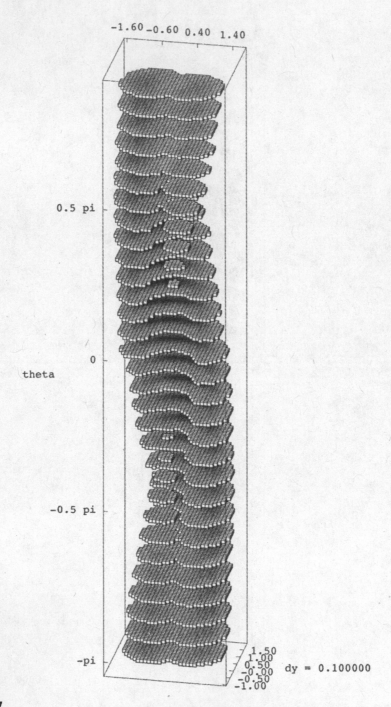

FIGURE 12.7

(Cont.) **(b) Workspace density for an 8-module manipulator with kinematic parameters** $q_2 =$ **0.25. Scale is** $\Delta x = \Delta y = 0.1$ **and** $\Delta\theta = \pi/15$. **(Chirikjian, G.S. and Ebert-Uphoff, I., Numerical convolution on the Euclidean group with applications to workspace generation,** *IEEE Trans. on Robotics and Autom.*, **14(1), 123–136, 1998. © 1998 IEEE.)**

Tables 12.1 and 12.2 show error, memory, and time, for the approximation of the manipulator of four modules described previously, if different discretizations are chosen for the representation of the workspace. In particular, the time listed is the net time to calculate workspace W4 by one numerical convolution of W2 with itself. The memory listed is the amount of memory needed to represent either W2 or W4 (whichever has more voxels). Since voxels are consolidated after convolution, it is possible for W4 to require fewer voxels than W2.

In all simulations the discretization is chosen as follows:

(a) choose Δx_2 for workspace W2 and Δx_4 for W4.

(b) calculate the angular discretization, $\Delta \theta_2$, of W2 according to Eq. (12.3).

(c) define the scale factor f as $f = \frac{\Delta x_4}{\Delta x_2}$, and determine $\Delta \theta_4$ such that f is also the factor for the angular discretization, i.e., $\Delta \theta_4 = f \cdot \Delta \theta_2$.

As a practical matter, integer scale factors are the easiest to work with. However, it is possible to consolidate voxels using more general scale factors if interpolation is used.

Results and Analysis

(1) For the case considered, W4 can be calculated from W2 with one convolution in less than 12 minutes with an error smaller than 10%. W8 is calculated from W4 with one convolution within another nine minutes. (No error is reported, since comparison with brute force results is impossible for W8).

(2) The best results in this particular example were obtained using a factor of $f \approx 2$, i.e., if the resolution is twice as fine before convolution than after (see Table 12.1). In general, an appropriate scale factor can be estimated even if there is no brute force data against which to compare the results of convolution. This may be achieved by examining the relative error of sequences of density functions generated using convolution and picking the scale factor for which the relative error is smallest.

(3) For a constant factor $f = 2$ the error decreases slowly with higher discretization (see Table 12.2).

(4) The result of the discrete convolution is closer to a brute force calculation if 8-point interpolation is used to generate the brute force results, as compared to no interpolation. Table 12.3 lists the error between brute force (BF) and discrete (numerical) convolution (DC) if different numbers of points are used in the interpolation, together with the corresponding computation time for either method. (1-point means no interpolation at all, 2-point means interpolation only in the angle, 8-point means interpolation in all three coordinate axes of the grid.)

Table 12.1 Results for Fixed Discretization (Δx_4, $\Delta \theta_4$) and Varying Factor f (Chirikjian, G.S. and Ebert-Uphoff, I., Numerical convolution on the Euclidean group with applications to workspace generation. *IEEE Trans. on Robotics and Autom.*, 14(1), 123–136, 1998. © 1998 IEEE.)

factor f	(Δx_2 , $\Delta \theta_2$)	(Δx_4 , $\Delta \theta_4$)	E_1	E_2	E_3	max. memory	Time
1	(0.0500, 0.101)	(0.050, 0.101)	18.95%	17.08%	96.42%	48 KB	2.4 min
2	(0.0250, 0.051)	(0.050, 0.101)	8.27%	9.64%	9.42%	442 KB	11.5 min
4	(0.0125, 0.025)	(0.050, 0.101)	21.82%	27.05%	30.18%	3546 KB	70.8 min

One would expect the error to reduce more or less monotonically with finer discretization. As can be seen in Table 12.2 this is true for the shape error, E_3. For error E_1 and E_2 the general tendency is to decrease for higher resolution, but this does not happen strictly monotonically. One likely reason for this is sensitivity of density to shifting of the grid by a tiny amount. This effect only appears if

Table 12.2 Results for Fixed Factor $f = 2$ and Varying Discretization (Δx_4, $\Delta \theta_4$) (Chirikjian, G.S. and Ebert-Uphoff, I., Numerical convolution on the Euclidean group with applications to workspace generation. *IEEE Trans. on Robotics and Autom.*, 14(1), 123–136, 1998. © 1998 IEEE.)

factor f	(Δx_2 , $\Delta \theta_2$)	(Δx_4 , $\Delta \theta_4$)	E_1	E_2	E_3	max. memory	Time
2	(0.0500, 0.101)	(0.1000, 0.203)	13.77%	15.37%	13.65%	54 KB	23 sec
2	(0.0375, 0.077)	(0.0750, 0.153)	7.48%	7.91%	13.58%	131 KB	91 sec
2	(0.0250, 0.051)	(0.0500, 0.101)	8.27%	9.64%	9.42%	442 KB	11.5 min
2	(0.0188, 0.038)	(0.0375, 0.076)	7.46%	9.05%	7.68%	1056 KB	47.1 min
2	(0.0125, 0.025)	(0.0250, 0.050)	5.20%	6.18%	6.47%	3546 KB	7hrs 11min

the density is distributed very unevenly and differs considerably in neighboring voxels. Hence we expect this effect also to disappear for manipulators of higher resolution, i.e., if the actuators have a higher resolution or a larger number of modules are considered as a single unit. This expectation is supported by the data in Table 12.3 which indicates a dramatic convergence of results obtained by brute force and convolution as the number of actuator states in each leg is doubled.

It is worth noting that for manipulators composed of P modules which each have a very large number of states, the $\mathcal{O}(P)$ or $\mathcal{O}(\log P)$ calculations required for P convolutions can be a significant savings over $\mathcal{O}(K^P)$. However, in the example presented here, the 64 frames reachable by each module are relatively sparse. This means that the actual time required to perform the $\log_2(8) = 3$ convolutions would far exceed the time required to generate the density function for half of the manipulator [which requires $(64)^{8/2}$ kinematic calculations] and it would perform one convolution. Thus in this case, $\mathcal{O}(K^{P/2})$ is better than $\mathcal{O}(\log P)$. In either case convolution plays a critical role in avoiding the $\mathcal{O}(K^P)$ calculations which cannot be performed in a reasonable amount of time.

Finally, we note that the purpose of this implementation is to show that the concept of numerical convolution on $SE(D)$ works and provides usable results. The error analysis resulting in Eq. (12.1) serves as a guide for balancing the number of discretizations in position and orientation. This direct numerical convolution also gives a sense for what kinds of errors are acceptable when using FFT algorithms. That is, if we know *a priori* that Euclidean group convolutions will not result in exact results, it becomes less important for the FFT to reproduce convolutions exactly. Some error is tolerable. For the spatial ($D = 3$) case, it is not possible to do brute force convolutions with the fine resolution we desire, and this indicates a need for $SE(3)$ FFT algorithms.

Table 12.3 Results for Different Types of Interpolation and Different Number of States per Actuator (Chirikjian, G.S. and Ebert-Uphoff, I., Numerical convolution on the Euclidean group with applications to workspace generation. *IEEE Trans. on Robotics and Autom.*, 14(1), 123–136, 1998. © 1998 IEEE.)

(Δx_2 , $\Delta \theta_2$)	(Δx_4 , $\Delta \theta_4$)	brute force	disc. conv.	E_1	E_2	E_3	T_{BF}	T_{DC}
		Four states per actuator:						
(0.0200, 0.041)	(0.040, 0.082)	1-point	2-point	10.56%	13.26%	13.58%	4.8 min	12.7 min
(0.0200, 0.041)	(0.040, 0.082)	1-point	8-point	10.90%	14.56%	9.83%	4.8 min	20.8 min
(0.0200, 0.041)	(0.040, 0.082)	8-point	8-point	7.72%	9.58%	8.27%	12.1 min	33.9 min
(0.0250, 0.051)	(0.050, 0.101)	1-point	2-point	11.91%	15.24%	16.74%	4.8 min	4.3 min
(0.0250, 0.051)	(0.050, 0.101)	1-point	8-point	9.78%	12.58%	9.60%	4.8 min	7.2 min
(0.0250, 0.051)	(0.050, 0.101)	8-point	8-point	8.27%	9.64%	9.42%	16.8 min	11.5 min
		Two states per actuator:						
(0.0250, 0.051)	(0.050, 0.101)	1-point	2-point	151.22%	235.52%	75.43%	0.1 sec	3.4 min
(0.0250, 0.051)	(0.050, 0.101)	1-point	8-point	121.50%	158.80%	62.51%	0.3 sec	12.5 min
(0.0250, 0.051)	(0.050, 0.101)	8-point	8-point	37.70%	21.64%	21.64%	0.7 sec	20.3 min

12.3 Inverse Kinematics of Binary Manipulators: the Ebert-Uphoff Algorithm

The concept of the manipulator workspace density function is useful in solving the inverse kinematics problem for discretely actuated manipulators with many states [21, 23]. If one were to try to solve the inverse kinematics problem by evaluating the manipulator forward kinematics for all K^P states, the computational cost would be prohibitive. If we have a cascade of manipulator workspace density functions corresponding to each of P sections of the manipulator, this exponential complexity is reduced to a problem that is linear in P.

Let $\rho_1, \ldots, \rho_P$ be the workspace density functions for each of the P sections of the manipulator. Instead of computing densities from the base to the distal end of the manipulator, the calculation can be performed in reverse order. That is,

$$\rho^P \triangleq \rho_P,$$

$$\rho^{P-1} \triangleq \rho_{P-1} * \rho_P,$$

$$\rho^{P-k} \triangleq \rho_{P-k} * \cdots * \rho_P,$$

$$\rho^1 \triangleq \rho_1 * \cdots * \rho_P.$$

ρ^1 is then the workspace density for the whole manipulator.

Now for $k \in \{1, \ldots, P\}$, let g_k be the transformation that relates the distal end of the kth segment to the base of the kth segment. The position of the top of the kth segment relative to the base of the manipulator then

$$g^{(k)} \triangleq g_1 \circ g_2 \circ \cdots \circ g_k$$

and the position and orientation of the distal end of the manipulator relative to the distal end of the kth segment is

$$\left(g^{(k)}\right)^{-1} \circ g^{(P)} = g_{k+1} \circ g_{k+2} \circ \cdots \circ g_P.$$

There are K possible states for each g_k. Using the information in the cascade of density functions $\rho^{(2)}, \ldots, \rho^{(P-1)}$, we can sequentially choose states of each section which, at each instant, maximize the probability density around the particular frame of reference which we seek to reach.

In other words, given that we want the end of the manipulator to reach $g_{des} \in SE(D)$, we start at the base of the manipulator and ask which state of segment 1 maximizes $\rho^{(P-1)}((g^{(1)})^{-1} \circ g_{des})$. After searching through all K possible values of g_1, and the optimal $g^{(1)} = g_1$ is fixed, we proceed up the manipulator one unit. That is, we next calculate $\rho^{(P-2)}((g^{(2)})^{-1} \circ g_{des})$. Since g_1 is fixed, K values of $g^{(2)} = g_1 \circ g_2$ are searched until the value of g_2 that maximizes is found. This procedure is performed by sequentially maximizing $\rho^{(P-k)}((g^{(k)})^{-1} \circ g_{des})$ for all $k \in \{1, 2, \ldots, P-1\}$. When $k = P$, the one out of K values of g_P that minimizes some measure of distance

$$C = d\left(g_{des}, g_1 \circ g_2 \circ \cdots \circ g_P\right)$$

is chosen.

The procedure we have described here is a way of specifying a state of the whole manipulator in $\mathcal{O}(P)$ arithmetic operations such that the distal end reaches g_{des} approximately.

12.4   Symmetries in Workspace Density Functions

When a manipulator arm possesses geometrical symmetries, this will induce certain symmetries in the manipulator's density function. As an example, consider the planar 3-bit platform in Fig. 12.1. This manipulator can be rotated by 180 degrees about its center through the axis pointing out of the plane of the figure, and if all the legs have the same two length states, the result will be the same manipulator with the same eight states. The only difference is that the configurations will be labeled differently, e.g., 110 might become 011. Hence, performing the same operations on the corresponding density function (which in this case is a sum of eight Dirac delta functions) should, in some sense, preserve the density function as long as we take into account the fact that frames which were originally moving freely at the top are now fixed, and frames originally fixed at the base are now free to move. In the following section we quantify how such discrete symmetries in the arm induce symmetries in the density function. Then in Section 12.4.2 we show how continuous symmetries reduce the density function of a spatial manipulator to a form which does not depend on all six parameters of $SE(3)$.

12.4.1   Discrete Manipulator Symmetries

Let us now consider in greater detail the density function of the symmetric 3-bit planar binary platform manipulator. This is a platform where all the states are uniform, i.e., the zero (one) state corresponds to the same length for all of the legs, and so rotating this platform about its center by 180 degrees results in functionally the same device. The density function is of the form

$$f(g) = \sum_{i=0}^{1} \sum_{j=0}^{1} \sum_{k=0}^{1} \delta\left(g_{(ijk)}^{-1} \circ g\right)$$

where ijk is a binary number describing one of the eight frames.

Now imagine shifting each of the original eight configurations of the platform so that the top plate in the platform now resides where the fixed base was previously. If a delta function is placed at the frame attached to what used to be the base of the manipulator for each of the eight configurations we get the new density function

$$\tilde{f}(g) = \sum_{i=0}^{1} \sum_{j=0}^{1} \sum_{k=0}^{1} \delta\left(\left(g_{(ijk)}^{-1}\right)^{-1} \circ g\right).$$

The second inverse is introduced because instead of the top of the manipulator reaching $g_{(ijk)}$, the bottom is now reaching $g_{(ijk)}^{-1}$ since the top is fixed at the identity frame. In general for a group G, and any $h \in G$, one finds $(h^{-1})^{-1} = h$. Also, the delta function has the special properties $\delta(g^{-1}) = \delta(g)$ and $\delta(h \circ g) = \delta(g \circ h)$. Applying these three rules we observe that

$$\tilde{f}(g) = \sum_{i=0}^{1} \sum_{j=0}^{1} \sum_{k=0}^{1} \delta\left(g_{(ijk)}^{-1} \circ g^{-1}\right) = f\left(g^{-1}\right). \tag{12.4}$$

This is true independent of any symmetry in the manipulator. Furthermore, the relationship $\tilde{f}(g) = f(g^{-1})$ holds for a cascade of platforms since the density function for the cascade will be a convolution, and it is easy to check that

$$\left(\tilde{f}_2 * \tilde{f}_1\right)(g) = (f_1 * f_2)\left(g^{-1}\right).$$

Hence, what holds for a single platform holds for a cascade due to the fact that $SE(n)$ is unimodular, and the resulting properties of convolution. The function $\tilde{f}(g)$ in Eq. (12.4) is shown in Fig. 12.8.

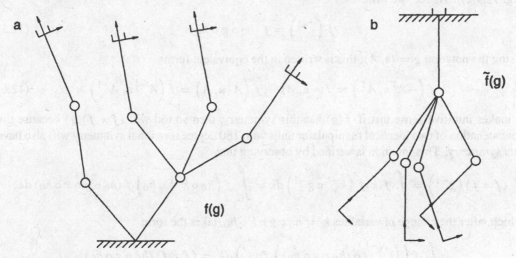

FIGURE 12.8
Workspace density of a manipulator with (a) proximal end fixed and distal end free (b) distal end fixed and proximal end free.

In the case when rotational symmetry about the center of the manipulator exists, we can say more. In this case, we observe that if any of the original configurations of the platform had been rotated by 180 degrees about the e_1 axis pointing out of the page *at the center of its base,* the resulting configuration would coincide with one of the shifted configurations discussed earlier. However, the frames attached to the platform would differ by an orientation of 180 degrees about e_1.

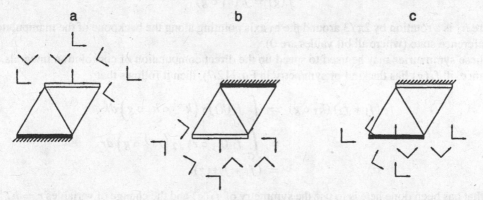

FIGURE 12.9
Frames reachable by a 3-bit manipulator with C_2 rotational symmetry: (a) proximal end fixed, (b) proximal end rotated by 180 degrees and fixed, (c) distal end fixed.

Let $f(g)$ be the sum of eight Dirac delta functions on $SE(2)$ which constitute the workspace of the 3-bit manipulator in Fig. 12.1 as depicted in Fig. 12.9(a). Let h_0 be a 180° rotation about e_1 (the axis pointing out of the plane) in Fig. 12.9. Then $h_0 = h_0^{-1}$. The version of the density function corresponding to the case when the original platform is rotated at its base (Fig. 12.9(b)) is $f(h_0 \circ g)$. An additional rotation by h_0 of the frames reachable at the distal end of the platform changes this to $f(h_0 \circ g \circ h_0)$. In the case of a manipulator with rotational symmetry about its center, this twice rotated density function will be the same as $\tilde{f}(g)$ (which is shown for this example platform in Fig. 12.9(c)). Hence, we write

$$f\left(g^{-1}\right) = f(h_0 \circ g \circ h_0). \tag{12.5}$$

Using the notation $g = (\mathbf{a}, A)$, this is written in the equivalent forms

$$f\left(-A^T \mathbf{a}, A^T\right) = f(-\mathbf{a}, A) \quad f\left(A^{\frac{1}{2}}\mathbf{a}, A\right) = f\left(A^{-\frac{1}{2}}\mathbf{a}, A^{-1}\right). \tag{12.6}$$

It makes intuitive sense that if $f(g)$ has this symmetry, then so too will $(f * f)(g)$ because the concatenation of two identical manipulator units with 180 degree rotational symmetry will also have this symmetry. This intuition is verified by observing that

$$(f * f)\left(g^{-1}\right) = \int_G f(k) f\left(k^{-1} \circ g^{-1}\right) dk = \int_G f\left(h_0 \circ k^{-1} \circ h_0\right) f(h_0 \circ g \circ k \circ h_0) dk,$$

which, after the change of variables $k' = h_0 \circ g \circ k \circ h_0$, takes the form

$$\int_G f\left(\left(k'\right)^{-1} \circ (h_0 \circ g \circ h_0)\right) f\left(k'\right) dk' = (f * f)(h_0 \circ g \circ h_0).$$

Thus the intuition is correct. A direct consequence of this is that when generating the density function for a manipulator which is a concatenation of identical modules, we only need to calculate the density function for half of its support, since the other half can be reconstructed from Eqs. (12.5) or (12.6).

Other discrete symmetries can exist as well. For instance, if a manipulator is composed of a cascade of binary Stewart/Gough platforms with three-fold symmetry about the $\mathbf{e}_3$ axis pointing from the center of the base plate through the center of the top one (as in Fig. 12.2), the density function will reflect this symmetry with the constraint

$$f(g) = f(h_1 \circ g), \tag{12.7}$$

where h_1 is a rotation by $2\pi/3$ around the $\mathbf{e}_3$ axis pointing along the backbone of the manipulator in its reference state (where all bit values are 0).

These symmetries may be used to speed up the direct computation of convolution integrals. For instance, if $f_1(g)$ has the kind of symmetry in Eq. (12.7), then it follows that

$$
\begin{aligned}
(f_1 * f_2)(h_1 \circ g) &= \int_G f_1(k) f_2\left(k^{-1} \circ h_1 \circ g\right) dk \\
&= \int_G f_1(h_1 \circ r) f_2\left(r^{-1} \circ g\right) dr \\
&= (f_1 * f_2)(g).
\end{aligned}
$$

All that has been done here is to use the symmetry of $f_1(g)$ and the change of variables $r = h_1^{-1} \circ k$, and from this it is clear that $(f_1 * f_2)(g)$ inherits the symmetry of $f_1(g)$ [apparently regardless of any symmetries of $f_2(g)$].

More generally, if H is any cyclic subgroup of $G = SE(3)$ and h_1 is the generator of H with Eq. (12.7) holding, then the computation of the convolution integral can be sped up by a factor of $|H|$ by evaluating $f_1(g)$ and $(f_1 * f_2)(g)$ only at one representative of each of the cosets $\sigma \in H \backslash G$ instead of evaluating them for all $g \in G$. To see this, use

$$\int_G f(g) dg = \sum_{h \in H} \int_{\sigma \in H \backslash G} f(h \circ g_\sigma) d_{H \backslash G}(\sigma),$$

for any $g_\sigma \in \sigma$ where $G = SE(3)$ and H a finite subgroup of G, to rewrite the convolution integral

$$(f_1 * f_2)(g_2) = \int_G f_1(g_1) f_2\left(g_1^{-1} \circ g_2\right) dg_1$$

as

$$(f_1 * f_2)(g_2) = \sum_{h \in H} \int_{\sigma \in H \backslash G} f_1(h \circ g_\sigma) f_2\left((h \circ g_\sigma)^{-1} \circ g_2\right) d_{H \backslash G}(\sigma). \tag{12.8}$$

Since $H \backslash G$ has the same dimension as G, and can be viewed as a measurable portion of G, it makes sense that $d_{H \backslash G}(\sigma) = d(g_\sigma)$ where $d(g) = dg$ is just the integration measure for $SE(3)$. And the only difference between integrating over G and $H \backslash G$ in this case is the bounds of integration for the parameters.

Using the fact that both $f_1(g)$ and $(f_1 * f_2)(g)$ have symmetry in Eq. (12.7), we can calculate Eq. (12.8) at $g_2 = g_{\sigma'}$ where $\sigma' \in H \backslash G$ as

$$(f_1 * f_2)(g_{\sigma'}) = \int_{\sigma \in H \backslash G} f_1(g_\sigma) \left(\sum_{h \in H} f_2\left(g_\sigma^{-1} \circ h^{-1} \circ g_{\sigma'}\right)\right) d(g_\sigma).$$

The calculation of

$$f_2'(g_\sigma, g_{\sigma'}) = \sum_{h \in H} f_2\left(g_\sigma^{-1} \circ h^{-1} \circ g_{\sigma'}\right)$$

can be performed in $\mathcal{O}(|H| \cdot (N/|H|)^2)$ computations where N is the number of discretizations of G.

Integrating over $H \backslash G$ is $|H|$ times faster than integrating over G. The combination of faster integration (due to a reduction in the domain of integration by a factor of $|H|$) and evaluation of the product on a domain reduced by a factor of $|H|$ leads to a speed-up by a factor of $|H|^2$ in all computations once $f_2'(g_\sigma, g_{\sigma'})$ is known. Hence, the limiting calculation is $f_2'(g_\sigma, g_{\sigma'})$, and a speed-up by a factor of $|H|$ is realized. This savings is not nearly as helpful as using the FFT for the motion group.

12.4.2 3D Manipulators with Continuous Symmetries: Simplified Density Functions

Consider a manipulator with a rotary actuator at the base as shown in Fig. 12.10. As with all manipulators, the workspace can be viewed as the support of a workspace density function, $f(g)$. However, if position is described using polar or spherical coordinates, the density function for this manipulator will have no dependence on the azimuthal angle ϕ. Likewise, if a rotary actuator is placed at the distal end of the arm, it will have the effect of averaging over the γ Euler angle, and so instead of considering a density function on the six-dimensional space $\mathbb{R}^3 \times SO(3)$, we can instead deal with the four-dimensional space $\mathbb{R}^2 \times S^2$. This leads to savings in the direct computation of convolutions for reasons analogous to those seen in the case of discrete symmetries.

It also forces many of the Fourier transform matrix elements to zero. From Eq. (10.34) it is clear for this case that only elements of the form

$$U_{l',m';l,0}^s(\mathbf{a}, A; p) = [l', m' \mid p, s \mid l, m'](\mathbf{a}) \, \tilde{U}_{m'0}^l(A)$$

contribute to the Fourier expansion of functions on $SE(3)$ that are independent of ϕ and γ. The

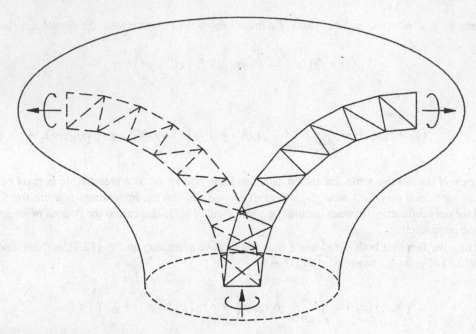

FIGURE 12.10
A 3D manipulator with continuous symmetries.

translation matrix elements take on the following simplified form for this special case

$$[l', m' \mid p, s \mid l, m'] (\mathbf{a})$$

$$= \sqrt{\frac{(2l' + 1)}{(2l + 1)}} \cdot \sum_{k=|l'-l|}^{l'+l} i^k (2k + 1) j_k(p a) C (k, 0; l', s \mid l, s)$$

$$\cdot C (k, 0; l', m' \mid l, m') P_k(\cos\theta). \tag{12.9}$$

From the perspective of calculating workspaces using the $SE(3)$ Fourier transform and convolution theorem, the independence of workspace density functions of ϕ and γ gives special structure to the corresponding Fourier matrices. It is yet to be determined what computational advantages can be gained from this special structure. From the perspective of direct numerical convolution, these special density functions are functions of a smaller number of variables, and the convolutions can be performed more efficiently.

We note that if the manipulator consists of planar sections connected with rotary joints that allow the manipulator to twist out of the plane, then the workspace density function will be independent of α. This means that only $m' = 0$ elements will contribute to the Fourier expansion. While such a manipulator possesses the freedom to reach points in space, its workspace density function only has three degrees of freedom. In this case

$$U_{l',0;l,0}^s(\mathbf{a}, A; p) = [l', 0 \mid p, s \mid l, 0] (\mathbf{a}) P_l(\cos\beta).$$

By defining $U_{l',l}^s(r, \theta, \beta; p) \triangleq U_{l',0;l,0}^s(\mathbf{a}, A; p)$ and $\hat{f}_{l,l'}^s(p) \triangleq \hat{f}_{l,0;l',0}^s(p)$, the Fourier expansion for workspace density functions on $SE(3)$ with this high degree of symmetry reduces to

$$f(\mathbf{a}, A) = \frac{1}{2\pi^2} \sum_{s=-\infty}^{\infty} \sum_{l'=|s|}^{\infty} \sum_{l=|s|}^{\infty} \int_0^{\infty} p^2 \, dp \, \hat{f}_{l,l'}^s(p) U_{l',l}^s(r, \theta, \beta; p)$$

where

$$\hat{f}^s_{l,l'}(p) = (2\pi)^3 \int_0^{\pi} \int_0^{\pi} \int_0^{\infty} f(r, \theta, \beta) \overline{U^s_{l',l}(r, \theta, \beta; p)} r \sin\theta \sin\beta \, dr d\theta d\beta.$$

The convolution theorem becomes

$$(\mathcal{F}(f_1 * f_2))^s_{l',l}(p) = \sum_{j=|s|}^{\infty} \left(\hat{f}_2\right)^s_{l',j}(p) \left(\hat{f}_1\right)^s_{j,l}(p).$$

12.4.3 Another Efficient Case: Uniformly Tapered Manipulators

We have seen how the concept of convolution can reduce a problem which would take K^P evaluations of the forward kinematics to one which requires P convolutions. In the special case when all the units are the same, this is reduced further to $\log_2 P$ convolutions. When discrete or continuous symmetries exist, each of these convolutions can be computed more efficiently than when using direct integration. Fast Fourier transforms also provide a tool which will speed up the computation of each convolution. But by far the largest savings in computational time results when the K^P problem is reduced to a $\log_2 P$ problem. We therefore examine a case here, in addition to the manipulator with uniform modules, for which this dramatic savings is possible.

Consider the case when each of the modules in a manipulator is a scaled version of each other, so the others corresponding density function for the ith module is

$$f_i(\mathbf{a}, A) = f(\mathbf{a}/s_i, A)$$

where s_i is the scale factor for the ith module, and $f(\mathbf{a}, A)$ is the density function for the base module where $s_1 = 1$.

The convolution of two adjacent density functions of this kind is written as

$$(f_i * f_{i+1})(\mathbf{a}, A) = \int_{SO(n)} \int_{\mathbb{R}^n} f(\mathbf{r}/s_i, R) f\left(R^T((\mathbf{a}-\mathbf{r})/s_{i+1}), R^T A\right) d\mathbf{r} dR.$$

Making the change of variables $\mathbf{r}' = \mathbf{r}/s_i$, this integral is transformed to the following form when $s_i = (s_0)^i$ for some scalar s_0

$$(s_i)^n \int_{SO(n)} \int_{\mathbb{R}^n} f(\mathbf{r}', R) f\left(R^T\left((\mathbf{a}-s_i\mathbf{r}')/s_{i+1}\right), R^T A\right) d\mathbf{r} dR = (s_i)^n (f * f_1)(\mathbf{a}/s_i, A).$$

This means that the density function for all pairs of modules for $i = 0, 2, 4, \ldots$ can be calculated with one convolution, and can be scaled appropriately to yield $f_{i,i+1} \triangleq f_i * f_{i+1}$. Then, using the same logic, $f_{i,i+2}$ is calculated by convolving appropriately scaled versions of $f_{i,i+1}$ with itself. This too is found by performing a single convolution and rescaling the output.

12.5 Inverse Problems in Binary Manipulator Design

In the previous sections of this chapter we have observed that the density function for the workspace of a binary manipulator is generated by the convolution of the density functions for two halves

$$(f_1 * f_2)(h) = \int_{SE(3)} f_1(g) f_2\left(g^{-1} \circ h\right) d(g) = f_3(h).$$

In addition to the *forward* problem, where $f_1(g)$ and $f_2(g)$ are given and we seek $f_3(g)$ for a manipulator with known geometry, certain *inverse* problems arise in the kinematic design of manipulators. One problem in the design of binary manipulators is to set kinematic parameters (e.g., actuator stroke length stops) so that a prescribed workspace density function is attained. This problem can be solved more easily if the manipulator is broken into subunits and each is designed separately. Another problem is to design the distal end of the manipulator when the proximal end is already built so that the whole workspace density comes as close as possible to the desired one. The latter problem reduces to the solution of a linear convolution equation, i.e., find $f_2(g)$ for given $f_1(g)$ so that the workspace of the combination comes as close as possible to the specified function $f_3(g)$. A nonlinear convolution problem (find $f_1(g)$ so that $(f_1 * f_1)(g) = f_3(g)$ for given $f_3(g)$) arises when we seek to design the sections of the manipulator separately [10, 36, 37].

The most natural way to solve these problems is to use the Fourier transform of functions on $SE(3)$. Using the convolution theorem, the convolution equation may be written in the form

$$\hat{f}_2 \hat{f}_1 = \hat{f}_3, \tag{12.10}$$

where $\hat{f}_i$ denotes the $SE(3)$ Fourier transform of function f_i.

In general, the Fourier transform can contain an infinite number of harmonics l, l', and block elements s. We assume that the contribution of the higher (rapidly oscillating) harmonics can be neglected, and truncate the Fourier transform at some $l = l'$ for each block and take all nonzero blocks for $| s | \le l$. Thus the problem may be reduced to the solution of the matrix Eq. (12.10). If the functions are "band limited" (i.e., only a finite number of harmonics give contributions to the Fourier transform), the Fourier transform matrices are finite.

For nonsingular matrix $\hat{f}_1$ the inverse Fourier transform is used to generate the solution

$$f_2(g) = \mathcal{F}^{-1}\left(\hat{f}_3(p)\hat{f}_1^{-1}(p)\right). \tag{12.11}$$

However, in practice, $\hat{f}_1(p)$ is usually singular for most or all values of p, and so a means of regularization is required.

This is a perfect application of the operational properties derived in Chapter 10. One can extend the Tikhonov regularization technique [26], used for integral equations of real-valued argument, to the case of $SE(3)$. That is, instead of solving the original problem, we seek an approximate solution which minimizes the cost function (this is a particular example of first order Tikhonov regularization)

$$C = \int_{SE(3)} \left(|(f_1 * f_2)(g) - f_3(g)|^2 + \epsilon |f_2(g)|^2 + \nu |\nabla_{\mathbf{a}} f_2(g)|^2 \right.$$
$$\left. + \eta \left(f_2(g), -\nabla_A^2 f_2(g)\right)\right) d(g) \tag{12.12}$$

for small parameters ϵ, ν, and η (higher order derivatives may be added for higher order regularization). Here $g = (\mathbf{a}, A)$.

Using the operational properties in Eqs. (10.60) and (10.63) together with the Plancherel equality in Eq. (10.50), one can convert this cost function into an algebraic expression in the dual space, do algebraic manipulations, and convert back using the inverse transform.

The Fourier transform converts Eq. (12.12) into

$$C = \frac{1}{2\pi^2} \int_0^\infty c\left(\hat{f}_2^\dagger(p), \hat{f}_2(p)\right) p^2 dp = \frac{1}{2\pi^2} \int_0^\infty \left(\| \hat{f}_2(p)\hat{f}_1(p) - \hat{f}_3(p)\right) \|_2^2$$
$$+ \epsilon \| \hat{f}_2(p) \|_2^2 + \nu \| p \hat{f}_2(p) \|_2^2 + \eta Tr\left(\hat{f}_2^\dagger(p) \mathcal{A}\hat{f}_2(p)\right) p^2 dp, \tag{12.13}$$

where $\mathcal{A}_{l',m';l,m} = l'(l'+1)\delta_{l'l}\delta_{m'm}$.

The equation for $\hat{f}_2(p)$, which minimizes the functional C, may be found by differentiating $c(\hat{f}_2^\dagger(p), \hat{f}_2(p))$ with respect to $\hat{f}_2^\dagger(p)$ (or $\hat{f}_2(p)$)

$$\frac{\partial c}{\partial \hat{f}_2^\dagger} = 0,$$

[differentiation with respect to $\hat{f}_2(p)$ gives a Hermitian conjugate equation]. This equation is written explicitly as

$$\hat{f}_2 \left(\hat{f}_1 \hat{f}_1^\dagger + \left(\epsilon + \nu p^2 \right) \mathbb{I} \right) + \eta \mathcal{A} \hat{f}_2 = \hat{f}_3 \hat{f}_1^\dagger, \tag{12.14}$$

where $\mathbb{I}$ is an appropriately dimensioned identity matrix. After truncating the Fourier transforms and contraction of indices according to Eq. (10.53) this equation is analogous to the matrix equation

$$\hat{f}_2 \mathcal{B} + \mathcal{A} \hat{f}_2 = \mathcal{D} \tag{12.15}$$

for given matrices $\mathcal{A}, \mathcal{B}, \mathcal{D}$. Methods for solving this equation can be found in [3, 24, 25].

We have to solve Eq. (12.14) for smaller and smaller values of the parameters ϵ, ν, η. When the solution starts to exhibit unpleasant behavior (the solution shows singular-like growth in some regions and starts to oscillate) the calculations must be stopped. "Physical" arguments may be used in the choice of the particular values of the parameters, i.e., we want to pay more attention to the restriction on the magnitude of derivatives of the solution (parameters ν) or to the restriction just on the magnitude of the solution (parameter ϵ). The solution for these values of the parameters ϵ, ν, η is an approximation to the solution of the convolution Eq. (10.4).

For more on this linear inverse problem, see [10]. For the nonlinear inverse problem, see [36].

12.6   Error Accumulation in Serial Linkages

Error accumulation in serial linkages and cascades of platform manipulators is another problem that is easily quantified as a group-theoretic convolution. Intuitively, the errors due to manufacturing inaccuracies, backlash, and flexibility of the constituent components "add up" as one traverses the length of an open kinematic chain. This intuitive notion is quantified in the following subsections using the concepts of convolution of functions on the group of rigid body motions. But first, we review an approach that assumes infinitesimal (as opposed to finite) error.

12.6.1   Accumulation of Infinitesimal Spatial Errors

Consider two homogeneous transforms (positions and orientations) that are measured and are hence known to some error. A natural question to ask is what the error of the concatenation of these transformations will look like. This question was addressed in detail in the case of planar motions by Smith and Cheeseman [55]. We now present a more general formulation of the same ideas.

Let H_1 be the "exact" proximal homogeneous transform, and let H_2 be the "exact" distal homogeneous transform. Relative to these transforms are measured homogeneous transforms $H_1(\mathbb{I} + \Sigma_1)$ and $H_2(\mathbb{I} + \Sigma_2)$. Here

$$\Sigma_i = \sum_{k=1}^{6} \epsilon_k^{(i)} X_k$$

is an element of the Lie algebra $se(3)$, and $\epsilon_k^{(i)}$ are error probability density functions. This model assumes the errors are very small. For larger errors, the discussion of errors in Subsection 12.6.2 is more appropriate.

While the exact concatenation of transformations is $H_1 H_2 = H_3$, when measurement error or uncertainty are introduced, it becomes

$$H_1 \left(\mathbb{I} + \Sigma_1 \right) H_2 \left(\mathbb{I} + \Sigma_2 \right) = H_3 \left(\mathbb{I} + \Sigma_3 \right).$$

We now determine Σ_3 from the given information.

Expanding the preceding products and using the smallness of the error to neglect second order effects, one finds

$$H_3 \left(\mathbb{I} + \Sigma_3 \right) = H_1 H_2 \left[\mathbb{I} + \Sigma_2 + H_2^{-1} \Sigma_1 H_2 \right].$$

Using the 6-dimensional screw description of the small motions, one writes

$$(\Sigma_3)^{\vee} = (\Sigma_2)^{\vee} + Ad\,(H_2)\,(\Sigma_1)^{\vee},$$

where by definition

$$Ad\,(H_2)\,(\Sigma_1)^{\vee} = \left(H_2^{-1} \Sigma_1 H_2 \right)^{\vee}.$$

Since the errors are small enough, they can be considered to evolve in the Lie algebra, $se(3)$, rather than on the group $SE(3)$, and the standard tools used in statistics on $\mathbb{R}^N$ are applicable here. For instance, one can calculate what the covariance of the resulting error is, given the covariances of the errors of the two constituent homogeneous transforms.

12.6.2   Model Formulation for Finite Error

Suppose we are given a manipulator consisting of two six degree-of-freedom sub-units. These units could be Stewart-Gough platforms or six serial links connected with revolute joints. One unit is stacked on top of the other. The proximal unit will be able to reach each frame $h_1 \in SE(3)$ with some error. This error may be different for each different frame h_1. This is expressed mathematically as a real-valued function of $g \in SE(3)$ which has a peak in the neighborhood of h_1 and decays rapidly away from h_1. If the unit could reach h_1 exactly, this function would be a delta function. Explicitly this function may have many forms depending on what error model is used. However, it will always be the case that it is of the form $\rho_1(h_1, g_1)$ for $h_1, g_1 \in SE(3)$. That is, the error will be a function on $g_1 \in SE(3)$ for each frame h_1 that the top of the module tries to attain relative to its base. Likewise, the second module will have an error function $\rho_2(h_2, g_2)$ for $h_2, g_2 \in SE(3)$ that describes the distribution of frames around h_2 that might be reached when h_2 is the actual goal.

The error that results from the concatenation of two modules with errors $\rho_1(\cdot)$ and $\rho_2(\cdot)$ results from sweeping the error of the second module by that of the first. This is written mathematically as

$$\rho\,(h_1 \circ h_2, g) = (\rho_1 \otimes \rho_2)\,(h_1 \circ h_2, g) \triangleq \int_{SE(3)} \rho_1\,(h_1, g_1)\,\rho_2\left(h_2, g_1^{-1} \circ g \right) d\,(g_1). \quad (12.16)$$

In the case of no error, the multiplication of homogeneous transforms h_1 and h_2 as $h_1 \circ h_2$ represents the composite change in position and orientation from the base of the lower unit to the interface between units, and from the interface to the top of the upper unit. In the case of inexact kinematics, the error function for the upper unit is shifted by the lower unit $[\rho_2(h_2, g_1^{-1} \circ g)]$, weighted by the error of the lower unit $[\rho_1(h_1, g_1)]$ and integrated over the support of the error function of the lower unit [which is the same as integrating over all of $SE(3)$ since outside of the support of the error function the integral is zero]. The result of this integration is by definition the error function around the frame $h_1 \circ h_2$, and this is denoted as $(\rho_1 \otimes \rho_2)(h_1 \circ h_2, g)$. We illustrate Eq. (12.16) in Fig. 12.11.

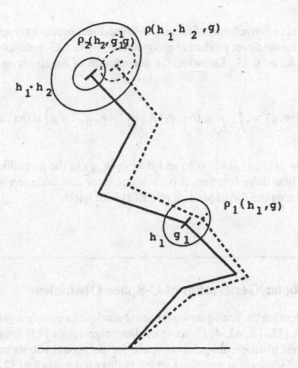

FIGURE 12.11
Error propagation in serial linkages.

12.6.3 Related Mathematical Issues

To test this formulation, consider the case of exact kinematics. In this case, the error functions have a very special form: they are Dirac delta functions on $SE(3)$. Completely analogous to the usual Dirac delta function on the real line, we have the properties

$$\int_{SE(3)} \delta(g)d(g) = 1 \qquad \delta\left(h^{-1} \circ g\right) = \delta\left(g^{-1} \circ h\right)$$

and

$$\int_{SE(3)} f(h)\delta\left(h^{-1} \circ g\right) d(h) = f(g).$$

Using these properties, the error functions for both units may be written as

$$\rho_i(h_i, g) = \delta\left(h_i^{-1} \circ g\right).$$

Then in this special case, Eq. (12.16) reduces to

$$
\begin{aligned}
(\delta \otimes \delta)(h_1 \circ h_2, g) &= \int_{SE(3)} \delta\left(h_1^{-1} \circ g_1\right) \delta\left(h_2^{-1} \circ g_1^{-1} \circ g\right) d(g_1) \\
&= \delta\left(h_2^{-1} \circ h_1^{-1} \circ g\right) \\
&= \delta\left((h_1 \circ h_2)^{-1} \circ g\right).
\end{aligned}
\tag{12.17}
$$

In other words, in the case of exact kinematics, we have exactly the result which is expected. Moreover, the delta function is a tool that also allows us to reexamine the general error propagation equation in Eq. (12.16) as a convolution of functions on the direct product group $SE(3) \times SE(3)$. Recall that the group law for the direct product of group with itself, $G \times G$, is simply $(h_1, g_1) \hat{o} (h_2, g_2) = (h_1 \circ h_2, g_1 \circ g_2)$ for $h_i, g_i \in G$. Likewise, the convolution of functions on the direct product is defined as

$$(\alpha * \beta)(h, g) = \int_G \int_G \alpha(h_3, g_1) \beta\left(h_3^{-1} \circ h, g_1^{-1} \circ g\right) d(h_3) d(g_1). \tag{12.18}$$

By letting $\alpha(h, g) = \rho_1(h, g) \delta(h_1^{-1} \circ h)$ and $\beta = \rho_2(h, g)$ in the preceding equation and using the properties of the Dirac delta function, it is clear that error accumulation in Eq. (12.16) may be written as a convolution on the direct product of $SE(3)$ with itself.

12.7   Mobile Robots: Generation of C-Space Obstacles

The question of computing the configuration-space obstacles of a mobile robot has been considered in a number of papers [27, 28, 42, 45]. Some of these algorithms [42] compute analytically the boundary of the regions of free configuration space for polygonal robots and obstacles. In this section we develop and implement a method which builds on the work of [27, 28], which may be applied to both polygonal and non-polygonal shapes. We suggest faster implementation of this method to compute the configuration space, which also allows one to solve inverse problems, i.e., design of a robot for given static obstacles and desired configuration space (or computation of the maximal size of the obstacles for a given robot and configuration space).

The mathematical formulation of the method is given below. We compute a function on the configuration space (the space of translations and rotations of the robot) which has nonzero values only in the regions where the robot hits the obstacle. The magnitude of this "density function" is the ratio of the overlapping volume (area) of the robot to the total volume (area), i.e., it changes from 0 to 1 in the overlapping regions. To calculate this value we compute the integral

$$c(\mathbf{x}, A) = \frac{\int_{\mathbb{R}^n} f_1(\mathbf{y}) f_2(A^{-1}(\mathbf{y} - \mathbf{x})) d^n y}{\int_{\mathbb{R}^n} f_2(\mathbf{y}) d^n y} \tag{12.19}$$

where $f_{1,2}(\mathbf{x})$ are equal to 1 if the vector $\mathbf{x}$ is inside or on the boundary of obstacles (robot) and zero otherwise, $n = 2, 3$ for 2D (3D) coordinate space and $d^n y = dy_1 \ldots dy_n$ is the usual integration measure for $\mathbb{R}^n$. The function $c(\mathbf{x}, A)$ is normalized to have maximal value of 1, i.e., it is divided by the volume (area) of the robot, $\mathbf{x} \in \mathbb{R}^n$ and $A \in SO(n)$. The geometrical meaning of this function is that it is zero when the obstacle and robot do not intersect, and it has increasing positive value as the area of intersection increases.

To compute this integral directly, or simply to check pixel by pixel, that the obstacle $f_1(\mathbf{y})$ and the rotated and translated robot $f_2(A^{-1}(\mathbf{y} - \mathbf{x}))$ do not overlap, we need to perform $O(N_R N_r^2)$ computations, where N_R is the number of sampled orientations and N_r is the number of sampled points in a bounded region of $\mathbb{R}^n$. For a large 3D array of values the computation of this integral by direct summation may be quite slow. The "overlap function" $c(\mathbf{x}, A)$, however, may be computed in $O(N_R N_r \log N_r)$ computations using Fourier methods on the discrete motion group and FFT methods. We discuss applications of this method to mobile robot configuration space generation in detail in the following section.

12.7.1 Configuration-Space Obstacles of a Mobile Robot

Here we apply the numerically implemented Fourier transform on the discrete motion group of the plane to the computation of the free configuration space of a rigid mobile robot moving among static 2D obstacles. To find the configuration space we compute the integral in the numerator of Eq. (12.19) using the Fourier methods for the discrete motion group described in Chapter 11. Because the obstacles and the robot are functions of only Cartesian position [i.e., to evaluate Eq. (12.19) we need to compute the Fourier transform of $f_2(\mathbf{x})$ only for one fixed orientation], the direct Fourier transform is performed faster than for an arbitrary function on the motion group.

A function of position may be considered as a function on the discrete motion group which does not depend on the orientation, i.e., $f(\mathbf{x}, A_i) = f(\mathbf{x})$. Then the "overlap" function in Eq. (12.19) may be formally written as the integral

$$c(\mathbf{x}, A_j) = \frac{1}{N_R} \frac{\sum_{i=0}^{N_R-1} \int_{\mathbb{R}^2} f_1(\mathbf{y}, A_i) f_2(A_j^{-1}(\mathbf{y} - \mathbf{x}), A_j^{-1} A_i) d^2 y}{\int_{\mathbb{R}^2} f_2(\mathbf{y}) d^2 y}. \tag{12.20}$$

Because the functions are real, the integral in the numerator may be written as

$$\frac{1}{N_R} \sum_{i=0}^{N_R-1} \int_{\mathbb{R}^2} \overline{f_1(\mathbf{y}, A_i)} f_2(A_j^{-1}(\mathbf{y} - \mathbf{x}), A_j^{-1} A_i) d^2 y$$

$$= \int_{G_{N_R}} \overline{f_1(h)} f_2\left(g^{-1} h\right) d(h)$$

where we denote $d(h)$ for the discrete motion group, G_{N_R}, to mean integration with respect to $\mathbb{R}^2$ and the summation through the A_i, and the group elements are of the form $g = (\mathbf{x}, A_j)$. Using the orthogonality properties of the Fourier matrix elements, this integral may be written as

$$\frac{1}{N_R} \sum_{q=0}^{N_R-1} \sum_{n=0}^{N_R-1} \int_0^\infty \int_v \sum_{m=0}^{N_R-1} \left(\overline{(\hat{f}_1)_{mn}}(\hat{f}_2)_{mq}\right) U_{qn}\left(g^{-1}; p, \phi\right) p\, dp\, d\phi$$

$$= \frac{1}{N_R} \sum_{q=0}^{N_R-1} \sum_{n=0}^{N_R-1} \int_0^\infty \int_v \sum_{m=0}^{N_R-1} \left(\overline{(\hat{f}_2)_{mq}}(\hat{f}_1)_{mn}\right) U_{nq}(g; p, \phi) p\, dp\, d\phi \tag{12.21}$$

where integration with respect to ϕ is integration on the circle in the angular interval $F_q = [2\pi q/N_R, 2\pi (q-1)/N_R]$. For the second expression we used the unitarity of the matrix elements and the fact that the expression is real (i.e., we take a complex conjugate of the integral). The matrices $(\hat{f}_{1,2})_{mn}$ are the Fourier transforms of the functions $f_{1,2}(\mathbf{x}, A_i)$.

Because functions $f_{1,2}(\mathbf{x}, A_i) = f_{1,2}(\mathbf{x})$ do not depend on the orientations A_i, matrix elements of the Fourier transforms in the same column are the same, i.e.,

$$\left(\hat{f}_{1,2}\right)_{mn} = \left(\hat{f}_{1,2}\right)_{qn}$$

for any m, q. This may be observed from the expression

$$U\left(g^{-1}; p, \phi\right)_{mn} = e^{-i\, p\, \mathbf{u}_n^\phi \cdot \mathbf{r}} \delta_{A_i^{-1} \mathbf{u}_n, \mathbf{u}_m}$$

(the exponent depends only on the n-index), the definition of the direct discrete-motion-group Fourier transform, and the fact that the functions do not depend on the orientation. Thus we compute a row of the Fourier matrix for a particular orientation (for example $A_0 = \mathbb{I}$)

$$\left(\hat{f}_{1,2}\right)_n = \left(\hat{f}_{1,2}\right)_{nn}.$$

This may be done using the 2D FFT for the functions $f_{1,2}(\mathbf{x})$ and interpolating the Fourier values to points on a polar coordinate grid. This requires $O(N_r \log(N_r))$ computations. Thus the "overlap" function in Eq. (12.19) is written as

$$c(\mathbf{x}, A_j) = C \frac{\sum_q \sum_n \int_0^\infty \int_\phi \left(\hat{\bar{f}}_{2q} \hat{f}_{1n} \right) U_{nq}(g; p, \phi) \, p \, dp \, d\phi}{\int_{\mathbb{R}^2} f_2(\mathbf{y}) \, d^2 y} \tag{12.22}$$

where $C = \frac{1}{N_R}$. The product of the column $\hat{\bar{f}}_2$ and the row $\hat{f}_1$ may be performed in $O(N_r N_R)$ computations, and the inverse transform may be performed in $(N_R N_r \log(N_r))$ computations. The normalization of the function $f_2(\mathbf{x})$ may be computed by direct integration in $O(N_r)$. Thus the inverse transform is the largest time-consuming computation. We note that the direct Fourier transform is performed only for one orientation, this is N_R times faster (for discrete rotation group C_{N_R}) than performing the FFT for each orientation as is done in [28], but the inverse transform is of the same order of computations. Thus, as N_R becomes large, there is a built-in factor of two speed increases using our method. We also mention that inverse problems, i.e., problems of finding the allowed shape of the robot for the given desired configuration space and the shape of the obstacles, may be solved using Fourier methods on the motion group, because it is reduced to a functional matrix equation (regularization methods for the solution of singular functional matrix equations are discussed in [36]).

When we discretize the coordinate regions and use the FFT to compute Eq. (12.19), the coefficient C in Eq. (12.22) must be defined as $C = A/N_R$, where A is the area of the compact region of $\mathbf{x}$ values where the FFT is computed (the functions must have a support inside the area A, the functions are considered periodic outside of this region). The area factor arises because the discrete Fourier transform can be considered as an approximation of the continuous case using

$$x_1 \to \frac{L}{N_r} i; \quad p_1 \to \frac{2\pi}{L} i,$$

(and analogously for the 2nd component). Here L is the length of the compact region in the x (y) direction. While the factor L is canceled out of equations if we apply direct and inverse discrete Fourier transform to the function, it appears in the convolution-like integral in Eq. (12.19).

Thus we compute the 2D FFT of $f_{1,2}(\mathbf{x})$, interpolate the Fourier elements to the polar grid, arrange them into the Fourier column and row, multiply column and row $\hat{f}_{mn} = \hat{\bar{f}}_{2m} \hat{f}_{1n}$, and take the inverse Fourier transform (the corresponding elements $m = n - i$ from the Fourier matrix $\hat{f}_{mn}$ must be taken for each orientation A_i and interpolated back to the Cartesian grid to take the 2D inverse FFT).

We implemented the computation of Eq. (12.22) using the FFT in the C programming language. Time to compute $c(\mathbf{x}, A_j)$ was 30 sec (on a 250 MHz SGI workstation) for a 256×256 square grid in $\mathbb{R}^2$ for the C_{10} group ($N_R = 10$, and we subdivide each segment into $N_\phi = 20$ subsegments).

Because for small values of $c(\mathbf{x}, A_j)$ the function exhibits oscillations (due to finite discretization of the integration area in the Fourier transform), we depict the boundary of the configuration space, defined by the contour line where $c(\mathbf{x}, A_j) = \epsilon$. The smallest possible choice of ϵ was in our examples in the region $0.005 - 0.035$. To increase accuracy, we use the following method. We may increase the value of $f_1(\mathbf{x})$ in the region near the border of the robot. For a convex robot this may be done by scaling down the robot and increasing the value of $f_1(\mathbf{x})$ between the scaled boundary of the robot and the original boundary. There is also another way to produce a rim for both concave and convex robots using the concept of an "offset curve." If the original boundary is the curve $\mathbf{x}(t)$, the offset curve is $\mathbf{O}_\delta(t) = \mathbf{x}(t) - \delta \mathbf{n}(t)$, where $\mathbf{n}(t)$ is the unit normal (pointing out) from the boundary curve.

While the change of the function values in the "rim" region does not change the shape and the size of the robot, the overlapping area of robot and obstacle is, effectively, smaller than ϵ for given

ϵ. If the overlap is completely in the "rim" of the increased values near the boundary, the following estimate may be received for the intersection area ϵ' for given value of ϵ

$$\epsilon' = \epsilon/q,$$

where $q = \frac{V}{V-1/k^2(V-1)}$, (the function f_1 is equal to one in the area scaled down by the factor k, and increased to the value V in the "rim" near the boundary of the robot).

In addition, we may increase the size of the robot and depict the configuration space for the scaled robot. The area between the two configuration boundaries (of the scaled robot and the original robot) is the "near-collision" area of the robot and the obstacle. This is an important region for motion planning. We note that the direct method [i.e., direct integration of Eq. (12.19) or the direct check if the robot and the obstacle overlap] may be used to find a precise boundary in the "near-collision" area. This may be performed in $O(N^2 A/A_{tot})$ computations, where A/A_{tot} is the ratio of the "near-collision" region to the total region of the coordinate space and $N = N_r N_R$. Because this ratio is generally small this gives very considerable (hundreds of times) savings in computations in the direct method.

We note that this method might not give good results if a very narrow and long object is attached to the robot (i.e., when the intersection area is not sensitive to the overlap). However, the value of $f_1(\mathbf{x})$ can be increased in these regions and this method may be useful for some of these shapes.

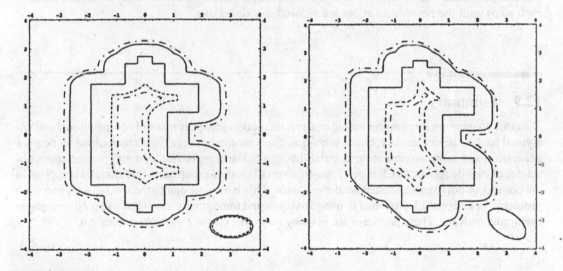

FIGURE 12.12
Configuration-space obstacle boundaries generated using the Fourier transform for the discrete motion group.

As an example we depict slices of configuration space in Fig. 12.12. The obstacle has a polygonal shape (the Fourier method can be applied both for polygonal and for "smooth" obstacles. The results are better for "smoother" obstacles, i.e., the value of ϵ can be chosen smaller) and the robot has an elliptic shape (the size and the orientation of the robot are depicted in the lower right corner on the figures). The solid line outside the obstacle depicts the boundary of the configuration space ($\epsilon \approx 0.030 - 0.035$). We increase the value of the function f_1 to 10 in the rim depicted in Fig. 12.12 (between the original boundary and the boundary scaled down by factor $k = 1.1$). We also depict in the pictures the boundary (dashed line) which describes the configuration space of the robot when it is located completely inside the obstacle ($g(\mathbf{x}, A) \approx 0.95 - 0.97$).

The figures also depict the corresponding boundaries for the scaled robot (enlarged by the factor 1.27). This is the region where the exact position of the configuration space boundary should be found by direct integration.

12.8   Mobile Robot Localization

Consider a rigid mobile robot navigating in a known environment where the location of the robot in the environment is initially unknown. The robot determines its pose (position and orientation) using range sensors. This often provides a number of noisy distance measurements, and the question is how to localize (determine pose) of the robot from these measurements.

One approach is similar to the C-space obstacle generation problem. We construct a polygon with vertices corresponding to the measured distances. Define a function equal to unity on the polygon and zero otherwise. Then convolve this function (for each orientation) with the function defined to be zero in free space and one on the obstacles. The result is a function which will be zero for all poses of the robot when the "sensor polygon" occupies free space.

Of course, it is possible that one set of measurements may not be sufficient to localize the robot. But this procedure can be repeated after the robot moves (storing the most likely candidate poses at each step) until the possible locations are reduced to a single one.

12.9   Summary

In this chapter we saw how harmonic analysis on motion groups is a useful computational and analytical tool in the context of robotics problems. Such problems range from manipulator workspace generation and error accumulation in serial linkages to the computation of configuration-space obstacles of mobile robots. In Chapter 17, applications of motion-group harmonic analysis is applied in the context of polymer-chain statistical mechanics. This is very similar to the workspace generation problem. In Chapter 13, problems in image analysis and tomography are posed using motion-group harmonic analysis. These problems are in many ways like C-space obstacle generation.

References

[1] Alciatore, D.G. and Ng, C.-C.D., Determining manipulator workspace boundaries using the Monte Carlo method and least square segmentation, *ASME Robotics: Kinematics, Dyn. and Controls*, DE-72, 141–146, 1994.

[2] Angeles, J., The design of isotropic manipulator architectures in the presence of redundancies, *Int. J. of Robotics Res.*, 11(3), 1992.

[3] Bartels, R.H. and Stewart, G.W., Solution of the matrix equation $AX + XB = C$, *Comm. ACM*, 15, 820–826, 1972.

[4] Basavaraj, U. and Duffy, J., End-effector motion capabilities of serial manipulators, *Int. J. of Robotics Res.*, 12(2), 132–145, 1993.

[5] Blackmore, D. and Leu, M.C., Analysis of swept volume via Lie groups and differential equations, *Int. J. of Robotics Res.*, 11(6), 516–537, 1992.

[6] Canny, J.F., *The Complexity of Robot Motion Planning*, MIT Press, Cambridge, MA, 1988.

[7] Ceccarelli, M. and Vinciguerra, A., On the workspace of general 4R manipulators, *Int. J. of Robotics Res.*, 14(2), 152–160, 1995.

[8] Chirikjian, G.S., A binary paradigm for robotic manipulators, *Proc. of the 1994 IEEE Int. Conf. on Robotics and Automation*, San Diego, CA, 1994.

[9] Chirikjian, G.S., Kinematic synthesis of mechanisms and robotic manipulators with binary actuators, *ASME J. of Mech. Design*, 117, 573–580, 1995.

[10] Chirikjian, G.S., Fredholm integral equations on the Euclidean motion group, *Inverse Probl.*, 12, 579–599, 1996.

[11] Chirikjian, G.S. and Ebert-Uphoff, I., Numerical convolution on the Euclidean group with applications to workspace generation, *IEEE Trans. on Robotics and Autom.*, 14(1), 123–136, 1998.

[12] Chirikjian, G.S., Synthesis of discretely actuated manipulator workspaces via harmonic analysis, in *Recent Advances in Robot Kinematics*, Lenarcic, J., and Parenti-Castelli, V., Eds., Kluwer Academic, Boston, 1996.

[13] Chirikjian, G.S., Group theoretical synthesis of binary manipulators, in *Ro. Man. Sy. '11: Theory and Practice of Robots and Manipulators*, 107–114, Springer, New York, 1997.

[14] Chirikjian, G.S. and Lees, D.S., Inverse kinematics of binary manipulators with applications to service robotics, *Proc. of IROS'95*, 3, 65–71, Pittsburgh, PA, 1995.

[15] Chirikjian, G.S., Inverse kinematics of binary manipulators using a continuum model, *J. Intelligent and Robotic Systems*, 19, 5–22, 1997.

[16] Choset, H. and Burdick, J.W., Sensor based motion planning: the hierarchical generalized Voronoi graph, *Int. J. of Robotics Res.*, in press.

[17] Craig, J.J., *Introduction to Robotics, Mechanics and Control*, Addison-Wesley, Reading MA, 1986.

[18] Desai, J. and Kumar, V., Nonholonomic motion planning of multiple mobile manipulators, IEEE Int. Conf. on Robotics and Autom., Albuquerque, New Mexico, April 20–24, 1997.

[19] Duffy, J. and Crane, C.D., III, *Kinematic Analysis of Robot Manipulators*, Cambridge University Press, New York, 1998.

[20] Durrant-Whyte, H.F., Uncertain geometry in robotics, *IEEE J. of Robotics and Autom.*, 4(1), 23–31, 1988.

[21] Ebert-Uphoff, I., On the development of discretely actuated hybrid-serial-parallel manipulators, Ph.D. dissertation, Dept. of Mechanical Engineering, Johns Hopkins University, Baltimore, 1997.

[22] Ebert-Uphoff, I. and Chirikjian, G.S., Efficient workspace generation for binary manipulators with many actuators, *J. of Robotic Systems*, 12(6), 383–400, 1995.

[23] Ebert-Uphoff, I. and Chirikjian, G.S., Inverse kinematics of discretely actuated hyper-redundant manipulators using workspace densities, Proc. 1996 IEEE Int. Conf. on Robotics and Automat., 139–145, 1996.

[24] Gantmacher, F.R., *The Theory of Matrices*, Chelsea, New York, 1959.

[25] Golub, G.H., Nash S., and Van Loan, C., A Hessenburg-Shur method for the problem $AX + XB = C$, *IEEE Trans. Automat. Control*, AC-24, 909–913, 1979.

[26] Groetsch, C.W., *The Theory of Tikhonov Regularization for Fredholm Equations of the First Kind*, Pitman, Boston, 1984.

[27] Guibas, L., Ramshaw, L., and Stolfi, J., A kinetic framework for computational geometry, *Proc. of IEEE Symp. on Found. of Comput. Sci.*, 100–111, 1983.

[28] Kavraki, L., Computation of configuration-space obstacles using the fast Fourier transform, *IEEE Trans. on Robotics and Autom.*, 11, 408–413, 1995.

[29] Khatib, O., Mobile manipulation: the robotic assistant, *J. of Robotics and Autonomous Syst.*, 26, 175–183, 1999.

[30] Klein, C.A. and Blaho, B.E., Dexterity measures for the design and control of kinematically redundant manipulators, *Int. J. of Robotics Res.*, 6(2), 1987.

[31] Koliskor, A., The l-coordinate approach to the industrial robots design, in *Information Control Problems in Manufacturing Technology 1986, Proc. of the 5th IFAC/IFIP/IMACS/IFORS Conf.*, 225–232, Suzdal, USSR, preprint.

[32] Korein, J.U., A geometric investigation of reach, MIT Press, Cambridge, MA, 1985.

[33] Kumar, A. and Waldron, K.J., Numerical plotting of surfaces of positioning accuracy of manipulators, *Mech. Mach. Theory*, 16(4), 361–368, 1980.

[34] Kwon, S.-J., Youm, Y., and Chung, K.C., General algorithm for automatic generation of the workspace for n-link planar redundant manipulators, *ASME Trans.*, 116, 967–969, 1994.

[35] Kyatkin, A.B. and Chirikjian, G.S., Fourier methods on groups: applications in robot kinematics and motion planning, Proc. 3rd Workshop on Algorithmic Foundations of Robotics, Houston, Texas, March 5–8, 1998.

[36] Kyatkin, A.B. and Chirikjian, G.S., Regularization of a nonlinear convolution equation on the Euclidean group, *Acta. Appl. Math*, 53, 89–123, 1998.

[37] Kyatkin, A.B. and Chirikjian, G.S., Synthesis of binary manipulators using the Fourier transform on the Euclidean group, *ASME J. Mech. Design*, 121, 9–14, 1999.

[38] Latombe, J.-C., *Robot Motion Planning*, Kluwer Academic, Boston, 1991.

[39] Lees, D.S. and Chirikjian, G.S., An efficient trajectory planning method for binary manipulators, *ASME Mech. Conf.*, 96-DETC/MECH-1161, Irvine, CA, August, 1996.

[40] Lees, D.S. and Chirikjian, G.S., An efficient method for computing the forward kinematics of binary manipulators, Proc. IEEE Int. Conf. of Robotics and Autom., 1012–1017, Minneapolis, MN, 1996.

[41] Lees, D.S. and Chirikjian, G.S., A combinatorial approach to trajectory planning for binary manipulators, Proc. IEEE Int. Conf. of Robotics and Autom., 2749–2754, Minneapolis, MN, 1996.

[42] Lozano-Perez, T., Spatial planning: a configuration space approach, *IEEE Trans. on Comput.*, 32, 108–120, 1983.

[43] Merlet, J.-P., *Les Robots Parallèles,* Traité des nouvelles Technologies, Série Robotique, Hermes, 1990.

[44] Murray, R.M., Li, Z., and Sastry, S.S., *A Mathematical Introduction to Robotic Manipulation*, CRC Press, Boca Raton, FL, 1994.

[45] Newman, W. and Branicky, M., Real-time configuration space transforms for obstacle avoidance, *Int. J. of Robotics Res.*, 10, 650–667, 1991.

[46] Park, F.C. and Brockett, R.W., Kinematic dexterity of robotic mechanisms, *Int. J. of Robotics Res.*, 13(1), 1–15, 1994.

[47] Pieper, D.L., The kinematics of manipulators under computer control, Ph.D. dissertation, Stanford University, Stanford, CA, 1968.

[48] Popplestone, R.J., Group theory and robotics, in *Robotics Res.: The First Int. Symp.*, Brady, M., and Paul, R., Eds., MIT Press, Cambridge, MA, 1984.

[49] Rastegar, J. and Deravi, P., Methods to determine workspace, its subspaces with different numbers of configurations and all the possible configurations of a manipulator, *Mech. Mach. Theory*, 22(4), 343–350, 1987.

[50] Rastegar, J. and Deravi, P., The effect of joint motion constraints on the workspace and number of configurations of manipulators, *Mech. Mach. Theory*, 22(5), 401–409, 1987.

[51] Roth, B., Rastegar, J., and Scheinman, V., On the design of computer controlled manipulators, First CISM-IFTMM Symp. on Theor. and Practice of Robots and Manipulators, 93–113, 1973.

[52] Sciavicco, L. and Siciliano, B., *Modeling and Control of Robot Manipulators*, McGraw-Hill, New York, 1996.

[53] Selig, J.M., *Geometrical Methods in Robotics*, Springer, New York, 1996.

[54] Sen, D. and Mruthyunjaya, T.S., A discrete state perspective of manipulator workspaces, *Mech. Mach. Theor.*, 29(4), 591–605, 1994.

[55] Smith, R.C. and Cheeseman, P., On the representation and estimation of spatial uncertainty, *Int. J. of Robotics Res.*, 5(4), 56–68, 1986.

[56] Spong, M.W. and Vidyasagar, M., *Robot Dynamics and Control*, John Wiley & Sons, New York, 1989.

[57] Tsai, L.-W., *Robot Analysis: The Mechanics of Serial and Parallel Manipulators*, John Wiley & Sons, New York, 1999.

[58] Yoshikawa, T., Manipulability of robotic mechanisms, *Int. J. of Robotics Res.*, 4(2), 3–9, 1985. (Also see *Foundations of Robotics: Analysis and Control*, MIT Press, Cambridge, MA, 1990.)

Chapter 13

Image Analysis and Tomography

In this chapter we formulate two kinds of problems in the language of noncommutative harmonic analysis: the template matching problem from two-dimensional pattern recognition, and the forward and inverse tomography problems. Both can be formulated as convolution equations on the group of rigid-body motions. In the former, the fast evaluation of $SE(2)$-convolutions is required. In the latter, the accurate inversion of convolution equations with a particular kind of kernel is sought. In Section 13.1 we address template matching. In Section 13.2 we address the forward tomography (imaging) problem. In Section 13.3 we address the inverse tomography (radiotherapy planning) problem.

13.1 Image Analysis: Template Matching

For a given template object we want to find if this template object is present in a given image, and, if it is found, determine its position and orientation. We use a correlation method (see [32] and references therein) for this purpose, which is extended to include rotations and dilations of the template object in addition to translations.[1] Essentially, we translate, rotate, and dilate the template object, overlap it with the image, and compute an overlap area (weighted by the intensity value at each pixel) with the proper normalization.

The correlation method is implemented using the Fourier transform on the discrete motion group. Fourier methods on the discrete motion group also provide a fast method to distinguish "identical" images (up to possible translations and rotations of the image) from "different" ones.

The discrete motion group (see Chapter 11) can be viewed as the set of matrices of the form

$$g = \begin{pmatrix} A_i & \mathbf{a} \\ \mathbf{0}^T & 1 \end{pmatrix}, \tag{13.1}$$

where

$$A_i = \begin{pmatrix} \cos 2\pi i/N_R & -\sin 2\pi i/N_R \\ \sin 2\pi i/N_R & \cos 2\pi i/N_R \end{pmatrix}, \tag{13.2}$$

for fixed natural number N_R and $i \in [0, N_R - 1]$ [A_i is replaced by an arbitrary element of $SO(2)$ for the continuous motion group $SE(2)$]. The group law is simply matrix multiplication.

---

[1] The material in this section was first presented in [38].

The problem of template matching is quite old, and has been approached in a number of different ways. Perhaps the most common (and oldest) approach is that of "matched filters" [68]. In this approach the Fourier transform of the image and template are taken, they are multiplied, and a peak is sought. This method can be implemented via digital computer or analog optical computation [36]. The drawback of this standard approach is that rotations are handled in a very awkward manner. Several works have considered rotation-invariant approaches (e.g., [3]). In such approaches, polar coordinates are used and images are expanded in series of Zernike polynomials (e.g., [6]) or the Hankel transform is used. The problem with such approaches is that rotational invariance is usually gained at the expense of the translational invariance offered by the usual (Abelian) Fourier transform.

A number of works have discussed considered using invariants of images for recognition (e.g., [1]). When one begins discussing invariants, the most natural analytical tool is group theory. In this chapter we apply group theory (in particular noncommutative harmonic analysis) to the template matching problem. In particular, if we are given a function $f(\mathbf{x})$, the Euclidean-group Fourier transform from Chapters 10 and 11 is a matrix function which has the property

$$\mathcal{F}\left(f\left(A^T(\mathbf{x} - \mathbf{a})\right)\right) = \mathcal{F}(f(\mathbf{x}))U(A, \mathbf{a})$$

where U is a unitary representation matrix that depends on rotation A and translation $\mathbf{a}$, and $\mathcal{F}$ denotes the non-Abelian Fourier transform. The preceding expression cannot be written as a matrix product for the usual Abelian Fourier transform for $A \neq \mathbb{I}_{2 \times 2}$, though it is completely analogous to the behavior of the Abelian Fourier transform applied to translated functions. In other words, noncommutative harmonic analysis provides a tool for both translation *and* rotation invariant pattern matching. Furthermore, since U is unitary $\|U\mathcal{F}(f)\|_2 = \|\mathcal{F}(f)\|_2$, and so this generalized Fourier transform provides a tool for generating a whole continuum of pattern invariants under rigid-body motion.

The connection between group theory and the theory of wavelets (which has become a very popular tool in image analysis) has been well established. In essence, expanding a function in a wavelet basis is achieved by starting with a mother wavelet and superposing affine-transformed versions of the mother wavelet to best approximate a given function. The interested reader is pointed to [2, 42, 49, 63] for further reading on the subject of wavelets, their applications in image analysis, and their connection with group theory.

The approach presented in this chapter is to use the non-Abelian Fourier transform and generalized concepts of convolution and correlation. This is very different than wavelet approaches, which have become very popular in the image analysis context in recent years. While wavelets typically allow one to efficiently approximate functions (or images), they have the drawback of not behaving well under operations such as convolution, which is the most natural tool in matched filtering.

In Section 13.1.1 we describe the correlation method. Section 13.1.2 shows why Fourier analysis on the discrete motion group is a useful tool in this context. In Section 13.1.3 we describe the numerical implementation of the correlation method using the Fourier transform on the discrete motion group.

13.1.1   Method for Pattern Recognition

In this section we extend the correlation method for pattern recognition [32] to include rotations and dilations (in addition to translations) as the allowed transformations of the image. To find if the template object is present in the image we take a section from the image and compare it with a rotated, translated, and dilated version of the template pattern. Taking a section from the image is equivalent to multiplication of the image by a "window" function, which is rotated, translated, and dilated the same way as the template pattern. Mathematically, the correlation function is written as

$$q(\mathbf{a}, A, k) =$$

$$\frac{\int_{\mathbb{R}^2} f_1(\mathbf{x}) f_2(A^{-1}(k\,\mathbf{x} - \mathbf{a}))\,d^2x}{[\int_{\mathbb{R}^2} (f_1(\mathbf{x}))^2\,(W(A^{-1}(k\,\mathbf{x} - \mathbf{a})))^2\,d^2x]^{1/2}\,[\int_{\mathbb{R}^2} (f_2((A^{-1}(k\,\mathbf{x} - \mathbf{a})))^2\,d^2x]^{1/2}} \tag{13.3}$$

where $A \in SO(2)$, $\mathbf{a} \in \mathbb{R}^2$, $k \in \mathbb{R}^+$ is close to one and $W(\mathbf{x})$ is a window function. $f_1(\mathbf{x})$ is the image function and $f_2(\mathbf{x})$ is the template function. Often f_1 and f_2 are normalized and zeroed in advance, in which case $-1 \leq q \leq 1$. For similar template pattern and windowed image the value of the correlation coefficient should be close to one. We note that for $k = 1$ the integral

$$\int_{\mathbb{R}^2} \left(f_2 \left(A^{-1}(k\,\mathbf{x} - \mathbf{a}) \right) \right)^2 d^2x \tag{13.4}$$

is just the square of the norm of function f_2

$$\int_{\mathbb{R}^2} (f_2(\mathbf{x}))^2\,d^2x.$$

According to the Cauchy-Schwarz inequality,

$$\int_{\mathbb{R}^2} f_1(\mathbf{x})\,f_2(\mathbf{x})\,d^2x \leq \left[\int_{\mathbb{R}^2} (f_1(\mathbf{x}))^2\,d^2x \int_{\mathbb{R}^2} (f_2(\mathbf{x}))^2\,d^2x \right]^{1/2},$$

the correlation coefficient in Eq. (13.3) is always smaller than or equal to one, and it is equal to one for identical pattern and windowed image. We note that the value of correlation coefficient does not change if we change the overall intensity of the original image or template object.

For the dilation coefficient $k = 1$ we observe that the correlation function $q = q(\mathbf{a}, A)$ is a function on the Euclidean motion group $SE(2)$. It appears that this group has not been used extensively in applications to the image processing; the authors are aware of only a few previous works using this group, (e.g., [26, 35, 43]).

Using Fourier methods on the motion group, we can compute the correlation coefficient in a much more efficient way than using direct integration. Indeed, the direct computation of integral Eq. (13.3) is very costly (we consider for simplicity the $k = 1$ case). For $N_r = N_x \cdot N_y$ samples of the image (and template) on an $N_x \times N_y$ rectangular grid, and for N_R samples of orientation, we need to perform $O(N_r^2 N_R)$ computations [and we need to compute the convolution-like integrals twice, in the denominator and numerator of Eq. (13.3)]. For $N_r = 256 \times 256$ and $N_R = 60$, the computations require $5 \cdot 10^{11}$ operations, which requires a day of computer work on a 250 MHZ workstation. In this section we use advantages of Fourier methods on the discrete motion group [i.e., subgroup of $SE(2)$, where the orientation angle has discrete values from the C_{N_R} subgroup of $SO(2)$, $\theta = 2\pi\,i/N_R$ for $i = 0, \ldots, N_R - 1$], and Fast Fourier Transform (FFT) methods [18, 25] to compute the correlation coefficient in $O(N_R N_r \log N_r)$ computations. In addition, Fourier methods on the discrete motion group provide a very fast method for comparison of two images which are translated and rotated relative to each other.

In the next section we briefly discuss applications of Fourier methods on the discrete motion group in the context of the template matching problem.

13.1.2 Application of the FFT on $SE(2)$ to the Correlation Method

The convolution-like integrals in the numerator and denominator of Eq. (13.3) may be formally written (for simplicity we consider first the $k = 1$ case; the case including dilations is considered in

Section 13.1.3) as integrals

$$c\left(\mathbf{x}, A_j\right) = \frac{1}{N_R} \sum_{i=0}^{N_R-1} \int_{\mathbb{R}^2} f_1\left(\mathbf{y}, A_i\right) f_2\left(A_j^{-1}\left(\mathbf{y}-\mathbf{x}\right), A_j^{-1} A_i\right) d^2y \tag{13.5}$$

where the functions $f_{1,2}$ are orientation-independent, i.e., $f_{1,2}(\mathbf{x}, A) = f_{1,2}(\mathbf{x})$. The correlation function, however, is a function on the discrete motion group G_{N_R}, so we may use the Fourier transform on G_{N_R} to write this integral as a product of Fourier transforms.

Because the functions are real the integral in the numerator of Eq. (13.3) may be written as

$$\frac{1}{N_R} \sum_{i=0}^{N_R-1} \int_{\mathbb{R}^2} \overline{f_1\left(\mathbf{y}, A_i\right)} f_2\left(A_j^{-1}\left(\mathbf{y}-\mathbf{x}\right), A_j^{-1} A_i\right) d^2y$$

$$= \int_{G_{N_R}} \overline{f_1(h)}\, f_2\left(g^{-1} \circ h\right) d(h) \tag{13.6}$$

where we denote integration over the discrete motion group to mean integration over $\mathbb{R}^2$ and summation through the A_i, and the group elements are of the form $g = (\mathbf{x}, A_j)$. Note that Eq. (13.6) is not a convolution of f_1 and f_2. Rather, it is a correlation.

Using the orthogonality and homomorphism properties of the Fourier matrix elements, Eq. (13.6) may be written as

$$\frac{1}{N_R} \sum_{q=0}^{N_R-1} \sum_{n=0}^{N_R-1} \int_0^\infty \int_0^{2\pi/N_R} \sum_m \left(\overline{\left(\hat{f}_1\right)_{mn}} \left(\hat{f}_2\right)_{mq}\right) U_{qn}\left(g^{-1}; p, \phi\right) p\, dp\, d\phi \tag{13.7}$$

$$= \frac{1}{N_R} \sum_{q=0}^{N_R-1} \sum_{n=0}^{N_R-1} \int_0^\infty \int_0^{2\pi/N_R} \sum_m \left(\overline{\left(\hat{f}_2\right)_{mq}} \left(\hat{f}_1\right)_{mn}\right) U_{nq}(g; p, \phi) p\, dp\, d\phi$$

where ϕ is measured from $2\pi q/N_R$. For the second expression we used the unitarity of the matrix elements U_{mn} and the fact that the expression is real. The matrices $(\hat{f}_{1,2})_{mn}$ are the G_{N_R}-Fourier transforms of the functions $f_{1,2}(\mathbf{x}, A_i)$. We note that this integral is the inverse Fourier transform of $\hat{f}_2^\dagger \cdot \hat{f}_1$, and thus the expression depends only on three indices.

Because functions $f_{1,2}(\mathbf{x}, A_i) = f_{1,2}(\mathbf{x})$ do not depend on the orientations A_i, matrix elements in the same column are the same, i.e.,

$$\left(\hat{f}_{1,2}\right)_{mn} = \left(\hat{f}_{1,2}\right)_{qn}$$

for any $m, q \in [0, N_R - 1]$. This may be observed from the expression

$$U_{mn}\left(g^{-1}; p, \phi\right) = e^{-i p \mathbf{u}_n^\phi \cdot \mathbf{a}} \delta_{A_i^{-1} \mathbf{u}_n, \mathbf{u}_m} \tag{13.8}$$

where $g = (\mathbf{a}, A_i) \in G_{N_R}$ [note that the exponent in Eq. (13.8) depends only on the n-index], the definition of the forward G_{N_R}-Fourier transform, and the fact that the functions do not depend on the orientation. Thus we compute a row of the Fourier matrix for a particular orientation (for example $A_0 = \mathbb{I}_{3\times3}$)

$$\left(\hat{f}_{1,2}\right)_n \triangleq \left(\hat{f}_{1,2}\right)_{nn}. \tag{13.9}$$

This can be done using the 2D FFT for the functions $f_{1,2}(\mathbf{x})$ and interpolating the Fourier values to points on a polar coordinate grid. The value of p is determined by $|\mathbf{p}|$ and the values of m and ϕ are determined by the angular part of $\mathbf{p}$. This requires $O(N_r \log(N_r))$ computations. Thus the integrals in Eq. (13.3) may be written as

$$c(\mathbf{x}, A_j) = C \sum_{q=0}^{N_R-1} \sum_{n=0}^{N_R-1} \int_0^\infty \int_0^{2\pi/N_R} \left(\left(\hat{f}_2 \right)_q (p, \phi) \left(\hat{f}_1 \right)_n (p, \phi) \right) \times$$
$$U_{nq}(g; p, \phi) \, p \, dp \, d\phi \tag{13.10}$$

where $C = \frac{1}{N_R}$.

We observe that the convolution-like integrals may be computed by taking the Fourier transform, computing the product of transforms, and taking the inverse Fourier transform on the discrete motion group.

Invariants of the Discrete Motion Group

Let us assume that it is desirable to compute properties of the image (object) which are invariant with respect to translations and rotations of the image. The Fourier transform on the discrete motion group provides a very efficient tool to compute these invariants. Let us construct a function with values in $\mathbb{R}^+$

$$\eta(p; \phi) = \sum_{m=0}^{N_R-1} \left[\overline{\hat{f}_m(p; \phi)} \, \hat{f}_m(p; \phi) \right] \tag{13.11}$$

for each fixed $\phi = 0, \ldots, N_\phi - 1$, where $\hat{f}_m(p; \phi)$ is the Fourier transform on the discrete motion group of $f(\mathbf{x})$. Then Eq. (13.11) is invariant with respect to rotations and translations of $f(\mathbf{x})$, i.e., $\eta(p; \phi)$ does not change if we compute Eq. (13.11) using the Fourier transform for the discrete motion group for $f'(\mathbf{x}) = f(A^{-1}(\mathbf{x} - \mathbf{a}))$.

We note that for orientation-independent functions (i.e., for functions on $\mathbb{R}^2$) the Fourier transform elements $\hat{f}_m$ can be arranged as a matrix which has the same matrix elements in the same column

$$\hat{f}_{qm} = \hat{f}_{rm} = \hat{f}_m.$$

Then Eq. (13.11) may be written also as a trace

$$\eta(p; \phi) = \text{Tr} \left[\hat{f}^\dagger(p; \phi) \, \hat{f}(p; \phi) \right] \tag{13.12}$$

where $\hat{f}^\dagger(p; \phi)$ is the Hermitian conjugate matrix.

Using the general definition of a group Fourier transform given in Chapter 8, $\hat{f}_{qm}$ for $f(\mathbf{x})$ may be written as

$$\hat{f}_{qm}(p; \phi) = \int_{G_{N_R}} f(h) \, U_{qm}^{-1}(h; p, \phi) \, d(h)$$

where the integral over G_{N_R} denotes integration with respect to $\mathbf{x}$ *and* summation through the elements of C_{N_R}, and $f(h) = f(\mathbf{x})$. The function $f'(\mathbf{x}) = f(A_i^{-1}(\mathbf{x} - \mathbf{a}))$ may be formally written as $f(g^{-1} \circ h)$, where $g = (\mathbf{a}, A_i) \in G_{N_R}$. Then $\hat{f}'_{qm}$ is written as

$$\hat{f}'_{qm}(p; \phi) = \int_{G_{N_R}} f\left(g^{-1} \circ h \right) U_{qm}^{-1}(h; p, \phi) \, d(h).$$

Using the invariance of the integration measure we can write this integral as

$$\hat{f}'_{qm}(p;\phi) = \int_{G_{N_R}} f\left(h'\right) U_{qm}^{-1}\left(g \circ h'; p, \phi\right) d\left(h'\right).$$

Using the homomorphism properties of U we may write it as

$$\hat{f}'_{qm}(p;\phi) = \sum_{r=0}^{N_R-1} \left[\int_{G_{N_R}} f\left(h'\right) U_{qr}^{-1}\left(h'; p, \phi\right) d\left(h'\right) \right] \cdot U_{rm}^{-1}(g; p, \phi)$$

$$= \sum_{r=0}^{N_R-1} \hat{f}_{qr}(p,\phi) U_{rm}^{\dagger}(g; p, \phi)$$

where we have used a unitarity property of U. Thus the Fourier matrix is transformed under rotations and translations $g \in G_{N_R}$ as

$$\hat{f}'(p,\phi) = \hat{f}(p,\phi) U^{\dagger}(g; p, \phi).$$

Using the cyclic property of Tr and unitarity of U it is clear that

$$\text{Tr}\left[\left(\hat{f}'\right)^{\dagger}(p,\phi)\hat{f}'(p,\phi) \right] = \text{Tr}\left[U(g; p, \phi) \hat{f}^{\dagger}(p,\phi) \hat{f}(p,\phi) U^{\dagger}(g; p, \phi) \right]$$

$$= \text{Tr}\left[\hat{f}^{\dagger}(p,\phi) \hat{f}(p,\phi) \right],$$

which proves the invariance of Eq. (13.11). We note that the invariant, written in the form Eq. (13.12) is valid also for orientation-dependent functions (i.e., for general functions on the discrete motion group). The use of invariants for pattern recognition was suggested in [26].

Efficiency of Computation of Convolution-like Integrals

As we mentioned before, the direct integration of Eq. (13.3) requires $O(N_r^2 N_R)$ computations for C_{N_R}, where N_r is the number of sampling points in an $\mathbb{R}^2$ region. Using the Fourier transform on the discrete motion group we have to compute direct Fourier transforms for image and template, compute the matrix product (in our case it is a column-row product) of the Fourier transform, which describes the Fourier transform of the convolved functions, and then calculate the inverse Fourier transform.

The calculation of direct Fourier transform and the "matrix" (column — row) product is a fast computation. The direct Fourier transform for $f_{1,2}(\mathbf{x})$ may be computed using a usual two-dimensional FFT [25] in $O(N_r \log N_r)$ computations. The FFT gives, however, values of Fourier elements computed on a Cartesian square (rectangular) grid of $\mathbf{p}$ values. To find the Fourier transform elements $\hat{f}_m(p, \phi)$ on the discrete motion group we have to interpolate values on the Cartesian grid to a polar coordinate grid; the p value is the magnitude of $\mathbf{p}$ and the m and ϕ indices are determined by the angular part of $\mathbf{p}$ (thus the constraint $N_p N_\phi N_R \approx N_r$ may be used). The linear interpolation requires $O(N_r)$ computations. The product of Fourier column $\hat{f}_m^T(p, \phi)$ and row $\hat{f}_n(p, \phi)$, which defines the matrix $\hat{F}_{mn}(p, \phi)$, can be performed in $O(N_R^2 N_p N_\phi) = O(N_R N_r)$ computations. We note that the trace in invariants in Eq. (13.11) can be computed in $O(N_R N_p N_\phi) = O(N_r)$ (for all ϕ-values). Thus the direct Fourier transform and the "matrix" product can be computed in $O(N_r \log N_r) + O(N_R N_r)$ computations for linear or spline interpolation.

The inverse Fourier transform calculation is a slower computation. One element from each row and column of $\hat{F}_{mn}(p, \phi)$ is used in computation of the inverse Fourier transform for each rotation element A_i. First, we interpolate the value of the Fourier transform on the square grid $N_r \times N_r$ of **p** to polar coordinates. The radial coordinate is $p = |\mathbf{p}|$; the polar angle is determined by the values of m and ϕ (the value of n is determined by m and the index of rotation i, $n = m + i$, thus we take $\hat{F}_{m,m+i}$ elements from the Fourier matrix to compute the inverse transform for fixed orientation A_i). After inverse interpolation to Cartesian coordinates [which can be done in $O(N_R N_r)$ computations], the inverse Fourier integration can be performed in $O(N_r \log(N_r))$ for each of the N nonzero matrix elements of U using the FFT. Thus in $O(N_R N_r \log(N_r))$ computations we reproduce the function for all A_i. We note that the inverse Fourier transform computation is $O(N_R)$ [or $O(\log N_r)$, depending on which is larger] times more time-consuming because we reproduce a function on the discrete motion group, rather than a function on $\mathbb{R}^2$.

Thus the total required is $O(N_R N_r \log(N_r))$ computations and these computations are, for the most part, calculations of the inverse Fourier transform. We also note that we have to perform calculations twice to compute convolution-like integrals in denominator and numerator of Eq. (13.3).

For a comparison with standard methods, see [38].

13.1.3   Numerical Examples

Correlation Method Including Rotations and Translations

In this section we compute the correlation function in Eq. (13.3) (for dilation $k = 1$) for some practical examples. We compute most examples for $N_r = 256 \times 256$ and $N_R = 60$ (C_{60} group), although the computing time for other arrays is also reported.

We consider the image depicted in Fig. 13.1. This is a 256×256 array of gray values (256 gray levels of intensity for each pixel). We note that because our implementation of the direct and inverse Fourier transforms on the motion group uses spline interpolations, the image after application of direct and inverse Fourier transforms on the discrete motion group is not reproduced with the same quality.

We choose a template, depicted in Fig. 13.2, which is a rotated (by angle $\theta = -\pi/3$) and translated pattern taken from the image. The arrow shown on the picture is used as a reference arrow to find the position and orientation of this template in the image. The correlation function depicted for the $\theta = \pi/3$ angle is depicted in Fig. 13.3. The highest value of the correlation function is at the original position and orientation of the pattern in the image. We also find positions and orientations of local maxima in each of $m \times m$ subregions of the original image. For $m = 8$ the positions and orientations of local maxima in each of the subregions are shown in Fig. 13.4. The highest value is depicted by the arrow, which is rotated and translated from the arrow in Fig. 13.2. Other local maxima (with a value of correlation which is greater than 0.85) are depicted by a white square; a small line attached to the square shows orientation.

We note that the precise values of correlation at the locations of 64 maxima can be found by direct integration, and the Fourier method may be used as a fast filter method to find locations of these maxima. It is especially important to compute precise values in the case when the template object does not match exactly the pattern in the image. For example, for a template extracted from a filtered version of Fig. 13.1 (which does not match exactly the corresponding pattern in Fig. 13.1), the position of the absolute maximum found by the Fourier method (0.97) is located at the "wrong" position, although the local maxima with the close value (0.91) of correlation coefficient is located at the precise position. Computation of precise values of correlation coefficients by direct integration at the locations of maxima finds the precise location of the pattern with correlation 0.954 (identical template and pattern in the image correlation was 0.999).

The approximate value of correlation coefficient, which distinguishes the case when the template object is "found" from the case when it is "not-found" in the image, depends on the desired level of accuracy and general properties of the image (such as if it contains a background or it is a "black

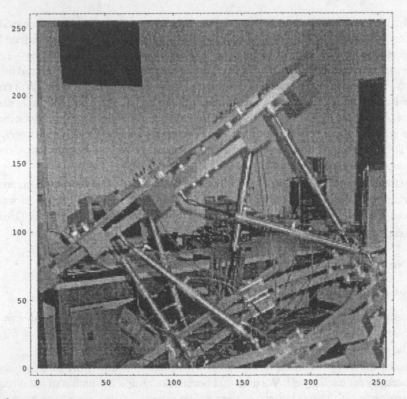

FIGURE 13.1

The image — 256 × 256 array of gray values. (Kyatkin, A.B. and Chirikjian, G.S., Pattern matching as a correlation on the discrete motion group, *Comput. Vision and Image Understanding*, 74(1), 22–35, 1999. © 1999 Academic Press.)

and white" contrast image). For images containing almost uniform background, like in the example discussed previously, the level of "threshold" correlation must be high and may be set around 0.95.

The accuracy of the method in our examples was around 0.5 pixel. For example, for a rotation of $\pi/7$ (4.29 rotational pixels) and translations of 41.2 and 68.8 pixels of the template relative to the similar pattern in the image (they do not match exactly, i.e., the template taken from a filtered version of Fig. 13.1), the found object was at the location 4 rotational pixels ($2\pi/15$) and translations of 41 and 69 pixels.

We note that the resolution of the Fourier method can be increased in some cases if we "precompute" the image to increase contrast. For images containing almost uniform background, application of the Laplacian operator (in Fourier space this is just a multiplication of Fourier transform by p^2) smoothed by a Gaussian weight function

$$\mathcal{F}(p) \to \text{const } p^2 \exp\left(-p^2/\left(2\sigma^2\right)\right)\mathcal{F}(p) \qquad (13.13)$$

(like in edge detection problems as in [12], but we do not compute "zero crossing" to get actual edges since we desire the overlap of the template and the image to be large), may increase the resolution. For example, the template shown in Fig. 13.5 produces a local maximum at the location of a similar (they do not match exactly) pattern in Fig. 13.1. This still gives an absolute maximum at the "right" location when we compute precise values of correlation using direct integration at the locations of local maxima. Application of the Laplacian operator gives the image shown in Fig. 13.6 for

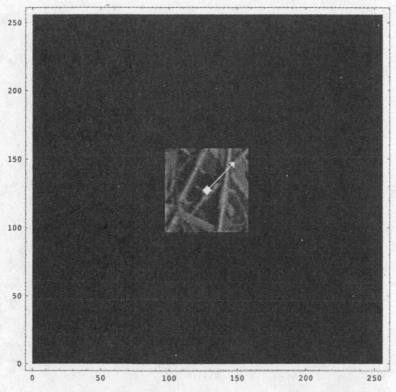

FIGURE 13.2
The template pattern: the arrow is used as a reference to find position and orientation of this
pattern in the image. (Kyatkin, A.B. and Chirikjian, G.S., Pattern matching as a correlation
on the discrete motion group, *Comput. Vision and Image Understanding,* 74(1), 22–35, 1999.
© 1999 Academic Press.)

$\sigma^2 = p_{max}^2/12$ where $p_{max} = N_x/2 = N_y/2$ is the maximal p value; the *const* in Eq. (13.13) was
set to normalize the gray values of the image to the interval [0, 1]. When we use the Fourier method
to find the position of the template (we also apply the Laplacian operator to the template), it produces
an absolute maximum with correlation 0.71. The next closest local maximum has the value 0.63.
Direct integration performed at the location of maxima gives the precise value of correlation 0.81.
The next closest local maxima has a value 0.61. For the "black and white" contrast pictures the
"threshold" correlation may be set around 0.8.

In the following table we listed the computing time of the method (given in minutes and seconds,
on a 250 MHz SGI workstation), implemented in the C programming language. N_R is listed along
the horizontal; the right column lists the time to compute correlation coefficients at 64 maxima using
direct integration. The N_r array size is given along the vertical.

	$N_R = 60$	$N_R = 30$	$N_R = 10$	Dir. int.
$N_r = 256 \times 256$	4:06	2:11	0:48	0:55
$N_r = 128 \times 128$	1:12	0:36	0:16	0:13
$N_r = 64 \times 64$	0:25	0:12	0:04	0:03

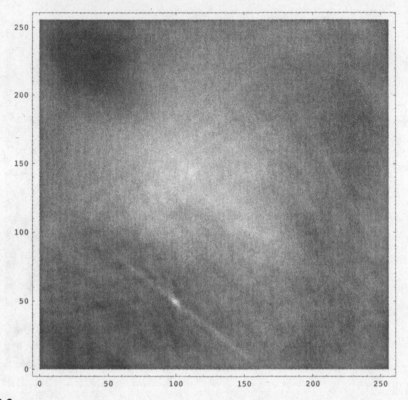

FIGURE 13.3
The correlation function depicted for the $\theta = \pi/3$ orientation angle. (Kyatkin, A.B. and Chirikjian, G.S., Pattern matching as a correlation on the discrete motion group, *Comput. Vision and Image Understanding*, 74(1), 22–35, 1999. © 1999 Academic Press.)

Using the Invariants on the Motion Group to Compare Images

As we have shown before, function in Eq. (13.11) is the same for images which are rotated and translated relative to each other. It may be used to compare images and determine if they are identical (similar) or not. Again, given image functions $f_{1,2}(\mathbf{x})$, we may compute the correlation coefficient of invariant functions $\eta_{1,2}(p, \phi)$ defined in Eq. (13.11). Then these can be compared as

$$\eta(\phi) = \frac{\int_0^\infty \eta_1(p, \phi)\,\eta_2(p, \phi)\,p\,dp}{\left(\int_0^\infty \eta_1(p, \phi)^2\,p\,dp\right)^{1/2} \left(\int_0^\infty \eta_2(p, \phi)^2\,p\,dp\right)^{1/2}}.$$

This is a fast computation of the order $O(N_p\,N_\phi) \approx O(N_r/N_R)$ (to compute N_ϕ coefficients) and it may be done using usual integration techniques. As we mentioned before, the direct Fourier transform may be computed in $O(N_r \log(N_r))$ computation; computation of the sum in Eq. (13.11) may be done in $O(N_r)$ computations.

We compare the images depicted in Figs. 13.7–13.8, which are just rotated and translated relative to each other. We use the value $\nu = (1.0 - \eta) \cdot 10^3$ to compare images, which is more convenient to use for η values which are close to 1.0. The greater ν is, the worse the correlation is. In the following table we show ν for $\phi = 0, \ldots, 5$.

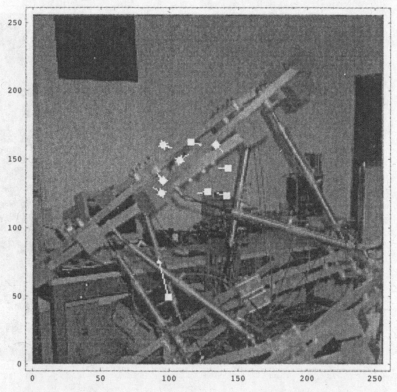

FIGURE 13.4

The image: positions and orientations of absolute maximum (shown by arrow) and local maxima of correlation function are depicted. (Kyatkin, A.B. and Chirikjian, G.S., Pattern matching as a correlation on the discrete motion group, *Comput. Vision and Image Understanding,* **74(1), 22–35, 1999. © 1999 Academic Press.)**

ϕ	0	1	2	3	4	5
ν	0.016	0.017	0.017	0.017	0.016	0.016

We see that values are very close for different ϕ, thus we may use ν for any one of the ϕ values. The values are small which indicates a very strong correlation.

The exact level to separate "identical" from "different" images depends on the array size and desired level of accuracy, and it must be set "by hand." For images of type depicted in Figs. 13.7–13.8 the value $\nu \approx 0.5$ separates "identical" from "different" images. We note that when we compare images of different size, the ν value becomes large if we compare Figs. 13.1 and 13.7 $\nu = 520$.

The time to compute direct Fourier transforms and correlations ν was around 3 sec on a 250 MHZ workstation.

Correlation Method Including Dilations

It may be shown that the convolution-like integral which includes dilations

$$c\left(\mathbf{x}, A_j, k\right) = \int_{\mathbb{R}^2} f_1(\mathbf{y}) \, f_2 \left(A_j^{-1} \left(k\mathbf{y} - \mathbf{x}\right)\right) \, d^2y$$

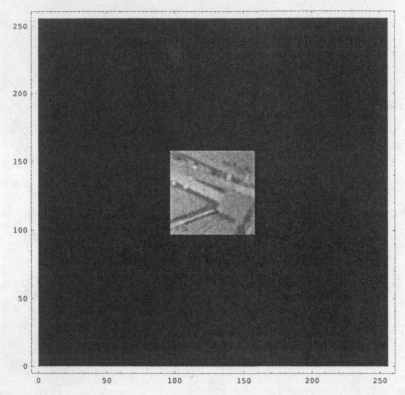

FIGURE 13.5
The template object. (Kyatkin, A.B. and Chirikjian, G.S., Pattern matching as a correlation on the discrete motion group, *Comput. Vision and Image Understanding,* **74(1), 22–35, 1999. © 1999 Academic Press.)**

may be written, analogous to the expression in Eq. (13.10), as

$$c\left(\mathbf{x}, A_j, k\right) = \frac{1}{N_R} \sum_{q=0}^{N_R-1} \sum_{n=0}^{N_R-1} \int_0^\infty \int_0^{2\pi/N_R} \left(\overline{\left(\hat{f}_2\right)_q (p, \phi)} \left(\hat{f}_1\right)_n (kp, \phi) \right) \times$$
$$U_{nq}(g; p, \phi)\, p dp\, d\phi. \tag{13.14}$$

The derivation of Eq. (13.14) is analogous to the derivation of Eq. (13.10). We also have to use the property of matrix elements of U

$$U(k\,\mathbf{r}, A; p, \phi) = U(\mathbf{r}, A; k\,p, \phi)$$

which may be easily observed from the expression in Eq. (13.8).
We also note that the integral in Eq. (13.4) is equal to

$$\frac{1}{k^2} \int_{\mathbb{R}^2} (f_2(\mathbf{x}))^2 \ d^2 x.$$

Thus to compute the correlation function in the case when dilations are allowed, we have to compute the direct Fourier transforms for $f_{1,2}(\mathbf{x})$ [which can be done in $O(N_r \log N_r)$ operations] to multiply the Fourier transforms for each k from the discrete set of values N_k [which can be done

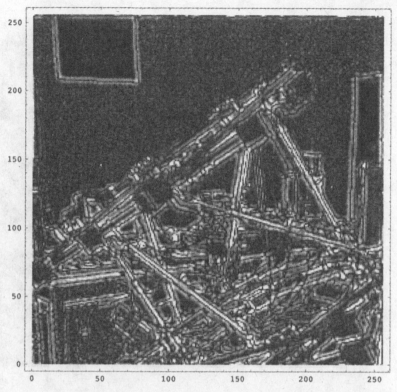

FIGURE 13.6
The image of Fig. 13.1 after application of "smoothed" Laplacian operator. (Kyatkin, A.B. and Chirikjian, G.S., Pattern matching as a correlation on the discrete motion group, *Comput. Vision and Image Understanding*, 74(1), 22–35, 1999. © 1999 Academic Press.)

in $O(N_k N_R N_r)$ operations] and to take the inverse Fourier transform on the discrete motion group for each k [which can be done in $O(N_k N_R N_r \log N_r)$ operations]. In total, the computations are approximately N_k times more time-consuming.

We note that we also have to increase the number of subregions where we look for local maxima of correlation functions (because the number of local maxima is increased).

In Fig. 13.9 we depict the template which is enlarged in size by the factor of 1.2 and compare with the similar pattern in the image (they are not identical). The scaling may be observed by comparing the pictures of Figs. 13.2 and 13.9. We computed the correlation function on a 256 × 256 × 60 array for possible scaling (reduction) by factor 1.0, 1.1, 1.2 (computing time is around 12 minutes). Local maxima were found in a 12 × 12 grid of subregions of the image. The template object produces local maxima at the locations with similar pattern in the image (with correct scaling factor). We compute the correlation coefficient at the locations of local maxima. The "right" location and scaling factor gives the highest correlation, as depicted in Fig. 13.10 (for correlation greater than 0.9 it is a single maximum).

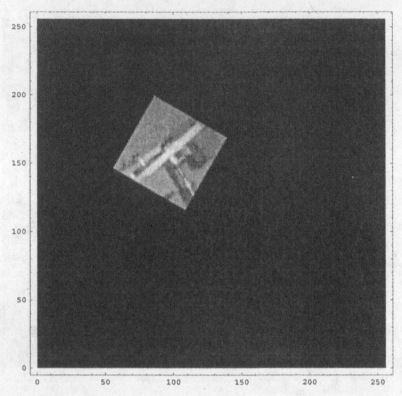

FIGURE 13.7
A window of the original image. (Kyatkin, A.B. and Chirikjian, G.S., Pattern matching as a correlation on the discrete motion group, *Comput. Vision and Image Understanding,* **74(1), 22–35, 1999. © 1999 Academic Press.)**

13.2   The Radon Transform and Tomography

The Radon transform and its generalizations play an important role in fields as diverse as medical imaging [4, 23, 34, 64], electron microscopy, and radio astronomy [9]. The reconstruction of images from data collected in the form of a Radon transform is an inverse problem which has received much attention over the past 25 years. New methods for the inversion of the Radon transform remain popular today (see, for example, [31, 37, 62]).

In this section, it is shown how the Radon transform can be viewed as a convolution on the Euclidean motion group (group of rigid-body motions) where the convolution product is defined relative to the group operation. While this has been observed before in abstract settings (see, for example, [27, 30, 55]), we go two steps further to show: (1) how the Euclidean-group Fourier transform is used to rewrite the Radon transform as a matrix product; and (2) how the regularized inversion of the Radon transform is achieved with efficient and accurate numerical codes based on this group-theoretical approach. The key to this approach is the use of "fast" Fourier transforms for motion groups.

We show how this approach handles the case of the usual Radon transform, as well as the case of finite-width beams. We note that only one other work, [20], was found which deals with this kind of generalization of the Radon transform. In contrast, a number of works have dealt with the

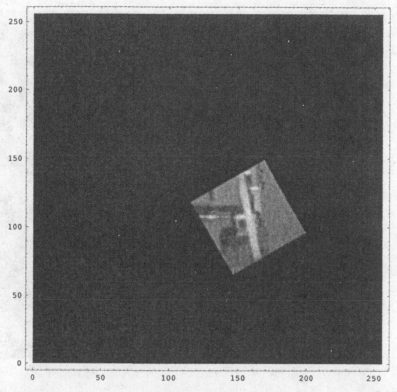

FIGURE 13.8
A rotated and translated version of the image in Fig. 13.7. (Kyatkin, A.B. and Chirikjian, G.S., Pattern matching as a correlation on the discrete motion group, *Comput. Vision and Image Understanding*, 74(1), 22–35, 1999. © 1999 Academic Press.)

exponential/attenuated Radon transforms in the context of single photon emission tomography (see, for example, [5, 15, 29, 48, 50, 67]) and their applications in radiation treatment planning [7] and inverse source problems [53]. While, in principle, exponential/attenuated Radon transforms can also be written as a matrix product upon application of the Euclidean group Fourier transform, we do not handle this case in this chapter for lack of an appropriate regularization technique.

In Subsection 13.2.1 the relationship between the Radon transform and motion-group convolutions is established. In Subsection 13.2.2 the derivation of the Radon transform for finite beam width is given. The application of Fourier transforms on motion groups to computed tomography algorithms is given in Subsection 13.2.3.

13.2.1 Radon Transforms as Motion-Group Convolutions

The Radon transform and its generalizations, such as the attenuated and exponential Radon transforms, play a central role in medical imaging and computed tomography (see, for example, [23] and references therein). This is because when an X-ray passes through an object it loses energy. Sensors on the other side of the object which measure the final energy in the rays effectively provide a means of determining the integral of the mass density of the object along the path of the X-ray. For recent works that address applications and techniques for inversion of the Radon transform see [10, 19, 24, 39].

In this subsection, we show how the Radon transform can be viewed as a convolution on the Euclidean motion group (group of rigid-body motions) where the convolution product is defined

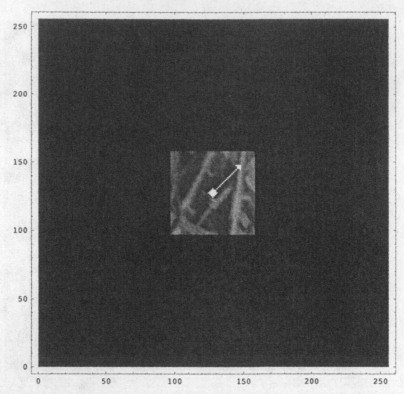

FIGURE 13.9
The template object enlarged by the factor 1.2 compared to the similar pattern in the image.

relative to the group operation. We then show how the Fourier transform of functions on the Euclidean motion group is used to rewrite the Radon transform as a matrix product. The regularized inversion of the Radon transform then reduces to matrix computations, followed by the inverse Fourier transform for the Euclidean motion group.

Given a function $f(\mathbf{x})$ on $\mathbb{R}^N$ which is integrable on each $(N-1)$-dimensional hyperplane in $\mathbb{R}^N$, the corresponding Radon transform is

$$\hat{f}(p, \boldsymbol{\xi}) = \mathcal{R}\{f(\mathbf{x})\} = \int_{\mathbb{R}^N} f(\mathbf{x})\delta(p - \boldsymbol{\xi} \cdot \mathbf{x})d\mathbf{x} \tag{13.15}$$

where $d\mathbf{x} = dx_1 \ldots dx_N$ is the usual integration measure for $\mathbb{R}^N$, and $\delta(\cdot)$ is the Dirac delta function on the real line. In principle, if $f(\mathbf{x})$ is differentiable, and the value of Radon transform is known for all of these hyperplanes (each hyperplane in $\mathbb{R}^N$ is determined by the equation $\boldsymbol{\xi} \cdot \mathbf{x} = p$ where $\boldsymbol{\xi}$ is its unit normal), then the original function can be reconstructed using the inverse Radon transform.

The Radon transform has certain properties under affine transformations of the form: $\mathbf{x} \to A\mathbf{x} + \mathbf{b} \overset{\triangle}{=} a \circ \mathbf{x}$ where $A \in GL(N, \mathbb{R})$ (i.e., A is an invertible $N \times N$ matrix). Like the Euclidean motion group, the group of affine transformations of $\mathbb{R}^N$ can be expressed using homogeneous transform notation. While $a \circ \mathbf{x}$ denotes the action of the affine transformation a on $\mathbf{x} \in \mathbb{R}^N$, the same affine transform applied to a function on $\mathbb{R}^N$ yields

$$f(\mathbf{x}) \to f\left(a^{-1} \circ \mathbf{x}\right) = f\left(A^{-1}(\mathbf{x} - \mathbf{b})\right).$$

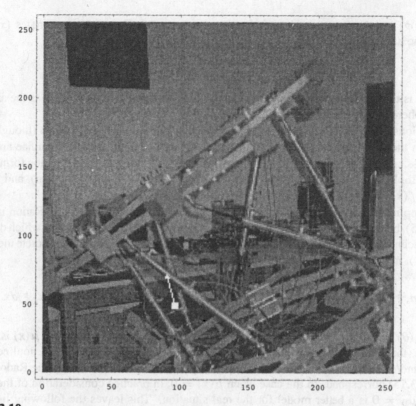

FIGURE 13.10
The absolute maximum found by direct integration at the location of local maxima.

Applying the Radon transform to an affinely transformed function, we see that

$$\mathcal{R}\left\{f\left(A^{-1}(\mathbf{x}-\mathbf{b})\right)\right\} = \int_{\mathbb{R}^N} f\left(A^{-1}(\mathbf{x}-\mathbf{b})\right)\delta(p-\boldsymbol{\xi}\cdot\mathbf{x})d\mathbf{x}.$$

Making the change of coordinates $\mathbf{y} = A^{-1}(\mathbf{x}-\mathbf{b})$, or equivalently $\mathbf{x} = A\mathbf{y} + \mathbf{b}$, we see that $d\mathbf{x} = |\det(A)|d\mathbf{y}$ and

$$\mathcal{R}\left\{f\left(A^{-1}(\mathbf{x}-\mathbf{b})\right)\right\} = |\det(A)|\int_{\mathbb{R}^N} f(\mathbf{y})\delta\left(p-\boldsymbol{\xi}\cdot(A\mathbf{y}+\mathbf{b})\right)d\mathbf{y}$$

$$= |\det(A)|\hat{f}\left(p-\boldsymbol{\xi}\cdot\mathbf{b}, A^T\boldsymbol{\xi}\right).$$

In the planar case when $\boldsymbol{\xi} = [\cos\theta, \sin\theta]^T$ we write explicitly

$$\hat{f}(p,\theta) = \hat{f}(p,\boldsymbol{\xi}(\theta)) = \int_{-\infty}^{\infty}\int_{-\infty}^{\infty} f(x_1, x_2)\delta(p - x_1\cos\theta - x_2\sin\theta)\,dx_1dx_2.$$

Johann Radon proved in 1917 that for the case when $N = 2$, the following inversion formula holds [57]

$$f(r,\phi) = \frac{1}{2\pi^2}\int_0^{\pi}\int_{-\infty}^{\infty}\frac{\partial\hat{f}}{\partial p}\frac{dp}{r\sin(\phi-\theta)-p}d\theta$$

where (r, ϕ) are the polar coordinates for the corresponding Cartesian coordinates (x_1, x_2). For $N = 3$ one has

$$f(\mathbf{x}) = -\frac{1}{8\pi^2}\nabla^2\left(\int_{S^2}\hat{f}(\boldsymbol{\xi}\cdot\mathbf{x}, \boldsymbol{\xi})ds(\boldsymbol{\xi})\right)$$

where S^2 is the unit sphere in three dimensions, ∇^2 is the Laplacian, and $ds(\boldsymbol{\xi})$ is the area element on this sphere (see, for example, [23, 27, 30]).

Extensions of the inversion formula to other dimensions are also well known (though, as can be seen from the above formulae, for even and odd dimensions the inversion formulae are different). Regardless, in many practical applications, one is unable to simply use this inversion formula because data is sampled discretely (i.e., only a finite number of measurements are taken), and the original function $f(\mathbf{x})$ may in fact not be continuous, much less differentiable.

To complicate matters further, in many practical applications, it is not the Radon transform in Eq. (13.15), but rather a *generalized, attenuated,* or *exponential* Radon transform which describes the physical scenario of interest [15, 48, 50, 67]. For example, instead of obtaining discrete measurements of $\hat{f}(p, \boldsymbol{\xi})$, in an experimental setting one may observe

$$\hat{f}_\mu(p, \boldsymbol{\xi}) = \int_{\mathbb{R}^2} f(\mathbf{x})\exp\left(-\int_{0\leq\mathbf{x}'\cdot\boldsymbol{\xi}_\perp\leq\mathbf{x}\cdot\boldsymbol{\xi}_\perp}\mu(\mathbf{x}')\delta(p-\mathbf{x}'\cdot\boldsymbol{\xi})d\mathbf{x}'\right)\delta(p-\mathbf{x}\cdot\boldsymbol{\xi})d\mathbf{x} \qquad (13.16)$$

where $\boldsymbol{\xi}_\perp(\theta) = (-\sin\theta, \cos\theta)$ satisfies $\boldsymbol{\xi}_\perp\cdot\boldsymbol{\xi} = 0$. In the general case when $\mu(\mathbf{x})$ is completely unknown, this problem cannot be solved. In the cases when $\mu(\cdot)$ is a known nonlinear function of $\rho(\mathbf{x})$ and/or $\mathbf{x}$, the inversion is still a hopelessly difficult problem. While the Radon transform in Eq. (13.15) corresponds to the case when $\mu(\mathbf{x}) = 0$, in practice consideration of the case when $\mu(\mathbf{x}) = \mu_0 \neq 0$ is a better model for the real situation. This leaves the following linear inverse problem to solve for $f(\mathbf{x})$:

$$\hat{f}_{\mu_0}(p, \boldsymbol{\xi}) = \int_{\mathbb{R}^2} f(\mathbf{x})\exp(-\mu_0\mathbf{x}\cdot\boldsymbol{\xi}_\perp)\delta(p-\mathbf{x}\cdot\boldsymbol{\xi})d\mathbf{x}. \qquad (13.17)$$

Our main observation in this section is that the Radon transforms and its generalizations such as in Eq. (13.17) can be viewed as a Euclidean group convolution written in the form

$$\int_{SE(N)} k\left(g_2\circ g_1^{-1}\right)f(g_1)d(g_1) = (k*f)(g_2). \qquad (13.18)$$

If we view a function on the motion group, $f(g)$, as a function on the Cartesian product of the translation and rotation subgroups, $f(\mathbf{r}, R)$, the convolution as written in Eq. (13.18) can be rewritten as

$$(f_2*f_1)(\mathbf{r}, R) = \int_{SO(N)}\int_{\mathbb{R}^N} f_1(\mathbf{x}, Q)f_2\left(-RQ^T\mathbf{x}+\mathbf{r}, RQ^T\right)dQd\mathbf{x}$$

where dQ is the invariant integration measure on the rotation group and $d\mathbf{x}$ is the usual differential volume element in $\mathbb{R}^N$. By making the choice

$$f_1(\mathbf{x}, Q) = f(\mathbf{x})\delta(Q) \quad \text{and} \quad f_2(\mathbf{x}, Q) = \delta(\mathbf{e}_1\cdot\mathbf{x})$$

where $\delta(x_1)$ is the usual Dirac delta function on the x_1 coordinate axis in $\mathbb{R}^N$ and $\delta(Q)$ is the Dirac delta function on the rotation group which has the properties

$$\int_{SO(N)}\delta(Q)dQ = 1 \qquad \delta(Q) = \delta(Q^T) \qquad \int_{SO(N)} f(Q)\delta\left(Q^T R\right)dQ = f(R),$$

one sees that

$$(f_1 * f_2)(\mathbf{q}, R) = \int_{\mathbb{R}^N} f(\mathbf{x}) \left[\int_{SO(N)} \delta(Q) \delta\left(\mathbf{e}_1 \cdot \left(\mathbf{r} - RQ^T\mathbf{x}\right)\right) dQ \right] d\mathbf{x}$$

$$= \int_{\mathbb{R}^N} f(\mathbf{x}) \delta\left(\mathbf{e}_1 \cdot \mathbf{r} - \mathbf{x} \cdot \left(R^T\mathbf{e}_1\right)\right) d\mathbf{x} = \hat{f}\left(\mathbf{r} \cdot \mathbf{e}_1, R^T\mathbf{e}_1\right).$$

While the above choice of functions $f_1(\cdot)$ and $f_2(\cdot)$ yields the Radon transform, this choice is not the only one that will work. For example, recalling that the *conjugate* of a complex-valued function $f(g)$ on a group G is defined as $f^*(g) = f(g^{-1})$, we observe that

$$\left(f_1 * f_2^*\right)(g) = \int_G f_1(h) f_2\left(g^{-1} \circ h\right) d(h) = \int_G f_1(g \circ h) f_2(h) d(h).$$

This is useful in the current context, because when $G = SE(N)$ we can write

$$\left(f_1 * f_2^*\right)(\mathbf{r}, R) = \int_{SO(N)} \int_{\mathbb{R}^N} f_1(R\mathbf{x} + \mathbf{r}, RQ) f_2(\mathbf{x}, Q) d\mathbf{x} dQ.$$

Choosing $f_1(\cdot)$ and $f_2(\cdot)$ as follows

$$f_1(\mathbf{r}, R) = \delta\left(\mathbf{r} \cdot \mathbf{e}_1\right) \quad \text{and} \quad f_2(\mathbf{r}, R) = f(-\mathbf{r}) \tag{13.19}$$

[where $f(\cdot)$ is real-valued], then

$$\left(f_1 * f_2^*\right)(\mathbf{r}, R) = \text{Vol}(SO(N)) \cdot \int_{\mathbb{R}^N} \delta\left(\left(R^T\mathbf{e}_1\right) \cdot \mathbf{x} + \mathbf{r} \cdot \mathbf{e}_1\right) f(-\mathbf{x}) d\mathbf{x}.$$

By proper normalization of the volume element for $SO(N)$, $\text{Vol}(SO(N)) = 1$. Making the change of coordinates $\mathbf{x} \to -\mathbf{x}$ we see that the choice in Eq. (13.19) yields

$$\left(f_1 * f_2^*\right)(\mathbf{r}, R) = \hat{f}\left(\mathbf{r} \cdot \mathbf{e}_1, R^T\mathbf{e}_1\right). \tag{13.20}$$

Thus we have a choice in defining the functions f_1 and f_2 in such a way that their convolution results in the Radon transform. Similar manipulations can be performed for the $\mu_0 \neq 0$ case.

13.2.2 The Radon Transform for Finite Beam Width

In this subsection we consider the 2D-computed tomography problem for the case when finite-width X-ray beams are used for taking parallel projections, i.e., when several detectors (that are each several pixels wide) are used for taking the parallel projection values. We assume that the beam width is Δ and, as in ordinary computed tomography approaches for parallel beams, the parallel projections are taken for angles $0 \leq \theta \leq \pi$ at distances ξ from the origin to the center line of the beam. In order to get an expression for the ratio of the incident and transmitted beam intensity we consider the $\theta = 0$ beam, shifted at distance $x = \xi$ from the origin. We denote $I(y)$ to be the total energy flow along the direction of propagation at coordinate y. The intensity of the beam $i(x, y)$ (the energy flow per unit length, in the direction of propagation) can be written as

$$I(y) = \int_{-\infty}^{\infty} i(x, y) \, dx$$

where the integral (the beam width may be infinite) denotes the integral along the cross-section of the beam, centered at $x = \xi$. Then, considering two close cross-sections of the beam located at coordinates y and $y + dy$, taking the small "volume" element of length dy and cross-section (which is just a length in the planar case) dx and using the conservation of energy, the equation for $i(x, y)$ can be written as

$$(i(x, y + dy) - i(x, y)) \, dx = -(\mu(x, y) \, i(x, y)) \, dx \, dy.$$

The right part of the equation shows the energy loss (mostly due to scattering and photoelectric absorption at small photon energies [33]) which is effectively taken into account by the linear attenuation coefficient $\mu(x, y)$. This equation is illustrated in Fig. 13.11. Integrating with respect to x along the cross-section of the beam, the equation is written as

$$\int_{-\infty}^{\infty} \frac{\partial i(x, y)}{\partial y} \, dx = -\int_{-\infty}^{\infty} \mu(x, y) \, i(x, y) \, dx. \tag{13.21}$$

We write the intensity of the beam in the form

$$i(x, y) = j(y) \, f^{\xi}(x, y),$$

where $f^{\xi}(x, y)$ is a dimensionless function which describes the intensity "profile" along the cross-section of the beam, located at ξ.

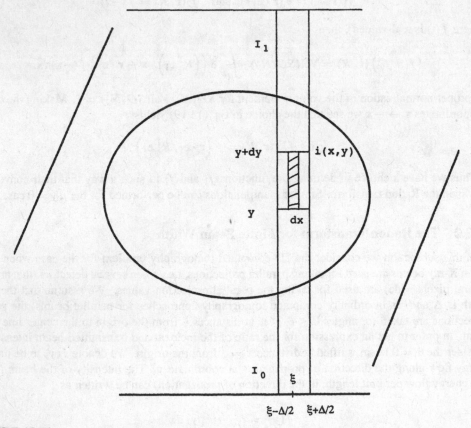

FIGURE 13.11
Illustration of the radiation transfer equation.

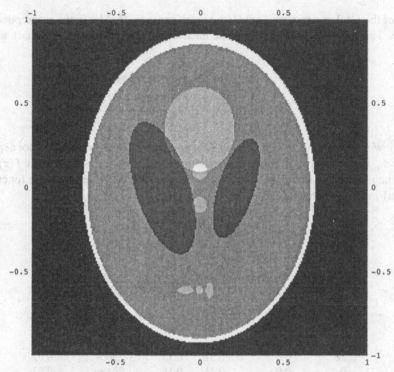

FIGURE 13.12
The Shepp-Logan image phantom.

We assume also that the intensity beam profile is not changing along the propagation path of the beam, i.e., we are looking for solutions to Eq. (13.21) of the form

$$i(x, y) = j(y) f^{\xi}(x).$$

We assume also that the profile of the beam located at ξ is just a shifted version of the beam located at $\xi = 0$

$$f^{\xi}(x) = f^0(x - \xi) = f(x - \xi).$$

Then, Eq. (13.21) takes the form

$$\frac{dj(y)}{dy} \int_{-\infty}^{\infty} f(x - \xi)\, dx = -\left(\int_{-\infty}^{\infty} \mu(x, y) f(x - \xi)\, dx \right) j(y).$$

This equation has a solution of the form

$$j(y) = j(y_0) \exp\left(-\frac{\left(\int_{y_0}^{y} \int_{-\infty}^{\infty} \mu(x, y) f(x - \xi)\, dx\, dy \right)}{\int_{-\infty}^{\infty} f(x)\, dx} \right) \qquad (13.22)$$

where y_0 is the coordinate of the source and y is the detector coordinate. We used the invariance of integration of functions on the real line under shifts to simplify

$$\int_{-\infty}^{\infty} f(x - \xi)\, dx = \int_{-\infty}^{\infty} f(x)\, dx.$$

The ratio of the total energy flow $I(y)/I(y_0)$ is the same as $j(y)/j_0$ in the assumption of constant beam profile. The logarithm of this ratio is the measured quantity (projection value), and it is equal to

$$
\begin{aligned}
\rho(0, \xi) &= -\log\left(I(y)/I(y_0)\right) = -\log\left(j(y)/j_0\right) \\
&= \frac{\left(\int_{y_0}^{y} \int_{-\infty}^{\infty} \mu(x, y) \, f(x - \xi) \, dx \, dy\right)}{\int_{-\infty}^{\infty} f(x) \, dx}.
\end{aligned}
$$

We may view the beam profile function as a function on the plane, which does not depend on the y coordinate, i.e., $f(\mathbf{r}) = f(x)$. This also may be written using the scalar product as $f(\mathbf{r}) = f(\mathbf{e}_1 \cdot \mathbf{r})$, where $\mathbf{e}_1$ is the unit vector along the x direction. For the shift $\mathbf{a}$ the beam profile function centered at $\xi = (\mathbf{e}_1 \cdot \mathbf{a})$ may be written as

$$
f(x - \xi) = f(\mathbf{e}_1 \cdot (\mathbf{r} - \mathbf{a})).
$$

Table 13.1   Parameters for the Shepp-Logan Head Phantom

Center x Coordinate	Center y Coordinate	Major Axis	Minor Axis	Rotation Angle	Refractive Index
0.0	0.0	0.92	0.69	$\pi/2$	2.0
0.0	−0.0184	0.874	0.6624	$\pi/2$	−0.98
0.22	0.0	0.31	0.11	1.25664	−0.02
−0.22	0.0	0.41	0.16	1.88496	−0.02
0.0	0.35	0.25	0.21	$\pi/2$	0.01
0.0	0.1	0.087	0.087	0.0	0.01
0.0	−0.1	0.087	0.087	0.0	0.01
−0.08	−0.605	0.09	0.087	0.0	0.01
0.08	−0.605	0.09	0.087	$\pi/2$	0.01

For arbitrary orientation angle of the beam θ and shift ξ, the beam profile function can be described by the rotated and shifted version of $f(\mathbf{r})$

$$
f\left(\mathbf{e}_1 \cdot \left(R^{-1}\mathbf{r} - \mathbf{a}\right)\right),
$$

where

$$
R^{-1} = \begin{pmatrix} \cos\theta & \sin\theta \\ -\sin\theta & \cos\theta \end{pmatrix}, \tag{13.23}
$$

describes the planar $SO(2)$ rotation. Then, the projection $\rho(\theta, \xi)$ is written as

$$
\rho(\theta, \xi) = \frac{\left(\int_{-\infty}^{\infty} \int_{-\infty}^{\infty} \mu(\mathbf{r}) \, f(\mathbf{e}_1 \cdot (R^{-1}\mathbf{r} - \mathbf{a})) \, d^2r\right)}{\int_{-\infty}^{\infty} f(x) \, dx} \tag{13.24}
$$

where we have used the relation

$$
\int_{-\infty}^{\infty} f(x) \, dx = \int_{-\infty}^{\infty} f\left(\mathbf{e}_1 \cdot \left(R^{-1}\mathbf{r} - \mathbf{a}\right)\right) d\mathbf{n}_{\perp}
$$

for integration along the direction $\mathbf{n}_{\perp}$ perpendicular to the propagation path. Equation (13.24) can also be derived writing an energy transfer equation for the rotated beam.

Equation (13.24) is written explicitly in (x, y) components as

$$\rho(\theta, \xi) = \frac{\left(\int_{-\infty}^{\infty} \int_{-\infty}^{\infty} \mu(x, y) f(x \cos\theta + y \sin\theta - \xi) \, dx \, dy\right)}{\int_{-\infty}^{\infty} f(x) \, dx}.$$

For $f(x) = \delta(x)$ this equation reproduces the ordinary Radon transform (see [34] and references therein).

13.2.3 Computed Tomography Algorithm on the Motion Group

We observe that the numerator of the finite-beam-width Radon transform in Eq. (13.24) may be viewed as a correlator on the 2D Euclidean motion group of the form

$$c(\mathbf{x}, R) = \int_{\mathbb{R}^2} f_1(\mathbf{y}) f_2(R \mathbf{y} + \mathbf{x}) \, d^2 y \qquad (13.25)$$

where R is an $SO(2)$ rotation matrix and $\mathbf{x}$ is the translation vector. As with the template-matching problem, this is a correlation.

We observe that the correlation integral in Eq. (13.25) is written in Fourier space as[2]

$$A \, \overline{\hat{f}_n^{(1)}(p)} \, \hat{f}_m^{(2)}(p) = \hat{c}_{nm}(p). \qquad (13.26)$$

Here A is the area of the compact region in $\mathbb{R}^2$ where the FFT is computed.

Then, the solution for $\hat{f}_n^{(1)}$ is written as

$$\hat{f}_n^{(1)}(p) = \frac{\overline{\hat{c}_{nm}(p)}}{A \, \hat{f}_m^{(2)}(p)} \qquad (13.27)$$

where m is some fixed value of the matrix index, and we also take $\phi = 0$. We note that in cases when $\hat{f}_m^{(2)}(p) = 0$ for some p, regularization is needed. We discuss regularization methods below.

For the case of the Radon transform the left part of Eq. (13.25) is a function of $\mathbf{e}_1 \cdot \mathbf{a} = \xi$ rather than a function of $\mathbf{a}$. This means that only the $\hat{c}_{n,0}(p)$ and $\hat{c}_{n,N_R/2}(p)$ Fourier elements are nonzero. The $m = N_R/2$ index corresponds to the $\phi = \pi$ slice of the Fourier transform. We assume, for simplicity, that N_R is an even number. For odd N_R the $\phi \neq 0$ index corresponding to $\phi = \pi$ has to be taken. This can be observed from the fact that

$$\frac{1}{(2\pi)^2} \int_{-\infty}^{\infty} \int_{-\infty}^{\infty} c(x) \exp(-ip_1 x) \exp(-ip_2 y) \, dx \, dy = \hat{c}(p_1) \, \delta(p_2). \qquad (13.28)$$

The Fourier transform on the group is found [using the discrete version of Eq. (13.28)] by interpolating the values on the Cartesian grid to the polar grid. It is clear from Eq. (13.28) that only the values for $p_2 = 0$ (i.e., along the p_1 axis) are nonzero, which means that only $m = 0$ and $m = N_R/2$ components give contributions.

For the discrete Fourier transform taking the 2D Fourier Transform of the projection slice $c(x, \theta)$ corresponds to computation of

$$\mathcal{F}(c(x, \theta)) = \hat{c}_i \, \delta_{k0} \qquad (13.29)$$

---

[2] In this section we use the notation $\hat{f}_n^{(i)}(p)$ to denote the discrete-motion-group Fourier transform of a function $f_i(\mathbf{x})$ as in Eq. (13.9). Here $n \in [0, N_R - 1]$ where N_R is the number of orientations and A is a scalar constant.

where i, k correspond to the first and second component of the Fourier vector. The values in Eq. (13.29) correspond to $\hat{c}_{n0}(p)$ for $i \geq 0$ and $\hat{c}_{n+N_R/2,N_R/2}(p)$ for $i < 0$ motion group Fourier transform matrix elements, where $n = -j$ for $\theta = (2\pi/N_R) j$ rotation angle, and $p > 0$.

We compute for convenience the correlation

$$\eta(\theta, \xi) = \frac{\left(\int_{-\infty}^{\infty} \int_{-\infty}^{\infty} \mu(\mathbf{r}) \, f(\mathbf{e}_1 \cdot (R\mathbf{r} + \mathbf{a})) \, d^2r\right)}{\int_{-\infty}^{\infty} f(x) \, dx}, \tag{13.30}$$

which is related with the measured quantity in Eq. (13.24) as

$$\eta(\theta, \xi) = \rho(-\theta, -\xi).$$

We note that due to the relation

$$\rho(\theta, \xi) = \rho(\theta + \pi, -\xi) \tag{13.31}$$

the relation

$$\eta(\theta, \xi) = \rho(-\theta + \pi, \xi) \tag{13.32}$$

is also valid.

The Radon Transform as a Particular Case of Correlation on the Motion Group

For the case of the ordinary Radon transform, i.e., for intensity profile function $f(x) = f_2(x) = \delta(x)$, the solution of the motion group convolution equation takes the form

$$\hat{f}_n^{(1)}(p) = \frac{\overline{\hat{\eta}_{n0}(p)}}{A \, \hat{f}_0^{(2)}(p)} \tag{13.33}$$

in Fourier space [the function $\eta(\theta, \xi)$ can be computed from the measured projections $\rho(\theta, \xi)$ via Eq. (13.32)], and the corresponding equation

$$\hat{f}_{n+N_R/2}^{(1)}(p) = \frac{\overline{\hat{\eta}_{n+N_R/2,N_R/2}(p)}}{A \, \hat{f}_{N_R/2}^{(2)}(p)} \tag{13.34}$$

for index $N_R/2$ in the right part of the equation. Equation (13.34) is equivalent to considering $p < 0$ values. $\eta_n(p)$ are radial slices in Radon transform with finite width beam, and $f_{N_R/2}(p)$ is Fourier transform of the beam profile function. $f_n^1(p)$ corresponds to the slices in the ordinary Radon transform [i.e., it is the Fourier transform of $f_1(r)$].

In fact, because for the rotation A_j the relations between indices m and n in $\hat{\eta}_{mn}$ is $m = n - j$, for the rotation θ ($\theta = 2\pi j/N_R$) we find that Eq. (13.33) takes the form

$$\hat{f}_n^{(1)}(p) = \frac{\overline{\hat{\eta}_{n0}(p)}}{L} \tag{13.35}$$

(and the corresponding equation for $N_R/2$ index), where $n = -j$ for the projection enumerated by j ($\theta = 2\pi j/N_R$ rotation angle) and L is a linear dimension of the discrete Fourier transform. Here we use the fact that the discrete 2D Fourier transform of $\delta(x)$ may be written as

$$\hat{\delta} = \frac{1}{L} \delta_{k0}$$

where k corresponds to the second component (dual to y) of the Fourier vector [as we mentioned before it means that $\hat{f}_0^{(2)}(p) = 1/L$ for the motion group Fourier transform]. The $\delta(x)$ is approximated as $(L/N_R)\,\delta_{i0}$ (i corresponds to the x component) on a discrete 2D grid with $L/N_R \times L/N_R$ pixel size.

Thus, Eq. (13.35) provides an algorithm for image reconstruction. Using $j = 0, \ldots, N_R/2$ rotation angles (for 0 and $N_R/2$ indices) we can reproduce the Fourier transform of $\hat{f}^{(1)}$ on the polar grid. Interpolation in Fourier space to a Cartesian grid and, then, the 2D FFT may be used. In this case a technique more complicated than linear interpolation has to be used, such as spline interpolation [17, 40, 65], or as described in [66]. Otherwise, the backprojection algorithm (see [34] and references therein) may be used, in which case linear interpolation in the spatial domain gives good results. Recall that the backprojection algorithm is based on the expression [34]

$$f_1(\mathbf{r}) = \int_0^\pi \left[\int_{-\infty}^\infty \hat{f}_1(p, \theta)|p| \exp(i2\pi p\,t)\,dp \right] d\theta \qquad (13.36)$$

where $\hat{f}_1(p, \theta)$ is a radial slice of the usual Abelian Fourier transform of $f_1(\mathbf{r})$ where $\mathbf{r} \in \mathbb{R}^2$, and $t = x\cos\theta + y\sin\theta$. Due to the symmetry relation in Eq. (13.31), the backprojection may be also computed through the $(\pi, 2\pi)$ interval

$$f_1(\mathbf{r}) = \int_\pi^{2\pi} \left[\int_{-\infty}^\infty \hat{f}_1(p, \theta)|p| \exp(i2\pi p\,t)\,dp \right] d\theta.$$

The Fourier slice $\hat{f}_1(p, \theta)$ at angle $\theta = 2\pi n/N_R$ corresponds to the motion group Fourier transform matrix elements

$$\hat{f}_1(p, \theta) = \hat{f}_n^{(1)}(p)$$

for $p \geq 0$, and

$$\hat{f}_1(p, \theta) = \hat{f}_{n+N_R/2}^{(1)}(|p|)$$

for $p < 0$.

We also note that for real function $f(\mathbf{x})$,

$$\overline{\hat{f}_n(p)} = \hat{f}_{n+N_R/2}(p)$$

which reflects the fact that $\hat{f}(\mathbf{p}) = \overline{\hat{f}(-\mathbf{p})}$ for the ordinary 2D Fourier transform of a real function. For convenience we formally allow p to take $p < 0$ values and denote

$$\hat{f}_{n+N_R/2}^{(1)}(|p|) = \hat{f}_n^{(1)}(p)$$

for $p < 0$.

Thus Eq. (13.35) provides

$$\hat{f}_{-j}^{(1)}(p) = \hat{f}_1(p, -2\pi j/N_R),$$

Which is the Fourier slice which can be used in Eq. (13.36) for filtered backprojection.

Using motion group notation we write the backprojection algorithm as

$$f_1(\mathbf{r}) = 2\pi/N_R \sum_{j=0}^{N_R/2-1} \left[\int_{-\infty}^\infty \hat{f}_{-j}^{(1)}(p)|p| \exp(i2\pi p\,t)\,dp \right] \qquad (13.37)$$

where $t = x\cos(2\pi j/N_R) - y\sin(2\pi j/N_R)$, and $\hat{f}_{-j}^{(1)}(p)$ are computed using Eq. (13.35).

We have shown in Fig. 13.13 the backprojection reconstruction in Eq. (13.36) for the Shepp and Logan "head phantom" [60] (depicted in Fig. 13.12) using expression Eq. (13.37). The analytical expression for the projections was used, $N_R/2 = 60$ rotation angles were used, and the image was

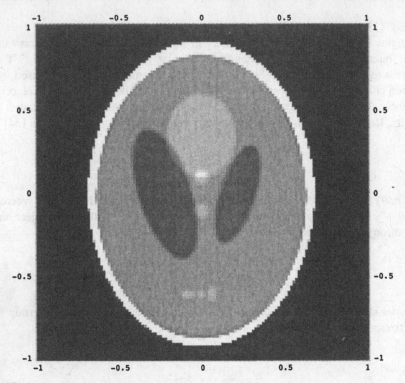

FIGURE 13.13
The Shepp-Logan image reconstruction.

located on a 128×128 grid, but the projection data were zero padded to a larger 256×256 grid to avoid the effect of periodic approximation to the continuous Fourier transform [34]. The $|p|$ multiplier in Eq. (13.36) was replaced by the Fourier transform of

$$h(0) = N_x N_x/4; \quad h(i) = 0; \ \text{for even } i;$$

$$h(i) = -\frac{N_x N_x}{(\pi^2 i^2)} \ \text{for odd } i;$$

($N_x = 256$ and $-N_x/2 < i \leq N_x/2$) in order to improve accuracy, and low pass filtering was used to reduce high-frequency "noise" [34]. We have used the Gaussian filter

$$g(p) = \exp\left(-\frac{p^2}{2\sigma^2}\right)$$

which is measured in units of maximal "momentum" $p_{max} = 2\pi N_R/(2L)$, $\sigma = c_1 p_{max}$. Using the notation $p = 2\pi i/L$, where $-N_R/2 < i \leq N_R/2$, we write the filter in the discrete form as

$$g(i) = \exp\left(-\sigma_1 \frac{i^2}{N_R^2}\right) \tag{13.38}$$

where $\sigma_1 = 2/c_1^2$. The reconstruction image depicted in Fig. 13.13 corresponds to the $\sigma_1 = 4$ value, i.e., for $\sigma = p_{max}/\sqrt{2}$.

The Backprojection Algorithm for Finite-Width Beams

It is clear from the expression in Eq. (13.33) that the backprojection algorithm can be easily modified to include the finite beam case. For finite beams a non-constant $\hat{f}_0^{(2)}(p)$ has to be used in the denominator of Eqs. (13.33) and (13.34) to correctly reproduce the left side of Eq. (13.35). The $\hat{f}_0^{(2)}(p)$ Fourier element is just the ordinary Fourier transform of the intensity profile function $f_2(x) = f(x)$.

Below we consider the case when the profile function is a step-like function

$$f(x) = \begin{cases} 1 & \text{if } |x| \le \Delta/2 \\ 0 & \text{otherwise} \end{cases} .$$

Then the denominator in Eq. (13.24) is just the width of the beam Δ. We assume that Δ is measured in an integer number of pixels.

In this case the analytical expression for the projection values can be easily found so we may test the accuracy of the algorithms on an analytically defined model.

The analytical expressions for the projections of an ellipse

$$x^2/a^2 + y^2/b^2 = 1$$

with constant linear attenuation coefficient q are given by

$$\rho(\theta, \xi) = \frac{q\,a\,b}{d(\theta)\,\Delta} \left((\xi + \Delta/2)\sqrt{d(\theta) - (\xi + \Delta/2)^2} \right.$$
$$- (\xi - \Delta/2)\sqrt{d(\theta) - (\xi - \Delta/2)^2}$$
$$\left. + d(\theta) \left(\arcsin\frac{((\xi + \Delta/2)}{d(\theta)} - \arcsin\frac{((\xi - \Delta/2)}{d(\theta)} \right) \right)$$

if $-\sqrt{d(\theta)} \le (\xi - \Delta/2)$ and $(\xi + \Delta/2) \le \sqrt{d(\theta)}$. Here $d(\theta) = a^2(\cos\theta)^2 + b^2(\sin\theta)^2$. For $(\xi - \Delta/2) \le \sqrt{d(\theta)} < (\xi + \Delta/2)$ we find the expression

$$\rho(\theta, \xi) = \frac{q\,a\,b}{d(\theta)\,\Delta} \left(-(\xi - \Delta/2)\sqrt{d(\theta) - (\xi - \Delta/2)^2} \right.$$
$$\left. + d(\theta)\left(\pi/2 - \arcsin\frac{(\xi - \Delta/2)}{d(\theta)} \right) \right).$$

For $(\xi - \Delta/2) < -\sqrt{d(\theta)} \le (\xi + \Delta/2)$ the expression is

$$\rho(\theta, \xi) = \frac{q\,a\,b}{d(\theta)\,\Delta} \left((\xi + \Delta/2)\sqrt{d(\theta) - (\xi + \Delta/2)^2} \right.$$
$$\left. + d(\theta)\left(\arcsin\left(\frac{(\xi + \Delta/2)}{d(\theta)} \right) + \pi/2 \right) \right).$$

The projection values are zero otherwise. We assume for simplicity that $\Delta < a$ and $\Delta < b$.
The Fourier transform of the profile functions is calculated as

$$f_0^{(2)}(i) = L\, \frac{\sin(\pi/L\,\Delta i)}{\pi\,\Delta i} \tag{13.39}$$

where $-N_R/2 < i < N_R/2$. Here L is a linear dimension of the image, $\delta = L/N_R$ is the pixel size.

Numerical Results

We observe that the expression in Eq. (13.39) may be equal to zero for some values of i (i.e., for some values of p). For the odd number $N_1 = \Delta/\delta$ (for power-of-two number of pixels N_R) Eq. (13.39) is nonzero, while for even number of pixels regularization is required.

Recall that our goal is to invert Eq. (13.26) for given A, $\hat{f}_m^{(2)}(p)$ and $\hat{c}_{nm}(p)$. When $\hat{f}_m^{(2)}(p) \approx 0$ then Eq. (13.34) is not a valid expression, and we must seek regularized solutions. One common method of regularizing expressions of the form $\hat{f}_i\,\hat{g}_i = \hat{h}_i$ for known $\hat{g}_i$ and $\hat{h}_i$ is Tikhonov regularization [28]

$$\hat{f}_i = \frac{\hat{h}_i\,\overline{\hat{g}_i}}{\epsilon + \hat{g}_i\,\overline{\hat{g}_i}}.$$

However this technique does not yield reconstructions with acceptable accuracy in this context. Instead, we use spline interpolation to find $\hat{f}_i$ when $\hat{g}_i \approx 0$. Neighboring values of $\hat{f}$ can be found by Eq. (13.34), and the value at a singular point may be found using spline interpolation.

We used the spline interpolation technique of degree three [22] to get the Fourier value at the singular points. The Fourier transform was interpolated by a piecewise cubic function with continuous second order derivatives. A fixed number of interpolating points was used. For the 12-point interpolating interval, the value of the Fourier transform at the singular point enumerated by i was interpolated as

$$\hat{f}_i = m_{i-1}/12 + m_{i+1}/12 + (\hat{f}_{i-1} - 2/3\,m_{i-1})/2 + \left(\hat{f}_{i+1} - 2/3\,m_{i+1}\right)/2.$$

Here $-N_R/2 < i \le N_R/2$ enumerates discrete p values, and values of the second order derivatives m_{i-1}, m_{i+1} were found from the solution of a system of linear equations with tridiagonal matrix [22]. The boundary conditions for the interpolating intervals were specified; we used natural splines (zero second order derivatives) and splines with given slope $\hat{f}'$ at the ends of the interval. Accuracy of the same order was achieved for both of these boundary conditions.

In Fig. 13.14 the Shepp-Logan-like computed simulation model is reconstructed using the ordinary backprojection algorithm, while the beam width is 5 pixels. We have listed the parameters of ellipses for simulations in Table 13.1. It is assumed that the image is the sum of ellipses with the parameters given previously.

Figure 13.15 depicts the reconstructed image using the algorithm in Eq. (13.33) which takes into account the beam width. The Gaussian filter in Eq. (13.38) with $\sigma_1 = 8$ was used.

The improvement in the image quality can be clearly observed.

Taking into account the beam width in the singular case also improves quality when the object size is comparable with the beam width. In Fig. 13.16(a) we depicted the reconstructed projection of the Shepp-Logan-like image for a beam width of 24 pixels, reconstructed using the ordinary backprojection algorithm. Figure 13.16(b) depicts the reconstruction (singular case) which takes into account beam width ($\sigma_1 = 20$ was used). The image is reconstructed with better accuracy, even though the deviation of order 0.4% may be observed. The non-singular case (23 pixels width) reproduces the image with much better accuracy, [see Fig. 13.16(c)] ($\sigma_1 = 12$ was used).

13.3   Inverse Tomography: Radiation Therapy Treatment Planning

Radiation therapy treatment planning (which also goes by other related names such as radiotherapy planning or radiation treatment planning) is the field concerned with delivering the best dose of

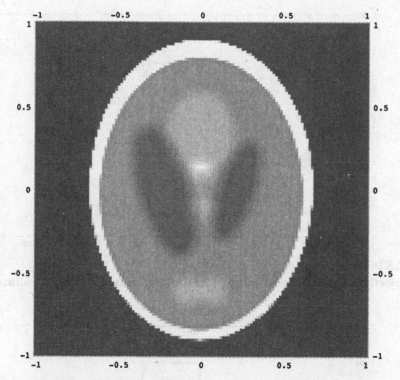

FIGURE 13.14
Image reconstruction for the finite beam case using the ordinary backprojection algorithm.

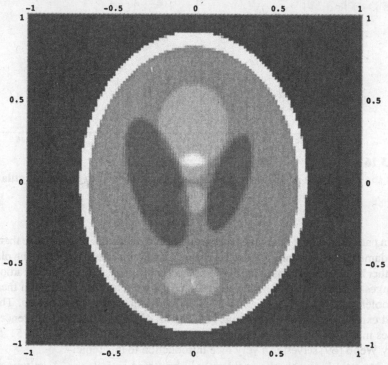

FIGURE 13.15
Image reconstruction which takes into account the beam width.

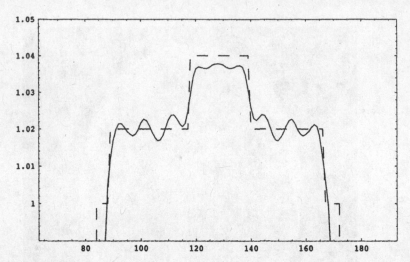

FIGURE 13.16

(a) The projection of reconstructed image without taking into account beam width; *(Continued).*

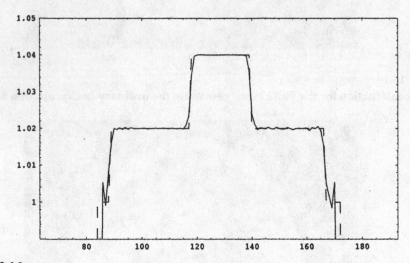

FIGURE 13.16

(Cont.). **(b) the projection of reconstructed image for finite width beam (singular case); *(Continued).***

radiation to a patient in order to damage or destroy a tumor, while causing no more than an acceptable amount of damage to surrounding healthy tissue. Radiation therapy treatment planning can be delivered either by external beams using a columnator, or by internal placement of radioactive material (either by ingestion or by a surgical procedure). When we refer to the radiation therapy treatment planning problem, we will mean radiation being delivered through external beams. This problem has been studied extensively (see, for example, [13, 14, 44, 45, 59, 58, 61] and references therein). For recent studies in medical physics and therapy planning see [8, 11, 21, 47, 41, 46, 51, 52, 56, 54, 70]. The book by Webb [69] serves as a very nice introduction to this topic.

In group-theoretic language, the problem may be stated as follows. A columnator produces a beam at position and orientation g. Associated with this beam is a certain normalized profile (shape)

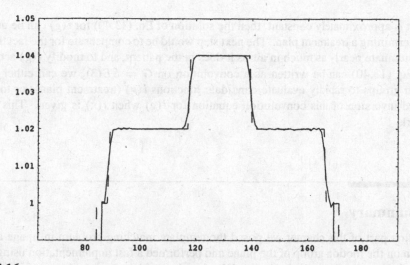

FIGURE 13.16

(Cont.). **(c) the projection of reconstructed image for finite width beam (non-singular case).**

that can be described as a function $b(\mathbf{x}, g)$ where $\mathbf{x} \in \mathbb{R}^3$. The form of the function $b(\mathbf{x}, g)$ (which describes the relative damage that the beam does to point $\mathbf{x}$ as compared to other positions in space when the beam originates from the origin of the frame g and is directed along an axis) depends on how tissue attenuates and/or distorts the cross-sectional profile of the beam, and what the beam profile is as it exits the columnator.

In the highly idealized case when it is assumed that there is no attenuation or beam spreading, and the beam profile is taken to have the same pin-point support for all positions and orientations of the columnator, then

$$b(\mathbf{x}, g) = b_0 \left(g^{-1} \circ \mathbf{x} \right)$$

where $b_0(\mathbf{x}) = \delta(x_1)\delta(x_2)$ is a function that has support on the x_3 axis, and constant values on this axis. In this idealized model, the question becomes how to specify the intensity, $i(g)$, of the beam for each position and orientation of the columnator so that the total dose is as close as possible to the desired one, $t(\mathbf{x})$. This is written as

$$t(\mathbf{x}) = \int_G i(g) b_0 \left(g^{-1} \circ \mathbf{x} \right) dg. \tag{13.40}$$

While our explanation of Eq. (13.40) was for a highly idealized function $b_0(\cdot)$, there are a number of ways in which $b_0(\cdot)$ can be made more realistic. For instance, in the reference frame $g = e$, the axis of the columnator is x_3, the beam originates at $x_3 = 0$, and it travels in the positive x_3 direction. The beam profile can be taken to be a Gaussian function with intensity that decreases with increasing value of x_3 and width that increases with x_3. That is,

$$b_0(\mathbf{x}) = c_1 (x_3) \exp \left(-c_2 (x_3) \left(x_1^2 + x_2^2 \right) \right)$$

where $c_1(x)$ and $c_2(x_3)$ decrease with x_3. This is a model that has been used to describe the evolution of a profile of a beam of high energy photons through water. Other kinds of profiles have been developed for other kinds of beams (e.g., protons).

The point is, Eq. (13.40) can be used to describe a wide variety of superposition problems involving beams of radiation emanating from different frames of reference provided the attenuation throughout the environment is constant. If $c_2(\cdot)$ is approximately constant and the attenuation constant within the patient is approximately constant, then the solution of Eq. (13.40) for $i(g)$ can be used as a first step in determining a treatment plan. The next step would be to compensate for the fact that the beam does not attenuate nearly as much in air as it does in the patient, and to modify $i(g)$ accordingly.

Since Eq. (13.40) can be written as a convolution on $G = SE(3)$, we can either use an FFT for motion groups to rapidly evaluate candidate functions $i(g)$ (treatment plans) or to attempt the regularized inversion of this convolution equation for $i(g)$ when $t(\mathbf{x})$ is given. This remains for future work.

13.4   Summary

In the first part of this chapter we posed the template matching problem in image analysis as a correlation on the motion group of the plane and performed a fast implementation using the Fourier transform for the discrete motion group. We also showed how a comparison of image invariants based on this group-theoretic Fourier transform yields a method to determine how alike two patterns are, regardless of their position and orientation.

In the second part of this chapter we showed how the Radon transform, and its generalization to the case of finite-width beams, can be viewed as a convolution (or correlation) of functions on the group of rigid-body motions. By discretizing the rotational part of this group into equal increments, the discrete motion group results. The Fourier transform of functions on this discrete motion group is used as a tool to reduce the tomographic reconstruction problem to a scalar algebraic equation in a generalized Fourier space. Numerical results using this approach were presented and appear to be promising.

In the third part of this chapter we discussed how some radiotherapy treatment planning problems can be posed in the language of noncommutative harmonic analysis.

References

[1] Abu-Mostafa, Y.S. and Psaltis, D., Recognition aspects of moment invariants, *IEEE Trans. Pattern Anal. Machine Intell.*, PAMI-6(6), 698–706, 1984.

[2] Antonini, M., Barlaud, M., Mathieu, P., and Daubechies, I., Image coding using wavelet transform, *IEEE Trans. on Image Process.*, 40(2), 205–220, 1992.

[3] Arsenault, H.H., Hsu, Y.N., and Chalasinsk-Macukow, K., Rotation-invariant pattern recognition, *Opt. Eng.*, 23, 705–709, 1984.

[4] Barrett, H.H. and Swindell, W., *Radiological Imaging: The Theory of Image Formation, Detection and Processing*, Academic Press, New York, 1981.

[5] Beylkin, G., The inversion problem and applications of the generalized Radon transform, *Commun. on Pure and Appl. Math.*, 37, 577–599, 1984.

[6] Bhatia, A.B. and Wolf, E., On the circle polynomials of Zernike and related orthogonal sets, *Proc. of the Cambridge Philos. Soc.*, 50, 40–48, 1954.

[7] Bortfeld, T.R. and Boyer, A.L., The exponential Radon transform and projection filtering in radiotherapy planning, *Int. J. of Imaging Syst. and Technol.*, 6, 62–70, 1995.

[8] Bortfeld, T., An analytical approximation of the Bragg curve for therapeutic proton beams, *Med. Phys.*, 24, 2024–2033, 1997.

[9] Bracewell, R.N. and Riddle, A.C., Inversion of fan-beam scans in radio astronomy, *Astrophys. J.*, 150, 427–434, 1967.

[10] Brady, M.L., A fast discrete approximation algorithm for the Radon transform, *SIAM J. of Comput.*, 27(1), 107–119, 1998.

[11] Brugmans, M.J.P., van der Horst, A., Lebesque, J.V., and Mijnheer, B.J., Dosimetric verification of the 95% isodose surface for a conformal irradiation technique, *Med. Phys.*, 25, 424–434, 1998.

[12] Canny, J., A computational approach to edge detection, *IEEE Trans. Patt. Anal. Mach. Intell.*, 8, 679–698, 1986.

[13] Censor, Y., Altschuler, M.D., and Powlis, W.D., A computational solution of the inverse problem in radiation-therapy treatment planning, *Appl. Math. and Comput.*, 25, 57–87, 1988.

[14] Censor, Y., Altschuler, M.D., and Powlis, W.D., On the use of Cimmino's simultaneous projections method for computing a solution of the inverse problem in radiation therapy treatment planning, *Inverse Probl.*, 4, 607–623, 1988.

[15] Chang, L.-T., Attenuation correction and incomplete projection in single photon emission computed tomography, *IEEE Trans. on Nuclear Sci.*, NS-26(2), 2780–2789, 1979.

[16] Chirikjian, G.S., Fredholm integral equations on the Euclidean motion group, *Inverse Probl.*, 12, 579–599, 1996.

[17] Choi, H. and Munson, D.C., Direct-Fourier reconstruction in tomography and synthetic aperture radar, *Int. J. of Imaging Syst. and Technol.*, 9, 1–13, 1998.

[18] Cooley, J.W. and Tukey, J., An algorithm for the machine calculation of complex Fourier series, *Math. of Comput.*, 19, 297–301, 1965.

[19] Cools, O.F., Herman, G.C., van der Weiden, R.M., and Kets, F.B., Fast computation of the 3-D Radon transform, *Geophysics*, 62(1), 362–364, 1997.

[20] Cormack, A.M., Sampling the Radon transform with beams of finite width, *Phys. Med. Biol.*, 23(6), 1141–1148, 1978.

[21] Das, I.J., McGee, K.P., and Cheng, C., Electron-beam characteristics at extended treatment distances, *Med. Phys.*, 22, 1667–1674, 1995.

[22] Davis, P. and Rabinowitz, P., *Methods of Numerical Integration*, Academic Press, New York, 1975.

[23] Deans, S.R., *The Radon Transform and Some of its Applications*, John Wiley & Sons, New York, 1983.

[24] Defrise, M., Clack, R., and Townsend, D.W., Image reconstruction from truncated, two-dimensional, parallel projections, *Inverse Probl.*, 11, 287–313, 1995.

[25] Elliott, D.F. and Rao, K.R., *Fast Transforms: Algorithms, Analyses, Applications*, Academic Press, New York, London, 1982.

[26] Gauthier, J.P., Bornard, G., and Sibermann, M., Motion and pattern analysis: harmonic analysis on motion groups and their homogeneous spaces, *IEEE Trans. Syst. Man Cybern.*, 21, 159–172, 1991.

[27] Gel'fand, I.M., Graev, M.I., and Vilenkin, N.Ya., *Generalized Functions*, Academic Press, New York, 1966.

[28] Groetsch, C.W., *The Theory of Tikhonov Regularization for Fredholm Equations of the First Kind*, Pitman, Boston, 1984.

[29] Heike, U., Single-photon emission computed tomography by inverting the attenuated Radon transform with least-squares collocation, *Inverse Probls.*, 2, 307–330, 1986.

[30] Helgason, S., *The Radon Transform*, Progress in Mathematics, Vol. 5, Birkhauser, 1980.

[31] Holschneider, M., Inverse Radon transforms through inverse wavelet transforms, *Inverse Probls.*, 7, 853–861, 1991.

[32] Jahne, B., *Spatio-Temporal Image Processing: Theory and Scientific Applications*, Springer-Verlag, Berlin, 1993.

[33] Johns, H.E. and Cunningham, J.R., *The Physics of Radiology*, Thomas, Springfield, IL, 1974.

[34] Kak, A. and Slaney, M., *Principles of Computerized Tomographic Imaging*, IEEE Press, New York, 1988.

[35] Kanatani, K., *Group-Theoretical Methods in Image Understanding*, Springer-Verlag, Berlin, Heidelberg, New York, 1990.

[36] Karim, M.A. and Awwal, A.A.S., *Optical Computing an Introduction*, John Wiley & Sons, New York, 1992.

[37] Kuchment, P., Lancaster, K., and Mogilevskaya, L., On local tomography, *Inverse Probls.*, 11, 571–589, 1995.

[38] Kyatkin, A.B. and Chirikjian, G.S., Pattern matching as a correlation on the discrete motion group, *Comput. Vision and Image Understanding*, 74(1), 22–35, 1999.

[39] Lanzavecchia, S. and Bellon, P.L., Fast computation of 3D Radon transform via a direct Fourier method, *Bioinformatics*, 14(2), 212–216, 1998.

[40] La Riviere, P.J. and Pan, X., Spline-based inverse Radon transform in two and three dimensions, *IEEE Trans. on Nuclear Sci.*, 45(4), 2224–2231, 1998.

[41] Llacer, J., Inverse radiation treatment planning using the dynamically penalized likelihood method, *Med. Phys.*, 24, 1751–1764, 1997.

[42] Leduc, J.-P., Spatio-temporal wavelet transforms for digital signal analysis, *Signal Process.*, 60, 23–41, 1997.

[43] Lenz, R., *Group Theoretical Methods in Image Processing*, lecture notes in Computer Science, Springer-Verlag, Berlin, Heidelberg, New York, 1990.

[44] Lind, B., Properties of an algorithm for solving the inverse problem in radiation therapy, *Inverse Probls.*, 6, 415–426, 1990.

[45] Lind, B. and Brahme, A., Development of treatment techniques for radiotherapy optimization, *Int. J. of Imaging Syst. and Technol.*, 6, 33–42, 1995.

[46] Lovelock, D.M.J., Chui, C.S., and Mohan, R., A Monte Carlo model of photon beams used in radiation therapy, *Med. Phys.*, 22, 1387–1394, 1995.

[47] Liu, H.H., Mackie, T.R., and McCullough, E.C., Correcting kernal tilting and hardening in convolution/superposition dose calculations for clinical divergent and polychromatic photon beams, *Med. Phys.*, 24, 1729–1741, 1997.

[48] Markoe, A., Fourier inversion of the attenuated X-ray transform, *SIAM J. Math. Anal.*, 15(4), 718–722, 1984.

[49] Murenzi, R., Wavelet transforms associated to the n-dimensional Euclidean group with dilations: signals in more than one dimension, in *Wavelets: Time-Frequency Methods and Phase Space*, Combes, J.M., Grossmann, A., and Tchamitchian, Ph., Eds., 239–246, 1990.

[50] Natterer, F., On the inversion of the attenuated Radon transform, *Numer. Math.*, 32, 431–438, 1979.

[51] Nizin, P.S. and Mooij, R.E., An approximation of central-axis absorbed dose in narrow photon beams, *Med. Phys.*, 24, 1775–1780, 1997.

[52] Ostapiak, O.Z., Zhu, Y., and Van Dyk, J., Refinements of the finite-size pencil beam model of three-dimensional photon dose calculations, *Med. Phys.*, 24, 743–750, 1997.

[53] Panchenko, A.N., Inverse source problem of radiative transfer: a special case of the attenuated Radon transform, *Inverse Probls.*, 9, 321–337, 1993.

[54] Perry, D., Wollin, M., Olch, A., and Buffa, A., Range spectra in electron penetration problems, *Med. Phys.*, 25, 43–55, 1998.

[55] Pintsov, D.A., Invariant pattern recognition, symmetry, and the Radon transforms, *J. Opt. Soc. Am.*, 6(10), 1545–1554, 1989.

[56] Olivares-Pla, M., Podgorsak, E.B., and Pla, C., Electron arc dose distributions as a function of beam energy, *Med. Phys.*, 24, 127–132, 1997.

[57] Radon, J., Über die bestimmung von funktionen durch ihre integralwerte längs gewisser mannigfaltigkeiten, *Berichte Sächsische Akademie der Wissenschaften. Leipzig, Math.-Phys. Kl.*, 69, 262–267, 1917.

[58] Raphael, C., Mathematical modelling of objectives in radiation therapy treatment planning, *Phys. Med. Biol.*, 37(6), 1293–1311, 1992.

[59] Raphael, C., Radiation therapy treatment planning: an $\mathcal{L}^2$ approach, *Appl. Maths. and Comput.*, 52, 251–277, 1992.

[60] Rosenfeld, A. and Kak, A., *Digital Picture Processing*, Academic Press, New York, 1982.

[61] Sandham, W.A., Yuan, Y., and Durrani, T.S., Conformal therapy using maximum entropy optimization, *Int. J. of Imaging Syst. and Technol.*, 6, 80–90, 1995.

[62] Sahiner, B. and Yagle, A.E., Iterative inversion of the Radon transform, *IEEE Eng. in Med. and Biol.*, 15(5), 112–117, 1996.

[63] Segman, J. and Zeevi, Y., Image analysis by wavelet-type transforms: group theoretical approach, *J. of Math. Imaging and Vision*, 3, 51–77, 1993.

[64] Shepp, L.S. and Logan, B.F., The Fourier reconstruction of a head section, *IEEE Trans. Nucl. Sci.*, NS-21, 21–43, 1974.

[65] Späth, H., *Two Dimensional Spline Interpolation Algorithms*, AK Peters, Wellesley, MA, 1995.

454   *ENGINEERING APPLICATIONS OF NONCOMMUTATIVE HARMONIC ANALYSIS*

[66] Stark, H., Woods, J.W., Paul, I., and Hingorani, R., Direct Fourier reconstruction in computed tomography, *IEEE Trans. Acoustics, Speech, Signal Process.*, ASSP-29, 237–245, 1981.

[67] Tretiak, O.J. and Metz, C., The exponential Radon transform, *SIAM J. Applied Math.*, 39, 341–354, 1980.

[68] Turin, G.L., An introduction to matched filters, *IRE Trans. Inf. Theory*, IT-6, 311–329, 1960.

[69] Webb, S., *The Physics of Three-Dimensional Radiation Therapy: Conformal Radiotherapy, Radiosurgery and Treatment Planning*, Medical Science Series, Institute of Physics Publishing, Bristol, UK, 1993.

[70] Wong, E. and Van Dyk, J., Lateral electron transport in FFT photon dose calculations, *Med. Phys.*, 24, 1992–2000, 1997.

Chapter 14

Statistical Pose Determination and Camera Calibration

In this chapter we examine the following problems: (1) finding the best rigid-body motion (pose) to fit to noisy measured data; and (2) performing camera calibration in vision systems mounted on robotic manipulators. In both cases it is assumed that only inexact measurements with known statistical properties are provided. We therefore begin this chapter with sections on basic probability theory and its extension to data on spheres, the rotation group, and the motion group.

First, Section 14.1 reviews basic definitions from probability theory. Section 14.2 discusses probability and statistics on the circle. Section 14.3 reviews the corresponding concepts for probability density functions (PDFs) on groups. Section 14.4 introduces PDFs on $SO(3)$ that have special properties. Section 14.5 provides definitions of mean and variance of PDFs on rotation and motion groups. Section 14.6 addresses the problem of finding the best rigid-body transformation to fit a set of measured data. Finally, Section 14.7 discusses the problem of camera calibration and the statistical aspects of this problem.

14.1 Review of Basic Probability

Given a probability density function (PDF) with real argument, $\rho(x)$, the classical apparatus of probability theory is used to analyze $\rho(x)$ in terms of its moments. The expected value is the center of mass, or mean, $E(x) = x_{cm}$ defined as the solution of

$$\int_{\mathbb{R}} (x - x_{cm}) \rho(x) dx = 0.$$

The median value corresponding to $\rho(x)$ is the value $x = x_{med}$ for which there is as much "mass" on one side as on the other

$$\int_{-\infty}^{x_{med}} \rho(x) dx = \int_{x_{med}}^{\infty} \rho(x) dx.$$

But since

$$\int_{-\infty}^{\infty} \rho(x) dx = 1,$$

this indicates that x_{med} satisfies the equation

$$\int_{-\infty}^{x_{med}} \rho(x) dx = \frac{1}{2}.$$

The median may also be viewed as the value of y that minimizes

$$F(y) = \int_{-\infty}^{\infty} |y - x| \rho(x) dx = -\int_{y}^{\infty} (y - x) \rho(x) dx + \int_{-\infty}^{y} (y - x) \rho(x) dx. \quad (14.1)$$

It is easy to see by setting $dF/dy = 0$ and solving for y that $F(x_{med})$ is the minimal value for any $\rho(x)$.

The nth moments about the expected value $E(x) = x_{cm}$ for all integers $n \geq 0$ are given by

$$M_n = \int_{\mathbb{R}} (x - x_{cm})^n \rho(x) dx.$$

$M_0 = 1$ is the "mass" of the probability density function and $M_1 = 0$ by definition. $M_2 = \sigma^2$ is the variance and σ is the standard deviation. Knowing all of the moments from $n = 2, \ldots, N$ for some finite N means that the important properties of the distribution $\rho(x)$ are known to some degree without regard to all of its details. As $N \to \infty$, the properties of $\rho(x)$ become more determined as more of the moments are known. Usually only the first few moments are of concern.

Analogous to the definition of the median as the value which minimizes the functional in Eq. (14.1), it is easy to show by direct calculation that $E(x) = x_{cm}$ may be calculated as the value of y which minimizes the functional

$$C(y) = \int_{\mathbb{R}} (x - y)^2 \rho(x) dx. \quad (14.2)$$

Hence, the value of y that minimizes $C(y)$ is the mean, and $C(x_{cm})$, the function C evaluated at the mean, is the variance.

The moments of probability density functions have nice properties under the operation of convolution. Recall that on the line

$$(\rho_1 * \rho_2)(x) = \int_{-\infty}^{\infty} \rho_1(\xi) \rho_2(x - \xi) d\xi.$$

Since $\rho_1(x) \geq 0$ for all $x \in \mathbb{R}$, it follows from a change of variables $z = x - \xi$ that the mass of the convolution of two PDFs is unity

$$\int_{-\infty}^{\infty} (\rho_1 * \rho_2)(x) dx = \left(\int_{-\infty}^{\infty} \rho_1(\xi) d\xi \right) \left(\int_{-\infty}^{\infty} \rho_2(z) dz \right) = 1 \cdot 1 = 1. \quad (14.3)$$

That is, the convolution of two PDFs results in a PDF. Similarly, one has that the higher moments of convolutions of PDFs can be generated from the moments of the PDFs being convolved

$$x_{cm}^{1*2} = \int_{-\infty}^{\infty} x \cdot (\rho_1 * \rho_2)(x) dx = \int_{-\infty}^{\infty} x \cdot \rho_1(x) dx + \int_{-\infty}^{\infty} x \cdot \rho_2(x) dx = x_{cm}^1 + x_{cm}^2 \quad (14.4)$$

and

$$\begin{aligned}
\left(\sigma^2 \right)_{1*2} &= \int_{-\infty}^{\infty} \left(x - x_{cm}^{1*2} \right)^2 (\rho_1 * \rho_2)(x) dx \\
&= \int_{-\infty}^{\infty} \left(x - x_{cm}^1 \right)^2 \rho_1(x) dx + \int_{-\infty}^{\infty} \left(x - x_{cm}^2 \right)^2 \rho_2(x) dx \\
&= \left(\sigma^2 \right)_1 + \left(\sigma^2 \right)_2.
\end{aligned} \quad (14.5)$$

As with Eq. (14.3), we find Eqs. (14.4) and (14.5) by substituting $z = x - \xi$ and using the invariance of integration under shifts.

14.1.1 Bayes' Rule

Let

$$P(x) = \int_{x}^{x+\Delta x} \rho(\xi)d\xi$$

be the probability that $\xi \in [x, x + \Delta x]$ corresponding to the PDF $\rho(x)$. Let $\rho(x, y)$ be the *joint probability* density describing the possibility that x and y occur simultaneously, and let

$$P(x, y) = \int_{x}^{x+\Delta x} \int_{y}^{y+\Delta y} \rho(\xi, \eta)d\xi d\eta.$$

The *conditional probability* that $\xi \in [x, x + \Delta x]$ given $\eta \in [y, y + \Delta y]$ is denoted $P(x|y)$, and is calculated as

$$P(x|y) = P(x, y)/P(y).$$

Similarly,

$$P(y|x) = P(x, y)/P(x).$$

Combining these two we get *Bayes' rule*[1]

$$P(x|y) = \frac{P(y|x)P(x)}{P(y)}. \tag{14.6}$$

Analogous to the definition of $P(x|y)$, let

$$\rho(x|y) = \rho(x, y)/\rho(y)$$

be the *conditional probability density function* of x given y. In the limit as Δx and Δy are allowed to approach zero, Eq. (14.6) is rewritten as

$$\rho(x|y)\Delta x = \frac{\rho(y|x)\Delta y \rho(x)\Delta x}{\rho(y)\Delta y}$$

which is simplified to

$$\rho(x|y) = \frac{\rho(y|x)\rho(x)}{\rho(y)}, \tag{14.7}$$

which is called *Bayes' rule for PDFs*.

14.1.2 The Gaussian Distribution

We have already seen the Gaussian distribution

$$\rho_G(x; m, \sigma) = \frac{1}{\sqrt{2\pi}\sigma}e^{-(x-m)^2/2\sigma^2}$$

in earlier chapters. The expected value of this distribution is $x_{cm} = m$, and its standard deviation is σ. The multi-dimensional Gaussian distribution is of the form

$$\rho_G(\mathbf{x}; \mathbf{m}, C) = \frac{\exp\left[-\frac{1}{2}\sum_{i,j=1}^{n} C_{ij}^{-1}(x_i - m_i)(x_j - m_j)\right]}{[(2\pi)^n \det C]^{\frac{1}{2}}}$$

---

[1] Thomas Bayes (1702–1761) established the foundations of statistical inference.

where $x_i = \mathbf{e}_i \cdot \mathbf{x}$, and m_i are the components of the mean vector

$$\mathbf{m} = \int_{\mathbb{R}^n} \mathbf{x} \cdot \rho_G(\mathbf{x}; \mathbf{m}, C) d\mathbf{x}.$$

C is an $n \times n$ covariance matrix with elements

$$C_{ij} = \int_{\mathbb{R}^n} (x_i - m_i)(x_j - m_j) \rho_G(\mathbf{x}; \mathbf{m}, C) d\mathbf{x},$$

and the normalization

$$\int_{\mathbb{R}^n} \rho_G(\mathbf{x}; \mathbf{m}, C) d\mathbf{x} = 1$$

is observed.

In the context of statistics, Gaussian distributions have some extremely useful properties. These include

- Closure under convolution, i.e., (an analogous result holds for $\mathbb{R}^n$) $\rho_G(x; m_1, \sigma_1) * \rho_G(x; m_2, \sigma_2) = \rho_G(x; m_3, \sigma_3)$ for appropriate values of m_3 and σ_3.

- Closure under conditioning [i.e., the product of two Gaussians is a Gaussian (to within a scale factor), and when the quotient can be normalized to be a PDF, this PDF will be a Gaussian].

- The central limit theorem [i.e., the convolution of a large number of well-behaved PDFs tends to the Gaussian distribution (see, for example, [23] and references therein)].

- Gaussians are solutions to the heat equation with δ-function as initial conditions (See Chapter 2).

- The Fourier transform of a Gaussian is a Gaussian.

- The Gaussian is an even function of its argument.

In the following sections we show how the well-known definitions and properties reviewed here are generalized in ways unfamiliar to most engineers. But first we review some non-Gaussian PDFs.

14.1.3   Other Common Probability Density Functions on the Line

While the Gaussian distribution and its generalizations play an important role in this chapter and those that follow, a number of other probability density functions on the line are worth noting.

The *Cauchy* distribution has the density

$$\rho_C(x; m, a) = \frac{a}{(\pi a)^2 + (x - m)^2} \tag{14.8}$$

for $a \in \mathbb{R}^+$ and mean $m \in \mathbb{R}$. This density has the property that it is closed under convolution, but it does not have many of the other nice properties of the Gaussian distribution listed in the previous section.

The *Laplace* distribution has the density

$$\rho_L(x; m, \sigma) = \frac{1}{\sqrt{2}\sigma} \exp\left(-\frac{\sqrt{2}|x - m|}{\sigma}\right).$$

The *rectangular* distribution has a density defined by the value $1/(b-a)$ on the interval $[a, b]$. For all values of $x \in \mathbb{R}$ outside of $[a, b]$ it takes the value zero. Clearly this is not closed under convolution. The result of the convolution of two rectangular distributions is one that is triangular.

In contrast, the sinc function defined in Chapter 2 has a Fourier transform that is (to within a constant) a rectangular distribution. The product of two rectangular distributions (in Fourier space) results in a rectangular distribution, and so sinc functions are closed under convolution. But a sinc function is not a PDF because it is not a strictly non-negative function.

14.2   Probability and Statistics on the Circle

We note that there are major differences between PDFs on the line and on the circle. Since the line is infinite in extent, it is not possible to define a PDF on the line that takes constant values. On the circle, the PDF $\rho(\theta) = 1/2\pi$ is perfectly valid. On the line, a PDF $\rho(x)$ always has a well-defined center of mass (the expected value of x). On the circle, it can be possible to have a single, multiple, or even an infinite number of "mean" values [e.g., for $\rho(\theta) = 1/2\pi$ all values can be called mean values]. Of course, when the density function is concentrated on one small portion of the circle, the tools developed for statistics on the line can be used without much difficulty. However, for distributions that are more spread out, the tools that are introduced in Section 14.3 are required to define concepts of mean and variance.

Probability density functions for the circle can be generated by "wrapping" (or "folding") PDFs defined on the line. Given that $\rho(x)$ is a PDF on the line, the function

$$\rho_W(\theta) = \sum_{n=-\infty}^{\infty} \rho(\theta - 2\pi n) \tag{14.9}$$

will be a PDF on the unit circle [or equivalently, $SO(2)$].

It is easy to verify that

$$\int_0^{2\pi} \sum_{n=-\infty}^{\infty} \rho(\theta - 2\pi n) d\theta = \int_{-\infty}^{\infty} \rho(x) dx = 1.$$

Likewise, since $\rho(x) \geq 0$ for all $x \in \mathbb{R}$, it follows that $\rho_W(\theta) \geq 0$ for all $\theta \in [0, 2\pi]$. Also, if $\rho(x, a)$ is a member of a family of density functions on the line such that

$$\rho(x, a) * \rho(x, b) = \rho(x, c)$$

(closure under convolution), then the convolution of wrapped versions of these functions will be closed under convolution on the circle. We verify this as follows:

$$\rho_W(\theta, a) * \rho_W(\theta, b) = \int_0^{2\pi} \left(\sum_{n=-\infty}^{\infty} \rho(\xi - 2\pi n, a) \right) \left(\sum_{m=-\infty}^{\infty} \rho(\theta - \xi - 2\pi m, b) \right) d\xi$$

$$= \int_{-\infty}^{\infty} \rho(\xi, a) \left(\sum_{m=-\infty}^{\infty} \rho(\theta - \xi - 2\pi m, b) \right) d\xi$$

$$= \sum_{m=-\infty}^{\infty} \int_{-\infty}^{\infty} \rho(\xi, a) \rho(\theta - \xi - 2\pi m, b) d\xi$$

$$= (\rho(x, a) * \rho(x, b))_W = \rho_W(x, c).$$

In other words, the operation of convolution on the line and wrapping on the circle can be replaced with wrapping first, then convolution on the circle, or vice versa. Hence, whenever functions are closed under convolution on the line, their wrapped versions will be closed under convolution on the circle.

If the PDF is very distributed, a Fourier series expansion will capture its shape well using few terms. If the PDF is very concentrated, then Eq. (14.9) can be truncated with good accuracy at small values of n.

For instance, the heat equation on the circle is

$$\frac{\partial f}{\partial t} = K \nabla^2 f$$

where

$$\nabla^2 = \frac{\partial^2}{\partial \theta^2}.$$

It is well known that the Fourier series solution of this heat equation under the initial condition $f(\theta, 0) = \delta(\theta - 0)$ is of the form

$$f(\theta, t) = \frac{1}{2\pi} \sum_{n=-\infty}^{\infty} e^{-n^2 K t} e^{in\theta} = \frac{1}{2\pi} + \frac{1}{\pi} \sum_{n=1}^{\infty} e^{-n^2 K t} \cos n\theta \qquad (14.10)$$

when the integration measure on the circle is $d\theta$. Another well-known form of the solution to the heat equation on the circle is what results from "wrapping" the solution of the heat equation on the line,

$$\rho(x, t) = \frac{1}{\sqrt{2\pi K t}} e^{-x^2/2Kt},$$

around the circle. It may be shown that the solution in Eq. (14.10) is related to this as

$$f(\theta, t) = \sum_{n=-\infty}^{\infty} \rho(\theta - 2\pi n, t).$$

The density of the Cauchy distribution in Eq. (14.8) can also be wrapped around the circle, the result of which can be expanded in the Fourier series [61]

$$\rho_C(\theta, b) = (2\pi)^{-1} \left(1 + 2 \sum_{n=1}^{\infty} b^n \cos n\theta \right)$$

where $b = e^{-a}$. Since

$$\sum_{n=1}^{\infty} b^n e^{-in\theta} = \sum_{n=1}^{\infty} \left(b e^{-i\theta} \right)^n$$

is a geometric series, it follows from the fact that the real part of this series is $\sum_{n=1}^{\infty} b^n \cos n\theta$ that [61]

$$\rho_C(\theta, b) = \frac{1}{2\pi} \frac{1 - b^2}{1 + b^2 - 2b \cos \theta}$$

for $0 \le b \le 1$.

Other PDFs for the circle have been defined without wrapping PDFs on the line around the circle. For instance, the *von Mises* distribution has a PDF

$$\rho_{VM}(\theta; m, \kappa) = \frac{1}{2\pi I_0(\kappa)} e^{\kappa \cos(\theta - m)}$$

where m is the mode value of θ where the PDF has its greatest value, κ dictates how spread out the density is ($\kappa = 0$ is the uniform density), and

$$I_0(\kappa) = \sum_{n=0}^{\infty} \frac{1}{(n!)^2} \left(\frac{\kappa}{2}\right)^{2r}$$

is the modified Bessel function of the first kind of order zero. Another PDF used in the context of statistics on the circle is the *cardioid* density

$$\rho_{CD}(\theta) = \frac{1}{2\pi}\left[1 + \frac{1}{2}\kappa\cos(\theta - m)\right].$$

This is a degenerate case of the von Mises density for small values of κ. See [61, 94] for a complete treatment. See [6, 31] for applications of orientational statistics (and in particular, statistics on the circle) in biology.

14.3 Metrics as a Tool for Statistics on Groups

In previous chapters we have seen a number of applications in which PDFs on groups and metrics on groups arise separately. Now we consider how these two concepts can synergistically interact.

14.3.1 Moments of Probability Density Functions on Groups and their Homogeneous Spaces

Clearly it would be useful to reduce the information in a probability density function on a group to the specification of its lowest moments. For example, prior to explicitly performing numerical convolutions on the Euclidean motion group (with or without the help of an FFT), it would be of value to know how moments of PDFs interact under convolution. But how should this be done? It is tempting to consider, as a direct analogy with the Abelian case reviewed in the previous section, a definition such as $E(g) = \int_G g\rho(g)d(g)$ for the definition of expected value. However, this definition would really not make any sense because while integration of functions on a group is well defined, integration of group elements is not. In fact, even the product of a scalar and a group element is not well defined. Even when we can get around these problems, such as when G is a matrix group, the issue remains that $E(g)$ as defined above is not even a group element in general. And while there are ways to adjust this definition to make sense in some cases (e.g., see [94]), we will use a different definition in this section.

Probability Density Functions on Groups

In this section we view the difference $x - y$ as a signed measure of distance between points on the line [instead of as the group product $x + (-y)$]. From this point of view, Eq. (14.2) is written as

$$C(y) = \int_{\mathbb{R}} d^2(x, y)\rho(x)dx$$

where $d(x, y) = |x - y|$ and $E(x)$ is the value of y which minimizes $C(y)$. Similarly, we may write Eq. (14.1) as

$$F(y) = \int_{\mathbb{R}} d(x, y)\rho(x)dx.$$

The straightforward generalization of these to PDFs on groups is that the expected value $E_d(g_1) = g_{cm}^d \in G$ is the group element which minimizes the function

$$C(g_1) = \int_G [d(g_1, g_2)]^2 \rho(g_2) d(g_2). \tag{14.11}$$

Here $d(g_1, g_2)$ is a metric [not to be confused with the integration measure $d(g)$], and clearly the center of mass in this case depends on how this metric is defined. Hence $E_d(g)$ is called the d-mean, and the value $C(g_{cm}^d)$ is what we will refer to as the d-variance. Likewise, the group element $g_{med}^d \in G$ which minimizes the function

$$F(g_1) = \int_G d(g_1, g_2) \rho(g_2) d(g_2) \tag{14.12}$$

is called the d-median. These definitions are known in the theoretical statistics literature (see, for example, [23]), but are not part of what is generally considered to be standard engineering mathematics. Other issues relating to probability density functions on groups in general [and $SO(N)$ in particular] are addressed in [48, 58, 82].

In traditional statistics in $\mathbb{R}^N$, $N \times N$ covariance matrices with entries of the form

$$C_{ij} = \int_{\mathbb{R}^N} (x_i - E[x_i])(x_j - E[x_j]) \rho(\mathbf{x}) dx$$

play an important role. Several possible extensions of this concept in the context of the rotation and motion groups are clear when using the matrix exponential. Assuming an arbitrary element is of the form

$$g(\alpha_1, \ldots, \alpha_N) = \exp\left(\sum_{i=1}^N \alpha_i X_i\right)$$

where $\{X_i\}$ is a basis for the Lie algebra, then $g(0, \ldots, \alpha_i, \ldots, 0)$ is a one parameter subgroup of G,

$$g(0, \ldots, \alpha_i, \ldots, 0) \circ g\left(0, \ldots, \alpha_i', \ldots, 0\right) = g\left(0, \ldots, \alpha_i + \alpha_i', \ldots, 0\right).$$

The logarithm of a group element is defined as

$$\log g(\alpha_1, \ldots, \alpha_N) = \sum_{i=1}^N \alpha_i X_i.$$

The inner product of two basis elements X_i and X_j is

$$(X_i, X_j) = \frac{1}{2}\text{trace}\left(X_i W X_j^T\right)$$

for an appropriate weighting matrix when the Lie algebra consists of real elements (see Chapter 7). Then we can define

$$C_{ij}^1 = \int_G \text{sgn}\left[(\alpha_i - E[\alpha_i])(\alpha_j - E[\alpha_j])\right] d\left(g(0, \ldots, \alpha_j, \ldots, 0), g(0, \ldots, E[\alpha_j], \ldots, 0)\right) \cdot$$

$$d\left(g(0, \ldots, \alpha_i, \ldots, 0), g(0, \ldots, E[\alpha_i], \ldots, 0)\right) \rho(g) dg,$$

where we define $E[\alpha_i]$ as

$$E[\alpha_i] \triangleq (\log(E[g]), X_i).$$

Using the left invariance of the metric, the exponential notation, and the shorthand $\Delta_i = \alpha_i - E[\alpha_i]$, one finds

$$C^1_{ij} = \int_G \operatorname{sgn}\left[\Delta_i \Delta_j\right] d\left(e, \exp\left(\Delta_i X_i\right)\right) d\left(e, \exp\left(\Delta_j X_j\right)\right) \rho(g) dg.$$

A second possible extension of the concept of covariance arises by using the same definitions as previously given, but with $E[\alpha_i]$ defined as the value of α_i that minimizes

$$\int_G d^2(g(0,\ldots,0,\alpha_i,0,\ldots,0))\rho(g)dg.$$

If this definition is used, we denote the corresponding covariance matrix as C^2_{ij}.

A third possibility is to first left shift the PDF so that the mean value is at the identity, and then do the calculations as in the C^1_{ij} case, hence resulting in

$$C_{ij} = \int_G \operatorname{sgn}\left[\alpha_i \alpha_j\right] d\left(e, \exp\left(\alpha_i X_i\right)\right) d\left(e, \exp\left(\alpha_j X_j\right)\right) \rho(E[g] \circ g)dg. \tag{14.13}$$

While there is no single correct choice, we use C_{ij} in Eq. (14.13) as the definition of covariance since it does not require the introduction of a choice for $E[\alpha_i]$. In the case when $G = \mathbb{R}^N$ all the definitions previously given degenerate to the classical one.

Statistics on the Homogeneous Space $SO(3)/SO(2)$

Analogous definitions hold for PDFs on homogeneous spaces G/H endowed with an integration measure invariant under G-shifts. The only difference is that integration is with respect to this measure [denoted as $d_{G/H}(g)$ or $d(gH)$ in Chapter 8] instead of $d_G(g)$, and the metric used is generated from a right-invariant metric on G as

$$d_{G/H}(g_1 H, g_2 H) \stackrel{\Delta}{=} \min_{h_1, h_2 \in H} d\left(g_1 \circ h_1, g_2 \circ h_2\right) = \min_{h \in H} d\left(g_1, g_2 \circ h\right).$$

For instance, a substantial amount of literature deals with statistics on the homogeneous space $SO(3)/SO(2) \cong S^2$. Applications include the orientation of fibers in composite materials, the orientation of pebbles in soil, and a number of other interesting problems referred to in [94]. As with the circle, it is possible to have multiple (or even an infinite number) of means and median values of a PDF on the sphere.

In the case of statistics on the sphere, there appears to be two natural ways to formulate statistics. The first is to use the concepts outlined previously, where $du = (1/4\pi)\sin\theta d\theta d\phi$ is the normalized integration measure, and $d(\mathbf{x}, \mathbf{y}) = |\cos^{-1}(\mathbf{x} \cdot \mathbf{y})|$ is the geodesic distance between two points $\mathbf{x}$ and $\mathbf{y}$ on the unit sphere S^2 written as vectors in $\mathbb{R}^3$. The second approach is to work in $\mathbb{R}^3$ without trying to "stay on the sphere" as is done in [94]. In that approach, the average of a set of points on the sphere is defined as the normalized spatial average of the points. Normalization ensures that the result is "put back" on the sphere after the average is calculated. There are some conveniences to both approaches. The first approach has the benefit of being part of a more general theory, and hence the results of general theorems can be localized to problems on the sphere. The benefit of the second approach is that a number of application areas already use it and have developed highly specialized tools in this context. For instance, in the context of paleomagnetic data the *Fisher distribution* [29]

$$f_F(\theta) = f(\mathbf{x}(\theta, \phi)) = \frac{\kappa}{\sinh \kappa} e^{\kappa \mathbf{e}_3 \cdot \mathbf{x}(\theta,\phi)} = \frac{\kappa}{\sinh \kappa} e^{\kappa \cos \theta} \tag{14.14}$$

is used to capture the distribution of magnetic poles in rocks. Here $\mathbf{e}_3$ is the direction of the mode and $\mathbf{x}$ is parameterized as in Eq. (4.19). A rotation of this PDF by $R \in SO(3)$ changes $\mathbf{x}$ to $R^T \mathbf{x}$ in the PDF (which is the same as changing $\mathbf{e}_3$ to $R\mathbf{e}_3$ and leaving $\mathbf{x}$ unchanged).

Another PDF commonly used to fit statistical data on spheres is the *Bingham density* [10]

$$b(\mathbf{x}) = \frac{e^{\mathbf{x}^T K \mathbf{x}}}{\int_{S^2} e^{\mathbf{x}^T K \mathbf{x}} d\mathbf{x}}, \tag{14.15}$$

where again $\mathbf{x} = \mathbf{x}(\theta, \phi) \in S^2$ and $K \in \mathbb{R}^{3 \times 3}$.

Clearly, when one is interested in the orientational statistics of sets of objects without an axis of symmetry, statistics on $SO(3)$ become important for exactly the same reason as statistics on S^2. The densities in Eqs. (14.14) and (14.15) have natural extensions to the quaternion sphere S^3, and densities on S^3 with antipodal symmetry are ideal for statistics on $SO(3)$.

A natural metric between elements $R_1, R_2 \in SO(3)$ can be interpreted as the absolute value of the angle of rotation, θ, found by solving $e^{\theta N} = R_1^T R_2$, or by viewing rotations as points on the upper hemisphere of S^3, and calculating $d(\mathbf{x}, \mathbf{y}) = |\cos^{-1}(\mathbf{x} \cdot \mathbf{y})|$ where now $\mathbf{x}, \mathbf{y} \in \mathbb{R}^4$ are unit vectors corresponding to rotations.

Questions regarding the average of a set of orientations or frames of reference are natural ones to ask in the field of robotics. In this context, one is often interested in how repeatably a robot arm can reach the same position and orientation in space. If, for instance, the same pick-and-place task were performed several hundred times, there would be some distribution of actual frames of reference reached. Ideally, all the reached frames would correspond to the desired one. However in practice, flexibility in the arm, backlash in gears, and even thermal effects in the motors and sensors due to the use of the arm can mean that the arm will not reach the desired frame, but rather a cloud of frames around it. Clearly then, having a measure of the average frame (position and orientation) together with the variance provides tools for describing the repeatability of a robotic arm. The methods presented in this subsection provide the vocabulary for analysis in the context of this very applied problem.

14.4 PDFs on Rotation Groups with Special Properties

In this section we review PDFs on the rotation groups $SO(2)$ and $SO(3)$ that have the special property of being closed under convolution. Since $SO(2)$ is Abelian, convolutions and shifts naturally commute. However, some effort is required in the three-dimensional case to find class functions that commute and are closed under convolution.

14.4.1 The Folded Gaussian for One-Dimensional Rotations

The Gaussian function "spread on the circle" is what results from dividing up the real line into increments of 2π, shifting (in multiples of 2π) the Gaussian function to the interval $[-\pi, \pi]$, and adding up all the shifted values

$$\psi(\theta; \beta) = \beta \sum_{k=-\infty}^{\infty} e^{-\pi \beta^2 (\theta - 2k\pi)^2}. \tag{14.16}$$

It is easy to check by direct calculation that this set of functions is closed under convolution, with

$$\psi(\theta; \beta) * \psi(\theta; \alpha) = \psi\left(\theta; \alpha'\right)$$

where

$$\frac{1}{\alpha'^2} = \frac{1}{\alpha^2} + \frac{1}{\beta^2}. \tag{14.17}$$

The Fourier series description of the same functions requires relatively few terms when β is small, and many terms when β is large. In contrast, the description of the same function as a sum of shifted Gaussians can be truncated at one or two terms with great accuracy for large values of β. Hence, for all intents and purposes, in engineering applications in which accurate sensors are used, the methods of classical probability theory on the line can be used for rotational processes in one dimension.

14.4.2 Gaussians for the Rotation Group of Three-Dimensional Space

In order to do statistics on $SO(3)$ it is useful to have a concept of Gaussian functions that have the properties enumerated at the end of Section 14.1.2.

The issue of closure under convolution is straightforward. As we have already seen, the convolution of any two band-limited functions on $SO(3)$ will be band limited, and the Fourier transform of the result will be the product of the Fourier transform matrices (in reverse order) of the original functions. In the special case when the Fourier transforms of two functions f and h are of the form

$$\hat{f}^l = \exp A_l \qquad \hat{h}^l = \exp B_l$$

[where l enumerates IURs of $SO(3)$], then it is clear that the result of the convolution will be of the form $\exp B_l \exp A_l$. If for all values of l, A_l and B_l commute ($[A_l, B_l] = 0_{(2l+1)\times(2l+1)}$) then

$$\mathcal{F}(f * h)_l = \exp(A_l + B_l) \qquad \text{and} \qquad f * h = h * f.$$

As a special case,

$$\rho(g, t_1) * \rho(g, t_2) = \rho(g, t_1 + t_2)$$

when

$$\hat{\rho}^l = \exp(B_l t).$$

In the special case when $B_l = b_l \mathbb{I}$ for all values of l, then $\rho(g, t)$ will be a class function for any value of t.

In Chapter 16, many of the solutions that result from rotational Brownian motion will have Fourier transforms of the form $\hat{f}^l(t) = \exp(A_l t)$, and hence $\hat{f}^l(t_1)\hat{f}^l(t_2) = \exp(A_l(t_1 + t_2))$. That is, solutions of certain PDEs on $SO(3)$ evaluated at different times will be closed under convolution.

We now consider a specific property of class functions under convolution.

THEOREM 14.1

The convolution of any function and a class function on a unimodular group is commutative, and the convolution of two class functions results in a class function.

PROOF By the definition of convolution, and the bi-invariance of the integration measure, a change of variables $k = h^{-1} \circ g$ allows one to write the convolution as

$$(f_1 * f_2)(g) = \int_G f_1\left(g \circ k^{-1}\right) f_2(k) d(k)$$

regardless of whether f_i are class functions or not. Assuming f_1 is a class function, then $f_1(g \circ k^{-1}) = f_1(k^{-1} \circ g)$, and since multiplication of scalar functions is commutative, we have $f_1 * f_2 = f_2 * f_1$. The same follows if f_2 had been a class function instead of f_1.

To see that the convolution of two class functions is a class function, one observes that

$$(f_1 * f_2)\left(k^{-1} \circ g \circ k\right) = \int_G f_1(h) f_2\left(h^{-1} \circ k^{-1} \circ g \circ k\right) d(h).$$

If f_2 is a class function, then $f_2(h^{-1} \circ k^{-1} \circ g \circ k) = f_2(k \circ h^{-1} \circ k^{-1} \circ g)$. Making the change of variables $s^{-1} = k \circ h^{-1} \circ k^{-1}$, one finds

$$(f_1 * f_2)\left(k^{-1} \circ g \circ k\right) = \int_G f_1\left(k^{-1} \circ s \circ k\right) f_2\left(s^{-1} \circ g\right) d(s) = \int_G f_1(s) f_2\left(s^{-1} \circ g\right) d(s),$$

where the last equality follows when f_1 is a class function. Hence we have

$$(f_1 * f_2)\left(k^{-1} \circ g \circ k\right) = (f_1 * f_2)(g),$$

which means $f_1 * f_2$ is a class function. ∎

In addition to closure under convolution, a much stronger condition is required for statistics on the rotation group to be of practical use in applications. Namely, the convolution of left-shifted versions of PDFs in a given family should be closed in the sense that they result in left-shifted PDFs in the same family, with resulting shifts equal to the composition of the originals. That is, we desire the property

$$f_1\left(g_1^{-1} \circ g\right) * f_2\left(g_2^{-1} \circ g\right) = (f_1 * f_2)\left((g_1 \circ g_2)^{-1} \circ g\right)$$

with f_1, f_2, and $f_1 * f_2$ in the same set. The next theorem describes when this is true.

THEOREM 14.2
The convolution of two left (right)-shifted functions on a non-Abelian unimodular group is equal to a left (right)-shifted version of the convolution of the unshifted functions if the original functions are class functions.

PROOF Let $\tilde{f}_i(g) = f_i(g_i^{-1} \circ g)$. Then $\tilde{f}_2(h^{-1} \circ g) = f_2(g_2^{-1} \circ (h^{-1} \circ g))$ and

$$\left(\tilde{f}_1 * \tilde{f}_2\right)(g) = \int_G f_1\left(g_1^{-1} \circ h\right) f_2\left(g_2^{-1} \circ h^{-1} \circ g\right) d(h).$$

Since f_1 is a class function, then

$$f_1\left(g_1^{-1} \circ h\right) = f_1\left(h \circ g_1^{-1}\right).$$

Letting $s = h \circ g_1^{-1}$, and using the fact that f_2 is a class function, then

$$\left(\tilde{f}_1 * \tilde{f}_2\right)(g) = \int_G f_1(s) f_2\left(g_2^{-1} \circ g_1^{-1} \circ s^{-1} \circ g\right) d(s) = (f_1 * f_2)\left(g \circ g_2^{-1} \circ g_1^{-1}\right).$$

From Theorem 14.1, $f_1 * f_2$ is a class function, and so

$$(f_1 * f_2)\left(g \circ g_2^{-1} \circ g_1^{-1}\right) = (f_1 * f_2)\left((g_1 \circ g_2)^{-1} \circ g\right).$$

A similar proof follows for right shifts. ∎

For $SO(3)$ we saw in Chapter 7 that all class functions are even functions of the angle of rotation. An analog of the folded normal distribution for the case of $SO(3)$ is [19]

$$\varphi_\beta(A) = \varphi\left(\theta'; \beta\right) = \frac{\beta^3}{\sin\theta'} \sum_{k=-\infty}^{\infty} \left(\theta' - 2k\pi\right) e^{-\pi\beta^2\left(\theta' - 2k\pi\right)^2} \tag{14.18}$$

for each fixed value of β where $\theta' = \theta/2$ and θ is the angle of rotation defined by the formula $A = e^{\theta N}$ and $\|\text{vect}(N)\| = 1$. It may be shown [19] that

$$\left(\varphi_\beta * \varphi_\alpha\right)(A) = \left(\varphi_\alpha * \varphi_\beta\right)(A) = \varphi_{\alpha'}(A)$$

where

$$\frac{1}{\alpha'^2} = \frac{1}{\alpha^2} + \frac{1}{\beta^2}. \tag{14.19}$$

We note that since $\varphi(\theta; \beta)$ is closed under convolution *and* is an even function of only θ for each fixed β (and hence a class function), shifted versions of these functions are closed under convolution as well (from Theorem 14.2).

14.5 Mean and Variance for $SO(N)$ and $SE(N)$

We now demonstrate the definitions of mean and variance in the context of the rotation and motion groups. For the rotation groups there are a variety of metrics we can use (see Chapters 5 and 6). Here we use the metric

$$d_{SO(N)}(A, R) = \|A - R\|_2 = \sqrt{\sum_{i,j=1}^{N} (A_{ij} - R_{ij})^2}$$

for all $A, R \in SO(N)$. For $SE(N)$ we use the metric

$$d_{SE(N)}(g_1, g_2) = \sqrt{\|\mathbf{a}_1 - \mathbf{a}_2\|_2^2 + L^2\|A_1 - A_2\|_2^2}$$

where $g_i = (\mathbf{a}_i, A_i)$ and $L \in \mathbb{R}^+$ is a length scale to put orientational and positional quantities in the same units.

14.5.1 Explicit Calculation for $SO(3)$

Using the metric $d_{SO(N)}$ described previously, the mean (expected value) associated with a PDF $\rho \in \mathcal{L}^2(SO(N))$ is $A_{cm} \in SO(N)$ that minimizes the function

$$C(A_1) = \int_{SO(N)} \|A_1 - A_2\|_2^2 \rho(A_2) \, dA_2.$$

In order to minimize with respect to A_1, we can differentiate the modified cost function

$$C'(A_1) = C(A_1) + \text{trace}\left(\Lambda\left(A_1^T A_1 - \mathbb{I}\right)\right)$$

with respect to the elements of A_1 and the elements of the Lagrange multiplier matrix $\Lambda = \Lambda^T \in \mathbb{R}^{N \times N}$. The solution to this problem can be written in closed form.

We note that when $N = 3$, instead of using the elements of the matrix A_1 as the variables and using six constraint equations, we could use Euler angles as the variables and no constraints, or the Euler parameters (or axis-angle parameterization) and one constraint. When $N = 2$, it is natural to use the absolute values of the differences of angles of rotation as a metric instead of $d_{SO(2)}$. Then the calculation of the mean is much like the case of statistics on $\mathbb{R}$.

The $d_{SO(N)}$-variance of $\rho(A)$ is found by evaluating $C(A_{cm})$. For closed-form formulas for the $d_{SO(N)}$-mean and variance of the convolution $(\rho_1 * \rho_2)(A)$ in terms of the $d_{SO(N)}$-means of ρ_1 and ρ_2 see [95].

14.5.2   Explicit Calculation for $SE(2)$ and $SE(3)$

The mean (expected value) associated with a PDF $\rho \in \mathcal{L}^2(SE(N))$ is the pair $(\mathbf{a}_{cm}, A_{cm}) \in SE(N)$ that minimizes the function

$$C(\mathbf{a}_1, A_1) = \int_{\mathbb{R}^N} \int_{SO(N)} \left\{ \|\mathbf{a}_1 - \mathbf{a}_2\|_2^2 + L \|A_1 - A_2\|_2^2 \right\} \rho(\mathbf{a}_2, A_2) \, d\mathbf{a}_2 dA_2.$$

The minimization with respect to $\mathbf{a}_1$ follows exactly like the case of a function on $\mathbb{R}^N$, and we find the value to be

$$\mathbf{a}_{cm} = \int_{\mathbb{R}^N} \mathbf{a}_2 \left(\int_{SO(N)} \rho(\mathbf{a}_2, A_2) \, dA_2 \right) d\mathbf{a}_2. \qquad (14.20)$$

In order to minimize with respect to A_1, we can differentiate the modified cost function

$$C'(\mathbf{a}_1, A_1) = C(\mathbf{a}_1, A_1) + \text{trace} \left(\Lambda \left(A_1^T A_1 - \mathbb{I} \right) \right)$$

with respect to the elements of A_1 and the elements of the Lagrange multiplier matrix $\Lambda = \Lambda^T \in \mathbb{R}^{N \times N}$. The solution to this problem is essentially like that for $SO(3)$.

As with the case of $SO(N)$, closed-form solutions for the orientational means and variances of the convolution of two PDFs on $SE(N)$ in terms of the means and variances of the original PDFs can be found in [95]. In particular, we are able to write the translational mean of the convolution of two PDFs in terms of the translational means of each PDF with relative ease. In order to see this, evaluate Eq. (14.20) with the PDF $(\rho_1 * \rho_2)(\mathbf{a}, A)$. We denote the result as

$$\mathbf{a}_{cm}^{1*2} = \int_{\mathbb{R}^N} \mathbf{a} \left(\int_{SO(N)} (\rho_1 * \rho_2)(\mathbf{a}, A) dA \right) d\mathbf{a}.$$

Substitution of the explicit form of $(\rho_1 * \rho_2)(\mathbf{a}, A)$ into the preceding equation yields

$$\mathbf{a}_{cm}^{1*2} = \int_{\mathbb{R}^N} \mathbf{a} \int_{SE(3)} \rho_1(\mathbf{r}, R) \left(\int_{SO(N)} \rho_2(R^{-1}(\mathbf{a} - \mathbf{r}), R^{-1}A) dA \right) d\mathbf{r} dR d\mathbf{a}.$$

Due to the invariance of integration on $SO(N)$, we may simplify the quantity inside the parenthesis by defining

$$F_2(\mathbf{a}) \overset{\Delta}{=} \int_{SO(N)} \rho_2(\mathbf{a}, A) dA.$$

Making the substitution $\mathbf{x} = R^{-1}(\mathbf{a} - \mathbf{r})$, we then have

$$\mathbf{a}_{cm}^{1*2} = \int_{SE(3)} \int_{\mathbb{R}^N} (R\mathbf{x} + \mathbf{r}) \rho_1(\mathbf{r}, R) F_2(\mathbf{x}) d\mathbf{r} dR d\mathbf{x}.$$

Passing integrals through terms that are invariant under the integral and using the fact that

$$\int_{SE(3)} \rho_i(g)dg = 1,$$

we find

$$\mathbf{a}_{cm}^{1*2} = \mathbf{a}_{cm}^1 + M\mathbf{a}_{cm}^2 \qquad (14.21)$$

where $\mathbf{a}_{cm}^i$ is the translational part of the mean of ρ_i, and

$$M = \int_{SO(N)} R \left(\int_{\mathbb{R}^N} \rho_1(\mathbf{r}, R)d\mathbf{r} \right) dR.$$

As an example of the usefulness of Eq. (14.21), consider the mean of the convolution of a PDF $\rho(\mathbf{a}, A)$ with itself n times. If $\mathbf{a}_{cm}$ is the translational mean of $\rho(\mathbf{a}, A)$, then the translational mean of $(\rho * \rho * \cdots * \rho)(\mathbf{a}, A)$ will be

$$\mathbf{a}_{cm}^{(n)} = \left(\mathbb{I} + \sum_{k=1}^n M^k \right) \mathbf{a}_{cm}.$$

But since

$$\left(\mathbb{I} + \sum_{k=1}^n M^k \right) (\mathbb{I} - M) = \left(\mathbb{I} - M^{n+1} \right),$$

it follows that if M has no eigenvalues equal to unity, then we can write

$$\mathbf{a}_{cm}^{(n)} = (\mathbb{I} - M)^{-1} \left(\mathbb{I} - M^{n+1} \right) \mathbf{a}_{cm}.$$

If all of the eigenvalues $|\lambda_i(M)| < 1$, then as $n \to \infty$ we have

$$\mathbf{a}_{cm}^{(n \to \infty)} \to (\mathbb{I} - M)^{-1} \mathbf{a}_{cm}.$$

In other words, for such PDFs on $SE(N)$, after an infinite number of convolutions, the translational mean will remain finite even though the translational mean of the original PDF is not zero. This does not happen with PDFs on $\mathbb{R}^N$.

14.6   Statistical Determination of a Rigid-Body Displacement

The problem of finding the rigid-body motion that will optimally match two sets of N points is of fundamental importance in a number of fields. Two variants on this static (time-independent) problem are described in the following sections.

14.6.1   Pose Determination without *A Priori* Correspondence

In the first variant, the one-to-one correspondence between points in each of the sets is assumed not to be known and must be determined. For each rigid-body motion that one of the sets undergoes relative to the other there are $|S_N| = N!$ possible correspondences where S_N is the group of premutations on N letters. However, all $N!$ need not be investigated, and algorithms exist that will calculate the optimal assignment of two sets of points in $O(N^3)$ arithmetic operations. The *value* of the resulting optimal assignment is the sum of distances between points whose correspondence is established by the optimal assignment algorithm. A measure of the congruence of the two sets of points comes from determining the value of the optimal assignment as one of the sets of points is displaced over all rigid-body motions and a minimum is found. Particular applications of this problem include the correspondence problem in computer vision [55] and the motion planning of modular self-reconfigurable robots [71]. While this problem makes use of the group of permutations, we are not aware of applications of noncommutative harmonic analysis in this context.

In the case when two objects (or patterns) are given, and one wants to determine how alike the objects are, measures of distance between shape are required. See [33, 71] for methods of defining distance between shapes and finding the best fit of shapes under rigid-body motion. Likewise, if it is assumed that the two patterns are the same (to within a rigid-body motion), it can still be the case that the correspondence between points in the two patterns is not known. One way to establish a correspondence is to have a metric that distinguishes how different the collection of coordinates is for two spatial patterns. That is, if the two patterns A and B occupy the same exact points in space, the distance between the patterns should be zero: $D(A, B) = 0$. $D(\cdot, \cdot)$ should satisfy all the properties of a metric. There are many pattern metrics used in computational geometry and discrete mathematics. One such metric is to assume the two point patterns are replaced with a sum of Gaussian functions centered at the points, and the $\mathcal{L}^2$ distance between the functions is calculated. That is, if $A = \{\mathbf{a}_1, \ldots, \mathbf{a}_n\}$ and $B = \{\mathbf{b}_1, \ldots, \mathbf{b}_n\}$, we define

$$f_A(\mathbf{x}) = \sum_{i=1}^n \frac{\exp\left[-\frac{1}{2\sigma}\|\mathbf{x} - \mathbf{a}_i\|^2\right]}{(2\pi\sigma)^{\frac{n}{2}}}$$

and

$$D_1(A, B) = \sqrt{\int_{\mathbb{R}^N} |f_A(\mathbf{x}) - f_B(\mathbf{x})|^2 d\mathbf{x}}.$$

Another metric, called the *Hausdorff distance* is defined as follows. Let

$$h(A, B) = \max_{a \in A} \min_{b \in B} d(a, b)$$

where $d(\cdot, \cdot)$ is any metric on $\mathbb{R}^N$. For example, the metric $d(\cdot, \cdot)$ could be Euclidean distance, or we could assign to each point a Gaussian function and calculate the $\mathcal{L}^2$ distance. It is well known that the triangle inequality holds for $h(A, B)$. For completeness, we illustrate this below

$$h(A, C) = \max_{a \in A} \min_{c \in C} d(a, c) \leq \max_{a \in A} \min_{c \in C} \left(d\left(a, b'\right) + d\left(b', c\right) \right)$$

$$= \max_{a \in A} d\left(a, b'\right) + \min_{c \in C} d\left(b', c\right) \quad \forall b' \in B.$$

This must be true for all $b' \in B$. Therefore it must be true for the choice $b' \in B$ such that $\max_{a \in A} d(a, b') = \max_{a \in A} \min_{b \in B} d(a, b)$. Therefore,

$$h(A, C) \leq \max_{a \in A} \min_{b \in B} d(a, b) + \min_{b \in B} \min_{c \in C} d(b, c) \leq h(A, B) + \max_{b \in B} \min_{c \in C} d(b, c)$$

$$= h(A, B) + h(B, C).$$

Since in general $h(A, B) \neq h(B, A)$ we may "repair" this by defining the undirected Hausdorff distance as

$$D_2(A, B) = \max(h(A, B), h(B, A)). \tag{14.22}$$

Another metric that has been shown to be useful in applications is the *optimal assignment metric* [71]. Again we begin with a measure of distance between points, $d(\cdot, \cdot)$. Then we define

$$D_3(A, B) = \min_{\pi \in \Pi_n} \sum_{i=1}^{n} d\left(\mathbf{a}_i, \mathbf{b}_{\pi(i)}\right) \tag{14.23}$$

where Π_n is the group of permutations on n letters.

Given a metric $D(\cdot, \cdot)$ that distinguishes between point patterns [e.g., any of the $D_i(\cdot, \cdot)$ mentioned previously], the rigid-body transformation with the best fit will be the one that minimizes the cost

$$C(g) = D(A, g \cdot B)$$

where $g \cdot B$ denotes each $\mathbf{b}_i \in B$ being moved as $g \cdot \mathbf{b}_i = R\mathbf{b}_i + \mathbf{r}$ where $g = (\mathbf{r}, R)$.

The optimal assignment metric in Eq. (14.23) is particularly attractive from the perspective that in addition to generating the optimal rigid-body match, it also generates a correspondence between points in A and B. This correspondence is defined by the permutation (optimal assignment) that performs the minimization in Eq. (14.23).

14.6.2   Pose Determination with *A Priori* Correspondence

In the second variant of the pose determination problem, it is assumed that the correspondence between points has been established *a priori*, and one seeks to find the relative rigid-body motion that will best match the two sets of points in the sense of least squared error. In this subsection, we review the work presented in [2, 45, 46, 68, 91]. This is what we call the static rigid-body motion estimation problem with *a priori* correspondence. In the context of computer vision and image analysis, many other related pose determination/estimation techniques have been studied extensively. See [27, 37, 50, 70, 74, 83]. In particular, the problem of rotation/orientation estimation has been addressed in [32, 47, 69].

Consider two sets of N points (positions) in $\mathbb{R}^3$ denoted as $\{\mathbf{p}_i\}$ and $\{\mathbf{q}_i\}$ where the correspondence $\mathbf{q}_i \leftrightarrow \mathbf{p}_i$ is assumed for all $i = 1, \ldots, N$. The set $\{\mathbf{q}_i\}$ can be thought of as exact points originating from a database (e.g., the vertices of a polygonal object). The set $\{\mathbf{p}_i\}$ can be thought of as measurements (e.g., if the corners of an object are detected experimentally). The goal is to find the rigid-body motion $(\mathbf{a}, A)$ such that $\{\mathbf{q}_i\}$ is moved to fit in the "best" way to $\{\mathbf{p}_i\}$. One way to define the best fit is as the minimization of the weighted mean-square error

$$E^2(A, \mathbf{a}) = \sum_{i=1}^{N} \|\mathbf{p}_i - (A\mathbf{q}_i + \mathbf{a})\|_{W_i}^2 . \tag{14.24}$$

Here we use the notation

$$\|\mathbf{x}\|_{W_i}^2 = \mathbf{x}^T W_i \mathbf{x}.$$

The 3×3 weighting matrices W_i are chosen based on how noisy the measurement of $\mathbf{p}_i$ is assumed to be. If the measurement error of all the points is assumed to be the same, then $W_i = W$ for all $i = 1, \ldots, N$.

Assuming the errors in position measurements are Gaussian distributed, we define the *likelihood function* to within a constant as

$$L(A, \mathbf{a}) = \sqrt{\det K} \exp\left(-\frac{1}{2}\epsilon^T K \epsilon\right)$$

where

$$\epsilon(A, \mathbf{a}) = \begin{pmatrix} \mathbf{p}_1 - (A\mathbf{q}_1 + \mathbf{a}) \\ \vdots \\ \mathbf{p}_N - (A\mathbf{q}_N + \mathbf{a}) \end{pmatrix} \in \mathbb{R}^{3N}$$

is a concatenated error vector.

When K is block diagonal of the form

$$K = \begin{pmatrix} W_1 & 0 & \ldots & & 0 \\ 0 & W_2 & 0 & & \ldots \\ 0 & \ldots & W_{N-1} & 0 & \\ & & & & \\ 0 & \ldots & & 0 & W_N \end{pmatrix},$$

then

$$-2 \log L = - \log \left(\prod_{i=1}^{N} \det W_i \right) + E^2(A, \mathbf{a})$$

where $E^2(A, \mathbf{a}) = \epsilon^T K \epsilon$.

In general the minimum of a function $f(x_1, \ldots x_n)$ and the minimum of $F(f(x_1, \ldots x_n))$ result in the same answer when $F(\cdot)$ is monotonically increasing. Hence the maximization of the likelihood function L is the same as the maximization of $\log L$. This, in turn, is the same as the minimization of the mean square error $E^2(A, \mathbf{a})$.

The problem of finding the optimal rotation matrix when $\mathbf{a} = \mathbf{0}$ is addressed in [12, 92, 93]. Namely, if we are given $\mathbf{p}_i = A\mathbf{q}_i + \epsilon_i$ then the $A \in SO(3)$ that minimizes $\sum_{i=1}^{n} \|\epsilon_i\|^2$ is

$$A = \left(P Q^T Q P^T \right)^{\frac{1}{2}} \left(Q P^T \right)^{-1}$$

where

$$P = M(\mathbf{p}) \stackrel{\Delta}{=} [\mathbf{p}_1, \ldots, \mathbf{p}_n] \; ; \qquad Q = M(\mathbf{q}) \stackrel{\Delta}{=} [\mathbf{q}_1, \ldots, \mathbf{q}_n].$$

Alternatively, one can compute the optimal estimate A in two stages by first computing

$$\hat{A} = \left[\sum_{k=1}^{n} \mathbf{p}_k \mathbf{q}_k^T \right] \left[\sum_{i=1}^{n} \mathbf{q}_i \mathbf{q}_i^T \right]^{-1} \tag{14.25}$$

(which solves the unconstrained least squares problem) and then finding the closest $A \in SO(3)$ to $\hat{A} \in \mathbb{R}^{3 \times 3}$ in the sense of the Frobenius (Hilbert-Schmidt) norm. The solution is the orthogonal matrix in the polar decomposition of $\hat{A}$

$$A = \hat{A} \left(\hat{A}^T \hat{A} \right)^{-\frac{1}{2}}. \tag{14.26}$$

Nádas [68] considers the case when all the individual error vectors $\epsilon_i = \mathbf{p}_i - (A\mathbf{q}_i + \mathbf{a})$ are assumed to be uncorrelated to each other, but have the same covariance matrices (hence $W_i = W$). That is, all the errors are assumed to be sampled from the same zero-mean normal distribution.

Define

$$M(\mathbf{x}) = [\mathbf{x}_1, \ldots, \mathbf{x}_N], \qquad \mathbf{x}_{cm} = \frac{1}{N} \sum_{i=1}^{N} \mathbf{x}_i, \quad \text{and} \quad \mathbf{x}_i^{rel} = \mathbf{x}_i - \mathbf{x}_{cm}. \tag{14.27}$$

Using the definition in Eq. (14.27) to define $\mathbf{p}^{rel}$ and $\mathbf{q}^{rel}$, the matrices $M(\mathbf{p}^{rel})$, $M(\mathbf{q}^{rel})$, and $M(\epsilon)$ then play an important role. In particular, the 3×3 matrix

$$C = M\left(\mathbf{p}^{rel}\right)\left[M\left(\mathbf{q}^{rel}\right)\right]^T \tag{14.28}$$

plays a special role. In the case considered by Nádas [68], it is straightforward to show that the optimal value of translation is determined by

$$\frac{\partial E^2}{\partial a_i} = 0 \quad \Rightarrow \quad \mathbf{a}_{opt} = \mathbf{p}^{rel} - WC\left(C^T W^2 C\right)^{-\frac{1}{2}} \mathbf{q}^{rel}.$$

This allows one to decouple the solution for the translational and rotational parts of the problem.

The solution for the optimal $A \in SO(3)$ can be approached in a number of ways. Perhaps the most intuitive is to formulate the problem as a constrained optimization problem, with the constraint of orthonormality of A imposed using Lagrange multipliers. This approach can be implemented in a variety of ways. For instance the six independent constraints given by the definition $AA^T = \mathbb{I}$ can be used [in which case the condition $\det(A) = 1$ must be checked after the fact], or if we define $A = [\mathbf{a}_1, \mathbf{a}_2, \mathbf{a}_3]$, then the six constraints

$$\mathbf{a}_1 \times \mathbf{a}_2 = \mathbf{a}_3, \quad \mathbf{a}_1 \cdot \mathbf{a}_1 = \mathbf{a}_2 \cdot \mathbf{a}_2 = 1, \quad \mathbf{a}_1 \cdot \mathbf{a}_2 = 0$$

ensure that a rotation will result. One could parameterize A with unit quaternions (Euler parameters) and minimize over these parameters subject to a single scalar constraint equation, or minimize over Euler angles subject to no constraint. Below we minimize over $A \in \mathbb{R}^{3 \times 3}$ subject to the constraint $AA^T = \mathbb{I}_{3 \times 3}$.

After substitution of $\mathbf{a}_{opt}$ into the cost function, and ignoring the constant $\log(\prod_{i=1}^N \det W_i)$, we then have the constrained cost function

$$F = \sum_{i=1}^N \left\|\mathbf{p}_i^{rel} - A\mathbf{q}_i^{rel})\right\|_{W_i}^2 + tr\left(\Lambda\left(A^T A - 1\right)\right),$$

where Λ is a symmetric Lagrange multiplier matrix with elements Λ_{ij} (and therefore has six independent Lagrange multipliers). Setting

$$\frac{\partial F}{\partial A_{ij}} = 0 \quad \text{and} \quad \frac{\partial F}{\partial \Lambda_{ij}} = 0$$

and assuming $W_i = W$ results in the system of equations [68]

$$W\left(M\left(\mathbf{p}^{rel}\right) - AM\left(\mathbf{q}^{rel}\right)\right)\left[M\left(\mathbf{q}^{rel}\right)\right]^T + A\Lambda = 0 \quad \text{and} \quad A^T A = \mathbb{I}_{3 \times 3}.$$

One can verify by inspection that when $W = w\mathbb{I}$,

$$A = WC\left(C^T W^2 C\right)^{-\frac{1}{2}} = C\left(C^T C\right)^{-\frac{1}{2}}$$

where the matrix C is defined in Eq. (14.28).

14.6.3 Problem Statement Using Harmonic Analysis

If $f(h)$ is the error probability density function when trying to fit rigid-body motion, then the measurements are

$$\int_{SE(N)} f(h)\delta_{\mathbf{q}_i}\left(h^{-1} \cdot \mathbf{x}\right) dh = p_i(\mathbf{x}). \tag{14.29}$$

Here $\delta_{\mathbf{q}_i}(\mathbf{x}) \overset{\triangle}{=} \delta(\mathbf{x} - \mathbf{q}_i)$ and $p_i(\mathbf{x})$ is the error PDF for point i. If the measurements were exact, then we would have $p_i(\mathbf{x}) = \delta(\mathbf{x} - \mathbf{p}_i)$ and $f(h) = \delta(g^{-1} \circ h)$ where $g \in SE(N)$ is the motion for which $g \cdot \mathbf{q}_i = \mathbf{p}_i$ for all i.

In the non-ideal case of inexact measurements we can write the system of Eq. (14.29) for all sample measurements $i \in [1, M]$ as

$$f * \Delta = P$$

where

$$\Delta(\mathbf{x}) = \left[\delta_{\mathbf{q}_1}(\mathbf{x}), \ldots, \delta_{\mathbf{q}_M}(\mathbf{x}) \right]$$

and

$$P(\mathbf{x}) = [p_1(\mathbf{x}), \ldots, p_M(\mathbf{x})],$$

and the convolution of f with each of these row vectors is defined to be the convolution with each element. The Fourier transform for $SE(N)$ can be used to reduce these convolutions to matrix products.

In practice, we will choose in advance an ansatz for $f(h)$ and seek to fit a few parameters to the data. In particular, we can define a Gaussian function on $SE(N)$ to be the product of Gaussians on $\mathbb{R}^N$ and $SO(N)$ centered around the respective identities. We denote

$$f(\mathbf{a}, A) = f_t(\mathbf{a}) f_r(A).$$

The convolution of two such functions results in a function of the same form when the translational Gaussian function is spherically symmetric.

Explicitly,

$$(f_1 * f_2)(\mathbf{a}, A) = \int_{SO(N)} \int_{\mathbb{R}^N} f_1(\mathbf{r}, R) f_2 \left(R^{-1}(\mathbf{a} - \mathbf{r}), R^{-1}A \right) d\mathbf{r} \, dR.$$

Substitution of $f_i(\mathbf{a}, A) = f_{t_i}(\mathbf{a}) f_{r_i}(A)$ and using the spherical symmetry $f_{t_i}(Q\mathbf{a}) = f_{t_i}(\mathbf{a})$ for all $Q \in SO(3)$, it may be shown that

$$(f_1 * f_2)(\mathbf{a}, A) = \left(f_{t_1} * f_{t_2} \right)(\mathbf{a}) \cdot \left(f_{r_1} * f_{r_2} \right)(A).$$

On the right side of the preceding equation, the convolutions are respectively those for $\mathbb{R}^N$ and $SO(N)$. Since the Gaussians for $\mathbb{R}^N$ and $SO(N)$ are closed under convolutions (in the sense that the convolution of two Gaussians is a Gaussian), it follows that the Gaussian defined for $SE(N)$ is also closed under convolution. Furthermore, if f_1 and f_2 are both Gaussians on $SE(N)$ then

$$(f_1 * f_2)(\mathbf{a}, A) = (f_2 * f_1)(\mathbf{a}, A).$$

However, while convolutions of such Gaussians commute with each other, they do not commute with arbitrary functions on $SE(N)$. That is, these Gaussians are not class functions.

14.7 Robot Sensor Calibration

This section reviews the robot sensor calibration problem solved by Shiu and Ahmad [88], Chou and Kamel [21], and Park and Martin [72]. We then extend the problem in a new direction, and show how noncommutative harmonic analysis may be applicable.

The problem is stated as follows. A robot arm which has already been calibrated has a frame of reference attached to its distal end. For any given movement the arm away from a home position, let this distal frame be denoted as g_A. When the arm is in the home position, we take g_A to be the identity frame. We now attach to this distal end (which can be considered the wrist of the robot arm) a new sensor (e.g., camera, or hand with tactile sensors). It is assumed that the kinematic parameters describing the position and orientation of a frame of reference attached to the new sensors, g_X, relative to the wrist frame, g_A, is unknown. However, the relative displacement of the new sensor location to its own position and orientation when the robot arm is in the home position can be measured. Hence, the composition of motions $g_A \circ g_X$ results in the position and orientation of the sensor frame relative to the identity frame. Likewise, $g_X \circ g_B$ describes the concatenation of motions from the identity frame to sensor frame when the arm is in the home position, and then the motion of this frame. Hence the equality [21, 72, 88]

$$g_A \circ g_X = g_X \circ g_B. \tag{14.30}$$

The solution to this problem for g_X in the case when it is assumed that one exact value of g_A and one exact value of g_B are known is underdetermined (i.e., there is a continuum of solutions) [88]. The case when multiple calibrating motions, (g_{A_i}, g_{B_i}) for $i = 1, \ldots, N$, are performed is addressed in [72]. We review both of these cases, and extend the analysis to answer the question: given two sets of measurements, where each measurement of g_A and g_B are assumed to have known error distributions, how is this reflected in the distribution of error in g_X? Figure 14.1 illustrates Eq. (14.30).

14.7.1 Solution with Two Sets of Exact Measurements

Using homogeneous transform notation, let $g = (\mathbf{q}, Q) \in SE(3)$ be represented as

$$H(g) = \begin{pmatrix} Q & \mathbf{q} \\ \mathbf{0}^T & 1 \end{pmatrix}.$$

Then Eq. (14.30) is rewritten as

$$\begin{pmatrix} Q_A & \mathbf{q}_A \\ \mathbf{0}^T & 1 \end{pmatrix} \begin{pmatrix} Q_X & \mathbf{q}_X \\ \mathbf{0}^T & 1 \end{pmatrix} = \begin{pmatrix} Q_X & \mathbf{q}_X \\ \mathbf{0}^T & 1 \end{pmatrix} \begin{pmatrix} Q_B & \mathbf{q}_B \\ \mathbf{0}^T & 1 \end{pmatrix}.$$

Performing the matrix multiplication results in two equations of the form

$$Q_A Q_X = Q_X Q_B \tag{14.31}$$

and

$$Q_A \mathbf{q}_X + \mathbf{q}_A = Q_X \mathbf{q}_B + \mathbf{q}_X. \tag{14.32}$$

The strategy to solve Eq. (14.30) would appear to reduce to first solving Eq. (14.31), and then rearranging Eq. (14.32) so as to find acceptable values of $\mathbf{q}_X$.

$$(Q_A - \mathbb{I}_{3\times 3}) \mathbf{q}_X = Q_X \mathbf{q}_B - \mathbf{q}_A.$$

However, there are some problems with this naive approach. As pointed out in [21, 72, 88], there is a one-parameter set of solutions to Eq. (14.31), and the matrix $Q_A - \mathbb{I}_{3\times 3}$ in general has rank 2. Hence, there are two unspecified degrees of freedom to the problem, and it cannot be solved uniquely unless additional measurements are taken.

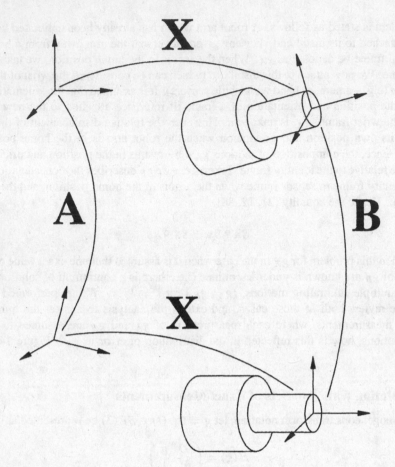

FIGURE 14.1
Robot sensor calibration: $AX = XB$.

This situation is rectified by considering two sets of exact measurements of the form in Eq. (14.30), i.e.,

$$g_{A_1} \circ g_X = g_X \circ g_{B_1} \quad \text{and} \quad g_{A_2} \circ g_X = g_X \circ g_{B_2}.$$

We now consider the technical details of the problem. To begin, we observe that Eq. (14.31) may be rewritten as

$$Q_A = Q_X Q_B Q_X^{-1}. \tag{14.33}$$

Regardless of what Q_X is, the trace is invariant under similarity transformation [trace$(Q_X Q_B Q_X^{-1}) =$ trace(Q_B)], and so

$$\text{trace}\,(Q_A) = \text{trace}\,(Q_B). \tag{14.34}$$

This means that the angle of rotation of Q_A and Q_B must be the same. For the present discussion, we will assume that trace$(Q_A) \neq -1$ (and the same for Q_B).

Using the vect$(\cdot)$ operation[2] on both sides of Eq. (14.33), observing that vect$(Q_X Q_B Q_X^{-1})$ $= Q_X$vect(Q_B), and dividing both sides by the magnitude $\|$vect$(Q_A)\| = \|$vect$(Q_B)\|$, results

---

[2]For matrices that are not skew symmetric the vect$(\cdot)$ operation acts on the skew-symmetric part.

in the equation

$$Q_X \mathbf{b} = \mathbf{a}$$

where $\mathbf{a}$ and $\mathbf{b}$ are the unit vectors pointing along the axes of rotation of Q_A and Q_B, respectively.

A particular solution to this equation for given $\mathbf{a}$ and $\mathbf{b}$ is

$$Q_Y = R(\mathbf{b}, \mathbf{a}), \qquad (14.35)$$

which is the rotation defined in (5.24), that takes $\mathbf{b}$ into $\mathbf{a}$.

It is also clear that if Q_Y is a solution to Eq. (14.31), then so must

$$Q_X = \mathrm{ROT}[\mathbf{a}, \theta_1] \, Q_Y \mathrm{ROT}[\mathbf{b}, \theta_2] \qquad (14.36)$$

for arbitrary $\theta_1, \theta_2 \in [0, 2\pi]$. Using the notation $\exp(\theta\hat{\mathbf{a}}) = \mathrm{ROT}[\mathbf{a}, \theta]$, Eq. (14.36) is written as[3]

$$Q_X = Q_Y \exp\left(\theta_1 Q_Y^T \hat{\mathbf{a}} Q_Y\right) \mathrm{ROT}[\mathbf{b}, \theta_2]$$

where $\mathrm{vect}(\hat{\mathbf{a}}) = \mathbf{a}$. This is simplified to

$$Q_X = Q_Y \mathrm{ROT}\left[Q_Y^T \mathbf{a}, \theta_1\right] \mathrm{ROT}[\mathbf{b}, \theta_2] = Q_Y \mathrm{ROT}[\mathbf{b}, \theta_1 + \theta_2].$$

Hence, despite the appearance of Eq. (14.36), there is a one parameter set of solutions to the rotational part of the problem.

To solve for a unique Q_X, two sets of measurements (Q_{A_1}, Q_{B_1}) and (Q_{A_2}, Q_{B_2}) are required. For each we have the condition

$$Q_X \mathbf{b}_i = \mathbf{a}_i.$$

One also observes that

$$\mathbf{a}_1 \times \mathbf{a}_2 = (Q_X \mathbf{b}_1) \times (Q_X \mathbf{b}_2) = Q_X (\mathbf{b}_1 \times \mathbf{b}_2). \qquad (14.37)$$

Combining all this into one equation, one has

$$Q_X [\mathbf{b}_1, \mathbf{b}_2, \mathbf{b}_1 \times \mathbf{b}_2] = [\mathbf{a}_1, \mathbf{a}_2, \mathbf{a}_1 \times \mathbf{a}_2].$$

This is easily inverted to find Q_X provided $\mathbf{b}_1 \times \mathbf{b}_2 \neq \mathbf{0}$ [which, by Eq. (14.37) means $\mathbf{a}_1 \times \mathbf{a}_2 \neq \mathbf{0}$ as well].

Having solved the rotational part of the problem, the translation $\mathbf{q}_X$ is determined by simultaneously solving the two equations

$$\left(Q_{A_i} - \mathbb{I}_{3\times3}\right) \mathbf{q}_X = Q_X \mathbf{q}_{B_i} - \mathbf{q}_{A_i} \qquad (14.38)$$

for $i = 1, 2$.

This can be done in closed form as follows. Fix the value of i and write

$$\mathbf{q}_X = x_i \mathbf{n}_i + y_i \mathbf{v}_i + z_i (\mathbf{n}_i \times \mathbf{v}_i)$$

where $\mathbf{n}_i$ is the unit vector pointing in the direction of the axis of rotation of Q_{A_i} and $\mathbf{v}_i = [-n_2, n_1, 0]^T$. Then

$$Q_{A_i} \mathbf{v}_i = \mathbf{v}_i \cos\theta_i + (\mathbf{n}_i \times \mathbf{v}_i) \sin\theta_i$$
$$Q_{A_i} (\mathbf{n}_i \times \mathbf{v}_i) = -\mathbf{v}_i \sin\theta_i + (\mathbf{n}_i \times \mathbf{v}_i) \cos\theta_i.$$

---

[3] Recall that for any $B, X \in \mathbb{R}^{N \times N}$ it holds that $B e^X B^{-1} = e^{BXB^{-1}}$.

Using this fact, and taking the dot product of Eq. (14.38) with $\mathbf{v}_i$ and $(\mathbf{n}_i \times \mathbf{v}_i)$, two scalar equations result that can be arranged as

$$\begin{pmatrix} \cos\theta_i - 1 & -\sin\theta_i \\ \sin\theta_i & \cos\theta_i + 1 \end{pmatrix} \begin{pmatrix} y_i \\ z_i \end{pmatrix} = \begin{pmatrix} (Q_X\mathbf{q}_{B_i} - \mathbf{q}_{A_i}) \cdot \mathbf{v}_i \\ (Q_X\mathbf{q}_{B_i} - \mathbf{q}_{A_i}) \cdot (\mathbf{n}_i \times \mathbf{v}_i) \end{pmatrix}. \quad (14.39)$$

For $\theta_i \neq 0$ we can invert this to find y_i and z_i for each $i = 1, 2$. What remains is to find x_1 and x_2 such that

$$\mathbf{q}_X = x_1\mathbf{n}_1 + y_1\mathbf{v}_1 + z_1(\mathbf{n}_1 \times \mathbf{v}_1) = x_2\mathbf{n}_2 + y_2\mathbf{v}_2 + z_2(\mathbf{n}_2 \times \mathbf{v}_2).$$

This is nothing more than the problem of finding where two lines (each parameterized by x_i) intersect. This problem can be solved by finding the values of x_1 and x_2 that minimize the square of the distance between points on the two lines. This leads to the equations

$$\begin{pmatrix} 1 & -\mathbf{n}_1 \cdot \mathbf{n}_2 \\ -\mathbf{n}_1 \cdot \mathbf{n}_2 & 1 \end{pmatrix} \begin{pmatrix} x_1 \\ x_2 \end{pmatrix} = \begin{pmatrix} \mathbf{n}_1 \cdot (y_2\mathbf{v}_2 + z_2(\mathbf{n}_2 \times \mathbf{v}_2)) \\ \mathbf{n}_2 \cdot (y_1\mathbf{v}_1 + z_1(\mathbf{n}_1 \times \mathbf{v}_1)) \end{pmatrix}.$$

We note that this solution is particularly simple if we can choose $\mathbf{n}_1$ and $\mathbf{n}_2$ such that $\mathbf{n}_1 \cdot \mathbf{n}_2 = 0$. Things simplify even more when $\mathbf{v}_1 = \mathbf{n}_2$ and $\mathbf{v}_2 = \mathbf{n}_1$, in which case $x_1 = y_2$ and $x_2 = y_1$.

The problem of solving for Q_X and $\mathbf{q}_X$ for $N > 2$ measurements is done in [21, 72].

14.7.2   Calibration with Two Noisy Measurements

As with the analysis of accumulation of kinematic error presented in Chapter 12, we reformulate the calibration problem by assuming the measured values g_A and g_B are random variables with probability density functions $\rho_A(g)$ and $\rho_B(g)$. One way to generate these PDFs would be to have the robot go through a series of motions, always returning to the same calibration configuration. In this calibration pose, multiple values of g_A and g_B can be recorded, and the best-fit Gaussian distributions on $SE(3)$ that describe the scatter of g_A and g_B can be generated. Assuming that g_X will have a Gaussian PDF associated with it, the problem becomes one of finding the correct shift (mean) and variance for $\rho_X(g)$ for given $\rho_A(g)$ and $\rho_B(g)$. This means solving equations of the form

$$\left(\rho_{A_i} * \rho_X\right)(g) = \left(\rho_X * \rho_{B_i}\right)(g)$$

for $i = 1, 2$. This task would, at first sight, appear to be related to the inverse convolution equations examined in Chapter 12. However, it is fundamentally different in the sense that here the form of the functions is completely assumed in advance, and only a few fitting parameters are sought. These are determined by solving algebraic equations, without having to invoke the Fourier transform on $SE(3)$. In fact, the generally small variances in this kind of problem would make Fourier methods rather unattractive.

14.8   Summary

In this chapter we have shown how statistical tools can be defined in a group-theoretic setting and applied in the context of pose determination and camera calibration. In addition, review material on statistics on the circle and spheres was presented, and we introduced the concept of a folded normal distribution for $SO(3)$, as well as concepts of mean and variance for $SO(N)$ and $SE(N)$. We computed the translational part of the mean of the convolution of two PDFs on $SE(N)$ in terms of the means of each function.

References

[1] Arnold, K.J, *On Spherical Probability Distributions,* Ph.D. thesis, MIT, Cambridge, MA, 1941.

[2] Arun, K.S., Huang, T.S., and Blostein, S.D., Least-squares fitting of two 3-D point sets, *IEEE Trans. on Pattern Anal. and Mach. Intelligence,* 9(5), 698–700, 1987.

[3] Baghi, P. and Guttman, I. Theoretical considerations of the multivariate von Mises-Fisher distribution, *J. of Appl. Stat.,* 15, 149–169, 1988.

[4] Bai, Z.D., Rao, C.R., and Zhao, L.C. Kernel estimators of density function of directional data, *J. of Multivariate Anal.,* 27, 24–39, 1988.

[5] Ball, K.A. and Pierrynowski, M.R., Classification of errors in locating a rigid body, *J. Biomechanics,* 29(9), 1213–1217, 1996.

[6] Batschelet, E., *Circular Statistics in Biology,* Academic Press, New York, 1981.

[7] Beran, R.J., Testing for uniformity on a compact homogeneous space, *J. Appl. Prob.,* 5, 177–195, 1968.

[8] Beran, R. Exponential models for directional data, *The Ann. of Stat.,* 7(6), 1162–1178, 1979.

[9] Besl, P.J. and McKay, N.D., A method for registration of 3-D shapes, *IEEE Trans. on Pattern Anal. and Mach. Intelligence,* 14(2), 239–256, 1992.

[10] Bingham, C., An antipodally symmetric distribution on the sphere, *The Ann. of Stat.,* 2(6), 1201–1225, 1974. (See also *Distributions on the sphere and projective plane,* Ph.D. dissertation, Yale University, New Haven, CT, 1964.)

[11] Breitenberger, E., Analogues of the normal distribution on the circle and sphere, *Biometrika,* 50, 81–88, 1963.

[12] Brock, J.E., Optimal matrices describing linear systems, *AIAA J.,* 6, 1292–1296, 1968.

[13] Brunner, L.J. and Lo, A.Y., Bayes methods for a symmetric unimodal density and its mode, *The Ann. of Stat.,* 17(4), 1550–1566, 1989.

[14] Chang, T., Spherical regression, *Ann. of Stat.,* 14, 907–924, 1986.

[15] Chang, T., On statistical properties of estimated rotations, *J. of Geophys. Res.,* 92(B7), 6319–6329, 1987.

[16] Chang, T., Spherical regression with errors in variables, *Ann. of Stat.,* 17(1), 293–306, 1989.

[17] Chang, T., Stock, J., and Molnar, P., The rotation group in plate tectonics and the representation of uncertainties of plate reconstructions, *Geophys. J. Int.,* 101, 649–661, 1990.

[18] Chapman, G.R., Chen, G., and Kim, P.T., Assessing geometric integrity through spherical regression techniques, *Stat. Sinica,* 5, 173–220, 1995.

[19] Chételat, O. and Chirikjian, G.S., Sampling and convolution on motion groups using generalized Gaussian functions, preprint.

[20] Chevalley, C., *Theory of Lie Groups I,* Princeton University Press, Princeton, NJ, 1946.

[21] Chou, J.C.K. and Kamel, M., Finding the position and orientation of a sensor on a robot manipulator using quaternions, *Int. J. of Robotics Res.*, 10(3), 240–254, 1991.

[22] Creamer, G., Spacecraft attitude determination using gyros and quaternion measurements, *J. of the Astronautical Sci.*, 44(3), 357–371, 1996.

[23] Diaconis, P., *Group Representations in Probability and Statistics*, Lecture Notes-Monogr. Ser., Gupta, S.S., Ed., Institute of Mathamatical Statistics, Hayward, CA, 1988.

[24] Diaconis, P. and Shahshahani, M., On square roots of the uniform distribution on compact groups, *Proc. Am. Math. Soc.*, 98, 341–348, 1986.

[25] Diggle, P.J. and Hall, P., A Fourier approach to nonparametric deconvolution of a density estimate, *J. of the Royal Stat. Soc. B*, 55(2), 523–531, 1993.

[26] Downs, T.D., Orientation statistics, *Biometrika*, 59, 665–676, 1972.

[27] Eggert, D.W., Lorusso, A., and Fischer, R.B., Estimating 3-D rigid body transformations: a comparison of four major algorithms, *Mach. Vision and Appl.*, 9(5-6), 272–290, 1997.

[28] Falkowski, B.-J., Infinitely divisible positive functions on $SO(3) \times \mathbb{R}^3$, in Lecture Notes in Math. No. 706, Dold, A. and Eckmann, B., Springer-Verlag, New York, 1979.

[29] Fisher, S.R., Dispersion on a sphere, *Proc. of the Royal Soc.*, A217, 295–305, 1953.

[30] Fisher, N.I., Lewis, T., and Embleton, B.J., *Statistical Analysis of Spherical Data*, Cambridge University Press, Cambridge, MA, 1987.

[31] Giske, J., Huse, G., and Fiksen, O., Modelling spatial dynamics of fish, *Rev. in Fish Biol. and Fisheries*, 8, 57–91, 1998.

[32] Goryn, D. and Hein, S., On the estimation of rigid-body rotation from noisy data, *IEEE Trans. on Pattern Anal. and Mach. Intelligence*, 17(12), 1219–1220, 1995.

[33] Goodrich, M.T., Mitchell, J.S.B., and Orletsky, M.W., Approximate geometric pattern matching under rigid motions, *IEEE Trans. on Pattern Anal. and Mach. Intelligence*, 21(4), 371–379, 1999.

[34] Gorostiza, L.G., The central limit theorem for random motions of d-dimensional Euclidean space, *The Ann. of Probab.*, 1(4), 603–612, 1973.

[35] Grenander, U., Miller, M.I., and Srivastava, A., Hilbert-Schmidt lower bounds for estimators on matrix Lie groups for ATR, *IEEE Trans. on Pattern Anal. and Mach. Intelligence*, 20(8), 790–802, 1998.

[36] Halmos, P.R., *Measure Theory*, Van Nostrand and Co., Princeton, NJ, 1950.

[37] Haralick, R.M., Joo, H., Lee, C.-N., Zhuang, X., Vaidya, V.G., and Kim, M.B., Pose estimation from corresponding point data, *IEEE Trans. on Syst., Man, and Cybernetics*, 19(6), 1426–1446, 1989.

[38] Hartman, P. and Watson, G.S., 'Normal' distribution functions on spheres and the modified Bessel function, *The Ann. of Probability*, 2(4), 593–607, 1974.

[39] Healy, D.M., Jr., Hendriks, H., and Kim, P.T., Spherical deconvolution, *J. of Multivariate Anal.*, 67, 1–22, 1998.

[40] Healy, D.M., Jr. and Kim, P.T., An empirical Bayes approach to directional data and efficient computation on the sphere, *The Ann. of Stat.*, 24(1), 232–254, 1996.

[41] Hendriks, H. Nonparametric estimation of a probability density on a Riemannian manifold using Fourier expansions, *The Ann. of Stat.*, 18(2), 832–849, 1990.

[42] Herz, C.S., Bessel functions of matrix argument, *Ann. Math.*, 61, 474–523, 1955.

[43] Heyer, H., Moments of probability measures on a group, *Int. J. of Math. and the Math. Sci.*, 4(1), 1–37, 1981.

[44] Heyer, H., *Probability Measures on Locally Compact Groups*, Springer-Verlag, New York, 1977.

[45] Horn, B.K.P., Closed-form solution of absolute orientation using unit quaternions, *J. of the Opt. Soc. of Am.*, 4(4), 629–642, 1987.

[46] Horn, B.K.P., Hilden, H.M., and Negahdaripour, S., Closed-form solution of absolute orientation using orthonormal matrices, *J. of the Opt. Soc. of Am.*, 5(7), 1127–1135, 1988.

[47] Kanatani, K., Analysis of 3-D rotation fitting, *IEEE Trans. on Pattern Anal. and Mach. Intelligence*, 16(5), 543–549, 1994.

[48] Kim, P.T., Deconvolution density estimation on SO(N), *Ann. of Stat.*, 26(3), 1083–1102, 1998.

[49] James, A.T., Normal multivariate analysis and the orthogonal group, *Ann. Math. Statist.*, 25, 40–75, 1954.

[50] Joseph, S.H., Optimal pose estimation in two and three dimensions, *Comput. Vision and Image Understanding*, 73(2), 215–231, 1999.

[51] Jupp, P.E. and Mardia, K.V., Maximum likelihood estimation for the matrix von Mises-Fisher and Bingham distributions, *Ann. of Stat.*, 7, 599–606, 1979.

[52] Kanatani, K., Analysis of 3-D rotation fitting, *IEEE Trans. on Pattern Anal. and Mach. Intelligence*, 16(5), 543–549, 1994.

[53] Khatri, C.G. and Mardia, K.V., The von Mises-Fisher matrix distribution in orientation statistics, *J. of the Royal Stat. Soc. B*, 39, 95–106, 1977.

[54] Krumbein, W.C., Preferred orientation of pebbles in sedimentary deposits, *J. of Geol.*, 47, 673–706, 1939.

[55] Lee, S.Y. and Chirikjian, G.S., Matching sets of points under perspective transformation, in preparation.

[56] Levy, P., L'addition des variables aléatoires définies sur une circonference, *Bull. Soc. Math. France*, 67, 1–41, 1939.

[57] Littlewood, D.E., Invariant theory under orthogonal groups, *Proc. London Math. Soc. 2*, 50, 349–379, 1948.

[58] Lo, J.T.-Y. and Ng, S.K., Characterizing Fourier-series representation of probability-distributions on compact Lie-groups, *SIAM J. on Appl. Math.*, 48(1), 222–228, 1988.

[59] MacKenzie, J.K., The estimation of an orientation relationship, *Acta Crystallographica*, 10, 61–62, 1957.

[60] Madsen, C.B., A comparative study of the robustness of two pose estimation techniques, *Mach. Vision and Appl.*, 9, 291–303, 1997.

[61] Mardia, K.V., *Statistics of Directional Data*, Academic Press, New York, 1972.

[62] Markley, F.L., Attitude determination using vector observations and the singular value decomposition, *J. of the Astronautical Sci.*, 36(3), 245–258, 1988.

[63] Maximov, V.M., Local theorems for Euclidean motions. I, *Zeitschrift für Wahrscheinlichkeitstheorie und verwandte Gebiete*, 51, 27–38, 1980.

[64] Miles, R.E., On random rotations in $\mathbb{R}^3$, *Biometrica*, 52, 636–639, 1965.

[65] Moran, P.A.P., Quaternions, Haar measure and the estimation of a paleomagnetic rotation, in *Perspectives in Probability and Statistics*, Gani, J., Ed., 295–301, Academic Press, Orlando, FL, 1976.

[66] Murnaghan, F.D., *The Unitary and Rotation Groups*, Spartan Books, Washington, DC, 1962.

[67] Nachbin, L., *The Haar Integral*, Van Nostrand and Co., Princeton, NJ, 1965.

[68] Nádas, A., Least squares and maximum likelihood estimates of rigid motion, *IBM Res. Rep.* RC 6945 (#29783), Mathematics, 1978.

[69] Ohta, N. and Kanatani, K., Optimal estimation of three-dimensional rotation and reliability evaluation, *IEICE Trans. on Inf. and Syst.*, E81D(11), 1247–1252, 1998.

[70] Or, S.H., Luk, W.S., Wong, K.H., and King, I., An efficient iterative pose estimation algorithm, *Image and Vision Comput.*, 16(5), 353–362, 1998.

[71] Pamecha, A., Ebert-Uphoff, I., and Chirikjian, G.S., Useful metrics for modular robot motion planning, *IEEE Trans. on Robotics and Autom.*, 13(4), 531–545, 1997.

[72] Park, F.C. and Martin, B.J., Robot sensor calibration: solving $AX = XB$ on the Euclidean group, *IEEE Trans. Robotics and Autom.*, 10(5), 717–721, 1994.

[73] Parthasarathy, K.R., The central limit theorem for the rotation group, *Probab. Theor. and its Appl.*, 9(2), 248–257, 1964.

[74] Pennec, X. and Thirion, J.P., A framework for uncertainty and validation of 3-D registration methods based on points and frames, *Int. J. of Comput. Vision*, 25(3), 203–229, 1997.

[75] Prentice, M.J., On invariant tests of uniformity for directions and orientations, *Ann. Stat.*, 6, 169–176, 1978.

[76] Prentice, M.J., Antipodally symmetric distributions for orientation statistics, *Stat. Plann. and Inference*, 66, 205–214, 1982.

[77] Prentice, M.J., Orientation statistics without parametric assumptions, *J. R. Stat. Soc. B*, 48(2), 214–222, 1986.

[78] Raffenetti, R.C. and Ruedenberg, K., Parameterization of an orthogonal matrix in terms of generalized Eulerian angles, *Int. J. of Quant. Chem.*, III S, 625–634.

[79] Rivest, L.-P., Spherical regression for concentrated Fisher-von Mises distributions, *The Ann. of Stat.*, 17(1), 307–317, 1989.

[80] Rivest, L.P., Some linear models for estimating the motion of rigid bodies with applications to geometric quality assurance, *J. of the Am. Stat. Assoc.*, 93(442), 632–642, 1998

[81] Roberts, P.H. and Ursell, H.D., Random walk on a sphere and on a Riemannian manifold, *Philos. Trans. of the Royal Soc. of London*, A252, 317–356, 1960.

[82] Rosenthal, J.S., Random rotations — characters and random-walks on $SO(N)$, *Ann. of Probab.*, 22(1), 398–423, 1994.

[83] Rosin, P.L., Robust pose estimation, *IEEE Trans. on Syst., Man and Cybernetics Part B-Cybernetics*, 29(2), 297–303, 1999.

[84] Rukhin, A.L., Estimation of a rotation parameter on a sphere, *Zapiski Nauchnykh Seminarov Leningradskogo Otdeleniya Matematicheskogo Instituta im. V.A. Steklova Akademii Nauk SSSR*, 29, 74–91, 1972 (in Russian).

[85] Saw, J.G., A family of distributions on the m-sphere and some hypothesis tests, *Biometrika*, 65, 69–73, 1978.

[86] Schaub, H., Junkins, J.L., and Robinett, R.D., New penalty functions and optimal control formulation for spacecraft attitude control problems, *J. of Guidance, Control, and Dynamics*, 20(3), 428–434, 1997.

[87] Selby, B., Girdle distributions on a sphere, *Biometrika*, 51, 381–392, 1964.

[88] Shiu, Y.C. and Ahmad, S., Calibration of wrist-mounted robotic sensors by solving homogeneous transform equations of the form $AX = XB$, *IEEE Trans. Robotics and Autom.*, 5(1), 16–29, 1989.

[89] Stephens, M.A., Vector correlation, *Biometrika*, 66, 41–48, 1979.

[90] Tutubalin, V.N., The central limit theorem for random motions of a Euclidean space, *Selected Transl. in Math.: Stat. and Probab.*, 12, 47–57, 1973.

[91] Umeyama, S., Least-squares estimation of transformation parameters between two point patterns, *IEEE Trans. on Pattern Anal. and Mach. Intelligence*, 13(4), 376–380, 1991.

[92] Wahba, G., Problem 65-1, a least squares estimate of spacecraft attitude, *SIAM Rev.*, 7(3), 409, 1965.

[93] Wahba, G., et al., Problem 65-1 (Solution) *SIAM Rev.*, 8, 384–386, 1966.

[94] Watson, G.S., *Statistics on Spheres*, Wiley-InterScience, New York, 1983.

[95] Suthakorn, J., Chirikjian, G.S., A New Inverse Kinematics Algorithm for Binary Manipulators with many actuators, preprint.

Chapter 15

Stochastic Processes, Estimation, and Control

In this chapter we consider estimation of systems in motion and control of certain kinds of dynamical systems. A dynamical system could be something as simple as the furnace in a building. In this case the controller might consist of a thermostat which measures the current temperature, compares to the desired, and either turns the furnace on or off to push the actual temperature toward the desired. The dynamical system of interest could also be something as complicated as a high performance airplane, robot, or even the human body. The physical mechanisms, by which the system evolves over time may be completely different. The system could be purely electronic, electromechanical, chemical, or even physiological. One may desire to control the rate of the system (as in chemical processes) or the position (as in a robot).

The field of feedback control seeks to modify existing dynamical systems in order for the modified system to evolve in time in a desired way. Pose determination is a prerequisite for the control of rigid bodies in the context of an imprecise world. Hence, the developments presented here build on those of Chapter 14.

In the first few sections of this chapter, we briefly review basic nonlinear models of mechanical systems, classical linear systems theory, recursive least-squares fitting and stochastic processes, together with the associated control system design techniques. Many of the most popular stochastic control techniques use the properties of a Gaussian probability density function. We extend these ideas of recursive least-squares fitting to systems which have states in the rotation and motion groups. Finally, we extend the results of linear control and estimation theory to systems that evolve in these Lie groups.

15.1 Stability and Control of Nonlinear Systems

Regardless of the particular application, there are several broad categories of models into which most dynamical systems fall, and corresponding analytical/computational techniques for control system design have been developed for each of these categories. In the most general context, the "system" can be described with the mathematical model

$$\frac{d\mathbf{x}}{dt} = \mathbf{f}[\mathbf{x}, \mathbf{u}, \mathbf{w}, t]. \tag{15.1}$$

Here $\mathbf{x}(t)$ is the actual state of the system at time t. For a purely mechanical system, the state can be viewed as the set of all generalized coordinates that describe the geometry of the system together with the rates of these generalized coordinates. $\mathbf{u}(t)$ is the control input to the system. For a mechanical system this may be a set of generalized forces which alter the behavior of the system. $\mathbf{w}(t)$ is some kind of external random disturbance acting on the system at time t. It is assumed within this model that the system itself is known exactly (i.e., that it has already been calibrated). If this were not the case, then the parameters characterizing the system would have to be identified. The area of *system identification* is a large area of research in itself, which is not addressed here.

In practice, one is often not able to directly measure the state vector $\mathbf{x}$. Instead, indirect measurements are observed

$$\mathbf{y} = \mathbf{h}[\mathbf{x}, \mathbf{v}, t], \tag{15.2}$$

where $\mathbf{v}(t)$ is a vector describing measurement error at time t.

The goal behind *stochastic control* is to derive a *feedback law*

$$\mathbf{u}(t) = \mathbf{g}(\mathbf{y}, \mathbf{w}(t), \mathbf{v}(t))$$

based on the observation $\mathbf{y}$ and the statistics of the system and measurement noise $\mathbf{w}$ and $\mathbf{v}$ so that, within acceptable error bounds, the system executes a desired behavior $\mathbf{x}_{des}(t)$. In contrast, in *deterministic control* it is assumed that there is no uncertainty, and $\mathbf{v}$ and $\mathbf{w}$ are zero. We discuss the basics of deterministic control below as background for the stochastic case (which is where noncommutative harmonic analysis is applicable).

In the study of systems it is natural to make several classifications. In addition to the labels "deterministic" or "stochastic," system models are treated differently depending on whether they are *linear* or *nonlinear*. The most developed theory is for linear systems (see, for example, [47, 79] for complete treatments). Nonlinear systems come in many varieties. One important class of nonlinear systems for which globally stable control laws can be applied is classical finite-dimensional mechanical systems. The configuration of such a system is specified by n *generalized coordinates* that are often written in the form of an array $\mathbf{q} = [q_1, \dots, q_n]^T$. These generalized coordinates specify the "shape" of the system as well as its location in space. The *state* of the system is the pair $(\mathbf{q}, \dot{\mathbf{q}})$.

For example, Fig. 15.1 shows a two-link planar robot arm with revolute (hinge-like) joints. This is a two-degree-of-freedom mechanical system. The first generalized coordinate can be taken to be the counterclockwise-measured angle that the first link makes with respect to the horizontal, and the second generalized coordinate can be taken to be the counterclockwise-measured angle that the second link makes with respect to the first, i.e., $\mathbf{q} = [q_1, q_2]^T$. These two angles completely specify the geometry of such an arm with fixed link lengths L_1 and L_2. If we assume that the mass in this arm is concentrated at the hinges, the positions to the masses with respect to a frame of reference fixed in space at the base of the arm are

$$\mathbf{x}_1 = \begin{pmatrix} L_1 \cos q_1 \\ L_1 \sin q_1 \end{pmatrix}$$

and

$$\mathbf{x}_2 = \begin{pmatrix} L_1 \cos q_1 + L_2 \cos(q_1 + q_2) \\ L_1 \sin q_1 + L_2 \sin(q_1 + q_2) \end{pmatrix}$$

where q_1 is the angle of the first link, and q_2 is the relative angle of the second link with respect to the first.

In general, the kinetic energy (energy of motion) of a finite-dimensional conservative mechanical system can be written as

$$T = \frac{1}{2} \dot{\mathbf{q}}^T M(\mathbf{q}) \dot{\mathbf{q}} \tag{15.3}$$

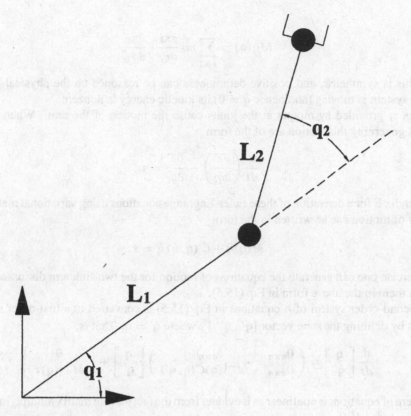

FIGURE 15.1
A planar two-link revolute manipulator.

where $M(\mathbf{q})$ is a symmetric positive definite matrix (called the *mass matrix*) and the potential energy is typically a function $V = V(\mathbf{q})$. Henceforth when referring to mechanical systems we will mean finite-dimensional conservative systems.

For the above example of a two-link revolute planar manipulator arm (operating in the plane normal to gravity) $V = 0$ and

$$T = \frac{1}{2}m_1 \dot{\mathbf{x}}_1 \cdot \dot{\mathbf{x}}_1 + \frac{1}{2}m_2 \dot{\mathbf{x}}_2 \cdot \dot{\mathbf{x}}_2 \qquad (15.4)$$

where m_1 and m_2 are the masses at the hinges. Explicitly, the kinetic energy is written as

$$T = \frac{1}{2}m_1 L_1^2 \dot{q}_1^2 + \frac{1}{2}m_2 \left\{ L_1^2 \dot{q}_1^2 + (\dot{q}_1 + \dot{q}_2)^2 L_2^2 + 2\dot{q}_1 (\dot{q}_1 + \dot{q}_2) L_1 L_2 \cos q_2 \right\}.$$

We can rearrange this to read

$$T = \frac{1}{2} \begin{bmatrix} \dot{q}_1 & \dot{q}_2 \end{bmatrix} \begin{pmatrix} (m_1 + m_2)L_1^2 + m_2(L_2^2 + 2L_1 L_2 \cos q_2) & m_2(L_1 L_2 \cos q_2 + L_2^2) \\ m_2(L_1 L_2 \cos q_2 + L_2^2) & m_2 L_2^2 \end{pmatrix} \begin{bmatrix} \dot{q}_1 \\ \dot{q}_2 \end{bmatrix}.$$

The 2×2 matrix in this equation is the mass matrix $M(\mathbf{q})$ for this example.

Alternatively, a straightforward application of the chain rule in Eq. (15.4) allows us to rewrite the elements of this mass matrix as

$$M_{ij}(\mathbf{q}) = \sum_{k=1}^{2} m_k \frac{\partial \mathbf{x}_k}{\partial q_i} \cdot \frac{\partial \mathbf{x}_k}{\partial q_j}.$$

Clearly this is symmetric, and positive definiteness can be reasoned on the physical grounds that when the system is moving (and hence $\dot{\mathbf{q}} \neq \mathbf{0}$) its kinetic energy is nonzero.

Torques τ_i provided by motors at the joints cause the motion of the arm. When $V = 0$, the equations governing this motion are of the form

$$\frac{d}{dt}\left(\frac{\partial T}{\partial \dot{q}_i}\right) - \frac{\partial T}{\partial q_i} = \tau_i.$$

See Appendix E for a derivation of these Euler-Lagrange equations using variational methods. These equations of motion can be written in the form

$$M(\mathbf{q})\ddot{\mathbf{q}} + C(\mathbf{q}, \dot{\mathbf{q}})\dot{\mathbf{q}} = \tau. \tag{15.5}$$

As an exercise one can generate the equations of motion for the two-link arm discussed previously, and write them in the above form in Eq. (15.5).

The second-order system of n equations in Eq. (15.5) is converted to a first-order system of $2n$ equations by defining the state vector $[\mathbf{q}^T, \mathbf{q}'^T]^T$ where $\mathbf{q}' = \dot{\mathbf{q}}$. That is,

$$\frac{d}{dt}\begin{bmatrix} \mathbf{q} \\ \mathbf{q}' \end{bmatrix} = \begin{pmatrix} \mathbb{0}_{n \times n} & \mathbb{I}_{n \times n} \\ \mathbb{0}_{n \times n} & -M^{-1}(\mathbf{q})C(\mathbf{q}, \mathbf{q}') \end{pmatrix} \begin{bmatrix} \mathbf{q} \\ \mathbf{q}' \end{bmatrix} + \begin{bmatrix} \mathbf{0} \\ M^{-1}(\mathbf{q})\tau \end{bmatrix}.$$

This system of equations is nonlinear as is evident from the fact that the matrix multiplying $[\mathbf{q}^T, \mathbf{q}'^T]^T$ depends on $[\mathbf{q}^T, \mathbf{q}'^T]^T$.

These equations of motion (as well as those for all conservative mechanical systems) exhibit the *passivity* condition when $\tau_i = 0$. That is, substitution of to Eq. (15.5) into the expression

$$\dot{E} = \frac{d}{dt}(T + V)$$

yields

$$\dot{E} = 0.$$

In our case $V = 0$. This is just the condition that the total energy is conserved. For conservative mechanical systems with $V = 0$, the following proportional-derivative (PD) control law is guaranteed to drive the system to the desired $\mathbf{q}_d$ as $t \to \infty$

$$\tau = -K(\mathbf{q} - \mathbf{q}_d) - D\dot{\mathbf{q}}.$$

Here K and D are symmetric positive definite matrices.

The first term is like a linear spring with equilibrium at $\mathbf{q}_d$ and the second term is like a viscous damping. If we write $V' = \frac{1}{2}(\mathbf{q} - \mathbf{q}_d)^T K(\mathbf{q} - \mathbf{q}_d)$, and define an "artificial energy" as $E = T + V'$, then we see that substitution on of Eq. (15.5) into the expression for $\dot{E}$ yields

$$\dot{E} = -\dot{\mathbf{q}}^T D \dot{\mathbf{q}} \leq 0$$

for all $(\mathbf{q}, \dot{\mathbf{q}}) \neq (\mathbf{0}, \mathbf{0})$, with equality holding only when $\dot{\mathbf{q}} = \mathbf{0}$. From physical arguments it is reasonable to assume that as $t \to \infty$ the kinetic energy dissipates to zero since $E \geq 0$ and $\dot{E} \leq 0$. The only way $E \to 0$ is if $\dot{\mathbf{q}} \to \mathbf{0}$ and $\mathbf{q} \to \mathbf{q}_d$. What we have just reasoned can be formalized as an example of *Lyapunov stability theory*, and in particular, La Salle's Invariance Theorem.

More generally, control laws for which we can write

$$\tau_i = -\frac{\partial R}{\partial \dot{q}_i} - \frac{\partial V'}{\partial q_i}$$

[where $V'(\mathbf{q})$ has one critical point which is a global minimum at $\mathbf{q}_d$] will cause $(\mathbf{q}, \dot{\mathbf{q}}) \to (\mathbf{q}_d, 0)$ as $t \to \infty$ when

$$R = \frac{1}{2}\dot{\mathbf{q}}^T D(\mathbf{q})\dot{\mathbf{q}}$$

and $D(\mathbf{q})$ is a positive definite matrix. R is called a *Rayleigh dissipation function*, and in general

$$\frac{d}{dt}(T + V') = -2R$$

means that as $t \to \infty$ then $T + V' \to 0$, and so $\mathbf{q} \to \mathbf{q}_d$.

Of course, this assumes that we can exactly measure the current state of the system $(\mathbf{q}, \dot{\mathbf{q}})$ at each time t, and that we can respond instantaneously. It also assumes that the actuators (in the case of the two-link robot these are motors) are able to provide these desired torques without any time delay.

While this PD control law is powerful in its generality, some natural questions arise. For instance: how should the gain matrices K and D be chosen so that $\mathbf{q}$ approaches $\mathbf{q}_d$ rapidly? What if we want to track a desired trajectory $\mathbf{q}_d(t)$ instead of moving from point to point? How do we handle the issues of limited actuator capability, time delays, and uncertainty in the measurement of the system's state?

Some of these questions have been answered completely for highly restricted classes of systems. The next section reviews results for one of these classes: linear systems. Later in this chapter we come back to a class of nonlinear systems that can be viewed as evolving on Lie groups and $SO(3)$ in particular.

15.2 Linear Systems Theory and Control

The linear form, of Eqs. (15.1) and (15.2) are respectively written as

$$\frac{d\mathbf{x}}{dt} = F(t)\mathbf{x} + G(t)\mathbf{u} + L(t)\mathbf{w} \qquad (15.6)$$

and

$$\mathbf{y} = H(t)\mathbf{x} + \mathbf{v}. \qquad (15.7)$$

If $G(t) = 0$ for all values of time t, the problem of determining how Eqs. (15.6) and (15.7) evolve over time is the *linear estimation* problem. If there are no dynamics [i.e., $F(t)$, $G(t)$, and $L(t)$ are all zero in Eq. (15.6)] the determination of Eq. (15.7) for constant $H(t)$ is the *static linear estimation* problem. This was addressed in Chapter 14 in the context of least squares pose determination. Section 15.3 reviews the recursive computation of this solution.

On the other hand, if $\mathbf{v}$ and $\mathbf{w}$ are taken to be appropriately dimensioned zero vectors, the pair of Eqs. (15.6) and (15.7) represents a linear deterministic system. The stability, controllability, and observability of such systems have been treated extensively in the literature and in a number of textbooks. See e.g., [47, 79]. Given a deterministic linear system for which the state can be observed and for which a finite control effort can direct the system to a desired state, the question becomes what control law will achieve this result with the least effort.

In this deterministic context the *linear quadratic regulator* (LQR) provides an optimal solution. In the remainder of this section we explain the LQ regulator, with an eye towards its generalization to systems on Lie groups, and the stochastic analogs of the deterministic theory.

Given the system

$$\frac{d\mathbf{x}}{dt} = F(t)\mathbf{x} + G(t)\mathbf{u} \tag{15.8}$$

with exact observation of the state $\mathbf{x}$, the question becomes how to select the control input $\mathbf{u}$ optimally to drive $\mathbf{x}$ to $\mathbf{0}$ as $t \to T < \infty$. Optimality is defined in the LQR setting as the $\mathbf{u}$ that minimizes the cost functional [36, 56, 87]

$$C = \frac{1}{2}\mathbf{x}^T(T)S\mathbf{x}(T) + \frac{1}{2}\int_0^T \left[\mathbf{x}^T(t)Q(t)\mathbf{x}(t) + \mathbf{u}^T(t)R(t)\mathbf{u}(t)\right] dt \tag{15.9}$$

where $t \in [0, T]$. The matrices S, $Q(t)$, and $R(t)$ are all symmetric and positive definite. The vector $\mathbf{x}(t)$ is defined so that $\mathbf{x}(T) = \mathbf{0}$ is the desired state.

It may be shown (see [17]) that the optimal control law in this context is

$$\mathbf{u} = -\mathcal{C}\mathbf{x} \tag{15.10}$$

where $\mathcal{C}$ is the matrix function of time defined by

$$\mathcal{C} = R^{-1}G^T S$$

where S is a matrix that satisfies the *Riccati equation*

$$\dot{S} = -SF - F^T S + SGR^{-1}G^T S - Q$$

subject to the end constraint $\mathcal{S}(T) = S$.

In practice, control laws are implemented using digital computers by sampling continuous systems at discrete times. If τ' is the sample period with $N\tau' = T$, and if the control law keeps $\mathbf{u}$ constant during each sample period, then Eq. (15.8) is replaced with the approximation

$$\mathbf{x}_{k+1} = \mathcal{F}_k\mathbf{x}_k + \mathcal{G}_k\mathbf{u}_k \tag{15.11}$$

where $\mathbf{x}_k = \mathbf{x}(k\tau')$, $\mathbf{u}_k = \mathbf{u}(k\tau')$,

$$\mathcal{F}_k = \Phi\left((k+1)\tau', k\tau'\right)$$

and

$$\mathcal{G}_k = \int_{k\tau'}^{(k+1)\tau'} \Phi\left((k+1)\tau', t\right) G(t)\, dt$$

where $\Phi(t, t_0)$ is the *state transition matrix*, which is the solution to the matrix differential equation

$$\frac{d}{dt}\Phi(t, t_0) = F(t)\Phi(t, t_0) \quad \text{with} \quad \Phi(t_0, t_0) = \mathbb{I}.$$

The matrix function $\Phi(t, t_0)$ has the property that solutions to $\dot{\mathbf{y}} = F(t)\mathbf{y}$ with $\mathbf{y}(t_0)$ given can be written as $\mathbf{y}(t) = \Phi(t, t_0)\mathbf{y}(t_0)$. In the special case when $F(t) = F_0$ (a constant matrix), we write

$$\Phi(t, t_0) = \exp\left((t - t_0) F_0\right).$$

The cost function in Eq. (15.9) is modified in the case of digital control to

$$C = \frac{1}{2}\mathbf{x}_N^T S_N \mathbf{x}_N + \frac{1}{2}\tau' \sum_{k=0}^{N-1} \left[\mathbf{x}_k^T Q_k \mathbf{x}_k + \mathbf{u}_k^T R_k \mathbf{u}_k \right]$$

where $Q_k = Q(k\tau')$, $R_k = R(k\tau')$ and $S_N = S$. In what follows, we assume time is measured in units such that

$$\tau' = 1.$$

The optimal solution of this discretized problem is found by minimizing $\mathbf{x}_k^T Q_k \mathbf{x}_k + \mathbf{u}_k^T R_k \mathbf{u}_k$ at each value of k subject to the constraint in Eq. (15.11). This is achieved recursively by defining [16, 23, 36, 56]

$$S_k = \mathcal{F}_k^T \left[S_{k+1} - S_{k+1}\mathcal{G}_k \left(\mathcal{G}_k^T S_{k+1}\mathcal{G}_k + R_k \right)^{-1} \mathcal{G}_k^T S_{k+1} \right] \mathcal{F}_k + Q_k$$

and

$$\mathbf{u}_k = -\left(\mathcal{G}_k^T S_{k+1}\mathcal{G}_k + R_k \right)^{-1} \mathcal{G}_k^T S_{k+1}\mathcal{F}_k \mathbf{x}_k$$

for $k < N$. This provides an optimal digital control law for deterministic linear systems.

The next section shows how estimation of a static linear transformation subject to measurement noise can also be found recursively.

15.3   Recursive Estimation of a Static Quantity

Consider the linear observer equation

$$\mathbf{y} = H\mathbf{x} + \mathbf{v},$$

where $\mathbf{x} \in \mathbb{R}^n$, $H \in \mathbb{R}^{k \times n}$, and $\mathbf{y}, \mathbf{v} \in \mathbb{R}^k$. $\mathbf{x}$ can be viewed as either a static vector with k fixed, or as a series of k scalar measurements taken over time. Here $\mathbf{v}$ represents the measurement noise.

We seek an estimate of $\mathbf{x}$ (denoted $\hat{\mathbf{x}}$) such that a quadratic cost function of the form

$$C_{\mathbf{x}}^{(1)}(\hat{\mathbf{x}}) = \frac{1}{2}(\mathbf{x} - \hat{\mathbf{x}})^T \mathcal{W}(\mathbf{x} - \hat{\mathbf{x}})$$

is minimized for a symmetric positive-definite weighting matrix $\mathcal{W}$. Since $\mathbf{x}$ is not known, we instead seek $\hat{\mathbf{x}}$ which minimizes a cost function of the observed quantity $\mathbf{y}$

$$C_{\mathbf{y}}^{(2)}(\hat{\mathbf{x}}) = \frac{1}{2}(\mathbf{y} - H\hat{\mathbf{x}})^T \mathcal{W}(\mathbf{y} - H\hat{\mathbf{x}}).$$

Minimization of $C_{\mathbf{y}}^{(2)}(\hat{\mathbf{x}})$ with respect to $\hat{\mathbf{x}}$ yields

$$\hat{\mathbf{x}} = \left(H^T \mathcal{W}^{-1} H \right)^{-1} H^T \mathcal{W}^{-1} \mathbf{y}. \tag{15.12}$$

A suitable choice for the weighting matrix $\mathcal{W}$ that incorporates properties of noise into the formulation is the *measurement error covariance*

$$\mathcal{W} = E\left(\mathbf{v}\mathbf{v}^T \right).$$

In cases when k (the dimension of the noise and observation vectors) is very large, or if new observation data is rapidly streaming in, it makes sense to consider *recursive least-squares estimators*. If, for instance, the observation data is gathered sequentially over time, it often makes sense to calculate the best estimate at a given time for $k' < k$ observations, and recursively update the estimate thereafter instead of recalculating Eq. (15.12) for successively large values of k.

Let us assume that for some number of measurements, k_{i-1}, the estimation problem

$$\mathbf{y}_{i-1} = H_{i-1}\mathbf{x} + \mathbf{v}_{i-1}$$

has been solved to yield

$$\hat{\mathbf{x}}_{i-1} = \left(H_{i-1}^T \mathcal{W}_{i-1}^{-1} H_{i-1}\right)^{-1} H_{i-1}^T \mathcal{W}_{i-1}^{-1} \mathbf{y}_{i-1}.$$

Now assume a new set of measurements,

$$\mathbf{y}_i = H_i\mathbf{x} + \mathbf{v}_i,$$

are made. The goal is to use the old information together with this newly observed information to arrive at the best estimate $\hat{\mathbf{x}}_2$. This is achieved by minimizing the quadratic cost function [87]

$$C_{\mathbf{y}_i}^{(3)}\left(\hat{\mathbf{x}}_i\right) = \left[(\mathbf{y}_{i-1} - H_{i-1}\hat{\mathbf{x}}_i)^T, (\mathbf{y}_i - H_i\hat{\mathbf{x}}_i)^T\right] \begin{pmatrix} \mathcal{W}_{i-1}^{-1} & 0_{(i-1)\times(i)} \\ 0_{(i)\times(i-1)} & \mathcal{W}_i^{-1} \end{pmatrix} \begin{bmatrix} \mathbf{y}_{i-1} - H_{i-1}\hat{\mathbf{x}}_i \\ \mathbf{y}_i - H_i\hat{\mathbf{x}}_i \end{bmatrix}.$$

The zero blocks appear in the weighting matrix because it is assumed that the noise is uncorrelated in the two different sequences of measurements.

The value of $\hat{\mathbf{x}}_i$ that minimizes $C_{\mathbf{y}_i}^{(3)}$ is of the form

$$\hat{\mathbf{x}}_i = \left(P_{i-1}^{-1} + H_i^T \mathcal{W}_i^{-1} H_i\right)^{-1} \left(H_{i-1}^T \mathcal{W}_{i-1}^{-1} \mathbf{y}_{i-1} + H_i^T \mathcal{W}_i^{-1} \mathbf{y}_i\right) \qquad (15.13)$$

where (see [87, p. 310])

$$P_{i-1}^{-1} = H_{i-1}^T \mathcal{W}_{i-1}^{-1} H_{i-1} = E\left[(\mathbf{x} - \hat{\mathbf{x}})(\mathbf{x} - \hat{\mathbf{x}})^T\right]$$

is the *residual covariance matrix.*

One calculates analytically (see [36, p. 558]) that

$$\left(P_{i-1}^{-1} + H_i^T \mathcal{W}_i^{-1} H_i\right)^{-1} = P_{i-1} - P_{i-1} H_i^T \left(H_i P_{i-1} H_i^T + \mathcal{W}_i\right)^{-1} H_i P_{i-1}.$$

Making this substitution and using the definitions of $\hat{\mathbf{x}}_{i-1}$ and the *recursive weighted least-squares estimator gain matrix*

$$K_i = P_{i-1} H_i^T \left(H_i P_{i-1} H_i^T + \mathcal{W}_i\right)^{-1}, \qquad (15.14)$$

Eq. (15.13) simplifies to

$$\hat{\mathbf{x}}_i = \hat{\mathbf{x}}_{i-1} + K_i\left(\mathbf{z}_i - H_i\hat{\mathbf{x}}_{i-1}\right). \qquad (15.15)$$

The matrix P_k is updated recursively by observing that

$$P_i^{-1} = P_{i-1}^{-1} + H_i^T \mathcal{W}_i^{-1} H_i.$$

For more on estimation (of dynamic as well as static quantities) see [28, 48, 49].

15.4 Determining a Rigid-Body Displacement: Recursive Estimation of a Static Orientation or Pose

In this section we consider recursive formulations of algorithms presented in Chapter 14 for least-squares fitting of rotations and rigid-body motions to measured data. Various iterative and recursive approaches to least-squares fitting of a rotation matrix have been introduced in the satellite attitude estimation and control literature over the years (see [1], [3]–[8], [10, 21, 22, 25, 26, 53, 58, 73, 74, 76, 77, 96]).

A recursive formulation of the batch least-squares fitting procedure for finding the optimal rotation matrix to fit measured vector data was derived in [21]. A Kalman filtering technique was presented in [5].

One method to compute $A \in SO(3)$ that provides the best least-squares fit between two sets of measured vector data $\{\mathbf{p}_i\}$ and $\{\mathbf{q}_i\}$ related as $\mathbf{p}_i = A\mathbf{q}_i$ is to recursively calculate a series of matrices $\hat{A}_k$ with $\hat{A}_N = \hat{A}$ [where $\hat{A}$ is defined in Eq. (14.25)] and use Eq. (14.26) at the end of the process. This is addressed in [21]. The basic idea is that the unconstrained least-squares estimate based on the first k pairs of data vectors $\{\mathbf{p}_i, \mathbf{q}_i\}$ for $i = 1, \ldots, k$ can be written as

$$\hat{A}_k = \left[R_{k-1} + \mathbf{p}_k \mathbf{q}_k^T \right] S_k^{-1}$$

where

$$R_{k-1} = \sum_{i=1}^{k-1} \mathbf{p}_i \mathbf{q}_i^T$$

and

$$S_k = \sum_{i=1}^{k} \mathbf{q}_i \mathbf{q}_i^T.$$

A recursion can be started because S_k^{-1} can be written in terms of S_{k-1}^{-1}, $\mathbf{p}_k$, and $\mathbf{q}_k$. Explicitly,

$$S_k^{-1} = S_{k-1}^{-1} - \left[1 + \mathbf{q}_k^T S_{k-1}^{-1} \mathbf{q}_k \right]^{-1} \cdot S_{k-1}^{-1} \mathbf{q}_k \mathbf{q}_k^T S_{k-1}^{-1}.$$

The matrix $\hat{A}_k$ is then found recursively as

$$\hat{A}_k = \hat{A}_{k-1} + \left[1 + \mathbf{q}_k^T S_{k-1}^{-1} \mathbf{q}_k \right]^{-1} \cdot \left[\mathbf{p}_k - \hat{A}_{k-1} \mathbf{q}_k \right] \mathbf{q}_k^T S_{k-1}^{-1}.$$

Note that in the formulations of this section the group structure of rigid-body motions was not used. And further, the concepts of probability density functions on groups were not required. We show in later sections of this chapter how the group structure can be used in different kinds of estimation problems.

Before examining more sophisticated estimation problems, we need to have a language for describing noise in systems that leads to measurement errors. This is the subject of the next section.

15.5 Gaussian and Markov Processes and Associated Probability Density Functions and Conditional Probabilities

In this section we review basic stochastic processes. See [39, 43, 55, 70, 88] for in-depth treatments. Our brief review closely follows [52, 69, 94].

Let $p_1(\mathbf{x}, t)d\mathbf{x}$ be the probability that the random process $\mathbf{X}(t)$ is contained in the d-dimensional voxel with volume $d\mathbf{x} = dx_1 \ldots dx_d$ centered at $\mathbf{x} \in \mathbb{R}^d$. Likewise, let $p_n(\mathbf{x}_1, t_1; \mathbf{x}_2, t_2; \ldots \mathbf{x}_n, t_n)$ $d\mathbf{x}_1 \ldots d\mathbf{x}_n$ be the probability that for each time t_i, each $\mathbf{X}(t_i)$ is in the voxel centered at $\mathbf{x}_i$ for each $i = 1, \ldots, n$. Hence, $p_1(\mathbf{x}, t)$ is a probability density function on $\mathbb{R}^d$ for each fixed t, while $p_n(\mathbf{x}_1, t_1; \mathbf{x}_2, t_2; \ldots \mathbf{x}_n, t_n)$ is a PDF on $\mathbb{R}^{d \cdot n} = \mathbb{R}^d \times \mathbb{R}^d \times \cdots \times \mathbb{R}^d$ for each fixed choice of $(t_1, \ldots, t_n)^T \in \mathbb{R}^n$.

By integrating the PDF $p_n(\cdot)$ over the last $n - k$ copies of $\mathbb{R}^d$, one observes the general relationship

$$p_k(\mathbf{x}_1, t_1; \mathbf{x}_2, t_2; \ldots \mathbf{x}_k, t_k) = \int_{\mathbb{R}^d} \cdots \int_{\mathbb{R}^d} p_n(\mathbf{x}_1, t_1; \mathbf{x}_2, t_2; \ldots \mathbf{x}_n, t_n) \, d\mathbf{x}_{k+1} \cdots d\mathbf{x}_n. \quad (15.16)$$

For the case when $d = 1$, a closed-form example of Eq. (15.16) is easily verified for the *Gaussian process*

$$p_n(x_1, t_1; x_2, t_2; \ldots x_n, t_n) = \frac{\exp\left[-\frac{1}{2} \sum_{i,j=1}^{n} C_{ij}^{-1}(x_i - m_i)(x_j - m_j)\right]}{[(2\pi)^n \det C]^{\frac{1}{2}}}$$

where C is an $n \times n$ covariance matrix with elements C_{ij} and $m_i = \langle X_i(t) \rangle$ are the components of the mean of p_n. The extension to the case of higher dimensions is straightforward.

The two PDFs, p_k and p_n, are also related by definition of conditional probability density function $\Pi_{n,k}$ as

$$p_n(\mathbf{x}_1, t_1; \mathbf{x}_2, t_2; \ldots \mathbf{x}_n, t_n) = p_k(\mathbf{x}_1, t_1; \mathbf{x}_2, t_2; \ldots \mathbf{x}_k, t_k) \, \Pi_{n,k}(\mathbf{x}_1, t_1; \mathbf{x}_2, t_2; \ldots \mathbf{x}_n, t_n). \quad (15.17)$$

This definition means that $\Pi_{n,k}(\mathbf{x}_1, t_1; \mathbf{x}_2, t_2; \ldots \mathbf{x}_n, t_n)d\mathbf{x}_{k+1} \cdots d\mathbf{x}_n$ is the conditional probability that each $\mathbf{X}(t_i)$ is in the voxel centered at $\mathbf{x}_i$ at time t_i for all $i = k + 1, \ldots, n$ given that each $\mathbf{X}(t_j)$ is in the voxel centered at $\mathbf{x}_j$ at time t_j for $j = 1, \ldots, k$.

A direct consequence of the definition in Eq. (15.17) and the observation in Eq. (15.16) is that

$$\int_{\mathbb{R}^d} \cdots \int_{\mathbb{R}^d} \Pi_{n,k}(\mathbf{x}_1, t_1; \mathbf{x}_2, t_2; \ldots \mathbf{x}_n, t_n) \, d\mathbf{x}_{k+1} \cdots d\mathbf{x}_n = 1.$$

A *Markov process*[1] is one which satisfies the condition

$$\Pi_{n,k}(\mathbf{x}_1, t_1; \mathbf{x}_2, t_2; \ldots \mathbf{x}_n, t_n) \overset{\Delta}{=} \Pi(\mathbf{x}_{n-1}, t_{n-1}; \mathbf{x}_n, t_n). \quad (15.18)$$

That is, it is a process with memory limited to only the preceding step. For a Markov process, the *Chapman-Kolmogorov equation*

$$\Pi(\mathbf{x}_1, t_1; \mathbf{x}_2, t_2) = \int_{\mathbb{R}^d} \Pi(\mathbf{x}_1, t_1; \boldsymbol{\xi}, t) \, \Pi(\boldsymbol{\xi}, t; \mathbf{x}_2, t_2) \, d\boldsymbol{\xi} \quad (15.19)$$

is satisfied for $t_1 \leq t \leq t_2$. It also follows that

$$p_n(\mathbf{x}_1, t_1; \mathbf{x}_2, t_2; \ldots \mathbf{x}_n, t_n) = p_{n-1}(\mathbf{x}_1, t_1; \mathbf{x}_2, t_2; \ldots \mathbf{x}_{n-1}, t_{n-1}) \, \Pi(\mathbf{x}_{n-1}, t_{n-1}; \mathbf{x}_n, t_n),$$

which is generally not true of non-Markovian processes.

A *stationary* Markov process is one for which

$$\Pi(\mathbf{x}_{i-1}, t_{i-1}; \mathbf{x}_i, t_i) = \Pi(\mathbf{x}_{i-1}, 0; \mathbf{x}_i, t_i - t_{i-1}). \quad (15.20)$$

---

[1] This is named after the Russian mathematician Andrei Andreyevich Markov (1856–1922).

In this context, one sometimes uses the notational shorthand

$$\Pi(\mathbf{x}_{i-1}, 0; \mathbf{x}_i, t) = \Pi(\mathbf{x}_{i-1}|\mathbf{x}_i, t).$$

Hence, for this case the Chapman-Kolmogorov equation is written as [52]

$$\Pi(\mathbf{x}_1|\mathbf{x}_2, t) = \int_{\mathbb{R}^d} \Pi(\mathbf{x}_1|\boldsymbol{\xi}, s)\, \Pi(\boldsymbol{\xi}|\mathbf{x}_2, t - s)\, d\boldsymbol{\xi}. \tag{15.21}$$

Finally, a stationary Markov process with the property [69]

$$\lim_{t \to \infty} \Pi(\mathbf{x}_0|\mathbf{x}, t) = p_1(\mathbf{x}, t) \tag{15.22}$$

is called a *Brownian motion*.

15.6 Wiener Processes and Stochastic Differential Equations

Consider the system of d stochastic differential equations

$$dx_i = f_i(x_1, \ldots, x_d, t)\, dt + \sum_{j=1}^{m} F_{ij}(x_1, \ldots, x_d, t)\, dW_j(t) \tag{15.23}$$

where $\mathbf{W}(t) = [W_1, \ldots, W_m]^T$ is a *Wiener process*. That is, each of the components W_j have zero ensemble average, are taken to be zero at time zero, and are stationary and independent processes. Denoting an ensemble average as $\langle \cdot \rangle$, these properties are written as [69]: (a) $\langle W_j(t) \rangle = 0$; (b) $W_j(0) = 0$; (c) $W_j(t_1 + t) - W_j(t_2 + t) = W_j(t_1) - W_j(t_2)$ for $t_1 \geq 0, t_2 \geq 0, t_1 + t \geq 0$ and $t_2 + t \geq 0$; and (d) $\langle (W_j(t_i) - W_j(t_n))(W_j(t_k) - W_j(t_l)) \rangle = 0$ for $t_i > t_n \geq t_k > t_l \geq 0$.

From these defining properties, it is clear that for the Wiener process, $W_j(t)$, one has

$$\begin{aligned}
\left\langle \left[W_j(t_1 + t_2) \right]^2 \right\rangle &= \left\langle \left[W_j(t_1 + t_2) - W_j(t_1) + W_j(t_1) - W_j(0) \right]^2 \right\rangle \\
&= \left\langle \left[W_j(t_1 + t_2) - W_j(t_1) \right]^2 + \left[W_j(t_1) - W_j(0) \right]^2 \right\rangle \\
&= \left\langle \left[W_j(t_1) \right]^2 \right\rangle + \left\langle \left[W_j(t_2) \right]^2 \right\rangle.
\end{aligned}$$

For the equality

$$\left\langle \left[W_j(t_1 + t_2) \right]^2 \right\rangle = \left\langle \left[W_j(t_1) \right]^2 \right\rangle + \left\langle \left[W_j(t_2) \right]^2 \right\rangle \tag{15.24}$$

to hold for all values of time t_1, t_2, it must be the case that [69]

$$\left\langle \left[W_j(t - s) \right]^2 \right\rangle = \sigma_j^2 |t - s| \tag{15.25}$$

for some positive real number σ_j^2. This, together with the absolute value signs, ensures that $\langle [W_j(t - s)]^2 \rangle > 0$.

One calculates the correlation of a scalar-valued Wiener process with itself at two different times t and s with $0 \leq s \leq t$ as

$$\begin{aligned}
\langle W_j(s) W_j(t) \rangle &= \langle W_j(s) \left(W_j(s) + W_j(t) - W_j(s) \right) \rangle \\
&= \left\langle \left[W_j(s) \right]^2 \right\rangle + \langle \left(W_j(s) - W_j(0) \right) \left(W_j(t) - W_j(s) \right) \rangle = \sigma_j^2 s.
\end{aligned}$$

The notation dW_j is defined by

$$dW_j(t) = W_j(t + dt) - W_j(t).$$

Hence, from the definitions and discussion above,

$$\langle dW_j(t) \rangle = \langle W_j(t + dt) \rangle - \langle W_j(t) \rangle = 0$$

and

$$
\begin{aligned}
\left\langle \left[dW_j(t) \right]^2 \right\rangle &= \langle (W_j(t + dt) - W_j(t)) (W_j(t + dt) - W_j(t)) \rangle \\
&= \left\langle \left[W_j(t + dt) \right]^2 \right\rangle - 2 \langle W_j(t) W_j(t + dt) \rangle + \left\langle \left[W_j(t) \right]^2 \right\rangle \\
&= \sigma_j^2 (t + dt - 2t + t) = \sigma_j^2 dt.
\end{aligned}
$$

Finally, we note that for the m-dimensional Wiener process $\mathbf{W}(t)$, each component is uncorrelated with the others for all values of time. This is written together with what we already know from the above discussion as

$$\langle W_i(s) W_j(t) \rangle = \sigma_j^2 \delta_{ij} \min(s, t)$$

and

$$\langle dW_i(t_k) dW_j(t_l) \rangle = \sigma_j^2 \delta_{ij} \delta(t_k - t_l) dt_k \, dt_l. \tag{15.26}$$

From these properties and Eq. (15.23) one observes that[2]

$$\langle dx_i \rangle = \langle f_i(x_1, \ldots, x_d, t) \, dt \rangle + \sum_{j=1}^m F_{ij}(x_1, \ldots, x_d, t) \langle dW_j(t) \rangle = f_i(x_1, \ldots, x_d, t) \, dt \tag{15.27}$$

and

$$
\begin{aligned}
\langle dx_i dx_k \rangle &= \left\langle \left(f_i(x_1, \ldots, x_d, t) \, dt + \sum_{j=1}^m F_{ij}(x_1, \ldots, x_d, t) \, dW_j(t) \right) \left(f_k(x_1, \ldots, x_d, t) \, dt \right. \right. \\
&\qquad \left. \left. + \sum_{l=1}^m F_{kl}(x_1, \ldots, x_d, t) \, dW_l(t) \right) \right\rangle \\
&= \sum_{j=1}^m \sum_{l=1}^m F_{ij}(x_1, \ldots, x_d, t) F_{kl}(x_1, \ldots, x_d, t) \langle dW_j(t) dW_l(t) \rangle + O\left((dt)^2 \right).
\end{aligned}
$$

Substitution of Eq. (15.26) into the preceding equation, and neglecting the higher order terms in dt, results in

$$\langle dx_i dx_k \rangle = \sum_{j=1}^m \sigma_j^2 F_{ij}(x_1, \ldots, x_d, t) F_{kj}(x_1, \ldots, x_d, t) \, dt. \tag{15.28}$$

---

[2]The assumptions $< F_{ij}(x_1, \ldots, x_d, t) dW_j(t) >= F_{ij}(x_1, \ldots, x_d, t) < dW_j(t) >$ and $< F_{ij} F_{kl} dW_j dW_l >= F_{ij} F_{kl} < dW_j \, dW_l >$ are equivalent to interpreting Eq. (15.23) as an Itô equation.

15.7 Stochastic Optimal Control for Linear Systems

For a linear system with Gaussian noise, we have

$$\frac{d\mathbf{x}}{dt} = F(t)\mathbf{x} + G(t)\mathbf{u} + L(t)\mathbf{w} \tag{15.29}$$

where $\mathbf{w}dt = d\mathbf{W}$ is a zero-mean Gaussian white noise process. An observation with noise is of the form

$$\mathbf{y} = H(t)\mathbf{x} + \mathbf{v}. \tag{15.30}$$

We assume $\mathbf{v}$ is also a zero-mean Gaussian white noise process that is uncorrelated with $\mathbf{w}$.

The LQG regulator is a control system based on the minimization of the expected value of the same cost function as in the deterministic case, but subject to the fact that measurements of the state are observed indirectly through Eq. (15.30). The main result of LQG control theory is that the *certainty-equivalence principle* holds (see [36, 87]). That is, the best controller-estimator combination is the same as the optimal estimator of the uncontrolled system and the deterministic LQ regulator based on the mean values of the observed state.

The optimal estimator for a discretized system is called a *Kalman filter* [48]. For a continuous system it is the *Kalman-Bucy filter* [49]. This subject has received a great deal of attention over the past 40 years, and the reader interested in learning more about this subject can draw on any of a number of sources. (See the reference list for this chapter.) As it is a rather involved discussion to explain Kalman filtering, we will not address this subject here. Rather, we will examine the effect of noise in the system of Eq. (15.29) on the quadratic cost function used in the deterministic controller case under the assumption of perfect state measurement.

Following [17, 87], the expected value of the quadratic cost function in the stochastic case is

$$C = E\left[\frac{1}{2}\mathbf{x}^T(T)S\mathbf{x}(T) + \frac{1}{2}\int_0^T \left[\mathbf{x}^T(t)Q(t)\mathbf{x}(t) + \mathbf{u}^T(t)R(t)\mathbf{u}(t)\right]dt\right]. \tag{15.31}$$

Using the control law in Eq. (15.10) (which would be optimal if $\mathbf{x}$ were known exactly), Eq. (15.29) then becomes

$$\frac{d\mathbf{x}}{dt} = (F - GC)\mathbf{x} + L\mathbf{w}. \tag{15.32}$$

Define the matrix

$$X(t) \triangleq E\left[(\mathbf{x}(t) - E[\mathbf{x}(t)])(\mathbf{x}(t) - E[\mathbf{x}(t)])^T\right].$$

Then a straightforward application of the product rule for differentiation gives

$$\frac{d}{dt}X(t) = E\left[\frac{d}{dt}(\mathbf{x}(t) - E[\mathbf{x}(t)])(\mathbf{x}(t) - E[\mathbf{x}(t)])^T\right] + E\left[(\mathbf{x}(t) - E[\mathbf{x}(t)])\frac{d}{dt}(\mathbf{x}(t) - E[\mathbf{x}(t)])^T\right].$$

Substitution of Eq. (15.32) into the right side of this equation together with appropriate changes of order in taking expectations gives

$$\dot{X} = (F - GC)X + X(F - GC)^T + LQ'L^T$$

where

$$E\left[\mathbf{w}(t)\mathbf{w}^T(\tau)\right] = Q'\delta(t - \tau)$$

is another way to write Eq. (15.26).

This gives a deterministic set of equations that describe the evolution of covariances of the stochastic process $\mathbf{x}(t)$.

15.8   Deterministic Control on Lie Groups

The dynamics and control of rotating systems such as satellites have been addressed in a large number of works in the guidance and control literature (see [41, 44, 50, 59, 82, 91, 97, 98, 103]). Group-theoretic and differential-geometric approaches to this problem have been explored in [9, 11, 12, 13, 14, 15, 20, 27, 46, 67, 90, 93, 95]. Similar problems have been studied in the context of underwater vehicles (see [35, 54]).

In these works, it is assumed that the state of the system is known completely, and the control torques used to steer the system implement the control law exactly. Below we review this deterministic problem in the hope that a sub-optimal (though conceptually easy) solution to the stochastic version of this problem can be implemented by solving the estimation and control problems separately. This is analogous to the LQG case, except for the fact that optimality is not retained.

We begin by noting a difference between the conventions used throughout this book and those used in the attitude control literature. In attitude control, rotation matrices called *direction cosine matrices* describe a frame fixed in space relative to one fixed in a satellite. That is, whereas we use $R \in SO(3)$ to denote the orientation of a satellite with respect to a space-fixed frame, in the attitude control literature everything would be formulated in terms of $R' = R^T$. This means that when comparing the equations that follow with those in the literature, many things will appear to be "backwards." For example, if ω is the angular velocity vector of the satellite with respect to a fixed frame (with components described in satellite-fixed coordinates) we would write

$$\omega = \text{vect}\left(R^T \dot{R}\right) \tag{15.33}$$

or

$$\dot{R} = R\Omega \tag{15.34}$$

where $\omega = \text{vect}(\Omega)$. In contrast, in the attitude control literature they often write everything in terms of R' as

$$\dot{R}' = \Omega' R'$$

where $\Omega' = -\Omega$.

15.8.1   Proportional Derivative Control on $SO(3)$ and $SE(3)$

There are many possible variants of the idea of PD control on the matrix Lie groups $SO(3)$ and $SE(3)$. When we make statements that hold for both of these groups we will refer to G rather than $SO(3)$ or $SE(3)$ in particular. Perhaps the conceptually simplest PD control scheme on G is to define an artificial potential based on any of the metrics discussed in Chapters 5 and 6

$$V'(g) = d^2(g, g_d)$$

where $g, g_d \in G$. A Rayleigh dissipation function of the form

$$R = \frac{1}{2}\text{trace}\left(\dot{g} W \dot{g}^T\right)$$

for appropriate positive-definite symmetric damping matrix W can also be chosen when g is interpreted as a matrix. When doing this, the general principle behind PD control of mechanical systems ensures that controllers of this type will work.

However there is some question as to whether or not such controllers are "natural" in the sense of utilizing all the structure that the properties of the group G has to offer. Such issues are addressed by Bullo and Murray [18]. We summarize some of their results below.

PD Control on $SO(3)$

Bullo and Murray [18] show that if $(\cdot, \cdot)$ is the usual Ad-invariant inner product on $SO(3)$ (see Chapter 7) and if $\theta(t)$ is the angle of rotation of $R(\mathbf{n}, \theta)$ then

$$\frac{1}{2}\frac{d(\theta^2)}{dt} = \left(\log R, R^T \dot{R}\right) = \left(\log R, \dot{R}R^T\right),$$

and hence, given the system $\dot{R} = R\Omega$ where it is assumed that we have the ability to specify Ω, the control law

$$\Omega = -k_p \log R, \quad k_p > 0$$

provides a feedback law that sends $R \to \mathbb{I}$ as $t \to \infty$. This is shown by using the Lyapunov function

$$\mathcal{V}(R) = \frac{1}{2}(\log R, \log R) = \frac{1}{2}\theta^2$$

and calculating

$$\dot{\mathcal{V}} = \left(\log R, -k_p \log R\right) = -2k_p\mathcal{V}$$

indicating that $\mathcal{V}$ is proportional to $e^{-2k_p t}$. Similarly, for a control torque that sends the system composed of Euler's equations of motion and the kinematic constraint $\dot{R} = R\Omega$ to the state $(R, \omega) = (\mathbb{I}, 0)$, Bullo and Murray [18] choose

$$\tau = \omega \times (I\omega) - \text{vect}\left(k_p \log R - k_d R^T \dot{R}\right). \tag{15.35}$$

Here $k_p, k_d \in \mathbb{R}^+$ are gains that provide an exponentially stable feedback law. We see this by constructing the Lyapunov function

$$\mathcal{V} = \frac{1}{2}(\log R, \log R) + k_p \left(R^T \dot{R}, R^T \dot{R}\right).$$

Differentiation with respect to time and substitution of the controlled equations of motion results in $\dot{\mathcal{V}} \leq 0$ in the usual way, with $\dot{\mathcal{V}} = 0$ at a set of measure zero.

Of course this is not the only control law that will stabilize the system as $t \to \infty$. For instance, Koditschek [51] introduced the feedback law

$$\tau = -k_p \log R - K_d \text{vect}\left(R^T \dot{R}\right)$$

where $K_d = K_d^T$ is a positive definite matrix (not to be confused with the scalar k_d previously mentioned). This control law is an example of the general approach described at the beginning of this section. In contrast, Eq. (15.35) uses a feedforward term to cancel the nonlinearities, which, strictly speaking, no longer makes it a pure PD controller.

PD Control on $SE(3)$

Recall that if X is an arbitrary element of $se(3)$ it can be written as

$$X = \begin{pmatrix} \theta N & \mathbf{h} \\ 0^T & 0 \end{pmatrix}$$

where $N \in so(3)$ and $\|\text{vect}(N)\| = 1$. Then

$$\exp X = \begin{pmatrix} R(\mathbf{n}, \theta) & J_L(\theta \mathbf{n})\mathbf{h} \\ 0^T & 1 \end{pmatrix} \tag{15.36}$$

where $R(\mathbf{n}, \theta) = \exp \theta N$ is the axis-angle parameterization of $SO(3)$ and

$$J_L(\theta \mathbf{n}) = \mathbb{I} + \frac{1 - \cos \theta}{\theta^2} N + \left(\frac{1}{\theta^2} - \frac{\sin \theta}{\theta^3} \right) N^2$$

where $J_L(\mathbf{x})$ is the left Jacobian defined in Eq. (5.44) for the parameterization $R = \exp X$. Bullo and Murray observe that $J_L(\theta \mathbf{n})$ and $R(\mathbf{n}, \theta)$ are related by expressions such as [18]

$$J_L(\theta \mathbf{n}) R(\mathbf{n}, \theta) = R(\mathbf{n}, \theta) J_L(\theta \mathbf{n}) = 2 J_L(2\theta \mathbf{n}) - J_L(\theta \mathbf{n})$$

and

$$\frac{d}{d\theta} J_L(\theta \mathbf{n}) = \frac{1}{\theta} (R(\mathbf{n}, \theta) - J_L(\theta \mathbf{n})).$$

These relationships are useful in constructing control systems on $SE(3)$. For instance, if $g \in SE(3)$ is parameterized as $g = \exp X$, consider the simple system

$$\frac{d}{dt} \exp X = (\exp X) \begin{pmatrix} \Omega & \mathbf{v} \\ \mathbf{0}^T & 0 \end{pmatrix}.$$

This is equivalent to the system

$$\mathcal{J}_R(\mathbf{x}) \dot{x} = \begin{pmatrix} \boldsymbol{\omega} \\ \mathbf{v} \end{pmatrix}$$

where $\mathbf{x} = (X)^\vee$ are exponential coordinates for $SE(3)$ and $\mathcal{J}_R(\mathbf{x})$ is the right Jacobian for $SE(3)$ using these coordinates.

Here $\boldsymbol{\omega} = \text{vect}(\Omega)$ and $\mathbf{v}$ are the quantities that are assumed to be specified by the controller. The preceding system can be controlled with the PD law [18]

$$\begin{pmatrix} \Omega & \mathbf{v} \\ \mathbf{0}^T & 0 \end{pmatrix}^\vee = - \begin{pmatrix} k_\omega \mathbb{I}_{3 \times 3} & 0_{3 \times 3} \\ 0_{3 \times 3} & (k_\omega + k_v) \mathbb{I}_{3 \times 3} \end{pmatrix} (X)^\vee$$

where $g = \exp X$ as in Eq. (15.36). Bullo and Murray show that the closed-loop system with this control law can be written as

$$\frac{d}{dt} (\theta N) = -k_\omega \theta N$$

$$\frac{d\mathbf{h}}{dt} = -k_\omega \mathbf{h} - k_v \left[J_L(\theta \mathbf{n}) \right]^{-T} \mathbf{h}.$$

[Here we use the notation $J^{-T} \triangleq (J^{-1})^T = (J^T)^{-1}$.] Clearly, the first of these equations indicates that $\theta N \to 0_{3 \times 3}$ exponentially as $t \to \infty$, and substitution of this information into the second equation indicates that $\mathbf{h} \to \mathbf{0}$ exponentially as $t \to \infty$ as well. This indicates that g goes to the identity.

Similar control laws for second-order systems can also be devised. Our point in addressing this material is to show that the Lie-group approach to kinematics presented in earlier chapters has direct applicability in deterministic control of mechanical systems.

15.8.2 Deterministic Optimal Control on $SO(3)$

A straightforward engineering way to state the optimal attitude control problem is as follows. We seek to find the control torques $\boldsymbol{\tau}(t)$ (defined in satellite-fixed coordinates) that will bring a satellite from state $(R_1, \boldsymbol{\omega}_1)$ at time t_1 to state $(R_2, \boldsymbol{\omega}_2)$ at time t_2 while minimizing the cost

$$C = \int_{t_1}^{t_2} \left\{ \boldsymbol{\omega}^T K(t) \boldsymbol{\omega} + \boldsymbol{\tau}^T L(t) \boldsymbol{\tau} \right\} dt$$

(for given weighting matrices $k(t) and L(t)$) subject to the kinematic constraint in Eq. (15.33) and the governing equation of motion

$$I\dot{\omega} + \omega \times (I\omega) = \tau. \tag{15.37}$$

Such problems have been addressed in the attitude control literature (see [45]).

In the geometric control literature, different variations on this problem have been addressed (see [93]). To our knowledge, the case when $K(t) = I$ (so that kinetic energy is the cost) and $L(t) \neq 0_{3\times3}$ does not have a closed-form solution. However, a number of numerical techniques exist for approaching this problem (see [45]).

A number of subcases with simple solutions do, however exist. For instance, if $I = a\mathbb{I}$, then Eq. (15.37) becomes a linear equation. Other cases when I is not a multiple of the identity also have solutions. When $L(t) = 0_{3\times3}$ and $K(t) = k(t)\mathbb{I}$, and assuming that angular velocity can be controlled exactly, an optimal solution will be [85, 86]

$$\omega^* = \mathbf{c}/k(t)$$

where if $R_0 \stackrel{\Delta}{=} R_1^{-1} R_2$ then

$$\mathbf{c} = \frac{\theta(R_0)}{2(\sin \theta(R_0)) \int_{t_1}^{t_2} 1/k(t) dt} \text{vect}\left(R_0 - R_0^T\right).$$

This value of ω^* is substituted for ω in Eq. (15.37) to find the control torques.

15.8.3 Euler's Equations as a Double Bracket

In recent years, it has become fashionable to rewrite Eq. (15.37) in Lie-theoretic notation as [2]

$$\dot{L} = [L, \Omega] \tag{15.38}$$

when $\tau = 0$. Here $\text{vect}(L) = I\omega$ is the angular momentum vector written in body-fixed coordinates and $\omega = \text{vect}(\Omega)$. One can go a step further and define the linear operator $\Lambda(L)$ such that

$$\text{vect}(\Lambda(L)) = I^{-1}\text{vect}(L).$$

This allows us to write Eq. (15.38) as $\dot{L} = [L, \Lambda(L)]$. One may show that

$$\Lambda(L) = \{\Lambda, L\} \stackrel{\Delta}{=} \mathcal{I}L + L\mathcal{I}$$

for an appropriate symmetric matrix $\mathcal{I} = \mathcal{I}(I)$. The bracket $\{\cdot, \cdot\}$ is called the anti-commutator, in contrast to $[\cdot, \cdot]$ which has a minus sign instead of a plus. This allows one to write Euler's equations of motion in a "double bracket" form

$$\dot{L} = [L, \{\mathcal{I}, L\}].$$

See [12] and references therein for quantitative results concerning the behavior of other double bracket equations.

15.8.4 Digital Control on $SO(3)$

In a digital control approach, where thrusters are modeled as instantaneous torques, the angular velocity can be modeled as being piecewise continuous. This is possible if a small torque of the form $\tau. = \omega \times (I\omega)$ is applied by the thrusters (so that $I\dot{\omega} = 0$). As a practical matter, a sub-optimal solution can be obtained by three "bang-bang" rotations (one about each of the principle axes of the satellite). When the satellite is starting and ending at rest, this reduces the attitude control problem to three one-dimensional linear equations of the form $I_i\ddot{\theta}_i = \tau_i$. The solution for the torque components τ_i in terms of the desired orientation relative to the initial then reduces to the kinematics problem of finding the Euler angles of the desired relative rotation.

If the time interval $[t_1, t_2]$ is divided into N segments each of duration τ', and the angular velocity in the ith subinterval is the constant $\omega_i = \text{vect}(\Omega_i)$, then the solution to Eq. (15.34) at any $t \in [t_1, t_2]$ will be a product of exponentials of the form

$$R = \exp\left(\tau'\Omega_1\right) \cdots \exp\left(\tau'\Omega_n\right) \exp\left((t - n\tau') \Omega_{n+1}\right) \tag{15.39}$$

when $n\tau' \le t \le (n + 1)\tau'$. If over each of these intervals the torque $\tau = \omega_n \times (I\omega_n)$ (which is small when $\|\omega_n\|$ is small) is replaced with $\tau = 0$, then Eq. (15.39) is no longer exact.

In this discretized problem, an optimality condition is the minimization of

$$C' = \sum_{n=1}^{N} \omega_n^T I \omega_n = \sum_{n=1}^{N} \text{trace}\left(\Omega_n^T J \Omega_n\right)$$

subject to the constraint that

$$R_1^{-1} R_2 = \exp\left(\tau'\Omega_1\right) \cdots \exp\left(\tau'\Omega_N\right).$$

Since the duration of each of these intervals is small, it makes sense to approximate

$$\exp\left(\tau'\Omega_n\right) \approx \mathbb{I} + \tau'\Omega_n.$$

However, in doing so, small errors will make

$$R_1^{-1} R_2 \approx \mathbb{I} + \tau' \sum_{n=1}^{N} \Omega_n$$

a bad approximation unless R_1 and R_2 are close.

15.8.5   Analogs of LQR for $SO(3)$ and $SE(3)$

One generalization of LQR to the context of matrix Lie groups is reviewed in this section. To begin, assume that we have complete control over velocities (angular and/or translational). Then for $g(t) \in G = SO(3)$ or $SE(3)$,

$$\dot{g} = gU \quad \text{or} \quad \left(g^{-1}\dot{g}\right)^{\vee} = \mathbf{u}$$

where $\mathbf{u} = (U)^{\vee}$ is the control input. Variations on this "kinematic control" where we assume the ability to specify velocities (rather than forces or torques) have been studied in [35, 37, 54, 93]. If the desired trajectory is $g_d(t)$, the control is simply $\mathbf{u} = \text{vect}(g_d^{-1}\dot{g}_d)$. However, if it is desirable to minimize control effort while tracking a trajectory in G, then we seek $\mathbf{u}$ such that the cost functional

$$C(\mathbf{u}) = d^2\left(g(T), g_d(T)\right) + \frac{1}{2} \int_0^T \left\{ \alpha \left\| \left(g_d^{-1}\dot{g}_d\right)^{\vee} - \mathbf{u} \right\|^2 + \beta\|\mathbf{u}\|^2 \right\} dt$$

is minimized where $d(\cdot, \cdot)$ is a metric on G and $\| \cdot \|$ is a norm on the Lie algebra $\mathcal{G}$. Here α and $\beta \in \mathbb{R}^t$ weight the relative importance of accurately following the trajectory and minimizing control effort.

Han and Park [40] address a modification of the preceding problem subject to the constraint that $f(g(t)) = 0$ is satisfied for given $f : G \to \mathbb{R}$ where $G = SE(3)$, $\beta = 0$, and they choose $d(g_1, g_2) = \| \log(g_1^{-1} g_2) \|$. They give a closed-form solution to this problem.

In the following section we examine the other half of the satellite attitude problem: estimation. When we cannot assume precise measurements of orientation, methods for attitude estimation become critical if we are to have any hope of developing stochastic versions of the deterministic control approaches presented in this section.

The estimation problem on Lie groups is a direct true application of noncommutative harmonic analysis (as opposed to the group theory or differential geometry used in this chapter thus far). Recursive estimation on Lie groups is a prerequisite for analogs of LQG on groups.

15.9   Dynamic Estimation and Detection of Rotational Processes

In this section we explore some of the techniques used in the estimation of time-varying rotational processes. The estimation of rotational processes has been addressed extensively in the satellite attitude estimation and navigation literature using a number of different sensing modalities and algorithms. See [24, 38, 42, 65, 66, 71, 72, 75, 78, 80, 81, 89, 92] for recent works in this area. One of the more popular kinds of rotation sensors is the fiber-optic gyroscope. See [19, 84] for a description of the operating principles of these devices. See [104] for descriptions of traditional mechanical gyros.

In contrast to the applied works just mentioned, a number of theoretical works on estimation on Lie groups and Riemannian manifolds have been presented over the years. For instance, see the work of Duncan [29]–[34], Willsky [99]–[100], Willsky and Marcus [102], Lo and Willsky [60, 61], and Lo and Eshleman [62]–[64]. Our discussions in the sections that follow will review some aspects of estimation that use harmonic analysis on $SO(3)$.

15.9.1   Attitude Estimation Using an Inertial Navigation System

A vast literature exists on the attitude (orientation) estimation problem in the satellite guidance and control literature. In addition to those works mentioned previously, see [1, 6, 7, 22, 53, 73, 74, 83], or almost any recent issue of the *Journal of Guidance Control and Dynamics* for up-to-date approaches to this problem.

In this section we consider an estimation problem on $SO(3)$ that was introduced in [99]–[100].

Consider a rigid object with orientation that changes in time, e.g., a satellite or airplane. It is a common problem to use on-board sensors to estimate the orientation of the object at each instant in time. We now examine several models.

Let $R(t) \in SO(3)$ denote the orientation of a frame fixed in the body relative to a frame fixed in space at time t. From the perspective of the rotating body, the frame fixed in space appears to have the orientation $R'(t) \stackrel{\Delta}{=} R^T(t)$. The angular velocity of the body with respect to the space-fixed frame at time t as seen in the space-fixed frame will be $\omega_L = \text{vect}(\dot{R}R^T)$, and the same angular velocity as seen in the rotating frame will be $\omega_R = R^T \omega_L = \text{vect}(R^T \dot{R})$ where $\dot{R} = dR/dt$. These are also easily rewritten in terms of R'.

If $\{X_i\}$ denotes the set of basis elements of the Lie algebra $SO(3)$

$$X_1 = \begin{pmatrix} 0 & 0 & 0 \\ 0 & 0 & -1 \\ 0 & 1 & 0 \end{pmatrix} ; \quad X_2 = \begin{pmatrix} 0 & 0 & 1 \\ 0 & 0 & 0 \\ -1 & 0 & 0 \end{pmatrix} ; \quad X_3 = \begin{pmatrix} 0 & -1 & 0 \\ 1 & 0 & 0 \\ 0 & 0 & 0 \end{pmatrix},$$

then

$$dR(t) = \left(\sum_{i=1}^{3} (\boldsymbol{\omega}_L \cdot \mathbf{e}_i dt) X_i \right) R(t)$$

and

$$dR(t) = R(t) \left(\sum_{i=1}^{3} (\boldsymbol{\omega}_R \cdot \mathbf{e}_i dt) X_i \right).$$

In practice, the increments $du_i = \boldsymbol{\omega}_R \cdot \mathbf{e}_i dt$ are measured indirectly. For instance, an *inertial platform* (such as a gyroscope) within the rotating body is held fixed (as best as possible) with respect to inertial space. This cannot be done exactly, i.e., there is some drift of the platform with respect to inertial space. Let $A \in SO(3)$ denote the orientation of the space-fixed frame relative to the inertial platform in the rotating body. If there were no drift, A would be the identity $\mathbb{I}$. However, in practice there is nonnegligible drift. This has been modeled as [99][3]

$$dA(t) = \left(\sum_{i=1}^{3} X_i dW_i(t) \right) A(t) \qquad (15.40)$$

where the three-dimensional Wiener process $\mathbf{W}(t)$ defines the noise model.

The orientation that is directly observable is the relative orientation of the inertial platform with respect to the frame of reference fixed in the rotating body. We denote this as Q^{-1}. In terms of R and A,

$$A^{-1} = RQ^{-1}.$$

Therefore Q, which is the orientation of the rotating body with respect to the inertial platform, is

$$Q = AR.$$

Taking the inverse of both sides gives

$$Q' = R'A'. \qquad (15.41)$$

From the chain rule this means

$$dQ' = dR'A' + R'dA'.$$

Substitution of Eq. (15.40) and the corresponding expression for dR gives

$$dQ' = \left(\sum_{i=1}^{3} -du_i X_i \right) R'A' + R'A' \left(\sum_{i=1}^{3} X_i dW_i(t) \right).$$

Using Eq. (15.41) and defining $x_i dt = -du_i$, this is written as [99]

$$dQ' = \left(\sum_{i=1}^{3} x_i dt X_i \right) Q' + Q' \left(\sum_{i=1}^{3} X_i dW_i(t) \right). \qquad (15.42)$$

---

[3]We are not viewing this Itô equation, but rather as a Stratonovich equation (see, e.g., [39] for clarification).

In Chapter 16 we discuss how to derive a partial differential equation (a Fokker-Planck equation) corresponding to the stochastic differential Eq. (15.42). That Fokker-Planck equation can be reduced to a system of simpler equations in Fourier space using techniques from noncommutative harmonic analysis.

In a series of works, other applications of noncommutative harmonic analysis have been explored in the context of estimation of processes on the rotation group. The following section examines one such technique in detail.

15.9.2   Exponential Fourier Densities

In this section we review the work of Lo and Eshleman [64] and discuss extensions of their results. Consider a random rotation, S_0, and the resulting discrete sequence of rotations

$$S_{k+1} = R_k S_k$$

where $\{R_k\}$ is a sequence of known deterministic rotations. It is assumed that S_k is not known directly, but rather, is measured through some noisy measurement process as

$$M_k = V_k S_k$$

where $\{V_k\}$ is a sequence of independent rotations describing the noise. The estimation problem is then that of finding each S_k given the set of measurements $\{M_1, \ldots, M_k\}$. In order to do this an analogous way to the case of vector measurements, both an appropriate error function (analog of the square of a matrix norm) and class of PDFs describing the noise process (analog of the Gaussian distribution) are required. We already discussed metrics on $SO(3)$ in Chapter 5, and, in principle, the square of any of these can be used. For the sake of concreteness, the function

$$d(R, Q) = \|R - Q\|^2$$

is used here.

Ideally, the PDF describing the noise process should be closed under the operations of conditioning and convolution. Of course, band-limited PDFs on $SO(3)$ will be closed under convolution. Lo and Eshleman chose instead to focus on a set of PDFs that are closed under conditioning. This is achieved by exponentiating a band-limited function. These *exponential Fourier densities* for the random initial rotation, measurement noise, and conditional probability $p(S_{k-1}|M_{k-1}, \ldots, M_1)$ are all assumed to be of the exponential Fourier form

$$\rho(S_0) = \exp\left(\sum_{l=0}^{N} \sum_{m,n=-l}^{l} a_{mn}^{l0} U_{mn}^l(S_0)\right),$$

$$\rho(V_k) = \exp\left(\sum_{l=0}^{N} \sum_{m,n=-l}^{l} b_{mn}^{lk} U_{mn}^l(V_k)\right),$$

and

$$\rho(S_{k-1}|M_{k-1}, \ldots, M_1) = \exp\left(\sum_{l=0}^{N} \sum_{m,n=-l}^{l} a_{mn}^{l,k-1} U_{mn}^l(S_{k-1})\right).$$

We now show that PDFs of this form are closed under conditioning, i.e., they are closed under operations of the form

$$\rho(S_k|M_k, \ldots, M_1) = c_k \rho(M_k|S_k) \rho(S_{k-1}|M_{k-1}, \ldots, M_1) \tag{15.43}$$

where $c_k = \frac{1}{\rho(M_k/M_{k-1})}$ is a normalization constant.

Substitution of $S_{k-1} = R_{k-1}^{-1} S_k$ into the previous expression for $\rho(S_{k-1}|M_{k-1}, \ldots, M_1)$ yields

$$\rho(S_{k-1}|M_{k-1}, \ldots, M_1) = \exp\left(\sum_{l=0}^{N} \sum_{m,n=-l}^{l} a_{mn}^{l,k-1} U_{mn}^l \left(R_{k-1}^{-1} S_k\right)\right)$$

$$= \exp\left(\sum_{l=0}^{N} \sum_{m,n=-l}^{l} \left[\sum_{j=-l}^{l} a_{mn}^{l,k-1} U_{mj}^l \left(R_{k-1}^{-1}\right)\right] U_{jn}^l(S_k)\right).$$

Similarly, observing that $\rho(M_k|S_k) = \rho(V_k)$ and making the substitution $V_k = M_k S_k^{-1}$ yields

$$\rho(M_k|S_k) = \exp\left(\sum_{l=0}^{N} \sum_{m,n=-l}^{l} b_{mn}^{l,k} U_{mn}^l \left(M_k S_k^{-1}\right)\right)$$

$$= \exp\left(\sum_{l=0}^{N} \sum_{m,n=-l}^{l} \left[(-1)^{m+n} \sum_{j=-l}^{l} b_{j,-m}^{lk} U_{j,-n}(M_k)\right] U_{mn}^l(S_k)\right).$$

By making the definition $\rho(S_0|M_0) = \rho(S_0)$, it is clear that the coefficients a_{mn}^{lk} defining

$$\rho(S_k|M_k, \ldots, M_1) = \exp\left(\sum_{l=0}^{N} \sum_{m,n=-l}^{l} a_{mn}^{l,k} U_{mn}^l(S_{k-1})\right)$$

can be recursively updated as

$$a_{mn}^{lk} = \sum_{j=-l}^{l} \left[a_{mn}^{l,k-1} U_{jm}\left(R_{k-1}^{-1}\right) + (-1)^{m+n} \sum_{j=-l}^{l} b_{j,-m}^{lk} U_{j,-n}(M_k)\right]$$

where $l, k > 0$. For $l = 0$, the coefficients a_{00}^{0k} serve to enforce the constraint that ρ is a PDF on $SO(3)$.

We note that one of the main features of the exponential Fourier densities introduced by Lo and Eshleman is closure under conditioning. The folded normal distribution for $SO(3)$ discussed in Chapter 14 effectively has this property when the variance is small (which is what we would expect from good sensors). In addition, the folded normal for $SO(3)$ is closed under convolution.

15.10   Towards Stochastic Control on $SO(3)$

In the previous section we reviewed several estimation problems on $SO(3)$ based on different noise models that have been presented in the literature. Here we review some models of noise in the dynamical system to be controlled under the assumption of perfect observation of the state variables. This separation of estimation without regard to control, and control without regard to estimation, is motivated by the linear-system case in which the two problems can be decoupled. To our knowledge, the problem of combined estimation and optimal stochastic control has not been solved in the Lie-group-theoretic setting. However it intuitively makes sense that a feasible (though suboptimal) stochastic controller could be constructed based on suboptimal estimation [using the $SO(3)$-Gaussian function] and stochastic control based on the expected value of the state variables. A first step toward generating such controllers is to have reasonable stochastic system models.

Perhaps the most natural stochastic control model is one where Euler's equations of motion are the governing equations with the torque vector consisting of a deterministic control part and a noise vector

$$\tau = \mathbf{u} + \mathbf{w}.$$

The question then becomes how to specify $\mathbf{u}(t)$ such that the rotation $R(t)$ and angular velocity $\omega(t)$ are driven to desired values in an optimal way given known statistical properties of the noise $\mathbf{w}dt = d\mathbf{W}$. This appears to be a difficult problem that has not been addressed in the literature.

Another noise model assumes that the rigid body to be controlled is bombarded with random impacts that make the angular momentum (rather than the torque) the noisy quantity. The model we discuss here is a variant of that presented by Liao [57].

Let X_1, X_2, X_3 be the usual basis elements of $SO(3)$ normalized so that $(X_i, X_j) = \delta_{ij}$ where $(\cdot, \cdot)$ is the Ad-invariant inner product discussed in Chapter 7. The kinematic equation $\dot{R} = R\Omega$ can be written as

$$dR = R\Lambda Ad\left(R^{-1}\right)M\,dt$$

where $\text{vect}(M) = R\text{vect}(L)$ is the angular velocity vector of the body as it appears in the space-fixed frame and Λ is the operator defined in subsection 15.8.3. In this model, the angular momentum will consist of three parts: the initial angular momentum; the component due to the random disturbances; and the component due to the control (where it is assumed that control implements any desired angular velocity). Then

$$M = \sum_{i=1}^{3} X_i \left(u_i + dW_i\right) + M_0$$

where $\text{vect}(M_0)$ is the initial angular momentum, u_i for $i = 1, 2, 3$ are the controls, and W_i is a Wiener process representing noise in the system. Then the full system model is

$$dR = R\Lambda Ad\left(R^{-1}\right)\sum_{i=1}^{3} X_i \left(u_i + dW_i\right) + R\Lambda Ad\left(R^{-1}\right)M_0\,dt. \tag{15.44}$$

To our knowledge, it remains to find an optimal stochastic control law $\{u_i\}$ for such a model.

15.11 Summary

In this chapter we have reviewed a number of basic results from the fields of linear systems theory, control, and stochastic processes. Many of these results will be used in subsequent chapters in a variety of contexts. A second goal of this chapter was to compile a number of results from the estimation theory literature in which ideas from noncommutative harmonic analysis are used. In particular, it appears that the statistical ideas presented in Chapter 14 may have a place in estimation problems on the rotation group.

References

[1] Aramanovitch, L.I., Quaternion nonlinear filter for estimation of rotating body attitude, *Math. Methods in the Appl. Sci.*, 18(15), 1239–1255, 1995.

[2] Arnol'd, V.I., *Mathematical Methods of Classical Mechanics*, Springer-Verlag, New York, 1978.

[3] Bar-Itzhack, I., Orthogonalization techniques of a direction cosine matrix, *IEEE Trans. Aerosp. and Electron. Syst.*, AES-5(5), 798–804, 1969.

[4] Bar-Itzhack, I., Iterative optimal orthogonalization of the strapdown matrix, *IEEE Trans. Aerosp. and Electron. Syst.*, AES-11(1), 30–37, 1975.

[5] Bar-Itzhack, I.Y., and Reiner, J., Recursive attitude determination from vector observations: direction cosine matrix identification, *J. Guidance*, 7(1), 51–56, 1984.

[6] Bar-Itzhack, I.Y. and Oshman, Y., Attitude determination from vector observations — quaternion estimation, *IEEE Trans. on Aerosp. and Electron. Syst.*, 21(1), 128–136, 1985.

[7] Bar-Itzhack, I.Y. and Idan, M., Recursive attitude determination from vector observations — Euler angle estimation, *J. of Guidance Control and Dyn.*, 10(2), 152–157, 1987.

[8] Bar-Itzhack, I., Montgomery, P.Y., and Gerrick, J.C., Algorithms for attitude determination using the global positioning system, *J. of Guidance, Control, and Dyn.*, 22(2), 193–201, 1999.

[9] Bharadwaj, S., Osipchuk, M., Mease, K.D., and Park, F.C., Geometry and inverse optimality in global attitude stabilization, *J. of Guidance, Control, and Dyn.*, 21(6), 930–939, 1998.

[10] Björck, A. and Bowie, C., An iterative algorithm for computing the best estimate of an orthogonal matrix, *SIAM J. Numerical Anal.*, 8(2), 358–364, 1971.

[11] Bloch, A.M., Krishnaprasad, P.S., Marsden, J.E., and de Alverez, G.S., Stabilization of rigid body dynamics by internal and external torques, *Autom.*, 28(4), 745–756, 1992.

[12] Bloch, A.M., Brockett, R.W., and Crouch, P.E., Double bracket equations and geodesic flows on symmetric spaces, *Commun. in Math. Phys.*, 187(2), 357–373, 1997.

[13] Brockett, R.W., Lie theory and control systems on spheres, *SIAM J. Applied Math.*, 25(2), 213–225, 1973.

[14] Brockett, R.W., System theory on group manifolds and coset spaces, *SIAM J. Control*, 10(2), 265–284, 1972.

[15] Brockett, R.W., Lie algebras and Lie groups in control theory, in *Geometric Methods in System Theory*, Mayne, D.Q., and Brockett, R.W., Eds., Reidel, Dordrecht-Holland, 1973.

[16] Brown, R.G. and Hwang, P.Y.C., *Introduction to Random Signals and Applied Kalman Filtering*, 2nd ed., John Wiley & Sons, New York, 1992.

[17] Bryson, A.E., Jr. and Ho, Y.-C., *Applied Optimal Control: Optimization, Estimation, and Control*, Hemisphere, reprinted by Taylor and Francis, Bristol, PA, 1975.

[18] Bullo, F. and Murray, R.M., Proportional derivative (PD) control on the Euclidean group, *Proc. of the 3rd Eur. Control Conf.*, 1091–1097, Rome, 1995. (See also Caltech CDS Technical Report 95-010).

[19] Burns, W.K., Ed., *Optical Fiber Rotation Sensing,* Academic Press, Boston, 1994.

[20] Byrnes, C.I. and Isidori, A., On the attitude stabilization of rigid spacecraft, *Autom.,* 27(1), 87–95, 1991.

[21] Carta, D.G. and Lackowski, D.H., Estimation of orthogonal transformations in strapdown inertial systems, *IEEE Trans. on Autom. Control,* AC-17, 97–100, 1972.

[22] Chaudhuri, S. and Karandikar, S.S., Recursive methods for the estimation of rotation quaternions, *IEEE Trans. on Aerosp. and Electron. Syst.,* 32(2), 845–854, 1996.

[23] Chui, C.K. and Chen, G., *Kalman Filtering with Real-Time Applications,* 3rd ed., Springer, Berlin, 1999.

[24] Comisel, H., Forster, M., Georgescu, E., Ciobanu, M., Truhlik, V., and Vojta, J., Attitude estimation of a near-earth satellite using magnetometer data, *Acta Astron.,* 40(11), 781–788, 1997.

[25] Crassidis, J.L., Lightsey, E.G., and Markley, F.L., Efficient and optimal attitude determination using recursive global positioning system signal operations, *J. of Guidance, Control, and Dyn.,* 22(2), 193–201, 1999.

[26] Crassidis, J.L. and Markley, F.L., Predictive filtering for attitude estimation without rate sensors, *J. of Guidance, Control, and Dyn.,* 20(3), 522–527, 1997.

[27] Crouch, P.E., Spacecraft attitude control and stabilization: applications of geometric control theory to rigid body models, *IEEE Trans. on Autom. Control,* AC-29(4), 321–331, 1984.

[28] Davis, M.H.A., *Linear Estimation and Stochastic Control,* Chapman & Hall, London, 1977.

[29] Duncan, T.E., Stochastic systems in Riemannian manifolds, *J. of Optimization Theor. and Applic.,* 27, 175–191, 1976.

[30] Duncan, T.E., Some filtering results in Riemann manifolds, *Inf. and Control,* 35(3), 182–195, 1977.

[31] Duncan, T.E., Estimation for jump processes in the tangent bundle of a Riemann manifold, *Appl. Math. and Optimization,* 4, 265–274, 1978.

[32] Duncan, T.E., An estimation problem in compact Lie groups, *Syst. & Control Lett.,* 10, 257–263, 1998.

[33] Duncan, T.E., Some solvable stochastic control problems in noncompact symmetric spaces of rank one, *Stochastics and Stochastics Rep.,* 35, 129–142, 1991.

[34] Duncan, T.E. and Upmeier, H., Stochastic control problems and spherical functions on symmetric spaces, *Trans. of the Am. Math. Soc.,* 347(4), 1083–1130, 1995.

[35] Egeland, O., Dalsmo, M., and Sordalen, O.J., Feedback control of a nonholonomic underwater vehicle with a constant desired configuration, *Int. J. of Robotics Res.,* 15(1), 24–35, 1996.

[36] Elbert, T.F., *Estimation and Control of Systems,* Van Nostrand Reinhold, New York, 1984.

[37] Enos, M.J., Optimal angular velocity tracking with fixed-endpoint rigid body motions, *SIAM J. on Control and Optimization,* 32(4), 1186–1193, 1994.

[38] Fujikawa, S.J. and Zimbelman, D.F., Spacecraft attitude determination by Kalman filtering of global positioning system signals, *J. of Guidance, Control, and Dyn.,* 18(6), 1365–1371, 1995.

[39] Gardiner, C.W., *Handbook of Stochastic Methods,* 2nd ed., Springer-Verlag, Berlin, 1985.

[40] Han, Y. and Park, F.C., Least squares tracking on the Euclidean group, *ASME Design Technical Meetings,* Baltimore, MD, September 2000.

[41] Hughes, P.C., *Spacecraft Attitude Dynamics,* John Wiley & Sons, New York, 1986.

[42] Idan, M., Estimation of Rodrigues parameters from vector observations, *IEEE Trans. on Aerosp. and Electron. Syst.,* 32(2), 578–586, 1996.

[43] Itô, K. and McKean, H.P., *Diffusion Processes and their Sample Paths,* Springer-Verlag, 1974.

[44] Joshi, S.M., Kelkar, A.G., and Wen, J.T.-Y., Robust attitude stabilization of spacecraft using nonlinear quaternion feedback, *IEEE Trans. on Autom. Control,* 40(10), 1800–1803, 1995.

[45] Junkins, J.L. and Turner, J.D., *Optimal Spacecraft Rotational Maneuvers,* Elsevier, New York, 1986.

[46] Jurdjevic, V. and Sussmann, H.J., Control systems on Lie groups, *J. of Differential Eq.,* 12, 313–329, 1972.

[47] Kailath, T., *Linear Systems,* Prentice-Hall, Englewood Cliffs, NJ, 1980.

[48] Kalman, R.E., A new approach to linear filtering and prediction problems, *Trans. of the ASME, J. of Basic Eng.,* 82, 35–45, 1960.

[49] Kalman, R.E. and Bucy, R.S., New results in linear filtering and prediction theory, *Trans. of the ASME, J. of Basic Eng.,* 83, 95–108, 1961.

[50] Kane, T.R., Likins, P.W., and Levinson, D.A., *Spacecraft Dynamics,* McGraw-Hill, New York, 1983.

[51] Koditschek, D.E., The application of total energy as a Lyapunov function for mechanical control systems, in *Dynamics and Control of Multibody Systems,* Krishnaprasad, P.S., Marsden, J.E., and Simo, J.C., Eds., 97, 131–157, AMS, Providence, RI, 1989.

[52] Kolmogorov, A., Uber die analytischen methoden in der wahrscheinlichkeitsrechnung, *Math. Ann.,* 104, 415–458, 1931.

[53] Lefferts, E.J., Markley, F.L. and Shuster, M.D., Kalman filtering for spacecraft attitude estimation, *J. of Guidance Control and Dyn.,* 5(5), 417–429, 1982.

[54] Leonard, N.E. and Krishnaprasad, P.S., Motion control on drift-free, left-invariant systems on Lie groups, *IEEE Trans. on Autom. Control,* 40(9), 1539–1554, 1995.

[55] Lévy, P., *Processus Stochastiques et Mouvement Brownien,* Gauthier-Villars, Paris, 1948.

[56] Lewis, F.L., *Optimal Control,* Wiley-InterScience, New York, 1986.

[57] Liao, M., Random motion of a rigid body, *J. of Theor. Probab.* 10(1), 201–211, 1997.

[58] Lin, C.-F., *Modern Navigation, Guidance, and Control Processing,* Prentice Hall, Englewood Cliffs, NJ, 1991.

[59] Lizarralde, F. and Wen, J.T., Attitude control without angular velocity measurement: a passivity approach, *IEEE Trans. on Autom. Control,* 41(3), 468–472, 1996.

[60] Lo, J.T.-H. and Willsky, A.S., Estimation for rotational processes with one degree of freedom (parts I-III), *IEEE Trans. on Automatic Control,* AC-20(1), 10–33, 1975.

[61] Lo, J.T.-H. and Willsky, A.S., Stochastic control of rotational processes with one degree of freedom, *SIAM J. Control,* 13(4), 886–898, 1975.

[62] Lo, J.T.-H. and Eshleman, L.R., Exponential Fourier densities and optimal estimation for axial processes, *IEEE Trans. on Inf. Theor.*, IT-25(4), 463–470, 1979.

[63] Lo, J.T.-H. and Eshleman, L.R., Exponential Fourier densities on S^2 and optimal estimation and detection for directional processes, *IEEE Trans. on Inf. Theor.*, IT-23, 321–336, 1977.

[64] Lo, J.T.-H. and Eshleman, L.R., Exponential Fourier densities on $SO(3)$ and optimal estimation and detection for rotational processes, *SIAM J. Appl. Math.*, 36(1), 73–82, 1979.

[65] Lo, J.T.-H., Optimal estimation for the satellite attitude using star tracker measurements, *Autom.*, 22(4), 477–482, 1986.

[66] Malyshev, V.V., Krasilshikov, M.N. and Karlov, V.I., *Optimization of Observation and Control Processes*, AIAA Educ. Ser., Przemieniecki, J.S., Ed., Washington, D.C., 1992.

[67] Mayne, D.Q. and Brockett, R.W., Eds., *Geometric Methods in System Theory*, Reidel, The Netherlands, 1973.

[68] M'Closkey, R. and Morin, P., Time-varying homogeneous feedback: design tools for the exponential stabilization of systems with drift, *Int. J. of Control*, 71(5), 837–869, 1998.

[69] McConnell, J., *Rotational Brownian Motion and Dielectric Theory*, Academic Press, New York, 1980.

[70] McKean, H.P., Jr., *Stochastic Integrals*, Academic Press, New York, 1969.

[71] Mortari, D., Moon-sun attitude sensor, *J. of Guidance, Control, and Dyn.*, 34(3), 360–364, 1997.

[72] Mortari, D., Euler-q algorithm for attitude determination from vector observations, *J. of Guidance, Control, and Dyn.*, 34(3), 328–334, 1998.

[73] Oshman, Y. and Markley, F.L., Minimal-parameter attitude matrix estimation from vector observations, *J. of Guidance Control and Dyn.*, 21(4), 595–602, 1998.

[74] Oshman, Y. and Markley, F.L., Sequential attitude and attitude-rate estimation using integrated-rate parameters, *J. of Guidance Control and Dyn.*, 22(3), 385–394, 1999.

[75] Park, F.C., Kim, J., and Kee, C., Geometric descent algorithms for attitude determination using the global positioning system, *J. of Guidance Control and Dyn.*, 23(1), 26–33, 2000.

[76] Pisacane, V.L. and Moore, R.C., Eds., *Fundamentals of Space Systems*, Oxford University Press, New York, 1994.

[77] Psiaki, M.L., Martel, F., and Pail, P.K., Three-axis attitude determination via Kalman filtering of magnetometer data, *J. Guidance, Control, and Dyn.*, 13(3), 506–514, 1990.

[78] Reynolds, R.G., Quaternion parameterization and a simple algorithm for global attitude estimation, *J. of Guidance Control and Dyn.*, 21(4), 669–671, 1998.

[79] Rugh, W.J., *Linear System Theory*, 2nd ed., Prentice Hall, Englewood Cliffs, NJ, 1996.

[80] Savage, P.G., Strapdown inertial navigation integration algorithm design part 1: attitude algorithms, *J. of Guidance, Control, and Dyn.*, 21(1), 19–28, 1998.

[81] Savage, P.G., Strapdown inertial navigation integration algorithm design part 2: velocity and position algorithms, *J. of Guidance, Control, and Dyn.*, 21(2), 208–221, 1998.

[82] Shrivastava, S.K. and Modi, V.J., Satellite attitude dynamics and control in the presence of environmental torques — a brief survey, *J. of Guidance, Control, and Dyn.*, 6(6), 461–471, 1983.

[83] Shuster, M.D. and Oh, S.D., Three-axis attitude determination form vector observations, *J. of Guidance, Control, and Dyn.*, 4(1), 70–77, 1981.

[84] Smith, R.B., Ed., *Fiber Optic Gyroscopes*, SPIE Milestone Ser., MS 8, SPIE Optical Engineering Press, Washington, D.C., 1989.

[85] Spindler, K., Optimal attitude control of a rigid body, *Appl. Math. and Optimization*, 34(1), 79–90, 1996.

[86] Spindler, K., Optimal control on Lie groups with applications to attitude control, *Math. of Control, Signals, and Syst.*, 11, 197–219, 1998.

[87] Stengel, R.F., *Optimal Control and Estimation*, Dover, New York, 1994.

[88] Stratonovich, R.L., *Topics in the Theory of Random Noise: Vols. I and II*, Gordon and Breach Science, New York, 1963.

[89] Tekawy, J.A., Precision spacecraft attitude estimators using an optical payload pointing system, *J. of Spacecraft and Rockets*, 35(4), 480–486, 1998.

[90] Tsiotras, P., Stabilization and optimality results for the attitude control problem, *J. Guidance, Control, and Dyn.*, 19(4), 772–779, 1996.

[91] Vadali, S.R. and Junkins, J.L., Optimal open-loop and stable feedback control of rigid spacecraft attitude maneuvers, *J. of the Astron. Sci.*, 32(2), 105–122, 1984.

[92] Vedder, J.D., Star trackers, star catalogs, and attitude determination: probabilistic aspects of system design, *J. of Guidance, Control, and Dyn.*, 16(3), 498–504, 1993.

[93] Walsh, G.C., Montgomery, R., and Sastry, S.S., Optimal path planning on matrix Lie groups, Proc. 33rd Conf. on Decision and Control, 1258–1263, December 1994.

[94] Wang, M.C. and Uhlenbeck, G.E., On the theory of Brownian motion II, *Rev. Modern Physics.*, 7, 323–342. 1945.

[95] Wen, J.T.-Y. and Kreutz-Delgado, K., The attitude control problem, *IEEE Trans. on Automatic Control*, 36(10), 1148–1162, 1991.

[96] Wertz, J.R., Ed., *Spacecraft Attitude Determination and Control*, Kluwer Academic, Boston, 1978.

[97] Wie, B. and Barba, P.M., Quaternion feedback for spacecraft large angle maneuvers, *J. of Guidance, Control, and Dyn.*, 8(3), 360–365, 1985.

[98] Wie, B. and Weiss, H., Arapostathis, A., Quaternion feedback regulator for spacecraft eigenaxis rotations, *J. of Guidance, Control, and Dyn.*, 12(3), 375–380, 1989.

[99] Willsky, A.S., Some estimation problems on Lie groups, in *Geometric Methods in System Theory*, Mayne, D.Q. and Brockett, R.W., Eds., Reidel, Dordrecht-Holland, 1973.

[100] Willsky, A.S., *Dynamical Systems Defined on Groups: Structural Properties and Estimation*, Ph.D. dissertation, Dept. of Aeronautics and Astronautics, MIT, Cambridge, MA, 1973.

[101] Willsky, A.S. and Marcus, S.I., Estimation for bilinear stochastic systems, in *Lecture Notes in Economics and Mathematical Systems, No. 111*, 116–137, Springer Verlag, New York, 1975.

[102] Willsky, A.S. and Marcus, S.I., The use of harmonic analysis in suboptimal estimator design, *IEEE Trans. Autom. Control*, AC-23(5), 911–916, 1978.

[103] Wisniewski, R., *Magnetic Attitude Control for Low Earth Orbit Satellites*, Springer-Verlag, New York, 1999.

[104] Wrigley, W., Hollister, W., and Denhard, W., *Gyroscopic Theory, Design, and Instrumentation*, MIT Press, Cambridge, MA, 1969.

Chapter 16

Rotational Brownian Motion and Diffusion

Brownian motion is a special kind of random motion driven by white noise. In this chapter we examine Brownian motion of a rigid body. Such problems have been studied extensively in the context of molecular motion in liquids (see [21, 37, 40], [45]–[49], [91, 76, 111]), as well as in more abstract settings (see [28, 43, 52, 53, 77, 86, 87, 92, 96, 98, 107, 113, 114]). In the context of molecular motion in liquids, it is usually assumed that viscosity effects and random disturbances are the dominant forces. Random motion of a rigid body at the macro scale in the absence of a fluid medium is a problem in which inertial terms dominate and viscous ones are often negligible. Potential applications in this area include the motion of a satellite subjected to random environmental torques. In addition, some problems contain viscous and inertial effects which are on the same order of importance. One can imagine the motion of an underwater robot or high performance airplanes subjected to turbulent currents as examples of this. In these different scenarios the translational and rotational motion often decouples, and so these problems may be solved separately.

In Section 16.1 the classical derivations of translational Brownian motion are reviewed in the context of physical problems. The results of modern mathematical treatments of Brownian motion (i.e., the Itô and Stratonovich calculus) are briefly mentioned here. The interested reader can see [50, 54, 59, 64, 85, 104] for more in depth treatments. In Section 16.2 we review the derivation of the classical Fokker-Planck equation and show how this derivation extends in a straightforward way to processes on manifolds. In Section 16.3 we review rotational Brownian motion from a historical perspective. When inertial effects are important, the diffusion process evolves in both orientation and angular velocity. This is addressed in Section 16.4. Section 16.5 reviews how Fourier expansions on $SO(3)$ are useful in solving PDEs governing the motion of molecules in the non-inertial theory of liquid crystals. Section 16.6 examines a diffusion equation from polymer science that evolves on $SO(3)$.

It is well known that the Abelian Fourier transform is a useful tool for solving problems in translational Brownian motion and heat conduction on the line (see Chapter 2). We review this fact, and show also that the Fourier transform on $SO(3)$ presented in Chapter 9 is a useful tool for rotational Brownian motion when viscous terms dominate. The Fourier techniques presented in Chapter 11 for $\mathbb{R}^3 \times SO(3)$ are relevant in the context of simplifying the form of some PDEs in Section 16.4 where inertial terms cannot be neglected, and rotational displacement and angular velocity are considered to be independent state variables.

16.1   Translational Diffusion

In this section we review the classical physical theories of translational Brownian motion that were developed at the beginning of the 20th century. In Section 16.1.1 we review Albert Einstein's theory of translational Brownian motion. In Section 16.1.2 the non-inertial Langevin and Smoluchowski equations are reviewed. Section 16.1.3 reviews the classical translational Fokker-Planck equation, which describes how a PDF on position and velocity evolves over time.

16.1.1   Einstein's Theory

In 1828, the botanist Robert Brown observed the random movement of pollen suspended in water (hence the name *Brownian motion*). At the beginning of the 20th century, Albert Einstein, in addition to developing the theory of special relativity, was instrumental in the development of the mathematical theory of Brownian motion. See [27] for the full development of Einstein's treatment of Brownian motion.

In this section we review Einstein's approach to the problem. The following section reviews the other traditional approaches to translational Brownian motion.

Consider the case of one-dimensional translation. Let $f(x, t)dx$ denote the number of particles on the real line in the interval $[x, x + dx]$ at time t. That is, $f(x, t)$ is a number density per unit length. Einstein reasoned that the number density of particles per unit x-distance at time t and at time $t + \tau$ for small τ must obey

$$f(x, t + \tau)dx = dx \int_{-\infty}^{\infty} f(x + \Delta, t)\phi(\Delta)d\Delta \qquad (16.1)$$

where $\phi(\Delta)$ is a symmetric probability density function

$$\int_{-\infty}^{\infty} \phi(\Delta)d\Delta = 1 \quad \text{and} \quad \phi(\Delta) = \phi(-\Delta).$$

Einstein assumed the PDF $\phi(\Delta)$ to be supported only in a small region around the origin. The PDF $\phi(\Delta)$ describes the tendency for a particle to be pushed from side to side.

Since τ is assumed to be small, the first-order Taylor series expansion

$$f(x, t + \tau) \approx f(x, t) + \tau \frac{\partial f}{\partial t}$$

is valid. Likewise, the second-order Taylor series expansion

$$f(x + \Delta, t) \approx f(x, t) + \Delta \frac{\partial f}{\partial x} + \frac{\Delta^2}{2!} \frac{\partial^2 f}{\partial x^2}$$

is taken. Substitution of these expansions in Eq. (16.1) yields the heat equation

$$\frac{\partial f}{\partial t} = D \frac{\partial^2 f}{\partial x^2}$$

where

$$D = \frac{1}{\tau} \int_{-\infty}^{\infty} \frac{\Delta^2}{2}\phi(\Delta)d\Delta.$$

The Fourier transform for $\mathbb{R}$ is used to generate the solution, as in the case of the heat equation (see Chapter 2).

Approaches that came after Einstein's formulation attempted to incorporate physical parameters such as viscosity and inertia. These are described in the next section.

16.1.2 The Langevin, Smoluchowski, and Fokker-Planck Equations

In this section the relationship between stochastic differential equations governing translational Brownian motion of particles in a viscous fluid and the partial differential equation governing the probability density function of the particles is examined. We note that for the translational cases discussed here, the stochastic differential equations can be viewed either as being of Itô or Stratonovich type. Regardless, the same Fokker-Planck equations result. Our treatment closely follows that of McConnell [76].

The Langevin Equation

In 1908 Langevin [65] considered translational Brownian motion of a spherical particle of mass m and radius R in a fluid with viscosity η with an equation of the form

$$m\frac{d^2x}{dt^2} = -6\pi R\eta \frac{dx}{dt} + X(t). \tag{16.2}$$

Here $X(t)$ is a random force function. It was not until 1930 when Uhlenbeck and Ornstein [108, 110] made assumptions about the behavior of $X(t)$ [in particular that $X(t)dt = mdW$ where $W(t)$ is a Wiener process] that a PDE governing the behavior of the probability density function, $f(x, t)$, governing the behavior of the particle could be derived directly from the Langevin equation in Eq. (16.2).

In particular, if we neglect the inertial term on the left side of Eq. (16.2), the corresponding PDE is

$$\frac{\partial f}{\partial t} = \frac{k_B T}{6\pi R\eta}\frac{\partial^2 f}{\partial x^2},$$

where the coefficient results from the assumption that the molecules are Boltzmann distributed in the steady state (k_B is the Boltzmann constant and T is temperature).

This is the same result as Einstein's, derived by a different means.

The Smoluchowski Equation

Extending this formulation to the three-dimensional case and including a deterministic force, one has the stochastic differential equation

$$m\frac{d^2\mathbf{x}}{dt^2}dt = -6\pi R\eta d\mathbf{x} + \mathbf{F}dt + md\mathbf{W}. \tag{16.3}$$

The Smoluchowski equation [101] is the PDE governing the PDF corresponding to the above SDE (stochastic differential equation) in the case when the inertial term on the left is negligible. It is derived using the apparatus of Section 16.2 as

$$\frac{\partial f}{\partial t} = \frac{m^2\sigma^2}{2(6\pi R\eta)^2}\nabla^2 f - \frac{1}{6\pi R\eta}\nabla \cdot (\mathbf{F}f).$$

Where σ is a measure of the strength of the noise $d\mathbf{W}$. Assuming the molecules are Boltzmann-distributed in the steady state, it may be shown that $\frac{m^2\sigma^2}{2(6\pi R\eta)^2}$ $= k_B T/(6\pi R\eta)$ and hence in the case when $\mathbf{F} = \mathbf{0}$, the heat equation

$$\frac{\partial f}{\partial t} = \frac{k_B T}{6\pi R\eta}\nabla^2 f$$

results.

Fourier methods are applicable here for solving this diffusion equation in direct analogy with the one-dimensional case.

16.1.3   A Special Fokker-Planck Equation

Consider classical Brownian motion as described by the Langevin equation

$$m\ddot{x}dt = -\xi\dot{x}dt + md\mathbf{W} \tag{16.4}$$

where $\xi = 6\pi\nu a$ is the viscosity coefficient. When the inertial effects are negligible in comparison to the viscous ones, one can write this as

$$d\mathbf{x} = Fd\mathbf{W}$$

where $F = \frac{m}{\xi}\mathbb{I}$. If inertial terms are not negligible, and in addition a deterministic force is applied to the particle, then the Smoluchowski equation

$$m\ddot{x}dt = -\xi\dot{x}dt + \mathbf{h}dt + md\mathbf{W} \tag{16.5}$$

results. By defining the vector

$$\mathbf{X} = \begin{pmatrix} \mathbf{x} \\ \mathbf{v} \end{pmatrix}$$

where $\mathbf{v} = \dot{\mathbf{x}}$, this equation is then put in the form of Eq. (16.8) as

$$\frac{d}{dt}\begin{pmatrix} \mathbf{x} \\ \mathbf{v} \end{pmatrix}dt = \begin{pmatrix} \mathbf{v} \\ -\frac{\xi}{m}\mathbf{v} + \frac{1}{m}\mathbf{h} \end{pmatrix}dt + \begin{pmatrix} 0_{3\times3} \\ \mathbb{I}_{3\times3} \end{pmatrix}d\mathbf{W}. \tag{16.6}$$

A PDE governing the PDF, $F(\mathbf{x}, \mathbf{v}, t)$, of the particle in phase space may be generated. In the one dimensional case this is

$$\frac{\partial f}{\partial t} = -\frac{\partial}{\partial x}(vf) + \gamma\frac{\partial}{\partial v}(vf) + \frac{\gamma k_B T}{m}\frac{\partial^2 f}{\partial v^2} \tag{16.7}$$

where $\gamma = 6\pi R\eta$ is the damping coefficient and the random forcing (Wiener process) has strength $2\gamma k_B T/m$ where T is the temperature and k_B is the Boltzmann constant. See [76, 111] for details.

A general methodology for generating partial differential equations for PDFs of random variables described by stochastic differential equations is described in the next section. Equation (16.7) can be derived from Eq. (16.6) using this technique. This methodology was first employed by Fokker [39] and Planck [94].

16.2   The General Fokker-Planck Equation

In this section we review Fokker-Planck equations, which are partial differential equations that describe the evolution of probability density functions associated with random variables governed by certain kinds of stochastic differential equations.

That is, we begin with stochastic differential equations of the form

$$d\mathbf{X}(t) = \mathbf{h}(\mathbf{X}(t), t)dt + H(\mathbf{X}(t), t)d\mathbf{W}(t) \tag{16.8}$$

where $\mathbf{X} \in \mathbb{R}^p$ and $\mathbf{W} \in \mathbb{R}^m$. Here the deterministic dynamical system $\dot{\mathbf{X}} = \mathbf{h}(\mathbf{X}, t)$ is perturbed at every value of time by noise, or random forcing described by $\mathbf{W}(t)$ where $\mathbf{W}(t)$ is a Wiener process and $H \in \mathbb{R}^{p\times m}$. The Fokker-Planck equation is a partial differential equation that governs the time evolution of a probability density function for $\mathbf{X}$. Whereas $\mathbf{x} = \mathbf{x}(t)$ is a dependent variable, the

PDF that describes the behavior of $\mathbf{x}$ is a function of time and the range of all possible $\mathbf{x}$ values. This range is parameterized with Cartesian coordinates $\mathbf{x} = [x_1 \dots x_p]^T$.

In the first section we examine the classical Fokker-Planck equation, and in the second section we show how the Fokker-Planck equation is derived for processes on certain manifolds. Our presentation is in a style appropriate for physical/engineering scientists, and follows [18, 41, 69, 68, 76, 97, 99, 102, 109].

16.2.1   Derivation of the Classical Fokker-Planck Equation

The goal of this section is to review the derivation of the Fokker-Planck equation, which governs the evolution of the PDF $f(\mathbf{x}, t)$ for a given stationary Markov process, e.g., for a system of the form in Eq. (15.23) which is forced by a Wiener process. The derivation reviewed here has a similar flavor to the arguments used in classical variational calculus in the sense that functionals of $f(\mathbf{x})$ and its derivatives, $m(f(\mathbf{x}), f'(\mathbf{x}), \dots, \mathbf{x})$, are projected against an "arbitrary" function $\epsilon(\mathbf{x})$, and hence integrals of the form

$$\int_{\mathbb{R}^p} m(f(\mathbf{x}), f'(\mathbf{x}), \dots, \mathbf{x})\epsilon(\mathbf{x})d\mathbf{x} = 0$$

are localized to

$$m(f(\mathbf{x}), f'(\mathbf{x}), \dots, \mathbf{x}) = 0$$

using the "arbitrariness" of the compactly supported function $\epsilon(\mathbf{x})$. The details of this procedure are now examined.

To begin, let $\mathbf{x} = \mathbf{X}(t)$ and $\mathbf{y} = \mathbf{X}(t + dt)$ where dt is an infinitesimal time increment. Using the properties of $\Pi(\mathbf{x}|\mathbf{y}, t)$ from Chapter 15, it may be observed [when Eq. (16.8) is interpreted as Itô equation] that [76]

$$\int_{\mathbb{R}^p} (y_i - x_i)\, \Pi(\mathbf{x}|\mathbf{y}, dt)d\mathbf{y} = \langle y_i - x_i \rangle = h_i(\mathbf{x}, t)dt \tag{16.9}$$

and

$$\int_{\mathbb{R}^p} (y_i - x_i)\,(y_j - x_j)\, \Pi(\mathbf{x}|\mathbf{y}, dt)d\mathbf{y} = \langle (y_i - x_i)\,(y_j - x_j) \rangle = \sum_{k=1}^m \sigma_k^2 H_{ik} H_{kj}^T dt, \tag{16.10}$$

where σ_k^2 for $k = 1, \dots, m$ indicate the strengths of the Wiener process $\mathbf{W}(t)$.

Using the Chapman-Kolmogorov equation, Eq. (15.21), together with the definition of partial derivative, one writes

$$\frac{\partial \Pi(\mathbf{x}|\mathbf{y}, t)}{\partial t} = \lim_{\Delta t \to 0} \frac{1}{\Delta t} \left[\Pi(\mathbf{x}|\mathbf{y}, t + \Delta t) - \Pi(\mathbf{x}|\mathbf{y}, t) \right]$$

$$= \lim_{\Delta t \to 0} \frac{1}{\Delta t} \left[\int_{\mathbb{R}^p} \Pi(\mathbf{x}|\boldsymbol{\xi}, t)\Pi(\boldsymbol{\xi}|\mathbf{y}, \Delta t)d\boldsymbol{\xi} - \Pi(\mathbf{x}|\mathbf{y}, t) \right].$$

Let $\epsilon(\mathbf{x})$ be an arbitrary compactly supported function for which $\partial \epsilon / \partial x_j$ and $\partial^2 \epsilon_i / \partial x_j \partial x_k$ are continuous for all $i, j, k = 1, \dots, p$. Then the projection of $\partial \Pi / \partial t$ against $\epsilon(\mathbf{y})$ can be expanded as

$$\int_{\mathbb{R}^p} \frac{\partial \Pi(\mathbf{x}|\mathbf{y}, t)}{\partial t}\epsilon(\mathbf{y})d\mathbf{y} = \lim_{\Delta t \to 0} \frac{1}{\Delta t} \left[\int_{\mathbb{R}^p} \epsilon(\mathbf{y})d\mathbf{y} \int_{\mathbb{R}^p} \Pi(\mathbf{x}|\boldsymbol{\xi}, t)\Pi(\boldsymbol{\xi}|\mathbf{y}, \Delta t)d\boldsymbol{\xi} \right.$$

$$\left. - \int_{\mathbb{R}^p} \Pi(\mathbf{x}|\boldsymbol{\xi}, t)\epsilon(\boldsymbol{\xi})d\boldsymbol{\xi} \right].$$

Inverting the order of integration in the left term on the right side results in

$$\int_{\mathbb{R}^p} \frac{\partial \Pi(\mathbf{x}|\mathbf{y},t)}{\partial t} \epsilon(\mathbf{y}) d\mathbf{y} = \lim_{\Delta t \to 0} \frac{1}{\Delta t} \int_{\mathbb{R}^p} \Pi(\mathbf{x}|\xi,t) \left[\int_{\mathbb{R}^p} \Pi(\xi|\mathbf{y},\Delta t)\epsilon(\mathbf{y}) d\mathbf{y} - \epsilon(\xi) \right] d\xi.$$

Expanding the function $\epsilon(\mathbf{y})$ in its Taylor series about ξ

$$\epsilon(\mathbf{y}) = \epsilon(\xi) + \sum_{i=1}^{p} (y_i - \xi_i) \frac{\partial \epsilon_i}{\partial \xi_i} + \frac{1}{2} \sum_{i,j=1}^{p} (y_i - \xi_i)(y_j - \xi_j) \frac{\partial^2 \epsilon_i}{\partial \xi_j \partial \xi_k} + \cdots,$$

substituting this series into the previous equation, and observing Eqs. (16.9) and (16.10), results in

$$\int_{\mathbb{R}^p} \frac{\partial \Pi(\mathbf{x}|\mathbf{y},t)}{\partial t} \epsilon(\mathbf{y}) d\mathbf{y} = \int_{\mathbb{R}^p} \left[\frac{\partial \epsilon}{\partial y_i} h_i(\mathbf{y},t) + \frac{1}{2} \sum_{i,j=1}^{p} \frac{\partial^2 \epsilon}{\partial y_i \partial y_j} \sum_{k=1}^{m} \sigma_k^2 H_{ik} H_{kj}^T \right] \Pi(\mathbf{x}|\mathbf{y},t) d\mathbf{y}$$

after renaming the dummy variable ξ to $\mathbf{y}$.

The final step is to integrate the two terms on the right side of the above equation by parts to generate

$$\int_{\mathbb{R}^p} \left\{ \frac{\partial \Pi(\mathbf{x}|\mathbf{y},t)}{\partial t} + \frac{\partial}{\partial y_i} (h_i(\mathbf{y},t) \Pi(\mathbf{x}|\mathbf{y},t)) \right.$$

$$\left. - \frac{1}{2} \sum_{k=1}^{m} \sigma_k^2 \sum_{i,j=1}^{p} \frac{\partial^2}{\partial y_i \partial y_j} \left(H_{ik} H_{kj}^T \Pi(\mathbf{x}|\mathbf{y},t) \right) \right\} \epsilon(\mathbf{y}) d\mathbf{y} = 0. \qquad (16.11)$$

Using the standard localization argument that the term in brackets must be zero since the integral is zero, implies that

$$\frac{\partial f(\mathbf{x},t)}{\partial t} + \frac{\partial}{\partial y_i} (h_i(\mathbf{x},t) f(\mathbf{x},t)) - \frac{1}{2} \sum_{k=1}^{m} \sigma_k^2 \sum_{i,j=1}^{p} \frac{\partial^2}{\partial x_i \partial x_j} \left(H_{ik} H_{kj}^T f(\mathbf{x},t) \right) = 0 \qquad (16.12)$$

where Eq. (15.22) has been used (with subscript on f suppressed) to write the Fokker-Planck equation for the evolution of the PDF $f(\mathbf{x},t)$ corresponding to the process $\mathbf{x}(t)$ under the condition of steady state Brownian motion.

16.2.2   The Fokker-Planck Equation on Manifolds

The derivation of the Fokker-Planck equation governing the time-evolution of PDFs on a Riemannian manifold proceeds in an analogous way to the derivation of the previous section. This subject has been studied extensively in the mathematics literature. See, for example, the works of Yosida [112], Itô [51, 52, 53], McKean [77], and Emery [33]. Aspects of diffusion processes on manifolds and Lie groups remain of interest today (see [2, 3, 32, 33, 87, 96, 107]).

Unlike many derivations in the modern mathematics literature, our derivation of the Fokker-Planck equation for the case of a Riemannian manifold is strictly coordinate-dependent.

We start by taking Eq. (16.8) to describe a stochastic process in an n-dimensional Riemannian manifold, M, with metric tensor $g_{ij}(x_1, \ldots, x_n)$. Here x_i are coordinates which do not necessarily form a vector in $\mathbb{R}^n$ (e.g., spherical coordinates). The array of coordinates is written as $x = (x_1, \ldots, x_n)$. In this context $X_i(t)$ denotes a stochastic process corresponding to the coordinate x_i. The definition of the Wiener process is given in Chapter 15, and the derivation of the Fokker-Planck equation for manifolds proceeds much like the classical case presented in Section 16.2.1. It appears that the only difference now is that the integration-by-parts required to isolate the function $\epsilon(x)$ from the rest of the integrand in the final steps in the derivation in the previous section will be weighted by $\sqrt{\det G}$ due to the fact that the volume element in M is $dV(x) = \sqrt{\det G}\,dx$ where $dx = dx_1 \cdots dx_n$. In particular,

$$\int_M \frac{\partial \epsilon}{\partial y_i} h_i(y,t) \Pi(x|y,t) dV(y) = \int_{\mathbb{R}^p} \frac{\partial \epsilon}{\partial y_i} h_i(y,t) \Pi(x|y,t) \sqrt{\det G(y)}\,dy =$$
$$- \int_{\mathbb{R}^p} \epsilon(y) \frac{\partial}{\partial y_i} \left(h_i(y,t) \Pi(x|y,t) \sqrt{\det G(y)} \right) dy.$$

The integration over all $\mathbb{R}^p$ is valid since $\epsilon(y)$ and its derivatives vanish outside of a compact subset of $\mathbb{R}^p$ which can be assumed to be contained in the range of a single coordinate chart of M. Then integrating by parts twice yields

$$\int_M \frac{\partial^2 \epsilon}{\partial y_i \partial y_j} h_i(y,t) \Pi(x|y,t) dV(y) =$$
$$\int_{\mathbb{R}^p} \epsilon(y) \frac{\partial^2}{\partial y_i \partial y_j} \left(h_i(y,t) \Pi(x|y,t) \sqrt{\det G(y)} \right) dy.$$

Using the standard localization argument as before, extracting the functional which multiplies $\epsilon(y)$ and setting it equal to zero yields, after division by $\sqrt{\det G}$,

$$\frac{\partial f(x,t)}{\partial t} + \frac{1}{\sqrt{\det G}} \sum_{i=1}^p \frac{\partial}{\partial x_i} \left(\sqrt{\det G} h_i(x,t) f(x,t) \right) =$$
$$\frac{1}{2} \sum_{k=1}^m \sigma_k^2 \sum_{i,j=1}^p \frac{1}{\sqrt{\det G}} \frac{\partial^2}{\partial x_i \partial x_j} \left(\sqrt{\det G} H_{ik} H_{kj}^T f(x,t) \right). \tag{16.13}$$

The second term on the left side of the preceding equation can be written as $\mathrm{div}(fh)$, and this leads one to inquire about the differential-geometric interpretation of the right side. In many cases, of interest, the matrices H_{ik} will be the inverse of the Jacobian matrix, and hence in these cases $\sum_k H_{ik} H_{kj}^T = \sum_k ((J_{ik})^{-1}((J_{kj})^{-1})^T) = (g_{ij}(x))^{-1} = (g^{ij}(x))$. Therefore, when $\sigma_i = 2K$ for all i, one finds in such cases that the Fokker-Planck equation on M becomes[1]

$$\frac{\partial f(x,t)}{\partial t} + \frac{1}{\sqrt{\det G}} \sum_{i=1}^p \frac{\partial}{\partial x_i} \left(\sqrt{\det G} h_i(x,t) f(x,t) \right) =$$
$$\frac{K}{\sqrt{\det G}} \sum_{i,j=1}^p \frac{\partial^2}{\partial x_i \partial x_j} \left(\sqrt{\det G} \left(g^{ij}(x) \right) f(x,t) \right). \tag{16.14}$$

This equation is similar to, though not exactly the same as, the heat equation on M written in coordinate-dependent form. In fact, a straightforward calculation explained by Brockett [15] equates the Fokker-Planck and heat equation

$$\frac{\partial f(x,t)}{\partial t} = \frac{K}{\sqrt{\det G}} \sum_{i,j=1}^p \frac{\partial}{\partial x_i} \left(\sqrt{\det G} (g^{ij}(x)) \frac{\partial}{\partial x_j} f(x,t) \right) = K \nabla^2 f \tag{16.15}$$

---

[1]When Eq. (16.8) is interpreted as a Stratonovich rather than Itô equation, the Fokker-Planck equation on a manifold must be modified accordingly.

in the special case when

$$h_i(x) = K \sum_{j=1}^{n} \left(\frac{1}{\sqrt{\det G}} \left(g^{ij}(x) \right) \frac{\partial \sqrt{\det G}}{\partial x_j} + \frac{\partial g^{ij}}{\partial x_j} \right).$$

This is clear by expanding the term on the right side of Eq. (16.14) as

$$\frac{K}{\sqrt{\det G}} \sum_{i,j=1}^{p} \frac{\partial}{\partial x_i} \left[\frac{\partial}{\partial x_j} \left(\sqrt{\det G} (g^{ij}(x)) f(x,t) \right) \right],$$

and observing the chain rule

$$\frac{\partial}{\partial x_j} \left(\sqrt{\det G} \left(g^{ij}(x) \right) f(x,t) \right) = \left(\left(g^{ij}(x) \right) \frac{\partial \sqrt{\det G}}{\partial x_j} + \sqrt{\det G} \frac{\partial g^{ij}}{\partial x_j} \right) f$$

$$+ \left(g^{ij}(x) \right) \sqrt{\det G} \frac{\partial f}{\partial x_j}.$$

If $(1/K)\sqrt{\det G} h_i$ is equated to the term in the parentheses, multiplying f on the right hand side of the preceding equation then the heat equation results.

16.3   A Historical Introduction to Rotational Brownian Motion

In this section stochastic differential equations and associate partial differential equations describing rotational Brownian motion are reviewed. First, in Section 16.3.1, theories which do not take into account inertia effects are considered. Then in Section 16.3.3 the equations governing the case when inertia cannot be ignored and no viscous forces are present is examined. In Section 16.4, the most general case is examined, together with simplifications based on axial and spherical symmetry.

16.3.1   Theories of Debye and Perrin: Exclusion of Inertial Terms

In 1913, Debye developed a model describing the motion of polar molecules in a solution subjected to an electric field. See [21] for a complete treatment of Debye's work. Our treatment follows [76].

Debye assumed the molecules to be spheres of radius R, the dipole moment to be μ, and the viscosity of the solution to be η. The deterministic equation of motion governing a single particle is then found by an angular balance which reduces in this case to

$$I\ddot{\theta} = -\mu F \sin \theta - 8\pi R^3 \eta \dot{\theta}$$

where I is the scalar moment of inertia of the spherical molecule, F is the magnitude of the applied electric field, and θ is the angle between the axis of the dipole and the field.

In Debye's approximation, the inertial terms are assumed to be negligible, and so the equation of motion

$$\dot{\theta} = -\frac{\mu F \sin \theta}{8\pi R^3 \eta} \tag{16.16}$$

results.

He then proceeded to derive a partial differential equation describing the concentration of these polar molecules at different orientations. We use $f(\theta, t)$ to denote the orientational density, or concentration, of molecules at time t with axis making angle θ with the direction of the applied electric field. Debye reasoned that the flux of $f(\theta, t)$ consisted of two components—one of which is due to the electric field of the form $f(\theta, t)\dot{\theta}$, and the other of which is of the form $-K \partial f/\partial \theta$ due to Fick's law. Combining these, the flux in orientational density is

$$J(\theta, t) = -\frac{\mu F f \sin \theta}{8\pi R^3 \eta} - K \frac{\partial f}{\partial \theta}$$

where K is a diffusion constant that will be determined shortly.

The change in the number of polar molecules with orientation from θ to $\theta + d\theta$ in time interval dt is $(\partial f/\partial t)d\theta dt$. Equating this to the net orientational flux into the sector $(\theta, \theta + d\theta)$ during the interval dt, which is given by $(J(\theta, t) - J(\theta + d\theta, t))dt = -(\partial J/\partial \theta)d\theta dt$, we see that

$$\frac{\partial f}{\partial t} = \frac{\partial}{\partial \theta}\left(\frac{\mu F \sin \theta}{8\pi R^3 \eta} + K \frac{\partial f}{\partial \theta}\right). \tag{16.17}$$

The diffusion constant K is determined, analogous to the case of translational diffusion, by observing that in the special case when the electric field is static $F(t) = F_0$, and the orientational distribution f is governed by the Maxwell-Boltzmann distribution $f = C \exp(-E/k_B T)$ for some constant C where $E = T + V = \frac{1}{2}I\dot{\theta}^2 - \mu F \cos \theta$ is the total energy of one molecule, k_B is the Boltzmann constant and T is temperature. Substitution of this steady state solution back into Eq. (16.17) dictates that

$$K = \frac{k_B T}{8\pi R^3 \eta}.$$

See [76] for details.

Note that the same result can be obtained as the Fokker-Planck equation corresponding to the stochastic differential equation

$$8\pi R^2 \eta d\theta = -\frac{\mu F \sin \theta}{8\pi R^3 \eta}dt + dW$$

with white noise of appropriate strength.

When no electric field is applied, $F(t) = 0$, the result is the heat equation on the circle or, equivalently, $SO(2)$. The solution of the heat equation on the circle is calculated using the Fourier series on $SO(2)$ subject to the normality condition

$$\int_0^{2\pi} f(\theta, t)d\theta = 1.$$

Clearly, this is a low degree-of-freedom model to describe a phenomenon which is, by nature, higher dimensional. In 1929, Debye reported an extension of this one-dof model to allow for random pointing (though not rotation about the axis) of the polar molecule. Hence, the model becomes one of symmetric diffusion on $SO(3)/SO(2) \cong S^2$. Since (α, β) specify points on the unit sphere, it is natural to assume the heat equation on the sphere would arise. However, because of the inherent axial symmetry imposed on the problem by the direction of the electric field, $f(\alpha, \beta, t) = f(\beta, t)$ for this problem as well. Hence, while the volume element for the sphere is $d\mathbf{u}(\alpha, \beta) = \sin \beta d\beta d\alpha$, we can consider the hoop formed by all volume elements for each value of β. This has volume $\int_0^{2\pi} \sin \beta d\beta d\alpha = 2\pi \sin \beta d\beta$. Again performing a balance on the number of particles pointing with orientations in the hoop $(\beta, \beta + d\beta)$, and dividing both sides by $\sin \beta$ and canceling common factors, the result is

$$8\pi R^2 \eta \frac{\partial f}{\partial t} = \frac{1}{\sin \beta} \frac{\partial}{\partial \beta} \left[\sin \beta \left(k_B T \frac{\partial f}{\partial \beta} + \mu F f \sin \beta \right) \right].$$

The normalization condition in this case is

$$2\pi \int_0^\pi f(\beta, t) \sin \beta d\beta = 1.$$

Implicitly what Debye did was to consider all dipole molecules with polar axes pointing in the same cone to be considered as being in the same state. Hence, without explicitly stating this, what he was really doing was defining an equivalence relation on $SO(3)$ for all orientations with the same second Euler angle, i.e., the model has as its variable elements of the double coset $SO(2)\backslash SO(3)/SO(2)$. In the case when no electric field is applied, the axial symmetry in the problem disappears, and the heat equation on S^2 results. When the molecule itself lacks axial symmetry, a diffusion equation on $SO(3)$ must be considered.

Perrin [89, 90, 91] extended the same kind of model to include full rotational freedom of the molecules, including elliptic shape. Here the heat equation on $SO(3)$ results in some contexts, and we seek $f(A(\alpha, \beta, \gamma), t)$ such that

$$\frac{\partial f}{\partial t} = K \nabla_A^2 f.$$

Solutions to this $SO(3)$-heat equation are given in Chapter 14. Note that Perrin's formulation also ignored the inertial terms in the equation of motion. This is the most elementary case of a non-inertial theory of rotational Brownian motion. A general treatment of non-inertial rotational Brownian motion is given in the next section.

16.3.2   Non-Inertial Theory of Diffusion on $SO(3)$

Consider the Euler-Langevin equation (i.e., the rotational analog of the Langevin equation)

$$\omega_R dt = \frac{1}{c} d\mathbf{W},$$

which we consider to be a Stratonovich equation. This equation holds when viscous terms far outweigh inertial ones (such as in the case of a polar molecule) and the object is spherical, and there is no external field applied.

If our goal is to find out how the process develops on the rotation group, then it makes sense to substitute this equation with the system

$$\text{vect}\left(R^T \dot{R} \right) dt = \frac{1}{c} d\mathbf{W}.$$

In the case when ZXZ Euler angles α, β, γ are used, $\text{vect}(R^T \dot{R}) = J_R(\alpha, \beta, \gamma)(\dot{\alpha}, \dot{\beta}, \dot{\gamma})^T$ and so we write the system of equations

$$\begin{pmatrix} d\alpha \\ d\beta \\ d\gamma \end{pmatrix} = \frac{1}{c} \begin{pmatrix} \sin \beta \sin \gamma & \cos \gamma & 0 \\ \sin \beta \cos \gamma & -\sin \gamma & 0 \\ \cos \beta & 0 & 1 \end{pmatrix}^{-1} \begin{pmatrix} dW_1 \\ dW_2 \\ dW_3 \end{pmatrix}.$$

This is a stochastic differential equation that evolves on $SO(3)$.

If this were viewed as an Itô equation, the Fokker-Planck equation for the PDF $f(A(\alpha, \beta, \gamma), t)$ corresponding to this SDE would be

$$\frac{\partial f}{\partial t} = \frac{1}{2} \frac{\sigma^2}{c} \left[\nabla_A^2 f + \cot \beta \frac{\partial f}{\partial \beta} - f \right] \tag{16.18}$$

where σ^2 is the variance corresponding to $d\mathbf{W}$. If a constant moment of magnitude M_0 were to be applied in the spatial e_3 direction, then the equation would be modified as

$$\frac{\partial f}{\partial t} = -\frac{M_0}{c} \sin \beta \frac{\partial f}{\partial \alpha} + \frac{1}{2} \frac{\sigma^2}{c} \left[\nabla_A^2 f + \cot \beta \frac{\partial f}{\partial \beta} - f \right].$$

However, viewing the above SDE as being of Stratonovich type changes things. Regardless, the operational properties of the $SO(3)$ Fourier transform can be used to reduce these equations, to a simplified form in the Fourier coefficients.

16.3.3 Diffusion of Angular Velocity

In the case when viscous effects are not dominant, the PDF $f(\omega, t)$ for the evolution of angular velocity distribution corresponding to the equation

$$d\omega = -B\omega - I^{-1}(\omega \times I\omega) dt + d\mathbf{W}$$

as given by Steele [103], Hubbard [45, 46, 47, 48, 49] and Gordon [42] is

$$\frac{\partial f}{\partial t} = -\nabla_\omega \cdot \left[B\omega f + I^{-1}(\omega \times I\omega) f \right] + \sum_{i=1}^{3} \sigma_i^2 \frac{\partial^2 f}{\partial \omega_i^2}. \tag{16.19}$$

Here $\{\sigma_i\}$ are the strengths of the random forcing in each coordinate direction and I and B are assumed to be diagonal matrices. (I is the inertia and B is the viscosity tensor.)

If one assumes that initially the angular velocity of all the bodies is ω_0, then $f(\omega, 0) = \delta(\omega - \omega_0)$. Likewise, it can be assumed that as $t \to \infty$ the distribution of angular velocities will approach the Maxwell-Boltzmann distribution [46, 103]

$$\lim_{t \to \infty} f(\omega, t) = \frac{(I_1 I_2 I_3)^{\frac{1}{2}}}{(2\pi k_B T)^{\frac{3}{2}}} \exp\left[-\omega^T I \omega / 2k_B T \right].$$

In order for this to hold, the strengths σ_i^2 for $i = 1, 2, 3$ must be of the form

$$\sigma_i^2 = B_i k_B T / I_i.$$

In general Eq. (16.19) does not lend itself to simple closed-form analytical solutions. However, for the spherically symmetric case when $I_1 = I_2 = I_3 = I$, this reduces to

$$\frac{\partial f}{\partial t} = -\nabla_\omega \cdot [B\omega f] + \sum_{i=1}^{3} \sigma_i^2 \frac{\partial^2 f}{\partial \omega_i^2}. \tag{16.20}$$

Treating ω as a vector in $\mathbb{R}^3$, application of the Fourier transform for $\mathbb{R}^3$ reduces Eq. (16.20) to the following first-order PDE in Fourier space [46]

$$\left[\frac{\partial}{\partial t} + \sum_{j=1}^{3} \left(B_j p_j \frac{\partial}{\partial p_j} + \sigma_j^2 p_j^2 \right) \right] \hat{f} = 0$$

where

$$\hat{f}(\mathbf{p}, t) = \frac{1}{8\pi^3} \int_{\mathbb{R}^3} f(\boldsymbol{\omega}, t) e^{-i\mathbf{p} \cdot \boldsymbol{\omega}} d\boldsymbol{\omega}.$$

The above linear first-order PDE can be solved as [46, 103]

$$\hat{f}(\mathbf{p}, t) = \prod_{j=1}^{3} \exp - \left[\left(\frac{k_B T}{2I} \right) p_j^2 \left(1 - e^{-2B_j t} \right) + i \left(\boldsymbol{\omega}_0 \cdot \mathbf{e}_j \right) p_j e^{-B_j t} \right].$$

Fourier inversion yields

$$f(\boldsymbol{\omega}, t) = \prod_{j=1}^{3} \left(\frac{I}{2\pi k_B T \left(1 - e^{-2B_j t} \right)} \right)^{\frac{1}{2}} \exp \left(-\frac{I \left(\omega_j - (\boldsymbol{\omega}_0 \cdot \mathbf{e}_j) e^{-B_j t} \right)^2}{2 k_B T \left(1 - e^{-2B_j t} \right)} \right).$$

16.4   Brownian Motion of a Rigid Body: Diffusion Equations on $SO(3) \times \mathbb{R}^3$

Consider a rigid body immersed in a viscous fluid and subjected to random moments superimposed on a steady state moment due to an external force field. This model could represent a polar molecule in solution as studied in [21, 76, 111], an underwater vehicle performing station keeping maneuvers while being subjected to random currents, or in the case of zero viscosity, a satellite in orbit subjected to random bombardments of micrometeorites.

The stochastic differential equation of motion describing the most general situation is

$$I d\boldsymbol{\omega} + \boldsymbol{\omega} \times I \boldsymbol{\omega} \, dt = -C\boldsymbol{\omega} + \mathbf{N} dt + d\mathbf{W}. \tag{16.21}$$

$I = \text{diag}[I_1, I_2, I_3]$ is the moment of inertia matrix in a frame attached to the body at its center of mass and oriented with principal axes. $C = IB = \text{diag}[c_1, c_2, c_3]$ is assumed to be diagonal in this body-fixed frame, but may be otherwise if the geometry of the object is not reflected in I due to inhomogeneous mass density or variations in surface characteristics. $\mathbf{N}$ is the steady external field, and $d\mathbf{W}$ is the white noise forcing.

The corresponding Fokker-Planck equation is [46]

$$\frac{\partial F}{\partial t} = -\nabla_{\boldsymbol{\omega}} \cdot \left[B\boldsymbol{\omega} F + I^{-1} (\boldsymbol{\omega} \times I \boldsymbol{\omega}) F \right] + \sum_{i=1}^{3} \left(\sigma_i^2 \frac{\partial^2}{\partial \omega_i^2} - \omega_i X_i^R \right) F. \tag{16.22}$$

Here $F(A, \boldsymbol{\omega}, t)$ for $A \in SO(3)$ is related to $f(\boldsymbol{\omega}, t)$ of the previous section as

$$f(\boldsymbol{\omega}, t) = \int_{SO(3)} F(A, \boldsymbol{\omega}, t) \, dA.$$

Likewise, when $F(A, \boldsymbol{\omega}, t)$ is known, we can find

$$\mathcal{F}(A, t) = \int_{\mathbb{R}^3} F(A, \boldsymbol{\omega}, t) d\boldsymbol{\omega}.$$

When the body is spherical, one can calculate the PDF $F(A, \boldsymbol{\omega}, t)$ from the simplified equation

$$\frac{\partial F}{\partial t} = -\nabla_{\boldsymbol{\omega}} \cdot [B\boldsymbol{\omega} F] + \sum_{i=1}^{3} \left(\sigma_i^2 \frac{\partial^2}{\partial \omega_i^2} - \omega_i X_i^R \right) F. \tag{16.23}$$

In this context, the angular momentum space is identified with $\mathbb{R}^3$ and the Fourier transform on $SO(3) \times \mathbb{R}^3$ can be used to simplify this equation.

Steele [103] reasoned that the average of F over all values of angular momentum must satisfy the equation

$$\frac{\partial \mathcal{F}}{\partial t} = R_1(t) \left[-\frac{\partial^2 \mathcal{F}}{\partial \gamma^2} + \nabla_A^2 \mathcal{F} \right] + R_3(t) \frac{\partial^2 \mathcal{F}}{\partial \gamma^2} \tag{16.24}$$

where

$$R_i(t) = (k_B T / B_i I) \left[1 - \exp(-B_i t) \right].$$

The operational properties of the $SO(3)$ Fourier transform converts Eq. (16.24) to a system of linear time-varying ODEs that can be solved.

While our goal here has been primarily to illustrate how diffusion equations on $SO(3)$ and $SO(3) \times \mathbb{R}^3$ arise in chemical physics, we note that there are many other aspects to rotational Brownian motion that have been addressed in the literature. Classic papers include [1, 16, 20, 31, 34, 37, 38, 40, 61, 62, 75, 78, 79, 100]. Books that address this topic include [18, 76, 93, 102]. Recent works include [4, 12, 13, 55, 56, 57, 81, 105, 115].

16.5 Liquid Crystals

Liquid crystals are often considered to be a phase of matter that lies somewhere in between solid crystalline structure and an isotropic liquid. The special properties of liquid crystals result from the long-range orientational order induced by the local interactions of molecules. Liquid crystalline phases are usually categorized according to whether the constituent molecules are rod-like or disc-like.[2] In both cases, there is usually an "average direction" along which the molecules tend to line up (at least locally), called the *director,* and denoted as $\hat{n}$. In the most general case, the director will be a function of position, $\hat{n}(a)$ where $a \in R^3$. Without loss of generality one can take the director as seen in inertial frame to be $\hat{n}(0) = e_3$.

If we consider $F(\mathbf{a}, A)$ to be a probability density function on $SE(3)$ describing the position and orientation of all molecules in a liquid crystal phase, then one formal definition of the director is $\hat{n} = \mathbf{n}/|\mathbf{n}|$ where

$$\mathbf{n}(\mathbf{a}) = \int_{SO(3)} A e_3 F(\mathbf{a}, A) dA.$$

Another definition, which for engineering purposes is equivalent to this, is that for each $\mathbf{a} \in \mathbb{R}^3$ it is the mode (or "mean" in the sense defined in Chapter 14) of the following function on the sphere

$$k(\mathbf{u}(\alpha, \beta); \mathbf{a}) = \frac{1}{2\pi} \int_{\gamma \in SO(2)} A e_3 F(\mathbf{a}, A) d\gamma,$$

where α, β, and γ are Euler angles, and, by themselves, α and β can be interpreted as the azimuthal and polar angles forming spherical coordinates on S^2. Often, the director does not depend on spatial position, and so $\hat{n}(\mathbf{a}) = e_3$ is the same as what results from averaging $\hat{n}(\mathbf{a})$ over a finite cube of material. Sometimes, as discussed in the following subsections, there is a different director in different planar layers of the liquid crystal. In such cases, the director is a function of only one

---

[2]These terms refer to aspect ratios and do not necessarily imply symmetry.

component of translation, which is the component in the direction which the director rotates. We take this to be the component a_1.

Within the rod-like category are the *nematic* and several *smectic* phases. The nematic phase is characterized by the lack of translational order in the molecules. In this case, all the order is most simply described using an *orientational distribution function* (ODF), which is a probability density function on the rotation group. In the case of the various smectic phases, there is translational order, and the combination of orientational and translational order may be described as a function on the motion group.

In the following sections, we closely follow the analytical descriptions of the evolution of liquid crystal PDFs presented in [19, 24, 58, 70, 71], and recast these descriptions in the terminology of earlier chapters of this book. Other classic works on the evolution of liquid crystal PDFs include [29, 44]. Recent works include [6, 7, 9, 10, 26, 80, 106]. Books on the mechanics and applications of liquid crystals include [11, 17, 22, 23].

16.5.1   The Nematic Phase

In nematogens, the PDF describing the order in the system is only a function of rotation

$$F(\mathbf{a}, A) = f(A).$$

Given an orientation-dependant physical property of a single molecule, $X(A)$, the ensemble average of this property over all molecules in the nematic phase is calculated as

$$\langle X \rangle = \int_{SO(3)} X(A) f(A) dA$$

where it is usually assumed that the ODF $f(A)$ is related to an orientational potential $V(A)$ as

$$f(A) = \frac{\exp\left[-V(A)/k_B T\right]}{\int_{SO(3)} \exp\left[-V(A)/k_B T\right] dA}. \tag{16.25}$$

In the case when the molecules have cylindrical symmetry and no translational order, the functions $X(A)$ and $f(A)$ are constant over $SO(2)$ rotations along the axis of cylindrical symmetry, and all functions on $SO(3)$ (and integrations) are reduced to those on $SO(3)/SO(2) \cong S^2$. Most nematic phases consist of apolar molecules that possess C_2 symmetry about an axis normal to the axis of cylindrical symmetry, and so only the upper hemisphere S^2/C_2 need be considered.

If the medium has orientational order along the inertial $\mathbf{e}_3$ direction, the ODF will be a function on $SO(2) \backslash SO(3)/SO(2)$. In this case the phase is called *uniaxial*.

In the case of uniaxial nematics, the Maier-Saupe potential [72, 73, 74]

$$V(\beta) = -\epsilon S P_2(\cos \beta)$$

is a widely used approximation for the potential function $V(A) = V(A(0, \beta, 0))$. Here ϵ scales the intermolecular interaction, and

$$S = \langle P_2 \rangle = \int_0^1 P_2(\cos \beta) f(A(0, \beta, 0)) d(\cos \beta)$$

can be determined experimentally from nuclear magnetic resonance (NMR).

Other variants on the nematic phase are *biaxial* nematics and *chiral* nematics. In chiral nematics, the molecules have a handedness which imparts a helical twist to the orientation axis along a direction orthogonal to the major axis.

In the case of chiral nematics, one would expect the orientational PDF to have a spatial dependence of the form

$$F(\mathbf{a}, A) = f\left(\text{ROT}\left[\mathbf{e}_3, -2\pi a_1/p\right] \cdot A\right),$$

where $a_1 = \mathbf{e}_1 \cdot \mathbf{a}$ is distance along the axis of the helical motion, which is normal to the plane in which the director (which at $a_1 = 0$ points in the $\mathbf{e}_3$ direction) rotates. The variable p is the pitch, which is the distance along the a_1 direction required for the director to make one full counter clockwise rotation.

16.5.2 The Smectic Phases

In the smectic phases (labeled A through I), varying degrees of translational order persists in addition to orientational order. For instance, in the smectic A phase, the molecules possess spatial order in the direction of the director, but no order in the planes normal to the director. In the smectic C phase, there is order in planes whose normals are a fixed angle away from the director. We shall not address the other smectic phases, but rather show how the PDF for the smectic A phase (and by extension the other smectic phases) is viewed as a function on the motion group.

The fact that smectic A phase liquid crystals have a director and centers of mass that lie in planes normal to the director has led researchers in the field of liquid crystal research to describe this order with several different PDFs. One is of the form [24]

$$F_D(A(\alpha, \beta, \gamma)), \mathbf{a}) = \sum_{l=0}^{\infty} \sum_{n=0}^{\infty} A_{ln} P_l(\cos\beta) \cos\left(2\pi n a_3/d\right) \tag{16.26}$$

where $a_3 = \mathbf{a} \cdot \mathbf{e}_3$ and d is a characteristic distance between the parallel planes that is on the order of the length of the long axis of the rod-like molecules composing the smectic A phase. In this context, the normalization

$$\int_0^\pi \int_0^d F_D \sin\beta \, d\beta \, da_3 = 1$$

is imposed, and the ensemble average of a quantity, X, is calculated as

$$\langle X \rangle = \int_0^\pi \int_0^d X F_D \sin\beta \, d\beta \, da_3.$$

Usually three terms in the series in Eq. (16.26) are kept, and so this model is defined by the parameters [24]

$$\eta = \langle P_2(\cos\beta) \rangle \qquad \tau = \langle \cos(2\pi a_3/d) \rangle \qquad \sigma = \langle P_2(\cos\beta) \cos(2\pi a_3/d) \rangle.$$

Another model for the PDF of smectic A phase is [78, 79]

$$F_M(A(\alpha, \beta, \gamma)), \mathbf{a}) = \frac{\exp\left[-V_M(\beta, a_3)/k_B T\right]}{\int_0^\pi \int_0^d \exp\left[-V_M(\beta, a_3)/k_B T\right] \sin\beta \, d\beta \, da_3} \tag{16.27}$$

where

$$V_M(\beta, a_3) = -v\left[\delta\alpha_0\tau \cos(2\pi a_3/d) + (\eta + \alpha_0\sigma \cos(2\pi a_3/d)) P_2(\cos\beta)\right]$$

and v and δ are physical constants.

Clearly, both of these PDFs are functions on the motion group, $SE(3)$.

16.5.3   Evolution of Liquid Crystal Orientation

Experimentally, many properties of liquid crystals are determined when the time evolution of an orientational correlation function of the form [6, 7]

$$\phi^{ll'}_{mm',nn'} = \int_{SO(3) \times SO(3)} f(A_0) f(A_0, A, t) \overline{U^l_{nm}(A_0)} U^{l'}_{m'n'}(A) \, dA \, dA_0$$

is known. Here $U^l_{mn}(A)$ are matrix elements of $SO(3)$ IURs (or equivalently, Wigner D-functions), $f(A_0, A, t)$ is the probability that a molecule will be at orientation A at time t given that it was at A_0 at time $t_0 = 0$, and $f(A)$ is the equilibrium orientational probability density given by Eq. (16.25).

In order to calculate the correlation function $\phi^{ll'}_{mm',nn'}$, it is first necessary to determine $f(A_0, A, t)$ for a given model of the orientational potential, $V(A)$. In the most general case, this is written in terms of a band-limited Fourier series on $SO(3)$ as

$$\frac{V(A)}{k_B T} = \sum_{l=0}^{N} (2l+1) \text{trace} \left[\hat{C} U(A) \right],$$

where the Fourier coefficients $\hat{C}^l_{mn}$ are given.

The evolution of $f(A_0, A, t)$ is governed by the diffusion equation [24]

$$\frac{\partial}{\partial t} f(A_0, A, t) = \Gamma f(A_0, A, t), \tag{16.28}$$

where the operator Γ is defined as

$$\Gamma = \sum_{i,j=1}^{3} D_{ij} X^R_i \left[X^R_j + X^R_j \left(\frac{V(A)}{k_B T} \right) \right],$$

X^R_i are the differential operators defined in Chapter 9, and $D_{ij} = D_i \delta_{ij}$ are components of the diffusion tensor in the body-fixed molecular frame.

It is common to assume a band-limited solution of the form

$$f(A_0, A, t) = \sum_{l=0}^{M} (2l+1) \text{trace} \left[\hat{f}(A_0, t) U(A) \right] \tag{16.29}$$

and the initial conditions $f(A_0, A, 0) = \delta(A_0^T A)$, in which case Eq. (16.28) is transformed into the system of ODEs

$$\frac{d\hat{f}}{dt} = \mathcal{R} \hat{f}$$

with initial conditions $\hat{f}(A_0, 0) = U(A_0^{-1})$. The components of $\mathcal{R}$ are calculated as

$$\mathcal{R}^{l,l'}_{m',n';m,n}(A_0) = \int_{SO(3)} \overline{U^{l'}_{n'm'}(A)} \Gamma U^l_{mn}(A) \, dA. \tag{16.30}$$

The solution for the Fourier coefficients, in principle, can be thought of as the matrix exponential

$$\hat{f}(A_0, t) = \exp[\mathcal{R}t] U\left(A_0^{-1} \right). \tag{16.31}$$

However, in practice, the evaluation of this solution is greatly simplified when $\mathcal{R}$ is a normal matrix. This cannot always be guaranteed, and so a symmetrized version of the problem is solved, in which Γ is replaced by the self-adjoint operator [24]

$$\tilde{\Gamma} = f^{-\frac{1}{2}}(A)\Gamma f^{\frac{1}{2}}(A)$$

and one solves the equation

$$\frac{\partial}{\partial t}\tilde{f}(A_0, A, t) = \tilde{\Gamma}\tilde{f}(A_0, A, t)$$

where

$$\tilde{f}(A_0, A, t) = f^{-\frac{1}{2}}(A)f(A_0, A, t)f^{\frac{1}{2}}(A_0). \tag{16.32}$$

In this case $\tilde{\mathcal{R}}$ [the matrix with elements defined as in Eq. (16.30) with $\tilde{\Gamma}$ in place of Γ] is unitarily diagonalizable, and the matrix exponential solution [Eq. (16.31) with $\hat{\tilde{f}}$ and $\tilde{\mathcal{R}}$ in place of $\hat{f}$ and $\mathcal{R}$] is efficiently calculated. Then, using Eq. (16.32), $f(A_0, A, t)$ is recovered from $\tilde{f}(A_0, A, t)$ [which is calculated as in Eq. (16.29) with $\tilde{f}$ and $\hat{\tilde{f}}$ replacing f and $\hat{f}$].

16.5.4 Explicit Solution of a Diffusion Equation for Liquid Crystals

The PDF for the orientation distribution of liquid crystals is governed by a viscosity-dominated formulation (zero inertia) resulting in the equation [19, 82, 83, 84]

$$\frac{\partial f}{\partial t} = \left[D_1\left(\left(X_1^R\right)^2 + \left(X_2^R\right)^2\right) + D_2\left(X_3^R\right)^2\right]f - K\operatorname{div}(f\operatorname{grad}V) \tag{16.33}$$

where V is a mean potential field. This equation is based on the *Maier-Saupe model* of nematic liquid crystals [72, 73, 74].

D_1 and D_2 are viscosity constants (which are different from each other because of the axial symmetry of the molecules), and K is a constant depending on external field strength and temperature. $u = \cos^2\beta$ is the normalized field potential for their model, and div and grad are defined in the differential-geometric context described in Chapter 4.

After using properties such as Eq. (9.58), Kalmykov and Coffey [58] write the $SO(3)$-Fourier transform of Eq. (16.33) as a system of the form

$$\frac{d\mathbf{f}}{dt} = A\mathbf{f}$$

where $f_l = \hat{f}^l_{mn}$ and A is a banded infinite-dimensional matrix. If, however, the assumption of a band-limited approximation to the solution is sought, then A becomes finite, and the explicit solution

$$\mathbf{f}(t) = \exp(At)\mathbf{f}(0)$$

can be calculated for bandlimit L.

16.6 A Rotational Diffusion Equation from Polymer Science

In polymer science the following diffusion-type PDE arises (see Chapter 18 for derivation)

$$\frac{\partial f}{\partial t} = \mathcal{K}f \tag{16.34}$$

where the operator $\mathcal{K}$ is defined as

$$\mathcal{K} = \frac{1}{2} \sum_{i,j=1}^{3} D_{ij} X_i^R X_j^R + \sum_{k=1}^{3} d_k X_k^R$$

and X_i^R are the differential operators defined in Chapter 9, and d_k and D_{ij} are constants.

The operational properties of the $SO(3)$ Fourier transform can be used to convert this kind of PDE into a system of linear ODEs. In this section we take the IUR matrix elements of $SO(3)$ to be of the form

$$U_{mn}^l(g(\alpha, \beta, \gamma)) = (-1)^{n-m} e^{-i(m\alpha+n\gamma)} P_{mn}^l(\cos \beta). \tag{16.35}$$

This differs by a factor of i^{n-m} from those in Chapter 9.

By expanding the PDF in the PDE in Eq. (16.34) into a Fourier series on $SO(3)$, the solution can be obtained once one knows how the differential operators X_i^R transform the matrix elements $U_{m,n}^l(A)$. Adjusted for the definition of U_{mn}^l in Eq. (16.35), these operational properties are

$$X_1^R U_{mn}^l = \frac{1}{2} c_{-n}^l U_{m,n-1}^l - \frac{1}{2} c_n^l U_{m,n+1}^l; \tag{16.36}$$

$$X_2^R U_{mn}^l = \frac{1}{2} i c_{-n}^l U_{m,n-1}^l + \frac{1}{2} i c_n^l U_{m,n+1}^l; \tag{16.37}$$

$$X_3^R U_{mn}^l = -in U_{mn}^l; \tag{16.38}$$

where $c_n^l = \sqrt{(l-n)(l+n+1)}$ for $l \geq |n|$ and $c_n^l = 0$ otherwise. From this definition it is clear that $c_k^k = 0$, $c_{-(n+1)}^l = c_n^l$, $c_{n-1}^l = c_{-n}^l$, and $c_{n-2}^l = c_{-n+1}^l$). Equations (16.36)–(16.38) follow from Eq. (16.35) and the fact that

$$\frac{d}{dt} U_{mn}^l (\text{ROT}(e_1, t))|_{t=0} = \frac{1}{2} c_{-n}^l \delta_{m+1,n} - \frac{1}{2} c_n^l \delta_{m-1,n} \tag{16.39}$$

$$\frac{d}{dt} U_{mn}^l (\text{ROT}(e_2, t))|_{t=0} = \frac{i}{2} c_{-n}^l \delta_{m+1,n} + \frac{i}{2} c_n^l \delta_{m-1,n} \tag{16.40}$$

$$\frac{d}{dt} U_{mn}^l (\text{ROT}(e_3, t))|_{t=0} = -in \delta_{m,n}. \tag{16.41}$$

By repeated application of these rules one finds

$$\left(X_1^R\right)^2 U_{mn}^l = \frac{1}{4} c_{-n}^l c_{-n+1}^l U_{m,n-2}^l - \frac{1}{4} \left(c_{-n}^l c_{n-1}^l + c_n^l c_{-n-1}^l\right) U_{mn}^l + \frac{1}{4} c_n^l c_{n+1}^l U_{m,n+2}^l;$$

$$\left(X_2^R\right)^2 U_{mn}^l = -\frac{1}{4} c_{-n}^l c_{-n+1}^l U_{m,n-2}^l - \frac{1}{4} \left(c_{-n}^l c_{n-1}^l + c_n^l c_{-n-1}^l\right) U_{mn}^l - \frac{1}{4} c_n^l c_{n+1}^l U_{m,n+2}^l;$$

$$\left(X_3^R\right)^2 U_{mn}^l = -n^2 U_{mn}^l;$$

$$X_1^R X_2^R U_{m,n}^l = \frac{i}{4} c_{-n}^l c_{-n+1}^l U_{m,n-2}^l + \frac{i}{4} \left(-c_{-n}^l c_{n-1}^l + c_n^l c_{-n-1}^l\right) U_{m,n}^l - \frac{i}{4} c_n^l c_{n+1}^l U_{m,n+2}^l;$$

$$X_2^R X_1^R U_{m,n}^l = \frac{i}{4} c_{-n}^l c_{-n+1}^l U_{m,n-2}^l + \frac{i}{4} \left(c_{-n}^l c_{n-1}^l - c_n^l c_{-n-1}^l\right) U_{m,n}^l - \frac{i}{4} c_n^l c_{n+1}^l U_{m,n+2}^l;$$

$$X_1^R X_3^R U_{m,n}^l = i\frac{n}{2} \left(-c_{-n}^l U_{m,n-1}^l + c_n^l U_{m,n+1}^l\right);$$

$$X_3^R X_1^R U_{m,n}^l = -i\frac{n-1}{2}c_{-n}^l U_{m,n-1}^l + i\frac{n+1}{2}c_n^l U_{m,n+1}^l;$$

$$X_3^R X_2^R U_{m,n}^l = \frac{(n-1)}{2}c_{-n}^l U_{m,n-1}^l + \frac{(n+1)}{2}c_n^l U_{m,n+1}^l;$$

$$X_2^R X_3^R U_{m,n}^l = \frac{n}{2}\left(c_{-n}^l U_{m,n-1}^l + c_n^l U_{m,n+1}^l\right).$$

As a direct result of the definition of the $SO(3)$-Fourier inversion formula, one observes that if a differential operator X transforms U_{mn}^l as

$$XU_{m,n}^l = x(n)U_{m,n+p}^l,$$

then there is a corresponding operational property of the Fourier transform

$$\mathcal{F}(Xf)_{m,n}^l = x(m-p)\hat{f}_{m-p,n}^l. \tag{16.42}$$

We use this to write

$$\mathcal{F}\left(X_1^R f\right)_{mn}^l = \frac{1}{2}c_{-m-1}^l \hat{f}_{m+1,n}^l - \frac{1}{2}c_{m-1}^l \hat{f}_{m-1,n}^l;$$

$$\mathcal{F}\left(X_2^R f\right)_{mn}^l = \frac{1}{2}ic_{-m-1}^l \hat{f}_{m+1,n}^l + \frac{1}{2}ic_{m-1}^l \hat{f}_{m-1,n}^l;$$

$$\mathcal{F}\left(X_3^R f\right)_{mn}^l = -im\hat{f}_{mn}^l;$$

$$\mathcal{F}\left(\left(X_1^R\right)^2 f\right)_{mn}^l = \frac{1}{4}c_{m+1}^l c_{-m-1}^l \hat{f}_{m+2,n}^l - \frac{1}{4}\left(c_{-m}^l c_{m-1}^l + c_m^l c_{-m-1}^l\right)\hat{f}_{mn}^l + \frac{1}{4}c_{-m+1}^l c_{m-1}^l \hat{f}_{m-2,n}^l;$$

$$\mathcal{F}\left(\left(X_2^R\right)^2 f\right)_{mn}^l = -\frac{1}{4}c_{m+1}^l c_{-m-1}^l \hat{f}_{m+2,n}^l - \frac{1}{4}\left(c_{-m}^l c_{m-1}^l + c_m^l c_{-m-1}^l\right)\hat{f}_{mn}^l - \frac{1}{4}c_{-m+1}^l c_{m-1}^l \hat{f}_{m-2,n}^l;$$

$$\mathcal{F}\left(\left(X_3^R\right)^2 f\right)_{mn}^l = -m^2 \hat{f}_{mn}^l;$$

$$\mathcal{F}\left(\left(X_1^R X_2^R + X_2^R X_1^R\right) f\right)_{m,n}^l = \frac{i}{2}c_{m+1}^l c_{-m-1}^l \hat{f}_{m+2,n}^l - \frac{i}{2}c_{-m+1}^l c_{m-1}^l \hat{f}_{m-2,n}^l;$$

$$\mathcal{F}\left(\left(X_1^R X_3^R + X_3^R X_1^R\right) f\right)_{m,n}^l = -i\frac{2m+1}{2}c_{-m-1}^l \hat{f}_{m+1,n}^l + i\frac{2m-1}{2}c_{m-1}^l \hat{f}_{m-1,n}^l;$$

$$\mathcal{F}\left(\left(X_3^R X_2^R + X_2^R X_3^R\right) f\right)_{m,n}^l = \frac{(2m+1)}{2}c_{-m-1}^l \hat{f}_{m+1,n}^l + \frac{(2m-1)}{2}c_{m-1}^l \hat{f}_{m-1,n}^l.$$

Collecting everything together we have

$$\mathcal{F}\left(\left(\frac{1}{2}\sum_{i,j=1}^{3} D_{ij} X_i^R X_j^R + \sum_{i=1}^{3} d_i X_i^R\right) f\right)_{mn}^l = \sum_{k=\max(-l,m-2)}^{\min(l,m+2)} \mathcal{A}_{m,k}^l \hat{f}_{k,n}^l,$$

where

$$\mathcal{A}_{m,m+2}^l = \left[\frac{(D_{11}-D_{22})}{8} + \frac{i}{4}D_{12}\right]c_{m+1}^l c_{-m-1}^l;$$

$$\mathcal{A}_{m,m+1}^l = \left[\frac{(2m+1)}{4}(D_{23}-iD_{13}) + \frac{1}{2}(d_1+id_2)\right]c_{-m-1}^l;$$

$$\mathcal{A}_{m,m}^l = \left[-\frac{(D_{11} + D_{22})}{8} \left(c_{-m}^l c_{m-1}^l + c_m^l c_{-m-1}^l \right) - \frac{D_{33}m^2}{2} - id_3 m \right];$$

$$\mathcal{A}_{m,m-1}^l = \left[\frac{(2m-1)}{4} (D_{23} + iD_{13}) + \frac{1}{2}(-d_1 + id_2) \right] c_{m-1}^l;$$

$$\mathcal{A}_{m,m-2}^l = \left[\frac{(D_{11} - D_{22})}{8} - \frac{i}{4} D_{12} \right] c_{-m+1}^l c_{m-1}^l.$$

Hence, application of the $SO(3)$-Fourier transform to Eq. (16.34) and corresponding initial conditions reduces Eq. (16.34) to a set of linear time-invariant ODEs of the form

$$\frac{d\hat{f}^l}{dL} = \mathcal{A}^l \hat{f}^l \quad \text{with} \quad \hat{f}^l(0) = \mathbb{I}_{(2l+1)} \times (2l+1). \tag{16.43}$$

Here $\mathbb{I}_{(2l+1)} \times (2l+1)$ is the $(2l+1) \times (2l+1)$ identity matrix and the banded matrix $\mathcal{A}^l$ are of the following form for $l = 0, 1, 2, 3$

$$\mathcal{A}^0 = \mathcal{A}_{0,0}^0 = 0; \quad \mathcal{A}^1 = \begin{pmatrix} \mathcal{A}_{-1,-1}^1 & \mathcal{A}_{-1,0}^1 & \mathcal{A}_{-1,1}^1 \\ \mathcal{A}_{0,-1}^1 & \mathcal{A}_{0,0}^1 & \mathcal{A}_{0,1}^1 \\ \mathcal{A}_{1,-1}^1 & \mathcal{A}_{1,0}^1 & \mathcal{A}_{1,1}^1 \end{pmatrix};$$

$$\mathcal{A}^2 = \begin{pmatrix} \mathcal{A}_{-2,-2}^2 & \mathcal{A}_{-2,-1}^2 & \mathcal{A}_{-2,0}^2 & 0 & 0 \\ \mathcal{A}_{-1,-2}^2 & \mathcal{A}_{-1,-1}^2 & \mathcal{A}_{-1,0}^2 & \mathcal{A}_{-1,1}^2 & 0 \\ \mathcal{A}_{0,-2}^2 & \mathcal{A}_{0,-1}^2 & \mathcal{A}_{0,0}^2 & \mathcal{A}_{0,1}^2 & \mathcal{A}_{0,2}^2 \\ 0 & \mathcal{A}_{1,-1}^2 & \mathcal{A}_{1,0}^2 & \mathcal{A}_{1,1}^2 & \mathcal{A}_{1,2}^2 \\ 0 & 0 & \mathcal{A}_{2,0}^2 & \mathcal{A}_{2,1}^2 & \mathcal{A}_{2,2}^2 \end{pmatrix};$$

$$\mathcal{A}^3 = \begin{pmatrix} \mathcal{A}_{-3,-3}^3 & \mathcal{A}_{-3,-2}^3 & \mathcal{A}_{-3,-1}^3 & 0 & 0 & 0 & 0 \\ \mathcal{A}_{-2,-3}^3 & \mathcal{A}_{-2,-2}^3 & \mathcal{A}_{-2,-1}^3 & \mathcal{A}_{-2,0}^3 & 0 & 0 & 0 \\ \mathcal{A}_{-1,-3}^3 & \mathcal{A}_{-1,-2}^3 & \mathcal{A}_{-1,-1}^3 & \mathcal{A}_{-1,0}^3 & \mathcal{A}_{-1,1}^3 & 0 & 0 \\ 0 & \mathcal{A}_{0,-2}^3 & \mathcal{A}_{0,-1}^3 & \mathcal{A}_{0,0}^3 & \mathcal{A}_{0,1}^3 & \mathcal{A}_{0,2}^3 & 0 \\ 0 & 0 & \mathcal{A}_{1,-1}^3 & \mathcal{A}_{1,0}^3 & \mathcal{A}_{1,1}^3 & \mathcal{A}_{1,2}^3 & \mathcal{A}_{1,3}^3 \\ 0 & 0 & 0 & \mathcal{A}_{2,0}^3 & \mathcal{A}_{2,1}^3 & \mathcal{A}_{2,2}^3 & \mathcal{A}_{2,3}^3 \\ 0 & 0 & 0 & 0 & \mathcal{A}_{3,1}^3 & \mathcal{A}_{3,2}^3 & \mathcal{A}_{3,3}^3 \end{pmatrix}.$$

As is well known in systems theory, the solution to Eq. (16.43) is of the form of a matrix exponential

$$\hat{f}^l(L) = e^{L\mathcal{A}^l}. \tag{16.44}$$

Since $\mathcal{A}^l$ is a band-diagonal matrix for $l > 1$, the matrix exponential can be calculated much more efficiently (either numerically or symbolically) for large values of l than for general matrices of dimension $(2l+1) \times (2l+1)$. One also gains efficiencies in computing the matrix exponential of $L\mathcal{A}^l$ by observing the symmetry

$$\mathcal{A}_{m,n}^l = (-1)^{m-n} \overline{\mathcal{A}_{-m,-n}^l}.$$

Matrices with this kind of symmetry have eigenvalues which occur in conjugate pairs, and if x_m are the components of the eigenvector corresponding to the complex eigenvalue λ, then $(-1)^m \overline{x_{-m}}$ will be the components of the eigenvector corresponding to $\bar{\lambda}$ (see Theorem 10.1).

In general, the numerically calculated values of $\hat{f}^l(L)$ may be substituted back into the $SO(3)$-Fourier inversion formula to yield the solution for $f(A; L)$ to any desired accuracy. When $D_{11} =$

$D_{22} = 1/\alpha_0$ and $D_{33} \to \infty$, and every other parameter in D and $\mathbf{d}$ is zero, the matrices $\mathcal{A}^l$ are all diagonal. This implies that the nonzero Fourier coefficients are of the form $\hat{f}^l_{m,m}(L) = \exp(L\mathcal{A}^l_{m,m})$. However, for $m \neq 0$ the value of D_{33} causes $\hat{f}^l_{m,m}(L)$ to be zero and what remains is a series in l with $m = 0$

$$f(A; L) = \sum_{l=0}^{\infty} (2l + 1)e^{-l(l+1)L/2\alpha_0} U^l_{0,0}(A) = \sum_{l=0}^{\infty} (2l + 1)e^{-l(l+1)L/2\alpha_0} P_l(\cos\beta).$$

This special case corresponds to the "Kratky-Porod" model discussed in Chapter 17.

All of what is described in this section is well known in the literature, though it is rarely expressed using group-theoretical notation and terminology.

16.7 Other Models for Rotational Brownian Motion

While problems in chemical physics are a rich source of rotational Brownian motion problems, there are other fields in which such problems arise. In this section we examine two non-physics-based models in which diffusions on $SO(3)$ occur.

16.7.1 The Evolution of Estimation Error PDFs

In Chapter 15, we encountered the estimation Eq. (15.42). This may be written in the form

$$(Q')^T dQ' = (Q')^T \left(\sum_{i=1}^{3} x_i dt X_i \right) Q' + \left(\sum_{i=1}^{3} X_i dW_i(t) \right). \tag{16.45}$$

Using the tools of this chapter, it is possible to write a corresponding Fokker-Planck equation. First, we take the vect($\cdot$) of both sides, resulting in

$$J_R(\phi)d\phi = Ad\left(Q'\right)\mathbf{x}dt + d\mathbf{W}$$

where ϕ is the array of ZXZ Euler angles corresponding to Q'. If $\mathbf{x} = \mathbf{0}$, the Fokker-Planck equation for the PDF of Q' is the same as given in Eq. (16.18) with $c = 1$. Otherwise, the drift term is different.

16.7.2 Rotational Brownian Motion Based on Randomized Angular Momentum

A different model of the random motion of a rigid body is formulated by Liao [67]. In this formulation, the random forces acting on the body are defined such that they instantaneously change the body's angular momentum in random ways. The corresponding stochastic differential equation is [67]

$$dR = R\Lambda Ad\left(R^{-1}\right) \sum_{i=1}^{3} X_i dW_i + R\Lambda Ad\left(R^{-1}\right) M_0 dt$$

where $M_0 \in so(3)$ is the initial angular momentum of the body in the space-fixed frame and Λ is the inverse inertia operator described in Chapter 15. Multiplying both sides of the above equation by R^{-1} on the left, taking the vect($\cdot$) of both sides, and using the properties of Λ, we find

$$\omega_R dt = I^{-1}\text{vect}\left(Ad\left(R^{-1}\right) \sum_{i=1}^{3} X_i dW_i \right) + I^{-1}\text{vect}\left(Ad\left(R^{-1}\right) M_0 dt \right).$$

Using the fact that $\omega_R = J_R\dot{\phi}$ where $\phi = [\alpha, \beta, \gamma]^T$ is the array of Euler angles, and for $SO(3)$ we have from Chapter 5 that $[Ad(R)] = R$, we find

$$d\phi = J_R^{-1} I^{-1} R^{-1} d\mathbf{W} + J_R^{-1} I^{-1} R^{-1} \mathbf{M}_0$$

where $\mathbf{M}_0 = \text{vect}(M_0)$ and $\mathbf{W} = \text{vect}(W)$.

The Fokker-Planck equation for this case may be generated using a formulation similar to that presented in Sections 16.2.1 and 16.2.2 with the modification that the Stratonovich calculus be used. In the special case when $I = \mathbb{I}$ and $\mathbf{M}_0 = \mathbf{0}$ the fact that $RJ_R = J_L$ is used to simplify matters, and we get a "left version" of Eq. (16.18) in the sense that now $d\phi = J_L^{-1} d\mathbf{W}$ instead of $d\phi = J_R^{-1} d\mathbf{W}$.

16.7.3   The Effects of Time Lag and Memory

One can imagine versions of Eq. (16.34) in which there are either time lags or memory effects. In the first case we would have

$$\frac{\partial}{\partial t} f(A, t) = \mathcal{K} f(A, t - \tau)$$

and in the second case,

$$\frac{\partial}{\partial t} f(A, t) = \int_0^t h(t - \tau) \mathcal{K} f(A, \tau) d\tau.$$

In both of these cases the *Laplace transform* can be used to convert the time dependence of these equations into an algebraic dependence on the Laplace transform parameter. Recall that the Laplace transform of a function $x(t)$ where $t \in \mathbb{R}^+$ is defined as

$$X(s) = L[x] = \int_0^\infty x(t) e^{-st} dt$$

where $s \in \mathbb{C}$ has positive real part.

Much like the Fourier transform of functions on the line, the Laplace transform has some useful operational properties. Using integration by parts, one can show that

$$L\left[\frac{dx}{dt}\right] = s L[x] - x(0).$$

Hence, given a system of linear equations with constant coefficients of the form

$$\frac{d\mathbf{x}}{dt} = A\mathbf{x}, \tag{16.46}$$

we can write

$$s L[\mathbf{x}] - \mathbf{x}(0) = A L[\mathbf{x}].$$

With initial conditions $\mathbf{x}(0)$ given, we can write

$$(s\mathbb{I} - A) L[\mathbf{x}] = \mathbf{x}(0).$$

When $\det(s\mathbb{I} - A) \neq 0$ we can symbolically write

$$\mathbf{x}(t) = L^{-1}\left[(s\mathbb{I} - A)^{-1} \mathbf{x}(0)\right]$$

where L^{-1} is the inverse Laplace transform. See [5] for a description of the inversion process. From a practical perspective, it is convenient to write the solution to Eq. (16.46) as

$$\mathbf{x}(t) = \exp(tA)\mathbf{x}(0).$$

However, for equations with time lags or memory, the Laplace transform technique is useful. This is because of the two operational properties discussed below.

The Laplace transform also converts convolutions of the form

$$(x * y)(t) = \int_0^t x(\tau)y(t - \tau)d\tau$$

to products as

$$L[x * y] = L[x]L[y].$$

Note that the bounds of integration in this convolution are different than in the rest of the book.

An operational property for shifts is [5]

$$L[x(t - t_0)u(t - t_0)] = e^{-t_0 s}\mathcal{L}[x(t)],$$

where $u(t - t_0)$ is the unit step function equal to zero for $t < t_0$ and equal to unity for $t \geq t_0$.

Hence, the $SO(3)$ Fourier transform can be used in combination with the Laplace transform operational properties to handle the time-dependent part of the equations at the beginning of this section.

16.8   Summary

In this chapter we reviewed the classical theories of rotational Brownian motion and their generalizations. Diffusion processes on the rotation group are associated with a diverse array of scenarios ranging from chemical physics and the analysis of liquid crystals to estimation problems. Fokker-Planck equations describing the time evolution of probability density functions in rotational Brownian motions are derived. In the non-inertial theory of rotational Brownian motion, a PDF governed by the Fokker-Planck equation is a function of rotation. In the inertial theory of rotational Brownian motion these PDFs are functions of both rotation and angular velocity.

In addition to the references provided throughout this chapter, topics related to rotational Brownian motion can be found in [8, 14, 25, 30, 35, 36, 60, 88, 92, 95].

References

[1] Adelman, S.A., Fokker-Planck equations for simple non-Markovian systems, *J. of Chem. Phys.*, 64(1), 124–130, 1976.

[2] Albeverio, S., Arede, T., and Haba, Z., On left invariant Brownian motions and heat kernels on nilpotent Lie groups, *J. of Math. Phys.*, 31(2), 278–286, 1990.

[3] Applebaum, D. and Kunita, H., Lévy flows on manifolds and Lévy processes on Lie groups, *J. Math. Kyoto. Univ.*, 33/34, 1103–1123, 1993.

[4] Balabai, N., Sukharevsky, A., Read, I., Strazisar, B., Kurnikova, M., Hartman, R.S., Coalson, R.D., and Waldeck, D.H., Rotational diffusion of organic solutes: the role of dielectric friction in polar solvents and electrolyte solutions, *J. of Molecular Liquids*, 77, 37–60, 1998.

[5] Bellman, R.E. and Roth, R.S., *The Laplace Transform,* World Scientific, Singapore, 1984.

[6] Berggren, E., Tarroni, R., and Zannoni, C., Rotational diffusion of uniaxial probes in biaxial liquid crystal phases, *J. Chem. Phys.,* 99(8), 6180–6200, 1993.

[7] Berggren, E. and Zannoni, C., Rotational diffusion of biaxial probes in biaxial liquid crystal phases, *Mol. Phys.,* 85(2), 299–333, 1995.

[8] Bernassau, J.M., Black, E.P., and Grant, D.M., Molecular motion in anisotropic medium. I. The effect of the dipolar interaction on nuclear spin relaxation, *J. Chem. Phys.,* 76(1), 253–256, 1982.

[9] Blenk, S., Ehrentraut, H., and Muschik, W., Statistical foundation of macroscopic balances for liquid-crystals in alignment tensor formulation, *Physica A,* 174(1), 119–138, 1991.

[10] Blenk, S. and Muschik, W., Orientational balances for nematic liquid-crystals, *J. of Non-Equilibrium Thermodyn.,* 16(1), 67–87, 1991.

[11] Blinov, L.M., *Electro-Optical and Magneto-Optical Properties of Liquid Crystals,* Wiley-InterScience, New York, 1984.

[12] Blokhin, A.P. and Gelin, M.F., Rotation of nonspherical molecules in dense fluids: a simple model description, *J. Phys. Chem. B,* 101, 236–243, 1997.

[13] Blokhin, A.P. and Gelin, M.F., Rotational Brownian motion of spherical molecules: the Fokker-Planck equation with memory, *Physica A,* 229, 501–514, 1996.

[14] Brilliantov, N.V., Vostrikova, N.G., Denisov, V.P., Petrusievich, Yu.M., and Revokatov, O.P., Influence of dielectric friction and near-surface increase of viscosity on rotational Brownian motion of charged biopolymers in solution, *Biophys. Chem.,* 46, 227–236, 1993.

[15] Brockett, R.W., Notes on stochastic processes on manifolds, in *Systems and Control in the Twenty-First Century,* Byrnes, C.I., et al., Eds., Birkhäuser, Boston, 1997.

[16] Budó, A., Fischer, E., and Miyamoto, S., Einfluß der molekülform auf die dielektrische relaxation, *Physikalische Zeitschrift,* 40, 337–345, 1939.

[17] Chigrinov, V.G., *Liquid Crystal Devices: Physics and Applications,* Artech House, Boston, 1999.

[18] Coffey, W., Evans, M., and Grigolini, P., *Molecular Diffusion and Spectra,* John Wiley & Sons, New York, 1984.

[19] Coffey, W.T. and Kalmykov, Yu.P., On the calculation of the dielectric relaxation times of a nematic liquid crystal from the non-inertial Langevin equation, *Liquid Crystals,* 14(4), 1227–1236, 1993.

[20] Cukier, R.I., Rotational relaxation of molecules in isotropic and anisotropic fluids, *J. Chem. Phys.,* 60(3), 734–743, 1974.

[21] Debye, P., *Polar Molecules,* Dover, New York, 1929. (See also *The Collected Papers of Peter J.W. Debye,* Wiley InterScience, New York, 1954.)

[22] de Gennes, P.G. and Prost, J., *The Physics of Liquid Crystals,* 2nd ed., Clarendon Press, Oxford, 1998.

[23] de Jeu, W.H., *Physical Properties of Liquid Crystalline Materials,* Gordon and Breach, New York, 1980.

[24] Dong, R.Y., *Nuclear Magnetic Resonance of Liquid Crystals,* 2nd ed., Springer, New York, 1997.

[25] Edwards, D., Steady motion of a viscous liquid in which an ellipsoid is constrained to rotate about a principal axis, *Q. J. of Pure and Appl. Math.,* 26, 70, 1893.

[26] Ehrentraut, H., Muschik, W., and Papenfuss, C., Mesoscopically derived orientation dynamics of liquid crystals, *J. of Non-Equilibrium Thermodyn.,* 22(3), 285–298, 1997.

[27] Einstein, A., *Investigations on the Theory of the Brownian Movement,* Dover, New York, 1956.

[28] Epperson, J.B. and Lohrenz, T., Brownian motion and the heat semigroup on the path space of a compact Lie group, *Pac. J. of Math.,* 161(2), 233–253, 1993.

[29] Ericksen, J.L. and Kinderlehrer, D., Eds., *Theory and Applications of Liquid Crystals,* IMA Vols. in Math. and its Appl., 5, Springer-Verlag, New York, 1987.

[30] Edén, M. and Levitt, M.H., Computation of orientational averages in solid-state NMR by Gaussian spherical quadrature, *J. of Magn. Resonance,* 132, 220–239, 1998.

[31] Egelstaff, P.A., Cooperative rotation of spherical molecules, *J. of Chem. Phys.,* 53(7), 2590–2598, 1970.

[32] Elworthy, K.D., *Stochastic Differential Equations on Manifolds,* Cambridge University Press, Cambridge, 1982.

[33] Emery, M., *Stochastic Calculus in Manifolds,* Springer-Verlag, Berlin, 1989.

[34] Evans, G.T., Orientational relaxation of a Fokker-Planck fluid of symmetric top molecules, *J. of Chem. Phys.,* 67(6), 2911–2915, 1977.

[35] Evans, G.T., Momentum space diffusion equations for chain molecules, *J. of Chem. Phys.,* 72(7), 3849–3858, 1980.

[36] Evans, M.W., Ferrario, M., and Grigolini, P., The mutual interaction of molecular rotation and translation, *Mol. Phys.,* 39(6), 1369–1389, 1980.

[37] Favro, L.D., Theory of the rotational Brownian motion of a free rigid body, *Phys. Rev.,* 119(1), 53–62, 1960.

[38] Fixman, M. and Rider, K., Angular relaxation of the symmetrical top, *J. of Chem. Phys.,* 51(6), 2425–2438, 1969.

[39] Fokker, A.D., Die mittlere Energie rotierender elektrischer Dipole im Strahlungsfeld, *Ann. Physik.,* 43, 810–820, 1914.

[40] Furry, W.H., Isotropic rotational Brownian motion, *Phys. Rev.,* 107(1), 7–13, 1957.

[41] Gardiner, C.W., *Handbook of Stochastic Methods,* 2nd ed., Springer-Verlag, Berlin, 1985.

[42] Gordon, R.G., On the rotational diffusion of molecules, *J. of Chem. Phys.,* 44(5), 1830–1836, 1966.

[43] Gorman, C.D., Brownian motion of rotation, *Trans. of the Am. Math. Soc.,* 94, 103–117, 1960.

[44] Hess, S. Fokker-Planck equation approach to flow alignment in liquid-crystals, *Zeitschrift für Naturforschung Section A—A J. of Phys. Sci.,* 31(9), 1034–1037, 1976.

[45] Hubbard, P.S., Rotational Brownian motion. II. Fourier transforms for a spherical body with strong interactions, *Phys. Rev. A,* 8(3), 1429–1436, 1973.

[46] Hubbard, P.S., Rotational Brownian motion, *Phys. Rev. A,* 6(6), 2421–2433, 1972.

[47] Hubbard, P.S., Theory of nuclear magnetic relaxation by spin-rotational interactions in liquids, *Phys. Rev.,* 131(3), 1155–1165, 1963.

[48] Hubbard, P.S., Nuclear magnetic relaxation in spherical-top molecules undergoing rotational Brownian motion, *Phys. Rev. A,* 9(1), 481–494, 1974.

[49] Hubbard, P.S., Angular velocity of a nonspherical body undergoing rotational Brownian motion, *Phys. Rev. A,* 15(1), 329–336, 1977.

[50] Ikeda, N. and Watanabe, S., *Stochastic Differential Equations and Diffusion Processes,* 2nd ed., North-Holland, Amsterdam, 1989.

[51] Itô, K., Brownian motions in a Lie group, *Proc. Japan Acad.,* 26, 4–10, 1950.

[52] Itô, K., Stochastic differential equations in a differentiable manifold, *Nagoya Math. J.,* 1, 35–47, 1950.

[53] Itô, K., Stochastic differential equations in a differentiable manifold (2), *Sci. Univ. Kyoto Math., Ser. A,* 28(1), 81–85, 1953.

[54] Itô, K. and McKean, H.P., Jr., *Diffusion Processes and their Sample Paths,* Springer, 1996.

[55] Kalmykov, Y.P. and Quinn, K.P., The rotational Brownian motion of a linear molecule and its application to the theory of Kerr effect relaxation, *J. of Chem. Phys.,* 95(12), 9142–9147, 1991.

[56] Kalmykov, Yu. P., Rotational Brownian motion in an external potential field: a method based on the Langevin equation, *Chem. Phys. Rep.,* 16(3), 535–548, 1997.

[57] Kalmykov, Y.P., Rotational Brownian motion in an external potential: the Langevin equation approach, *J. of Mol. Liquids,* 69, 117–131, 1996.

[58] Kalmykov, Y.P. and Coffey, W.T., Analytical solutions for rotational diffusion in the mean field potential: application to the theory of dielectric relaxation in nematic liquid crystals, *Liquid Crystals,* 25(3), 329–339, 1998.

[59] Karatzas, I. and Shreve, S.E., *Brownian Motion and Stochastic Calculus,* 2nd ed., Springer, 1991.

[60] Kendall, D.G., Pole-seeking Brownian motion and bird navigation, *Royal Stat. Soc.,* 36, Ser. B, 365–417, 1974.

[61] Kobayashi, K.K., Theory of translational and orientational melting with application to liquid crystals. I, *J. of the Phys. Soc. of Japan,* 29(1), 101–105, 1970.

[62] Kobayashi, K.K., Theory of translational and orientational melting with application to liquid crystals, *Mol. Crystals and Liquid Crystals,* 13, 137–148, 1971.

[63] Kolmogorov, A., Uber die analytischen Methoden in der Wahrscheinlichkeitsrechnung, *Math. Ann.,* 104, 415–458, 1931.

[64] Kunita, H., *Stochastic Flows and Stochastic Differential Equations,* Cambridge University Press, Cambridge, 1997.

[65] Langevin, P., Sur la théorie du mouvement Brownien, *Comptes. Rendues Acad. Sci. Paris,* 146, 530–533, 1908.

[66] Lévy, P., *Processus stochastiques et Muvement Brownien,* Gauthier-Villars, Paris, 1948.

[67] Liao, M., Random motion of a rigid body, *J. of Theoretical Probability*, 10(1), 201–211, 1997.

[68] Lin, Y.K., *Probabilistic Theory of Structural Dynamics*, Robert E. Krieger, Malabar, FL, 1986.

[69] Luckhurst, G.R. and Gray, G.W., Eds., *The Molecular Physics of Liquid Crystals*, Academic Press, New York, 1979.

[70] Luckhurst, G.R. and Sanson, A., Angular dependent linewidths for a spin probe dissolved in a liquid crystal, *Mol. Phys.*, 24(6), 1297–1311, 1972.

[71] Luckhurst, G.R., Zannoni, C., Nordio, P.L., and Segre, U., A molecular field theory for uniaxial nematic liquid crystals formed by non-cylindrically symmetric molecules, *Mol. Phys.*, 30(5), 1345–1358, 1975.

[72] Maier, W. and Saupe, A., Eine einfache molekulare Theorie des nematischen kristallinflüssigen Zustandes, *Z. Naturforsch*, 13a, 564–566, 1958.

[73] Maier, W. and Saupe, A., Eine einfache molekular-statistische theorie der nematischen kristallinflüssigen phase. Teil I, *Z. Naturforsch*, 14a, 882–889, 1959.

[74] Maier, W. and Saupe, A., Eine einfache molekular-statistische theorie der nematischen kristallinflüssigen phase. Teil II, *Z. Naturforsch*, 15a, 287–292, 1960.

[75] McClung, R.E.D., The Fokker-Planck-Langevin model for rotational Brownian motion. I. General theory, *J. of Chem. Phys.*, 73(5), 2435–2442, 1980.

[76] McConnell, J., *Rotational Brownian Motion and Dielectric Theory*, Academic Press, New York, 1980.

[77] McKean, H.P., Jr., Brownian motions on the 3-dimensional rotation group, *Memoirs of the College of Science, University of Kyoto, Ser. A*, 33(1), 2538, 1960.

[78] McMillan, W.L., Simple molecular model for the smectic A phase of liquid crystals, *Phys. Rev. A*, 4(3), 1238–1246, 1971.

[79] McMillan, W.L., X-ray scattering from liquid crystals. I. Cholesteryl nonanoate and myristate, *Phys. Rev. A*, 6(3), 936–947, 1972.

[80] Muschik, W. and Su, B., Mesoscopic interpretation of Fokker-Planck equation describing time behavior of liquid crystal orientation, *J. of Chem. Phys.*, 107(2), 580–584, 1997.

[81] Navez, P. and Hounkonnou, M.N., Theory of the rotational Brownian motion of a linear molecule in 3D: a statistical mechanics study, *J. of Mol. Liquids*, 70, 71–103, 1996.

[82] Nordio, P.L. and Busolin, P., Electron spin resonance line shapes in partially oriented systems, *J. Chem. Phys.*, 55(12), 5485–5490, 1971.

[83] Nordio, P.L., Rigatti, G., and Segre, U., Spin relaxation in nematic solvents, *J. Chem. Phys.*, 56(5), 2117–2123, 1972.

[84] Nordio, P.L., Rigatti, G., and Segre, U., Dielectric relaxation theory in nematic liquids, *Mol. Phys.*, 25(1), 129–136, 1973.

[85] Øksendal, B., *Stochastic Differential Equations, An Introduction with Applications*, 5th ed., Springer, Berlin, 1998.

[86] Orsingher, E., Stochastic motions on the 3-sphere governed by wave and heat equations, *J. Appl. Prob.*, 24, 315–327, 1987.

[87] Pap, G., Construction of processes with stationary independent increments in Lie groups, *Arch. der Math.*, 69, 146–155, 1997.

[88] Perico, A., Guenza, M., and Mormino, M., Protein dynamics: rotational diffusion of rigid and fluctuating three dimensional structures, *Biopolymers*, 35, 47–54, 1995.

[89] Perrin, P.F., Mouvement Brownien d'un ellipsoide (I). Dispersion diélectrique pour des molécules ellipsoidales, *Le J. de Phys. et Le Radium*, 7(10), 497–511, 1934.

[90] Perrin, P.F., Mouvement Brownien d'un ellipsoide (II). Rotation libre et dépolarisation des fluorescences. Translation et diffusion de molécules ellipsoidales, *Le J. de Phys. et Le Radium*, 7(1), 1–11, 1936.

[91] Perrin, P.F., Étude mathématique du mouvement Brownien de rotation, *Ann. Sci. de L' École Normale Supérieure*, 45, 1–51, 1928.

[92] Piccioni, M. and Scarlatti, S., An iterative Monte Carlo scheme for generating Lie group-valued random variables, *Adv. Appl. Prob.*, 26, 616–628, 1994.

[93] Pikin, S.A., *Structural Transformations in Liquid Crystals*, Gordon and Breach, New York, 1991.

[94] Planck, M., Uber einen satz der statistischen dynamik und seine erweiterung in der quantentheorie, *Sitz. ber. Berlin A Akad. Wiss.*, 324–341, 1917.

[95] Polnaszek, C.F., Bruno, G.V., and Freed, J.H., ESR line shapes in the slow-motional region: anisotropic liquids, *J. Chem. Phys.*, 58(8), 3185–3199, 1973.

[96] Rains, E.M., Combinatorial properties of Brownian motion on the compact classical groups, *J. of Theor. Prob.*, 10(3), 659–679, 1997.

[97] Risken, H., *The Fokker-Planck Equation, Methods of Solution and Applications*, 2nd ed., Springer-Verlag, Berlin, 1989.

[98] Roberts, P.H. and Ursell, H.D., Random walk on a sphere and on a Riemannian manifold, *Philos. Trans. of the Royal Soc. of London*, A252, 317–356, 1960.

[99] Rothschild, W.G., *Dynamics of Molecular Liquids*, John Wiley & Sons, New York, 1984.

[100] Sack, R.A., Relaxation processes and inertial effects I, II, *Proc. Phys. Soc.*, B, 70, 402–426, 1957.

[101] Smoluchowski, M.V., Über Brownsche Molekularbewegung unter einwirkung äußerer Kräfte und deren zusammenhang mit der veralgemeinerten diffusionsgleichung, *Ann. Physik.*, 48, 1103–1112, 1915.

[102] Steele, D. and Yarwood, J., Eds., *Spectroscopy and Relaxation of Molecular Liquids*, Elsevier, 1991.

[103] Steele, W.A., Molecular reorientation in liquids. I. Distribution functions and friction constants. II. Angular autocorrelation functions, *J. of Chem. Phys.*, 38(10), 2404–2418, 1963.

[104] Stratonovich, R.L., *Topics in the Theory of Random Noise: Vols. I and II*, Gordon and Breach Science, New York, 1963.

[105] Tang, S. and Evans, G.T., Free and pendular-like rotation: orientational dynamics in hard ellipsoid fluids, *J. of Chem. Phys.*, 103, 1553–1560, 1995.

[106] Tarroni, R. and Zannoni, C., On the rotational diffusion of asymmetric molecules in liquid crystals, *J. Chem. Phys.*, 95(6), 4550–4564, 1991.

[107] Tsoi, A.H., Integration by parts for a Lie group valued Brownian motion, *J. of Theor. Probab.*, 6(4), 693–698, 1993.

[108] Uhlenbeck, G.E. and Ornstein, L.S., On the theory of Brownian motion, *Phys. Rev.*, 36, 823–841, 1930.

[109] van Kampen, N.G., *Stochastic Processes in Physics and Chemistry*, North Holland, Amsterdam, 1981.

[110] Wang, M.C. and Uhlenbeck, G.E., On the theory of Brownian motion II, *Rev. Mod. Phys.*, 7, 323–342, 1945.

[111] Wyllie, G., Random motion and Brownian rotation, *Phys. Rep.*, 61(6), 327–376, 1980.

[112] Yosida, K., Integration of Fokker-Planck's equation in a compact Riemannian space, *Arkiv für Matematik*, 1(9), 71–75, 1949.

[113] Yosida, K., Brownian motion on the surface of the 3-sphere, *Ann. Math. Stat.*, 20, 292–296, 1949.

[114] Yosida, K., Brownian motion in a homogeneous Riemannian space, *Pacific J. Math.*, 2, 263–296, 1952.

[115] Zhang, Y. and Bull, T.E., Nonlinearly coupled generalized Fokker-Planck equation for rotational relaxation, *Phys. Rev. E*, 49(6), 4886–4902, 1994.

Chapter 17

Statistical Mechanics of Macromolecules

The macromolecules that we consider here are polymer chains consisting of simpler monomer units. These units can be identical (in which case a homopolymer results) or they may be different (in which case a heteropolymer results). Polymers exist in nature (including DNA and proteins), but usually when one refers to polymer theory, it is in the context of the man-made polymers.

Our presentation of the statistical mechanics of polymers is from a very kinematic perspective. We show how certain quantities of interest in polymer physics can be generated numerically using Euclidean group convolutions. We also show how for wormlike polymer chains, a partial differential equation governs a process that evolves on the motion group and describes the diffusion of end-to-end position and orientation. This equation can be solved using the $SE(3)$-Fourier transform.

Before proceeding to the group-theoretic issues, we present background material on polymer theory in the first few sections that follow. This review material is consolidated from the following books devoted to polymer theory [4, 6, 15, 16, 17, 18, 19, 21, 29, 30, 51, 66, 77, 83, 89].

17.1 General Concepts in Systems of Particles and Serial Chains

Given a collection of n identical particles which are connected by links, the vector from the initial particle to the final one as seen in a frame of reference $\mathcal{F}$ fixed to the initial particle is[1]

$$\mathbf{x}_n = \sum_{i=1}^{n-1} \mathbf{l}_i$$

where $\mathbf{l}_i$ is the vector from particle i to particle $i+1$. The square of the end-to-end distance is then $r^2 = |\mathbf{x}_n|^2$ which may be written as

$$r^2 = \sum_{i=1}^{n-1}\sum_{j=1}^{n-1} \mathbf{l}_i \cdot \mathbf{l}_j = \sum_{i=1}^{n-1} |\mathbf{l}_i|^2 + 2\sum_{i=1}^{n-1} \sum_{j=i+1}^{n-1} \mathbf{l}_i \cdot \mathbf{l}_j. \tag{17.1}$$

In the frame of reference $\mathcal{F}$, the center of mass of the collection of n particles is denoted $\mathbf{x}_{cm}$. The position of the ith particle is $\mathbf{x}_i$ as seen in $\mathcal{F}$ and its position relative to a frame parallel to $\mathcal{F}$ with origin at the center of mass of the system is denoted $\mathbf{r}_i$. Hence

$$\mathbf{x}_i = \mathbf{x}_{cm} + \mathbf{r}_i. \tag{17.2}$$

---

[1] In polymer science n usually denotes the number of bonds. Therefore replacing n with $n+1$ for the number of particles would make our presentation consistent with others in the field.

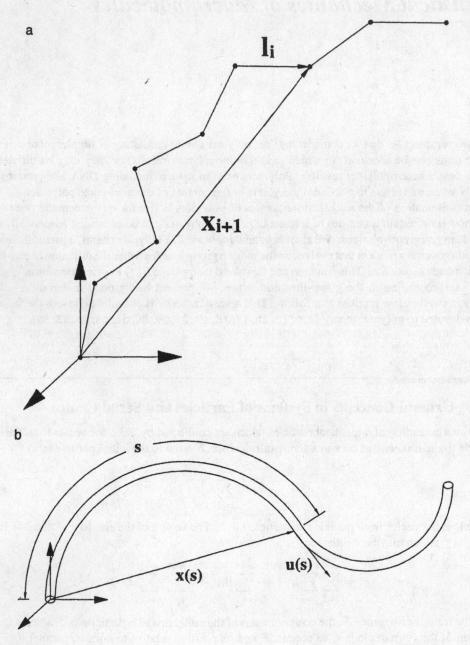

FIGURE 17.1
Simple polymer models: (a) a discrete chain and (b) a continuous (wormlike) chain.

The position of particle j relative to particle i is

$$x_{ij} = x_j - x_i = r_j - r_i.$$

Hence $x_{i,i+1} = l_i$ and $x_{1i} = x_i$. Furthermore, by definition of the center of mass we have

$$x_{cm} = \frac{1}{n} \sum_{i=1}^{n} x_i$$

and summing both sides of Eq. (17.2) over all values of $i \in [1, n]$ therefore yields

$$\sum_{i=1}^{n} r_i = 0. \tag{17.3}$$

The radius of gyration of the system of particles is defined as

$$S = \sqrt{\frac{1}{n} \sum_{i=1}^{n} r_i \cdot r_i}. \tag{17.4}$$

It is desirable to have a general statement relating the radius of gyration of a collection of particles to the relative positions of the particles with respect to each other. We now derive such an expression following [21, 43].

Substituting $r_i = -x_{cm} + x_i$ into this equation and observing Eq. (17.3) and the fact that

$$x_{cm} \cdot x_{cm} = \frac{1}{n^2} \sum_{i=1}^{n} \sum_{j=1}^{n} x_i \cdot x_j$$

allows us to write [21]

$$S^2 = \frac{1}{n} \sum_{i=1}^{n} x_i \cdot x_i - \frac{1}{n^2} \sum_{i=1}^{n} \sum_{j=1}^{n} x_i \cdot x_j.$$

In vector form, the law of cosines reads

$$x_{ij} \cdot x_{ij} = x_i \cdot x_i + x_j \cdot x_j - 2x_i \cdot x_j$$

or equivalently,

$$x_i \cdot x_j = \left(x_i \cdot x_i + x_j \cdot x_j - x_{ij} \cdot x_{ij} \right) /2.$$

Using this to substitute in for $x_i \cdot x_j$ in the previous expression for S^2, yields

$$S^2 = \frac{1}{n^2} \left(\sum_{i=1}^{n} x_i \cdot x_i + \frac{1}{2} \sum_{i=1}^{n} \sum_{j=1}^{n} x_{ij} \cdot x_{ij} \right).$$

Recalling that $x_i = x_{1i}$, and using the symmetry of the dot product operation allows us to finally write

$$S = \frac{1}{n} \sqrt{\sum_{1 \le i < j \le n} x_{ij} \cdot x_{ij}}. \tag{17.5}$$

The previous equation is *Lagrange's theorem*, and the derivation we presented is essentially the same as those in [21, 43].

17.2   Statistical Ensembles and Kinematic Modeling of Polymers

Many of the quantitative results in polymer science depend on the averaging of physical properties over statistical mechanical ensembles. Hence, instead of considering a single polymer chain, one considers the collection of all possible shapes attainable by the chain, each weighted by its likelihood of occurrence. Those with high energy are less likely than those with low energy. In cases when it can be reasoned that all conformations have the same energy, the problem of generating an ensemble is purely geometrical. This purely geometrical, or kinematical, analysis is the subject of this section. The next section reviews classical techniques for incorporating conformational energy in special cases.

Two of the most important quantities that describe the ensemble properties of a polymer chain are the distribution of end-to-end distances and the radius of gyration of the chain over all energy-weighted conformations. In fact, a simple computation shows that the distribution of radii of gyration can be found if the distribution of distances between all points in the chain is known. Hence the problem is really one of finding the distribution of distances between all points in the chain over all energy-weighted conformations. In the simplest models such as the Gaussian chain (or random walk), the freely-jointed chain, and the Kratky-Porod continuous chain where all conformations are assumed to have the same energy and the effects of excluded volume are neglected, closed-form expressions for this distribution (or at least the moments of the distribution) are known. In applications such as relating polymer structure to macroscopic quantities such as viscosity or elasticity, it is usually the statistical mechanical average of the square of the end-to-end distance or radius of gyration which are of importance [5, 48, 80]. In their most general form these follow from Eqs. (17.1) and (17.5) as

$$\left\langle r^2 \right\rangle = (n-1)l^2 + 2 \sum_{i=1}^{n-2} \sum_{j=i+1}^{n-1} \left\langle \mathbf{l}_i \cdot \mathbf{l}_j \right\rangle \tag{17.6}$$

and

$$\left\langle s^2 \right\rangle = \frac{1}{n^2} \sum_{1 \le i < j \le n} \left\langle \mathbf{x}_{ij} \cdot \mathbf{x}_{ij} \right\rangle, \tag{17.7}$$

whereas in Chapters 15 and 16, $< \cdot >$ means the average. In Eq. (17.6) it is assumed that $\langle |\mathbf{l}_i|^2 \rangle = l^2$ is the same for each bond.

The behavior of $\langle \mathbf{l}_i \cdot \mathbf{l}_j \rangle$ and $\langle \mathbf{x}_{ij} \cdot \mathbf{x}_{ij} \rangle$ depends heavily on the particular polymer model.

17.2.1   The Gaussian Chain (Random Walk)

Perhaps the most common model for the distribution of end-vectors is the Gaussian distribution

$$W_G(\mathbf{r}) = \left(\frac{3}{2\pi \left\langle r^2 \right\rangle} \right)^{\frac{3}{2}} \exp\left[-\frac{3r^2}{2 \left\langle r^2 \right\rangle} \right]. \tag{17.8}$$

This distribution is spherically symmetric (and hence depends only on $r = |\mathbf{r}|$ where r is spatial position as measured relative to one end of the chain). It is normalized so that it is a probability density function,

$$\int_{\mathbb{R}^3} W(\mathbf{r}) d_r^3 = 4\pi \int_0^\infty W(\mathbf{r}) r^2 dr = 1,$$

satisfying

$$\int_{\mathbb{R}^3} W(\mathbf{r}) |\mathbf{r}|^2 d_r^3 = 4\pi \int_0^\infty W(\mathbf{r}) r^4 dr = \left\langle r^2 \right\rangle.$$

17.2.2 The Freely-Jointed Chain

The freely-jointed chain model assumes that each link is free to move relative to the others with no constraint on the motion and no correlation between the motion of adjacent links. Hence, in this model

$$\left\langle \mathbf{l}_i \cdot \mathbf{l}_j \right\rangle = 0 \; \forall \; i \neq j \; \Rightarrow \; \left\langle r^2 \right\rangle = (n-1) l^2.$$

Assuming analogous behavior for each subchain in a freely jointed chain means

$$\left\langle \mathbf{x}_{ij} \cdot \mathbf{x}_{ij} \right\rangle = (j - i) l^2 \; \Rightarrow \; \left\langle S^2 \right\rangle = \frac{l^2}{2n^2} \sum_{j=1}^{n-1} j(j+1).$$

Using elementary summation formulae, this is written as

$$\left\langle S^2 \right\rangle = \frac{(n+1)(n-1) l^2}{6n}. \tag{17.9}$$

For large n, we then have

$$\left\langle S^2 \right\rangle \approx \left\langle r^2 \right\rangle / 6.$$

Flory [21] derives the statistical distribution of end positions of a freely jointed chain as a special case of his more general theory, and gives its form as

$$W_F(\mathbf{r}) = \frac{1}{2\pi^2 r} \int_0^\infty \sin(qr)[\sin(ql)/ql]^n q \, dq \tag{17.10}$$

where again $r = |\mathbf{r}|$. Other classical derivations can be found in [7, 67].

Application of the usual Abelian Fourier transform yields

$$\mathcal{F}(W_F(\mathbf{r})) = G(\mathbf{q}) = [\sin(ql)/ql]^n$$

where $q = |\mathbf{q}|$ and $\mathbf{q}$ is the vector of Fourier parameters. From this fact it is clear by the convolution theorem that

$$W_F^{(n_1)} * W_F^{(n_2)} = W_F^{(n_1+n_2)}$$

where the superscript (n_i) is the number of links in the chain.

Flory [21] derived the following approximate expression for the distribution of end positions

$$W(\mathbf{r}) = \left(A / r l^2 \right) \mathcal{L}^{-1}(r/nl) \exp \left[-\frac{1}{l} \int_0^r \mathcal{L}^{-1}(\sigma/nl) d\sigma \right] \tag{17.11}$$

where

$$\mathcal{L}(x) = \coth x - 1/x$$

is the *Langevin* function and $\mathcal{L}^{-1}$ is its inverse, which has the series expansion

$$\mathcal{L}^{-1}(a) \approx 3a + \frac{9}{5}a^3 + \frac{297}{175}a^5.$$

Substitution of this into Eq. (17.11) provides a quasi-closed-form solution.

Treloar [80] derives that

$$W(\mathbf{r}) = \left(Ab(r)/rl^2\right) \left(\frac{\sinh b(r)}{b(r)}\right)^n e^{-b(r)r/l}$$

where

$$b(r) = \mathcal{L}^{-1}(r/nl),$$

and A is the normalization ensuring that $W(\mathbf{r})$ is a probability density function.

17.2.3   The Freely Rotating Chain

In the freely rotating chain model of a polymer, each bond in the chain has the same fixed length, $l = |\mathbf{l}_i|$, and each bond vector $\mathbf{l}_{i+1}$ makes a bond angle θ with respect to $\mathbf{l}_i$. Hence $\mathbf{l}_i \cdot \mathbf{l}_{i+1} = l^2 \cos\theta$. Since this is true for each bond in any conformation, it is also true when averaging over all conformations: $\langle \mathbf{l}_i \cdot \mathbf{l}_{i+1}\rangle = l^2 \cos\theta$.

We may resolve the bond vector $\mathbf{l}_{i+1}$ in a direction along bond i and orthogonal to it. Then

$$\mathbf{l}_{i+1} = \cos\theta\, \mathbf{l}_i + \mathbf{v}_i \tag{17.12}$$

where $\mathbf{v}_i \cdot \mathbf{l}_i = 0$ by definition. If the unhindered rotation around bond i is parameterized as $\phi_i \in [0, 2\pi]$, then $\mathbf{v}_i$ is a function of all the angles $\{\phi_j\}$ for all $j \le i$, while $\mathbf{l}_i$ is a function of all angles $\{\phi_j\}$ for all $j < i$. Hence all conformational averages of the form

$$\langle \mathbf{l}_i \cdot \mathbf{v}_{i+n}\rangle = 0 \tag{17.13}$$

follow for all $n \ge 0$ because in this average is an integration over ϕ_{i+n}. That is, for unhindered rotation

$$\int_{S^1} \mathbf{v}_{i+n} d\phi_{i+n} = 0 \Rightarrow \int_{S^1} \mathbf{l}_i \cdot \mathbf{v}_{i+n} d\phi_{i+n} = 0$$

where the second equation above is true because $\mathbf{l}_i$ is not a function of ϕ_{i+n} and can be taken out of the integral. Using Eq. (17.12) recursively, one may write $\mathbf{l}_{i+n}$ as the sum of $\cos^n\theta\, \mathbf{l}_i$ and a series of terms of the form $a_{i+j}(\theta)\mathbf{v}_{i+j}$ for $j = 0, \ldots, n-1$. Calculating the conformational averages $\langle \mathbf{l}_i \cdot \mathbf{l}_{i+j}\rangle$ and observing Eq. (17.13) one finds that

$$\langle \mathbf{l}_i \cdot \mathbf{l}_{i+j}\rangle = l^2 (\cos\theta)^j. \tag{17.14}$$

It then follows from Eq. (17.6) that

$$\langle r^2\rangle = (n-1)l^2 + 2l^2 \sum_{i=1}^{n-1}\sum_{j=i}^{n-1} (\cos\theta)^{j-i}.$$

Manipulating this expression as in [21, 65], one finds

$$\langle r^2\rangle = (n-1)l^2 \frac{1 + \cos\theta}{1 - \cos\theta} - 2l^2 \frac{\cos\theta \left(1 - \cos^{n-1}\theta\right)}{(1 - \cos\theta)^2}.$$

Similarly, for this model the conformational average of the radius of gyration is

$$\langle S^2\rangle = \frac{l^2 (n-1)(n+1)(1 + \cos\theta)}{6n(1 - \cos\theta)} - \frac{2l^2 \cos\theta}{n^2(1 - \cos\theta)^2} \sum_{j=1}^{n-1} \left[j - \left(\frac{1 - \cos^j\theta}{1 - \cos\theta}\right)\cos\theta\right].$$

In the limit as $n \to \infty$, these become

$$\left(\langle r^2\rangle/(n-1)l^2\right)_n = \frac{1 + \cos\theta}{1 - \cos\theta} \quad \text{and} \quad \left(\langle S^2\rangle/(n-1)l^2\right)_n = \frac{1}{6}\left(\langle r^2\rangle/(n-1)l^2\right)_n.$$

17.2.4   Kratky-Porod Chain

The Kratky-Porod (or KP) model [40, 65] is for semi-flexible polymer chains. By imposing the condition that the polymer act like a continuous curve with random curvature, the evolution of distal-end position and orientation is more gradual than the other models considered previously in this chapter. This model intrinsically takes into account the properties of chains which are stiff over short segments, but are nonetheless flexible on the length scale of the whole chain due to their long and slender nature (e.g., DNA).

In this model, the position of a point on an individual polymer chain at an arclength s from the proximal end is

$$\mathbf{x}(s) = \int_0^s \mathbf{u}(\sigma)d\sigma,$$

and the relative position of two points on the chain is given by

$$\mathbf{x}(s_1, s_2) = \mathbf{x}(s_2) - \mathbf{x}(s_1).$$

$\mathbf{u}(s) \in S^2$ is the unit tangent vector to the curve at s. It is assumed that the proximal end is attached to the origin of our frame of reference with $\mathbf{u}(0) = \mathbf{e}_3$, and the polymer has total length L. Denoting the position of the distal end as $\mathbf{r} = \mathbf{x}(L)$, the expressions for $\langle |\mathbf{r}|^2 \rangle$ and $\langle S^2 \rangle$ derived for chains with discrete links become

$$\left\langle |\mathbf{r}|^2 \right\rangle = \langle \mathbf{r} \cdot \mathbf{r} \rangle = \int_0^L \int_0^L \langle \mathbf{u}(s_1) \cdot \mathbf{u}(s_2) \rangle \, ds_1 ds_2 \qquad (17.15)$$

and

$$\left\langle S^2 \right\rangle = \frac{1}{L^2} \int_0^L \left(\int_{s_1}^L \left\langle |\mathbf{x}(s_1, s_2)|^2 \right\rangle ds_2 \right) ds_1.$$

Yamakawa [89] reasons that $\langle |\mathbf{x}(s_1, s_2)|^2 \rangle = \langle |\mathbf{x}(s_2 - s_1)|^2 \rangle$ and hence

$$\left\langle S^2 \right\rangle = \frac{1}{L^2} \int_0^L (L - s) \left\langle |\mathbf{x}(s)|^2 \right\rangle ds. \qquad (17.16)$$

Ideally one would like to solve for the probability density function $W_C(\mathbf{u}, \mathbf{x}, s)$ for the orientation (modulo roll) and position of the frame of reference attached to the curve for each value of s (and $s = L$ in particular). In Section 17.2.4 we examine the moments $\langle |\mathbf{r}|^2 \rangle$ and $\langle S^2 \rangle$ for this model. In Section 17.2.4 we review methods for approximating the probability density function $W_C(\mathbf{u}, \mathbf{x}, s)$.

Moments of the Continuous Wormlike Chains

Let $\theta(s_1, s_2)$ denote the angle between the tangents to the curve at two different values of arclength, s_1 and s_2

$$\cos \theta(s_1, s_2) = \mathbf{u}(s_1) \cdot \mathbf{u}(s_2).$$

If one considers all possible conformations of a continuous curve, then it can be reasoned based on uniformity of behavior everywhere on the chain that the average $\langle \cos \theta(s_1, s_2) \rangle$ over all conformations is of the form $\langle \cos \theta(s_1 - s_2) \rangle$. Furthermore, as stated in [29], it is often a realistic assumption that this function has the property

$$\langle \cos \theta(s_1 - s_2) \rangle = \langle \cos \theta(s_1) \rangle \langle \cos \theta(-s_2) \rangle.$$

The solution of this functional equation is of the form

$$\langle \cos \theta(s) \rangle = e^{-s/a} \qquad (17.17)$$

where a is a quantity called the *persistence length* of the polymer. Essentially, it is a parameter which indicates how far one must travel along the polymer before the tangent directions are no longer correlated. Physically this describes how stiff the molecule is. For instance, if $a \to 0$ then $\langle \cos \theta(s) \rangle \approx 0$ for all $s \in [0, L]$, whereas if $a \to \infty$ then $\langle \cos \theta(s) \rangle \approx 1$ for all $s \in [0, L]$. These two situations correspond to completely flexible and rigid chains, respectively.

We note that there is a different way to view this same result. The orientation distribution, $w(\mathbf{u}; s)$ corresponding to the ensemble of tangent vectors $\{\mathbf{u}(s)\}$ [each $\mathbf{u}(s)$ measured at arclength s along its respective chain] may be viewed as a diffusion process on the sphere with arclength replacing time. Following this reasoning, one needs to solve the heat equation on the sphere

$$\frac{\partial w}{\partial s} = K \nabla_u^2 w,$$

where ∇_u^2 is the spherical Laplacian, subject to the initial conditions $w(\mathbf{u}; 0) = \delta(\mathbf{u})$ (the δ-function concentrated at the North Pole). When $\mathbf{u} = \mathbf{u}(\theta, \phi)$ is parameterized using spherical coordinates,

$$\delta(\mathbf{u}) = \sum_{l=0}^{\infty} \left(\frac{2l+1}{4\pi} \right) P_l(\cos \theta).$$

The solution to the heat equation is therefore independent of ϕ due to the axial symmetry of the initial conditions around the $\mathbf{e}_3$ axis. The solution to the heat equation is written as

$$w(\mathbf{u}; s) = w(\theta; s) = \sum_{l=0}^{\infty} \left(\frac{2l+1}{4\pi} \right) e^{-l(l+1)s/K} P_l(\cos \theta).$$

From this distribution we may calculate

$$\langle \mathbf{u}(s) \cdot \mathbf{u}(0) \rangle = \int_{S^2} \mathbf{u} \cdot \mathbf{e}_3 w(\mathbf{u}; s) d\mathbf{u} = \int_0^\pi \int_0^{2\pi} \cos \theta w(\theta, \phi; s) \sin \theta d\phi d\theta = e^{-s/K}.$$

Hence the denominator of the exponential in this description is the persistence length and so we write the diffusion constant as $K = a$.

Substituting the function in Eq. (17.17) into Eq. (17.15) one finds [17, 29]

$$\langle r^2 \rangle = 2aL \left[1 - (a/L) \left(1 - e^{-L/a} \right) \right].$$

Substituting this into Eq. (17.16) one finds [17, 29]

$$\langle s^2 \rangle = (aL/3) \left(1 - (3a/L) \left[1 - 2(a/L) + 2(a/L)^2 \left(1 - e^{-L/a} \right) \right] \right).$$

In the limit as $L \to \infty$ one finds that as with the Gaussian chain

$$\langle s^2 \rangle \approx \langle r^2 \rangle / 6.$$

The PDF for the Continuous Chain

As with the Gaussian and freely jointed chains, it is desirable to have the full distribution of end positions and orientations associated with the model. As has been shown [14, 35, 89], the Fokker-Planck equation for the evolution of the probability density function $W_C(\mathbf{u}, \mathbf{x}; L)$ is

$$\frac{\partial W_C}{\partial L} = K \nabla_u^2 W_C - \mathbf{u} \cdot \nabla_x W_C. \tag{17.18}$$

The initial conditions corresponding to the PDF at the proximal end are

$$W_C(\mathbf{u}, \mathbf{x}; 0) = \delta\left(\mathbf{u}\delta(\mathbf{u})\right)\delta(\mathbf{x}).$$

A series solution to this partial differential equation is given in [28, 89]. In the context of the current work, it suffices to say that while the distribution is not spherically symmetric for nonzero finite values of L and K, one has nonetheless

$$\lim_{L\to\infty}\int_{S^2} W_C(\mathbf{u}, \mathbf{r}; L)d\mathbf{u} = W_G(|\mathbf{r}|).$$

While we know of no closed-form solution for the distribution of end-vector position in the continuous curve model, certain approximations can be found in the literature. These include expansions in the ratio a/L (see [88, 89]).

In very recent work, Thirumalai and Ha [79] relaxed the local constraint $\mathbf{u}(s) \cdot \mathbf{u}(s) = 1$ in the Kratky-Porod model and instead enforced the global constraint $\langle\mathbf{u}(s) \cdot \mathbf{u}(s)\rangle = 1$. They then derived the distribution

$$W_T(r) = \frac{A}{\left(1 - r^2/L^2\right)^{\frac{9}{2}}} \exp\left[-\frac{\alpha_0}{\left(1 - r^2/L^2\right)^{\frac{9}{2}}}\right] \tag{17.19}$$

where $\alpha_0 = -9L/8a$ (a is the persistence length) and

$$A = \frac{4}{\pi^{1/2}e^{-\alpha_0}\alpha_0^{-3/2}\left(1 + 3/\alpha_0 + 15/4\alpha_0^2\right)}$$

is the constant required to normalize $W_T(\cdot)$ so that it is a probability distribution.

17.3 Theories Including the Effects of Conformational Energy

17.3.1 Chain with Hindered Rotations

The hindered rotation model is the simplest of several classical models of polymer chains which explicitly include the effects of conformational energy. It assumes that the *bond angles*, $\{\theta_i\}$, are fixed, and each *torsion angle*, ϕ_i, has preferred values dictated by the energy function $E_i(\phi_i)$. The total conformational energy is

$$E = \sum_i E_i(\phi_i).$$

In this model, the probability density function $p(\phi_i)$ describing the distribution of attainable torsion angles is given by

$$p(\phi_i) = \frac{\exp\left[-E_i(\phi_i)/k_BT\right]}{\int_0^{2\pi}\exp\left[-E_i(\phi)/k_BT\right]d\phi} \tag{17.20}$$

where k_B is the Boltzmann constant and T is the temperature measured in degrees Kelvin. Hence, for high temperatures this model reduces to the freely rotating chain model. In any case, the conformational average of any scalar, vector, or matrix-valued function of ϕ_i will be calculated as

$$\langle f\rangle = \int_0^{2\pi} f(\phi_i)\,p(\phi_i)\,d\phi_i.$$

The probability density function on the whole conformation space of the chain is the product of those for each torsion angle ϕ_i. This means that the conformational average of the product of functions of individual torsion angles will be the product of conformational averages of these functions. In particular, consider the important example given below.

Following the notation in [21], the rotation matrix describing the orientation of a frame attached to bond vector $i + 1$ relative to the frame attached to bond i is written as

$$T_i = \begin{pmatrix} \cos\theta_i & \sin\theta_i & 0 \\ \sin\theta_i \cos\phi_i & -\cos\theta_i \cos\phi_i & \sin\phi_i \\ \sin\theta_i \sin\phi_i & -\cos\theta_i \sin\phi_i & -\cos\phi_i \end{pmatrix}. \tag{17.21}$$

In the current context $\theta_i = \theta = const$, and so $T_i = T_i(\phi_i)$. Assuming that the energy function is symmetric, $E_i(\phi_i) = E_i(-\phi_i)$, then $p(\phi_i)$ is also symmetric and so $\langle \sin\phi_i \rangle = 0$. Hence

$$\langle T_i \rangle = \begin{pmatrix} \cos\theta_i & \sin\theta_i & 0 \\ \sin\theta_i \langle\cos\phi_i\rangle & -\cos\theta_i \langle\cos\phi_i\rangle & 0 \\ 0 & 0 & -\langle\cos\phi_i\rangle \end{pmatrix}.$$

Since the bond vector $\mathbf{l}_i = l\mathbf{e}_1$ in its own coordinate system, and in this same coordinate system $\mathbf{l}_j$ for $j > i$ appears as $lT_jT_{j-1}\cdots T_i\mathbf{e}_1$, it follows that

$$\langle \mathbf{l}_i \cdot \mathbf{l}_j \rangle = l^2 \mathbf{e}_1^T \langle T_jT_{j-1}\cdots T_i \rangle \mathbf{e}_1 = l^2 \mathbf{e}_1^T \langle T \rangle^{j-i} \mathbf{e}_1, \tag{17.22}$$

where the last product results from the assumption that all of the bond angles and torsional energies are independent of i (and so $\langle T_i \rangle = \langle T \rangle$), and the conformational average of a product of functions of single torsion angles is the product of their conformational averages.

Substitution of Eq. (17.22) into Eq. (17.1) results in the average of the square of the end-to-end distance given in [21]

$$\langle r^2 \rangle = (n-1)l^2 \mathbf{e}_1^T \left[(\mathbb{I} + \langle T \rangle)(\mathbb{I} - \langle T \rangle)^{-1} - (2\langle T \rangle/(n-1))\left(\mathbb{I} - \langle T \rangle^n\right)(\mathbb{I} - \langle T \rangle)^{-2} \right] \mathbf{e}_1.$$

The average of the square of the radius of gyration follows in a similar way.

In the limit as $n \to \infty$ one finds the result from Oka's 1942 paper [62]

$$\left(\langle r^2 \rangle /(n-1)l^2 \right)_{n\to\infty} = \left(\frac{1+\cos\theta}{1-\cos\theta} \right) \left(\frac{1+\langle\cos\phi\rangle}{1-\langle\cos\phi\rangle} \right) = 6 \left(\langle S^2 \rangle /(n-1)l^2 \right)_{n\to\infty}.$$

This reduces to the case of the freely rotating chain when $\langle \cos\phi \rangle = 0$.

17.3.2   Rotational Isomeric State Model

Let the homogeneous transform (element of $SE(3)$) which describes the position and orientation of the frame of reference attached to a serial chain molecule at the intersection of the ith and $i + 1$st bond vectors relative to the frame attached at the intersection of the $i - 1$st and ith bond vectors be given by

$$g_i = (R_i, \mathbf{l}_i).$$

$\mathbf{l}_i = [l_i, 0, 0]^T$ is the ith bond vector as seen in this frame i (which is attached at the intersection of the $i - 1$st and ith bond vectors).[2]

The vector pointing to the tip of this bond vector from the origin of frame j is $\mathbf{r}_j^i$.

---

[2] In contrast to Section 17.1, here n is the number of bonds and $n + 1$ is the number of backbone atoms.

The end-to-end vector for the chain, $\mathbf{r} = \mathbf{r}_0^n$, is the concatenation of these homogeneous transforms applied to $\mathbf{l}_n$

$$\mathbf{r} = g_1 \circ g_2 \circ \cdots \circ g_n \circ \mathbf{l}_n.$$

Since

$$g_1 \circ g_2 \circ \cdots \circ g_n = \left(\prod_{i=1}^{n} R_i, \sum_{i=1}^{n-1} \left(\prod_{j=0}^{i-1} R_j \right) \mathbf{l}_i \right)$$

(where $R_0 = \mathbb{I}$) one finds that

$$\mathbf{r} = \left(\sum_{i=1}^{n} \left(\prod_{j=0}^{i-1} R_j \right) \mathbf{l}_i \right).$$

From this, the square of the end-to-end distance, $r^2 = \mathbf{r} \cdot \mathbf{r}$, is easily calculated. It is relatively straightforward to show that this can be put in the form [51]

$$r^2 = \begin{bmatrix} 1 & 2\mathbf{l}_1^T R_1 & l_1^2 \end{bmatrix} G_2 G_3 \cdots G_{n-1} \begin{bmatrix} l_n^2 \\ \mathbf{l}_n \\ 1 \end{bmatrix} \tag{17.23}$$

where

$$G_i = \begin{pmatrix} 1 & 2\mathbf{l}_i^T R_i & l_i^2 \\ 0 & R_i & \mathbf{l}_i \\ 0 & \mathbf{0}^T & 1 \end{pmatrix}.$$

Similarly, the square of the radius of gyration for a chain of $n + 1$ identical atoms is

$$S^2 = \frac{1}{(n+1)^2} \sum_{k=0}^{n-1} \sum_{j=k+1}^{n} \left| \mathbf{r}_k^j \right|^2.$$

This can be written as the matrix product [51]

$$S^2 = \begin{bmatrix} 1 & 1 & 2\mathbf{l}_1^T R_1 & l_1^2 & l_1^2 \end{bmatrix} H_2 H_3 \cdots H_{n-1} \begin{bmatrix} l_n^2 \\ l_n^2 \\ \mathbf{l}_n \\ 1 \\ 1 \end{bmatrix}, \tag{17.24}$$

where

$$H_i = \begin{pmatrix} 1 & 1 & 2\mathbf{l}_i^T R_i & l_i^2 & l_i^2 \\ 0 & 1 & 2\mathbf{l}_i^T R_i & l_i^2 & l_i^2 \\ 0 & 0 & R_i & \mathbf{l}_i & \mathbf{l}_i \\ 0 & 0 & \mathbf{0}^T & 1 & 1 \\ 0 & 0 & \mathbf{0}^T & 1 & 1 \end{pmatrix}.$$

The forms for r^2 and S^2 in Eqs. (17.23) and (17.24) make it easy to perform conformational averages. These averages are weighted by conformation partition functions which reflect the relative likelihood of one conformation as compared to another based on the energy of each conformation. The rotational isomeric state (RIS) model [21, 69] is perhaps the most widely known method to generate the statistical information needed to weight the relative occurrence of conformations in a statistical mechanical ensemble. There are two basic assumptions to the RIS model developed by Flory [21, 22]

- The conformational energy function for a chain molecule is dominated by interactions between each set of three groups of atoms at the ends of two adjacent bond vectors. Hence the conformational energy function can be written in the form

$$E(\phi) = \sum_{i=2}^{n-1} E_i (\phi_{i-1}, \phi_i)$$

where $\phi = (\phi_1, \phi_2, \ldots, \phi_{n-1})$ is the set of all rotation angles around bond vectors.

- The value of the conformational partition function $e^{-E(\phi)/k_B T}$ is negligible except at the finite number of points where $E(\phi)$ is minimized, and hence averages of conformation-dependent functions may be calculated in the following way

$$\langle f \rangle = \frac{\int_{T^{n-1}} f(\phi) e^{-E(\phi)/k_B T} d\phi_1 \cdots d\phi_{n-1}}{\int_{T^{n-1}} e^{-E(\phi)/k_B T} d\phi_1 \cdots d\phi_{n-1}} \approx \frac{\sum_{\phi_\eta \in T^{n-1}} f(\phi_\eta) e^{-E(\phi_\eta)/k_B T}}{\sum_{\phi_\eta \in T^{n-1}} e^{-E(\phi_\eta)/k_B T}}.$$

Here the finite set conformations $\{\phi_\eta\}$ are local minima of the energy function $E(\phi)$ and T^{n-1} is the $(n-1)$-dimensional torus.

As a consequence of the preceding two assumptions, conformational statistics are generated in the context of the RIS model using statistical weight matrices. A detailed explanation of how this technique is used to generate $\langle r^2 \rangle$ and $\langle S^2 \rangle$ can be found in Flory's classic text [21] (in particular, see Chapter 4 of that book).

One drawback of this technique, as pointed out in [17, 51], is that the actual distribution of end-to-end distance is not calculated using the RIS method. The traditional technique for generating statistical distributions is to use Monte Carlo simulations. While this is a quite effective technique, it has drawbacks, including the fact that it will, by definition, fail to pick up the tails of a distribution. In contrast, the Euclidean group convolution technique discussed later in this chapter guarantees that the tails (which give the border of the distribution) will be captured.

17.3.3   Helical Wormlike Models

The helical wormlike (HW) chain model developed by Yamakawa [89] is a generalization of the Kratky-Porod model, in which the effects of torsional stiffness are included in addition to those due to bending. Hence, in this model, if one desires to find $\langle r^2 \rangle$ it is not sufficient to consider only the diffusion of the unit tangent vector $\mathbf{u}(s) \in S^2$ as s progresses from 0 to L, but also how a frame attached to the curve twists about this tangent. This is a diffusion process on $SO(3)$, and the techniques of noncommutative harmonic analysis play an important role in solving, or at least simplifying, the governing equations.

Let $R(\alpha(s), \beta(s), \gamma(s))$ be the rotation matrix describing the orientation of the frame of reference attached to the curve at s relative to the base frame $R(\alpha(0), \beta(0), \gamma(0)) = \mathbb{I}$ where α, β, γ are ZYZ Euler angles. The unit tangent to the curve is $\mathbf{u}(s) = R(\alpha(s), \beta(s), \gamma(s))\mathbf{e}_3$. The angular velocity matrix

$$\dot{R}R^T = \begin{pmatrix} 0 & -\omega_3 & \omega_2 \\ \omega_3 & 0 & -\omega_1 \\ -\omega_2 & \omega_1 & 0 \end{pmatrix},$$

where

$$\omega_1 = \dot{\beta} \sin \gamma - \dot{\alpha} \sin \beta \cos \gamma \qquad (17.25)$$
$$\omega_2 = \dot{\beta} \cos \gamma + \dot{\alpha} \sin \beta \sin \gamma \qquad (17.26)$$
$$\omega_3 = \dot{\alpha} \cos \beta + \dot{\gamma}, \qquad (17.27)$$

describes the instantaneous evolution in orientation for a small change in s. A dot represents differentiation with respect to s.

In the HW model, Yamakawa assumes the preferred state of the curve (which is assumed to be a solid tube of uniform material) is a helix, which is the result when the following potential energy function is minimized at each point on the curve

$$V = \frac{1}{2}\alpha_0 \left[\omega_1^2 + (\omega_2 - \kappa_0)^2 \right] + \frac{1}{2}\beta_0 (\omega_3 - \tau_0)^2.$$

The equality $\beta_0 = \alpha_0/(1+\sigma)$ holds for the stiffnesses α_0 and β_0 in the case when the cross-section of the tube is a circular disc and σ is the Poisson's ratio of the material representing the curve $(0 \leq \sigma \leq 1/2)$ [89].

Yamakawa generated equations for the probability density function, $P(R, s)$, governing the evolution of orientations of the frame attached to the curve at each value of s

$$\frac{\partial P(R; s)}{\partial s} = \left(\lambda \nabla_R^2 + \lambda \sigma L_3^2 - \kappa_0 L_2 + \tau_0 L_3 \right) P(R, s) \text{ where } P(\mathbf{r}, R; 0) = \delta(R)\delta(r). \quad (17.28)$$

He derived a similar equation for $\mathcal{P}(\mathbf{r}, R; s)$ as

$$\frac{\partial \mathcal{P}(\mathbf{r}, R; s)}{\partial s} = \left(\lambda \nabla_R^2 + \lambda \sigma L_3^2 - \kappa_0 L_2 + \tau_0 L_3 - \mathbf{e}_3^T R^T \nabla_\mathbf{r} \right) \mathcal{P}(\mathbf{r}, R; s)$$

$$\mathcal{P}(\mathbf{r}, R; 0) = \delta(R)\delta(\mathbf{r}). \quad (17.29)$$

The operators L_i in the preceding equations are the derivatives

$$L_1 = \sin\gamma \frac{\partial}{\partial\beta} - \frac{\cos\gamma}{\sin\beta}\frac{\partial}{\partial\alpha} + \cot\beta\cos\gamma\frac{\partial}{\partial\gamma} \quad (17.30)$$

$$L_2 = \cos\gamma \frac{\partial}{\partial\beta} + \frac{\sin\gamma}{\sin\beta}\frac{\partial}{\partial\alpha} - \cot\beta\sin\gamma\frac{\partial}{\partial\gamma} \quad (17.31)$$

$$L_3 = \frac{\partial}{\partial\gamma} \quad (17.32)$$

(which are the same as X_i^R in ZYZ Euler angles) and $\nabla_R^2 = L_1^2 + L_2^2 + L_3^2$ is the Laplacian for $SO(3)$. The solutions of Eqs. (17.28) and (17.29) are related as

$$\int_{\mathbb{R}^3} \mathcal{P}(\mathbf{r}, R; s)d\mathbf{r} = P(R; s).$$

In fact, Eq. (17.28) is derived from Eq. (17.29) by integrating over $\mathbb{R}^3$ and observing that

$$\int_{\mathbb{R}^3} \nabla_\mathbf{r} \mathcal{P}(\mathbf{r}, R; s)d\mathbf{r} = 0.$$

Knowing $P(R, s)$ allows one to find $\langle \mathbf{u}(s) \cdot \mathbf{e}_3 \rangle$ as in the Kratky-Porod model, but in this context it is calculated as

$$\langle \mathbf{u}(s) \cdot \mathbf{e}_3 \rangle = \int_{SO(3)} \mathbf{e}_3^T Re_3 P(R; s)d(R).$$

This may then be substituted into Eqs. (17.15) and (17.16) to obtain $\langle r^2 \rangle$ and $\langle S^2 \rangle$. For the case when $\sigma = 0$, Yamakawa [89] has calculated these quantities. In particular,

$$\langle r^2 \rangle = \frac{(4 + \tau_0^2)L}{4 + \nu^2} - \frac{\tau_0^2}{2\nu^2} - \frac{2\kappa_0^2(4 - \nu^2)}{\nu^2(4 + \nu^2)^2}$$

$$+ \frac{e^{-2L}}{\nu^2}\left(\frac{\tau_0^2}{2} + \frac{2\kappa_0^2}{(4 + \nu^2)^2}\left[(4 - \nu^2)\cos\nu L - 4\nu\sin\nu L \right] \right),$$

where

$$\nu = \left(\kappa_0^2 + \tau_0^2 \right)^{\frac{1}{2}}.$$

When $\sigma = \tau_0 = \kappa_0 = 0$, we see that Eq. (17.28) reduces to the heat equation on $SO(3)$ and has the closed-form series solution analogous to the KP-model. When $\sigma \neq 0$ and $\tau_0 = \kappa_0 = 0$ the diffusion equation for the axially symmetric rotor results.

It is also possible to integrate Eq. (17.29) over $SO(3)$ to solve for $W_{HW}(\mathbf{r}; s)$. In particular, the solution may be expanded in a series of Hermite polynomials as [21, 89]

$$W_{HW}(\mathbf{r}; L) = (á/\pi)^3 e^{-(ár)^2} \sum_{k=0}^{\infty} c_{2k} H_{2k+1}(ár)/(ár) \qquad (17.33)$$

where,

$$á = \left(\frac{3}{2\langle R^2 \rangle} \right)^{\frac{1}{2}},$$

$H_j(x)$ is the jth Hermite polynomial, and

$$c_{2k} = \frac{\langle H_{2k+1}(ár)/(ár) \rangle}{2^{2(k+1)}(2k+1)!}.$$

17.3.4   Other Computational and Analytical Models of Polymers

One of the most widely used traditional computational techniques for determining statistical properties of macromolecular conformations is the Monte Carlo method [70]. The Monte Carlo method is very general, but it may not capture the "tails" of slowly decreasing probability density functions.

The long-range effects of excluded volume are often modeled using self-avoiding walks [1, 25, 29, 36, 47]. The self-avoiding walk is more sophisticated than the traditional random walk (see [74] for a discussion of random walks), since self-interpenetration is not allowed. Polymer models based on self-avoiding walks are usually limited to rather short chains ($N << 100$) and the polymer is assumed to have backbone atoms that sit on a regular lattice. Other techniques for modeling long-range interactions can be found in [2, 3, 11, 60]. This requires models of the forces of interaction, such as those presented in [44, 76] (and references therein).

In contrast, renormalization group methods [24, 68, 71] are usually limited to very long ones ($N \to \infty$).

Recently, statistics of polymer intertwining have been studied using the concept of a braid group from knot theory [8, 12, 55, 56, 57].

An altogether different approach is to do direct molecular dynamic simulations (see [34, 63] for recent examples). The primary drawback of molecular simulations is the intensive computational requirements.

In the following section, we review an approach introduced in [9] which, in a sense, combines ideas of the self-avoiding walk and renormalization group. Essentially, the statistics of a polymer chain are generated using a weighted convolution on $SE(3)$, where the weighting reflects a bias against interpenetration of the molecule with itself.

17.4   Mass Density, Frame Density, and Euclidean Group Convolutions

In this section we define three statistical properties of macromolecular ensembles. Later it will be shown how these quantities are used to model the interactions of segments which may be far apart as measured along the chain, but are proximal in space. Before introducing these definitions, we review some notations introduced in earlier chapters.

Recall that $\mathbb{R}^3$ denotes 3-dimensional Euclidean space, and $SE(3)$ denotes the group of rigid-body motions. $\mathbf{x} \in \mathbb{R}^3$ is a position vector, and $g = (R, \mathbf{r}) \in SE(3)$ is a rigid body transformation (or frame of reference). The terminology $SE(3)$ stands for "special Euclidean" group of 3-dimensional space. This group is the semi-direct product of the rotation group $SO(3)$ and the translation group $(\mathbb{R}^3, +)$. The group law is $g_1 \circ g_2 = (R_1 R_2, R_1 \mathbf{r}_2 + \mathbf{r}_1)$ and geometrically represents the concatenation of rigid-body motions, or equivalently, a sequential change of reference frames. The inverse of any element $g \in SE(3)$ and the group identity element are given respectively as $g^{-1} = (R^T, -R^T \mathbf{r})$ and $e = (\mathbb{I}, \mathbf{0})$ where $\mathbb{I}$ is the 3×3 identity. The action of elements of $SE(3)$ on elements of $\mathbb{R}^3$ is defined as $g \circ \mathbf{x} = R\mathbf{x} + \mathbf{r}$.

The three statistical quantities of importance in the present formulation are: (1) the ensemble mass density for the whole chain $\rho(\mathbf{x})$; (2) the ensemble tip frame density $f(g)$ (where g is the frame of reference of the distal end of the chain relative to the proximal); and (3) the function $\mu(g, \mathbf{x})$, which is the ensemble mass density of all configurations which grow from the identity frame fixed to one end of the chain and terminate at the relative frame g at the other end. These quantities are illustrated in Fig. 17.2.

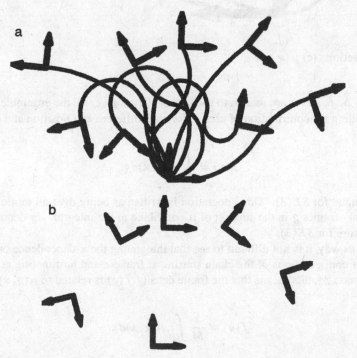

FIGURE 17.2
The functions (a) $\rho(\mathbf{x})$, (b) $f(g)$ *(Continued)*.

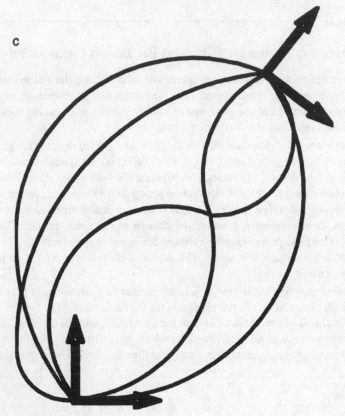

FIGURE 17.2
(Cont.) **The functions (c) $\mu(g, \mathbf{x})$.**

The functions ρ, f, and μ are related to each other. Given $\mu(g, \mathbf{x})$, the ensemble mass density is calculated by adding the contribution of each μ for each different end position and orientation

$$\rho(\mathbf{x}) = \int_G \mu(g, \mathbf{x}) dg. \tag{17.34}$$

Here G is shorthand for $SE(3)$. This integration is written as being over all motions of the end of the chain, but only frames g in the support of μ contribute to the integral. dg denotes the invariant integration measure for $SE(3)$.

In an analogous way, it is not difficult to see that integrating the $\mathbf{x}$-dependence out of μ provides the total mass of configurations of the chain starting at frame e and terminating at frame g. Since each chain has mass M, this means that the frame density $f(g)$ is related to $\mu(g, \mathbf{x})$ as

$$f(g) = \frac{1}{M} \int_{\mathbb{R}^3} \mu(g, \mathbf{x}) d\mathbf{x}. \tag{17.35}$$

We note the total number of frames attained by one end of a discrete-state chain relative to the other is

$$F = K^N = \int_G f(g) dg$$

when each of the chains N degrees of freedom has K preferred states. It then follows that

$$\int_{\mathbb{R}^3} \rho(\mathbf{x})d\mathbf{x} = F \cdot M.$$

If the functions $\rho(\mathbf{x})$ and $f(g)$ are known for the whole chain, then a number of important thermodynamic and mechanical properties of the polymer can be determined. For instance, the moments of any positive integer power of the end-to-end distance, $\langle|\mathbf{r}|^m\rangle$, can be calculated from $f(g) = f(R, \mathbf{r})$ by first integrating out the orientational dependence

$$\tilde{\rho}(\mathbf{r}) = \int_{SO(3)} f(R, \mathbf{r})dR,$$

where dR is the normalized invariant integration measure for $SO(3)$. Then

$$\langle|\mathbf{r}|^m\rangle = \int_{\mathbb{R}^3} |\mathbf{r}|^m \tilde{\rho}'(\mathbf{r})d\mathbf{r}$$

where $d\mathbf{r} = dr_1 dr_2 dr_3$ is the usual integration measure on $\mathbb{R}^3$ and $\tilde{\rho}' = \tilde{\rho}/K^n$ is the normalized version of $\tilde{\rho}$. The PDF of end-to-end distances is given as

$$d(r) = r^2 \int_{S^2} \tilde{\rho}'(\mathbf{r})d\mathbf{u} \qquad (17.36)$$

where in spherical coordinates $\mathbf{r} = r\mathbf{u}$. Here $\mathbf{u} \in S^2$ is a point on the unit sphere and $d\mathbf{u}$ is the area element for the unit sphere. In the special case when $\tilde{\rho}'(\mathbf{r})$ is spherically symmetric $d(r) = 4\pi r^2 \tilde{\rho}'(r)$. More generally, it follows that

$$\int_{\mathbb{R}^3} \tilde{\rho}'(\mathbf{r})d\mathbf{r} = \int_0^\infty d(r)dr.$$

The functions $\rho(\mathbf{x})$ and $\tilde{\rho}(\mathbf{r})$ are also related in the following way. Imagine the chain is made up of n units,[3] each of which has K preferred energy states, and the ith of which has mass m_i. Then the function $\tilde{\rho}_i(\mathbf{r})$ relating the distribution of points reachable by the distal end of the ith unit relative to the proximal end of the first unit is related to the mass density of the collection of all units as

$$\rho(\mathbf{x}) \approx \frac{1}{K^n} \sum_{i=0}^{n-1} m_i K^{n-i} \tilde{\rho}_i(\mathbf{x}).$$

The weights K^{n-i} reflect the fact that units at the base of the chain will be counted more times than those at the top since they contribute to all subsequent units.

From an analytical viewpoint, all the previous information can be calculated if $\mu(g, \mathbf{x})$ is known for the whole chain. If, however, one views the problem from a computational perspective, it is difficult to rationalize storing numerical values of the function $\mu(g, \mathbf{x})$ because $SE(3)$ is a six-dimensional Lie group, $\mathbb{R}^3$ is three-dimensional, and the computational storage required for accurate approximation of a function on the nine-dimensional space $SE(3) \times \mathbb{R}^3$ is prohibitive. For example, 100 sample points in each coordinate direction would require 10^{18} memory locations. However, calculating $\rho(\mathbf{x})$ and $f(g)$ separately with the same sampling makes the problem literally a million times easier to handle.

---

[3]These could be either segments of the chain or single monomer units.

Moreover, there are two major problems that must be addressed in order to accurately and efficiently calculate the functions $\rho(\mathbf{x})$ and $f(g)$. The first is that for a chain with $N \in [100, 10000]$ units, each neighboring pair of which has K potential wells, it is impossible with current technology to enumerate the K^N conformations of the chain corresponding to all possible combinations of potential wells. The second is that the long-range interactions of units that are distant in the chain but proximal in space cannot be modeled using only the functions $\rho(\mathbf{x})$ and $f(g)$. Both of these problems are addressed using the recursive procedure outlined in the following sections.

17.5   Generating Ensemble Properties for Purely Kinematical Models

This section explores mathematical and computational methods for generating statistical ensembles of macroscopically-serial chain macromolecules. In particular, given ensemble properties of subchains, we explore how the properties of the whole chain can be generated if the effects of conformational energy between segments are ignored. While this can be a valid approximation in situations when the energy of interaction is considered negligible, our goal in neglecting energy effects in this section is purely a matter of pedagogy. For example, introducing useful syntax in the context of a purely kinematical model makes the definitions and concepts used in subsequent sections (where energy effects are incorporated) easier to understand.

Because of the exponential growth in the number of conformations as a function of the number of backbone atoms, it is not possible to generate the functions $\rho(\mathbf{x})$ and $f(g)$ for the whole chain by explicitly enumerating all configurations. This is well known in the polymer science literature. As reviewed in the introduction, the standard approaches to avoiding this exponential growth are: (1) asymptotic approximations in which the ensemble properties are assumed to be Gaussian as $N \to \infty$; and (2) Monte Carlo simulations which choose a small portion of the conformations and approximate overall behavior based on appropriate sampling.

In this section a completely different approach is taken. At the core of this approach remains the philosophy of considering ensemble properties of subsegments of the chain. This idea is not new; its roots go back more than half of a century [23, 42]. However, the way in which properties of the whole chain are calculated based on the properties of segments of the chain is quite different than traditional approaches.

The basic idea is that we imagine dividing the chain up into P statistically significant segments. P is chosen large enough so that the continuum approximations to the ensembles $\rho(\mathbf{x})$ and $f(g)$ meet acceptable measures of accuracy. For example, if $N \approx 1000$ and $K = 3$ (as might be the case for torsion angles in a polyethylene molecule) we might choose $P \approx 40$. In this way $K^{N/P}$ is a umber which can be managed with a personal computer.

17.5.1   The Mathematical Formulation

For each of the P statistical segments in the chain we can calculate $\rho_i(\mathbf{x})$ and $f_i(g)$ where g is the *relative* frame of reference of the distal end of the ith segment with respect to the proximal end. For a homogeneous chain, such as polyethylene, these functions are the same for each value of $i = 1, \ldots, P$.

In the general case of a heterogenous chain, we can calculate the functions $\rho_{i,i+1}(\mathbf{x})$ and $f_{i,i+1}(g)$ for the concatenation of segments i and $i + 1$ from those of segments i and $i + 1$ separately in the following way

$$\rho_{i,i+1}(\mathbf{x}) = F_{i+1}\rho_i(\mathbf{x}) + \int_G f_i(h)\rho_{i+1}\left(h^{-1} \circ \mathbf{x}\right) dh \tag{17.37}$$

and

$$f_{i,i+1}(g) = (f_i * f_{i+1})(g) = \int_G f_i(h)f_{i+1}\left(h^{-1} \circ g\right) dh. \tag{17.38}$$

In these expressions $h \in G = SE(3)$ is a dummy variable of integration. The meaning of Eq. (17.37) is that the mass density of the ensemble of all conformations of two concatenated chain segments results from two contributions. The first is the mass density of all the conformations of the lower segment (weighted by the number of different upper segments it can carry, which is $F_{i+1} = \int_G f_{i+1} dg$). The second contribution results from rotating and translating the mass density of the ensemble of the upper segment and adding the contribution at each of these poses (positions and orientations). This contribution is weighted by the number of frames that the distal end of the lower segment can attain relative to its base. Mathematically $L(h)\rho_{i+1}(\mathbf{x}) = \rho_{i+1}(h^{-1} \circ \mathbf{x})$ is a left-shift operation which geometrically has the significance of rigidly translating and rotating the function $\rho_{i+1}(\mathbf{x})$ by the transformation h. The weight $f_i(h)dh$ is the number of configurations of the ith segment terminating at frame of reference h.

The meaning of Eq. (17.38) is that the distribution of frames of reference at the terminal end of the concatenation of segments i and $i + 1$ is the group-theoretical *convolution* of the frame densities of the terminal ends of each of the two segments relative to their respective bases.

Equations (17.37) and (17.38) can be iterated with $\rho_{i,i+1}$ and $f_{i,i+1}$ taking the places of ρ_i and f_i, and $\rho_{i,i+2}$ and $f_{i,i+2}$ taking the places of $\rho_{i,i+1}$ and $f_{i,i+1}$. This will be described in detail shortly.

17.5.2   Generating $\mu_{i,i+1}$ from μ_i and μ_{i+1}

Later in the chapter the effects of interaction between distal segments of a macromolecule are approximated in an average sense[4] by considering how the functions $\mu_{i,i+j}(g, \mathbf{x})$ and $\mu_{i+j+1,k}(g, \mathbf{x})$ overlap for $k \geq i + j + 1$. In order to calculate the interaction of these functions, one must first calculate the functions $\mu_{i,i+j}(g, \mathbf{x})$ from the set of functions $(\mu_{i,i+1}(g, \mathbf{x}), \ldots, \mu_{i+j-1,i+j}(g, \mathbf{x}))$. Analogous to the way $\rho_{i,i+j}(\mathbf{x})$ and $f_{i,i+j}(g)$ are calculated by repeating the computations in Eqs. (17.37) and (17.38), $\mu_{i,i+j}(g, \mathbf{x})$ is constructed recursively by first calculating $\mu_{i,i+1}(g, \mathbf{x})$ from $\mu_i(g, \mathbf{x})$ and $\mu_{i+1}(g, \mathbf{x})$.

This is achieved by observing that

$$\mu_{i,i+1}(g, \mathbf{x}) = \int_G \left(\mu_i(h, \mathbf{x})f_{i+1}\left(h^{-1} \circ g\right) + f_i(h)\mu_{i+1}\left(h^{-1} \circ g, h^{-1} \circ \mathbf{x}\right)\right) dh. \tag{17.39}$$

This equation says that there are two contributions to $\mu_{i,i+1}(g, \mathbf{x})$. The first comes from adding up all the contributions due to each $\mu_i(h, \mathbf{x})$. This is weighted by the number of upper segment conformations with distal ends that reach the frame g given that their base is at frame h. The second comes from adding up all shifted (translated and rotated) copies of $\mu_{i+1}(g, \mathbf{x})$, where the shifting is performed by the lower distribution and the sum is weighted by the number of distinct configurations of the lower segment that terminate at h. This number is $f_1(h)dh$.

The veracity of this derivation may be confirmed by integrating the resulting function $\mu_{i,i+1}(g, \mathbf{x})$ over $\mathbb{R}^N$ and $SE(3)$ and comparing with Eqs. (17.37) and (17.38). Note that segment i has mass M_i, segment $i + 1$ has mass M_{i+1}, and so $M_{i,i+1} = M_i + M_{i+1}$ is the value of M in Eq. (17.35) in the context of the present discussion.

---

[4]The meaning of this "average sense" is made precise later in the chapter.

17.5.3 Computationally Efficient Strategies for Calculating $f(g)$ and $\rho(x)$

Given the previous means for calculating the functions $\rho_{i,i+1}(x)$ and $f_{i,i+1}(g)$ from the functions $f_i(g)$, $f_{i+1}(g)$, $\rho_i(x)$, and $\rho_{i+1}(x)$, it is now possible to formulate algorithms for generating these functions for the whole chain: $\rho(x) = \rho_{1,P}(x)$ and $f(g) = f_{1,P}(g)$.

On a serial processor the most straightforward way to do this is to sequentially start at one end of the chain and repeatedly perform the required integrations. It does not matter at which end we begin. Starting at the base and working toward the distal end, we would calculate the sequence of functions $(\rho_{1,2}(x), f_{1,2}(g))$, ... , $(\rho_{1,i}(x), f_{1,i}(g))$, ... , $(\rho_{1,P}(x), f_{1,P}(g))$. Starting from the other end we would calculate $(\rho_{P-1,P}(x), f_{P-1,P}(g))$, ... , $(\rho_{P-i,P}(x), f_{P-i,P}(g))$, ... , $(\rho_{1,P}(x), f_{1,P}(g))$. In either case, viewing the number of computations required to calculate Eqs. (17.37) and (17.38) as a constant, the recursive computation of these convolution-like integrals requires $\mathcal{O}(P)$ calculations. This is after each of the functions $\rho_i(x)$ and $f_i(g)$ have been calculated, which can be achieved in $\mathcal{O}(P \cdot K^{(N/P)})$ calculations, i.e., $\mathcal{O}(K^{(N/P)})$ calculations to explicitly enumerate configurations of each of the P segments.

The computational speed of the previous approach on a parallel computer with P processors is much faster than on a single processor. Clearly, in this case the enumeration of segment conformations is reduced to an $\mathcal{O}(K^{(N/P)})$ time calculation since each of the P ensembles can be calculated separately. Furthermore, instead of explicitly computing a P-fold convolution requiring $\mathcal{O}(P)$ time, convolutions of adjacent functions can be calculated in a pairwise fashion on different processors. This reduces the running time to $\mathcal{O}(\log_2 P)$. For example, if $P = 8$, then the convolutions $f_{1,2} = f_1 * f_2$, $f_{3,4} = f_3 * f_4$, $f_{5,6} = f_5 * f_6$, and $f_{7,8} = f_7 * f_8$ are all performed at the same time using $4 = P/2$ processors. Then $f_{1,4} = f_{1,2} * f_{3,4}$ and $f_{5,8} = f_{5,6} * f_{7,8}$ are calculated at the next level using two processors. Finally, $f_{1,8} = f_{1,4} * f_{5,8}$ is performed using a single processor. Thus we have in this example an 8-fold convolution calculated in the same time as $3 = \log_2(8)$ convolutions on a serial processor.

It is worth noting that the same speed-up achieved for a heterogeneous chain calculated on a parallel processor is valid in the case of a homogeneous chain calculated on a single processor. In this special case $\mathcal{O}(K^{(N/P)})$ time is required for brute force enumeration of one segments of length N/P. Similarly, a P-fold convolution of the same function with itself only requires $\mathcal{O}(\log_2 P)$ distinct convolutions, and thus this order of time. For example, the three convolutions $f_1 = f * f$, $f_2 = f_1 * f_1$, and $f_3 = f_2 * f_2$ generate the same result as an eight-fold convolution of f with itself.

17.6 Incorporating Conformational Energy Effects

Including the effects of conformational energy, the density function describing the distribution of tip-to-base positions and orientations of a macromolecule may be written in the form

$$f(g)\,\mathrm{Vol}(\Delta(g)) = \int_{\phi \in Im(\Delta(g))} e^{-E(\phi)/k_B T}\,d\phi \qquad (17.40)$$

where in the case of a serial chain $\phi = (\phi_1, \phi_2, \ldots, \phi_{n-1})$ is the set of all torsion angles, and $d\phi = d\phi_1 \cdots d\phi_{n-1}$.

$\Delta(g)$ is a small 6-dimensional voxel (box) in $SE(3)$ containing the group element g, and $\mathrm{Vol}(\Delta(g))$ is the volume of this voxel. Since the support of f is finite, it can be divided into a finite number of voxels, $\mathcal{M}$. $g(\phi)$ is the end frame of reference of the chain relative to its base for given torsion angles ϕ, and $Im(\Delta(g))$ is the set of all torsion angles such that $g(\phi) \in \Delta(g)$.

Each torsion angle takes its values from the unit circle, T^1, and so the whole collection of angles takes its values from the $n - 1$-dimensional torus $T^{n-1} = T^1 \times \cdots \times T^1$.

One can normalize $f(g)$ in Eq. (17.40) by observing that the sum

$$\sum_{i=1}^{M} f(g_i) \, \text{Vol}(\Delta(g_i)) = \sum_{i=1}^{M} \int_{\phi \in Im(\Delta(g_i))} e^{-E(\phi)/k_B T} d\phi$$

becomes

$$\int_G f(g) dg = \int_{\phi \in T^{n-1}} e^{-E(\phi)/k_B T} d\phi$$

for sufficiently small voxels.

We note that as $k_B T \to \infty$, $f(g)$ reduces to the purely kinematic model discussed earlier. In the following sections we examine how the density function in Eq. (17.40) is related to the Euclidean group convolution model under a wide variety of conditions.

17.6.1 Two-State and Nearest-Neighbor Energy Functions

One of the simplest kinds of conformational energy functions is one of the form

$$E(\phi) = \sum_{i=1}^{n-1} E_i(\phi_i). \tag{17.41}$$

This kind of conformational energy function models nearest-neighbor interactions and leads to a separable partition function.

The frame density function for the concatenation of two chain molecules with this kind of energy function is again given by a Euclidean group convolution. That is, since the energy function is additive and the partition function is separable, the PDFs of two concatenated segments are multiplied and integrated as

$$(f_1 * f_2)(g) = f(g).$$

This is derived by substituting the energy function in Eq. (17.41) into Eq. (17.40) for a chain with $n = n_1 + n_2$ torsion angles. The functions ρ and μ are calculated analogously.

17.6.2 Interdependent Potential Functions

When interdependent potential functions are used, it is not possible to completely separate the conformational partition function and perform straight Euclidean group convolutions. Instead, one can write the conformational energy function as

$$E(\phi) = E_1(\phi_1, \ldots, \phi_i) + E_{i+1}(\phi_i, \phi_{i+1}) + E_2(\phi_{i+1}, \ldots, \phi_{n-1}). \tag{17.42}$$

Then frame densities for the lower and upper segments are generated as before. However, unlike the previously discussed cases, these frame densities are not only functions on $SE(3)$, but also dependant on the bond angles contributing to the energy of interaction between the two chain segments. Hence we define f_1 and f_2 by the equalities

$$f_1\left(g', \phi_i\right) \text{Vol}\left(\Delta\left(g'\right)\right) = \int_{\{T^{i-1}|g(\phi_1,\ldots,\phi_i) \in Im(\Delta(g'))\}} e^{-E_1(\phi_1,\ldots,\phi_i)/k_B T} d\phi_1 \ldots d\phi_{i-1}$$

and

$$f_2\left(g', \phi_{i+1}\right) \text{Vol}\left(\Delta\left(g'\right)\right)$$
$$= \int_{\{T^{n-i-2}|g(\phi_{i+1},...,\phi_{n-1})\in Im(\Delta(g'))\}} e^{-E_2(\phi_{i+1},...,\phi_{n-1})/k_BT} d\phi_{i+2}...d\phi_{n-1}.$$

The functions f_1 and f_2 are written more cleanly in the limit of very small voxels using the Dirac delta as

$$f_1\left(g', \phi_i\right) = \int_{T^{i-1}} e^{-E_1(\phi_1,...,\phi_i)/k_BT} \delta\left(g^{-1}(\phi_1,...,\phi_i) \circ g'\right) d\phi_1...d\phi_{i-1}$$

and

$$f_2\left(g', \phi_{i+1}\right) = \int_{T^{n-i-2}} e^{-E_2(\phi_{i+1},...,\phi_{n-1})/k_BT} \delta\left(g^{-1}(\phi_{i+1},...,\phi_{n-1}) \circ g'\right) d\phi_{i+2}...d\phi_{n-1}$$

where

$$\delta(g) = \begin{cases} 0 & g \notin \Delta(e) \\ \frac{1}{\text{Vol}(\Delta(e))} & g \in \Delta(e) \end{cases}$$

where e is the identity of G.

These two contributions add to give the composite frame density function using a combination of convolution and weighted integration over the last torsion angle of the first chain segment and first torsion angle of the second chain segment

$$f(g) = \int_{T^2} \int_{SE(3)} f_1(h, \phi_i) f_2\left(h^{-1} \circ g, \phi_{i+1}\right) e^{-E_{i+1}(\phi_i,\phi_{i+1})/k_BT} dh d\phi_i d\phi_{i+1}.$$

This follows from the evaluation of Eq. (17.40) with Eq. (17.42). From the forms given previously for f_1 and f_2, and the properties of the Dirac delta, it is clear that the normalization

$$\int_G f(g)dg = \int_{T^{n-1}} e^{-E(\phi)/k_BT} d\phi$$

holds.

The assumption of discrete sampling of most probable values of torsion angles, as in the RIS model, reduces the integrations over T^2 to summations. Hence, from a computational perspective, interdependent energy functions require K^2 convolutions instead of one, where K is the number of sample points for each torsion angle. Typically for an organic chain molecule $K = 3$.

This procedure is iterated analogous to the algorithm developed in the previous section for the purely kinematic model. Namely, instead of breaking a chain molecule into two imaginary pieces, it can be broken into an arbitrary number of statistically significant segments, and an energy-weighted convolution of each pair of adjacent segments can be performed.

Hence convolution-like integrals of the form

$$f_{i,l}(g, \phi_i, \phi_l) \tag{17.43}$$
$$= \int_{T^2} \int_{SE(3)} f_1(h, \phi_i, \phi_k) f_{k+1,l}\left(h^{-1} \circ g, \phi_{k+1}, \phi_l\right) e^{-E_{i+1}(\phi_k,\phi_{k+1})/k_BT} dh d\phi_k d\phi_{k+1}$$

are performed, where now both the values of base and distal torsion angles must be recorded so that the process can be iterated. The drawback of this is that if the T^2 integral is approximated as a sum over K^2 values, then the K^2 convolutions on $SE(3)$ must be performed for each of the K^2 values of the pairs (ϕ_i, ϕ_l). Hence, K^4 convolutions on $SE(3)$ are required at each step. While this is troublesome, we note that these calculations are independent for each (ϕ_i, ϕ_l) and hence can be distributed over a parallel computer.

17.6.3 Ensemble Properties Including Long-Range Conformational Energy

In this section we model the longe-range interactions in a macroscopically serial chain using an averaging approach which builds on the formulation of the previous sections. The model uses the functions $\mu(g, \mathbf{x})$ which were not explicitly needed in the case when long-range interactions were not important.

Accurately modeling interactions of atoms which are distal in the chain but proximal in space (due to bending of the chain) is one of the most difficult problems in the study of macromolecules, independent of whether they are man-made polymers, proteins, or DNA. Explicitly accounting for all such interactions for all possible configurations by brute force enumeration requires a mind-boggling amount of computational time. At the other extreme, the simplified closed-form analytical models such as the Gaussian random walk do not explicitly account for these interactions. Furthermore, models based on self-avoiding walks and renormalization group methods are respectively limited to rather short chains ($N << 100$) or very long ones ($N \to \infty$).

A number of different approaches are considered here to incorporate the effects of energy. Perhaps the most straightforward is to penalize contributions in Eq. (17.39) so that when the support of appropriately shifted functions μ_i and μ_{i+1} intersect, these functions would be disallowed from contributing to the computation of $\mu_{i,i+1}$. Clearly this would generate ensemble statistical distributions which would be lower estimates of those generated from the self-avoiding walk model. A slightly more sophisticated model would calculate the energy of interaction of the distributions μ_i and μ_{i+1} and use this information in an appropriate conformational partition function. We now quantify this discussion. The interaction of segment i and $i + 1$ is approximated by considering the interaction of the corresponding functions μ_i and μ_{i+1} as

$$E_{i,i+1}(h, g) = \int_{\mathbb{R}^3} \int_{\mathbb{R}^3} \mu_i(h, \mathbf{x}) \mu_{i+1}\left(h^{-1} \circ g, h^{-1} \circ \mathbf{y}\right) V(\mathbf{x} - \mathbf{y}) d\mathbf{x} d\mathbf{y}. \tag{17.44}$$

We assume here that the potential between any two atoms located at positions $\mathbf{x}$ and $\mathbf{y}$ is $V(\mathbf{x} - \mathbf{y})$. Then Eq. (17.44) is an approximation of the interaction of all configurations of segment i which terminate at frame h and all configurations of segment $i + 1$ with distal end at g and proximal end at h (hence the relative displacement $h^{-1} \circ g$). In the case when the potential function is used to represent pure hard-sphere repulsion and no attraction, a reasonable model for $V(\cdot)$ is

$$V(\mathbf{x}) = E_0 \delta(\mathbf{x}).$$

In this case one writes

$$E_{i,i+1}(h, g) = E_0 \int_{\mathbb{R}^3} \mu_i(h, \mathbf{x}) \mu_{i+1}\left(h^{-1} \circ g, h^{-1} \circ \mathbf{x}\right) d\mathbf{x}. \tag{17.45}$$

This may then be used to approximate $\mu_{i,i+1}$ as

$$\mu_{i,i+1}(g, \mathbf{x}) \tag{17.46}$$
$$= \int_G \left(\mu_i(h, \mathbf{x}) f_{i+1}\left(h^{-1} \circ g\right) + f_i(h) \mu_{i+1}\left(h^{-1} \circ g, h^{-1} \circ \mathbf{x}\right)\right) e^{-E_{i,i+1}(h,g)/k_B T} dh.$$

By definition, it follows that

$$f_{i,i+1}(g) = \frac{1}{M_i + M_{i+1}} \int_{\mathbb{R}^3} \mu_{i,i+1}(g, \mathbf{x}) d\mathbf{x}$$
$$= \int_G f_i(h) f_{i+1}\left(h^{-1} \circ g\right) e^{-E_{i,i+1}(h,g)/k_B T} dh. \tag{17.47}$$

In this way, the purely kinematical model is modified so as to take into account the energy of interaction of two adjacent segments. As E_0/k_BT becomes large, the only contributions to $\mu_{i,i+1}$ are from the shifted versions of the functions μ_i and μ_{i+1} which do not overlap at all. This extreme case is a lower bound on the $\mu_{i,i+1}$ that the self-avoiding walk would generate. This follows for large E_0/k_BT because the model in Eq. (17.46) not only disallows the intersection of two adjacent segments, but also disallows any contribution if Eq. (17.45) is nonzero, i.e., if μ_i and μ_{i+1} overlap. On the other hand, for smaller values of E_0/k_BT the present model might provide more realistic statistics than lattice self-avoiding walks since there is no artificial restriction on bond and torsion angles that limit conformations to conform to a lattice in the present model, and when $E_0/k_BT \to 0$ the model becomes purely kinematic.

17.7   Statistics of Stiff Molecules as Solutions to PDEs on $SO(3)$ and $SE(3)$

Experimental measurements of the stiffness constants of DNA and other stiff (or semi-flexible) macromolecules have been reported in a number of papers, as well as the statistical mechanics of such molecules. See [32, 37, 38, 45, 49, 52, 53, 54, 58, 59, 61, 64, 72, 73, 75, 82, 86, 87, 88].

As is often the case in theoretical polymer science, analogies between the motion of a particle along a path and the motion of an observer traversing a polymer chain allow for tools from classical and quantum mechanics to be applied. We review one such analogy in the following subsection.

17.7.1   Model Formulation

A number of authors have derived potential energies of bending and/or twisting of a stiff chain that are of the form

$$E = \int_0^L U(\omega(s))ds$$

where L is the length of the macromolecule and

$$U = \frac{1}{2}\omega^T B\omega - b^T\omega + \beta'. \tag{17.48}$$

Here $B = B^T \in \mathbb{R}^{3\times 3}$ is a positive semi-definite matrix, $b \in \mathbb{R}^3$ and $\beta' \in \mathbb{R}$. ω is the "angular velocity" of a frame of reference which traverses the macromolecule, coinciding with each frame $(a(s), A(s))$ affixed to the backbone of the molecule for each value of arclength s. This "angular velocity" is the dual vector of the skew symmetric matrix $A^T\dot{A}$, where $(\dot{\ }) = d/ds$, i.e., $\omega \times x = A^T\dot{A}x$ for all $x \in \mathbb{R}^3$. This is completely analogous to the definition of angular velocity of a rigid body as seen in the body fixed frame with s taking the place of time. Henceforth, we will use the notation $U = U(\omega) = U(A, \dot{A})$ to denote the fact that the bending energy is a function of the rotation matrix and its derivative through the definition of ω.

As well-known examples of Eq. (17.48) from the polymer science literature, consider

The Kratky-Porod Model [17]

$$B = \begin{pmatrix} \alpha_0 & 0 & 0 \\ 0 & \alpha_0 & 0 \\ 0 & 0 & 0 \end{pmatrix}; \quad b = \begin{pmatrix} 0 \\ 0 \\ 0 \end{pmatrix}; \quad \beta' = 0.$$

The Yamakawa Model [89]

$$B = \begin{pmatrix} \alpha_0 & 0 & 0 \\ 0 & \alpha_0 & 0 \\ 0 & 0 & \beta_0 \end{pmatrix}; \quad b = \begin{pmatrix} 0 \\ \alpha_0\kappa_0 \\ \beta_0\tau_0 \end{pmatrix}; \quad \beta' = \frac{1}{2}\left(\beta_0\tau_0^2 + \alpha_0\kappa_0^2\right).$$

The Marko-Siggia DNA Model [49]

$$B = \begin{pmatrix} A' & 0 & B \\ 0 & A & 0 \\ B & 0 & C \end{pmatrix}; \quad b = \begin{pmatrix} B\omega_0 \\ 0 \\ C\omega_0 \end{pmatrix}; \quad \beta' = \frac{1}{2}C\omega_0^2.$$

The Revised Marko-Siggia Model [33]

$$B = \begin{pmatrix} A + B^2/C & 0 & B \\ 0 & A & 0 \\ B & 0 & C \end{pmatrix}; \quad b = \begin{pmatrix} B\omega_0 \\ 0 \\ C\omega_0 \end{pmatrix}; \quad \beta' = \frac{1}{2}C\omega_0^2.$$

Other modifications of these models may be made to include stretching effects, though the current presentation is restricted to the inextensible case.

We now generate the diffusion equation that governs the evolution of the positional and orientation probability density function $F(a, A; s)$ for all values of $0 \le s \le L$. Assuming that the proximal end is fixed at the frame $(0, \mathbb{I})$, then $F(a, A; 0) = \delta(a)\delta(A)$. Here the Dirac delta function on the motion group is written as the product of those for $\mathbb{R}^3$ and $SO(3)$.

Under the constraint that the molecule is inextensible, and all the frames of reference are attached to the backbone with their local z-axis pointing in the direction of the next frame, one observes

$$\mathbf{a}(L) = \int_0^L \mathbf{u}(s)ds \quad \text{and} \quad \mathbf{u}(s) = A(s)\mathbf{e}_3. \tag{17.49}$$

Hence, the PDF of interest can be formulated as the following path integral over the rotation group

$$F(\mathbf{a}, A; L) = \int_{A(0)=I}^{A(L)=A} \delta\left(\mathbf{a}(L) - \int_0^L \mathbf{u}(s)ds\right) \exp\left[-\int_0^L U(A, \dot{A})ds\right] \mathcal{D}[A(s)], \tag{17.50}$$

where it is assumed that the bending energy U is measured in units of $k_B T$. Path integration over the rotation group has been studied extensively in the literature in the context of quantum mechanics (see Appendix E). Our notation and formulation follows that in [89].

Using the classical Fourier transform pair

$$\hat{f}(\mathbf{k}) = \int_{\mathbb{R}^3} f(\mathbf{a})e^{-i\mathbf{k}\cdot\mathbf{a}}d^3a \quad \leftrightarrow \quad f(\mathbf{a}) = \frac{1}{(2\pi)^3}\int_{\mathbb{R}^3} \hat{f}(\mathbf{k})e^{i\mathbf{k}\cdot\mathbf{a}}d^3k, \tag{17.51}$$

one writes

$$\hat{F}(\mathbf{k}, A; L) = \int_{A(0)=I}^{A(L)=A} \exp\left[-\int_0^L (i\mathbf{k}\cdot\mathbf{u} + U)ds\right] \mathcal{D}[A(s)].$$

Treating the innermost integrand as i times a Lagrangian with kinetic and potential energies,

$$T = \frac{1}{2}i\omega^T B\omega$$

and

$$V = i\left[\mathbf{b}\cdot\boldsymbol{\omega} - \beta'\right] + \mathbf{k}\cdot\mathbf{u},$$

(the constant β' can be ignored) one calculates the momenta and Hamiltonian in the usual way, which for this case means

$$p_k = \frac{\partial L}{\partial \omega_k} \rightarrow H = -i \left[\frac{1}{2} \mathbf{p}^T B^{-1} \mathbf{p} \right] + \mathbf{b}^T B^{-1} \mathbf{p} + \mathbf{k} \cdot \mathbf{u}. \tag{17.52}$$

Here and henceforth B is assumed to be positive definite (and hence invertible).

The quantization

$$p_i = -i X_i^R \tag{17.53}$$

is used, where the differential operators X_i^R acting on functions on the rotation group are defined as

$$X_i^R f(A) = \left. \frac{df(A \cdot \text{ROT}[\mathbf{e}_i, t])}{dt} \right|_{t=0} = \left. \frac{df(A(I + tX_i))}{dt} \right|_{t=0}, \tag{17.54}$$

where

$$X_1 = \begin{pmatrix} 0 & 0 & 0 \\ 0 & 0 & -1 \\ 0 & 1 & 0 \end{pmatrix}; \quad X_2 = \begin{pmatrix} 0 & 0 & 1 \\ 0 & 0 & 0 \\ -1 & 0 & 0 \end{pmatrix}; \quad X_3 = \begin{pmatrix} 0 & -1 & 0 \\ 1 & 0 & 0 \\ 0 & 0 & 0 \end{pmatrix}.$$

The superscript R in X_i^R denotes the fact that the infinitesimal rotation $I + tX_i$ is applied on the right of the argument of the function. This corresponds to an infinitesimal motion relative to the body-fixed frame in a rigid body.

Using the ZXZ Euler angles (α, β, γ) these operators have the explicit form

$$\begin{aligned} X_1^R &= -\cot\beta \sin\gamma \frac{\partial}{\partial \gamma} + \frac{\sin\gamma}{\sin\beta} \frac{\partial}{\partial \alpha} + \cos\gamma \frac{\partial}{\partial \beta}; \\ X_2^R &= -\cot\beta \cos\gamma \frac{\partial}{\partial \gamma} + \frac{\cos\gamma}{\sin\beta} \frac{\partial}{\partial \alpha} - \sin\gamma \frac{\partial}{\partial \beta}; \\ X_3^R &= \frac{\partial}{\partial \gamma}. \end{aligned} \tag{17.55}$$

The Schrödinger-like equation corresponding to the Hamiltonian in Eq. (17.52) and quantization in Eq. (17.53) is

$$i \frac{\partial \hat{F}}{\partial L} = H \hat{F}.$$

This takes the explicit form

$$\left(\frac{\partial}{\partial L} - \frac{1}{2} \sum_{k,l=1}^{3} \left(B_{lk}^{-1} X_l^R X_k^R - 2 B_{lk}^{-1} b_k X_l^R \right) + i\mathbf{k} \cdot \mathbf{u} \right) \hat{F} = 0. \tag{17.56}$$

Henceforth we will use the quantities $D = B^{-1}$ and $\mathbf{d} = -B^{-1}\mathbf{b}$.

The classical Fourier inversion formula in Eq. (17.51) then converts Eq. (17.56) to

$$\left(\frac{\partial}{\partial L} - \frac{1}{2} \sum_{k,l=1}^{3} D_{lk} X_l^R X_k^R - \sum_{l=1}^{3} d_l X_l^R + \mathbf{u} \cdot \nabla_{\mathbf{a}} \right) F = 0, \tag{17.57}$$

or equivalently

$$\left(\frac{\partial}{\partial L} - \frac{1}{2} \sum_{k,l=1}^{3} D_{lk} \tilde{X}_l^R \tilde{X}_k^R - \sum_{l=1}^{3} d_l \tilde{X}_l^R + \tilde{X}_6^R \right) F = 0, \tag{17.58}$$

which is a PDE on the motion group, $SE(3)$. The initial conditions are $F(\mathbf{a}, A; 0) = \delta(\mathbf{a})\delta(A)$.

Integrating F over all positions, $\mathbf{a} \in \mathbb{R}^3$, results in a purely orientational density function

$$f(A; s) = \int_{\mathbb{R}^3} F(\mathbf{a}, A; s)d^3a.$$

Performing this integration over the initial conditions and Eq. (17.57) results in the $SO(3)$-diffusion equation

$$\left(\frac{\partial}{\partial L} - \frac{1}{2} \sum_{k,l=1}^{3} D_{lk}X_l^R X_k^R - \sum_{l=1}^{3} d_l X_l^R \right) f = 0 \qquad (17.59)$$

with initial conditions $f(A; 0) = \delta(A)$.

Equation (17.59) is a partial differential equation which governs the evolution of the function f on the rotation group, $SO(3)$. It can be solved in series form using techniques from noncommutative harmonic analysis.

We note that as a direct result of the structure of IUR matrix elements for $SE(3)$ and the Fourier inversion formula (Chapter 10) that

$$\int_{SO(3)} F(\mathbf{a}, A)dA = \frac{1}{2\pi^2} \sum_{l'=0}^{\infty} \sum_{m'=-l'}^{l'} \int_0^{\infty} p^2 dp\, \hat{F}_{0,0;l',m'}^0(p) \left[l', m' \,|p, 0| 0, 0 \right](\mathbf{a}).$$

If this distribution of end positions is then integrated over the surface of a sphere with radius $a = |\mathbf{a}|$, the result is the end-to-end distance distribution

$$\frac{a^2}{2\pi^2} \int_{S^2} \int_{SO(3)} F(a\mathbf{u}, A)d\mathbf{u}dA = \frac{2}{\pi}a^2 \int_0^{\infty} p^2 dp\, \hat{F}_{0,0;0,0}^0(p)[0, 0\,|p, 0| 0, 0](\mathbf{a}).$$

It is easy to verify that $[0, 0\,|p, 0| 0, 0](\mathbf{a}) = \sin(pa)/pa$.

These expressions provide a means of addressing PDFs of end-to-end relative position and end-to-end distance when knowledge of orientation is not critical.

17.7.2 Operational Properties and Solutions of PDEs

By the definition of the $SE(3)$-Fourier transform $\mathcal{F}[\cdot]$ and operators $\tilde{X}_i^R$ reviewed in Chapter 10 one observes that

$$\mathcal{F}\left[\tilde{X}_i^R F \right] = \int_{SE(3)} \frac{d}{dt} \left(F \left(g \circ \exp\left(t\tilde{X}_i \right) \right) \right) \Big|_{t=0} U^s \left(g^{-1}, p \right) dg. \qquad (17.60)$$

Here g can be thought of as $H(g)$ and $\exp(t\tilde{X}_i)$ is an element of the subgroup of $SE(3)$ generated by $\tilde{X}_i$, which for small values of t is approximated as $I + t\tilde{X}_i$. By performing the change of variables $h = g \circ \exp(t\tilde{X}_i)$ and using the homomorphism property of the representations $U^s(\cdot)$, one finds

$$\mathcal{F}\left[\tilde{X}_i^R F \right] = \int_{SE(3)} F(h)\frac{d}{dt} \left(U^s \left(\exp\left(t\tilde{X}_i \right) \circ h^{-1}, p \right) \right) \Big|_{t=0} dh \qquad (17.61)$$

$$= \frac{d}{dt} \left(U^s \left(\exp\left(t\tilde{X}_i \right), p \right) \right) \Big|_{t=0} \int_{SE(3)} F(h)U^s \left(h^{-1}, p \right) dh. \qquad (17.62)$$

By defining

$$u^s \left(\tilde{X}_i, p \right) = \frac{d}{dt} \left(U^s \left(\exp\left(t\tilde{X}_i \right), p \right) \right) \Big|_{t=0},$$

we write

$$\mathcal{F}\left[\tilde{X}_i^R F\right] = u^s\left(\tilde{X}_i, p\right) \hat{F}^s(p).$$

Hence, Eq. (17.58) can be transformed to the infinite system of linear differential equations

$$\frac{d\hat{F}^s}{dL} = B^s \hat{F}^s, \tag{17.63}$$

where

$$B^s = \frac{1}{2} \sum_{k,l=1}^{3} D_{lk} u^s\left(\tilde{X}_l, p\right) u^s\left(\tilde{X}_k, p\right) + \sum_{l=1}^{3} d_l u^s\left(\tilde{X}_l, p\right) - u^s\left(\tilde{X}_6, p\right).$$

In principle, $F(\mathbf{a}, A; L)$ is then found by simply substituting $\hat{F}^s(p; L) = \exp(LB^s)$ into the $SE(3)$ Fourier inversion formula. In practice, however, exponentiation of a non-diagonal infinite-dimensional matrix poses some difficulties.

Explicitly, for $i = 1, 2, 3$ we have

$$u^s\left(\tilde{X}_i, p\right) = \frac{d}{dt} U^s_{l',m';l,m}\left(0, \exp\left[tX_i\right]; p\right)|_{t=0} = \delta_{l,l'} \frac{d}{dt} U^l_{m',m}\left(\exp\left[tX_i\right]\right)|_{t=0}.$$

The second equality above follows easily from the structure of the matrix elements $U^s_{l',m';l,m}$, and $U^l_{m,n}(\exp[tX_i])|_{t=0}$ are given explicitly in Chapter 9. This, together with the operational property (Chapter 10)

$$u^s_{l',m';l,m}\left(\tilde{X}_6, p\right) = \frac{d}{dt} U^s_{l',m';l,m}\left(t\mathbf{e}_3, I; p\right)|_{t=0}$$

$$= ip\kappa^s_{l',m'}\delta_{l'-1,l}\delta_{m',m} + ip\frac{m's}{l'(l'+1)}\delta_{l'l}\delta_{m',m} + ip\kappa^s_{l,m}\delta_{l',l-1}\delta_{m',m}$$

where

$$\kappa^s_{l',m'} = \left(\frac{(l'^2 - m'^2)(l'^2 - s^2)}{(2l'+1)(2l'-1)l'^2}\right)^{1/2}$$

allows us to write the elements of $B^s(p)$ as

$$B^s_{l',m';l,m} = A^l_{m',m}\delta_{l',l} - ip\kappa^s_{l',m'}\delta_{l'-1,l}\delta_{m',m} - ip\frac{m's}{l'(l'+1)}\delta_{l'l}\delta_{m',m} - ip\kappa^s_{l,m}\delta_{l',l-1}\delta_{m',m}.$$

This explicit form of the elements of the matrix B^s, together with the Fourier inversion formula for $SE(3)$, allows us to numerically solve for the probability density function $F(\mathbf{a}, A; L)$ for each L, and more specifically the probability density of end-to-end distances. From this, any desired moments can be computed. Numerical implementation by truncating and exponentiating the infinite-dimensional matrices B^s is described in [10].

17.8 Summary

In this chapter we illustrated how noncommutative harmonic analysis can be used in theoretical polymer science. We reviewed classical approaches for describing statistical mechanics of macromolecules. We also showed how Euclidean-group convolutions can be used as a numerical tool for generating conformational statistics. In the special case of stiff macromolecules, we showed how the Fourier transform for $SE(3)$ converts a diffusion-type partial differential equation evolving on $SE(3)$ into a system of linear ordinary differential equations with constant coefficients. For more on the numerical implementation of these ideas see [9, 10].

References

[1] Alexandrowicz, Z., Monte Carlo of chains with excluded volume: a way to evade sample attrition, *J. of Chem. Phys.*, 51(2), 561–565, 1969.

[2] Allegra G., Ganazzoli F., and Bontempelli S., Good- and bad-solvent effect on the rotational statistics of a long chain molecule, *Comput. and Theor. Polymer Sci.*, 8(1-2), 209–218, 1998.

[3] Batoulis, J. and Kremer, K., Statistical properties of biased sampling methods for long polymer chains, *J. of Phys. A: Math. and General*, 21, 127–146, 1988.

[4] Birshtein, T.M. and Ptitsyn, O.B., *Conformations of Macromolecules*, Wiley InterScience, New York, 1966.

[5] Bohdanecký, M. and Kovár, J., *Viscosity of Polymer Solutions*, Elsevier, New York, 1982.

[6] Boyd, R.H. and Phillips, P.J., *The Science of Polymer Molecules*, Cambridge Solid State Science Series, Cambridge University Press, Cambridge, 1993.

[7] Chandrasekhar, S., Stochastic problems in physics and astronomy, *Rev. of Mod. Phys.*, 15(1), 1–89, 1943.

[8] Chen, S.-J. and Dill, K.A., Symmetries in proteins: a knot theory approach, *J. Chem. Phys.*, 104(15), 5964–5973, 1996.

[9] Chirikjian, G.S., Conformational statistics of macromolecules using generalized convolution, *Comput. and Theor. Polymer Sci.*, in press.

[10] Chirikjian, G.S. and Wang, Y., Conformational statistics of stiff macromolecules as solutions to PDEs on the rotation and motion groups, *Phys. Rev. E.*, July, 2000.

[11] Collet, O. and Premilat, S., Calculation of conformational free energy and entropy of chain molecules with long-range interactions, *Macromol.*, 26, 6076–6080, 1993.

[12] Comtet, A. and Nechaev, S., Random operator approach for word enumeration in braid groups, *J. Phys. A: Math. Gen.*, 31, 5609–5630, 1998.

[13] Cush, R., Russo, P.S., Kucukyavuz, Z., Bu, Z., Neau, D., Shih, D., Kucukyavuz, S., and Ricks, H., Rotational and translational diffusion of a rodlike virus in random coil polymer solutions, *Macromol.*, 30(17), 4920–4926, 1997.

[14] Daniels, H.E., The statistical theory of stiff chains, *Proc. Roy. Soc.*, A63, 290–311, Edinburgh, Scotland, 1952.

[15] de Gennes, P.G., *Introduction to Polymer Dynamics*, Cambridge University Press, Cambridge, 1990.

[16] de Gennes, P.G., *Scaling Concepts in Polymer Physics*, Cornell University Press, Ithaca, NY, 1979.

[17] des Cloizeaux, J. and Jannink, G., *Polymers in Solution: Their Modelling and Structure*, Clarendon Press, Oxford, 1990.

[18] Doi, M. and Edwards, S.F., *The Theory of Polymer Dynamics*, Clarendon Press, Oxford, 1986.

[19] Doi, M., *Introduction to Polymer Physics*, Oxford University Press, New York, 1996.

[20] Fisher, M.E., Shape of a self-avoiding walk or polymer chain, *J. of Chem. Phys.*, 44(2), 616–622, 1966.

[21] Flory, P.J., *Statistical Mechanics of Chain Molecules*, John Wiley & Sons, 1969, reprint, Hanser, Munich, 1989.

[22] Flory, P.J., Foundations of rotational isomeric state theory and general methods for generating configurational averages, *Rotational Isomeric State Theor. and Methods*, 7(3), 381–392, 1974.

[23] Flory, P.J., The configuration of real polymer chains, *J. of Chem. Phys.*, 17(3), 303–310, 1949.

[24] Freed, K.F., *Renormalization Group Theory of Macromolecules*, John Wiley & Sons, New York, 1987.

[25] Freed, K.F., Polymers as self-avoiding walks, *Ann. of Probab.*, 9(4), 537–556, 1981.

[26] Freire, J.J. and Horta, A., Mean reciprocal distances of short polymethylene chains. Calculation of the translational diffusion coefficient of n-alkanes, *J. of Chem. Phys.*, 65(10), 4049–4054, 1976.

[27] Fuller, F.B., Decomposition of the linking number of a closed ribbon: a problem from molecular biology, *Proc. Nat. Acad. Sci. USA*, 75(8), 3557–3561, 1978.

[28] Gobush, W., Yamakawa, H., Stockmayer, W.H., and Magee, W.S., Statistical mechanics of wormlike chains. I. Asymptotic behavior, *J. of Chem. Phys.*, 57(7), 2839–2843, 1972.

[29] Grosberg, A. Yu. and Khokhlov, A.R., *Statistical Physics of Macromolecules*, American Institute of Physics, New York, 1994.

[30] Grosberg, A., Ed., *Theoretical and Mathematical Models in Polymer Research*, Academic Press, Boston, 1998.

[31] Haas, E., Wilchek, M., Katchalski-Katzir, E., and Steinberg, I.Z., Distribution of end-to-end distances of oligopeptides in solution as estimated by energy transfer, *Proc. of the Nat. Acad. of Sci.*, 72(5), 1807–1811, 1975.

[32] Hagerman, P.J., Analysis of the ring-closure probabilities of isotropic wormlike chains: application to duplex DNA, *Biopolymers*, 24, 1881–1897, 1985.

[33] Haijun, Z. and Zhong-can, O., Bending and twisting elasticity: a revised Marko-Siggia model on DNA chirality, *Phys. Rev. E*, 58(4), 4816–4819, 1998.

[34] He, S. and Scheraga, H.A., Macromolecular conformational dynamics in torsional angle space, *J. Chem. Phys.*, 108(1), 271–286, 1998.

[35] Hermans, J.J. and Ullman, R., The statistics of stiff chains, with applications to light scattering, *Physica*, 18(11), 951–971, 1952.

[36] Honeycutt, J.D., A general simulation method for computing conformational properties of single polymer chains, *Comput. and Theor. Polymer Sci.*, 8(1/2), 1–8, 1998.

[37] Horowitz, D.S. and Wang, J.C., Torsional rigidity of DNA and length dependence of the free energy of DNA supercoiling, *J. of Mol. Biol.*, 173, 75–91, 1984.

[38] Klenin, K., Merlitz, H., and Langowski, J., A Brownian dynamics program for the simulation of linear and circular DNA and other wormlike chain polyelectrolytes, *Biophys. J.*, 74, 780–788, 1998.

[39] Kloczkowski, A. and Jernigan, R.L., Computer generation and enumeration of compact self-avoiding walks within simple geometries on lattices, *Comput. and Theor. Polymer Sci.*, 7(3/4), 163–173, 1997.

[40] Kratky, O. and Porod, G., Röntgenuntersuchung gelöster fadenmoleküle, *Recueil des Travaux Chimiques des Pays-Bas*, 68(12), 1106–1122, 1949.

[41] Krishnaswami, S., Ramkrishna, D., and Caruthers, J.M., Statistical-mechanically exact simulation of polymer conformation in an external field, *J. of Chem. Phys.*, 107, 5929–5944, 1997.

[42] Kuhn, W., Über die gestalt fadenförmiger moleküle in lösungen, *Kolloid-Zeitschrift*, 68, 2–15, 1934.

[43] Lagrange, J.L., Sur une nouvelle propriété du centre de gravité, *Oeuvres de Lagrange*, 5, 535–540, 1870.

[44] Lennard-Jones, J.E., The equation of state of gases and critical phenomena, *Physica*, 4(10), 941–956, 1937.

[45] Levene, S.D. and Crothers, D.M., Ring closure probabilities for DNA fragments by Monte Carlo simulation, *J. of Molec. Biol.*, 189, 61–72, 1986.

[46] Lodge, A.S., *Body Tensor Fields in Continuum Mechanics, with Applications to Polymer Rheology*, Academic Press, New York, 1974.

[47] Madras, N. and Slade, G., *The Self-Avoiding Walk*, Birkhäuser, Boston, 1996.

[48] Mark, J.E. and Erman, B., *Elastomeric Polymer Networks*, Prentice Hall, Englewood Cliffs, NJ, 1992.

[49] Marko, J.F. and Siggia, E.D., Bending and twisting elasticity of DNA, *Macromol.*, 27, 981–988, 1994.

[50] Mattice, W.L., Asymmetry of flexible chains, macrocycles, and stars, *Macromol.*, 13, 506–511, 1980.

[51] Mattice, W.L. and Suter, U.W., *Conformational Theory of Large Molecules, The Rotational Isomeric State Model in Macromolecular Systems*, John Wiley & Sons, New York, 1994.

[52] Mondescu, R.P. and Muthukumar, M., Brownian motion and polymer statistics on certain curved manifolds, *Phys. Rev. E*, 57(4), 4411–4419, 1998.

[53] Moroz, J.D. and Nelson, P., Torsional directed walks, entropic elasticity, and DNA twist stiffness, *Proc. of the Nat. Acad. of Sci. of the USA*, 94(26), 14418–14422, 1997.

[54] Moroz, J.D. and Nelson, P., Entropic elasticity of twist-storing polymers, *Macromol.*, 31(18), 6333–6347, 1998.

[55] Nechaev, S.K., Grosberg, A. Yu., and Vershik, A.M., Random walks on braid groups: Brownian bridges, complexity and statistics, *J. Phys. A: Math. Gen.*, 29, 2411–2433, 1996.

[56] Nechaev, S. and Desbois, J., Statistical mechanics of braided Markov chains: I. Analytic methods and numerical simulations, *J. of Stat. Phys.*, 88(1/2), 201–229, 1997.

[57] Nechaev, S.K., *Statistics of Knots and Entangled Random Walks*, World Scientific, Singapore, 1996.

[58] Nelson, P., Sequence-disorder effects on DNA entropic elasticity, *Phys. Rev. Lett.*, 80(26), 5810-5812, 1998.

[59] Nelson, P., New measurements of DNA twist elasticity, *Biophys. J.*, 74(5), 2501-2503, 1998.

[60] Norisuye, T., Tsuboi, A., and Teramoto, A., Remarks on excluded-volume effects in semiflexible polymer solutions, *Polymer J.*, 28(4), 357–361, 1996.

[61] Odijk, T., Physics of tightly curved semiflexible polymer chains, *Macromol.*, 26, 6897–6902, 1993.

[62] Oka, S., Zur theorie der statistischen molekülgestalt hochpolymerer kettenmoleküle unter berücksichtigung der behinderung der freien drehbarkeit, *Proc. Phys.-Math. Soc. of Japan*, 24, 657–672, 1942.

[63] Plimpton, S. and Hendrickson, B., A new parallel method for molecular dynamics simulation of macromolecular systems, *J. of Comput. Chem.*, 17(3), 326–337, 1996.

[64] Podtelezhnikov, A.A., Cozzarelli, N.R., and Vologodskii, A.V., Equilibrium distributions of topological states in circular DNA: interplay of supercoiling and knotting, *Proc. of the Nat. Acad. of Sci. of the USA*, 96(23), 12974–12979, 1999.

[65] Porod, G., X-ray and light scattering by chain molecules in solution, *J. Polymer Sci.*, 10(2), 157–166, 1953.

[66] Prohofsky, E., *Statistical Mechanics and Stability of Macromolecules*, Cambridge University Press, Cambridge, 1995.

[67] Rayleigh, L., On the problem of random vibrations, and of random flights in one, two, or three dimensions, *Phil. Mag.*, 37(6), 321–347, 1919.

[68] Redner, S. and Reynolds, P.J., Position-space renormalisation group for isolated polymer chains, *J. of Phys. A: Math. and General*, 14, 2679–2703, 1981.

[69] Rehahn, M., Mattice, W.L., and Suter, U.W., Rotational isomeric state models in macromolecular systems, *Adv. in Polymer Sci.*, 131/132, 1–6, 1997.

[70] Rosenbluth, M.N. and Rosenbluth, A.W., Monte Carlo calculation of the average extension of molecular chains, *J. of Chem. Phys.*, 23(2), 356–359, 1955.

[71] Shalloway, D., Application of the renormalization group to deterministic global minimization of molecular conformation energy functions, *J. of Global Optimization*, 2, 281–311, 1993.

[72] Shi, Y., He, S., and Hearst, J.E., Statistical mechanics of the extensible and shearable elastic rod and of DNA, *J. of Chem. Phys.*, 105(2), 714–731, 1996.

[73] Shimada, J. and Yamakawa, H., Statistical mechanics of DNA topoisomers, *J. of Mol. Biol.*, 184, 319–329, 1985.

[74] Solc, K. and Stockmayer, W.H., Shape of a random-flight chain, *J. of Chem. Phys.*, 54(6), 2756–2757, 1971.

[75] Shore, D. and Baldwin, R.L., Energetics of DNA twisting, *J. of Molec. Biol.*, 170, 957–981, 1983.

[76] Stellman, S.D. and Gans, P.J., Efficient computer simulation of polymer conformation. I. Geometric properties of the hard-sphere model, *Macromol.*, 5(4), 516–526, 1972.

[77] Sun, S.F., *Physical Chemistry of Macromolecules: Basic Principles and Issues,* John Wiley & Sons, New York, 1994.

[78] Theodorou, D.N. and Suter, U.W., Shape of unperturbed linear polymers: polypropylene, *Macromol.*, 18, 1206–1214, 1985.

[79] Thirumalai, D. and Ha, B.-Y., Statistical mechanics of semiflexible chains: a mean field variational approach, in *Theoretical and Mathematical Models in Polymer Research*, Grosberg, A., Ed., 1–35, Academic Press, New York, 1998.

[80] Treloar, L.R.G., *The Physics of Rubber Elasticity,* Oxford University Press, New York, 1958.

[81] Vaia, R.A., Dudis, D., and Henes, J., Chain conformation and flexibility of thorny rod polymers, *Polymer,* 39(24), 6021–6036, 1998.

[82] Vinograd, J., Lebowitz, J., Radloff, R., Watson, R., and Laipis, P., The twisted circular form of polyoma viral DNA, *Proc. Nat. Acad. Sci. USA,* 53(5), 1104–1111, 1965.

[83] Volkenstein, M.V., *Conformational Statistics of Polymeric Chains,* Wiley InterScience, New York, 1963 (originally published in Russian by IzdRtel'stvo Akad. Nauk SSSR, Moscow, 1959).

[84] Volkenstein, M.V., A linear polymer as a mix of rotational isomers, *Dokl. Akad. Nauk. SSSR,* 78(5), 879–882, 1951.

[85] Volkenstein, M.V. and Ptitsyn, O.B., On stretching short polymer chains, *Dokl. Akad. Nauk. SSSR,* 91(6), 1313–1316, 1953.

[86] Vologodskii, A.V., Anshelevich, V.V., Lukashin, A.V., and Frank-Kamenetskii, M.D., Statistical mechanics of supercoils and the torsional stiffness of the DNA double helix, *Nature,* 280, 294–298, 1979.

[87] White, J.H. and Bauer, W.R., Calculation of the twist and the Writhe for representative models of DNA, *J. of Mol. Biol.,* 189, 329–341, 1986.

[88] Wilhelm, J. and Frey, E., Radial distribution function of semiflexible polymers, *Phys. Rev. Lett.,* 77(12), 2581–2584, 1996.

[89] Yamakawa, H., *Helical Wormlike Chains in Polymer Solutions,* Springer-Verlag, Berlin, 1997.

Chapter 18

Mechanics and Texture Analysis

Classical mechanics is the study of the interplay between forces acting on matter and the resulting motion of matter. Quantitative results are based on four essential balances of quantities in any system composed of matter: (1) balancing the mass entering and leaving the system; (2) balancing the change in momentum of the system with applied force; (3) balancing the change in angular momentum with applied moments of force; and (4) balancing the total energy in the system with the net contributions from external sources. The laws of conservation of mass, momentum, angular momentum, and energy follow from these balances in the absence of external influences.

Traditionally, matter is assumed to exist in the solid, liquid, or gaseous states. Recently, the techniques of classical mechanics have also been applied to what are considered the new states of matter such as plasmas, liquid crystals, and gels. At the core of mechanics is the approximation of fluids (including liquids and gases) and solids as continuous distributions of matter, hence the term *continuum mechanics*. In the continuum model, properties of matter are averaged, or smoothed, over space instead of considering the huge collection of molecules that constitute the system. Recent research in mechanics has attempted to bridge the gap between the description of matter at the various length scales of engineering interest ranging from the atomic to microscopic to mesoscopic to macroscopic.

The material in this chapter illustrates why group theory, functions on the group of rotations, averaging over rotations, and diffusion on the rotation group have found applications in mechanics. We begin in Section 18.1 with a review of continuum mechanics applicable to both fluids and solids. In Section 18.2 we illustrate how the strength of polycrystalline solids and composites can be described as an orientational average of various constituents in a material. In Section 18.3 we show how material properties of a polycrystalline material can be found by averaging single crystal properties over orientations. In Section 18.4 we illustrate how the concept of convolution on groups enters in materials science. In Section 18.5 we show how symmetries in solids, such as the crystallographic point groups, can be used to reduce the complexity of the search for analytical relationships between stress and strain in solid materials. Section 18.6 points to literature on the mechanics of liquid crystals, and analogous descriptions of the dynamics of short fiber strands and ellipsoidal particles suspended in a flowing viscous liquid.

18.1 Review of Basic Continuum Mechanics

In this section we review the basic mechanics of solids and fluids. At the core of both discussions is the description of the relationship between external and internal forces (in the form of surface tractions and internal stresses) and the resulting deformation of a continuum. In the following sections we review the mathematical tools required to describe the interaction between forces and motions of solids and fluids.

18.1.1   Tensors and How They Transform

Continuum mechanics is based on the relationship between various Cartesian *tensors* describing deformation and force. A tensorial quantity is one which transforms in a prescribed way under rotations. A zeroth order tensor is simply a scalar quantity and has the same value under all changes in orientation. A first-order tensor is one which has components that transform as

$$x_i = \sum_{j=1}^{3} Q_{ij} x_j' \quad i, j = 1, 2, 3 \tag{18.1}$$

where Q_{ij} are the elements of a rotation matrix that transforms components defined in a rotated (body-fixed) frame to components defined in a space-fixed frame. We note that the convention in mechanics for the use of primes and how the rotation matrix is defined is different from that in rigid-body kinematics. In Chapter 5 we would write

$$\mathbf{x}_L = Q\mathbf{x}_R \tag{18.2}$$

where the meaning of Q is the rotation matrix describing motion from space-fixed to moving coordinates as seen from the perspective of the space-fixed coordinates, $\mathbf{x}_R$ is the position of a point before motion, and $\mathbf{x}_L$ is the position of the same point after motion (both described in space-fixed coordinates). In contrast, in continuum mechanics, a vector $\mathbf{x}$ (viewed as a first-order tensor) is independent of the reference frame. However its components are x_i' when viewed in the body-fixed frame and x_i are components of absolute position. That is, these components are how $\mathbf{x}$ *appears* if we change our frame of reference relative to a space-fixed frame. This means that *in continuum mechanics one would never write* Eq. (18.2) [since in Eq. (18.2) $\mathbf{x}_L = [x_i]$ and $\mathbf{x}_R = [x_i']$ are 3×1 arrays of numbers that represent the same vector quantity]. Rather Eq. (18.1) would be used.

In continuum mechanics we would not put a prime over the vector $\mathbf{x}$, but rather only its components

$$\mathbf{x} = x_1\mathbf{e}_1 + x_2\mathbf{e}_2 + x_3\mathbf{e}_3 = x_1'\mathbf{e}_1' + x_2'\mathbf{e}_2' + x_3'\mathbf{e}_3' \tag{18.3}$$

where $\mathbf{e}_i' \triangleq Q^T \mathbf{e}_i$ and $(\mathbf{e}_i)_j = \delta_{ij}$. Then,

$$x_i' = \mathbf{e}_i' \cdot \mathbf{x} = \left(Q^T \mathbf{e}_i\right)^T \left(\sum_{j=1}^{3} x_j\mathbf{e}_j\right) = \sum_{j=1}^{3} x_j \left(\mathbf{e}_i^T Q \mathbf{e}_j\right).$$

Since $\mathbf{e}_i' \cdot \mathbf{e}_j = \mathbf{e}_i^T Q \mathbf{e}_j = Q_{ij}$, this is the same as Eq. (18.1).

Hence, a first-order tensor is a vector, and the transformation law is the description of the elements of the vector in coordinate systems with different orientations and the same origin. Usually indicial notation, where summation is assumed over repeated indices, is used to write this more compactly as $x_i = Q_{ij} x_j'$. Using indicial notation, a second-order tensor is one which transforms as

$$X_{ij} = Q_{im} Q_{jn} X_{mn}' \quad i, j, m, n \in \{1, 2, 3\}. \tag{18.4}$$

The moment of inertia tensor for a rigid body is an example of a second-order tensor, i.e., the form of the moment-of-inertia matrix depends on the frame in which it is defined. In Chapter 5 we wrote the relationship between moment of inertia matrices $I = QI_0Q^T$ where I_0 was for the body-fixed frame. In the current notation, we would write $I = [I_{ij}]$ and $I_0 = [I_{ij}']$. Angular velocity, when viewed as a vector, transforms like a vector, and when viewed as a skew-symmetric matrix in a given coordinate system, transforms between different coordinate systems like a second-order tensor. In general an nth-order tensor is one which transforms as

$$X_{i_1, i_2, \ldots, i_n} = Q_{i_1 j_1} Q_{i_2 j_2} \cdots Q_{i_n, j_n} X'_{j_1, j_2, \ldots, j_n} \qquad i_1, \ldots, i_n, j_1, \ldots j_n \in \{1, 2, 3\}. \tag{18.5}$$

A tensor is completely defined by its components given in a frame of reference at a particular orientation. That is, the 3^n numbers $X_{j_1, j_2, \ldots, j_n}$ and the frame of reference $\{e_1, e_2, e_3\}$ completely define an nth-order tensor. The tensor is written in terms of its components and the basis vectors as

$$X = X_{j_1, j_2, \ldots, j_n} e_{j_1} \otimes e_{j_2} \otimes \cdots \otimes e_{j_n}$$

for $j_1, \ldots j_n \in \{1, 2, 3\}$. Here $\otimes$ denotes the tensor product of two vectors. For a second-order tensor, we can think of $e_{j_1} \otimes e_{j_2}$ as a matrix with (i, j)th element $\delta_{j_1, i} \delta_{j, j_2}$ when $(e_i)_j = \delta_{ij}$.

We note that Eq. (18.5) is also written as

$$X_{j_1, j_2, \ldots, j_n} e_{j_1} \otimes e_{j_2} \otimes \cdots \otimes e_{j_n} = X'_{j_1, j_2, \ldots, j_n} e'_{j_1} \otimes e'_{j_2} \otimes \cdots \otimes e'_{j_n}$$

where $e'_j = Q^T e_j$.

Another notation that says the same thing is

$$X = X_{j_1, j_2, \ldots, j_n}(Q) e_{j_1}(Q) \otimes e_{j_2}(Q) \otimes \cdots \otimes e_{j_n}(Q) \tag{18.6}$$

where $e_j(Q) = Q^T e_j$ and

$$X_{i_1, i_2, \ldots, i_n}(Q) = Q_{i_1 j_1} Q_{i_2 j_2} \cdots Q_{i_n, j_n} X'_{j_1, j_2, \ldots, j_n} \tag{18.7}$$

for any $Q \in SO(3)$ and $X_{j_1, j_2, \ldots, j_n}(\mathbb{I}) = X'_{j_1, j_2, \ldots, j_n}$. That is, when $Q = \mathbb{I}$, the components of a tensor in the space-fixed and body-fixed frames coincide.

The constraint imposed by Eq. (18.7) implies that a concatenation of changes of relative reference orientation by R then A results in the tensor elements

$$\left(R(A) X_{j_1, j_2, \ldots, j_n} \right)(Q) = X_{j_1, j_2, \ldots, j_n}(QA),$$

where $R(\cdot)$ is the right regular representation. When

$$X_{j_1, j_2, \ldots, j_n}(QA) = X_{j_1, j_2, \ldots, j_n}(Q), \tag{18.8}$$

for all A in a subgroup of $SO(3)$ for all $j_1, j_2, \ldots, j_n \in \{1, 2, 3\}$, this indicates a symmetry in the physical phenomenon under investigation. For instance, for the mechanics of a crystalline material, Eq. (18.8) should hold for all A in one of the 32 crystallographic point groups.

The form of the transformation law for the nth-order tensor in Eq. (18.6) makes it clear that tensor components $X_{j_1, j_2, \ldots, j_n}(Q)$ constitute a set of 3^n scalar functions on $SO(3)$.

In the previous discussion, we have considered tensors that are constant. That is, they are not functions of position in $\mathbb{R}^3$. In fact, they are not functions of anything (even though the components of a tensor in a given basis can be considered as functions of orientation). We now consider some issues regarding tensor functions of position, or *tensor fields*. In the same way that a scalar function $f(\mathbf{x})$ can be used to describe the relative density or temperature at a point $\mathbf{x}$, tensor quantities can vary over $\mathbb{R}^3$ as well. The components of a tensor field are then of the form $F_{j_1, j_2, \ldots, j_n}(\mathbf{x}, Q)$. Each of these components can be viewed as a function on $SE(3)$.

Something that will be useful later is to know how a tensor field transforms under rigid-body motion. That is, if we have a tensor field $F(\mathbf{x})$, and we rigidly move it by $g = (\mathbf{a}, A) \in SE(3)$, we want to know what the resulting tensor field looks like in the original frame of reference. For the scalar case we have seen in previous chapters that $f(\mathbf{x})$ transforms to $f(g^{-1} \cdot \mathbf{x}) = f(A^{-1}(\mathbf{x} - \mathbf{a}))$ under the motion $g \in SE(3)$. In the case of an nth-order tensor field, a motion by $g = (\mathbf{a}, A)$ results in the transformation of tensor components

$$F_{i_1, i_2, \ldots, i_n}(\mathbf{x}) \rightarrow A_{i_1 j_1} A_{i_2 j_2} \cdots A_{i_n, j_n} F_{j_1, j_2, \ldots j_n} \left(A^{-1}(\mathbf{x} - \mathbf{a}) \right), \tag{18.9}$$

where $F_{i_1, i_2, \ldots i_n}(\mathbf{x})$ are the components of $F(\mathbf{x})$ at $Q = \mathbb{I}$.

In the particular case of a vector field $\mathbf{v}(\mathbf{x})$ expressed in the frame $(\mathbf{0}, \mathbb{I})$ as the 3×1 array with components $v_i(\mathbf{x})$ we have[1]

$$v_i(\mathbf{x}) \rightarrow A_{ij} v_j \left(A^T(\mathbf{x} - \mathbf{a}) \right), \tag{18.10}$$

and for a 2-tensor field $F(\mathbf{x})$ expressed in the frame $(\mathbf{0}, \mathbb{I})$ as the 3×3 matrix with components $F_{ij}(\mathbf{x})$ we have

$$F_{ij}(\mathbf{x}) \rightarrow A_{im} A_{jn} F_{mn} \left(A^T(\mathbf{x} - \mathbf{a}) \right). \tag{18.11}$$

The following sections examine how tensors arise in various contexts in mechanics. Section 18.1.2 addresses solid mechanics, while Section 18.1.3 addresses fluids. These sections summarize results that can be found in greater detail in classic texts on mechanics such as [69, 77, 80].

18.1.2 Mechanics of Elastic Solids

The discussion here is limited to the statics of a continuous solid which responds elastically to applied load. Within this context the only important factors are the load applied to the solid, the resulting internal stresses, the material elastic properties, and how the material deforms as a result of the state of stress in the solid.

In Chapter 5 we already introduced aspects of the kinematics of continuous solids. Any point initially with coordinates $\mathbf{X} \in \mathbb{R}^3$ is moved to the point $\mathbf{x}(\mathbf{X}, t)$ at time t. This is called a referential, material, or Lagrangian description of the solid. The time t_0 when $\mathbf{x}(\mathbf{X}, t_0) = \mathbf{X}$ designates the referential state of the material. Usually this is when the material is not subjected to loads.

The mass density per unit volume at a point $\mathbf{X}$ in the referential configuration is $\rho(\mathbf{X})$. This is a scalar field, i.e., the value of density is independent of the orientation with which it is observed. The conservation of mass is automatically satisfied as $d\rho/dt = 0$ if the deformation $\mathbf{x}(\mathbf{X}, t)$ is *admissible*. Namely, if the Jacobian matrix with elements $\partial x_i/\partial X_j$ is not singular for all values of $\mathbf{x}$ and interpenetration of material is disallowed.

The velocity and acceleration of a material particle initially at referential position $\mathbf{X}$ are

$$\mathbf{v} = \frac{\partial \mathbf{x}}{\partial t} \quad \text{and} \quad \mathbf{a} = \frac{\partial \mathbf{v}}{\partial t} = \frac{\partial^2 \mathbf{x}}{\partial t^2}.$$

These definitions are valid for any value of t.

The relationship between the position of a particle and its near neighbors, and how this relationship changes when the material deforms, is quantified by the concept of strain. The *Lagrangian strain tensor* is essentially a constant multiple of the metric tensor for a deformed volume in $\mathbb{R}^3$ which is parameterized with Lagrangian coordinates,

$$E^L(\mathbf{X}, t) = \frac{1}{2} J^T(\mathbf{X}, t) J(\mathbf{X}, t),$$

---

[1] We often use the word vector to denote arrays such as $[v_i]$, but in this context it is important to make the distinction between the (physical) vector $\mathbf{v}$ and the array of numbers that describes this vector in a particular frame of reference.

where $J = \nabla \mathbf{x}$ is the Jacobian matrix with elements $J_{ij} = \partial x_i / \partial X_j$. Hence, E^L contains all the information regarding how each infinitesimal volume is distorted under the deformation.

The *displacement*, $\mathbf{u}(\mathbf{X}, t)$, corresponding to a deformation $\mathbf{x}(\mathbf{X}, t)$ indicates how much each point moves from its original position: $\mathbf{u}(\mathbf{X}, t) = \mathbf{x}(\mathbf{X}, t) - \mathbf{X}$. The *displacement gradient* is the matrix with elements $(\nabla \mathbf{u})_{ij} = \partial u_i / \partial X_j$. This is a useful definition in the context of small deformations, because in this case the Lagrangian strain tensor is replaced by the *infinitesimal strain tensor*

$$E = \frac{1}{2}\left((\nabla \mathbf{u}) + (\nabla \mathbf{u})^T\right).$$

That is, E is a first-order approximation to E^L which becomes exact for infinitesimally small deformations.

Given a force, $\mathbf{t}$, acting on the surface of any infinitesimal volume within a solid body, the *stress tensor*, T, within the volume relates the applied surface force and the normal to the surface as

$$\mathbf{t}(\mathbf{x}, \mathbf{n}, t) = T(\mathbf{x}, t)\mathbf{n}.$$

The vector $\mathbf{t}$ is called a surface traction acting on the infinitesimal volume. The linear relationship between $\mathbf{t}$ and $\mathbf{n}$ via the stress tensor is a result of *Cauchy's stress principle*.

Performing a balance of moments on any infinitesimal volume within a solid indicates that [69, 80]

$$T = T^T$$

as the volume shrinks to zero regardless of whether or not the material is in mechanical equilibrium. This is because inertial terms (which depend on the volume of the material element) become smaller much faster than area-dependent quantities such as T.

By performing a momentum balance on the infinitesimal volume, one arrives at the equilibrium equations

$$\nabla_{\mathbf{x}} \cdot T + \rho \mathbf{b} = \rho \mathbf{a}, \tag{18.12}$$

where $\rho(\mathbf{x})$ is the mass density per unit volume in the material at point $\mathbf{x}$, $\mathbf{a}$ is the acceleration, and $\mathbf{b}$ is body force per unit mass. In the case of static equilibrium $\mathbf{a} = \mathbf{0}$. For a solid undergoing small deformations,

$$\frac{\partial T_{ij}}{\partial x_j} \approx \frac{\partial T_{ij}}{\partial X_j} \quad \text{and} \quad \rho(\mathbf{x}) \approx \rho(\mathbf{X}). \tag{18.13}$$

Hence, if the stress tensor T can be related to the strain tensor E, then a partial differential equation in the displacements $u_i(\mathbf{X}, t)$ can be written with independent variables X_i and t.

In solid mechanics, the problem of finding the relationship $T = T(E)$ for particular materials (which is called a *constitutive equation*) is fundamental and significant because knowing this allows one to simulate the static and dynamic behavior of any solid material. The most common kind of constitutive law in solid mechanics is of the form

$$T_{ij} = C_{ijkl} E_{kl},$$

or equivalently,

$$E_{ij} = S_{ijkl} T_{kl},$$

where C is called the stiffness (or elasticity) tensor, and $S = C^{-1}$ is called the compliance tensor. Due to the symmetries $T_{ij} = T_{ji}$ and $E_{ij} = E_{ji}$, it follows that in general

$$C_{ijkl} = C_{jikl} = C_{ijlk}.$$

Using group theory to help make the search for constitutive laws (i.e., determining additional structure in the coefficients C_{ijkl}) more tractable is the subject of Section 18.5.

18.1.3   Fluid Mechanics

Whereas in the mechanics of solids the deformations are generally small and it is convenient to use referential descriptions, this is usually not the case in fluid mechanics. Instead, it is convenient to work in *spatial* or *Eulerian* coordinates. In this description a fluid particle is usually not tracked over time, but rather the velocity of fluid passing through the point $\mathbf{x}$ at time t is observed. The result is a velocity field $\mathbf{v}(\mathbf{x}, t)$ defined at each spatial coordinate and time. Since $\mathbf{x} = \mathbf{x}(\mathbf{X}, t)$, the observation of conservation of mass, $d\rho/dt = 0$, in the referential coordinates is no longer convenient as it is in solid mechanics. Instead, density at a particular instant in time must satisfy the equation

$$\frac{d\rho}{dt} = \frac{\partial \rho}{\partial t} + \mathbf{v} \cdot \nabla \rho = 0 \tag{18.14}$$

where $\mathbf{v}$ is the velocity of fluid passing through the spatial point $\mathbf{x}$ at time t, and $\nabla \rho$ is the gradient of ρ with respect to $\mathbf{x}$ (previously we denoted this as $\nabla_{\mathbf{x}}$). Equation 18.14, which is simply an application of the chain rule, may be viewed as a particular example of a general scalar conservation law in spatial coordinates.

The acceleration of a particle of fluid at $\mathbf{x}$ at time t is also computed using the chain rule

$$\mathbf{a}(\mathbf{x}, t) = \frac{d\mathbf{v}}{dt} = \frac{\partial \mathbf{v}}{\partial t} + (\nabla \mathbf{v})\mathbf{v}, \tag{18.15}$$

where the components of $\nabla \mathbf{v}$ are $\partial v_i / \partial x_j$.

The force balance in Eq. (18.12) is equally valid for fluid mechanics as it was in the case of solid mechanics. However, the assumptions in Eq. (18.13) are no longer valid since deformations are large. Furthermore, since we are no longer using a referential description, the acceleration must now be calculated as in Eq. (18.15).

Again, in order to write a partial differential equation which describes the fluid motion, a constitutive law is needed. In the context of fluid mechanics, laws of the form $T = T(\mathbf{v}, \nabla \mathbf{v})$ are common, the most common being the *Newtonian fluids* with

$$T = -p\mathbb{I} + \lambda \mathbb{I} \operatorname{trace}(\nabla \mathbf{v}) + \mu \left(\nabla \mathbf{v} + (\nabla \mathbf{v})^T \right),$$

where $p(\mathbf{x}, t)$ is the pressure, and λ, μ are constants specific to the fluid.

In the case of incompressible Newtonian fluids, the conservation of mass expression reduces to

$$\nabla \cdot \mathbf{v} = 0, \tag{18.16}$$

and Eq. (18.12) becomes the *Navier-Stokes* equations:

$$\rho \left(\frac{\partial \mathbf{v}}{\partial t} + (\nabla \mathbf{v})\mathbf{v} \right) = \rho \mathbf{b} - \nabla p + \mu \nabla^2 \mathbf{v}. \tag{18.17}$$

Hence, Eqs. (18.16) and (18.17) constitute four PDEs in the four variables v_1, v_2, v_3, and p.

In the special case when the density is constant throughout the fluid: $\rho(\mathbf{x}, t) = \rho_0$ and the body force $\mathbf{b}$ is conservative ($\mathbf{b} = -\nabla V$) the Navier-Stokes equations are transformed by taking the curl of both sides resulting in the form

$$\frac{\partial \boldsymbol{\xi}}{\partial t} + (\nabla \boldsymbol{\xi})\mathbf{v} = (\nabla \mathbf{v})\boldsymbol{\xi} + \nu \nabla^2 \boldsymbol{\xi} \tag{18.18}$$

where

$$\boldsymbol{\xi} = \nabla \times \mathbf{v} = \operatorname{vect}\left((\nabla \mathbf{v}) - (\nabla \mathbf{v})^T \right) \tag{18.19}$$

is called the *vorticity* and $\nu = \mu/\rho$.

In Section 18.6.1 we discuss how these equations are generalized in the literature to include the dynamics of fibers and elliptic particles embedded in a flow.

18.2 Orientational and Motion-Group Averaging in Solid Mechanics

In the first two sections that follow we describe techniques for averaging tensor properties over orientations. For these sections we follow the formulation in [68]. In the third subsection, we extend this formulation to include averages over rigid-body motion.

18.2.1 Tensor-Valued Functions of Orientation

Recall from Eq. (18.6) that every nth-order tensor quantity is defined by 3^n scalar functions on $SO(3)$ of the form in Eq. (18.7). However, a tensor quantity itself is, by definition, independent of the orientation from which it is viewed (although its components vary with changes in orientation). However, it is possible to define a tensor-valued function of orientation as

$$X(R) = X_{j_1, j_2, \ldots, j_n}(R, Q) e_{j_1}(Q) \otimes e_{j_2}(Q) \otimes \cdots \otimes e_{j_n}(Q), \tag{18.20}$$

where for each fixed R, the Q-dependence of $X_{j_1, j_2, \ldots, j_n}(R, Q)$ satisfies Eq. (18.7).

For instance, $X(R)$ may be a tensor where R denotes the orientation of the physical phenomenon relative to fixed space, and Q denotes the orientation relative to fixed space from which we observe the tensor quantity. If there is only one tensor quantity of interest, there is no reason to use both R and Q (a relative orientation of one with respect to the other will suffice). However, if many tensor quantities of the same kind occupy a localized region of space, then the average

$$\langle X \rangle \triangleq \int_{SO(3)} X(R)\rho(R)dR$$

can be a useful approximation of the average tensorial properties, where $\rho(R)$ is a PDF describing the distribution of physical quantities.

In component form we observe that

$$\langle X \rangle_{j_1, j_2, \ldots, j_n}(Q) = \int_{SO(3)} X_{j_1, j_2, \ldots, j_n}(R, Q)\rho(R)dR$$

defines a tensor whenever $X_{j_1, j_2, \ldots, j_n}(R, Q)$ does for each fixed R.

In [68] the case

$$X_{i_1, i_2, \ldots, i_n}(R, Q) = \left(QR^T\right)_{i_1 j_1} \left(QR^T\right)_{i_2 j_2} \cdots \left(QR^T\right)_{i_n, j_n} X_{j_1, j_2, \ldots, j_n}(R)$$

is considered. Using the notation

$$l_{ij}(Q) = Q_{ij} = e_i^T Q e_j = l_{ji}\left(Q^T\right)$$

to emphasize that the (i, j)th component of Q is a function of Q, and using the bi-invariance of integration on $SO(3)$, we see that [68]

$$\langle X \rangle_{i_1, i_2, \ldots, i_n}(Q) = \int_{SO(3)} l_{i_1 j_1}\left(QR^T\right) l_{i_2 j_2}\left(QR^T\right) \cdots l_{i_n, j_n}\left(QR^T\right) X_{j_1, j_2, \ldots, j_n}(R)dR$$

$$= \int_{SO(3)} l_{j_1 i_1}(R) l_{j_2 i_2}(R) \cdots l_{j_n, i_n}(R) X_{j_1, j_2, \ldots, j_n}(RQ)dR.$$

Also using the invariance of integration, and the fact that $l_{ij}(QR) = l_{ik}(Q)l_{kj}(R)$ (using summation notation), it may be shown that [68]

$$\langle X \rangle_{i_1, i_2, \ldots, i_n}(AQ) = A_{i_1 j_1} A_{i_2 j_2} \cdots A_{i_n, j_n} \langle X \rangle_{j_1, j_2, \ldots, j_n}(Q).$$

Hence $\langle X \rangle$ is a tensor.

18.2.2   Useful Formulas in Orientational Averaging of Tensor Components

When averaging components of an nth-order tensor over all orientations with $\rho(A) = 1$, it is natural to seek closed-form expressions for integrals of the form

$$I_{i_1,\dots i_n; j_1,\dots,j_n} = \int_{SO(3)} l_{i_1,j_1}(A) l_{i_2,j_2}(A) \cdots l_{i_n,j_n}(A) dA. \tag{18.21}$$

One way to approach this systematically is to rewrite Eq. (9.4) as

$$A = T_1 U^1(A) T_1^{-1},$$

and to sequentially rewrite products of rotation matrix elements in terms of sums of IUR matrix elements and Clebsch-Gordon coefficients. That is,

$$l_{i_1,j_1}(A) l_{i_2,j_2}(A) \cdots l_{i_n,j_n}(A) = \sum_{mnl} c_{i_1,j_1,\dots,i_n,j_n}^{mnl} U_{mn}^l,$$

where c_{mnl} is some combination of Clebsch-Gordon coefficients. Then the fact that

$$\int_{SO(3)} U_{mn}^l(A) dA = \delta_{l,0} \delta_{m,0} \delta_{n,0}$$

can be used to eliminate all contributions to the integral in Eq. (18.21) other than the coefficients of $U_{00}^0(A) = 1$.

While this is a systematic approach to the problem for a tensor of any order, the tensors in mechanics tend to be either of second order or fourth order. It is sufficient to evaluate

$$I_{i_1,i_2; j_1,j_2} = \int_{SO(3)} l_{i_1,j_1}(A) l_{i_2,j_2}(A) dA \tag{18.22}$$

at

$$\sum_{n=1}^{6} \sum_{m=1}^{n} 1 = \sum_{n=1}^{6} n = \frac{6 \cdot 7}{2} = 21 \tag{18.23}$$

independent values of the arguments of $I_{i_1,i_2; j_1,j_2}$.

Equation (18.23) comes from the symmetries in $I_{i_1,i_2; j_1,j_2}$. In particular, from the invariance of integration under inversion of the argument,

$$I_{i_1,i_2; j_1,j_2} = \int_{SO(3)} l_{i_1,j_1}\left(A^T\right) l_{i_2,j_2}\left(A^T\right) dA$$

$$= \int_{SO(3)} l_{j_1,i_1}(A) l_{j_2,i_2}(A) dA = I_{i_2,i_1; j_2,j_1}. \tag{18.24}$$

This is why the first sum on the left side of Eq. (18.23) is to 6 instead of 9. The second sum on the left side of Eq. (18.23) is to n instead of 6 because

$$I_{i_1,i_2; j_1,j_2} = I_{j_1,j_2; i_1,i_2}.$$

Similarly for the fourth-order case, due to symmetries there are not 9^4, but rather at most

$$\sum_{n=1}^{6} \sum_{m=1}^{n} \sum_{l=1}^{m} \sum_{k=1}^{l} 1 = \sum_{n=1}^{6} \sum_{m=1}^{n} \frac{m(m+1)}{2} = 126$$

integrals that need to be calculated.

We now enumerate some of the explicit results. Recall that ZXZ Euler angles express a double-coset decomposition of $SO(3)$, both $SO(3)/SO(2)$ and $SO(2)\backslash SO(3)$ can be identified with the unit sphere S^2, and the subgroups parameterized by $(\alpha, \beta, \gamma) = (\alpha, 0, 0)$ and $(0, 0, \gamma)$ are both isomorphic to $SO(2)$. With $G = SO(3)$ and $H = SO(2)$ we write

$$\int_{G/H} f(gH)d(gH) \triangleq \int_{S^2} f\left(\mathbf{u}'(\beta, \alpha)\right) d\mathbf{u}'(\beta, \alpha),$$

and

$$\int_{H\backslash G} f(Hg)d(Hg) \triangleq \int_{S^2} f\left(\mathbf{u}''(\beta, \gamma)\right) d\mathbf{u}''(\beta, \gamma).$$

For $SO(3)/SO(2)$ we take

$$\mathbf{u}'(\beta, \alpha) \triangleq A\mathbf{e}_3$$

and for $SO(2)\backslash SO(3)$ and we take

$$\mathbf{u}''(\beta, \gamma) \triangleq A^T\mathbf{e}_3.$$

We note that even though

$$\mathbf{u}(\theta, \phi) \neq \mathbf{u}'(\theta, \phi) \neq \mathbf{u}''(\theta, \phi),$$

it is nonetheless true that

$$d\mathbf{u}(\theta, \phi) = d\mathbf{u}'(\theta, \phi) = d\mathbf{u}''(\theta, \phi)$$

where

$$\mathbf{u}(\theta, \phi) = [\sin\theta\cos\phi, \sin\theta\sin\phi, \cos\theta]^T$$

is the standard parameterization of S^2.

A useful tool in evaluating integrals of products of rotation matrix elements over $SO(3)$ is to first determine integrals over $SO(2)$ and S^2. For instance[2] [68]

$$\frac{1}{2\pi} \int_0^{2\pi} l_{ki}(A)l_{kj}(A)d\alpha = \frac{1}{2}\mu_{ij}$$

for $k = 1, 2$ where

$$\mu_{ij} \triangleq \delta_{ij} - l_{3i}(A)l_{3j}(A)$$

and

$$\int_{S^2} l_{3i}(A)l_{3j}(A)d\mathbf{u}''(\beta, \gamma) = \frac{1}{3}\delta_{ij}$$

combine to yield

$$\int_{SO(3)} l_{ki}(A)l_{kj}(A)dA = \frac{1}{3}\delta_{ij}.$$

Similarly,

$$\frac{1}{2\pi} \int_0^{2\pi} l_{1,i}(A)l_{2,j}(A)d\alpha = -\frac{1}{2\pi} \int_0^{2\pi} l_{2,i}(A)l_{1,j}(A)d\alpha$$

$$= \frac{1}{2}\left(\epsilon_{1,i,j}l_{3,1}(A) + \epsilon_{2,i,j}l_{3,2}(A) + \epsilon_{3,i,j}l_{3,3}(A)\right)$$

---

[2]In this section repeated indices do not indicate summation.

(where ϵ_{ijk} is the alternating tensor) and the fact that

$$\int_{S^2} l_{3,i}(A) d\mathbf{u}''(\beta, \gamma) = 0$$

implies that

$$\int_{SO(3)} l_{1,i}(A) l_{2,j}(A) dA = \int_{SO(3)} l_{2,i}(A) l_{1,j}(A) dA = 0.$$

The decomposition of the $SO(3)$ integral into an integral over $SO(2)$ and $SO(3)/SO(2)$ yields a similar result. However, we only need to work with one decomposition because of the symmetry in Eq. (18.24).

For averaging over fourth-order tensor components, the following are useful [68]

$$\frac{1}{2\pi} \int_0^{2\pi} l_{mi}(A) l_{mj}(A) l_{mk}(A) l_{ml}(A) d\alpha = \frac{1}{8} \left(\mu_{ij}\mu_{kl} + \mu_{ik}\mu_{jl} + \mu_{il}\mu_{jk} \right)$$

for $m = 1, 2$;

$$\frac{1}{2\pi} \int_0^{2\pi} l_{1,i}(A) l_{2,j}(A) l_{2,k}(A) l_{2,l}(A) d\alpha = -\frac{1}{2\pi} \int_0^{2\pi} l_{2,i}(A) l_{1,j}(A) l_{1,k}(A) l_{1,l}(A) d\alpha;$$

$$\int_{S^2} l_{3,i}(A) l_{3,j}(A) l_{3,k}(A) l_{3,l}(A) d\mathbf{u}(\beta, \gamma) = \frac{1}{15} \left(\delta_{ij}\delta_{kl} + \delta_{ik}\delta_{jl} + \delta_{il}\delta_{jk} \right);$$

$$\int_{S^2} \left(l_{3,1}(A) \right)^{m_1} \left(l_{3,2}(A) \right)^{m_2} \left(l_{3,3}(A) \right)^{m_3} d\mathbf{u}''(\beta, \gamma) = 0$$

for $m_1 + m_2 + m_3 = 4$ and $m_i > 0$. Using the fact that

$$\int_{S^2} \mu_{ij}(\beta, \gamma) d\mathbf{u}''(\beta, \gamma) = \frac{2}{3}\delta_{ij},$$

the preceding equations combine to yield

$$\int_{SO(3)} l_{1,i}(A) l_{2,j}(A) l_{2,k}(A) l_{2,l}(A) dA = -\int_{SO(3)} l_{2,i}(A) l_{1,j}(A) l_{1,k}(A) l_{1,l}(A) dA$$

and

$$\int_{SO(3)} \left(l_{3,1}(A) \right)^{m_1} \left(l_{3,2}(A) \right)^{m_2} \left(l_{3,3}(A) \right)^{m_3} dA = 0$$

when $m_1 + m_2 + m_3 = 4$ and $m_i > 0$.

18.2.3 Motion-Group Averaging: Modeling-Distributed Small Cracks and Dilute Composites

Two problems for which ideas of noncommutative harmonic analysis such as integration and convolution on groups can be used in mechanics are: (1) modeling the effective properties of a solid body with a distribution of cracks with the same shape and size, and (2) modeling the effective properties of a solid body containing a dilute collection of identical inclusions.

A large body of literature in mechanics has considered the stress/strain tensor fields around cracks under various loading conditions, and the effective strength of a homogenized model that replaces the region containing the crack or inclusion with a continuum model that behaves similarly. If $F(\mathbf{x})$ is a tensor field quantity (such as effective stiffness, compliance, or elasticity) that replaces the crack or inclusion, then an approximation for the tensor field for the whole body can be written as

$$\langle F \rangle(\mathbf{x}) = \int_{SE(3)} \rho(g) F\left(g^{-1} \cdot \mathbf{x}\right) d(g) \qquad (18.25)$$

where $\rho(g)$ is the positional and orientational density of the cracks/inclusions at $g = (\mathbf{a}, A) \in SE(3)$. For instance, if $F(\mathbf{x})$ is a 2-tensor field with elements $F_{ij}(\mathbf{x})$ in the frame $(\mathbf{0}, \mathbb{I})$, then Eq. (18.25) is written in summation notation as

$$\langle F \rangle_{ij}(\mathbf{x}) = \int_{SE(3)} \rho(g) A_{ii_1} A_{jj_1} F_{i_1 j_1}\left(g^{-1} \cdot \mathbf{x}\right) d(g).$$

While in our discussion we have not assumed any particular form (or order) for the tensor fields, numerous references have considered the tensor fields around inclusions and cracks. For works that address the tensor fields in cracked media and those with rigid inclusions and techniques for homogenization (averaging) see [26, 27, 33, 37, 48, 52, 56, 71, 72, 73, 99, 123, 133].

We note that this formulation also holds for non-mechanical tensor fields such as in the optical properties of liquid crystals.

18.3 The Orientational Distribution of Polycrystals

Polycrystalline materials are composed of many identical crystals oriented in different ways. When it is the case that the crystals in any small volume element in $\mathbb{R}^3$ are evenly distributed in their orientation, the orientational distribution function (ODF) is assumed to be constant: $\rho(A) = 1$, and the overall material is treated as isotropic.

However, when $\rho(A) \neq 1$, the question arises as to how the mechanical properties of each crystal contribute to those properties of the bulk material. If each crystal in the aggregate has a rotational symmetry group $G_C < SO(3)$, then the ODF will satisfy

$$\rho(A) = \rho(AB)$$

for all $B \in G_C$. In contrast, it is also possible for the material sample to have symmetry, in which case

$$\rho(A) = \rho\left(B'A\right)$$

for all $B' \in G_S < SO(3)$. The symmetry groups G_C and G_S are respectively said to reflect crystal and sample symmetry [17]. For a deeper discussion of ODF symmetries see [6, 19, 42]. Our notation of using relative (as opposed to absolute) rotations gives the previous expressions a different form from those found in [17].

The analysis of texture [i.e., the form of the function $\rho(A)$] is used in the context of geology (see [29, 54, 95, 105, 124]) as well as in quantitative metallurgy (see [8, 20, 66, 82, 113, 126, 131]).

In addition to the orientation distribution function $\rho(A)$, it has been speculated that a second function, the *misorientation distribution function* (MODF), plays a role in the material properties of crystalline aggregates [24, 43, 97, 98, 142]. The use of ODFs to describe polycrystalline materials can also be found in [1, 2, 25].

Whereas the ODF describes the distribution of orientations of crystals in the aggregate, the MDOF describes the distribution of differences in orientation between adjacent crystals. That is, if one crystal has orientation A_1 and one of its neighbors has orientation A_2, then instead of summing the ODFs $\delta(A_i^{-1}A)$ over all crystals enumerated by i, the MODF sums $\delta(B_j^{-1}A)$ over all shared planes between crystals. B_j is a relative orientation such as $A_1^{-1}A_2$.

In Section 18.3.1 we examine the forward problem of approximating bulk properties from given single-crystal properties. In Section 18.3.2 we examine the inverse problem of approximating single-crystal properties from experimentally measured bulk material properties and a known orientation distribution.

18.3.1   Averaging Single-Crystal Properties over Orientations

A natural assumption is that the tensor components of the bulk material correspond to those of the crystal averaged over all orientations. However, the question immediately arises as to what tensor properties are to be averaged. That is, we must determine if it makes sense to average stiffness

$$\langle C \rangle_{ijkl} = C_{i_1,j_1,k_1,l_1} \int_{SO(3)} \rho(A) l_{i,i_1}(A) l_{j,j_1}(A) l_{k,k_1}(A) l_{l,l_1}(A) dA, \qquad (18.26)$$

or to average compliance

$$\langle S \rangle_{ijkl} = S_{i_1,j_1,k_1,l_1} \int_{SO(3)} \rho(A) l_{i,i_1}(A) l_{j,j_1}(A) l_{k,k_1}(A) l_{l,l_1}(A) dA. \qquad (18.27)$$

We note that while for a single crystal $S = C^{-1}$ and for the bulk material $S_{bulk} = C_{bulk}^{-1}$, in general

$$\langle S \rangle \neq \langle C \rangle^{-1}.$$

Hence, the computations in Eqs. (18.26) or (18.27) constitute a choice. If we choose to average stiffness as in Eq. (18.26) this is called a *Voigt* model [130], and if we choose to average compliance as in Eq. (18.27) this is called a *Reuss* model [103]).

Another approximation due to Hill [44] is to set the bulk properties as

$$C_{Hill} = \frac{1}{2}\left(\langle C \rangle + \langle S \rangle^{-1}\right); \qquad S_{Hill} = \frac{1}{2}\left(\langle S \rangle + \langle C \rangle^{-1}\right).$$

Apparently $C_{bulk} \approx C_{Hill}$ and $S_{bulk} \approx S_{Hill}$ can be useful approximations, even though mathematically $(C_{Hill})^{-1} \neq S_{Hill}$ and $(S_{Hill})^{-1} \neq C_{Hill}$.

In the remainder of this section we review another technique that others have proposed to circumvent this problem. In particular, we follow [4, 70, 87, 88, 90] and Chapter 7 of [61].

In general, C_{ijkl} can be decomposed as [70]

$$C_{ijkl} = C_{I,J} B_{ij}^{(I)} B_{kl}^{(J)}$$

where

$$B^{(1)} = \begin{pmatrix} 1 & 0 & 0 \\ 0 & 0 & 0 \\ 0 & 0 & 0 \end{pmatrix}; \qquad B^{(2)} = \begin{pmatrix} 0 & 0 & 0 \\ 0 & 1 & 0 \\ 0 & 0 & 0 \end{pmatrix}; \qquad B^{(3)} = \begin{pmatrix} 0 & 0 & 0 \\ 0 & 0 & 0 \\ 0 & 0 & 1 \end{pmatrix};$$

$$B^{(4)} = \frac{1}{\sqrt{2}}\begin{pmatrix} 0 & 1 & 0 \\ 1 & 0 & 0 \\ 0 & 0 & 0 \end{pmatrix}; \qquad B^{(5)} = \frac{1}{\sqrt{2}}\begin{pmatrix} 0 & 0 & 1 \\ 0 & 0 & 0 \\ 1 & 0 & 0 \end{pmatrix}; \qquad B^{(6)} = \frac{1}{\sqrt{2}}\begin{pmatrix} 0 & 0 & 0 \\ 0 & 0 & 1 \\ 0 & 1 & 0 \end{pmatrix}.$$

The matrices $B^{(I)}$ form an orthogonal basis for the set of all 3×3 symmetric matrices, in the sense that

$$\text{trace}\left(B^{(I)} B^{(J)}\right) = \delta_{IJ}$$

and so we can define the Ith element of a 6×1 array $\mathbf{X}$ as trace(XB^I) for any 2-tensor X. Likewise, a 6×6 stiffness matrix is defined as

$$C_{IJ} = B_{ij}^{(I)} C_{ijkl} B_{kl}^{(J)}.$$

This means that the relationship $T_{ij} = C_{ijkl} E_{kl}$ can be rewritten as the expression

$$T_I = C_{IJ} E_J.$$

Assume, as in [70], that in addition to those symmetries which are induced by the symmetries of T and E, that $C_{ijkl} = C_{klij}$ (so that the matrix C_{IJ} is symmetric). Then it is possible to diagonalize C_{IJ} by a proper change of basis. Let $\{X^{(I)}\}$ for $I = 1, 2, \ldots 6$ denote the basis for which $C_{IJ} = C^{(I)} \delta_{IJ}$ (no sum). That is,

$$C_{ijkl} X_{kl}^{(I)} = C^{(I)} X_{ij}^{(I)} \quad \text{(no sum on } I).$$

Then it is possible to write [61]

$$C_{ijkl} = \sum_{I=1}^{6} C^{(I)} X_{ij}^{(I)} X_{kl}^{(I)}. \tag{18.28}$$

Since $C^{(I)} > 0$ for all $I = 1, 2, \ldots 6$, it makes sense to define the logarithm of C component-by-component as

$$(\log C)_{ijkl} = \sum_{I=1}^{6} \log \left(C^{(I)}\right) X_{ij}^{(I)} X_{kl}^{(I)}.$$

In this way,

$$C = \exp(\log C).$$

The components of the average of $\log C$ over all orientations are then

$$\langle \log C \rangle_{ijkl} = \sum_{i=1}^{6} \log \left(C^{(I)}\right) X_{i_1, j_1}^{(I)} X_{k_1, l_1}^{(I)} \int_{SO(3)} \rho(A) l_{i,i_1}(A) l_{j,j_1}(A) l_{k,k_1}(A) l_{l,l_1}(A) dA.$$

An assumption that has been made in the literature for mathematical convenience is that the logarithm of the bulk stiffness tensor is approximately equal to the average of the logarithms of each crystal stiffness tensor [4, 61]

$$\tilde{C} \overset{\triangle}{=} \exp(\langle \log C \rangle).$$

Here $\tilde{C}$ is an approximation to C_{bulk}. In this way,

$$\left(\tilde{C}\right)^{-1} = \exp(-\langle \log C \rangle) = \exp\left(\left\langle \log C^{-1} \right\rangle\right) = \exp(\langle \log S \rangle) \overset{\triangle}{=} \tilde{S}.$$

While this is a mathematical assumption that allows one to average single-crystal properties over all orientations to yield the approximations $C_{bulk} \approx \tilde{C}$ and $S_{bulk} \approx \tilde{S}$ for a bulk polycrystalline material, and it has the desired property that $(\tilde{C})^{-1} = \tilde{S}$ (which is consistent with the requirement that $C_{bulk}^{-1} = S_{bulk}$), this is a purely mathematical construction. The only way to determine if this construction (or the whole hypothesis of orientational averaging) is valid is through experimentation. See [61] for discussions.

For other approaches to the calculation of mean (bulk) material properties (and bounds on these properties) using the geometric mean and other averaging techniques see [11, 12, 38, 57, 58, 63, 64, 72, 81, 83, 87, 91, 137].

18.3.2   Determining Single Crystal Strength Properties from Bulk Measurements of Polycrystalline Materials

In this section we examine the problem of finding the elasticity and compliance tensor components C_{ijkl} and S_{ijkl} when the ODF is known and either $\langle C \rangle_{ijkl}$ or $\langle S \rangle_{ijkl}$ are measured from experiment. Let

$$M_{ii'jj'kk'll'} = \int_{SO(3)} \rho(A)l_{i,i'}(A)l_{j,j'}(A)l_{k,k'}(A)l_{l,l'}(A)dA$$

be the eighth-order tensor that is known when the ODF $\rho(A)$ is known. In principle, if the Voigt model is used, the stiffness tensor is found from the corresponding bulk property by the simple inversion

$$C_{ijkl} = \langle C \rangle_{i'j'k'l'}M_{ii'jj'kk'll'}^{-1}.$$

If the Reuss model is used, the compliance tensor is found as

$$S_{ijkl} = \langle S \rangle_{i'j'k'l'}M_{ii'jj'kk'll'}^{-1}.$$

More complicated calculations are involved in inverting the Hill and geometric mean approaches. For approaches to solving for the single crystal properties from bulk measurements see [74, 139].

18.4   Convolution Equations in Texture Analysis

In Section 18.3, it is assumed that the orientational distribution is known and is a purely kinematic quantity, i.e., it does not explicitly depend on the magnitude of stress or strain in the material. In this section we examine two equations from the field of *texture analysis* where the ODF is either unknown or varies under load.[3] For developments in the field of texture analysis see [17, 34, 43, 54, 59, 61, 79, 86, 104, 108, 109, 119, 122, 129, 134, 135, 136, 141].

The two equations we examine respectively address the problem of determining the orientation distribution function from experimental measurements and modeling how the ODF changes under applied loads. Both can be formulated as convolutions on $SO(3)$.

Determination of the ODF from Experimental Measurements

In practice, the orientation distribution function in a material sample is not known prior to performing crystallographic experiments. These experiments use techniques in which the full orientations of crystals are not determined directly. Rather, a distribution of crystal axes is determined. This is called a *pole figure,* and can be represented as a function on the sphere S^2 for each orientation of the experiment. A subfield concerned with the reconstruction of structure from pole figures is called *stereology* (see [107, 125] for more on this topic).

By rotating the sample of material and taking an infinite number of projections, the ODF can be reconstructed from the distribution of crystal axes. In a sense, this is like a Radon transform for the rotation group. Mathematically, we can write the equation relating the orientational distribution $\rho(A)$ to experimental measurements as

$$\int_{SO(3)} \rho(A)\delta\left(\mathbf{y}^T A\mathbf{h} - 1\right)dA = p(\mathbf{h}, \mathbf{y}) \tag{18.29}$$

---

[3]This refers to the area of materials science concerned with modeling the bulk properties of crystal aggregates based on individual crystal properties.

where $\mathbf{h} \in S^2$ represents an axis in the crystal frame, $\mathbf{y} \in S^2$ represents an axis in the sample frame, and these two frames are related by the rotation A. We take the sample frame to be fixed and A to denote the orientation of the crystal frame relative to it. If the opposite convention is used, then Eq. (18.29) is changed by replacing $\rho(A)$ with $\tilde{\rho}(A) = \rho(A^{-1})$.

When a projection is taken with $\mathbf{h}$ fixed, the resulting function $p_\mathbf{h}(\mathbf{y}) \overset{\triangle}{=} p(\mathbf{h}, \mathbf{y})$ is a pole figure. When a projection is taken with $\mathbf{y}$ fixed, $p_\mathbf{h}(\mathbf{y}) \overset{\triangle}{=} p(\mathbf{h}, \mathbf{y})$ is called an *inverse pole figure*.

Usually in materials science Eq. (18.29) is not written in the form we have given, though it is equivalent (see [17], [61]). The issue of approximating the orientation distribution function from a finite number of pole figures has been addressed extensively in the literature [49, 50, 84, 85, 92, 115]. For other techniques to do pole figure inversion and ODF determination see [14, 35, 62].

We mention this problem here because by defining

$$\Delta_\mathbf{h}(\mathbf{u}) \overset{\triangle}{=} \delta\left(\mathbf{h}^T \mathbf{u} - 1\right)$$

for all $\mathbf{u} \in S^2$, we can write Eq. (18.29) as the convolution equation

$$\int_{SO(3)} \rho(A)\Delta_\mathbf{h}\left(A^T\mathbf{y}\right) dA = \rho * \Delta_\mathbf{h} = p_\mathbf{h}(\mathbf{y}).$$

The determination of $\rho(A)$ from a single equation like the preceding one is not possible [as we would expect from the fact that the $SO(3)$-Fourier transform matrix of the function $\Delta_\mathbf{h}(\mathbf{u})$ is singular].

We note that Eq. (18.29) can be written in the equivalent form [17]

$$p(\mathbf{h}, \mathbf{y}) = \frac{1}{2\pi} \int_0^{2\pi} \rho\left(R(\mathbf{e}_3, \mathbf{y}) \, \text{ROT}\,[\mathbf{e}_3, \theta]\, R(\mathbf{h}, \mathbf{e}_3)\right) d\theta$$

where $R(\mathbf{a}, \mathbf{b})$ is the rotation matrix in Eq. (5.24) that takes the vector $\mathbf{a}$ into $\mathbf{b}$. Expansion of $\rho(A)$ in a Fourier series on $SO(3)$ and using the property

$$U^l\left(R(\mathbf{e}_3, \mathbf{y}) \, \text{ROT}\,[\mathbf{e}_3, \theta]\, R(\mathbf{h}, \mathbf{e}_3)\right) = U^l\left(R(\mathbf{e}_3, \mathbf{y})\right) U^l\left(\text{ROT}\,[\mathbf{e}_3, \theta]\right) U^l\left(R(\mathbf{h}, \mathbf{e}_3)\right)$$

allows one to integrate out the θ in closed form and reduces the pole inversion problem to the solution of a set of linear algebraic equations. Sample and/or crystal symmetries are useful in reducing the dimension of the system of equations. A problem that has been observed is that the solution obtained by inverting pole figures can have negative values. This is called the *ghost* problem [78, 127, 128].

Modern experimental techniques exist that can more directly determine the orientation of single crystals (grains) in an aggregate polycrystalline material. These methods of imaging and orientation mapping are based on the measurement of Kikuchi patterns. See [3, 10, 30, 55, 76, 89, 110, 138] for descriptions of experimental methods.

Texture Transformation/Evolution

The ODF of a polycrystalline material, $f_1(A)$, can be changed by various mechanical processes to result in an ODF $f_2(A)$ (see [17, Chapter 8]). The change $f_1(A) \to f_2(A)$ is called texture transformation (or texture evolution). It has been postulated [17] that the process of texture transformation can be posed as the convolution

$$(w * f_1)(A) = f_2(A) \tag{18.30}$$

where, here, convolution is on the group $SO(3)$, and the function $w(A)$ is characteristic of the particular transformation process (e.g., phase transformation, recrystallization, etc.).

In this context, the three problems of interest that center around Eq. (18.30) as stated in [17] are (1) finding $f_2(A)$ for given $f_1(A)$ and $w(A)$, (2) finding the original $f_1(A)$ for given $f_2(A)$ and $w(A)$ and (3) finding $w(A)$ for given $f_2(A)$ and $f_1(A)$. The first of these problems is a convolution equation [which can be calculated efficiently numerically when $w(A)$ and $f_1(A)$ are band-limited functions]. The other two problems are inverse problems which generally require regularization techniques to solve. The issues in the regularization of convolution equations on $SO(3)$ are similar to those for $SE(3)$ (as described in Chapters 12 and 13).

For more on the idea of texture transformation, see [21, 22, 23, 53, 75, 106].

18.5   Constitutive Laws in Solid Mechanics

In this section we review a few of the simplest models for the relationship between stress and strain in solids. When the material is isotropic (has the same properties when a given point is viewed from all different orientations), this imposes severe restrictions on how stress and strain are related. Also, when the material is anisotropic, but has certain symmetries, this also imposes constraints that must be taken into account in the constitutive equations. Our presentation in this section is a summary of material that can be found in [7, 69, 114]

A constitutive law is a relationship between stress and strain. In solid mechanics, this is essentially a generalization of Hooke's law

$$F = \varphi(x) = kx \tag{18.31}$$

where F is the force applied to a spring and x is the amount of displacement or elongation of the spring. Hooke's law is a linear relationship, with the stiffness k serving as the constant of proportionality. However, for large displacements this model breaks down, and taking $\varphi(x) = kx + \epsilon x^3$ extends the range of applicability of the model for larger deformations. When $\epsilon > 0$ the spring is called *stiffening*, and for $\epsilon < 0$ it is called *softening*.

18.5.1   Isotropic Elastic Materials

For a three-dimensional element of material, a constitutive law relates stress and strain. The simplest kind of deformation is an elastic one

$$T = \varphi(E). \tag{18.32}$$

An elastic deformation is necessarily reversible, and does not depend on the history of strain states prior to the current time. The simplest of the elastic deformations is the linearly elastic model [which is a 3D analog of the 1D version of Eq. (18.31)]. This is written in indicial notation as

$$T_{ij} = C_{ijkl} E_{kl}, \tag{18.33}$$

where C_{ijkl} are the components of a fourth-order tensor called the *stiffness (or elasticity) tensor*. One can also write

$$E_{ij} = S_{ijkl} T_{kl},$$

where S_{ijkl} are components of the *compliance tensor* $S = C^{-1}$. In fact, we have seen these expressions earlier in this chapter. The tensorial nature of C with components C_{ijkl} follows from the fact that T and E are both second-order tensors, i.e., $T_{ij}(Q) = C_{ijkl}(Q)E_{kl}(Q)$ implies

$$Q_{i,i_1} Q_{j,j_1} T'_{i_1,j_1} = C_{ijkl}(Q) Q_{k,k_1} Q_{l,l_1} E'_{k_1,l_1}.$$

Then for Eq. (18.33) to hold in the primed coordinate system, $T'_{ij} = C'_{ijkl} E'_{kl}$, it must be that

$$Q_{i,i_1} Q_{j,j_1} C'_{i_1,j_1,k,l} E'_{kl} = C_{ijkl}(Q) Q_{k,k_1} Q_{l,l_1} E'_{k_1,l_1}.$$

Multiplying both sides by $Q_{i,m} Q_{j,n}$ and summing over i and j, yields

$$\delta_{i_1,m} \delta_{j_1,n} C'_{i_1,j_1,k,l} E'_{kl} = Q_{i,m} Q_{j,n} C_{ijkl}(Q) Q_{k,k_1} Q_{l,l_1} E'_{k_1,l_1}.$$

Using the localization argument that this must hold for all possible values of E'_{ij} (see [69]), and changing the indices of summation on the left from k, l to k_1, l_1, we find

$$C'_{m,n,k_1,l_1} = Q_{i,m} Q_{j,n} Q_{k,k_1} Q_{l,l_1} C_{ijkl}(Q).$$

This can also be written in the form

$$C_{ijkl}(Q) = Q_{i,i_1} Q_{j,j_1} Q_{k,k_1} Q_{l,l_1} C'_{i_1,j_1,k_1,l_1}.$$

This verifies that $C_{ijkl}(Q)$ transform as the components of a fourth-order tensor. A similar calculation shows that $S_{ijkl}(Q)$ transform as tensor components.

At first glance, it may appear that C_{ijkl} for $i, j, k, l \in \{1, 2, 3\}$ constitute a set of $3^4 = 81$ independent parameters. However, C_{ijkl} inherits symmetries from those of T and E. That is, $T_{ij} = T_{ji}$ and $E_{kl} = E_{lk}$ imply that

$$C_{ijkl} = C_{jikl} = C_{ijlk}. \tag{18.34}$$

Hence there are only 36 free parameters (6 for all i, j and each fixed k, l, and vice versa).

Further symmetry relations exist when the material has symmetry group $G \le SO(3)$ because then for all $A \in G$

$$C_{ijkl}(A) = \acute{C}_{ijkl}.$$

The most restrictive case is $G = SO(3)$. In this case, it may be shown (see [69, 114]) that

$$C_{ijkl} = \lambda \delta_{ij} \delta_{kl} + 2\mu \delta_{ik} \delta_{jl} \tag{18.35}$$

is the most general isotropic fourth-order tensor which also observes the symmetries Eq. (18.34). The two free constants λ and μ are called the Lamé constants. The corresponding elements of the compliance tensor are written as

$$S_{ijkl} = -\frac{\lambda}{2\mu} \delta_{ij} \delta_{kl} + \frac{1}{2\mu} \delta_{ik} \delta_{jl}. \tag{18.36}$$

For an isotropic linearly elastic solid, it is common to write the constitutive law with stiffness in Eq. (18.35) as [69, 77, 80]

$$\begin{pmatrix} T_{11} \\ T_{22} \\ T_{33} \\ T_{12} \\ T_{13} \\ T_{23} \end{pmatrix} = \begin{pmatrix} \lambda + 2\mu & \lambda & \lambda & 0 & 0 & 0 \\ \lambda & \lambda + 2\mu & \lambda & 0 & 0 & 0 \\ \lambda & \lambda & \lambda + 2\mu & 0 & 0 & 0 \\ 0 & 0 & 0 & \mu & 0 & 0 \\ 0 & 0 & 0 & 0 & \mu & 0 \\ 0 & 0 & 0 & 0 & 0 & \mu \end{pmatrix} \begin{pmatrix} E_{11} \\ E_{22} \\ E_{33} \\ 2E_{12} \\ 2E_{13} \\ 2E_{23} \end{pmatrix}. \tag{18.37}$$

The reason for the factor of 2 in front of the E_{ij} terms [as opposed to multiplying μ in the $(4, 4)$, $(5, 5)$, and $(6, 6)$ elements in the preceding matrix] is explained in the next section.

More generally, a constitutive law (which is elastic but not necessarily linear) must also preserve the fact that T and E are both tensors. This imposes the condition on Eq. (18.32) that [114]

$$Q_{i,i_1} Q_{j,j_1} \varphi_{i_1,j_1}(E_{kl}) = \varphi_{ij}(Q_{k,k_1} Q_{l,l_1} E_{k_1,l_1}).$$

18.5.2   Materials with Crystalline Symmetry

In the case when the symmetry group of a material is a finite subgroup $G < SO(3)$, the 36 independent constants that define the tensor elements C_{ijkl} do not reduce to only the two Lamé constants. The construction of C_{ijkl} for materials with various crystallographic symmetries can be found in [7, 114], and references therein.

Using the symmetries of T and E, and writing the equation $T_{ij} = C_{ijkl}E_{kl}$ in the matrix form

$$
\begin{pmatrix} T_{11} \\ T_{22} \\ T_{33} \\ T_{12} \\ T_{13} \\ T_{23} \end{pmatrix} = \begin{pmatrix} C_{1111} & C_{1122} & C_{1133} & C_{1112} & C_{1113} & C_{1123} \\ C_{2211} & C_{2222} & C_{2233} & C_{2212} & C_{2213} & C_{2223} \\ C_{3311} & C_{3322} & C_{3333} & C_{3312} & C_{3313} & C_{3323} \\ C_{1211} & C_{1222} & C_{1233} & C_{1212} & C_{1213} & C_{1223} \\ C_{1311} & C_{1322} & C_{1333} & C_{1312} & C_{1313} & C_{1323} \\ C_{2311} & C_{2322} & C_{2333} & C_{2312} & C_{2313} & C_{2323} \end{pmatrix} \begin{pmatrix} E_{11} \\ E_{22} \\ E_{33} \\ 2E_{12} \\ 2E_{13} \\ 2E_{23} \end{pmatrix}, \qquad (18.38)
$$

the problem of characterizing material properties when symmetries exist is the same as determining the structure of the 6×6 stiffness matrix above.[4] We will write Eq. (18.38) as

$$
\mathbf{T} = \hat{C}\mathbf{E}
$$

to distinguish it from Eq. (18.33).

Perhaps the most straightforward way to construct elasticity tensors for materials with symmetry is to enforce the equality

$$
C_{ijkl} = A_{i,i_1} A_{j,j_1} A_{k,k_1} A_{l,l_1} C_{i_1,j_1,k_1,l_1}
$$

for all $A \in G$. This results in a degenerate set of $36 \cdot |G|$ simultaneous equations which places a restriction on the form of the coefficients C_{ijkl}.

While this works for a finite group G, one cannot enumerate all the elements of a Lie group such as $SO(3)$.

Another way is to start with an arbitrary set of 36 elasticity constants C_{ijkl}, average over all elements of G by defining

$$
\langle C \rangle_{ijkl} = \frac{1}{|G|} \sum_{A \in G} A_{i,i_1} A_{j,j_1} A_{k,k_1} A_{l,l_1} C_{i_1,j_1,k_1,l_1},
$$

and determine the form of each $\langle C \rangle_{ijkl}$ as a function of all the C_{i_1,j_1,k_1,l_1}.

Clearly, for any $Q \in G$

$$
\begin{aligned}
Q_{i_2,i} Q_{j_2,j} Q_{k_2,k} Q_{l_2,l} \langle C \rangle_{ijkl} &= \frac{1}{|G|} \sum_{A \in G} Q_{i_2,i} Q_{j_2,j} Q_{k_2,k} Q_{l_2,l} A_{i,i_1} A_{j,j_1} A_{k,k_1} A_{l,l_1} C_{i_1,j_1,k_1,l_1} \\
&= \frac{1}{|G|} \sum_{A \in G} (QA)_{i_2,i_1} (QA)_{j_2,j_1} (QA)_{k_2,k_1} (QA)_{l_2,l_1} C_{i_1,j_1,k_1,l_1} \\
&= \langle C \rangle_{i_2,j_2,k_2,l_2}.
\end{aligned}
$$

This is because the sum of any function on a finite group over all group elements is invariant under shifts (See Chapters 7 and 8).

---

[4]Note the 2's in Eq (18.38) are required to make this matrix equation the same as Eq. (18.33) since E_{21}, E_{31}, E_{32} do not explicitly appear.

The same approach can be used for continuous symmetries, such as the isotropic case, with the summation being replaced with integration over the group. That is, we can calculate

$$\langle C \rangle_{ijkl} = \int_{SO(3)} A_{i,i_1} A_{j,j_1} A_{k,k_1} A_{l,l_1} C_{i_1,j_1,k_1,l_1} dA$$

and identify how many repeated sets of parameters from C_{i_1,j_1,k_1,l_1} are present in $\langle C \rangle_{ijkl}$.

While this method can be used in principle, it too can be quite tedious when performing calculations by hand.

Some special cases of importance are [80]

$$\hat{C} = \hat{C}^T,$$

in which case there are 21 rather than 36 free parameters;

$$\hat{C} = \begin{pmatrix}
C_{1111} & C_{1122} & C_{1133} & 0 & 0 & C_{1123} \\
C_{2211} & C_{2222} & C_{2233} & 0 & 0 & C_{2223} \\
C_{3311} & C_{3322} & C_{3333} & 0 & 0 & C_{3323} \\
0 & 0 & 0 & C_{1212} & C_{1213} & 0 \\
0 & 0 & 0 & C_{1312} & C_{1313} & 0 \\
C_{2311} & C_{2322} & C_{2333} & 0 & 0 & C_{2323}
\end{pmatrix},$$

which indicates a plane of elastic symmetry;

$$\hat{C} = \begin{pmatrix}
C_{1111} & C_{1122} & C_{1133} & 0 & 0 & 0 \\
C_{2211} & C_{2222} & C_{2233} & 0 & 0 & 0 \\
C_{3311} & C_{3322} & C_{3333} & 0 & 0 & 0 \\
0 & 0 & 0 & C_{1212} & 0 & 0 \\
0 & 0 & 0 & 0 & C_{1313} & 0 \\
0 & 0 & 0 & 0 & 0 & C_{2323}
\end{pmatrix},$$

which indicates orthotropic (three orthogonal planes of) symmetry.

18.6 Orientational Distribution Functions for Non-Solid Media

Our discussion in this chapter primarily addresses the application of techniques from noncommutative harmonic analysis in the mechanics of solids and quantitative materials science. In this short section we provide pointers to the literature for the reader with an interest in fluid mechanics and rheology.

18.6.1 Orientational Dynamics of Fibers in Fluid Suspensions

The problem of the evolution of orientations of fibers and rigid particles in fluid has been addressed extensively. See, for example, [9, 16, 28, 36, 40, 41, 45, 51, 60, 100, 101, 102, 111, 112, 117, 118]. Whereas problems in solid mechanics and quantitative materials science can be described well using the ODF, in fluid mechanics the ODF is not sufficient. Rather, a model of the interaction of the ODF (which becomes a function of time as well as in the context of fluids) and the fluid surrounding the collection of particles/fibers becomes critical. Therefore, the importance of harmonic analysis of the ODF alone would seem to be less relevant in the context of fluid mechanics under low viscosity conditions.

18.6.2 Applications to Liquid Crystals

In the analysis of optical and mechanical properties of liquid crystals a number of tensor quantities arise. See [31, 32, 47]. The ODF for liquid crystals is addressed in Chapter 16. Given this ODF, orientational averaging can be performed as in solid mechanics to determine bulk properties. The assumption of high viscosity makes a non-inertial model for liquid crystals appropriate, and the limitations in the application of harmonic analysis in fluid mechanics are not as severe.

18.7 Summary

In this chapter we have examined a variety of orientation-dependent issues in the mechanics of anisotropic solids and liquids containing rigid inclusions. The idea of averaging over orientations, as well as modeling the time evolution of certain processes using convolution on $SO(3)$ both naturally arose.

In our presentation, most of the phenomena were anisotropic but spatially uniform, and so ideas from harmonic analysis on $SO(3)$ entered. For properties that vary over both orientation and spatial position, a motion-group distribution function $f(\mathbf{a}, A)$ is a natural replacement for the ODF $\rho(A)$. Integration of such a function over orientation provides the spatial variability of properties, whereas integration over the translation (position) provides an average ODF for the whole sample.

While the ODF we have considered is a function on $SO(3)$, symmetries in the physical phenomena are represented in the form of symmetries of the ODF. In the case of fibers with axial symmetry, the ODF can be viewed as a function on S^2. For example, the problem of determining effective properties of fiber-reinforced composites using orientational averaging has been considered in a number of works including [5, 13, 39, 94, 116, 120, 121, 132]. The fact that such averages are usually taken over S^2 instead of $SO(3)$ reduces the need for knowledge of noncommutative harmonic analysis in that particular context.

References

[1] Adams, B.L., Description of the intercrystalline structure distribution in polycrystalline metals, *Metall. Trans.*, 17A, 2199–2207, 1986.

[2] Adams, B.L. and Field, D.P., A statistical theory of creep in polycrystalline materials, *Acta Metall. Mater.*, 39, 2405–2417, 1991.

[3] Adams, B.L., Wright, S.I., and Kunze, K., Orientation imaging: the emergence of a new microscopy, *Metallurgical Trans. A-Phys. Metallurgy and Materials Sci.*, 24(4), 819–831, 1993.

[4] Aleksandrov, K.S. and Aisenberg, L.A., Method of calculating the physical constants of poly-crystalline materials, *Soviet Phys.-Doklady*, 11, 323–325, 1966.

[5] Alwan, J.M. and Naaman, A.E., New formulation for elastic modulus of fiber-reinforced, quasibrittle matrices, *J. of Eng. Mech.*, 120(11), 2443–2461, 1994.

[6] Baker, D.W., On the symmetry of orientation distribution in crystal aggregates, *Adv. X-ray Anal.*, 13, 435–454, 1970.

[7] Bao, G., *Application of Group and Invariant-Theoretic Methods to the Generation of Constitutive Equations*, Ph.D. Dissertation, Lehigh University, Bethlehem, PA, 1987.

[8] Barrett, C.S. and Massalski, T.B., *Structure of Metals*, 3rd ed., McGraw-Hill, New York, 1980.

[9] Batchelor, G.K., The stress generated in a non-dilute suspension of elongated particles by pure straining motion, *J. of Fluid Mech.*, 46, 813–829, 1971.

[10] Baudin, T. and Penelle, R., Determination of the total texture function from individual orientation measurements by electron backscattering pattern, *Metallurgical Trans. A—Phys. Metallurgy and Materials Sci.*, 24(10), 2299–2311, 1993.

[11] Becker, R. and Panchanadeeswaran, S., Crystal rotations represented as Rodrigues vectors, *Textures and Microstruct.*, 10, 167–194, 1989.

[12] Beran, M.J., Mason, T.A., Adams, B.L., and Olsen, T., Bounding elastic constants of an orthotropic polycrystal using measurements of the microstructure, *J. of the Mech. and Phys. of Solids*, 44(9), 1543–1563, 1996.

[13] Boutin, C., Microstructural effects in elastic composites, *Int. J. of Solids and Struct.*, 33(7), 1023–1051, 1996.

[14] Bowman, K.J., Materials concepts using mathCAD. 1. Euler angle rotations and stereographic projection, *JOM-J. of the Miner. Met. and Mater. Soc.*, 47(3), 66–68, 1995.

[15] Brady, J.F., Phillips, R.J., Lester, J.C., and Bossis, G., Dynamic simulation of hydrodynamically interacting suspensions, *J. of Fluid Mech.*, 195, 257–280, 1988.

[16] Bretherton, F.P., The motion of rigid particles in a shear flow at low Reynolds number, *J. Fluid Mech.*, 14, 284–304, 1962.

[17] Bunge, H.-J., *Texture Analysis in Materials Science*, Butterworths, London, 1982. (See also *Mathematische Methoden der Texturanalyse*, Akademie-Verlag, Berlin, 1969.)

[18] Bunge, H.J., Esling, C., Dahlem, E., and Klein, H., The development of deformation textures described by an orientation flow field, *Textures and Microstruct.*, 6(3), 181–200, 1986.

[19] Bunge, H.-J., Zur Darstellung allgemeiner Texturen, *Z. Metallkunde*, 56, 872–874, 1965. (See also *Mathematische Methoden der Texturanalyse*, Akademie-Verlag, Berlin, 1969.)

[20] Bunge, H.J. and Esling, C., *Quantitative Texture Analysis*, Dtsch. Gesell. Metallkde., Oberursel, 1982.

[21] Bunge, H.J., Humbert, M., Esling, C., and Wagner, F., Texture transformation, *Memoires et études scientifiques de la revue de metallurgie*, 80(9), 518–518, 1983.

[22] Bunge, H.J., Humbert, M., and Welch, P.I., Texture transformation, *Textures and Microstruct.*, 6(2), 81–95, 1984.

[23] Bunge, H.J., 3-dimensional texture analysis, *Int. Mater. Rev.*, 32(6), 265–291, 1987.

[24] Bunge, H.J. and Weiland, H., Orientation correlation in grain and phase boundaries, *Textures and Microstruct.*, 7(4), 231–263, 1988.

[25] Bunge, H.J., Kiewel, R., Reinert, T., and Fritsche, L., Elastic properties of polycrystals— influence of texture and stereology, *J. of the Mech. and Phys. of Solids*, 48(1), 29–66, 2000.

[26] Castaneda, P.P., The effective mechanical properties of nonlinear isotropic composites, *J. of Mech. and Phys. of Solids*, 39(1), 45–71, 1991.

[27] Castaneda, P.P. and Willis, J.R., The effect of spatial distribution on the effective behavior of composite materials and cracked media, *J. of Mech. and Phys. of Solids*, 43(12), 1919–1951, 1995.

[28] Claeys, I.L. and Brady, J.F., Suspensions of prolate spheroids in Stokes flow. Part 2. Statistically homogeneous dispersions, *J. of Fluid Mech.*, 251, 443–477, 1993.

[29] Coulomb, P., *Les Textures dans les Metaux de Reseau Cubique*, Dunood, Paris, 1982.

[30] Davies, R.K. and Randle, V., Application of crystal orientation mapping to local orientation perturbations, *Eur. Phys. J.-Appl. Phys.*, 7(1), 25–32, 1999.

[31] de Gennes, P.G. and Prost, J., *The Physics of Liquid Crystals*, 2nd ed., Oxford University Press, New York, 1998.

[32] de Jeu, W.H., *Physical Properties of Liquid Crystalline Materials*, Gordon and Breach, New York, 1980.

[33] Eshelby, J., The determination of the elastic field of an ellipsoidal inclusion, and related problems, *Proc. Roy. Soc. Lond.*, A241, 376–396, 1957.

[34] Esling, C., Humbert, M., Philippe, M.J., and Wagner, F., H.J. Bunge's cooperation with Metz: initial work and prospective development, *Texture and Anisotropy of Polycrystals*, 273(2), 15–28, 1998.

[35] Ewald, P.P., The 'poststift'—a model for the theory of pole figures, *J. Less. Common Me.*, 28, 1–5, 1972.

[36] Folgar, F. and Tucker, C.L., Orientation behavior of fibers in concentrated suspensions, *J. Reinf. Plas. Compos.*, 3, 98–119, 1984.

[37] Forest, S., Mechanics of generalized continua: construction by homogenization, *J. of Phys. IV*, 8, 39–48, 1998.

[38] Frank, F.C., The conformal neo-Eulerian orientation map, *Phyl. Mag. A.*, A65, 1141–1149, 1992.

[39] Fu, S. and Lauke, B., The elastic modulus of misaligned short-fiber-reinforced polymers, *Composites Sci. and Technol.*, 58, 389–400, 1998.

[40] Ganani, E. and Powell, R.L., Rheological properties of rodlike particles in a Newtonian and a non-Newtonian fluid, *J. of Rheology*, 30, 995–1013, 1986.

[41] Graham, A.L., Mondy, L.A., Gottlieb, M., and Powell, R.L., Rheological behavior of a suspension of randomly oriented rods, *Appl. Phys. Lett.*, 50, 127–129, 1987.

[42] Guidi, M., Adams, B.L., and Onat, E.T., Tensorial representations of the orientation distribution function in cubic polycrystals, *Textures Microstruct.*, 19, 147–167, 1992.

[43] Haessner, F., Pospiech, J., and Sztwiertnia, K., Spatial arrangement of orientations in rolled copper, *Mater. Sci. Eng.*, 57, 1–14, 1983.

[44] Hill, R., The elastic behaviour of a crystalline aggregate, *Proc. Phys. Soc. Ser. A*, Part 5, 65(389), 349–354, 1952.

[45] Hinch, E.J. and Leal, L.G., The effect of Brownian motion on the rheological properties of a suspension of non-spherical particles, *J. of Fluid Mech.*, 52, 683–712, 1972.

[46] Hinch, E.J., Averaged-equation approach to particle interactions in a fluid suspension, *J. of Fluid Mech.*, 83, 695–720, 1977.

[47] Hsu, P., Poulin, P., and Weitz, D.A., Rotational diffusion of monodisperse liquid crystal droplets, *J. of Colloid and Interface Sci.*, 200, 182–184, 1998.

[48] Hutchinson, J.W., Elastic-plastic behaviour of polycrystalline metals and composites, *Proc. Roy. Soc. London*, A 319, 247–272, 1970.

[49] Imhof, J., Die ßestimmung einer näherung für die Funktion der Orientierungsverteilung aus einer Polfigur, *Z. Metallkunde*, 68, 38–43, 1977.

[50] Imhof, J., Texture analysis by iteration I. General solution of the fundamental problem, *Physica Status Solidi B—Basic Res.*, 119(2), 693–701, 1983.

[51] Jeffery, G.B., The motion of ellipsoidal particles immersed in a viscous fluid, *Proc. Roy. Soc. Lond.*, A102, 161–179, 1923.

[52] Kailasam, M. and Ponte Castaneda, P., A general constitutive theory for linear and nonlinear particulate media with microstructure evolution, *J. of Mech. and Phys. of Solids*, 46(3), 427–465, 1998.

[53] Kallend, J.S., Davies, G.J., and Morris, P.P., Texture transformations: the misorientation distribution function, *Acta Metall.*, 24, 361–368, 1976.

[54] Kallend, J.S., Kocks, U.F., Rollett, A.D., and Wenk, H.R., Operational texture analysis, *Mater. Sci. and Eng. A-Struct. Mater. Prop. Microstruct. and Process.*, 132, 1–11, 1991.

[55] Katrakova, D., Maas, C., Hohnerlein, D., and Mucklich, F., Experiences on contrasting microstructure using orientation imaging microscopy, *Praktische Metallographie-Practical Metallography*, 35(1), 4–20, 1998.

[56] Kassir, M.K. and Sih, G.C., Three-dimensional stress distribution around an elliptical crack under arbitrary loadings, *J. Appl. Mech.*, Ser. E, 88, 601–611, 1966.

[57] Kiewel, H. and Fritsche, L., Calculation of effective elastic moduli of polycrystalline materials including non-textured samples and fiber textures, *Phys. Rev. B*, B50, 5–16, 1994.

[58] Kneer, G., Über die ßerechnung der Elastizitätsmoduln vielkristalliner Aggregate mit Textur, *Phy. Stat. Sol.*, 9, 825, 1965.

[59] Knorr, D.B., Weiland, H., and Szpunar, J.A., Applying texture analysis to materials engineering problems, *JOM-J. of the Minerals Metals and Materials Soc.*, 46(9), 32–36, 1994.

[60] Koch, D.L., A model for orientational diffusion in fiber suspensions, *Phys. of Fluids*, 7(8), 2086–2088, 1995.

[61] Kocks, U.F., Tomé, C.N., and Wenk, H.-R., *Texture and Anisotropy: Preferred Orientations in Polycrystals and their Effect on Materials Properties*, Cambridge University Press, Cambridge, 1998.

[62] Krigbaum, W.R., A refinement procedure for determining the crystallite orientation distribution function, *J. Phys. Chem.*, 74, 1108–1113, 1970.

[63] Kröner, E., Berechnung der elastischen Konstanten des Vielkristalls aus den Konstanten der Einkristalle, *Z. Phys.*, 151, 504, 1958.

[64] Kröner, E., Bounds for effective elastic moduli of disordered materials, *J. Mech. Phys. Solids*, 25, 137, 1977.

[65] Krumbein, W.C., Preferred orientation of pebbles in sedimentary deposits, *J. of Geol.*, 47, 673–706, 1939.

[66] Kudriawzew, I.P., *Textures in Metals and Alloys,* Isdatel'stwo Metallurgija, Moscow, 1965 (in Russian).

[67] Lagzdiņš A. and Tamužs, V., Tensorial representation of the orientation distribution function of internal structure for heterogeneous solids, *Math. and Mech. of Solids*, 1, 193–205, 1996.

[68] Lagzdiņš A., Tamužs, V., Teters, G., and Krēgers, A., *Orientational Averaging in Solid Mechanics*, Pitman Res. Notes in Math. Ser., 265, Longman Scientific & Technical, London, 1992.

[69] Lai, W.M., Rubin, D., and Krempl, E., *Introduction to Continuum Mechanics*, Pergamon Press, New York, 1978.

[70] Leibfried, G. and Breuer, N., *Point Defects in Metals. I: Introduction to the Theory,* Springer-Verlag, Berlin, 1978.

[71] Li, B., Effective constitutive behavior of nonlinear solids containing penny-shaped cracks, *Int. J. of Plasticity,* 10(4), 405–429, 1994.

[72] Li, G. and Castaneda, P.P., The effect on particle shape and stiffness on the constitutive behavior of metal-matrix composites, *Int. J. of Solids and Struct.*, 30(23), 3189–3209, 1993.

[73] Li, G., Castaneda, P.P., and Douglas, A.S., Constitutive models for ductile solids reinforced by rigid spheroidal inclusions, *Mech. of Mater.*, 15, 279–300, 1993.

[74] Li, D.Y. and Szpunar, J.A., Determination of single-crystals elastic-constants from the measurement of ultrasonic velocity in the polycrystalline material, *Acta Metallurgica et Materialia,* 40(12), 3277–3283, 1992.

[75] Lindsay, R., Chapman, J.N., Craven, A.J., and McBain, D., A quantitative determination of the development of texture in thin films, *Ultramicrosc.*, 80(1), 41–50, 1999.

[76] Liu, Q., A simple and rapid method for determining orientations and misorientations of crystalline specimens in TEM, *Ultramicrosc.*, 60(1), 81–89, 1995.

[77] Love, A.E.H., *A Treatise on the Mathematical Theory of Elasticity,* 4th ed., Dover, New York, 1927.

[78] Lücke' K., Pospiech, J., Jura, J., and Hirsch, J., On the presentation of orientation distribution functions by model functions, *Z. Metallk.*, 77, 312–321, 1986.

[79] Luzin, V., Optimization of texture measurements. III. Statistical relevance of ODF represented by individual orientations, *Texture and Anisotropy of Polycrystals*, 273-275, 107–112, 1998.

[80] Malvern, L.E., *Introduction to the Mechanics of a Continuous Medium,* Prentice-Hall, Englewood Cliffs, NJ, 1969.

[81] Mason, T.A., Simulation of the variation of material tensor properties of polycrystals achieved through modification of the crystallographic texture, *Scripta Materialia*, 39(11), 1537–1543, 1998.

[82] Matthies, S., *Aktuelle Probleme der Texturanalyse,* Akad. Wiss. D.D.R., Zentralinstitut für Kernforschung, Rossendorf-Dresden, 1982.

[83] Matthies, S., Vinel, G.W., and Helming, K., *Standard Distributions in Texture Analysis: Maps for the Case of Cubic-Orthorhombic Symmetry,* Akademie Verlag, Berlin, 1987.

[84] Matthies, S., Reproducibility of the orientation distribution function of texture samples from pole figures (ghost phenomena), *Physica Status Solidi B-Basic Res.*, 92(2), K135–K138, 1979.

[85] Matthies, S., On the reproducibility of the orientation distribution function of texture samples from pole figures (VII), *Cryst. Res. Technol.*, 16, 1061–1071, 1981.

[86] Matthies, S., Helming, K., and Kunze, K., On the representation of orientation distributions by σ-sections -I. General properties of σ-sections; II. Consideration of crystal and sample symmetry, examples, *Phys. Stat. Sol. (b)*, 157, 71–83, 489–507, 1990.

[87] Matthies, S. and Humbert, M., The realization of the concept of the geometric mean for calculating physical constants of polycrystalline materials, *Phys. Stat. Sol. (b)*, 177, K47–K50, 1993.

[88] Matthies, S. and Humbert, M., On the principle of a geometric mean of even-rank symmetrical tensors for textured polycrystals, *J. of Appl. Crystallography*, Part 3, 28, 254–266, 1995.

[89] Morawiec, A., Automatic orientation determination from Kikuchi patterns, *J. of Appl. Crystallography*, Part 4, 32, 788–798, 1999.

[90] Morawiec, A., Calculation of polycrystal elastic constants from single crystal data, *Phys. Stat. Sol. (b)*, 154, 535–541, 1989.

[91] Morawiec, A., Review of deterministic methods of calculation of polycrystal elastic constants, *Textures and Microstruct.*, 22, 139, 1994.

[92] Muller, J., Esling, C., and Bunge, H.J., An inversion formula expressing the texture function in terms of angular distribution functions, *J. of Phys.*, 42, 161–165, 1981.

[93] Nikolayev, D.I., Savyolova, T.I., and Feldman, K., Approximation of the orientation distribution of grains in polycrystalline samples by means of Gaussians, *Textures and Microstruct.*, 19, 9–27, 1992.

[94] Ngollé A. and Péra, J., Microstructural based modelling of the elastic modulus of fiber reinforced cement composites, *Adv. Cement Based Mater.*, 6, 130–137, 1997.

[95] Owens, W.H., Strain modification of angular density distributions, *Tectonophys.*, 16, 249–261, 1973.

[96] Park, N.J., Bunge, H.J., Kiewel, H., and Fritsche, L., Calculation of effective elastic moduli of textured materials, *Textures and Microstruct.*, 23, 43–59, 1995.

[97] Pospiech, J., Lücke, K., and Sztwiertnia, J., Orientation distributions and orientation correlation functions for description of microstructures, *Acta Metall. Mater.*, 41, 305–321, 1993.

[98] Pospiech, J., Sztwiertnia, J., and Haessner, F., The misorientation distribution function, *Textures Microstruct.*, 6, 201–215, 1986.

[99] Pursey, H. and Cox, H.L., The correction of elasticity measurements on slightly anisotropic materials, *Philos. Mag.*, 45, 295–302, 1954.

[100] Rahnama, M. and Koch, D.L., The effect of hydrodynamic interactions on the orientation distribution in a fiber suspension subject to simple shear flow, *Phys. Fluids*, 7, 487–506, 1995.

[101] Ralambotiana, T., Blanc, R., and Chaouche, M., Viscosity scaling in suspensions of non-Brownian rodlike particles, *Phys. of Fluids*, 9(12), 3588–3594, 1997.

[102] Rao, B.N., Tang, L., and Altan, M.C., Rheological properties of non-Brownian spheroidal particle suspensions, *J. Rheology*, 38, 1–17, 1994.

[103] Reuss, A., Berechnung der Fließgrenze von Mischkristallen grund auf der Plastizitätsbedingung für Einkristalle, *Z. Angew. Math. Mech.*, 9, 49–58, 1929.

[104] Roe, R.-J., Description of crystallite orientation in polycrystalline materials III, general solution to pole figure inversion, *J. Appl. Phys.*, 36, 2024–2031, 1965.

[105] Sander, B., *An Introduction to the Study of Fabrics of Geological Bodies*, Pergamon, New York, 1970.

[106] Sargent, C.M., Texture transformation, *Scripta Metallica*, 8, 821–824, 1974.

[107] Saxl, I., *Stereology of Objects with Internal Structure*, Mater. Sci. Monogr. 50, Elsevier, Amsterdam, 1989.

[108] Schaeben, H., Parameterizations and probability distributions of orientations, *Textures Microstruct.*, 13, 51–54, 1990.

[109] Schaeben, H., A note on a generalized standard orientation distribution in PDF-component fit methods, *Textures and Microstruct.*, 23, 1–5, 1995.

[110] Schwarzer, R.A., Automated crystal lattice orientation mapping using a computer-controlled SEM, *Micron*, 28(3), 249–265, 1997.

[111] Shaqfeh, E.S.G. and Fredrickson, G.H., The hydrodynamic stress in a suspension of rods, *Phys. of Fluids*, 2, 7–24, 1990.

[112] Shaqfeh, E.S.G. and Koch, D.L., Orientational dispersion of fibers in extensional flows, *Phys. of Fluids*, 2, 1077–1093, 1990.

[113] Smallman, R.E., *Physical Metallurgy*, Pergamon, Oxford, 1970.

[114] Smith, G.F., *Constitutive Equations for Anisotropic and Isotropic Materials*, North-Holland, New York, 1994.

[115] Starkey, J., A crystallographic approach to the calculation of orientation diagrams, *Can. J. Earth Sci.*, 20, 932–952, 1983.

[116] Suquet, P. and Garajeu, M., Effective properties of porous ideally plastic or viscoplastic materials containing rigid particles, *J. of Mech. and Phys. of Solids*, 45(6), 873–902, 1997.

[117] Szeri, A.J. and Lin, J.D., A deformation tensor model of Brownian suspensions of orientable particles—the nonlinear dynamics of closure models, *J. of Non-Newtonian Fluid Mech.*, 64, 43–69, 1996.

[118] Szeri, A.J., Milliken, W.J. and Leal, L.G., Rigid particles suspended in time-dependent flows: irregular versus regular motion, disorder versus order, *J. of Fluid Mech.*, 237, 33–56, 1992.

[119] Tavard, C. and Royer, F., Indices de texture partiels des solides polycristallines et interprétation de l'approximation de Williams, *C.R. Acad. Sci. Paris*, 284, 247, 1977.

[120] Theocaris, P.S., Stavroulakis, G.E., and Panagiotopoulos, P.D., Calculation of effective transverse elastic moduli of fiber-reinforced composites by numerical homogenization, *Composites Sci. and Technol.*, 57, 573–586, 1997.

[121] Theocaris, P.S. and Stavroulakis, G.E., The homogenization method for the study of variation of Poisson's ratio in fiber composites, *Arch. of Appl. Mech.*, 68, 281–295, 1998.

[122] Tóth, L.S. and Van Houtte, P., Discretization techniques for orientation distribution functions, *Textures and Microstruct.*, 19, 229–244, 1992.

[123] Tresca, H., Mémoire sur l'écoulement des corps solides soumis à des fortes pressions, *Comptes Rendus Hébdom. Acad. Sci. Paris*, 59, 754–758, 1864.

[124] Turner, F.J. and Weiss, L.E., *Structural Analysis of Metamorphic Tectonics*, McGraw-Hill, New York, 1963.

[125] Underwood, E.E., *Quantitative Stereology*, Addison-Wesley, Reading, MA, 1970.

[126] Underwood, F.A., *Textures in Metal Sheets*, MacDonald, London, 1961.

[127] Van Houtte, P., The use of a quadratic form for the determination of non-negative texture functions, *Textures and Microstruct.*, 6, 1–20, 1983.

[128] Van Houtte, P., A method for the generation of various ghost correction algorithms—the example of the positivity method and the exponential method, *Textures and Microstruct.*, 13, 199–212, 1991.

[129] Viglin, A.S., A quantitative measure of the texture of a polycrystalline material—texture function, *It Fiz. Tverd. Tela*, 2, 2463–2476, 1960.

[130] Voigt, W., *Lehrbuch der Kristallphysik*, Teubner, Leipzig, 1928.

[131] Wassermann, G. and Grewen, J., *Texturen metallischer Werkstoffe*, Springer, Berlin, 1962.

[132] Watt, J.P., Davies, G.F., and O'Connell, R.J., Elastic properties of composite materials, *Rev. of Geophys.*, 14(4), 541–563, 1976.

[133] Weibull, W.A., A statistical theory of the strength of materials, *Proc. Roy. Swedish Inst. Eng. Research*, 151, 1939.

[134] Weiland, H., Microtexture determination and its application to materials science, *JOM-J. of the Miner. Met. and Mater. Soc.*, 46(9), 37–41, 1994.

[135] Weissenberg, K., Statistische Anisotropie in kristallinen Medien und ihre röntgenographische Bestimmung, *Ann. Phys.*, 69, 409–435, 1922.

[136] Wenk, H.-R., Ed., *Preferred Orientation in Deformed Metals and Rocks: An Introduction to Modern Texture Analysis*, Academic Press, New York, 1985.

[137] Wenk, H.-R., Matthies, S., Donovan, J., and Chateigner, D., BEARTEX: a Windows-based program system for quantitative texture analysis, *J. of Appl. Crystallography*, Part 2, 31, 262–269, 1998.

[138] Wright, S.I., Zhao, J.W., and Adams, B.L., Automated-determination of lattice orientation from electron backscattered Kikuchi diffraction patterns, *Textures and Microstruct.*, 13(2-3), 123–131, 1991.

[139] Wright, S.I., Estimation of single-crystal elastic-constants from textured polycrystal measurements, *J. of Appl. Crystallography*, Part 5, 27, 794–801, 1994.

[140] Yamane, Y., Kaneda, Y., and Dio, M., Numerical-simulation of semidilute suspensions of rodlike particles in shear-flow, *J. of Non-Newtonian Fluid Mech.*, 54, 405–421, 1994.

[141] Yashnikov, V.P. and Bunge, H.J., Group-theoretical approach to reduced orientation spaces for crystallographic textures, *Textures and Microstruct.*, 23, 201–219, 1995.

[142] Zhuo, L., Watanabe, T., and Esling, C., A theoretical approach to grain-boundary-character-distribution (GBCD) in textured polycrystalline materials, *Zeitschrift für Metallkunde*, 85(8), 554–558, 1994.

Appendix A

Computational Complexity, Matrices, and Polynomials

A.1 The Sum, Product, and Big-$\mathcal{O}$ Symbols

The symbols $\sum$ and $\prod$ are generally used to denote sums and products, respectively. Usually these symbols will have integer subscripts and superscripts

$$\sum_{i=m}^{n} a_i = a_m + a_{m+1} + \cdots + a_n$$

and

$$\prod_{i=m}^{n} a_i = a_m \cdot a_{m+1} \cdot \cdots \cdot a_n$$

where m and n are nonnegative integers and $m \leq n$.

As a special case, x raised to the power k is

$$x^k = \prod_{j=1}^{k} x_j \quad \text{for} \quad x_j = x. \tag{A.1}$$

In order to illustrate summation notation, consider the following example. An $M \times N$ *matrix* A is an array of numbers

$$A = \begin{pmatrix} a_{11} & a_{12} & \cdots & & a_{1N} \\ a_{21} & a_{22} & \cdots & & \vdots \\ \vdots & \vdots & \ddots & & a_{M-1,N} \\ a_{M1} & \cdots & a_{M,N-1} & & a_{MN} \end{pmatrix}.$$

The element in the ith row and jth column of A is denoted a_{ij} (and likewise, the elements of any matrix denoted with an upper case letter are generally written as subscripted lower case letters). Given an $N \times P$ matrix B the (i, j)th element of the product AB is defined as

$$(AB)_{ij} = \sum_{k=1}^{N} a_{ik} b_{kj}.$$

When $M = N$, the *trace* of A is defined as

$$\text{trace}(A) = \sum_{i=1}^{N} a_{ii}.$$

When a computational procedure (such as summing or multiplying n numbers, or multiplying two matrices) can be performed in $g(n)$ arithmetic operations, we say it has $\mathcal{O}(f(n))$ complexity if there exists a positive number N such that

$$|g(n)| \leq cf(n) \tag{A.2}$$

for some constant c and all $n \geq N$. The "$\mathcal{O}$" notation hides constants and lower order terms. For instance

$$10^{-10}n^2 + 10^6 n + 10^{10} = \mathcal{O}\left(n^2\right).$$

Similarly, when evaluating computational complexity, one often finds expressions like $\mathcal{O}\left(n^p(\log n)^r\right)$. The base of the logarithm need not be specified when only the order of computation is concerned since if $n = b^K$, then

$$n = \left(10^{\log_{10} b}\right)^K = 10^{K \log_{10} b},$$

and since b is a constant,

$$\mathcal{O}\left(\log_b n\right) = \mathcal{O}(K) = \mathcal{O}\left(K \log_{10} b\right) = \mathcal{O}\left(\log_{10} n\right).$$

As examples of the usefulness of this notation, consider the sums

$$\sum_{k=1}^{n} 1 = n \quad \text{and} \quad \sum_{k=1}^{n} k = \frac{n(n+1)}{2}.$$

Whereas these are easy to verify, it is often more convenient to write

$$\sum_{k=1}^{n} k^p = \mathcal{O}\left(n^{p+1}\right)$$

than to find the specific formula. (We know the above to be true because integrating x^p from 1 to n provides an approximation that grows in the same way as the sum for $p = 0, 1, 2, \ldots$.)

Often in computational procedures, loops are used to perform calculations. A common loop is to perform a sum of sums. In this regard we note the following convenient expression

$$\sum_{i=g}^{h} \left(\sum_{k=g}^{i} a_k\right) b_i = \sum_{i=g}^{h} a_i \left(\sum_{k=i}^{h} b_k\right) \tag{A.3}$$

where $g, h, i,$ and k are finite integers, and $g \leq h$.

A.2 The Complexity of Matrix Multiplication

Given $n \times n$ matrices $A = [a_{ij}]$ and $B = [b_{ij}]$, computing the product $C = AB$ by the definition

$$c_{ik} = \sum_{j=1}^{n} a_{ij} b_{jk}$$

uses n multiplications and $n-1$ additions for each fixed pair of (i, k). Doing this for all $i, k \in [1, n]$ then requires $\mathcal{O}(n^3)$ operations.

While this is the most straightforward way to compute matrix multiplications, theoretically, it is not the most efficient technique computationally. We now review a method introduced in [47] and discussed in detail in [2, 6]. Other methods for reducing the complexity of matrix multiplication can be found in [38, 39].

Strassen's algorithm [47] is based on the calculation of the product of 2×2 matrices $A = [a_{ij}]$ and $B = [b_{ij}]$ to yield $C = AB$ using the following steps. Let

$$
\begin{aligned}
d_1 &= (a_{12} - a_{22})(b_{21} + b_{22}) \\
d_2 &= (a_{11} + a_{22})(b_{11} + b_{22}) \\
d_3 &= (a_{11} - a_{21})(b_{11} + b_{12}) \\
d_4 &= (a_{11} + a_{12}) b_{22} \\
d_5 &= a_{11} (b_{12} - b_{22}) \\
d_6 &= a_{22} (b_{21} - b_{11}) \\
d_7 &= (a_{21} + a_{22}) b_{11}.
\end{aligned}
$$

From these intermediate calculations the entries in the 2×2 matrix $C = [c_{ij}]$ are calculated as

$$
\begin{aligned}
c_{11} &= d_1 + d_2 - d_4 + d_6 \\
c_{12} &= d_4 + d_5 \\
c_{21} &= d_6 + d_7 \\
c_{22} &= d_2 - d_3 + d_5 - d_7.
\end{aligned}
$$

If the only goal was to multiply 2×2 matrices efficiently, this algorithm (which uses 7 multiplications and 18 additions/subtractions) would be far less desirable than the straightforward way using the definition of matrix multiplication. However, the goal is to recursively use this algorithm for matrices that have dimensions $n = 2^K$ (since any matrix of dimensions not of this form can be embedded in a matrix of this form at little cost). This is possible because Strassen's algorithm outlined above holds not only when a_{ij} and b_{ij} are scalars, but also when these elements are themselves matrices.

It is not difficult to see from the 2×2 case that the number $T(n)$ of arithmetic operations required to compute the product of two $n \times n$ matrices using this algorithm recursively satisfies the recurrence [2]

$$
T(n) = 7T(n/2) + 18(n/2)^2.
$$

The complexity of any algorithm whose complexity satisfies such a recurrence relation will be of $\mathcal{O}(n^{\log 7})$ [2]. Hence as n becomes large, this, and other, recursive approaches to matrix multiplication become more efficient than direct evaluation of the definition.

Throughout the text, when we discuss matrix computations we use the complexity $\mathcal{O}(n^\gamma)$ keeping in mind that for direct evaluation of the definition $\gamma = 3$ and for Strassen's algorithm $\gamma \approx 2.81$. For other approaches and more up-to-date complexity bounds see [38, 39].

A.3 Polynomials

An nth-degree real (complex) polynomial is a sum of the form

$$p(x) = a_0 x^n + a_1 x^{n-1} + \cdots + a_{n-1} x + a_n = \sum_{k=0}^{n} a_k x^{n-k}$$

where $a_i \in \mathbb{R}$ (or $\mathbb{C}$), $a_0 \neq 0$ and $n \in \{0, 1, 2, \dots\}$.

A naive evaluation of $p(x)$ at a single value of x would calculate $z_k = x^k$ as a k-fold product for each $k \in [1, n]$. This would use $k - 1$ multiplications for each value of k, and hence a total of

$$\sum_{k=1}^{n} (k - 1) = \frac{n(n + 1)}{2} - n = \mathcal{O}\left(n^2\right)$$

multiplications. Then evaluating $\sum_{i=0}^{n} a_i z_{n-i}$ can be performed in $n - 1$ additions and n multiplications. The result is $n(n + 1)/2 + n - 1 = \mathcal{O}(n^2)$ arithmetic operations.

Better computational performance can be achieved by observing that $z_k = x \cdot z_{k-1}$. This is an example of a *recurrence relation*. Using this, the calculation of all x^k for $k = 1, \dots, n$ is achieved using n multiplications, and the evaluation of $p(x)$ at a particular value of x is reduced to $3n - 1 = \mathcal{O}(n)$ arithmetic operations.

This can be improved further using the *Newton-Horner* recursive scheme

$$p_0 = a_0 \qquad p_{i+1} = x \cdot p_i + a_{i+1}$$

for $i = 0, \dots, n - 1$. This uses one multiplication and one addition for each value of i, and is hence an algorithm requiring $2n = \mathcal{O}(n)$ computations. This algorithm also has the advantage of not having to store the values $z_1, \dots, z_n$.

Various algorithms for the efficient evaluation of a polynomial at a fixed value of x can be found in the literature. See, for example, [14, 37]. For algorithms that efficiently evaluate a polynomial at many points, and/or interpolate a polynomial to fit many points, see [2, 5, 6, 10, 26, 32, 34].

Naive approaches to finding the coefficients of products (by direct expansion) and quotients (using long division) of polynomials, as well as the evaluation of $p_n(x)$ at n distinct points $x_1, \dots, x_{n-1}$ lead to $\mathcal{O}(n^2)$ algorithms. The subsequent sections review more efficient algorithms to achieve the same results.

A.4 Efficient Multiplication of Polynomials

Consider the complexity of multiplying two nth-degree polynomials

$$\left(\sum_{i=0}^{n} a_i x^i\right)\left(\sum_{j=0}^{n} b_j x^j\right) = \sum_{k=0}^{2n} c_k x^k. \tag{A.4}$$

A naive computation as

$$\sum_{k=0}^{2n} c_k x^k = \sum_{i=0}^{n} \sum_{j=0}^{n} a_i b_j x^{i+j}$$

might lead one to believe that $\mathcal{O}(n^2)$ computations are required to find the set $\{c_k | k = 0, \ldots, 2n\}$. This, however, would be a wasteful computation because upon closer examination we see

$$c_k = \sum_{j=0}^{n} a_j b_{k-j}. \tag{A.5}$$

This is essentially the convolution of two sequences, and the set $\{c_k | k = 0, \ldots, 2n\}$ can therefore be calculated using the FFT in

$$M(n) \triangleq \mathcal{O}(n \log n)$$

arithmetic operations [2].

A.5 Efficient Division of Polynomials

Let p and q be polynomials of degree n and m, respectively, with $n > m$. Then we can always find two polynomials s and r such that

$$p = s \cdot q + r$$

where the degree of r is less than m, and the degree of s is $n - m$. In this case we say

$$p \equiv r \pmod{q}.$$

Finding r and s by long division can be performed in $\mathcal{O}(mn)$ computations. This is because there are $n - m + 1$ levels to the long division, each requiring at most $\mathcal{O}(m)$ arithmetic operations. For example, consider $p(x) = x^4 + 3x^3 + 2x^2 + x + 1$ and $q(x) = x^2 + 2x + 1$. Performing long division we observe that

$$x^4 + 3x^3 + 2x^2 + x + 1 = \left(x^2 + x - 1\right)\left(x^2 + 2x + 1\right) + 2x + 2.$$

Analogous to the way polynomial multiplication can be improved upon, so too can division. When $n = 2m$, division can be performed in

$$D(n) \triangleq \mathcal{O}(n \log n)$$

computations, as opposed to the $\mathcal{O}(n^2)$ required by long division [2].

A.6 Polynomial Evaluation

Given n distinct values of x labeled $x_0, \ldots, x_{n-1}$, the polynomial evaluation problem considered here is that of determining the values of $y_k = p(x_k)$ where $p(x)$ is a polynomial of degree $n - 1$ with coefficients $\{a_j\}$. This is written explicitly as

$$y_k = \sum_{j=0}^{n-1} b_{kj} a_j \quad \text{where} \quad b_{kj} = (x_k)^{n-j-1}. \tag{A.6}$$

A naive computation would be to simply perform the $\mathcal{O}(n^2)$ arithmetic operations indicated by the previous matrix-vector product. However this does not take advantage of the special structure of the matrix B comprised of the entries b_{kj}. This matrix, called a *Vandermonde* matrix [19, 24, 40], has the special property that its product with a vector of dimension n can be computed in $\mathcal{O}(n \log^2 n)$ arithmetic operations [13]. We now show why this result is so using arguments from [2, 5].

It is well known that $p(x_k)$ is equal to the remainder of the division $p(x)/(x - x_k)$

$$\frac{p(x)}{x - x_k} \equiv p(x_k) \pmod{(x - x_k)}. \tag{A.7}$$

For instance let $p(x) = x^3 + 2x + 1$. Then since

$$x^3 + 2x + 1 = (x - 4)\left(x^2 + 4x + 18\right) + 73$$

we calculate $p(4) = 73$.

Let $p_0, p_1, \ldots, p_{n-1}$ be monomials of the form $p_k = x - x_k$ for a distinct set of given values $x_k \in \mathbb{R}$ and let n be a power of 2. If p is a polynomial of degree n, then the set of n residues r_i defined by

$$p = s_i \cdot p_i + r_i$$

where s_i are polynomials can be computed in $\mathcal{O}(M(n) \log n) = \mathcal{O}(n \log^2 n)$ operations. This is better than performing the division algorithm n times [which would use $\mathcal{O}(n^2 \log n)$ computations].

This efficiency is achieved by first calculating the products $p_0 p_1,\ p_2 p_3,\ p_4 p_5,\ \ldots,\ p_{n-2} p_{n-1}$, then $p_0 p_1 p_2 p_3,\ p_4 p_5 p_6 p_7,\ \ldots,\ p_{n-4} p_{n-3} p_{n-2} p_{n-1}$, and so on as indicated by the upward-pointing arrow in Fig. A.1. Labeling the base of the tree as level zero, at the Kth level up from the bottom there are $n/2^K$ polynomials, each of degree 2^K. The product of all polynomials as described previously constitutes a total of

$$\mathcal{O}\left(\sum_{K=1}^{\log_2 n} \frac{n}{2^K}\left(2^K \log_2 2^K\right)\right) = \mathcal{O}\left(n \sum_{K=1}^{\log_2 n} K\right) = \mathcal{O}\left(n \log^2 n\right)$$

calculations.

Next, the $\log_2 n$ divisions corresponding to the downward-pointing arrow in Fig. A.1 are performed [each requiring $\mathcal{O}(n \log_2 n)$ operations]. The residues (remainders) that result at the end of the process are the desired values $p(x_k)$.

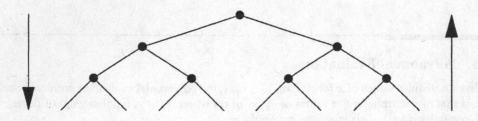

FIGURE A.1
Multiplications from bottom to top; divisions from top to bottom.

A.7 Polynomial Interpolation

We now review an efficient algorithm for polynomial interpolation. Given n values of x labeled $x_0, \ldots, x_{n-1}$, the polynomial interpolation problem is to find the n coefficients $a_0, \ldots, a_{n-1}$ defining an $n - 1$-degree polynomial, $p(x)$, that satisfies the equalities $p(x_k) = y_k$ for the given values $y_0, \ldots, y_{n-1}$.

In this context the values of x_k are given, and the problem of finding the vector $\mathbf{a} = [a_0, \ldots, a_{n-1}]^T$ could be written as an inversion of the matrix B in Eq. (A.6) followed by a product of B^{-1} with the vector $\mathbf{y}$. Using Gaussian elimination, $\mathcal{O}(n^3)$ arithmetic operations would be required to invert B. In principle, faster inversion methods exist for general matrices [47]. Furthermore, the special structure of B again allows us to efficiently find B^{-1} and perform interpolation in $\mathcal{O}(n \log^2 n)$ arithmetic computations.

Another way to achieve the same result is as follows. The *Lagrange interpolation polynomials* (discussed in Section 3.5.2 of Chapter 3) are defined as

$$L_n(x) = \sum_{k=0}^{n} f_k \frac{l^{n+1}(x)}{(x - \xi_k)\left(l^{n+1}\right)'(\xi_k)} \tag{A.8}$$

where

$$l^{n+1}(x) = \prod_{k=0}^{n} (x - \xi_k).$$

These have the property $L_n(\xi_j) = f_j$ for $j = 0, \ldots, n$ because for a polynomial $l^{n+1}(x)$ one finds from l'Hopital's rule that

$$\left.\frac{l^{n+1}(x)}{x - \xi_k}\right|_{x=\xi_k} = \left.\frac{dl^{n+1}}{dx}\right|_{x=\xi_k},$$

and when evaluated at $x = \xi_k$ all but one term in the sum in Eq. (A.8) vanish.

As a special case of Eq. (A.7), we observe that

$$\frac{l^{n+1}(x)}{x - \xi_k} \equiv \left.\frac{dl^{n+1}}{dx}\right|_{x=\xi_k} \mod (x - \xi_k).$$

Following [2], we use these facts to formulate an efficient polynomial interpolation procedure. First define

$$q_{ij} = \prod_{m=i}^{i+2^j-1} (x - \xi_m).$$

Then for $j \neq 0$ define

$$s_{ij} \doteq q_{ij} \sum_{m=i}^{i+2^j-1} \frac{f_m}{(x - \xi_m) \left.\frac{dl^{n+1}}{dx}\right|_{x=\xi_m}},$$

and

$$s_{i0} = \frac{f_i}{\left.\frac{dl^{n+1}}{dx}\right|_{x=\xi_i}}.$$

A recursive procedure defined as

$$s_{ij} = s_{i,j-1} q_{i+2^{j-1}, j-1} + s_{i+2^{j-1}, j-1} q_{i, j-1}$$

is then used to calculate $L_n(x) = s_{0,K}$ when $n = 2^K$. It starts by calculating s_{i0} for $i = 0, \ldots, n-1$ in unit increments. Then it runs through the values $j = 1, \ldots, K$ in unit increments and $i = 0, \ldots, n-1$ in increments of 2^j. This results in an $\mathcal{O}(n(\log_2 n)^2)$ algorithm.

The polynomial interpolation problem can be viewed as the dual to the evaluation problem. It therefore corresponds to the downward-pointing arrow in Fig. A.1.

Appendix B

Set Theory

B.1 Basic Definitions

The concept of a *set* is the foundation on which all of modern mathematics is built. A set is simply a collection of well-defined objects called *elements* or *points*. A set will be denoted here with capital letters. If S is a set, the notation $x, y \in S$ means x and y are elements of S. Sets can contain a finite number of elements or an infinite number. Sets can be discrete in nature or have a continuum of elements. The elements of the set may have an easily recognizable common theme, or the set may be composed of seemingly unrelated elements. Examples of sets include

$$
\begin{aligned}
S_1 &= \{a, 1, \triangle, Bob\}\,; \\
S_2 &= \{0, 1, \ldots, 11\}\,; \\
S_3 &= \mathbb{Z} = \{\ldots, -2, -1, 0, 1, 2, \ldots\} \\
S_4 &= \mathbb{R}\,; \\
S_5 &= \left\{A \in \mathbb{R}^{3\times 3} \,\middle|\, \det A = 1\right\}\,; \\
S_6 &= \{\{0\}, \{0, 1\}, \{0, 1, 2\}\}.
\end{aligned}
$$

S_1 is a finite set containing four elements that are explicitly enumerated. S_2 is a finite set with 12 elements (the numbers 0 through 11) that are not all explicitly listed because the pattern is clear. S_3 is the set of integers. This is a discrete set with an infinite number of elements. S_4 is the set of all real numbers. This is a set that possesses a continuum of values and is therefore infinite. S_5 is a set of 3×3 matrices with real entries that satisfy the condition of unit determinant. Often it is convenient to define sets as parts of other sets with certain conditions imposed. The vertical line in the definition of S_5 previously mentioned is read as "such that." In other words, the larger set in this case is $\mathbb{R}^{3\times 3}$, and S_5 is defined as the part of this larger set such that the condition on the right is observed. S_6 illustrates that a set can have elements that are themselves sets.

Unless otherwise specified, the order of elements in a set does not matter. Hence the following sets are all the same

$$\{0, 1, 2\} = \{1, 2, 0\} = \{2, 1, 0\} = \{0, 2, 1\}.$$

In the special case when the order of elements does matter, the set is called an *ordered set*. In general, two sets X and Y are equal if for every $x \in X$, it is also true that $x \in Y$, and for every $y \in Y$ it is also true that $y \in X$.

Two of the most basic concepts in set theory are those of intersection, $\cap$, and union, $\cup$. The intersection of two sets is the largest set that is shared by both sets

$$X \cap Y = \{x \,|\, x \in X \ \text{and} \ x \in Y\}.$$

The union of two sets is the smallest set that contains both sets

$$X \cup Y = \{x | x \in X \text{ or } x \in Y\}.$$

The order in which unions or intersections of two sets are performed does not matter. That is,

$$X \cup Y = Y \cup X \quad \text{and} \quad X \cap Y = Y \cap X.$$

The set-theoretic difference $X - Y$ of two sets X and Y is the set of elements in X but not in Y. For example,

$$
\begin{aligned}
S_1 \cap S_2 &= \{1\}; \\
S_2 \cap S_3 &= S_2; \\
S_1 \cup S_2 &= \{a, 0, 1, 2, 3, 4, 5, 6, 7, 8, 9, 10, 11, \triangle, Bob\}; \\
S_1 - S_1 \cap S_2 &= \{a, \triangle, Bob\}.
\end{aligned}
$$

Many concepts from set theory are demonstrated graphically with a *Venn diagram*. Figure B.1 illustrates union, intersection, and difference of sets.

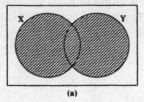 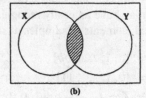 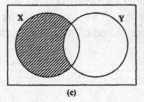

(a)   (b)   (c)

FIGURE B.1
Venn diagrams depicting: (a) $X \cup Y$; (b) $X \cap Y$; and (c) $X - Y$.

When two sets X and Y have no elements in common, one writes

$$X \cap Y = \emptyset.$$

The symbol $\emptyset$ is called the *empty set* and is the set with no elements. One immediately confronts the seemingly paradoxical statement that every set has the empty set as a subset, and therefore the intersection of any two sets contains the empty set. Hence every two sets have at least the empty set in common.[1] If the intersection of two sets is the empty set, the two sets are called *disjoint*. In the preceding examples, S_1 and S_6 are disjoint since none of the elements in both sets are the same (despite the fact that the element $1 \in S_1$ is contained in two of the elements of S_6).

Given a collection of sets $\{S_1, \ldots, S_n\}$ where n is a positive integer and $I = \{1, \ldots, n\}$ is a set of consecutive integers (called an *indexing set*), the union and intersection of the whole collection of sets are denoted, respectively, as

$$\bigcup_{i \in I} S_i = S_1 \cup S_2 \cup \cdots \cup S_n$$

and

$$\bigcap_{i \in I} S_i = S_1 \cap S_2 \cap \cdots \cap S_n.$$

---

[1] Other rules such as a set is not allowed to be an element of itself can be found in books on axiomatic set theory.

More generally, the indexing set can be any countable set (e.g., the integers) and need not consist of numbers (consecutive or otherwise).

If a set S_1 is contained in a set S_2 then S_1 is called a *subset* of S_2. The set X is contained in the set Y if all of the elements of X are elements of Y. The notation

$$X \subseteq Y$$

is used to denote that X is a subset of Y, and can possibly be equal to Y. When $X \subseteq Y$ and $X \neq Y$ we use the notation $X \subset Y$ and call X a *proper subset* of Y. We note that this notation is not standardized in mathematics. Often in mathematics the symbol $\subset$ is used to denote what we have defined as $\subseteq$. Our notational choice is so that the symbols $\leq$ and $<$ for subgroup and proper subgroup are consistent with the symbols for subset and proper subset.

Clearly

$$X \cap Y \subseteq Y \quad \text{and} \quad X \cap Y \subseteq X$$

whereas

$$X \subseteq X \cup Y \quad \text{and} \quad Y \subseteq X \cup Y.$$

If $A = \{a_1, \ldots, a_n\}$ and $B = \{b_1, \ldots, b_m\}$, then $|A| = n$ and $|B| = m$ are the number of elements in each. The number of elements in a set is called the *order* of the set. The *Cartesian product* of A and B is the set of ordered pairs $A \times B = \{(a, b) | a \in A, b \in B\}$. It follows that $|A \cup B| = |A| + |B| - |A \cap B|$ and $|A \times B| = |A| \cdot |B|$.

Given a set S, the concept of *equivalence* of elements $a, b \in S$ (denoted $a \sim b$) is any relationship between elements that satisfies the properties of reflexivity ($a \sim a$), symmetry (if $a \sim b$ then $b \sim a$), and transitivity (if $a \sim b$ and $b \sim c$, then $a \sim c$). An *equivalence class*, $[a] \subset S$ determined by $a \in S$ is defined as $[a] = \{b \in S | b \sim a\}$. The set S can be decomposed into disjoint equivalence classes since every element of the set is in one and only one equivalence class [36]. The set of equivalence classes is denoted as $S/\sim$. Hence for any $s \in S$, $[s] \in S/\sim$ and

$$[s_1] \cap [s_2] = \begin{cases} \emptyset & \text{for } [s_1] \neq [s_2] \\ [s_1] & \text{for } [s_1] = [s_2] \end{cases}.$$

An important example of equivalence classes is the one established by modular arithmetic. Two numbers a and b are said to be *congruent modulo n* if n divides $a - b$ without a remainder. This is denoted

$$a \equiv b \, (\text{mod} \, n).$$

It is easy to check that congruence defined in this way is an equivalence relation.

B.2   Maps and Functions

Another of the basic concepts of set theory is a *map* that assigns elements from one set to another. A map $f : X \to Y$ assigns to each element in X one and only one element in Y. Unless otherwise stated, the map $f : X \to Y$ need not "use up" all the elements in Y, and it may be the case that two or more elements in X are assigned the same element in Y. The subset of Y onto which all the elements of X map is called the *image* of f, and is denoted as $f(X)$ or $\text{Im}(f)$. The collection of elements in X that map to the same element $y \in Y$ is called the *preimage* or *fiber* over y in X. In the case when all the elements of Y are assigned to elements in X, the map is called *onto* or *surjective* and $f(X) = Y$. Another way to say this is that f is surjective when every point in Y has at least one preimage in X. A map that is not necessarily surjective, and satisfies the constraint for any $x_1, x_2 \in X$ that $f(x_1) = f(x_2)$ implies $x_1 = x_2$ is called *one-to-one* or *injective*. Another way to say this is that f maps to each element of $f(X)$ from only one element of X. In the case when f is both surjective and injective it is called *bijective* or *invertible*. In this case there is a unique map

$$f^{-1} : Y \to X$$

that is also bijective, such that $(f \circ f^{-1})(y) = y$ and $(f^{-1} \circ f)(x) = x$.

A *function* is a map from a set into either the real or complex numbers (or, more generally, any *field* as defined in Section C.1).

Sets with certain kinds of maps or properties associated with them are given special names. For instance, a *metric space* is a set S together with a real-valued function $d : S \times S \to \mathbb{R}$, called a *metric* or *distance function* such that $d(s_1, s_2) > 0$ if $s_1 \neq s_2$; $d(s_1, s_1) = 0$ for all $s_1 \in S$; $d(s_1, s_2) = d(s_2, s_1)$; and $d(s_1, s_3) \leq d(s_1, s_2) + d(s_2, s_3)$.

A *topological space* is a set S together with a collection of subsets of S (called a topology of S, and denoted T), such that: (1) the empty set, $\emptyset$, and S are contained in T; (2) the union of any number of subsets of T are in T; and (3) the intersection of any finite number of subsets of T are also in T.

Examples of topologies of an arbitrary set S include $\{S, \emptyset\}$ (called the indiscrete or trivial topology); the set of all subsets of S (called the discrete topology); and if the set is a metric space, the set of all subsets formed by the arbitrary union and finite intersection of all "ϵ-balls"

$$B_\epsilon(z) = \{s \mid d(s, z) < \epsilon\}$$

for $O < \epsilon \in \mathbb{R}$ and $s, z \in S$ is called the metric topology.

Given a topological space S with topology T, we say that the subset $U \subseteq S$ is *open* if $U \in T$. A subset $V \subseteq S$ is called *closed* if $S - V$ is open. Given a subset $A \subseteq S$, the *interior* of A [denoted $Int(A)$] is the union of all open sets contained in A. The closure of A [denoted $Cl(A)$] is the intersection of all closed sets in S that contain A. $Int(A)$ is an open set and $Cl(A)$ is a closed set, and the following relationship holds

$$Int(A) \subseteq A \subseteq Cl(A).$$

A topological space X is called a *Hausdorff space* if for each pair of distinct points $x_1, x_2 \in X$ there exist open sets $U_1, U_2 \subset X$ such that $x_1 \in U_1$, $x_2 \in U_2$, and $U_1 \cap U_2 = \emptyset$.

Any map $f : X \to Y$ induces equivalence relations on the two involved sets. For instance, if f is not surjective we can call two elements of Y equivalent if they are both in $f(X)$ or if they are both in $Y - f(X)$. Similarly, if f is not injective, all points in X which map to a given point in Y can be called equivalent. Finally, if f is bijective and an equivalence relation $x_1 \sim x_2$ exists between elements in X, then the elements $f(x_1)$ and $f(x_2)$ can be called equivalent as well.

For example, let $X = \mathbb{R}$, $Y = \{c \in \mathbb{C} \mid c\bar{c} = 1\}$, and $f(x) = e^{ix}$. Then all $x \in X$ that map to the same value $y \in Y$ can be called equivalent. The equivalence defined in this way is the same as congruence mod 2π, and the equivalence class is therefore $X/\sim = \mathbb{R}/2\pi\mathbb{R}$.

In general, since a set is divided into disjoint equivalence classes, we can write

$$\bigcup_{[s] \in S/\sim} [s] = S.$$

Another way to write this is by constructing an indexing set I consisting of one and only one representative from each equivalence class. Then

$$\bigcup_{s \in I} [s] = S.$$

We are assuming $S/\sim$ is a countable set. (This assumption can in fact be proved if we have the axion of choice [36].)

For example, consider

$$S = \{0, 1, 2, 3, 4, 5, 6, 7, 8, 9, 10, 11\}.$$

Using the equivalence relation $a \sim b \leftrightarrow a \equiv b \pmod 3$ we have the equivalence classes

$$[0] = \{0, 3, 6, 9\}$$
$$[1] = \{1, 4, 7, 10\}$$
$$[2] = \{2, 5, 8, 11\}.$$

Clearly $S = [0] \cup [1] \cup [2]$.

Given a complex-valued function of set-valued argument $f : S \to \mathbb{C}$, where S is a finite set, one can make the following decomposition

$$\sum_{s \in S} f(s) = \sum_{[s] \in S/\sim} \left(\sum_{x \in [s]} f(x) \right). \tag{B.1}$$

That is, using the disjointness property, we can first sum the values of the function evaluated for all the elements in each equivalence class, then sum these sums over all equivalence classes. An example of a function on a set is the *characteristic function* of a subset $X \subset S$. This function is defined for all $x \in S$ as

$$\chi_X(x) = \begin{cases} 1 & \text{for } x \in X \\ 0 & \text{for } x \notin X \end{cases}.$$

B.3   Invariant Measures and Metrics

In the case when S is not finite, one desires a tool for determining the net "mass" of a function evaluated over all of the elements $s \in S$. This is where measure theory is valuable. We follow the definitions given in [23, 46]. A *measure* is a real-valued, non-negative function $\mu(\cdot)$ with certain properties defined on a σ-*algebra* S. A σ-algebra S is defined as a family of subsets of a set S with the properties that $\emptyset, S \in S$, $S' \in S$ implies $S - (S \cap S') \in S$, and for any countable set $\{S_1, S_2, \dots\}$ with each $S_i \in S$ it holds that $\cup_{i=1}^{\infty} S_i \in S$ and $\cap_{i=1}^{\infty} S_i \in S$. A *measurable space* is the pair (S, S). A measure $\mu(\cdot)$ satisfies the properties $\mu(\emptyset) = 0$, $\mu(S') \geq 0$ for all $S' \in S$, and $\mu(\cup_{n=1}^{\infty} S_n) \leq \sum_{n=1}^{\infty} \mu(S_n)$ with equality holding if $S_i \cap S_j = \emptyset$ for all positive integers i, j.

A *Haar measure* is a measure on a locally compact group G (see Chapter 7) for which $\mu(g \circ A) = \mu(A)$ for any $g \in G$ and any compact subset $A \subset G$. The notation goA is defined as $goA \triangleq \{goA \forall a \in A\}$. If the subset A is divided into disjoint infinitesimal elements, $da \subset A$, then we define $\mu(da) = d\mu(a)$ where $a \in A$. Integration with respect to a given measure is then defined as $\mu(A) = \int_{a \in A} d\mu(a)$. The mathematician Haar proved that such a measure always exists on a locally compact group, and all such measures are scalar multiples of each other. Sometimes the term *Hurwitz* measure is used instead of Haar measure for the particular case of a Lie group. When a Haar measure is invariant under right as well as left shifts, it is called a *bi-invariant* measure.

Given any set S, on which G acts, a metric $d : S \times S \to \mathbb{R}$ is said to be G-*invariant* if $d(A, B) = d(goA, goB)$ $\forall g \in G$ for some group G and all $A, B \in S$ under the closure condition $goA, goB \in S$.

Two geometrical objects A and B (closed and bounded regions in $\mathbb{R}^n$) are said to be *similar*[2] (or equivalent under the action of a group G) if $d(A, g \circ B) = 0$ for some $g \in G$ and the G-invariant metric $d(\cdot, \cdot)$ [such as $d(A, B) = Vol(A \cap B)$ when G is locally volume preserving, e.g., rigid-body motion].

THEOREM B.1

The function $D_G([A], [B]) \stackrel{\Delta}{=} \min\limits_{g \in G} d(A, g \circ B)$ is a metric on the set of equivalence classes $S / \sim$ with elements of the form $[A] = \{A' \in S | d(A, g \circ A') = 0$ for some $g \in G\}$ when $d(A, B)$ is a G-invariant metric on S.

PROOF   Let

$$[A] = \left\{ A' \in S | d\left(A, g \circ A'\right) = 0 \quad \text{for some} \quad g \in G \right\}$$

denote the equivalence class containing A. Then $D_G([A], [B])$ is well defined, and is a metric because it satisfies:

Symmetry:

$$
\begin{aligned}
D_G([A], [B]) &= \min_{g \in G} d(A, g \circ B) \\
&= \min_{g \in G} d\left(g^{-1} \circ A, B\right) \\
&= \min_{g \in G} d(g \circ A, B) \\
&= \min_{g \in G} d(B, g \circ A) \\
&= D_G([B], [A]).
\end{aligned}
$$

Positive Definiteness:

$$D_G([A], [A]) = 0 \quad \text{(by definition since we may choose } g = e)$$
$$D_G([A], [B]) = 0 \Rightarrow \min_{g \in G} d(A, g \circ B) = 0 \rightarrow B \in [A] \quad \forall B \in [B].$$

Likewise, from symmetry, $A \in [B] \quad \forall A \in [A]$. Therefore $[A] = [B]$.

Triangle Inequality:

$$
\begin{aligned}
D_G([A], [C]) &= \min_{g \in G} d(A, g \circ C) \\
&\leq \min_{g \in G} \min_{h \in G} (d(A, h \circ B) + d(h \circ B, g \circ C)) \\
&= \min_{h \in G} d(A, h \circ B) + \min_{g \in G} \min_{h \in G} d(h \circ B, g \circ C) \\
&= D_G([A], [B]) + \min_{g, h \in G} d\left(B, \left(h^{-1} \circ g\right) \circ C\right) \\
&= D_G([A], [B]) + \min_{k \in G} d(B, k \circ C) \\
&= D_G([A], [B]) + D_G([B], [C]).
\end{aligned}
$$

---

[2]Note this definition is a generalization of the standard geometrical concept.

The inequality above follows from the fact that

$$\min_{g \in G} d(A, g \circ C) \leq \min_{g \in G} \left(d\left(A, h' \circ B\right) + d\left(h' \circ B, g \circ C\right) \right)$$

for *any* $h' \in G$ since $d(\cdot, \cdot)$ is a metric and thus satisfies the triangle inequality. Therefore all the metric properties are satisfied. ∎

Appendix C

Vector Spaces and Algebras

Consider a nonempty set $\mathcal{X}$, on which the operations of addition of elements $(x, y \in \mathcal{X} \rightarrow x+y \in \mathcal{X})$ and scalar multiplication with elements $(x \in \mathcal{X}$ and $\alpha \in \mathbb{C}$ (or $\mathbb{R}) \rightarrow \alpha \cdot x \in \mathcal{X})$ are defined. The triplet $(\mathcal{X}, +, \cdot)$ is said to be a *complex (or real) vector space*[1] if the following properties are satisfied

$$x + y = y + x \quad \forall \; x, y \in \mathcal{X} \tag{C.1}$$

$$(x + y) + z = x + (y + z) \quad \forall \; x, y, z \in \mathcal{X} \tag{C.2}$$

$$\exists \, 0 \in \mathcal{X} \quad \text{s.t.} \quad x + 0 = x \in \mathcal{X} \tag{C.3}$$

$$\exists \, (-x) \in \mathcal{X} \quad \text{for each} \quad x \in \mathcal{X} \quad \text{s.t.} \quad x + (-x) = 0 \tag{C.4}$$

$$\alpha \cdot (x + y) = \alpha \cdot x + \alpha \cdot y \quad \forall \; \alpha \in \mathbb{C} (\text{or} \, \mathbb{R}), x, y \in \mathcal{X} \tag{C.5}$$

$$(\alpha + \beta) \cdot x = \alpha \cdot x + \beta \cdot x \quad \forall \; \alpha, \beta \in \mathbb{C} (\text{or} \, \mathbb{R}), x \in \mathcal{X} \tag{C.6}$$

$$(\alpha\beta) \cdot x = \alpha \cdot (\beta \cdot x) \quad \forall \; \alpha, \beta \in \mathbb{C} (\text{or} \, \mathbb{R}), x \in \mathcal{X} \tag{C.7}$$

$$1 \cdot x = x \quad \forall \; x \in \mathcal{X}. \tag{C.8}$$

$(\mathbb{R}^N, +, \cdot)$ for any $N = 1, 2, \ldots$ is an example of a real vector space, and so is the set of all real-valued functions on $\mathbb{R}^N$.

The conditions (C.1)–(C.4) imply that every vector space is an Abelian group under addition. Let $V = (\mathcal{X}, +, \cdot)$ be a vector space. The elements of V (elements of $\mathcal{X}$) are called vectors. If a complex vector space V has an *inner* (or scalar) product $(\cdot, \cdot)$ defined between vectors such that $(x, y) \in \mathbb{C}$ for all $x, y \in V$, and satisfies the properties

$$(x, x) \geq 0, \quad \text{and} \quad (x, x) = 0 \Rightarrow x = 0 \tag{C.9}$$

$$(x, y + z) = (x, y) + (x, z) \tag{C.10}$$

$$(x, \alpha y) = \alpha(x, y) \quad \forall \alpha \in \mathbb{C} \tag{C.11}$$

$$(x, y) = \overline{(y, x)} \tag{C.12}$$

then the pair $(V, (\cdot, \cdot))$ is called an *inner product space*. A real inner product space can be viewed as the special case where $\alpha \in \mathbb{R}$ and Eq. (C.12) reduces to $(x, y) = (y, x)$.

Note that there is a difference between the way mathematicians and physicists define these properties. We have used the physicist's convention above whereas a mathematician would use the alternate defining property

$$(\alpha x, y) = \alpha(x, y)$$

---

[1] Even more generally one refers to a "vector space over a field." See Section C.1 for what is meant by a field.

in place of Eq. (C.11). Using Eqs. (C.10) and (C.12), the mathematician's convention amounts to linearity in the first argument rather than the second. We will use whichever convention is more convenient for a given application.

The *norm* of a vector is defined as

$$\|x\| = \sqrt{(x, x)}.$$

The usual dot (scalar) product of vectors $\mathbf{x}, \mathbf{y} \in \mathbb{R}^N$ defined by $\mathbf{x} \cdot \mathbf{y} = \sum_{i=1}^{N} x_i y_i$ is an example of an inner product, and $(\mathbb{R}^N, \cdot)$ is a real inner product space. Given $A, B \in \mathbb{C}^{N \times N}$, $(A, B) = \text{trace}(AB^\dagger)$ is an inner product in the mathematician's convention, where $\dagger$ denotes the complex conjugate transpose. Likewise, for $A, B \in \mathbb{R}^{N \times N}$, $(A, B) = \text{trace}(AB^T)$ is an inner product.

Since points in the plane can be identified with complex numbers, we can define the inner product between two complex numbers $a = a_1 + ia_2$ and $b = b_1 + ib_2$ as

$$(a, b) = \text{Re}\left(a\overline{b}\right) = a_1 b_1 + a_2 b_2.$$

This satisfies the definition of an inner product, and even though the vectors in this case are complex numbers, $\mathbb{C}$ in this example is a *real* vector space, and with the preceding definition of inner product, $(\mathbb{C}, (\cdot, \cdot))$ is a *real* inner product space.

An inner product of two real-valued functions $\alpha(\mathbf{x})$, $\beta(\mathbf{x})$ is defined as $(\mathcal{L}^1 \cap \mathcal{L}^2)(\mathbb{R}^N)$

$$(\alpha, \beta) = \int_{\mathbb{R}^N} \alpha(\mathbf{x})\beta(\mathbf{x})dx_1 \ldots dx_N.$$

A basis for a vector space is a set of linearly independent vectors $\{e_1, \ldots, e_N\}$ whose linear combination spans the vector space. That is, any $x \in V$ can be uniquely expressed as $x = \sum_{i=1}^{N} x^{(i)} \cdot e_i$ for appropriate scalars $\{x^{(i)}\}$. In the case when $V = \mathbb{R}^N$, $e_i = \mathbf{e}_i$ where $(\mathbf{e}_i)_j = \delta_{ij}$ is one possible basis. When $N < \infty$ elements are needed to span V, then V is called N-dimensional. If an infinite number of independent basis elements exist, then V is called an infinite-dimensional vector space. Function spaces, such as the set of all real-valued functions on $\mathbb{R}^N$, are usually infinite-dimensional.

Two general properties of inner-product spaces are: (1) that the *Cauchy-Schwarz inequality*

$$(x, y)^2 \leq (x, x)(y, y) \tag{C.13}$$

holds for any two vectors $x, y \in V$; and (2) given an arbitrary basis for V, an orthonormal one can be constructed using the *Gram-Schmidt orthogonalization process*. We now examine these in detail.

The Cauchy-Schwarz Inequality

To see that the Cauchy-Schwarz inequality holds, one need only observe that

$$f(t) = (x + ty, x + ty) = \|x + ty\|^2 \geq 0.$$

Expanding out, one finds the quadratic equation in t

$$f(t) = (x, x) + 2(x, y)t + (y, y)t^2 \geq 0.$$

Since the minimum of $f(t)$ occurs when $f'(t) = 0$ [i.e., when $t = -(x, y)/(y, y)$], the minimal value of $f(t)$ is

$$f(-(x, y)/(y, y)) = (x, x) - (x, y)^2/(y, y)$$

when $y \neq 0$. Since $f(t) \geq 0$ for all values of t, the Cauchy-Schwarz inequality follows. In the $y = 0$ case, Eq. (C.13) reduces to the equality $0 = 0$.

When the vector space is the space of square-integrable real-valued functions on the unit interval, the Cauchy-Schwartz inequality takes the form

$$\int_0^1 f_1(x)f_2(x)dx \le \left(\int_0^1 f_1^2(x)dx\right)^{\frac{1}{2}}\left(\int_0^1 f_2^2(x)dx\right)^{\frac{1}{2}}.$$

Choosing $f_1(x) = f(x)$ and $f_2(x) = \text{sign}(f(x))$ and squaring both sides, it follows that

$$\left(\int_0^1 |f(x)|dx\right)^2 \le \int_0^1 f^2(x)dx.$$

The Gram-Schmidt Orthogonalization Process

Let V be an n-dimensional inner product space, and let $\{v_1, \ldots, v_n\}$ be a basis for V. Then an orthonormal basis for V can be constructed as follows. First normalize one of the original basis vectors (e.g., v_1) and define

$$u_1 = v_1/\|v_1\|.$$

Then define u_2 by taking away the part of v_2 that is parallel to u_1 and normalizing what remains

$$u_2 = (v_2 - (v_2, u_1)u_1)/\|v_2 - (v_2, u_1)u_1\|.$$

It is easy to see that $(u_1, u_2) = 0$ and that they are both unit vectors. The process then is recursively performed by taking away the parts of v_i that are parallel to each of the $\{u_1, \ldots, u_{i-1}\}$. Then u_i is defined as the unit vector of what remains, i.e.,

$$u_i = \left(v_i - \sum_{k=1}^{i-1}(v_i, u_k)u_k\right)/\left\|v_i - \sum_{k=1}^{i-1}(v_i, u_k)u_k\right\|.$$

This process is repeated until a full set of orthonormal basis vectors $\{u_1, \ldots, u_n\}$ is constructed.

A real vector space V, for which an additional operation $\wedge$ is defined such that $x \wedge y \in V$ for every $x, y \in V$, is called a real *algebra* if for $x, y, z \in \mathcal{X}$ and $\alpha \in \mathbb{R}$,

$$(x + y) \wedge z = x \wedge z + y \wedge z \tag{C.14}$$

$$z \wedge (x + y) = z \wedge x + z \wedge y \tag{C.15}$$

$$(\alpha \cdot x) \wedge y = x \wedge (\alpha \cdot y) = \alpha \cdot (x \wedge y). \tag{C.16}$$

For every n-dimensional algebra, we can expand elements x and y in a basis $\{e_i\}$ so that $x = \sum_{i=1}^n x^{(i)}e_i$, $y = \sum_{i=1}^n y^{(i)}e_i$, and

$$x \wedge y = \sum_{i=1}^n x^{(i)}y^{(j)}\left(e_i \wedge e_j\right).$$

An algebra, $(V, \wedge)$, is called an *associative algebra* if for $x, y, z \in \mathcal{X}$

$$x \wedge (y \wedge z) = (x \wedge y) \wedge z.$$

$(V, \wedge)$ is called a *commutative algebra* if

$$x \wedge y = y \wedge x.$$

An example of an associative algebra is the set of $N \times N$ real matrices $\mathbb{R}^{N \times N}$, with $\wedge$ representing matrix multiplication. The subset of diagonal $N \times N$ real matrices is both an associative and commutative algebra under the operation of matrix multiplication.

A *Lie algebra* is an algebra where the operation $x \wedge y$ (denoted as $[x, y]$ in this special context) satisfies for $x, y, z \in (V, [, \,])$ the additional properties

$$[x, y] = -[y, x] \qquad [x, [y, z]] + [y, [z, x]] + [z, [x, y]] = 0. \qquad (C.17)$$

These properties are respectively called anti-symmetry and the Jacobi identity. The operation $[x, y]$ is called the *Lie bracket* of the vectors x and y. The property $[x, x] = 0$ follows automatically. Note that Lie algebras are not generally associative.

Examples of Lie algebras are: (1) $\mathbb{R}^3$ with the cross-product operation between two vectors and (2) $\mathbb{R}^{N \times N}$ with the matrix commutator operation: $[A, B] = AB - BA$ for $A, B \in \mathbb{R}^{N \times N}$.

C.1   Rings and Fields

In this section we follow the definitions given in Herstein [22].

An *associative ring* (or simply a *ring*) R is a nonempty set with two operations, denoted $+$ and $\cdot$, such that for all $a, b, c \in R$ and the special element $0 \in R$ the following properties are satisfied

$$a + b \in R \qquad (C.18)$$
$$a + b = b + a \qquad (C.19)$$
$$(a + b) + c = a + (b + c) \qquad (C.20)$$
$$\exists 0 \in R \quad s.t. \quad a + 0 = a \qquad (C.21)$$
$$\exists (-a) \in R \quad s.t. \quad a + (-a) = 0 \qquad (C.22)$$
$$a \cdot b \in R \qquad (C.23)$$
$$a \cdot (b \cdot c) = (a \cdot b) \cdot c \qquad (C.24)$$
$$a \cdot (b + c) = a \cdot b + a \cdot c \qquad (C.25)$$
$$(b + c) \cdot a = b \cdot a + c \cdot a. \qquad (C.26)$$

When, in addition to the additive identity element, 0, there is a multiplicative identity element $1 \in R$, R is called a *ring with unit element* (or *ring with unity*). If $a \cdot b = b \cdot a$, R is called a *commutative ring*.

A standard example of a ring is the set of integers with usual multiplication and addition (this is a commutative ring with unit element). A more exotic example of a ring is the set of all square and absolutely integrable functions on a locally compact group, with the operations of addition and convolution of functions.

A *field* is a commutative ring with unit element in which every nonzero element has a multiplicative inverse. A *skew field* is an associative (but not commutative) ring in which every nonzero element has a multiplicative inverse.

The real and complex numbers are both fields. The quaternions are an example of a skew field.

Appendix D

Matrices

D.1 Special Properties of Symmetric Matrices

The following properties of symmetric real matrices, and their generalizations, are very important in all areas of engineering mathematics. The proofs given below for real symmetric matrices follow for the case of Hermitian matrices[1] (and operators defined in Appendix E.2) as well when the appropriate inner product is used. In the real cases which follow, we take $(\mathbf{u}, \mathbf{v}) = \mathbf{u} \cdot \mathbf{v} = \mathbf{u}^T \mathbf{v}$ for $\mathbf{u}, \mathbf{v} \in \mathbb{R}^N$.

THEOREM D.1
Eigenvectors of symmetric matrices corresponding to different eigenvalues are orthogonal.

PROOF   Given a symmetric matrix A and two of its different eigenvalues λ_i and λ_j with corresponding eigenvectors $\mathbf{u}_i$ and $\mathbf{u}_j$, by definition the following is true: $A\mathbf{u}_i = \lambda_i\mathbf{u}_i$, and $A\mathbf{u}_j = \lambda_j\mathbf{u}_j$. Multiplying the first of these on the left by $\mathbf{u}_j^T$, multiplying the second one on the left by $\mathbf{u}_i^T$, and subtracting, we get

$$\left(\mathbf{u}_j, A\mathbf{u}_i\right) - \left(\mathbf{u}_i, A\mathbf{u}_j\right) = \left(\mathbf{u}_j, \lambda_i\mathbf{u}_i\right) - \left(\mathbf{u}_i, \lambda_j\mathbf{u}_j\right) = \left(\lambda_i - \lambda_j\right)\left(\mathbf{u}_i, \mathbf{u}_j\right).$$

Since A is symmetric, the left side of the equation is zero.[2] Since the eigenvalues are different, we can divide by their difference, and we are left with $\mathbf{u}_i \cdot \mathbf{u}_j = 0$.   ∎

THEOREM D.2
Eigenvalues of real symmetric matrices are real.

PROOF   To show that something is real, all we have to do is show that it is equal to its complex conjugate. Recall that given $a, b \in \mathbb{R}$, a complex number $c = a + b\sqrt{-1}$ has a conjugate $\overline{c} = a - b\sqrt{-1}$. If $c = \overline{c}$, then $b = 0$ and $c = a$ is real. The complex conjugate of a vector or matrix is just the complex conjugate of all its elements. Furthermore, the complex conjugate of a product is the product of the complex conjugates. Therefore, given $A\mathbf{u}_i = \lambda_i\mathbf{u}_i$, we can take the complex conjugate of both sides to get

$$\overline{(A\mathbf{u}_i)} = \overline{A}\,\overline{\mathbf{u}_i} = A\overline{\mathbf{u}_i} = \overline{(\lambda_i\mathbf{u}_i)} = \overline{\lambda}_i\overline{\mathbf{u}}_i.$$

---

[1] $H \in \mathbb{C}^{N \times N}$ such that $H = H^\dagger$ where by definition $H^\dagger = \overline{H^T}$.
[2] This follows from the transpose rule $(A\mathbf{u})^T = \mathbf{u}^T A^T$, and the fact that $(\mathbf{u}, \mathbf{v}) = (\mathbf{v}, \mathbf{u})$.

In this derivation $\overline{A} = A$ because the elements of A are real.

Then, using an argument similar to the one in Theorem D.1, we compute

$$(\overline{\mathbf{u}}_i, A\mathbf{u}_i) = (\overline{\mathbf{u}}_i, \lambda_i \mathbf{u}_i) \quad \text{and} \quad \left(\mathbf{u}_i, \overline{A}\overline{\mathbf{u}}_i\right) = \left(\mathbf{u}_i, \overline{\lambda}_i \overline{\mathbf{u}}_i\right).$$

Subtracting the second from the first, and using the fact that A is symmetric, we get

$$(\overline{\mathbf{u}}_i, A\mathbf{u}_i) - (\mathbf{u}_i, A\overline{\mathbf{u}}_i) = 0 = (\overline{\mathbf{u}}_i, \lambda_i \mathbf{u}_i) - \left(\mathbf{u}_i, \overline{\lambda}_i \overline{\mathbf{u}}_i\right) = \left(\lambda_i - \overline{\lambda}_i\right)(\mathbf{u}_i, \overline{\mathbf{u}}_i).$$

Dividing by $(\mathbf{u}_i, \overline{\mathbf{u}}_i)$, which is a positive real number, we get $\lambda_i - \overline{\lambda}_i = 0$ which means that the imaginary part of λ_i is zero, or equivalently $\lambda_i \in \mathbb{R}$. ∎

What these results mean is that given a real symmetric matrix with distinct eigenvalues (i.e., if all eigenvalues are different), we can write

$$A[\mathbf{u}_1, \ldots, \mathbf{u}_n] = [\lambda_1 \mathbf{u}_1, \ldots, \lambda_n \mathbf{u}_n] \quad \text{or} \quad AQ = Q\Lambda,$$

where

$$Q = [\mathbf{u}_1, \ldots, \mathbf{u}_n] \quad \text{and} \quad \Lambda = \begin{pmatrix} \lambda_1 & 0 & \ldots & 0 \\ 0 & \lambda_2 & 0 & \ldots \\ 0 & \ldots & \lambda_{n-1} & 0 \\ 0 & \ldots & 0 & \lambda_n \end{pmatrix}.$$

In fact, this would be true even if there are repeated eigenvalues, but we will not prove that here.

A *positive-definite* matrix $A \in \mathbb{R}^{N \times N}$ is one for which

$$\mathbf{x}^T A \mathbf{x} \geq 0$$

with equality holding only when $\mathbf{x} = \mathbf{0}$.

THEOREM D.3
All eigenvalues of a real, positive-definite symmetric matrix are positive.

PROOF   Using the information given previously, we can rewrite $(\mathbf{x}, A\mathbf{x})$ (whose sign must be positive for all possible choices of $\mathbf{x} \in \mathbb{R}^n$ if it is positive definite) as $(\mathbf{x}, Q\Lambda Q^T \mathbf{x}) = (Q^T \mathbf{x}, \Lambda(Q^T \mathbf{x}))$. Defining $\mathbf{y} = Q^T \mathbf{x}$, we see that

$$(\mathbf{x}, A\mathbf{x}) = (\mathbf{y}, \Lambda \mathbf{y}) = \sum_{i=1}^{n} \lambda_i y_i^2.$$

Since this must be positive for arbitrary vectors $\mathbf{x}$, and therefore for arbitrary vectors $\mathbf{y}$, each λ_i must be greater than zero. If they are not, it would be possible to find a vector for which the quadratic form is not positive. ∎

As a result of these theorems, we will always be able to write a symmetric positive definite matrix $A = A^T \in \mathbb{R}^{n \times n}$ in the form

$$A = Q\Lambda Q^T$$

where $Q \in SO(n)$ is a special orthogonal matrix, and Λ is diagonal with positive elements on the diagonal.

Matrix Functions

In the same way that it is possible to expand certain functions of a real argument x in a convergent Taylor series about $x = 0$

$$f(x) = \sum_{n=0}^{\infty} \frac{f^{(n)}(0)}{n!} x^n$$

[where $f^{(n)}(x)$ is the nth derivative of $f(x)$ with respect to x in some open interval about $x = 0$], it is also possible to define certain *matrix functions* using a Taylor series. In the same way that the Pth power of a matrix A is

$$A^P = \underbrace{AA \cdots A}_{P \ times},$$

the exponential of a matrix A is

$$\exp(A) = \sum_{n=0}^{\infty} \frac{1}{n!} A^n.$$

Similarly, $\cos(A)$ and $\sin(A)$ are well defined. Fractional powers of matrices cannot generally be represented using a Taylor series [e.g., try writing the Taylor series for $f(x) = \sqrt{x}$; it cannot be done because the slope of this function is not defined at $x = 0$]. In fact, it is not generally true that the root of a matrix can even be calculated. For example, try to find a matrix A such that

$$A^2 = \begin{pmatrix} 0 & 1 \\ 0 & 0 \end{pmatrix}.$$

However, in the case of symmetric real matrices, roots can always be taken.

As shown in the previous section, a symmetric matrix can always be represented using the form

$$A = Q \Lambda Q^T.$$

It follows naturally that any positive power P of the matrix A is of the form

$$A^P = Q \Lambda^P Q^T.$$

What may be surprising is that for fractional and negative powers this expression also holds. For instance, the inverse of a matrix A can be written as A^{-1} where

$$A^{-1} = \left(Q \Lambda Q^T \right)^{-1} = Q \Lambda^{-1} Q^T,$$

and Λ^{-1} is the diagonal matrix with diagonal elements $(\Lambda^{-1})_{ii} = 1/(\Lambda)_{ii}$. Clearly if any of the elements of Λ are zero, this is going to cause a problem (as we would expect, a matrix with a zero eigenvalue is not invertible).

We can define the square root of a positive definite matrix as

$$M^{\frac{1}{2}} = Q \Lambda^{\frac{1}{2}} Q^T$$

such that

$$M^{\frac{1}{2}} M^{\frac{1}{2}} = \left(Q \Lambda^{\frac{1}{2}} Q^T \right) \left(Q \Lambda^{\frac{1}{2}} Q^T \right) = Q \Lambda^{\frac{1}{2}} \left(Q^T Q \right) \Lambda^{\frac{1}{2}} Q^T = Q \Lambda^{\frac{1}{2}} \Lambda^{\frac{1}{2}} Q^T = Q \Lambda Q^T = M.$$

Note: Matrix functions defined as Taylor series or roots are *not* simply the functions or roots applied to matrix entries. In fact, for any square matrix, $A = [a_{ij}]$, of dimension greater than 1×1, it is always the case (even for diagonal matrices) that

$$\left(e^A \right)_{ij} \neq e^{a_{ij}}.$$

The matrix roots defined previously are only for symmetric matrices, and it is **not** simply the square root of the elements of the matrix. Also, the square root of a real symmetric matrix is symmetric, but it is only real if the original matrix is positive semi-definite.

D.2   Special Properties of the Matrix Exponential

Given the $n \times n$ matrices X and A where $X = X(t)$ is a function of time and A is constant, the solution to the differential equation

$$\frac{d}{dt}(X) = AX \tag{D.1}$$

subject to the initial conditions $X(0) = \mathbb{I}_{n \times n}$ is

$$X(t) = \exp(tA).$$

We now show that

$$\det(\exp A) = e^{\mathrm{trace}(A)} \tag{D.2}$$

where

$$\mathrm{trace}(A) = \sum_{i=1}^{n} a_{ii} \tag{D.3}$$

and using the permutation notation introduced in Chapter 7,

$$\det(X) = \sum_{\sigma \in S_n} (\mathrm{sign}\,\sigma) x_{\sigma(1),1} \cdots x_{\sigma(n),n} = \sum_{\sigma \in S_n} (\mathrm{sign}\,\sigma) x_{1,\sigma(1)} \cdots x_{n,\sigma(n)}. \tag{D.4}$$

Here $(\mathrm{sign}\,\sigma) \in \{-1, 1\}$ with the sign determined by how the permutation decomposes into basic *transpositions* (i.e., permutations of only two numbers, leaving the remaining $n - 2$ numbers unchanged). An even/odd number of transpositions corresponds to $+/-$, respectively.

Using the notation

$$\det X = \begin{vmatrix} x_{11} & x_{12} & \cdots & x_{1n} \\ x_{21} & x_{22} & \cdots & \vdots \\ \vdots & \vdots & \ddots & \vdots \\ x_{n1} & x_{n2} & \cdots & x_{nn} \end{vmatrix},$$

it follows from the product rule for differentiation and Eq. (D.4) that

$$\frac{d}{dt}(\det X) = \begin{vmatrix} \frac{dx_{11}}{dt} & \frac{dx_{12}}{dt} & \cdots & \frac{dx_{1n}}{dt} \\ x_{21} & x_{22} & \cdots & \vdots \\ \vdots & \vdots & \ddots & \vdots \\ x_{n1} & x_{n2} & \cdots & x_{nn} \end{vmatrix} + \begin{vmatrix} x_{11} & x_{12} & \cdots & x_{1n} \\ \frac{dx_{21}}{dt} & \frac{dx_{22}}{dt} & \cdots & \vdots \\ \vdots & \vdots & \ddots & \vdots \\ x_{n1} & x_{n2} & \cdots & x_{nn} \end{vmatrix} + \cdots$$

$$+ \begin{vmatrix} x_{11} & x_{12} & \cdots & x_{1n} \\ \vdots & \vdots & \ddots & \vdots \\ \frac{dx_{n-1,1}}{dt} & \frac{dx_{n-1,2}}{dt} & \cdots & \frac{dx_{n-1,n}}{dt} \\ x_{n1} & x_{n2} & \cdots & x_{nn} \end{vmatrix} + \begin{vmatrix} x_{11} & x_{12} & \cdots & x_{1n} \\ x_{21} & x_{22} & \cdots & \vdots \\ \vdots & \vdots & \ddots & \vdots \\ \frac{dx_{n1}}{dt} & \frac{dx_{n2}}{dt} & \cdots & \frac{dx_{nn}}{dt} \end{vmatrix}. \tag{D.5}$$

Equation (D.1) is written in component form as

$$\frac{dx_{ik}}{dt} = \sum_{j=1}^{N} a_{ij} x_{jk}.$$

After making this substitution, one observes that the ith term in Eq. (D.5) becomes

$$\begin{vmatrix} x_{11} & x_{12} & \cdots & x_{1n} \\ \vdots & \vdots & & \vdots \\ \sum_{j=1}^{N} a_{ij} x_{j1} & \sum_{j=1}^{N} a_{ij} x_{j2} & \cdots & \vdots \\ \vdots & \vdots & \ddots & \vdots \\ x_{n1} & x_{n2} & \cdots & x_{nn} \end{vmatrix} = \begin{vmatrix} x_{11} & x_{12} & \cdots & x_{1n} \\ \vdots & \vdots & & \vdots \\ a_{ii} x_{i1} & a_{ii} x_{i1} & \cdots & \vdots \\ \vdots & \vdots & \ddots & \vdots \\ x_{n1} & x_{n2} & \cdots & x_{nn} \end{vmatrix}.$$

This follows by subtracting a_{ij} multiplied times the jth row of X from the ith row of the left side for all $j \neq i$.

The result is then

$$\frac{d}{dt}(\det X) = \text{trace}(A)(\det X).$$

Since $\det X(0) = 1$, this implies

$$\det X = \exp(\text{trace}(A)t).$$

Evaluation of both sides at $t = 1$ yields Eq. (D.2).

The equalities

$$\exp(A + B) = \exp A \exp B = \exp B \exp A \qquad (D.6)$$

may be easily observed when

$$AB = BA$$

by expanding both sides in a Taylor series and equating term by term. What is perhaps less obvious is that sometimes the first and/or the second equality in Eq. (D.6) can be true when A and B do not commute. For example, Fréchet [17] observes that when

$$A = \begin{pmatrix} 0 & 2\pi \\ -2\pi & 0 \end{pmatrix} \qquad B = \begin{pmatrix} 1 & 0 \\ 0 & -1 \end{pmatrix},$$

$AB \neq BA$, but it is nonetheless true that

$$e^A e^B = e^B e^A.$$

In the same paper, he observed that if

$$A = \pi \begin{pmatrix} 0 & \lambda \\ -1/\lambda & 0 \end{pmatrix} \qquad B = \pi \begin{pmatrix} 0 & 10 + 4\sqrt{6}\lambda \\ \frac{-10+4\sqrt{6}}{\lambda} & 0 \end{pmatrix}$$

then for $\lambda \neq 0$,

$$e^A e^B = e^{A+B}$$

even though $AB \neq BA$. Another example of a choice for the matrices A and B can be found in [53] for which $AB \neq BA$ and $e^{A+B} = e^A e^B \neq e^B e^A$.

Having said this, $AB = BA$ is in fact a necessary condition for $\exp t(A + B) = \exp t A \exp t B = \exp t B \exp t A$ to hold for all values of $t \in \mathbb{R}^+$.

D.3   Matrix Decompositions

In this section we state without proof some useful definitions and theorems from matrix theory.

D.3.1   Decomposition of Complex Matrices

THEOREM D.4 (QR decomposition) [25]
For any $N \times N$ matrix, A, with complex entries, it is possible to find an $N \times N$ unitary matrix Q such that

$$A = QR$$

where R is upper triangular. In the case when A is real, one can take $Q \in O(N)$ and R real.

THEOREM D.5 (Cholesky decomposition) [25]
Any $N \times N$ complex matrix B that is decomposible as $B = A^\dagger A$ for some $N \times N$ complex matrix A can be decomposed as $B = LL^\dagger$ where L is lower triangular with nonnegative diagonal entries.

THEOREM D.6 (Schur's unitary triangularization theorem)[25]
For any $A \in \mathbb{C}^{N \times N}$, it is possible to find a matrix $U \in U(N)$ such that

$$U^\dagger A U = T$$

where T is upper (or lower) triangular with the eigenvalues of A on its diagonal.

Note: This does *not* mean that for real A that U will necessarily be real orthogonal. For example, if $A = -A^T$ then $Q^T A Q$ for $Q \in O(N)$ will also be skew-symmetric and hence cannot be upper triangular.

A *Jordon block* corresponding to a k-fold repeated eigenvalue is a $k \times k$ matrix with the repeated eigenvalue on its diagonal and the number 1 in the super diagonal. For example,

$$J_2(\lambda) = \begin{pmatrix} \lambda & 1 \\ 0 & \lambda \end{pmatrix},$$

$$J_3(\lambda) = \begin{pmatrix} \lambda & 1 & 0 \\ 0 & \lambda & 1 \\ 0 & 0 & \lambda \end{pmatrix},$$

etc.

The direct sum of two square matrices, $A \in \mathbb{C}^{N \times N}$ and $B \in \mathbb{C}^{M \times M}$, is the $(M + N) \times (M + N)$ block-diagonal matrix

$$A \oplus B = \begin{pmatrix} A & 0_{N \times M} \\ 0_{M \times N} & B \end{pmatrix}.$$

Clearly

$$\text{trace}(A \oplus B) = \text{trace}(A) + \text{trace}(B)$$

and

$$\det(A \oplus B) = \det(A) \cdot \det(B).$$

The notation $n_i J_i$ stands for the n_i-fold direct sum of J_i with itself

$$n_i J_i = J_i \oplus J_i \oplus \cdots \oplus J_i = \begin{pmatrix} J_i & 0 & 0 \\ 0 & \ddots & 0 \\ 0 & 0 & J_i \end{pmatrix}.$$

The notation

$$\sum_{i=1}^{m} \bigoplus A_i = A_1 \oplus A_2 \oplus \cdots \oplus A_m$$

is common, and we use it in Chapter 8, as well as below. Note that $\dim(J_i) = i$ and

$$\dim\left(\sum_{i=1}^{m} \bigoplus A_i\right) = \sum_{i=1}^{m} \dim A_i.$$

THEOREM D.7
Every matrix $A \in \mathbb{C}^{N \times N}$ can be written in the Jordan normal form

$$A = TJT^{-1}$$

where T is an invertible matrix and

$$J = \sum_{j=1}^{q} \bigoplus \left(\sum_{i=1}^{m} \bigoplus n_i^j J_i(\lambda_j)\right)$$

is the direct sum of a direct sum of Jordan blocks with m being the dimension of the largest Jordan block, q being the number of different eigenvalues, and n_i^j being the number of times $J_i(\lambda_j)$ is repeated in the decomposition of A.

Note that

$$\sum_{j=1}^{q} \sum_{i=1}^{m} i \cdot n_i^j = \dim(A).$$

For instance, if

$$J = J_1(\lambda_1) \oplus J_1(\lambda_2) \oplus J_2(\lambda_3) \oplus J_2(\lambda_3) \oplus J_3(\lambda_4) \oplus J_5(\lambda_5) \oplus J_6(\lambda_5),$$

then $m = 6$, $q = 5$, and all values of n_i^j are zero except for

$$n_1^1 = n_1^2 = n_3^4 = n_5^5 = n_6^5 = 1; \quad n_2^3 = 2.$$

In the special case when A is real *and* all of its eigenvalues are real, T can be taken to be real. A square matrix A with complex entries is called *normal* if

$$A^\dagger A = AA^\dagger.$$

Examples of normal matrices include Hermitian, skew-Hermitian, unitary, real orthogonal, and real symmetric matrices.

THEOREM D.8
Normal matrices are unitarily diagonalizable [25].

D.3.2   Decompositions of Real Matrices

In this subsection we review some numerical methods used for real matrices.

THEOREM D.9 (singular value decomposition) [19]

For any real $M \times N$ matrix A, there exist orthogonal matrices $U \in O(M)$ and $V \in O(N)$ such that

$$A = U \Lambda V^T$$

where Λ is an $M \times N$ matrix with entries $\Lambda_{ij} = \sigma_i \delta_{ij}$. The value σ_i is the ith largest singular value of A.

As a consequence of the above theorem, every real square matrix A can be written as the product of a symmetric matrix $S_1 = U \Lambda U^T$ and the orthogonal matrix

$$R = U V^T, \tag{D.7}$$

or as the product of R and $S_2 = V \Lambda V^T$. Hence we write

$$A = S_1 R = R S_2. \tag{D.8}$$

This is called the *polar decomposition*.

In the case when det $A \neq 0$, one calculates

$$R = A \left(A^T A \right)^{-\frac{1}{2}} \tag{D.9}$$

(the negative fractional root makes sense for a symmetric positive definite matrix). We note that Eq. (D.7) is always a stable numerical technique for finding R, whereas Eq. (D.9) becomes unstable as det(A) becomes small.

The Cholesky decomposition of a real symmetric positive definite matrix is

$$G = L L^T$$

where L is a lower triangular real matrix with nonnegative diagonal entries.

For $N \times N$ real matrices such that $\det(A_i) \neq 0$ for $i = 1, \ldots, N$,[3] it is possible to write

$$A = LU$$

where L is a unique lower triangular matrix and U is a unique upper triangular matrix. This is called an LU-decomposition [19].

See the classic references [18, 52] for other general properties of matrices and other decompositions.

---

[3]The notation A_i denotes the $i \times i$ matrix formed by the first i rows and i columns of the matrix A.

Appendix E

Techniques from Mathematical Physics

In this appendix we review the Dirac delta function, self-adjoint differential operators, path integration, and quantization. These methods from mathematical physics find their way into several of the chapters.

E.1 The Dirac Delta Function

In several chapters we use properties of the Dirac delta function that are presented here. Our treatment follows those in [29, 43].

The Dirac delta "function" is actually not a function in the most rigorous sense. It is an example of a class of mathematical objects called *generalized functions* or *distributions*. While the Dirac delta function has been used in the physical sciences to represent highly concentrated or localized phenomena for about a century, the development of a rigorous mathematical theory of distributions came much more recently. It was not until the middle of the 20th century that the French mathematician Laurent Schwartz articulated a precise mathematical theory of distributions, building on the foundations of the Russian mathematician Sergei Sobolev.

Intuitively, a Dirac delta function is a "spike" that integrates to unity and has the property that it "picks off" values of a continuous function when multiplied and integrated

$$\int_{-\infty}^{\infty} f(x)\delta(x)dx = f(0).$$

There are many ways to express the Dirac delta function as the limit of a sequence of functions substituted in the previous integral. That is, there are many sequences of functions $\{\delta_m^{(k)}(x)\}$ for $m = 1, 2, \ldots$ for which

$$\lim_{m \to \infty} \int_{-\infty}^{\infty} f(x)\delta_m^{(k)}(x)dx = f(0) \qquad \text{(E.1)}$$

is satisfied. Here we use k to specify each of a number of classes of such functions, and m enumerates individual functions for fixed k. For instance,

$$\delta_m^{(1)}(x) = \frac{\sin mx}{\pi x} = \frac{1}{2\pi} \int_{-m}^{m} e^{i\omega x} d\omega \; ;$$

$$\delta_m^{(2)}(x) = \frac{1}{\pi} \frac{m}{1 + m^2 x^2} \; ;$$

$$\delta_m^{(3)}(x) = \left(\frac{m}{\pi}\right)^{\frac{1}{2}} e^{-mx^2} \; ;$$

$$\delta_m^{(4)}(x) = \begin{cases} \dfrac{\exp\left(\frac{-1}{1-m^2 x^2}\right)}{\int_{-\frac{1}{m}}^{\frac{1}{m}} \exp\left(\frac{-1}{1-m^2 r^2}\right) dr} & \text{for } |x| < 1/m \\[6pt] 0 & \text{for } |x| \geq 1/m \end{cases} \; ;$$

$$\delta_m^{(5)}(x) = \begin{cases} 1/(2m) & \text{for } |x| < 1/m \\ 0 & \text{for } |x| \geq 1/m \end{cases} \; ;$$

$$\delta_m^{(6)}(x) = \begin{cases} 1/m & \text{for } 0 \leq x < 1/m \\ 0 & \text{elsewhere} \end{cases} \; ;$$

$$\delta_m^{(7)}(x) = \begin{cases} -m^2|x| + m & \text{for } |x| < 1/m \\ 0 & \text{elsewhere} \end{cases} .$$

Each of the sequences $\{\delta_m^{(k)}(x)\}$ has certain advantages. For instance, for $k = 1, \ldots, 4$, the functions are continuous for finite values of m. The cases $k = 5, 6, 7$ are intuitive definitions given in engineering texts. In the cases $k = 4 - 7$, the functions $\delta_m^{(4)}(x)$ all have finite support. The case $k = 1$ is used in the Shannon sampling theorem, and we use $k = 2$ (the Gaussian) in the proof of the Fourier inversion formula in Chapter 2. Others are possible as well, but our discussion is restricted to those mentioned above.

Regardless of which sequence of distributions is used to describe the Dirac delta function, we note that $\delta(x)$ is only rigorously defined when it is used inside of an integral, such as Eq. (E.1). Under this restriction, we use the notation

$$\delta(x) =_w \lim_{m \to \infty} \delta_m^{(k)}(x)$$

for any $k \in \{1, \ldots, 7\}$ to mean that Eq. (E.1) holds for any continuous and square integrable $f(x)$. We will call $=_w$ "weak equality." If $=_w$ is replaced with a strict equality sign, the limit does not rigorously exist. However, to avoid the proliferation of unwieldy notation throughout the book, *every equality involving delta functions on the outside of an integral will be assumed to be equality in the sense of $=_w$*. This is not such a terrible abuse, since $\delta(x)$ is only used inside of an integral anyway.

E.1.1   Derivatives of Delta Functions

Considering the "spike-like" nature of the Dirac delta function (which is so singular it is not even really a function), it may appear implausible to consider differentiating $\delta(x)$. However, since $\delta(x)$ is only used under an integral in which it is multiplied by a continuous (and in the present context, an n-times differentiable) function, we can use integration by parts to define

$$\int_{-\infty}^{\infty} f(x) \frac{d^n}{dx^n} \delta(x) dx = (-1)^n \int_{-\infty}^{\infty} \frac{d^n f}{dx^n} \delta(x) dx = (-1)^n \frac{d^n f}{dx^n}\bigg|_{x=0} .$$

In this context, we do not use the $k = 5, 6, 7$ definitions of the sequence $\delta_m^{(k)}(x)$ since they are not differentiable by definition. For the others, it is often possible to obtain analytical expressions for the integral of $\frac{d^n f}{dx^n} \delta_m^{(k)}(x)$, from which the limit can be calculated explicitly.

Other more intricate formulas involving derivatives of Dirac delta functions can be found in [29].

E.1.2 Dirac Delta Functions with Functional Arguments

In the context of orthogonal expansions and transforms in curvilinear coordinates, it is very useful to decompose Dirac delta functions evaluated at a function of the coordinates into delta functions in the coordinates themselves. For instance, in polar coordinates $x_1 = r \cos \theta$ and $x_2 = r \sin \theta$. Therefore,

$$\delta \left(\mathbf{x} - \mathbf{x}^0 \right) = \delta \left(x_1 - x_1^0 \right) \delta \left(x_2 - x_2^0 \right) = \delta \left(r \cos \theta - x_1^0 \right) \delta \left(r \sin \theta - x_2^0 \right).$$

It is useful to write this in terms of separate delta functions in r and θ. The first step in doing this is to write $\delta(y(x))$ in terms of $\delta(x)$ when $y(\cdot)$ is a differentiable and invertible transformation from the real line to itself with x_0 defined by the condition $y(x_0) = 0$. Using the standard change-of-coordinates formula

$$\int_{y(a)}^{y(b)} F(y) dy = \int_a^b F(y(x)) |dy/dx| dx$$

in reverse, we write

$$\int_{-\infty}^{\infty} \delta(y(x)) f(x) dx = \int_{-\infty}^{\infty} \delta(y) f(x(y)) |dx/dy| dy.$$

Since $\delta(y)$ "picks off" the value of the function in the second integral above at $y = 0$, we have

$$\int_{-\infty}^{\infty} \delta(y(x)) f(x) dx = [f(x(y)) |dx/dy|]_{y=0} = [f(x)/|dy/dx|]_{x=x_0}.$$

Since

$$[f(x)/|dy/dx|]_{x=x_0} = \int_{-\infty}^{\infty} \delta(x - x_0) f(x)/|dy/dx| dx,$$

we conclude that

$$\delta(y(x)) = \delta(x - x_0) /|dy/dx|.$$

Similarly, when $y(x)$ is not monotonic (and hence not invertible) with $y(x_i) = 0$ and $(dy/dx)_{x=x_i} \neq 0$ for $i = 0, \ldots, n - 1$, we have

$$\delta(y(x)) = \sum_{i=0}^{n-1} \frac{\delta(x - x_i)}{|y'(x)|} \tag{E.2}$$

where $y' = dy/dx$. As always, the equality sign in Eq. (E.2) is interpreted as equality under the integral, and this leads to multiple possible definitions of $\delta(y(x))$ that act in the same way. For instance, we can write $y'(x_i)$ in place of $y'(x)$ in Eq. (E.2) or multiply the right side of Eq. (E.2) by any continuous function $\phi(x)$ satisfying $\phi(x_i) = 1$. Hence, instead of considering the previous discussion as a derivation of Eq. (E.2), it can be considered as a justification for using Eq. (E.2) as a definition.

E.1.3   The Delta Function in Curvilinear Coordinates: What to do at Singularities

Given a multi-dimensional coordinate transformation $\mathbf{y} = \mathbf{y}(\mathbf{x})$ [which is written in components as $y_i = y_i(x_1, \ldots, x_n)$ for $i = 1, \ldots, n$], the following well-known integration rule holds

$$\int_{Im(Box(\mathbf{a},\mathbf{b}))} F(y_1, \ldots, y_n)\, dy_1 \cdots dy_n$$

$$= \int_{a_1}^{b_1} \cdots \int_{a_n}^{b_n} F(y_1(x_1, \ldots, x_n), \ldots, y_n(x_1, \ldots, x_n)) \,|\det J|\, dx_1 \cdots dx_n$$

where

$$J = \left[\frac{\partial \mathbf{y}}{\partial x_1}, \ldots, \frac{\partial \mathbf{y}}{\partial x_n} \right]$$

is the Jacobian matrix of the transformation and $|\det J|$ gives a measure of local volume change, and $Im(Box(\mathbf{a}, \mathbf{b}))$ is the "image" of the box $a_i \le x_i \le b_i$ for $i = 1, \ldots, n$ in y-coordinates.

Using the same reasoning as in the one-dimensional case and assuming the function $\mathbf{y}$ is differentiable and one-to-one, we write

$$\delta(\mathbf{y}(\mathbf{x})) = \delta(\mathbf{x} - \mathbf{x}_0)/|\det J(\mathbf{x})|$$

where $\mathbf{y}(\mathbf{x}_0) = \mathbf{0}$, and it is assumed that $|\det J(\mathbf{x}_0)| \ne 0$.

In the case when the independent variables are curvilinear coordinates $\mathbf{u} = (u_1, \ldots, u_n)$ and the dependent variable is the position $\mathbf{x} \in \mathbb{R}^n$, we have the parameterization $\mathbf{x}(\mathbf{u})$ and

$$
\begin{aligned}
\delta(\mathbf{x} - \mathbf{p}) &= \delta(\mathbf{x}(\mathbf{u}) - \mathbf{p}) \\
&= \delta(\mathbf{u} - \mathbf{q})/|\det J(\mathbf{u})| = \delta(u_1 - q_1) \cdots \delta(u_n - q_n)/|\det J(u_1, \ldots, u_n)| \quad \text{(E.3)}
\end{aligned}
$$

where $\mathbf{p} = \mathbf{x}(\mathbf{q})$, and we assume that $|\det J(\mathbf{q})| \ne 0$. For example, using the spherical coordinates described in Chapter 4, we write

$$\delta(\mathbf{x} - \mathbf{p}) = \frac{\delta(r - r_0)\, \delta(\theta - \theta_0)\, \delta(\phi - \phi_0)}{r^2 \sin\theta}$$

where $\mathbf{p} = \mathbf{x}(r_0, \theta_0, \phi_0)$ and $\mathbf{q} = [r_0, \theta_0, \phi_0]^T$ represents a point where the description is nonsingular.

At a singularity $\mathbf{x}(\sigma) = \mathbf{s}$, we have $|\det J(\sigma)| = 0$, and the calculation of the Dirac delta is a bit different. At a singularity, the curvilinear coordinate description breaks down in the sense that a continuum of coordinate values (as opposed to a single discrete value) map to one point. For instance, in spherical coordinates, when $\theta = 0$, all values of ϕ describe the same point on the x_3-axis, and when $r = 0$, all values of θ and ϕ describe the same point ($\mathbf{x} = \mathbf{0}$).

More generally, in curvilinear coordinates $\mathbf{u} = (u_1, \ldots, u_n)$ such that $\mathbf{x} = \mathbf{x}(\mathbf{u})$, we have as a special case that at a singular point

$$\mathbf{s} = \mathbf{x}(v_1, \ldots, v_k, \sigma_{k+1}, \ldots, \sigma_n)$$

where $\sigma_{k+1}, \ldots, \sigma_n$ are fixed parameter values that define $\mathbf{s}$, and $v_1, \ldots, v_k$ can take any values in a continuous set of values (denoted below as C_σ) for which they are defined. The definition of the Dirac delta function in curvilinear coordinates in Eq. (E.3) is then modified at singularities as

$$\delta(\mathbf{x} - \mathbf{s}) = \frac{\delta(u_{k+1} - \sigma_{k+1}) \cdots \delta(u_n - \sigma_n)}{\int_{C_\sigma} |\det J(v_1, \ldots, v_k, u_{k+1}, \ldots, u_n)|\, dv_1 \cdots dv_k}. \quad \text{(E.4)}$$

For a continuous function $f(\mathbf{x}) \in L^2(\mathbb{R})$ for $\mathbf{x} = \mathbf{x}(\mathbf{u})$, this has the desired effect of yielding the same result as the definition

$$\int_{\mathbb{R}^n} f(\mathbf{x})\delta(\mathbf{x} - \mathbf{s})d\mathbf{x} = f(\mathbf{s})$$

in Cartesian coordinates. However, Eq. (E.4) *only* holds at singularities.

The extension of these concepts to M-dimensional surfaces in $\mathbb{R}^N$ and Lie groups follows in exactly the same way, with the appropriate definition of the Jacobian. Examples are provided throughout the book.

E.2   Self-Adjoint Differential Operators

Linear second-order differential operators acting on complex-valued functions $u(x)$ defined on an interval $[0, 1]$ are written as

$$Lu = a_0(x)u'' + a_1(x)u' + a_2(x)u. \tag{E.5}$$

We will assume that the complex-valued functions $a_i(x)$ are at least twice differentiable, and $a_0(x) \neq 0$ for $0 < x < 1$.

Let $\{u_i(x)\}$ be a complete orthonormal basis for $\mathcal{L}^2([0, 1], \mathbb{C}, dx)$. The *matrix elements* of L in this basis are defined relative to the inner product

$$(u_i, u_j) = \int_0^1 \overline{u_1} u_2 \, dx$$

as

$$L_{ij} = (u_i, L u_j) = \int_0^1 \overline{u_i} L u_j \, dx.$$

The operator L is called *Hermitian* if [3]

$$(u_i, L u_j) = (L u_i, u_j). \tag{E.6}$$

That is, the operator L is called Hermitian when the matrix with elements $L_{ij} = (u_i, L u_j)$ is a Hermitian matrix. Since the basis in this case has an infinite number of elements, a Hermitian operator corresponds to an infinite-dimensional Hermitian matrix.

The *adjoint*, $L^\dagger$, of the operator L in Eq. (E.5) is defined by the equality

$$(u_i, L^\dagger u_j) = (L u_i, u_j). \tag{E.7}$$

In the current context, a self-adjoint operator ($L = L^\dagger$) is always Hermitian, though the concept of a self-adjoint operator also appears in more abstract settings (e.g., when $\{u_i\}$ are not complex valued).

Explicitly, the form of the adjoint of the operator in Eq. (E.5) is

$$L^\dagger u = \left[\overline{a_0(x)} u\right]'' - \left[\overline{a_1(x)} u\right]' + \overline{a_2(x)} u.$$

This is found by integrating $(L u_i, u_j)$ by parts and requiring the conditions

$$\left[\bar{a}_0 \bar{u}_i' u_j - \bar{u}_i (\bar{a}_0 u_j)' + \bar{a}_i \bar{u}_i u_j\right]\big|_0^1 = 0$$

The concepts of self-adjoint and Hermitian operators extend to functions on domains other than the unit interval. In the case of the circle and rotation groups, boundary conditions analogous to those above are satisfied automatically due to the continuity of the basis functions. The definitions in Eqs. (E.6) and (E.7) are exactly the same for these cases with the inner product being defined with the appropriate integration measures.

E.3   Functional Integration

The functional integral has been used extensively in theoretical physics to describe quantum and statistical phenomena [27, 42].

Functional integration may be considered as a generalization of multi-dimensional integration for the case of an infinite-dimensional domain. Let us first consider the integral

$$\int \exp\left(-f(x)\right) dx$$

(with finite or infinite interval of integration).

Now consider the n-dimensional integral

$$\int \cdots \int \exp\left(-f(x_1, \ldots, x_n)\right) dx_1 \ldots dx_n. \tag{E.8}$$

We assume that the integral in Eq. (E.8) is well defined. The functional integral is the generalization of Eq. (E.8) for the case of infinite dimensional domains (for real- or complex-valued functions). Let us consider $f(x)$ [we assume that $f(x) \in \mathcal{L}^2([x_-, x_+])$] on the one dimensional interval $[x_-, x_+]$ (which may be considered finite or infinite; in physical applications it is frequently infinite). We also consider some functional $S(f) = \int L(f(x))dx$. This functional has physical meaning (usually it is an action functional) and may contain differential operators. An expression of the type $\exp(-S)$ has practical importance for physical applications.

We discretize this interval into x_n, write $\int L(f(x))dx$ as the sum

$$\sum_n L\left(f(x_n)\right) \Delta x_n = \sum_n \tilde{L}\left(f(x_n)\right),$$

and integrate the product of function values $\exp S(f(x_n))$ for all possible values of $f_n = f(x_n)$

$$Z = \lim_{N \to \infty, \Delta x_n \to 0} \int \cdots \int \exp\left[-\sum_{n=0}^{N} \tilde{L}(f_n)\right] df_1 \ldots df_N. \tag{E.9}$$

We assume that the expression in Eq. (E.9) exists. In this case, this expression defines the functional integral

$$Z = \int \exp\left(-S(f(x))\right) Df \tag{E.10}$$

where Df stands for an infinite dimensional integration measure. It is assumed that the function $f(x)$ has well-defined limits f_-, f_+ at $x \to \pm\infty$ (or for $x = x_\pm$ on finite intervals). Specific boundary conditions may be imposed on $f(x)$ also at finite x.

The expression in Eq. (E.10) defines the one-dimensional functional integral [because $f(x)$ is a function of the one dimensional parameter x]; the functional integral may be also generalized for m-dimensional space $x_i, i = 1, \ldots, m$.

The term *path integration* is also frequently used for functional integration for the following reason. Let us consider a discrete set of values $f(x_n)$ for all possible values of x_n. The set of all possible values $f(x_n)$ can be depicted as the two dimensional grid in Fig. E.1. Then the functional integral in Eq. (E.9) can be considered as a sum of the products of weight factors $\exp(-\tilde{L}(f(x_n)))$ taken along all possible paths on the grid of Fig. E.1.

For example, for the path in Fig. E.1, we have to multiply factors $\exp(-\tilde{L}(f^i(x_n)))$ at each point x_n [where the function has some value $f^i(x_n)$] to get a contribution of this path to the functional integral. Summation of contributions along all possible paths satisfying boundary conditions $f(x) \to f_\pm$ (for $x \to \pm\infty$) defines the functional integral in Eq. (E.10) in the limit $\Delta x_n \to 0$, $N \to \infty$.

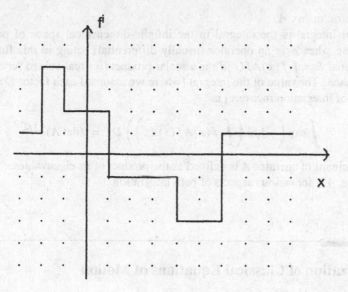

FIGURE E.1

Illustration of the path integral.

The contribution with the minimal value of product factors $\exp(-\tilde{L}(f^i(x_n)))$ [which defines the function $f(x)$ with the minimal action $S(f(x))$ for the case of a real-valued functional] gives the largest contribution to the functional integral, and other path contributions describe additional quantum or statistical corrections.

For quadratic (Gaussian) functionals $S(f(x))$ the integration may be performed exactly. Let us consider first the multi-dimensional integral

$$\int_{-\infty}^{\infty} \cdots \int_{-\infty}^{\infty} \exp\left(-1/2 \sum_{j=1}^{n} a_j x_j^2\right) dx_1 \ldots dx_n = \frac{(2\pi)^{n/2}}{\Pi_{j=1}^{n} a_j^{1/2}}.$$

We consider the set of coordinates $(x_1, \ldots, x_n)$ as a vector in n-dimensional space and write the sum as a scalar product of vectors $\mathbf{x}$ and $A\mathbf{x}$, where A is a diagonal matrix with diagonal elements a_n

$$\sum_{j=1}^{n} a_j x_j^2 = (\mathbf{x}, A\mathbf{x}).$$

We assume that a $1/(2\pi)^{1/2}$ factor is included in the definition of each integration measure. Then the result of the integration may be written as

$$\int \exp\left(-1/2(\mathbf{x}, A\mathbf{x})\right) d^n x / (2\pi)^{n/2} = (\det A)^{-1/2}.$$

This expression is also valid for the case of real positive definite symmetric matrices, and the Gaussian integration can also be performed for Hermitian matrices for the case of complex-valued functions. We also mention the expression

$$\int \exp\left(-[1/2(\mathbf{x}, A\mathbf{x}) + (\mathbf{b}, \mathbf{x}) + c]\right) d^n x / (2\pi)^{n/2}$$
$$= \exp\left(1/2\left(\mathbf{b}, A^{-1}\mathbf{b}\right) - c\right)(\det(A))^{-1/2}$$

for a real symmetric matrix A.

The functional integral is the integral in the infinite-dimensional space of possible values of $f(x)$. In the case when A is an operator (usually differential) acting in this function space, the quadratic functional $S = \int f(x) A f(x) dx$ is a scalar product (for real-valued functions) (f, Af) in the functional space. The value of the integral [where we assumed each factor $(2\pi)^{1/2}$ is absorbed in the definition of integration measure] is

$$\int \exp\left(-1/2\left(\int f(x) A f(x) dx\right)\right) Df = (\det A)^{-1/2}$$

where the determinant of operator A is defined as the product of its eigenvalues.

See [30, 44, 48, 49] for various aspects of path integration.

E.4   Quantization of Classical Equations of Motion

The functional integral of the form

$$Z = \int \exp\left(-S(f(s))\right) Df \tag{E.11}$$

where Df stands for an infinite-dimensional integration measure on the one-dimensional line (or segment) parameterized by s, describes a partition function of a statistical system in statistical physics. Similar integrals have been used to describe the amplitude of transitions and Green functions in quantum field theory. As we have discussed previously, the path with the smallest value of the action functional $S(f) = \int L(f(s), f'(s)) ds$ [where $L(f(s), f'(s))$ is the Lagrangian of the system] gives the largest contribution to the functional integral. The path with the minimal action gives a classical trajectory in classical mechanics. The variation of the action $S(f)$ with respect to f gives the well-known Euler-Lagrange equation (see Appendix F for derivation)

$$\frac{\partial L}{\partial f} = \frac{d}{ds}\left(\frac{\partial L}{\partial f'}\right).$$

An equivalent equation of motion can be found in the Hamiltonian formalism. We remind the reader that the Hamiltonian of the system is calculated from the Lagrangian L as

$$H = pf' - L$$

where $p = \partial L/\partial f'$ is the classical momentum, and $f' = df/ds$. The Hamiltonian for the system should be expressed as a function of p and f.[1] The classical equations of motion in the Hamiltonian formalism are written as

$$f' = \frac{\partial H(p, f)}{\partial p}$$

and

$$p' = -\frac{\partial H(p, f)}{\partial f}.$$

The Poisson bracket defined as

$$\{H, A\} = \frac{\partial H}{\partial p}\frac{\partial A}{\partial f} - \frac{\partial H}{\partial f}\frac{\partial A}{\partial p}$$

describes the evolution of a dynamical quantity with the "time"-coordinate, i.e.,

$$\frac{dA}{ds} = \{H, A\}.$$

We note

$$\{p, f\} = 1. \tag{E.12}$$

In the quantum case, momenta and coordinates become operators, which reflects the fact that they do not commute with each other as in the classical case. For a system with a single coordinate and momentum

$$\left[\hat{p}, \hat{f}\right] = \hat{p}\hat{f} - \hat{f}\hat{p} = -i. \tag{E.13}$$

This is a quantum analogy of Eq. (E.12).

This equation is valid when it acts on the wave functions describing a quantum system. Then the Hamiltonian $\hat{H}(\hat{p}, \hat{f})$ describing the system becomes an operator and the Schrödinger-like equation

$$i\frac{\partial \psi(f, t)}{\partial t} = \hat{H}\psi(f, t)$$

can be used to describe a non-relativistic one-particle quantum system. Thus, the main step in quantization consists of replacing classical momentum by the operator

$$p \rightarrow \hat{p} = -i\frac{\partial}{\partial f}.$$

The coordinate operators act as simple multiplication in this representation, i.e.,

$$\hat{f}\psi = f\psi.$$

The physical reason for quantization is the *uncertainty* principle, i.e., the principle stating that coordinates and momenta cannot be measured precisely at the same time. More rigorously, for the one dimensional case

$$\delta f \delta p \geq 1/2 \, h/(2\pi)$$

where δf and δp are the uncertainty in measurements of the coordinate and momentum. Here h is Planck's constant [units where $h/(2\pi)$ is set to 1 are used frequently; we assume this in this section].

---

[1] f is an analogy of the coordinate and s is an analogy of time in this notation.

Thus a quantum particle with precise value of the momentum p should have a completely uncertain coordinate f. The wave function

$$\psi = \exp(i \, p \, f)$$

satisfies this property. We also observe that the operator $\hat{p}$ acts on ψ as

$$\hat{p}4 = p4.$$

That is, this operator describes the momentum of a quantum particle. It can be easily checked that for operators $\hat{p}$ and $\hat{f}$ (where $\hat{f}\psi = f\psi$) that the equation

$$\left(\hat{p}\hat{f} - \hat{f}\hat{p}\right)\psi(f) = -i\frac{\partial(f\psi)}{\partial f} - \left(-i\frac{f\partial\psi}{\partial f}\right) = -i\psi$$

is valid, which is just the commutation relation in Eq. (E.13).

Another approach to look at quantization is a group theoretical approach. As we mentioned in Chapter 9, each invariant group of transformations of a system corresponds to a conservation law for the system (an m-parameter group corresponds to m conserving dynamical invariants for the system, which is a statement of Noether's theorem). The momentum of the system can be considered as an infinitesimal translation operator

$$\hat{p}\psi = -i\left.\frac{d\psi(f+\Delta)}{d\Delta}\right|_{\Delta=0} = -i\frac{\partial\psi}{\partial f}.$$

The eigenfunction of the momentum operator is $\psi = \exp ipf$ and the commutation relation in Eq. (E.13) is valid. The substitution of classical momentum p and coordinate f in the Hamiltonian by $\hat{p}$ and $\hat{f}$ is a quantization of the system. In the case of a polymer system, the infinitesimal operators X_i^R of rotations acting on functions on $SO(3)$ are used to quantize the classical angular momentum of the system.

Appendix F

Variational Calculus

From elementary calculus one knows that the extremal values (minima, maxima, or points of inflection) of a differentiable function $f(x)$ can be found by taking the derivative of the function and setting it equal to zero

$$\text{Solve:} \quad \left.\frac{df}{dx}\right|_{x=y} = 0 \quad \Leftrightarrow \quad y \in \mathcal{Y}$$

(i.e., $\mathcal{Y}$ is defined to be the set of extrema).

Determining whether or not the values $y \in \mathcal{Y}$ yield local minima, maxima, or points of inflection can be achieved by looking at whether the neighboring values of $f(y)$ for $y \in \mathcal{Y}$ are greater or less than these values themselves. This is equivalent to taking the second derivative of $f(y)$, when f is at least twice differentiable and checking its sign when evaluated at the points $y \in \mathcal{Y}$.

While engineers are familiar with the extension of elementary calculus to the case of several variables (called multivarible calculus), there is another very useful extension that some engineers may not be familiar with. Suppose that instead of finding the values of a variable y that extremizes a function $f(y)$, we are interested in extremizing a quantity of the form

$$J = \int_{x_1}^{x_2} f\left(y, \frac{dy}{dx}, x\right) dx$$

with respect to all possible *functions* $y(x)$. How can this be done? Two examples of when such problems arise are now given.

Example F.1

Suppose we want to find the curve $y = y(x)$ in the $x - y$ plane which connects the points (x_1, y_1) and (x_2, y_2), and which has minimal length. The length of the curve is

$$J = \int_{x_1}^{x_2} \sqrt{1 + \left(\frac{dy}{dx}\right)^2} \, dx.$$

While we already know the answer to this question (the straight line segment connecting the two points), this example illustrates the kinds of problems to which variational calculus can be applied.

Example F.2

As another example, we restate a problem of an economic/management nature found in [28]. Suppose that a small company produces a single product. The amount of the product in stock is $x(t)$ (it is a function of time because depending on sales and the rate at which it is produced, the product inventory changes). The rate with which the product can be made is $dx/dt = \dot{x}(t)$ (units per day).

Suppose the company receives an order to deliver an amount A in T days from the time of the order, i.e., $x(T) = A$. Assume nothing is in inventory when the order is placed, so $x(0) = 0$. Now, the decision as to what production schedule to use (plot of $\dot{x}$ as a function of time) depends on: (1) the cost of keeping inventory and (2) the cost associated with production. Suppose the cost of inventory is proportional to the amount of product that is currently being held in inventory: $C_I = c_2 x(t)$ (such as would be the case if the company pays rent by the square foot). Suppose the cost of producing one unit of product is linear in the rate at which it is produced: $C_P = c_1 \dot{x}(t)$. This can be a realistic scenario because if the company needs to produce more, it will have to pay overtime and hire part-time workers who would otherwise not be employed by the company. The cost per unit time for production will be the cost per unit produced times the rate of production: $C_P \dot{x}$. The total amount of money being expended at time t is $f(x, \dot{x}) = c_1(\dot{x})^2 + c_2 x$. The total expenditure between now and the time the shipment is due is $J = \int_0^T f(x, \dot{x}) dt$. Thus this problem is also of the general form of a variational calculus problem. $\square$

Note that nothing has been solved in the previous examples; we have only *formulated* the problem.

One possible approach to solve the problem is to assume the solution has a particular form. For instance if x is the dependent variable and we seek the extremal of $y(x)$, we could assume the function $y(x) = a_0 + a_1 x + a_2 x^2 + \cdots$, substitute this in, integrate, and then $J = J(a_0, \ldots)$. The optimal values for a_i could then be found using standard methods of multivariable calculus. Instead of a power series, we could have used a Fourier series, or any other complete series representation.

There is another way of solving the problem that is extremely useful in fields as different as classical mechanics and economics. It goes by many names including: variational calculus, calculus of variations, optimal control, Hamilton's principle, principle of least action, and dynamic optimization. We will now explore the basis of this method.

F.1 Derivation of the Euler-Lagrange Equation

Loosely speaking, a variational operator, denoted as δ, finds functions $y(x)$ that yield extremal values of the integral

$$J = \int_{x_1}^{x_2} f\left(y(x), y'(x), x\right) dx$$

for a given function $f(\cdot)$ in the same way that the operator d/dy finds the extremal values of a function $f(y)$. This new problem may be subject to boundary conditions $y(x_1) = y_1$ and $y(x_2) = y_2$, or the boundary conditions can be free. The way to formulate the problem is as follows.

If we were to *assume* that the optimal solution to the problem is $y(x)$, then the following will *not* be the optimal value of the integral

$$\hat{J}(\alpha) = \int_{x_1}^{x_2} f\left(Y, Y', x\right) dx,$$

where

$$Y = Y(x, \alpha) \stackrel{\Delta}{=} y(x) + \alpha \epsilon(x)$$

and $\epsilon(x)$ is any continuous function such that

$$\epsilon(x_1) = \epsilon(x_2) = 0. \tag{F.1}$$

The notation $\hat{J}$ is introduced to distinguish between the integral J resulting from the assumed function $y(x)$ and the value of the same integral evaluated with $Y(x, \alpha)$. That is, $J = \hat{J}(0)$ and $Y(x, 0) = y(x)$.

Note that $Y(x)$ satisfies the same boundary conditions as $y(x)$, but by definition it must be the case that $\hat{J}(\alpha) \geq J$. We can introduce the concept of a variational operator as follows

$$\delta \hat{J} = \frac{\partial \hat{J}}{\partial \alpha}\bigg|_{\alpha=0}. \tag{F.2}$$

α is a variable which is introduced into the calculus of variations problem to distinguish all functions "within the neighborhood" of the desired function and meeting the boundary conditions $Y(x_1, \alpha) = y_1$ and $Y(x_2, \alpha) = y_2$. This is shown in Fig. F.1.

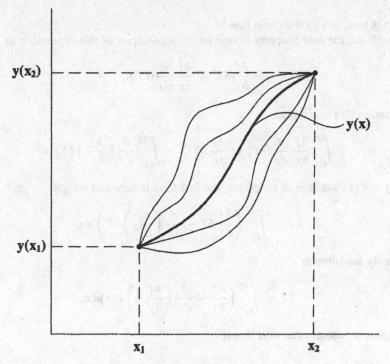

FIGURE F.1
The optimal path and surrounding candidate paths.

δ is nothing more than shorthand for the operation in Eq. (F.2). It is used like a derivative. There are four important properties of δ (which follow naturally from the previous equations). These are

- It commutes with integrals: $\delta \int_{x_1}^{x_2} f(Y, Y', x) dx = \int_{x_1}^{x_2} \delta f(Y, Y', x) dx$. This follows because δ is basically a derivative, and taking a derivative of an integral when the variables of integration and differentiation are different can be done in any order when the bounds of integration are finite and f is well behaved.

- It acts like a derivative on Y and Y' when it is applied to the function $f(Y, Y', x)$ but treats independent variables such as x like constants: $\delta f = \frac{\partial f}{\partial Y}\delta Y + \frac{\partial f}{\partial Y'}\delta Y'$. This is because Y depends on α by definition and x does not.

- It commutes with derivatives: $\delta \left(\frac{\partial Y}{\partial x} \right) = \frac{\partial}{\partial x} (\delta Y)$. This follows because δ is basically a derivative, and derivatives with respect to independent variables commute.

- The variation of $Y(x)$ vanishes at the endpoints: $\delta Y(x_1) = \delta Y(x_2) = 0$. This follows from Eq. (F.1).

We can use these properties to generate conditions which will yield the extremal solution $y(x)$ that we seek. Namely,

$$\delta \hat{J} = \delta \int_{x_1}^{x_2} f\left(Y, Y', x \right) dx = \int_{x_1}^{x_2} \delta f\left(Y, Y', x \right) dx = \int_{x_1}^{x_2} \left(\frac{\partial f}{\partial Y} \delta Y + \frac{\partial f}{\partial Y'} \delta Y' \right) dx$$

$$= \int_{x_1}^{x_2} \left(\frac{\partial f}{\partial y} \delta Y + \frac{\partial f}{\partial y'} \delta Y' \right) dx. \tag{F.3}$$

The last step is true, just by the chain rule.[1]

Now we will use the final property to rewrite the second part of this expression as

$$\frac{\partial f}{\partial y'} \delta Y' = \frac{\partial f}{\partial y'} \frac{d}{dx} (\delta Y).$$

Using integration by parts[2] means

$$\int_{x_1}^{x_2} \frac{\partial f}{\partial y'} \frac{d}{dx} (\delta Y) = \frac{\partial f}{\partial y'} \delta Y \Big|_{x_1}^{x_2} - \int_{x_1}^{x_2} \frac{d}{dx} \left(\frac{\partial f}{\partial y'} \right) \delta Y dx.$$

Since $\delta Y(x) = \epsilon(x)$ vanishes at endpoints, the first term is zero and we get

$$\delta \hat{J} = \int_{x_1}^{x_2} \left(\frac{\partial f}{\partial Y} \delta Y - \frac{d}{dx} \left(\frac{\partial f}{\partial Y'} \right) \delta Y \right) dx.$$

This is easily rewritten as

$$\delta \hat{J} = \int_{x_1}^{x_2} \left(\frac{\partial f}{\partial y} - \frac{d}{dx} \left(\frac{\partial f}{\partial y'} \right) \right) \epsilon(x) dx. \tag{F.4}$$

Using a general theorem that says that if

$$\int_a^b M(x) \epsilon(x) dx = 0$$

for all possible differentiable functions $\epsilon(x)$ for which $\epsilon(a) = \epsilon(b) = 0$, the function $M(x) = 0$ and the integrand in Eq. (F.4) can be set equal to zero, and so

$$\frac{\partial f}{\partial y} - \frac{d}{dx} \left(\frac{\partial f}{\partial y'} \right) = 0. \tag{F.5}$$

This is called the *Euler-Lagrange* equation. Note that if the function f does not depend on y', the second term vanishes, and we get the familiar minimization problem from Calculus.

---

[1] $\partial/\partial y(f(g(y), g'(y'), x)) = (\partial f/\partial g)(\partial g/\partial y)$. In our case, $g = Y$ and $\partial Y/\partial y = 1$. Therefore, $\partial f/\partial y = \partial f/\partial Y$. The same is true for $\partial f/\partial y' = \partial f/\partial Y'$.
[2] $\int_a^b u dv = uv|_a^b - \int_a^b v du.$

We now have the tools to solve the two example problems stated at the beginning of this appendix.

Solution for Example F.1: If $f = (1 + (y')^2)^{1/2}$, there are no y terms so the Euler-Lagrange equation reduces to

$$\frac{d}{dx}\left(\frac{\partial f}{\partial y'}\right) = 0 \qquad \Rightarrow \qquad \frac{\partial f}{\partial y'} = c.$$

In this particular problem, this gives: $(1 + (y')^2)^{-1/2} y' = c$. This can be rewritten as $c^2(1 + (y')^2) = (y')^2$. Solving for y' yields $y' = [c^2/(1 - c^2)]^{1/2} \triangleq a$. Since c was arbitrary, we have replaced the fraction with a, another arbitrary constant. Integration yields $y = ax + b$, where b is arbitrary. The boundary conditions $y(x_1) = y_1$ and $y(x_2) = y_2$ specify a and b completely.

Solution for Example F.2: Again we use the Euler-Lagrange equation, this time with $f = c_1(\dot{x})^2 + c_2 x$, and we get $\frac{\partial f}{\partial x} - \frac{d}{dt}\left(\frac{\partial f}{\partial \dot{x}}\right) = c_2 - 2c_1\ddot{x} = 0$. Rewriting as $\ddot{x} = \frac{1}{2}c_2/c_1$ and integrating twice, we get $x(t) = \frac{1}{4}c_2/c_1 t^2 + at + b$. We can solve for a and b given that $x(0) = 0$ and $x(T) = A$.

While we have presented the one-dimensional variational problem, the same methods are used for a functional dependent on many variables

$$J = \int_a^b f\left(y_1, \ldots, y_n, y'_1, \ldots, y'_n, x\right) dx.$$

In this case, a set of simultaneous Euler-Lagrange equations are generated

$$\frac{\partial f}{\partial y_i} - \frac{d}{dx}\left(\frac{\partial f}{\partial y'_i}\right) = 0. \tag{F.6}$$

Likewise, the treatment of problems where $f(\cdot)$ depends on higher derivatives of y is straightforward. The preceding derivation is simply extended by integrating by parts once for each derivative of y. See [7, 15] for further reading on the classical calculus of variations and [20, 28] for the more modern extensions of optimal control and dynamic optimization.

F.2 Sufficient Conditions for Optimality

The Euler-Lagrange equations provide *necessary* conditions for optimality, but there is usually no guarantee that solutions of the Euler-Lagrange equations will be optimal. However, in certain situations, the structure of the function $f(\cdot)$ will guarantee that the solution generated by the Euler-Lagrange equations is a globally optimal solution. We now examine one such case of relevance to the discussion in Chapter 6.

Suppose the integrand in the cost functional is of the form

$$f\left(y, y', x\right) = \left(y'\right)^2 g(y) \quad \text{with} \quad g(y) > 0.$$

The corresponding Euler-Lagrange equation is

$$2y'' g + \left(y'\right)^2 \frac{\partial g}{\partial y} = 0.$$

Multiplying both sides by y' and integrating yields the exact differential

$$\left(\left(y'\right)^2 g\right)' = 0.$$

Integrating both sides with respect to x and isolating y' yields

$$y' = cg^{-\frac{1}{2}}(y)$$

where c is the arbitrary constant of integration. With the boundary conditions $y(0) = 0$ and $y(1) = 1$, we may then write

$$\frac{1}{c}\int_0^y g^{\frac{1}{2}}(\sigma)d\sigma = x,$$

where

$$c = \int_0^1 g^{\frac{1}{2}}(\sigma)d\sigma.$$

Hence, we have an expression of the form $G(y) = x$ that can be inverted (G is monotonically increasing since $g > 0$) to yield $y = G^{-1}(x)$.

To see that this solution is optimal, one need only substitute

$$y' = g^{-\frac{1}{2}}(y)\int_0^1 g^{\frac{1}{2}}(\sigma)d\sigma \qquad (\text{F.7})$$

into the cost functional

$$J(Y) = \int_0^1 g(Y)\left(Y'\right)^2 dx$$

to see that

$$J(y) = \left(\int_0^1 g^{\frac{1}{2}}(y)dy\right)^2 = \left(\int_0^1 g^{\frac{1}{2}}(y(x))y'dx\right)^2.$$

Since in general

$$\int_0^1 [f(x)]^2 dx \ge \left(\int_0^1 f(x)dx\right)^2,$$

we see that by letting $f(x) = g^{\frac{1}{2}}(y(x))y'(x)$ that $J(y) \le J(Y)$ where y is the solution generated by the Euler-Lagrange equation, and $Y(x)$ is any other continuous function satisfying $Y(0) = 0$ and $Y(1) = 1$.

Appendix G

Manifolds and Riemannian Metrics

In this section we present definitions from differential geometry used throughout the text. Many works on differential geometry and its applications can be found. These include the very rigorous classic texts by Kobayashi and Nomizu [31], Abraham and Marsden [1], Guillemin and Pollack [21], and Warner [51]. We have found the classic text of Arnol'd [4] to be particularly readable, and the books by Choquet-Bruhat, DeWitt-Morette, and Dillard-Bleick [9], Schutz [45], and Burke [8] to be useful in connecting differential geometric methods to physical problems. We also have found the introductory texts of Millman and Parker [35] and Do Carmo [12] to serve as excellent connections between differential geometry in two and three dimensions and its abstraction to coordinate-free descriptions in higher dimensions. Finally, we have found the recent books by Lee [33] and Rosenberg [41] to be rigorous introductions written in a style appropriate for physicists and mathematically oriented engineers.

We note before proceeding that the use of differential forms has been avoided here in order to reduce the scope of this brief overview of basic differential geometry. The interested reader is referred to the books by Darling [11] and Flanders [16] (and references therein) for information on this subject.

We have already seen the definition of a metric, $d(\cdot, \cdot)$. Recall that when a metric is defined on a set X, the pair (X, d) is called a *metric space*. Henceforth it will be clear when the shorthand X is used to refer to the metric space (X, d).

An *open ball of radius* ϵ around $x \in X$ [denoted $B_\epsilon(x)$] is the set of all $y \in X$ such that $B_\epsilon(x) = \{y \in X | d(x, y) < \epsilon\}$. A subset $A \subset X$ is called *open* if for every point $x \in A$ it is possible to find an open ball $B_\epsilon(x) \subset A$, i.e., for every point $x \in A$, there must exist a $B_\epsilon(x)$ that is also in A for some value of $\epsilon \in \mathbb{R}^+$. Any open set containing a point $x \in X$ is called an *open neighborhood* of x. As a consequence of these definitions (or the more general properties of topological spaces) it follows that the union of two open sets is open, as is their intersection. It also may be shown that given any two distinct points in a metric space, it is possible to find neighborhoods of the points that do not intersect.

The concept of open sets is important in the definition of continuity of functions between metric spaces. For a continuous map from metric space X to metric space Y, $\phi : X \rightarrow Y$, every open set $V \subset Y$ has a preimage $\phi^{-1}(V) \subset X$ that is also open. The concept of continuity of mappings between metric spaces is of particular importance when one seeks an intrinsic description of higher-dimensional analogs of curves and surfaces. In this case, one of the metric spaces, Y, is taken to be $\mathbb{R}^n$ for some positive integer n, and X is denoted as M (for manifold). Since we have seen how various coordinates can be defined for the whole of $\mathbb{R}^n$ (and hence any of its open subsets), it follows that ϕ^{-1} provides a tool to map coordinates onto an open subset of M when a bijective mapping exists between that open subset and an open subset of $\mathbb{R}^n$, and both ϕ and ϕ^{-1} are continuous. More generally, a bijection f between two topological spaces X and Y is called a *homeomorphism* if both $f : X \rightarrow Y$ and $f^{-1} : Y \rightarrow X$ are continuous.

An *n-dimensional proper coordinate chart about* $x \in M$ is the pair (U, ϕ) where U is a neighborhood of x and $\phi : U \rightarrow \mathbb{R}^n$ is a homeomorphism. It then follows that $\phi(U)$ is open in $\mathbb{R}^n$. A collection of coordinate charts $\{U_i, \phi_i\}$ for $i \in I$ (I is a set that indexes the charts) is called an *atlas*. This allows us to rigorously define what is meant by a space that locally "looks like" $\mathbb{R}^n$ everywhere. A *manifold, M,* is a metric space for which there is an atlas with the properties

- $\{U_i, \phi_i\}$ exists so that for each $x \in M$, $x \in U_i$ for some $i \in I$.

- If (U_i, ϕ_i) and (U_j, ϕ_j) are any two coordinate charts in the atlas for which $(U_i, \phi_i) \cap (U_j, \phi_j) \neq \emptyset$, then the composed map

$$\phi_j \circ \phi_i^{-1} : \phi_i \left(U_i \cap U_j \right) \rightarrow \phi_j \left(U_i \cap U_j \right)$$

is continuous.

- All possible charts with the above two properties are contained in the atlas.

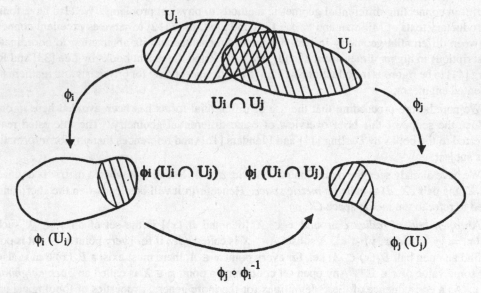

FIGURE G.1
Relationships between coordinate charts.

When all the mappings $\phi_j \circ \phi_i^{-1}$ are infinitely differentiable, the manifold is called a *differentiable manifold*. If, in addition, each of these mappings has a convergent Taylor series it is called an *analytic manifold*. Henceforth when referring to a manifold, we mean an analytic manifold. Complex analytic manifolds are defined in an analogous way (see [50]). In the context of our treatment it suffices to consider complex manifolds of complex dimension n as real manifolds of dimension $2n$ with additional properties.

In a manifold, one defines a curve, $x(t)$, as $x : \mathbb{R} \rightarrow M$. One can generally construct a set of curves $\{x_k(t)\}$ for $k = 1, \ldots, n$ that intersect at $x(0)$ such that $x_k(0) \in U_i$ for all $k = 1, \ldots, n$. Furthermore, the set of curves $\{x_k(t)\}$ can be constructed so that the tangents to the *images* of the curves $\{\phi(x_k(t))\}$ form a basis for $\mathbb{R}^n$. We call such a set of curves *nondegenerate*. An *orientation* is assigned to $x(0)$ by taking the sign of the determinant of the Jacobian matrix whose kth column is the image of the tangent to $\phi(x_k(t))$ at $t = 0$. A particular set of nondegenerate curves that is of importance is generated for U_i as $\phi_i^{-1}(u_k e_k)$ for $k = 1, \ldots, n$ and $u_k \in \mathbb{R}$. We will call this set of curves *coordinate curves*, since they take Cartesian coordinates in $\phi(U_i) \subset \mathbb{R}^n$ and map them to $U_i \subset M$. Clearly the orientation assigned to each set of intersecting coordinate curves is $\det(e_1, \ldots, e_n) = +1$.

The tangent vectors themselves are harder to visualize. Formally, a tangent vector is defined as an operator on the space of all smooth real-valued functions on M (denoted $C^\infty(M)$). A tangent vector at $x \in M$ is an operator

$$v_x : C^\infty(M) \to \mathbb{R}$$

satisfying linearity

$$v_x(af + bg) = a v_x(f) + b v_x(g)$$

and the condition

$$v_x(f \cdot g) = v_x(f)g(x) + f(x)v_x(f)$$

that is reminiscent of the product rule for differentiation. The set of all tangent vectors at $x \in M$ is denoted $T_x M$ and is called the *tangent space* of M at x.

One defines

$$TM \triangleq \{(x, v_x) \,|\, x \in M, v_x \in T_x M\}.$$

The *natural projection* is the mapping

$$\pi : TM \to M.$$

A *tangent bundle* consists of the *total space* TM, the *base space* M, the *fibers* $T_x M$, and the projection $\pi(x, v_x) = x$.

A manifold is called *orientable* if the sign of the Jacobian determinant of $\phi_j \circ \phi_i^{-1}$ is positive for all i, j for which $U_i \cap U_j \neq \emptyset$ where $\{U_i\}$ is an open covering. Since $\phi_j \circ \phi_i^{-1}$ is a mapping between two open sets in $\mathbb{R}^n$, the Jacobian is calculated in a similar way as for curvilinear coordinates in Chapter 4. Orientability essentially means that it is possible to define coordinate charts so that a set of coordinate curves of any point in one chart is positively oriented in any other chart containing the same point.

All the Lie groups considered in the text are orientable manifolds, as are all Lie subgroups of $GL(n, \mathbb{R})$ and $GL(n, \mathbb{C})$ [11].

A *Riemannian metric* on a manifold M is a family of smoothly varying positive definite inner products on $T_x M$ for each $x \in M$. For any two vectors $v_1, v_2 \in T_x M$, the value of the metric is denoted $g_x(v_1, v_2)$.

Note that a Riemannian metric might not exist for an arbitrary manifold. Hence the following definition is required.

A *Riemannian manifold* is the pair (M, g) where M is a smooth manifold and $g = g_x$ is a Riemannian metric.

An important result from differential topology that we use implicitly (without proof) in the text is that every N-dimensional differentiable manifold is homeomorphic to a subset of $\mathbb{R}^{2N+1}$. As an example, the manifold of $SO(3)$ has an image that is a subset of $\mathbb{R}^6$ by writing a rotation matrix as $R = [\mathbf{u}, \mathbf{v}, \mathbf{u} \times \mathbf{v}]$ where $\mathbf{u}$ and $\mathbf{v} \in \mathbb{R}^3$ are unit vectors, and the correspondence

$$R \leftrightarrow \left[\mathbf{u}^T, \mathbf{v}^T \right]^T$$

defines a homeomorphism between $SO(3)$ and a subset of $\mathbb{R}^6$. Since a subset of $\mathbb{R}^6$ is also one of $\mathbb{R}^7$, this example may be viewed as a demonstration of the theorem.

References

[1] Abraham, R. and Marsden, J.E., *Foundations of Mechanics*, 2nd ed., Benjamin/Cummings, San Mateo, CA, 1978.

[2] Aho, A.V., Hopcroft, J.E., and Ullman, J.D., *The Design and Analysis of Computer Algorithms*, Addison-Wesley, Reading, MA, 1974.

[3] Arfken, G.B. and Weber, H.J., *Mathematical Methods for Physicists*, 4th ed., Academic Press, San Diego, 1995.

[4] Arnol'd, V.I., *Mathematical Methods of Classical Mechanics*, Springer-Verlag, New York, 1978.

[5] Borodin, A. and Munro, I., *The Computational Complexity of Algebraic and Numerical Problems*, Elsevier, New York, 1975.

[6] Borodin, A.B. and Munro, I., Evaluating polynomials at many points, *Inf. Process. Lett.*, 1(2), 66–68, 1971.

[7] Brechtken-Manderscheid, U., *Introduction to the Calculus of Variations*, Chapman & Hall, New York, 1991.

[8] Burke, W.L., *Applied Differential Geometry*, Cambridge University Press, Cambridge, 1985.

[9] Choquet-Bruhat, Y., DeWitt-Morette, C., and Dillard-Bleick, M., *Analysis, Manifolds and Physics*, North-Holland, Amsterdam, 1982.

[10] Collins, G.E., Computer algebra of polynomials and rational functions, *Am. Math. Mon.*, 80(7), 725–754, 1973.

[11] Darling, R.W.R., *Differential Forms and Connections*, Cambridge University Press, Cambridge, 1994.

[12] Do Carmo, M., *Differential Geometry of Curves and Surfaces*, Prentice-Hall, Englewood Cliffs, NJ, 1976.

[13] Driscoll, J.R., Healy, D.M., Jr., and Rockmore, D.N., Fast discrete polynomial transforms with applications to data analysis for distance transitive graphs, *SIAM J. Comput.*, 26, 1066–1099, 1997.

[14] Eve, J., The evaluation of polynomials, *Numerische Math.*, 6, 17–21, 1964.

[15] Ewing, G.M., *Calculus of Variations With Applications*, W.W. Norton, New York, 1969.

[16] Flanders, H., *Differential Forms with Applications to the Physical Sciences*, Dover, New York, 1989.

[17] Fréchet, M., Les solutions non commutables de l'équation matricielle $e^X \cdot e^Y = e^{X+Y}$, *Rendiconti Del Circulo Matematico Di Palermo*, Ser. 2, 1, 11–21, 1952. (See also 2, 71–72, 1953).

[18] Gantmacher, F.R., *The Theory of Matrices*, Chelsea, New York, 1959.

[19] Golub, G.H. and Van Loan, C.F., *Matrix Computations*, 2nd ed., The Johns Hopkins University Press, Baltimore, 1989.

[20] Gruver, W.A. and Sachs, E., *Algorithmic Methods in Optimal Control*, Pitman Publishing, Boston, 1980.

[21] Guillemin, V. and Pollack, A., *Differential Topology*, Prentice-Hall, Englewood Cliffs, NJ, 1974.

[22] Herstein, I.N., *Topics in Algebra*, 2nd ed., John Wiley & Sons, 1975.

[23] Halmos, P.R., *Measure Theory*, Springer-Verlag, New York, 1974.

[24] Higham, N.J., Fast solution of Vandermonde-like systems involving orthogonal polynomials, *IMA J. Numerical Anal.*, 8, 473–486, 1988

[25] Horn, R.A. and Johnson, C.R., *Matrix Analysis*, Cambridge University Press, New York, 1985.

[26] Horowitz, E., A fast method for interpolation using preconditioning, *Inf. Process. Lett.*, 1(4), 157–163, 1972.

[27] Itzykson, C. and Zuber, J.B., *Quantum Field Theory*, McGraw-Hill, New York, 1980.

[28] Kamien, M.I. and Schwartz, N.L., *Dynamic Optimization: The Calculus of Variations and Optimal Control in Economics and Management*, 2nd ed., North-Holland, New York, 1991.

[29] Kanwal, R.P., *Generalized Functions, Theory and Technique*, 2nd ed., Birkhäuser, Boston, 1998.

[30] Kleinert, H., *Path Integrals in Quantum Mechanics, Statistics and Polymer Physics*, 2nd ed., World Scientific, Singapore, 1995.

[31] Kobayashi, S. and Nomizu, K., *Foundations of Differential Geometry Vols. I and II*, John Wiley & Sons, 1963. (Wiley Classics Library Ed., 1996).

[32] Kung, H.T., Fast evaluation and interpolation, Dept. of Computer Science, CMU, Pittsburgh, 1973.

[33] Lee, J.L., *Riemannian Manifolds: An Introduction to Curvature*, Springer-Verlag, New York, 1997.

[34] Lipson, J., Chinese remaindering and interpolation algorithms, Proc. 2nd Symp. of Symbolic and Algebraic Manipulation, 372–391, 1971.

[35] Millman, R.S. and Parker, G.D., *Elements of Differential Geometry*, Prentice-Hall, Englewood Cliffs, NJ, 1977.

[36] Munkres, J.R., *Topology, a First Course*, Prentice-Hall, Englewood Cliffs, NJ, 1975.

[37] Pan, V.Y., Methods of computing values of polynomials, *Russ. Math. Surv.*, 21(1), 105–136, 1966.

[38] Pan, V., How can we speed up matrix multiplication, *SIAM Rev.*, 26, 393–416, 1984.

[39] Pan, V., *How to Multiply Matrices Fast*, Springer-Verlag, Berlin, Heidelberg, 1984.

[40] Press, W.H., Flannery, B.P., Teukolsky, S.A., and Vetterling, W.T., *Numerical Recipes in C*, Cambridge University Press, New York, 1988.

[41] Rosenberg, S., *The Laplacian on a Riemannian Manifold: An Introduction to Analysis on Manifolds*, London Math. Soc. Student Texts, 31, Cambridge University Press, Cambridge, 1997.

[42] Ryder, L.H., *Quantum Field Theory*, Cambridge University Press, Cambridge, 1985.

[43] Saichev, A.I. and Woyczyński, W.A., *Distributions in the Physical and Engineering Sciences*, Appl. and Numerical Harmonic Anal. Ser., Benedetto, J.J., Ed., Birkhäuser, Boston, 1997.

[44] Schulman, L.R., *Techniques and Applications of Path Integration*, John Wiley & Sons, New York, 1981.

[45] Schutz, B., *Geometrical Methods of Mathematical Physics*, Cambridge University Press, Cambridge, 1980.

[46] Stokey, N.L., Lucas, R.E., Jr., and Prescott, E.C., *Recursive Methods in Economic Dynamics*, Harvard University Press, Cambridge, MA, 1989.

[47] Strassen, V., Gaussian elimination is not optimal, *Numerische Math.*, 13, 354-356, 1969.

[48] Swanson, M.S., *Path Integrals and Quantum Processes*, Academic Press, Boston, 1992.

[49] Tomé, W., *Path Integrals on Group Manifolds*, World Scientific, Singapore, 1998.

[50] Varadarajan, V.S., *Lie Groups, Lie Algebras, and their Representations*, Springer-Verlag, New York, 1984.

[51] Warner, F.W., *Foundations of Differentiable Manifolds and Lie Groups*, Springer-Verlag, New York, 1983.

[52] Wedderburn, J.H.M., *Lectures on Matrices*, Dover, New York, 1964.

[53] Wermuth, E.M.E., Brenner, J., Leite, F.S., and Queiro, J.F., Computing matrix exponentials, *SIAM Rev.*, 31(1), 125–126, 1989.

Index

2D Fourier transform in polar coordinates, 98
3D discrete motion groups, 366
3D Fourier transform in spherical coordinates, 101
3D manipulators with continuous symmetries, 403

absolutely integrable, 6
accumulation of infinitesimal spatial errors, 407
adjoint, 639
adjoint and tensor representations of $SO(3)$, 284
adjoint matrix, 157
adjoint of differential operators, 268
adjoint operator, 218
algebras, 623
algorithm, Ebert-Uphoff, 399
algorithms for fast $SE(3)$ convolutions using FFTs, 358
aliasing, 33, 34
alternative
    methods for computing $SE(D)$ convolutions, 371
    motion-group Fourier transform, 372
    to the intrinsic approach, 162
analogs of LQR for $SO(3)$ and $SE(3)$, 502
analysis
    and tomography, image, 419
    classical Fourier, 15
    convolution equations in texture, 592
    harmonic, 5
    mathematical texture, 13
    mechanics and texture, 579
    of materials, 12
    on groups, harmonic, 239
    on Lie groups, harmonic, 262
    on the Euclidean motion groups, harmonic, 321
    problem statement using harmonic, 473
    texture, 592
analytic function, 6, 281
analytic manifold, 652
analytical examples of Fourier transforms, 344

analytical models of polymers, 558
angular velocity, 127, 525, 568
anti-commutator, 501
application of the FFT on $SE(2)$ to the correlation method, 421
applications to liquid crystals, 598
approximate spline interpolation, 355
approximating rigid-body motions, 170
assembly planning, 1
assigning frames to curves and serial chains, 159
associated Legendre functions, 43, 100
associated probability density functions, 493
associative algebra, 625
associative ring, 626
asymptotic expansions, 60
atlas, 88, 652
attitude estimation, 11
attitude estimation using an inertial navigation system, 503
automorphism, 203
averaging single-crystal properties over orientations, 590
$ax + b$ group, 234
axis-angle and Cayley parameterizations, 300

backprojection algorithm for finite-width beams, 445
Baker-Campbell-Hausdorff formula, 219, 221
band-limited function, 16
band-pass filtered, 33
base space, 653
basic continuum mechanics, 579
basic definitions of sets, 615
basic probability review, 455
Bayes' rule, 457
Bessel functions, 55
Bessel polynomials, 58
Bessel's inequality, 42
bi-invariant measure, 619
biaxial nematics phase, 528

bijective map, 618
bilinear form, 232
bilinear transformations in the complex plane, 141
binary manipulator, 380, 399
binary manipulator design, 405
binary operation, 192
Bingham density, 464
binomial coefficient, 61
binormal vector, 159
biquaternions, 172
bond angle, 166
Bonner's recursion, 44
Brownian motion, 495, 516
Brownian motion based on randomized angular momentum, 535
Brownian motion of a rigid body, 526
bulk measurements of polycrystalline materials, 592
Burnside's formula, 249

calculating jacobians, 224
calculation for $SE(2)$ and $SE(3)$, 468
calculation for $SO(3)$, 467
calibration with two noisy measurements, 478
camera calibration, 455
cardioid density, 461
Cartan's criteria, 232
Cartesian coordinates in $\mathbb{R}^4$, 124
Cartesian product, 194, 617
Cartesian tensors, 580
Cauchy distribution, 458
Cauchy-Schwarz inequality, 624
Cauchy's stress principle, 583
Cauley/Rodrigues parameters, 123
Cayley table, 189
Cayley's formula, 115, 117
centered difference rule, 49
certainty-equivalence principle, 497
CGCs, 309
chain with hindered rotations, 553
Chapman-Kolmogorov equation, 494
characteristic function, 619
characters of finite groups, 245
Charlier polynomials, 63
Chebyshev polynomials, 46
checking irreducibility of representations of compact Lie groups, 272
checking irreducibility of representations of finite groups, 270
chiral nematics phase, 528
Christoffel-Darboux formula, 51
Christoffel symbols, 86
circle, 106
class

constants, 210
equation, 208
functions, 208, 222, 248
functions and sums, 208
products, 208
sum fuction, 209
classical equations of motion, 642
classical Fokker-Planck equation, 519
classical Fourier analysis, 15
classification of finite subgroups, 223
classification of quantum states according to representations of $SU(2)$, 315
Clebsch-Gordan coefficients, 309
Clebsch-Gordan coefficients and Wigner 3jm symbols, 309
closed binary operation, 192
colatitude, 91
common probability density functions on the line, 458
commutative and noncommutative convolution, 7
commutative ring, 626
commutativity of sampling and convolution, 35
compact Lie groups, 263, 272
complete in a set of functions S, 41
completeness of eigenfunctions, 41
completeness relation, 21
complex matrices, 632
complex spectrogram, 71
complex vector space, 623
complexity of matrix multiplication, 608
compliance tensor, 594
composing rotations, 121
composition of motions, 149
computation of convolution-like integrals, 424
computation of $SE(2)$ convolution integrals, 365
computation of the convolution product of functions on $SE(2)$, 386
computational and analytical models of polymers, 558
computational benefit, 385
computational complexity, 371, 607
computed tomography algorithm on the motion group, 441
computing $SE(D)$ convolutions, 371
computing $SE(D)$ convolutions using Fourier transforms, 373
conditional probability, 457
conditions for optimality, 649
configuration-space obstacles and localization, 9
configuration-space obstacles of a mobile robot, 411
confluent hypergeometric function, 61
conformational energy, 553
conformational energy effects, 564

conformational statistics of macromolecules, 12
congruent modulo, 617
conjugacy classes, 208, 211
conjugacy classes and class functions, 222
conjugacy classes and conjugate subgroups, 208
conjugate elements, 208
conjugate of a complex-valued function, 437
conjugate subgroup, 193, 208
conjugation, 192
constitutive equation, 583
constitutive laws in solid mechanics, 594
constructing invariant integration measures, 259
continuous
    Gabor transform, 71
    modulation-translation-based wavelet transforms, 71
    scale-translation-based wavelet transforms, 69
    symmetries, 403
    wavelet transforms, 69, 70
    wavelet transforms on the plane, 102
    wormlike chains, 551
continuum mechanics, 579
continuum model, 12
contraction of $SE(3)$ to $SO(4)$, 173
contraction of $SE(D)$ to $SO(D+1)$, 375
control of nonlinear systems, 485
converting to the Fourier transform, 59
convolution, 563
    commutative and noncommutative, 7
    equations in texture analysis, 592
    for workspace generation, 384
    like integrals, 424
    of functions, 337, 362
    of functions on $SE(D)$, 382
    product of functions on $SE(2)$, 386
    property, 17
    theorem, 248, 325
coordinate curves, 653
coordinates
    2D Fourier transform in polar, 98
    3D Fourier transform in spherical, 101
    chart, 88
    curvilinear, 81
    delta function in curvilinear, 638
    differential operators and laplacian for $SO(3)$ in spherical, 292
    Eulerian, 584
    examples of curvilinear, 90
    generalized, 486
    in $\mathbb{R}^4$, cartesian, 124
    jacobians for spherical, 131
    orthogonal curvilinear, 88
    orthogonal expansions and transforms in curvilinear, 98

orthogonal expansions in curvilinear, 81
    polar, 90
    spatial, 584
    spherical, 91, 125
correlation method including dilations, 429
correlation method including rotations and translations, 425
correlation on the motion group, 442
coset space, 201
cosets and orbits, 199
cosets spaces and quotient groups, 199
cost of interpolation, 353
counting formulas, 207
crystalline symmetry, 596
cube, Rubik's, 3
cubo-octahedral group, 224
curvature, 159
curvature of a surface, 86
curvilinear coordinates, 90, 98
curvilinear coordinates and surface parameterizations, 81
cyclic group, 214, 224

d'Alembert's solution, 25
DCT, 47
decomposition of complex matrices, 632
decomposition of functions on the circle, 15
decomposition of functions on the line, 19
deconvolution, 240
deformations of nonrigid objects, 111
delta function in curvilinear coordinates, 638
demonstration of theorems with $SO(3)$, 220
Denavit-Hartenberg parameters in robotics, 164
density function, 380
density function of a discretely actuated manipulator, 380
derivation of the classical Fokker-Planck equation, 519
derivation of the Euler-Lagrange equation, 646
derivations of delta functions, 636
derivative control on $SO(3)$ and $SE(3)$, 498
derivatives, 254
describing rotations in three-space, 135
determination of the ODF from experimental measurements, 592
determining a rigid-body displacement, 493
determining single crystal strength properties, 592
deterministic control, 486
deterministic control on Lie groups, 498
deterministic optimal control on $SO(3)$, 500
DFT, 15
differentiable manifold, 652
differential

geometry, 88
geometry of S^3 and $SU(2)$, 290
operators, 268
operators and laplacian for $SO(3)$ in spherical coordinates, 292
operators for $SE(3)$, 330
operators for $SO(3)$, 310
differentiation and integration of functions on Lie groups, 254
diffusion
    based wavelets for $SO(3)$, 305
    based wavelets for the sphere, 303
    coefficient, 23
    equation for liquid crystals, 531
    equations, 23
    equations on $SO(3) \times \mathbb{R}^3$, 526
    on angular velocity, 525
    on $SO(3)$, non-inertial theory of, 524
    rotational Brownian motion, 515
    translational, 516
digital control on $SO(3)$, 501
dihedral groups, 224
dilation, 274
dilations, correlation method including, 429
dilute composites, 588
Dirac delta function, 18, 635
Dirac delta functions with functional arguments, 637
direct convolution and the cost of interpolation, 353
direct Fourier transform, 358
direct product, 194
direct sum, 242
direction cosine matrices, 498
directional writhing number, 163
director, 527
discrete
    and fast Fourier transforms, 30
    cosine transformation, 47
    Fourier transform, 30
    manipulator symmetries, 400
    modulation-translation wavelet series, 74
    motion-group Fourier transform, 365
    motion group, invariants of the, 423
    motion group of the plane, 363
    polynomial transforms, 39, 49
    scale-translation wavelet series, 72
    wavelet transform, 72, 73
    wavelets and other expansions on the sphere, 107
discretely actuated manipulator, 380
discretization of parameter space, 388
disjoint, 616
displacement, 583
displacement gradient, 583

displacement, statistical determination of a rigid-body, 469
distance function, 618
distributed small cracks and dilute composites, 588
divergence, 88
division of polynomials, 611
double coset decomposition, 222
double cosets and their relationship to conjugacy classes, 211
drift coefficient, 23
dual
    numbers, 168
    of the group, 239
    orthogonal matrices, 169
    quaternions, 169
    unitary matrices, 169
    vector, 115
dynamic estimation and detection of rotational processes, 503
dynamics of fibers in fluid suspensions, 597

Ebert-Uphoff algorithm, 399
effects of conformational energy, 553
efficiency of computation of convolution-like integrals, 424
efficiency of computation of $SE(2)$ convolution integrals, 365
efficient division of polynomials, 611
efficient multiplication of polynomials, 610
eigenfunctions, 39
eigenfunctions of the Laplacian, 100, 294
eigenvalues, 39
eigenvalues and eigenvectors of rotation matrices, 114
Einstein's theory, 516
elastic solids, 582
elements, 615
empty set, 616
end effector, 9, 379
ensemble properties for purely kinematical models, 562
ensemble properties including long-range comformational energy, 567
epimorphism, 203
equation
    Chapman-Kolmogorov, 494
    class, 208
    constitutive, 583
    derivation of the classical Fokker-Planck, 519
    derivation of the Euler-Lagrange, 646
    diffusion, 23
    Euler's, 501
    Fokker-Planck, 517, 518

heat, 23
   in texture analysis, convolution, 592
   Langevin, 517
   of motion, 488
   of motion, Euler's, 134
   Riccati, 490
   rotational diffusion, 531
   scalar wave, 28
   Schrödinger, 23
   Smoluchowski, 517
   telegraph, 25
   wave, 24
   Wiener processes and stochastic differential, 495
   with convolution kernels, integral, 26
equatorial line, 83
equivalence class, 617
equivalence of elements, 617
equivalent representation, 241
error accumulation in serial linkages, 407
error measures, 391
estimation error PDFs, 535
estimation of a static orientation or pose, 493
estimation of a static quantity, 491
Euclidean distance, 112
Euclidean group convolutions, 559
Euclidean motion groups, 321
Euler
   angles, 125
   characteristic, 96
   characteristic of the surface, 96
   equations as a double bracket, 501
   equations of motion, 134
   Lagrange equation, 646, 648
   parameters, 122
   theorem, 118
Eulerian coordinates, 584
evolution of estimation error PDFs, 535
evolution of liquid crystal observation, 530
exact measurements, 475
examples of curvilinear coordinates, 90
examples of definitions, 214
examples of Lie groups, 220
examples of volume elements, 262
exclusion of inertial terms, 522
expansions on the sphere, 107
explicit calculation for $SE(2)$ and $SE(3)$, 468
explicit calculation for $SO(3)$, 467
explicit solution of a diffusion equation for liquid
   crystals, 531
exponential Fourier densities, 505
expressing rotations as 4×4 matrices, 137

faithful representation, 241

fast
   algorithm for the Fourier transform on the 2D
      motion group, 356
   exact Fourier interpolation, 354
   Fourier transforms, 32
     and convolution, 32
     discrete and, 30
     for finite groups, 252
     for motion groups, 353
   Legendre and associated Legendre transforms,
      45
   numerical transforms, 58
   polynomial transforms, 54
   $SE(3)$ convolutions using FFTs, 358
feedback law, 486
FFT, 15
FFT for $SO(3)$ and $SU(2)$, 301
FFTs for the sphere, 101
fiber, 617
fibers in fluid suspensions, 597
field, 626
filtering, 33
finite error, 408
finite groups, 189, 195, 269
   characters of, 245
   fast Fourier transforms for, 252
   Fourier transforms for, 240
   induced representations of, 269
   irreducibility of representations of, 270
   products of class sum functions for, 210
   representations of, 241
finite Haar transform, 66
finite subgroups, 223
finite Walsh transform, 68
finite-width beams, 445
Fisher distribution, 463
fluid mechanics, 584
fluid suspensions, 597
Fokker-Planck equation, 517, 518
Fokker-Planck equation on manifolds, 520
folded Gaussian for one-dimensional rotations, 464
formulas in orientational averaging of tensor com-
      ponents, 586
forward kinematics, 165, 379
forward tomography problem, 11
Fourier
   analysis, generalizations of, 39
   Bessel series, 56
   densities, exponential, 505
   inversion formula, 20
   optics, 27
   series, 15
   series expansion, 16
   transform, 19, 248

and convolution, fast, 32
discrete, 30
discrete and fast, 30
for 3D discrete motion groups, 366
for finite groups, 240
for $SE(2)$, 324
for $SE(3)$, 336
matrices for $SE(2)$ and $SE(3)$, 346
of a complex-valued function, 249
of functions on $SE(2)$, 325
on the 2D motion group, 356
on the discrete motion group, 364
on the discrete motion group of the plane, 363
to solve PDEs and integral equations, 23
frame density, 559
frames attached to serial linkages, 164
frames of least variation, 160
freely-jointed chain, 549
freely rotating chain, 550
Frenet frames, 159
Frenet-Serret apparatus, 159
Fresnel approximations, 29
function, 618
analytic, 6, 281
and the adjoint, modular, 261
associated Legendre, 43, 100
band-limited, 16
Bessel, 55
characteristic, 619
class, 208, 248
class sum, 209
confluent hypergeometric, 61
conjugacy classes and class, 222
conjugate of a complex-valued, 437
convolution of, 337, 362
derivations of delta, 636
Dirac delta, 18, 635, 637
distance, 618
Fourier transform of a complex-valued, 249
gamma, 61
generalized, 635
Haar, 66
Haar measures and shifted, 255
Hermite, 48
hypergeometric, 61
integration and convolution of rotation-dependent, 145
interdependent potential, 565
Langevin, 549
likelihood, 471
maps and, 617
misorientation distribution, 589
modular, 256

nearest-neighbor energy, 565
on Lie groups, 254
on the circle, 15
on the line, 19
on the line, sampling and reconstruction, 34
orientational distribution, 597
periodic, 15
piecewise constant orthogonal, 65
products of class sum, 210
properties of the δ, 343
Rademacher, 67
Rayleigh dissipation, 489
spectrum of the, 249
spherical Bessel, 55, 57
symmetries in workspace density, 400
tensor valued, 585
Walsh, 68
Wigner D, 299
window, 67, 420
functional integration, 640
fundamental period, 15

G-equivariance, 206
G-morphism, 206
Gabor transform on a unimodular group, 274
Gabor wavelets for the circle, 103
Gabor wavelets for the sphere, 105
Gabor wavelets on the circle and sphere, 103
Galilean invariance, 191
Galilean transformations, 191
Galilei group, 217
gamma function, 61
Gauss-Bonnet theorem, 96
Gauss integral, 163
Gauss map, 95
Gaussian
and Markov processes and associated probability density functions, 493
chain, 548
curvature, 86
distribution, 457
for the rotation group of three-dimensional space, 465
process, 494
quadrature, 52
Gegenbauer polynomials, 46
general concepts in systems of particles and serial chains, 545
general terminology, 192
generalized coordinates, 486
generalized functions, 635
generating ensemble properties for purely kinematical models, 562

generation of C-space obstacles, 410
generators of the group, 214
geometric interpretation of convolution of functions on $SE(D)$, 382
geometry-based metrics, 144
ghost problem, 593
Gibbs vector, 123
global properties of closed curves, 162
gradients, 88, 254
Gram-Schmidt orthogonalization process, 624
group
    $ax + b$, 234
    based on dilation and translation, 274
    characters of finite, 245
    checking irreducibility of representations of compact Lie, 272
    checking irreducibility of representations of finite, 270
    commutative, 192
    compact Lie, 263, 272
    computed tomography algorithm on the motion, 441
    cosets spaces and quotient, 199
    cyclic, 214, 224
    definition of, 192
    deterministic control on Lie, 498
    dihedral, 224
    examples of Lie, 220
    fast algorithm for the Fourier transform on the 2D motion, 356
    fast Fourier transforms for finite, 252
    finite, 189, 195, 240, 269
    Fourier transforms for finite, 240
    Fourier transforms on the discrete motion, 364
    Gabor transform on a unimodular, 274
    Galilei, 217
    generators of the, 214
    harmonic analysis on, 239
    harmonic analysis on Lie, 262
    harmonic analysis on the Euclidean motion, 321
    Heisenberg, 217
    hyper-octahedral, 199
    icosa-dodecahedral, 224
    induced representations of finite, 269
    induced representations of Lie, 271
    integration measures on Lie, 255
    intuitive introduction to Lie, 216
    irreducible unitary representations of the discrete motion, 363
    isomorphic, 203
    isotropy, 221
    Jacobians for the $ax + b$, 227

    Jacobians for the scale-Euclidean, 229
    Laplacians of functions on Lie, 254
    Lie, 216, 218, 254, 271
    little, 202
    locally compact, 194
    matrix Lie, 217, 218
    nilpotent, 232
    noncommutative, 192
    noncommutative unimodular, 266
    operational properties for unimodular Lie, 266
    probability density functions on, 461
    products of class sum functions for finite, 210
    proper Lorentz, 230
    quotient, 201
    Radon transform as a particular case of correlation on the motion, 442
    representations of finite, 241
    representations of Lie, 262
    semi-simple Lie, 232
    solvable, 232
    symmetric, 197
    tetrahedral, 224
    theory, 4, 187
    topological, 193
    unimodular, 256, 274
    wavelets and, 273
    wavelets on the sphere and rotation, 302
gyroscope, 504

Haar functions, 66
Haar measure, 619
Haar measures and shifted functions, 255
Haar system, 73
Hahn polynomials, 62
harmonic analysis, 5
harmonic analysis on groups, 239
harmonic analysis on Lie groups, 262
harmonic analysis on the Euclidean motion groups, 321
Hausdorff distance, 470
Hausdorff space, 618
head phantom, 444
heat equation, 23
Heisenberg group, 217
helical wormlike models, 556
helicity representations, 306
Hermite functions, 48
Hermite polynomials, 47
Hermitian operator, 639
histograms on $SE(2)$, 389
historical introduction to rotational Brownian motion, 522
homeomorphism, 651

homogeneous
    manifolds, 221
    polynomials, 281
    space $SO(3)/SO(2)$, 463
    spaces, 255, 461
    transformation matrices, 150
    transformation matrix, 150
homomorphism, 203
homomorphism theorem, 205
Hurwitz measure, 619
hyper-octahedral group, 199
hypergeometric function, 61
hypergeometric series, 61

icosa-dodecahedral group, 224
image, 617
image analysis and tomography, 419
image analysis: template matching, 419
image of the homomorphism, 203
image registration, 10
images of the curves, 652
images, using the invariants on the motion group
    to compare, 428
incorporating conformational energy effects, 564
indexing set, 616
induced representations, 308
    and test for irreducibility, 268
    of finite groups, 269
    of Lie groups, 271
inertial navigation system, 503
inertial platform, 504
infinitesimal motions, 156
infinitesimal rotations, 127
infinitesimal spatial errors, 407
infinitesimal strain tensor, 583
infinitesimal twist, 156
injective map, 618
inner product space, 623
integral equations, 23
integral equations with convolution kernels, 26
integration, 127
    and convolution of rotation-dependent func-
        tions, 145
    measures, invariant, 259
    measures on Lie groups, 255
    on G/H and $H\backslash G$, 256
    over rigid-body motions, 157
interdependent potential functions, 565
internal angular momentum, 316
interpolation, approximate spline, 355
interpolation, fast exact Fourier, 354
intrinsic approach, 162
introduction to robotics, 379

intuitive introduction to Lie groups, 216
invariant integration measures, 259
invariant measures and metrics, 619
invariants of the discrete motion group, 423
inverse
    Fourier transform, 361
    kinematics, 379
    kinematics of binary manipulators, 399
    pole figure, 593
    problem, 26
    problems in binary manipulator design, 405
    tomography problem, 11
    tomography: radiation therapy treatment plan-
        ning, 446
inversion, 248
inversion formula, proof of the, 327
inversion formulas for wavelets on spheres, 302
invertible map, 618
inverting Gabor wavelet transforms for the sphere,
    302
irreducibility, 268
    of representations of finite groups, 270
    of the representations $U(g, p)$ of $SE(2)$, 323
    of the representations $U_i(g)$ of $SU(2)$, 287
irreducible
    representations, 241
    representations from homogeneous polyno-
        mials, 281
    unitary representations of $SE(3)$, 331
    unitary representations of the discrete motion
        group, 363
isomorphic groups, 203
isomorphism, 195, 203
isotropic elastic materials, 594
isotropy groups, 221
isotropy subgroup, 202
IURs, 243
Iwasawa decomposition, 228

Jacobi identity, 219
Jacobi polynomials, 46
Jacobian matrix, 81
Jacobians
    associated with parameterized rotations, 128
    calculating, 224
    for $GL(2, \mathbb{R})$, 227
    for $SL(2, \mathbb{R})$, 228
    for spherical coordinates, 131
    for the $ax + b$ group, 227
    for the Cayley/Rodrigues parameters, 131
    for the Euler parameters, 132
    for the matrix exponential, 130
    for the modified Rodrigues parameters, 131

for the scale-Euclidean group, 229
for $Z \times Z$ Euler angles, 129
parameterization of motions and associated, 154
when rotations and translations are parameterized separately, 157
joint probability density, 457
Jordon block, 632

Kalman-Bucy filter, 497
Kalman filter, 497
kernel of the homomorphism, 203
Killing form, 232
kinematic modeling of polymers, 548
kinetic energy metric, 181
kinetic energy of a rotating rigid body, 135
Kratky-Porod chain, 551
Kratky-Porod model, 568
Krawtchouk polynomials, 63
kth Fourier coefficient, 16

L-periodic, 15
Lagrange fundamental polynomials, 52
Lagrange interpolation polynomial, 52, 613
Lagrange's theorem, 201, 547
Lagrangian
    description, 111
    description of a solid, 582
    strain tensor, 582
Laguerre polynomials, 47
Lamé constants, 595
Langevin equation, 517
Langevin function, 549
Laplace-Beltrami operator, 90
Laplace distribution, 458
Laplace transform, 536
Laplacian, 88
Laplacians of functions on Lie groups, 254
latin square, 188
left coset, 199
left invariant, 225
left-regular representations, 242
Legendre polynomials, 43
Lie algebra, 626
Lie bracket, 219, 626
Lie groups, 216, 218, 254, 271, 498
likelihood function, 471
linear algebraic properties of Fourier transform matrices, 346
linear estimation problem, 489
linear quadratic regulator, 490
linear systems, 497
linear systems theory and control, 489

linking number, 163
liquid crystal observation, 530
liquid crystals, 527, 531, 598
little group, 202
locally compact group, 194
locally compact space, 193
long-range comformational energy, 567
longitude, 91
LQR, 490
LQR for $SO(3)$ and $SE(3)$, 502
Lyapunov stability theory, 488

M-dimensional hyper-surfaces in $\mathbf{R}^N$, 85
Mackey machine, 269
macromolecule, 12
macromolecules, statistical mechanics of, 545
Maier-Saupe model, 531
manifolds, 88, 652
manifolds and Riemannian metrics, 651
manifolds, Fokker-Planck equation on, 520
mappings, 203
maps and functions, 617
Marko-Siggia DNA model, 569
Markov process, 494
mass density, 559
mass matrix, 487
matched filters, 420
materials, mechanics and texture analysis of, 12
materials with crystalline symmetry, 596
mathematical biology, 199
mathematical formulation, 367, 562
mathematical physics, 635
mathematical texture analysis, 13
matrices, 607, 627
matrices of $Ad(G)$, $ad(X)$, and $B(X, Y)$, 233
matrix
    decomposition, 632
    elements, 334, 639
        of IURs of $SE(2)$, 322
        of $SU(2)$ representations as eigenfunctions of the Laplacian, 294
        parameterized with Euler angles, 298
    exponential, 119, 155
    exponential, special properties of the, 630
    functions, 629
    Lie group, 217, 218
    mass, 487
    multiplication, 608
    normal, 633
    positive-definite, 628
    recursive weighted least-squares estimator gain, 492
    representation, 198

residual covariance, 492
state transition, 490
Vandermonde, 612
McCarthy's approach, 170
mean and variance for $SO(N)$ and $SE(N)$, 467
mean-square convergent, 41
measurable space, 619
measurement error covariance, 491
mechanics and texture analysis, 579
mechanics and texture analysis of materials, 12
mechanics of elastic solids, 582
mechanics of macromolecules, statistical, 545
Meixner polynomials, 63
method for pattern recognition, 420
methods for computing $SE(D)$ convolutions, 371
metric space, 618, 651
metric tensor, 81, 290
metrics
    and rotations, 141
    as a tool for statistics on groups, 461
    based on dynamics, 144
    examples and definitions, 142
    invariant measures and, 619
    manifolds and Riemannian, 651
    on motion, 174
    on motion induced by metrics on $\mathbb{R}^N$, 174
    on rotations, 143
    on $SE(3)$, 181
    on $SE(N)$ induced by norms, 178
    resulting from matrix norms, 143
misorientation distribution function, 589
mobile robot, 410, 411
mobile robot localization, 414
Möbius band, 94
model, continuum, 12
model formulation for finite error, 408
modeling-distributed small cracks and dilute composites, 588
MODF, 589
modified axis-angle parameterization, 127
modified Rodrigues parameters, 126
modular functions, 256
modular functions and the adjoint, 261
modulation-shift wavelet transform for the circle, 103
modulation-shift wavelet transform for the sphere, 105
moments of probability density functions on groups and their homogeneous spaces, 461
moments of the continuous wormlike chains, 551
monomorphism, 203
mother wavelet, 70
motion
    approximating rigid-body, 170

Brownian, 495, 516
composition of, 149
equations of, 488
group, 441
group averaging, 588
group averaging in solid mechanics, 585
historical introduction to rotational Brownian, 522
induced by metrics on $\mathbb{R}^N$, 174
infinitesimal, 156
integration over rigid-body, 157
metrics on, 174
parameterization of, 154
planar, 170
quantization of classical equations of, 642
rigid-body, 149
rotational Brownian, 515
screw, 152
spatial, 171
motion-group convolutions, 433
motion group to compare images, using the invariants on the, 428
motivational examples, 187
multi-dimensional continuous scale-translation wavelet transforms, 103
multi-dimensional Gabor transform, 102
multiplication of polynomials, 610
multiplication tables, 195

natural projection, 653
navigation system, 503
nearest-neighbor energy functions, 565
nematic phase, 528
Newton-Horner recursive scheme, 610
Newtonian fluids, 584
nilpotent group, 232
nilpotent matrices, 217
noisy measurements, 478
non-inertial theory of diffusion on $SO(3)$, 524
non-solid media, 597
noncommutative operations, 1
noncommutative unimodular groups in general, 266
nondegenerate set of curves, 652
nonisomorphic tables, 196
nonlinear systems, 485
norm of a vector, 624
normal spherical image, 95
normal vector, 159
numerical convolution of histograms on $SE(2)$, 389
numerical examples, 425
numerical methods for real matrices, 634
numerical results for planar workspace generation, 390

Nyquist criterion, 34

ODF from experimental measurements, 592
onto map, 617
open ball of radius, 651
operational calculus for $SU(2)$ and $SO(3)$, 281
operational properties, 313, 329
    and solutions of PDEs, 571
    for Lie group Fourier transforms, 339
    for unimodular Lie groups, 266
    other, 343
operations, noncommutative, 1
optimal assignment metric, 471
optimal reparameterization for least variation, 161
optimality, 649
optimally twisting frames, 160
orbits, 221
orbits and stabilizers, 201
order Hankel transform, 56
order of an element, 197
order of the set, 617
ordered set, 615
orientable manifold, 653
orientable surface, 95
orientational
    and motion-group averaging in solid mechan-
        ics, 585
    averaging, 13
    averaging of tensor components, 586
    distribution functions for non-solid media, 597
    distribution of polycrystals, 589
    dynamics of fibers in fluid suspensions, 597
orthogonal curvilinear coordinates, 88
orthogonal expansions and transforms in curvilin-
    ear coordinates, 98
orthogonal expansions in curvilinear coordinates,
    81
orthogonal expansions on the sphere, 99
orthogonality, 336
orthogonality of eigenfunctions, 39
other methods for describing rotations in three-
    space, 135
other models for rotational Brownian motion, 535
other operational properties, 343

parallel architecture, 379
parallel axis theorem, 134
parameterizations
    and Jacobians for $SL(2, \mathbb{C})$, 230
    axis-angle and Cayley, 300
    based on stereographic projection, 126
    curvilinear coordinates and surface, 81
    modified axis-angle, 127

    of motions and associated Jacobians, 154
    of rotation, 122
    of rotation as a solid ball, 125
    of the semi-circle, 82
    of the unit circle, 82
    of Tsiotras and Longuski, 127
    planar rotations, 82
    $SO(3)$ matrix representations in various, 298
parameterizing rotations, 117
Park's metric, 180
Parseval/Plancherel equality, 338
Parseval's equality, 18, 42, 328
partial sum, 16
passivity condition, 488
path integration, 641
pattern matching, 10
pattern recognition, 420
Pauli spin matrices, 139
PD control on $SE(3)$, 499
PD control on $SO(3)$, 499
PDEs, 23
PDEs, operational properties and solutions of, 571
PDF for the continuous chain, 552
PDFs on rotation groups with special properties,
    464
periodic boundary conditions, 40
periodic function, 15
periodic pulse train, 34
periodization, 105
permeability in the medium, 27
permittivity in the medium, 27
permutation matrices, 199
permutations and matrices, 197
persistence length of the polymer, 552
piecewise constant orthogonal functions, 65
planar manipulators, 387
planar motions, 170
planar workspace generation, 390
Plancherel formula, 328
platform architecture, 379
Pochhammer symbol, 61
points, 615
polar angle, 91
polar coordinates, 90
polar decomposition, 634
pole figure, 592
polycrystalline materials, 592
polycrystals, orientational distribution of, 589
polymer kinematics, 165
polymer, persistence length of the, 552
polymer science, rotational diffusion equation from,
    531
polymers, kinematic modeling of, 548
polynomials, 607, 610

Bessel, 58
Charlier, 63
Chebyshev, 46
  efficient division of, 611
  efficient multiplication of polynomials, 610
  evaluation, 611
Gegenbauer, 46
Hahn, 62
Hermite, 47
homogeneous, 281
interpolation, 613
irreducible representations from homogeneous, 281
Jacobi, 46
Krawtchouk, 63
Lagrange fundamental, 52
Lagrange interpolation, 52, 613
Laguerre, 47
Legendre, 43
Meixner, 63
recurrence relations for orthogonal, 50
Wilson, 65
with continuous and discrete orthogonalities, 65
Zernike, 64
pose determination with a priori correspondence, 471
pose determination without a priori correspondence, 469
positive-definite matrix, 628
preimage, 617
probability
  and statistics on the circle, 459
  conditional, 457
  density functions on groups, 461
  density functions on the line, 458
  density, joint, 457
  review, basic, 455
problem statement using harmonic analysis, 473
products of class sum functions for finite groups, 210
projectors, 83
proof of the inversion formula, 327
proper Lorentz group, 230
proper subset, 617
properties
  convolution and Fourier transforms of functions on $SE(2)$, 325
  of rotational differential operators, 342
  of the δ-function, 343
  of translation differential operators, 340
  operational, 329, 339
  other operational, 343
  symmetry, 323, 335

proportional derivative control on $SO(3)$ and $SE(3)$, 498
purely kinematical models, 562

quadrature rule, 49
quadrature rules and discrete polynomial transforms, 49
quantization of angular momentum, 316
quantization of classical equations of motion, 642
quantum states, 315
quaternion sphere, 137
quaternions, 135
quotient group, 199, 201

Rademacher functions, 67
radiation therapy treatment planning, 446
radon transform and tomography, 432
radon transform as a particular case of correlation on the motion group, 442
radon transform for finite beam width, 437
radon transforms as motion-group convolutions, 433
randomized angular momentum, 535
rapidly-decreasing, 6
Rayleigh dissipation function, 489
real algebra, 625
real matrices, 634
real vector space, 624
rectangular distribution, 459
recurrence relation, 610
recurrence relations for orthogonal polynomials, 50
recursive estimation of a static orientation or pose, 493
recursive estimation of a static quantity, 491
recursive least-squares estimators, 492
recursive weighted least-squares estimator gain matrix, 492
reducible representation, 241, 372
referential description, 111
regular tessellations, 95
regular tessellations of the plane and sphere, 97
regularization technique, 26
relationship between modular functions and the adjoint, 261
relationships between rotation and skew-symmetric matrices, 115
representations
  adjoint and tensor, 284
  checking irreducibility of, 270
  from homogeneous polynomials, irreducible, 281
  induced, 268
  irreducibility of the, 287

irreducible, 241
left-regular, 242
of finite groups, 241
of finite groups, induced, 269
of Lie groups, 262
of Lie groups, induced, 271
of $SE(3)$, irreducible unitary, 331
of $SO(4)$ and $SU(2) \times SU(2)$, 317
of $SU(2)$ and $SO(3)$, 281
of the discrete motion group, 363
reducible, 372
right-regular, 242
theory and operational calculus for $SU(2)$
    and $SO(3)$, 281
$U(g, p)$ of $SE(2)$, 323
representative of the double coset, 211
residual covariance matrix, 492
Reuss model, 590
review of basic continuum mechanics, 579
revised Marko-Siggia model, 569
ribbon, 163
Riccati equation, 490
Ricci and scalar curvature, 87
Riemannian curvature tensor, 87
Riemannian manifold, 653
Riemannian metric, 653
Riemannian metric tensor, 88
right-circular helix, 160
right invariant, 225
right-regular representations, 242
rigid-body
    Brownian motion of a, 526
    displacement, 469
    displacement, determining a, 493
    kinematics, 4
    mechanics, 132
    motion, 149
    rotations, 112
rigorous definitions, 218
ring, 626
ring with unit, 626
rings and fields, 626
robot sensor calibration, 474
robot workspaces, 9
robotic manipulator, 9
robotics, 379
Rodrigues formula, 43
Rodrigues vector, 123
rotation groups with special properties, 464
rotational
    Brownian motion, 11, 522
    Brownian motion based on randomized an-
        gular momentum, 535
    Brownian motion diffusion, 515

Brownian motion, other models for, 535
differential operators, 342
diffusion equation from polymer science, 531
isomeric state model, 554
processes, 503
rotations
    and skew-symmetric matrices, 115
    and translations, 425
    as 4×4 matrices, 137
    dependent functions, 145
    folded Gaussian for one-dimensional, 464
    group of three-dimensional space, 465
    in 3D as bilinear transformations in the com-
        plex plane, 141
    in $\mathbf{R}^4$, 171
    in three dimensions, 111
    in three-space, 135
    infinitesimal, 127
    Jacobians associated with parameterized, 128
    matrices, 114
    metrics and, 141
    metrics on, 143
    parameterizations of, 122
    parameterizing, 117
    rigid-body, 112
    rules for composing, 121
    transformation of cross products under, 117
Rubik's cube, 3
rules for composing rotations, 121

sampling, 33
    and FFT for $SO(3)$ and $SU(2)$, 301
    and reconstruction of functions on the line,
        34
    band-limited Sturm-Liouville expansions, 42
    of the Fourier-Bessel series, 59
scalar wave equation, 28
scale-Euclidean wavelet transform, 103
scale-translation wavelet transforms for the circle
    and sphere, 105
scaling property, 343
Schrödinger equation, 23
screw axis, 152
screw motions, 152
$SE(2)$, 235
$SE(2)$ convolution integrals, 365
$SE(2)$ revisited, 226
$SE(D)$ convolutions, 373
self-adjoint differential operators, 639
semi-direct product, 194
semi-simple Lie groups, 232
separable boundary conditions, 40
serial chain topology, 379

serial chains, 545

serial linkages, 407

set basic definitions, 615

set of special orthogonal, 113

set theory, 615

Shannon sampling theorem, 34

Shannon wavelet, 70

shift property, 343

shifted functions, 255

signed curvature, 162

simplified density functions, 403

Simpson's rule, 49

single-crystal properties over orientations, 590

single crystal strength properties, 592

skew field, 136, 626

small cracks and dilute composites, 588

smectic phase, 529

Smoluchowski equation, 517

$SO(3)$, 235

$SO(3)$ matrix representations in various parameterizations, 298

softening spring, 594

solid mechanics, 594

solution with two sets of exact measurements, 475

solutions of PDEs, 571

solvable group, 232

some differential geometry of S^3 and $SU(2)$, 290

spatial coordinates, 584

spatial motions, 171

special orthogonal, 113

special properties of symmetry matrices, 627

special properties of the matrix exponential, 630

special unitary 2×2 matrices, 138

spectrum of the function, 249

sphere, 106

spherical

    Bessel functions, 55, 57

    coordinates, 91, 125

    Hankel transform, 57

    harmonics, 100

    harmonics as eigenfunctions of the Laplacian, 100

square integrable, 6

stability and control of nonlinear systems, 485

stability subgroup, 202

stabilizer, 202

state of the system, 486

state transition matrix, 490

static linear estimation, 489

stationary Markov process, 494

statistical determination of a rigid-body displacement, 469

statistical ensembles and kinematic modeling of polymers, 548

statistical mechanics of macromolecules, 545

statistical pose determination and camera calibration, 455

statistics of stiff molecules as solutions to PDEs, 568

statistics on groups, 461

statistics on the circle, 459

statistics on the homogeneous space $SO(3)/SO(2)$, 463

stereographic projection, 83, 93

stereology, 592

stiff molecules as solutions to PDEs, 568

stiffening spring, 594

stiffness tensor, 594

stochastic control, 486

stochastic control on $SO(3)$, 506

stochastic differential equations, 495

stochastic optimal control for linear systems, 497

stochastic processes, estimation, and control, 485

stress tensor, 583

structure constants, 219, 234

Sturm-Liouville expansions, 39

Sturm-Liouville theory, 39

$SU(2)$ representations as eigenfunctions of the Laplacian, 294

subduction, 269

subgroup definition, 193

subset, 617

sufficient conditions for optimality, 649

sum, product, and Big-$\mathcal{O}$ symbols, 607

surface curvature, 86

surface, orientable, 95

surface parameterizations, 81

surface traction, 583

surjective map, 617

symmetric group, 197

symmetries, 338

symmetries in workspace density functions, 400

symmetry matrices, 627

symmetry operations on the equilateral triangle, 214

symmetry properties, 323, 335

system identification, 486

systems of particles and serial chains, 545

tangent, 159

tangent bundle, 653

tangent elements, 218

tangent space, 653

tangent vectors, 218

techniques from mathematical physics, 635

telegraph equation, 25

template matching, 419

template matching and tomography, 10

tensor components, 586

tensor fields, 581

tensor product, 285

tensor representation, 285

tensor-valued functions of orientation, 585

tensors and how they transform, 580

test for irreducibility, 268

tetrahedral group, 224

texture analysis, 592

texture analysis of materials, 12

texture transformation, 593

texture transformation/evolution, 593

the correlation method, 421

theorem

    convolution, 248, 325

    Euler's, 118

    Gauss-Bonnet, 96

    homomorphism, 205

    Lagrange's, 201

    parallel axis, 134

    Shannon sampling, 34

theories including the effects of conformational energy, 553

theories of Debye and Perrin, 522

three-term recurrence relations, 43

Tikhonov regularization, 26

tomography, 10, 419

tomography algorithm, 441

tomography, inverse, 446

tomography, radon transform and, 432

tomotherapy problem, 11

topological group, 193

topological space, 618

topology of surfaces and regular tessellations, 95

Torresani wavelets for the circle, 104

Torresani wavelets for the sphere, 105

torsion, 159

torsion angle, 166

torsion of the curve, 159

total space, 653

towards stochastic control on $SO(3)$, 506

transformation group, 193

transformation of cross products under rotation, 117

transformations of a ribbon, 194

transforms

    alternative motion-group Fourier, 372

    as motion-group convolutions, radon, 433

    continuous Gabor, 71

    continuous scale-translation-based wavelet, 69

    continuous wavelet, 69, 70

    converting to the Fourier, 59

    discrete and fast Fourier, 30

    discrete Fourier, 30

    discrete-motion-group Fourier, 365

    discrete polynomial, 39, 49

    discrete wavelet, 72, 73

    fast Fourier, 32, 252

    fast Legendre and associated Legendre, 45

    fast numerical, 58

    fast polynomial, 54

    finite Haar, 66

    finite Walsh, 68

    for finite beam width, radon, 437

    for finite groups, Fourier, 240

    Fourier, 19, 248

    Gabor, 274

    Hankel, 56

    in curvilinear coordinates, 98

    multi-dimensional continuous scale-translation wavelet, 103

    multi-dimensional Gabor, 102

    on the plane and R^n, 102

    operational properties for Lie group Fourier, 339

    quadrature rules and discrete polynomial, 49

    scale-euclidean wavelet, 103

    spherical Hankel, 57

    Walsh functions and, 67

    Walsh-Hadamard, 68

transitively, 193

translation, 274

translation differential operators, 340

translational diffusion, 516

transpositions, 630

trapezoid rule, 49

tree-like topology, 379

twist number, 163

two sets of exact measurements, 475

two state and nearest-neighbor energy functions, 565

uncertainty principle, 643

uniaxial phase, 528

uniformly tapered manipulators, 405

unimodular, 228

unimodular group, 256, 274

unimodular Lie groups, 266

unit quaternions, 137

unitary group representation, 241

unitary operations, 69

unitary representations of $SE(3)$, 331

use of convolution for workspace generation, 384

useful formulas in orientational averaging of tensor components, 586

using Fourier transform to solve PDEs and integral equations, 23

using the invariants on the motion group to compare
        images, 428
using the Laguerre series expansion, 60

Vandermonde matrix, 612
variance for $SO(N)$ and $SE(N)$, 467
variational calculus, 645
vector spaces and algebras, 623
Venn diagram, 616
Voigt model, 590
volume elements, 262
von Mises distribution, 460
vorticity, 584

Walsh function, 68
Walsh functions and transforms, 67
Walsh-Hadamard transform, 68
wave equation, 24
wave number, 28
wavelets, 39, 69, 70
        and groups, 273
        on spheres, 302
        on the circle and sphere, Gabor, 103
        on the sphere and rotation group, 302
        Torresani, 104
        transforms, 103
            on groups based on dilation and transla-
                tion, 274
            on the plane, continuous, 102
            scale-translation, 105
        Weil-Heisenberg, 103
Weil-Heisenberg wavelets, 103
Wiener process, 495
Wiener processes and stochastic differential equa-
        tions, 495
Wigner 3jm symbols, 309, 310
Wigner D-functions, 299
Wilson polynomials, 65
window function, 67, 420
workspace, 379
workspace density functions, 400
workspace generation, 379, 384
workspace generation for planar manipulators, 387
workspace of a manipulator, 9
wreath product, 199
writhe number, 163

Yamakawa model, 569

Zernike polynomials, 64
zero crossing, 426

Printed in the United States
by Baker & Taylor Publisher Services